THE GREEK ALPHABET

A	α	Alpha		N	ν	Nu
B	β	Beta		Ξ	ξ	Xi
Γ	γ	Gamma		O	ο	Omicron
Δ	δ	Delta		Π	π	Pi
E	ε	Epsilon		P	ρ	Rho
Z	ζ	Zeta		Σ	σ	Sigma
H	η	Eta		T	τ	Tau
Θ	θ	Theta		Υ	υ	Upsilon
I	ι	Iota		Φ	φ	Phi
K	κ	Kappa		X	χ	Chi
Λ	λ	Lambda		Ψ	ψ	Psi
M	μ	Mu		Ω	ω	Omega

CELL AND MOLECULAR BIOLOGY

CELL AND MOLECULAR BIOLOGY

THIRD EDITION

PHILLIP SHEELER

CALIFORNIA STATE UNIVERSITY, NORTHRIDGE

DONALD E. BIANCHI

CALIFORNIA STATE UNIVERSITY, NORTHRIDGE

JOHN WILEY & SONS, INC.

NEW YORK   **CHICHESTER**   **BRISBANE**   **TORONTO**   **SINGAPORE**

PHOTO CREDITS

PART 1: A cultured fibroblast cell just completing mitosis. (Courtesy of Drs. K. Feren and A. Reith. From *Scanning Electron Microscopy 1981*. Vol. II. p. 199. Copyright © 1981 by Scanning Electron Microscopy, Inc.)

PART 2: Circular DNA (i.e., plasmids) isolated from *Escherichia coli*. (Courtesy of Dr. E. Namork; in Electron microscopy of nucleic acids. *Micron 11*, 88, 1980. Copyright © by Pergamon Press.)

PART 3: A mitochondrion and surrounding endoplasmic reticulum and cytosol. (Courtesy of Dr. K. R. Porter.)

PART 4: An autoradiogram of radioactively-labeled proteins separated by polyacrylamide gel electrophoresis. (Courtesy of Drs. F. A. Eiserling and D. Coombs.)

PART 5: Transmission electron photomicrograph of intestinal epithelium. The cell on the far left is a mucus-secreting goblet cell. (Courtesy of R. Chao.)

PART 6: Sino atrial muscle tissue from frog heart. (Courtesy of Dr. A. Verna. From *Biol. Cellulaire 49*, 95, 1983. Copyright © Société Française de Microscopie Électronique.)

Cover Photo: Scanning electron photomicrograph of a human macrophage attempting to phagocytose a number of oil droplets. The cell, shown here at about 18,000 times its actual size, is an important member of the body's immune system. (Photo courtesy of Times (c) Boehringer Ingelheim International.)

Cover and text design: Karin Gerdes Kincheloe

Production: Elizabeth Meder

Library of Congress Cataloging in Publication Data:

Sheeler, Phillip.
    Cell and molecular biology.

    Bibliography: p.
    Includes index.
    1. Cytology. I. Bianchi, Donald E., 1938–
II. Title.

QH581.2.S53 1987   574.87   86-28113
ISBN 0-471-81758-9

Printed in the United States of America

10 9 8 7 6 5 4 3 2 1

To the memory of my father Barnett Sheeler,
to my mother Deborah,
to my wife Annette,
and to my children Wendy, Donna,
Lindsey, Paul, and Carly

P.S.

To my parents Ernest and Florence,
to my wife Georgia,
to my children Bill, Dave, Diana, and Jon,
and to my grandson Andy

D.E.B.

PREFACE

This book was written for sophomore and junior level courses in cell biology, molecular biology, and cellular physiology. We consider in detail the fine structure of eukaryotic and prokaryotic cells (and viruses), the chemical composition and organization of cells, cell metabolism, and bioenergetics and, for each major cell organelle or structural component, its particular molecular and supermolecular organization and its functions. Special attention has also been paid to a description of the major research tools used by cell biologists to increase our knowledge of cell structure, biochemistry, and function.

In preparing this third edition of *Cell and Molecular Biology*, we have drawn upon more than forty-five years of our own collective university teaching experience as well as the help of more than 25 university professors who have used the first and second editions of this book. We are hopeful that our early training in different areas of cell biology, the expertise of each reviewer's specialty, and our involvement in active research programs have helped to present an up-to-date, thorough and balanced approach.

We have necessarily made certain assumptions with regard to the background of students reading this book. For example, we have assumed that students have had courses in introductory biology and introductory chemistry. Most portions of the book dealing with physical or biochemical principles or concepts are preceded by a discussion of the requisite fundamentals.

We have revised, updated, and expanded this edition within the relatively short span of four years because of the rapid advances that have taken place in cell biology, especially in the fields of molecular genetics, genetic engineering, and immunology. Indeed, we have added a new chapter devoted specifically to the subjects of immunity and cancer. Concepts that are no longer tenable in the light of recent findings have been discarded, while newly developing ideas are presented and discussed. The clear and informative diagrams of the first and second editions of the book were among its major strengths; the best of these have been conserved, and a large number of new diagrams have been added. Once again, we have searched the literature to obtain the best and clearest electron photomicrographs depicting cellular organelles and other components, and this edition is replete with highly informative photomicrographs that have been obtained using the most recently evolved laboratory techniques.

Some illustrations are presented in stereoscopic form and, although individual members of each "stereo pair" may be viewed without an optical aid, the perspective effect is best obtained with the use of a stereo viewer. Plastic-frame viewers with adjustable lenses are available in most college and university campus bookstores or where graphic supplies are sold.

Many individuals provided invaluable help, suggestions, and guidance during the preparation, writing, and production of this book. We are indebted to them. Individual chapters in this or a prior edition were reviewed and critiqued by Betty D. Allamong (Ball State University), Jeffrey Beckner (University of Tennessee), Gerald Bergtrom (University of Wisconsin), James Bowman (Utah State University), Richard Bowmer (Idaho State University), Allyn Bregman (SUNY, New Paltz), Dennis E. Buetow (University of Illinois at Urbana-Champaign), Marvin H. Cantor (California State University, Northridge), George Card (University of Montana), Joseph U. Cassim (Ohio State University), Janice E. Chambers (Mississippi State University), George B. Cline (University of Alabama, Birmingham), Steve Free (State University of New York, Buffalo), Rosemary Grady (Loyola University), William J. Grimes (University of Arizona), Robert Hamilton (University of California, San Francisco Medical Center), Thomas Hejkal (Murray State University), Ray Holton (University of Tennessee), H. James Harmon (Oklahoma State University), Konrad Keck (University of Arizona), Bonnie Lamvermeyer (Denison University), Harvard Lyman (SUNY, Stony Brook), Ray Lynn (Utah State University), Janet L. Morgan (University of Kentucky), Eldon H. Newcomb (University of Wisconsin), Sally Nyquist (Bucknell University), Daniel G. Oldfield (DePaul University), William Oostonink (Colgate University), James Ownby (Oklahoma State University), Lansing Prescott (Augustana College), David Rooney (Saint Louis University), Peter Snustad (University of Minnesota),

Hugh Stanley (Utah State University), Philip Stukus (Denison University), Glenn Todd (Oklahoma State University), Gerald Vaughan (University of Tennessee), Eugene Vigil (University of Maryland), Ted Weinheimer (California State College, Bakersfield), Fred Weyter (Colgate University), and Gary Whitson (University of Tennessee).

In this edition, as in the first and second, we are especially grateful to those investigators who provided us with the unique electron photomicrographs from their collections. Our special thanks to Keith Porter (University of Colorado), Keiichi Tanaka (Tottori University School of Medicine), John Possingham (CSIRO, Adelaide, Australia), and Richard Chao, Daisy A. Kuhn, and Edward G. Pollock (California State University, Northridge).

The entire manuscript for this edition was prepared using word-processing equipment, which greatly facilitated our need to edit continuously sections that deal with topics in which progress is so rapid. We are especially grateful to Lindsey Deborah Myers for entering the text onto diskettes and for providing us with her word processing expertise.

The staff at Wiley has always been most helpful and we are once again indebted to our past editor Frederick C. Corey for his help and encouragement. Our thanks also to our current editor Katie Vignery for carrying the project to completion. The extraordinary talents of our artist John Balbalis are once again evident in this edition and we owe him a special thanks for his attention to detail and accuracy. We should like also to thank Safra Nimrod for helping us to find the photographic materials that we needed, Karin Kincheloe who designed the book, and Joshua Spieler and Elizabeth Meder who nurtured the book through production.

We apologize for any errors that we may have allowed into print and thank our readers in advance for bringing them to our attention.

Phillip Sheeler
Don Bianchi
Northridge, 1986

CONTENTS

CHAPTER 1

THE CELL: AN INTRODUCTION

Cell Biology as a science began within the progressive evolution of the **Cell Doctrine**. Briefly summarized, this doctrine states that cells are the fundamental units of both *structure* and *function* in all living things; that all forms of life (animal, plant, and microbial) are composed of cells and their secretions; and that cells arise only from *preexisting* cells, each cell having a life of its own in addition to its integrated role in multicellular organisms.

This statement seems both elementary and obvious to any student with some background in the biological sciences. Nevertheless, it took several centuries for this concept to be developed and accepted. The very existence of cells was not even suspected until the seventeenth century, because most cells are too small to be discerned with the naked eye, and because instruments for significantly magnifying small objects did not exist. However, with the introduction of the first crude light microscopes, investigators began to examine small organisms, tissues cut from plants or removed from animals, and the "animalcules" in pond water. The invention of the microscope and its gradual improvement went hand-in-hand with the development of the cell doctrine. It finally became apparent that a fundamental similarity existed in the structural organization of all the living things studied.

What follows is a brief description of a few of the historical highlights that culminated in the cell doctrine. Although a great many individuals made contributions of varying significance to the development of this concept, the works of a certain small number of people stand out as milestones.

Printed in 1558 were the results of *Conrad Gesner's* (Swiss, 1516–1565) studies on the structure of a group of protists called *foraminifera*. What is especially significant about this work is that Gesner's sketches included so much detail that they could only have been made if he had used some form of magnifying lens. This appears to be the earliest recorded use of a magnifying instrument in a biological study.

Francis and *Zacharias Janssen*, who manufactured eyeglasses in Holland, are generally credited with the construction of the first compound microscopes in 1590. Their microscopes had magnifying powers between $10\times$ and $30\times$ and were used primarily to examine small whole organisms such as fleas and other insects. The first microscopes were in fact referred to as "flea-glasses."

Although noted principally for his contributions in the fields of astronomy and physics, *Galileo Galilei* (Italian, 1564–1642) produced several important biological works. His own microscopes were constructed at about the same time as those of the Janssens (around 1610) and were used for several extensive studies on the arrangements of the *facets* in the compound eyes of insects.

Among the earliest descriptions of the microanatomy of tissues were those of *Marcello Malpighi* (Italian, 1628–1694), one of the first great animal and plant anatomists. He was the first to describe the existence of the capillaries, thereby completing the work on the circulation of the blood started by the great English physiologist William Harvey. Malpighi was among the first to use a microscope to examine and describe thin slices of animal tissues from such organs as the brain, liver, kidney, spleen, lungs, and tongue. His published works also include descriptions of the development of the chick embryo. In his later years, Malpighi turned to investigations of plant tissues and suggested that they were composed of structural units that he called "utricles" (later to be called "cells").

Antonie van Leeuwenhoek (Dutch, 1632–1723) was one of the most distinguished of all the early microscopists. Although it was only an avocation, Leeuwenhoek became an expert lens grinder and built numerous microscopes, some with magnifications approaching 300×. Leeuwenhoek was the first to describe microscopic organisms in rainwater collected from tubes inserted into the soil during rainfall. His sketches included numerous bacteria (bacilli, cocci, spirilla, etc.), protozoa, rotifers, and hydra. Leeuwenhoek was the first to describe sperm cells (of humans, dogs, rabbits, frogs, fish, and insects) and observe the movement of blood cells in the web capillaries of the frog's foot and the rabbit's ear. He described the blood cells of mammals, birds, amphibians, and fish, noting that those of fish and amphibians were oval in shape and contained a central body (i.e., the nucleus), while those of humans and other mammals were round. Leeuwenhoek's observations were recorded in a series of reports that he sent to the Royal Society of London.

Many of Leeuwenhoek's observations were confirmed in experiments conducted by *Robert Hooke* (English, 1635–1703), an architect and scientist employed by the Royal Society. Hooke popularized the use of microscopes among contemporary biologists in England and built several compound microscopes of his own. On one occasion, Hooke examined a thin slice cut from a piece of dried cork. In his description, Hooke wrote that he found the sections to be "all perforated and porous, much like a honeycomb" and referred to the boxlike structures as "cells." Thus, it is Hooke who introduced the term **cell** to biology. What

he observed, of course, were not cork cells but rather the empty spaces left behind after the living portion of the cells had disintegrated.

Nehemiah Grew (English, 1641–1712), together with Marcello Malpighi, is recognized as one of the founders of plant anatomy. His publications included accounts of the microscopic examination of sections through the flowers, roots, and stems of plants and clearly indicate that he recognized the cellular nature of plant tissue. Grew was also the first to recognize that flowers are the sexual organs of plants.

In 1824, *Rene Dutrochet* (French, 1776–1847) wrote that all animal and plant tissues were "aggregates of globular cells" and, in 1831, *Robert Brown* (English, 1773–1858) noted that the cells of plant epidermis, pollen grains, and stigmas contained certain "constant structures" that he called **nuclei**, thereby introducing this term to biology. Brown is also credited with the first description of the physical phenomenon now referred to as "Brownian motion." *Johannes E. Purkinje* (Czech, 1787–1869) coined the term **protoplasm** to describe the contents of cells.

Mathias J. Schleiden (German, 1804–1881) and *Theodor Schwann* (German, 1810–1882) are often credited, albeit incorrectly, with the first formal statement of a general cell theory. Their contributions to the development of the cell doctrine reside in the generalizations that they made based principally on the works of their predecessors. Schleiden and Schwann were particularly influential among their contemporaries and did, therefore, gain popular acceptance for the developing cell doctrine. Schleiden, a botanist, extended the studies begun by Robert Brown on the structure and function of the cell nucleus (which Schleiden called a "cytoblast") and was the first to describe **nucleoli**. Schleiden's writings clearly indicate his appreciation of the individual nature of cells. In 1838, he wrote that each cell leads a double life—one independent, pertaining to its own development, and another as an integral part of a plant. Schwann studied both plant and animal tissues. His work with connective tissues such as bone and cartilage led him to modify the evolving cell theory to include the notion that living things are composed of both cells *and the products of cells*. Schwann is also credited with introduction of the term **metabolism** to describe the activities of cells.

Rudolf Virchow (German, 1821–1902) was a pathologist and recognized the cellular basis of disease. His writings, often in Latin, also reveal his appreciation of the cellular basis of life's continuity, as summarized in his now famous expression *omnis cellula e cellula*, "all cells arise from [preexisting] cells."

In the last part of the 1800s and certainly by the turn of the century, the light microscope approached its limit in terms of magnification and resolving power, and nearly all major cellular structures had at least been described.

In this century, especially during the past 25 years, we have witnessed an unprecedented growth of our knowledge of the cell, its structural organization and diversity, its chemical organization, and the various functions of its component parts. This understanding is founded on the contributions of many thousands of scientists working in laboratories all over the world. Indeed, this textbook is a distillation of those contributions. To identify and credit all such contributors is, of course, beyond the scope of our concerns here, although many scientists and their particular contributions are identified in subsequent chapters of the book. Probably no symbol of recognition of the contributions made by scientists in this century has captured the imagination of the public (or for that matter scientists themselves) as has the Nobel Prize, an award recognizing specific contributions in diverse fields of human endeavor. Many such awards in the fields of *chemistry*, *physiology*, and *medicine* have been made for contributions that bear directly on cell biology (see Table 1–1), and we will discuss a number of these in subsequent chapters.

MICROSCOPY

Fundamentals of Light Microscopy and Transmission Electron Microscopy

Until the 1940s, most of our knowledge concerning the structure and organization of cells was obtained from studies conducted using light microscopy, and major structures and *organelles*, including the *cell wall*, *nuclei*, *chromosomes*, *chloroplasts*, *mitochondria*, *vacuoles*, *centrioles*, *flagella*, and *cilia*, had been described.

The smallest distance, *d*, between two points resolvable as *separate* points when viewed through lenses is given by the relationship

$$d = \frac{0.6\lambda}{n \sin \alpha} \qquad (1-1)$$

In this equation, λ is the **wavelength** of the light (radiation) that is used to illuminate the specimen; *n* is the

refractive index of the air or liquid between the specimen and the lens; and α is the **aperture angle**. The product $n \sin \alpha$ is called the lens **numerical aperture**, and for a good microscope lens its value is about 1.4.

Equation 1–1 also shows that the *resolving power* of a microscope varies with the wavelength of the source of illumination. The human eye cannot directly detect light having a wavelength of less than about 400 nm (see Table 1–2 for the units of metric measurement). Therefore, in the case of the light microscope, the maximum resolving power is about $0.6 \times 400/1.4$, or about $0.17\mu m$. That is, points less than about 0.2 μm apart cannot be distinguished as separate points by light microscopy (in practice, the limit is closer to 0.5 μm). From a practical standpoint, this means that even when glass optics of the finest quality are used, it is possible to observe cells at magnifications no greater than about $2000 \times$. Resolution is improved when sources emitting rays that have shorter wavelengths are employed. For example, the resolving power of the ultraviolet light microscope (which requires quartz optics because glass does not transmit ultraviolet light) is approximately double that of the light microscope.

Much greater resolution has been obtained with the *electron microscope*, developed in the 1930s, and with which magnifications of several hundred thousand are a practical possibility. The wavelength of radiations used with the electron microscope is typically about 0.005 nm (0.05 Å). Although resolution of an Angstrom or less is theoretically possible, the practical limit is about 5Å. This is many thousand times greater than that attainable using microscopes with glass optics. The basic features of a *transmission* electron microscope (abbreviated TEM) and shown in Figure 1–1, and a comparison between the component parts of the TEM and the light microscope is depicted diagrammatically in Figure 1–2. In the last decade, the *scanning* electron microscope (SEM) has become an increasingly important tool of the cell biologist. The SEM employs quite different principles than the TEM and will be considered separately later.

In both the light and electron microscopes, the source of radiation is an electrically heated tungsten filament. In the light microscope, the light emitted from the glowing filament is focused by a **condenser** onto the specimen to be observed. In the transmission electron microscope, the condenser focuses **electrons** emitted by the excited tungsten atoms into a beam, and electrodes accelerate the beam toward the specimen. Whereas the condenser of a light microscope

TABLE 1–1 NOBEL PRIZES FOR CELL BIOLOGY-RELATED CONTRIBUTIONS

Year	Division	Name	Contribution
1902	Chemistry	Emil Fischer	Pioneering studies in the field of biochemistry, especially studies of proteins
1906	Physiology & Medicine	Camillo Golgi S. Ramon y Cajal	Studies of the organization of the nervous system; contributions to histology, especially the structure of nerve cells
1908	Physiology & Medicine	Elie Metchnikoff Paul Ehrlich	Phagocytosis of bacteria during infection; staining procedures for bacteria; studies on immunity
1910	Physics	J. van der Waals	Physical properties of liquids
1915	Chemistry	Richard Wilstätter	Studies of chlorophyll and other plant pigments
1922	Physiology & Medicine	Archibald V. Hill Otto Meyerhof	Metabolism of muscle tissue; relationship between muscle metabolism and lactic acid
1926	Chemistry	Theodor Svedberg	Properties of colloids, especially proteins; development of analytical ultracentrifugation
1930	Physiology & Medicine	Karl Landsteiner	Discovery of human blood groups; studies of cellular agglutinins or antigens
1931	Physiology & Medicine	Otto Warburg	Studies on the nature and mode of action of respiratory enzymes; studies of oxidation and reduction in metabolism
1932	Chemistry	Irving Langmuir	Studies of surface chemistry, especially surface properties of liquids
1933	Physiology & Medicine	Thomas Hunt Morgan	Discoveries concerning the role of chromosomes in the transmission of heredity
1935	Physiology & Medicine	Hans Spemann	Discoveries concerning the role of the organizer during egg development
1936	Physiology & Medicine	Henry Dale Otto Loewi	Studies on the chemical transmission of nerve impulses
1937	Physiology & Medicine	Albert von Szent-Gyorgyi	Studies on biological oxidation and the involvement of vitamin C
1938	Chemistry	Richard Kuhn	Studies on the chemistry of carotenoids and vitamins
1943	Chemistry	George de Hevesy	Development of the radioisotopic tracer technique for following the progress of chemical reactions
1946	Physiology & Medicine	Hermann Joseph Muller	Studies of genetic mutations produced by X rays
1947	Physiology & Medicine	Carl F. Cori Gerty T. Cori	Studies of the metabolism of glycogen
1948	Chemistry	Arne Tiselius	Studies on the chemistry of proteins; development of electrophoresis
1952	Chemistry	Archer Morten Richard Synge	Development of chromatographic procedures for the separation of biological substances
1953	Chemistry	Hans A. Krebs	Elucidation of the tricarboxylic acid cycle (also called Krebs cycle)
		Fritz A. Lipman	Studies on coenzyme A
1954	Chemistry	Linus Pauling	Studies on the nature of chemical bonds, especially the peptide bond of proteins
1958	Physiology & Medicine	George W. Beadle Edward L. Tatum Joshua Lederberg	Studies of the organization and actions of genes in bacteria; the "one gene–one enzyme" concept
	Chemistry	Frederick Sanger	Analysis of protein structure, especially insulin
1959	Physiology & Medicine	Severo Ochoa Arthur Kornberg	Studies on the synthesis of RNA and DNA
1960	Physiology & Medicine	F. Macfarlane Burnet Peter Medawar	Studies of acquired immunological tolerance

TABLE 1-1 *continued*

Year	Division	Name	Contribution
1961	Chemistry	Melvin Calvin	Studies on the assimilation of CO_2 by plants, photosynthesis, the "Calvin cycle"
	Physiology & Medicine	Georg von Bekesy	Discoveries concerning the physiology of hearing
1962	Physiology & Medicine	James D. Watson Francis H. C. Crick Maurice Wilkins	Studies of the structure of genes, the double-helix model for DNA
	Chemistry	Max F. Perutz John C. Kendrew	Studies of the structure of globular proteins, especially myoglobin and hemoglobin
1963	Physiology & Medicine	John Eccles Alan Hodgkin Andrew Huxley	Role of sodium and potassium ions in the conduction of nerve impulses along the nerve cell membrane
1964	Physiology & Medicine	Konrad Bloch Feodor Lynen	Studies of the metabolism of cholesterol and fatty acids
1965	Physiology & Medicine	Francois Jacob Andre Lwoff Jacques Monod	Discovery of genes that regulate the actions of other genes; the "Operon" model
1966	Physiology & Medicine	Francis P. Rous	Studies on cellular transformation
1968	Physiology & Medicine	Robert W. Holley H. C. Khorana Marshall W. Nirenberg	Studies deciphering the genetic code
1969	Physiology & Medicine	Max Delbrück Alfred D. Hershey Salvador E. Luria	Studies of virus-mediated diseases
1970	Chemistry	Luis F. Leloir	Studies of the role of sugar nucleotides in the synthesis of carbohydrates
	Physiology & Medicine	Julius Axelrod Ulf von Euler Bernard Katz	Studies of the mechanisms of storage and release of neurohumors in nerve impulse transmission
1971	Physiology & Medicine	E. A. Sutherland	Mechanisms of action of hormones; role of cyclic AMP
1972	Physiology & Medicine	Maurice Edelman Rodney R. Porter	Chemical studies of immunoglobulins
	Chemistry	Christian B. Anfinsen Stanford Moore William H. Stein	Studies of ribonuclease Chemical studies of the amino acid composition of proteins
1974	Physiology & Medicine	Albert Claude Christian de Duve George Palade	Isolation and characterization of subcellular organelles and other particles
1975	Physiology & Medicine	Howard Temin Renato Dulbecco David Baltimore	Studies of the interaction between tumor viruses and cells; discovery of reverse transcription
1978	Physiology & Medicine	Werner Abner Hamilton O. Smith Daniel Nathens	Studies on genetic engineering; discovery of restriction enzymes; mapping of DNA sequences
	Chemistry	Peter Mitchell	Studies of bioenergetics
1980	Chemistry	Paul Berg Frederick Sanger	Studies on gene splicing Determination of nucleotide sequences of genes
1983	Physiology & Medicine	Barbara McClintock	Studies of movable genetic elements ("jumping genes")
1984	Physiology & Medicine	Cesar Milstein J. F. Kohler Niels K. Jerne	Molecular immunology

TABLE 1–2   METRIC MEASUREMENTS OF SIZE

1 meter (m)	=	39.4 inches (in.)
1 meter (m)	=	100 centimeters (cm)
1 centimeter (cm)	=	10 millimeters (mm)
1 millimeter (mm)	=	1000 micrometers (μm) or microns (μ)
1 micrometer (μm)	=	1000 nanometers (nm) or millimicrons (mμ)
1 nanometer (nm)	=	10 angstroms (Å)

FIGURE 1–1   A transmission electron microscope.
(Courtesy of Carl Zeiss, Inc., Thornwood, New York.)

LIGHT MICROSCOPE

ELECTRON MICROSCOPE

Tungsten filament

Condenser lens

Specimen (arrow)

Objective lens

Projector lens

Ocular lenses

Image seen by eye

Image on zinc sulfide screen

Zinc sulfide screen

Image captured on sheets of film in camera (screen lifted aside)

FIGURE 1–2 A comparison of the basic features of the light microscope and transmission electron microscope.

consists of one or a few glass lenses, the condenser of the electron microscope consists of several large, circular electromagnets. Indeed, all "lenses" of the electron microscope are electromagnets. In both microscopes, the radiation passes through the specimen and is then refocused by the **objective** lenses. The last lens of the light microscope is the **ocular**, through which the image may be viewed with the eye. The image of the electron microscope is viewed after its magnetic projection onto a zinc sulfide screen. The molecules of the screen are excited by the impinging electrons and emit visible light during their return to the ground state. Alternatively, the image may be captured on photographic film housed in a special camera mounted below the movable zinc sulfide screen.

The lenses of the light microscope have a fixed *focal length* and are focused by moving them nearer to or

further from the specimen. In the electron microscope, focusing is accomplished by manipulating the amount of current flowing through the windings of the series of electromagnetic lenses. This alters the electromagnetic field through which the electron beam must pass. The column through which the electron beam passes must be evacuated of air. If the vacuum is inadequate, the electrons will be scattered by collisions with residual gas molecules. Consequently, the specimen, the filament, the electromagnets, and the zinc sulfide screen are all mounted within a sealed compartment (i.e., a "column") that is connected to a vacuum pump. To avoid excessive scattering or absorption of electrons by the specimen itself, the material to be examined must be cut into extremely thin sections.

Two special forms of light microscopy warrant further description because of their widespread use and special application; these are **phase-contrast** microscopy and **fluorescence** microscopy.

Phase-Contrast Microscopy Although most regions of an unstained cell are transparent, they may have different densities and therefore different refractive indexes. Consequently, light rays travel through these regions at different velocities and may be refracted or bent to different extents. The phases of light rays that pass directly through an object and those that pass across its edges (i.e., at the interface where the refractive index changes) will necessarily be altered. The change increases the contrast between the object in focus and its surroundings. In the phase-contrast microscope, the phases of light rays entering the object are shifted by an annular diaphragm below the condenser. The phases of the rays passing through and around the object are shifted again by a phase plate in the objective lens. The result is a striking increase in the contrast of the object as certain regions appear much brighter (owing to additive effects of rays brought into phase) while other regions appear much darker (owing to the canceling effects of rays shifted further out of phase). The effect can be seen in Figure 1–39. Because phase-contrast microscopy produces added contrast in the material being studied without the need to employ stains, the technique is especially useful when examining living materials.

Fluorescence Microscopy Certain chemical substances emit visible light when they are illuminated with ultraviolet light. The effect is termed **fluorescence** and is put to use in the fluorescence microscope in which ultraviolet light rays are focused on the specimen. Some cellular components possess a natural fluorescence and appear in various colors. Other, nonfluorescing structures can be made to fluoresce by staining them with fluorescent dyes (**fluorochromes**). One of the most popular contemporary uses of fluorescence microscopy involves the preparation of **antibodies** that will bind to specific cellular proteins (see Chapter 4). The antibodies are first complexed with *fluorescein* (a fluorescent dye) and the fluorescein-labeled antibody is then applied to the cells. Cell structures containing the specific proteins capable of binding the fluorescein-labeled antibody are caused to fluoresce when examined with the fluorescence microscope, dramatically revealing their detail (Fig. 1–3).

Image intensifiers and microspectrofluorometers can also be coupled to fluorescence microscopes to provide images of or detect the presence of very small quantities of labeled molecules. Such adaptations provide sensitivities to as few as 10^5–10^6 molecules. In recent years, there have also been amazing advances in computer enhancement of the microscopic images.

Preparation of Materials for Microscopy

The preparation of biological material for examination with either the light microscope or the transmission electron microscope involves a series of physical and chemical manipulations that include (1) **fixation**, (2) **embedding**, (3) **sectioning**, and (4) **mounting.**

Fixation   One notable advantage of the light microscope is the capability to observe whole, **living** cells. It is also possible to employ "vital stains," which improve contrast but do not interfere with normal cell activity. More frequently, however, the cells are first killed and fixed. The fixation step is intended to preserve the structure of the material by preventing the growth of bacteria in the sample and by precluding postmortem changes. Formaldehyde and osmium tetroxide (OsO_4) are examples of fixatives most often employed for light microscopy. OsO_4 has a very high electron density, and because this gives contrast to the resulting image, OsO_4 has also found widespread use as a fixative in electron microscopy. Other popular fixatives include potassium permanganate and glutaraldehyde. After fixation for the required length of time, the samples are dehydrated by successive exposures to increasing concentrations of alcohol or acetone.

Embedding   Cells or tissues to be examined by light microscopy are usually embedded in warm, liquid paraffin wax. The wax, which both surrounds the tissue and infiltrates it, hardens on cooling, thereby supporting the tissue externally and internally. The resulting solid paraffin block is then trimmed to the appropri-

FIGURE 1–3   The network of *microtubules* present in this cell is clearly revealed using the combined techniques of fluorescent antibody labeling and fluorescence microscopy. (Photomicrograph courtesy of Drs. M. Osborn and K. Weber; copyright © 1977 by MIT Press, *Cell* **12**, 561.)

FIGURE 1–4   Plastic blocks of various shapes containing fixed and embedded tissue. (Courtesy of R. Chao.)

ate shape before being sectioned. The ultrathin sections required for electron microscopy necessitate the use of harder embedding and infiltrating materials such as epoxy plastics. These initially are in liquid form and are poured into small molds containing pieces of the fixed tissues; on heating, the liquid undergoes polymerization to form a hard plastic (Fig. 1–4).

Sectioning   The trimmed blocks containing the embedded samples are sectioned using a **microtome** (Fig. 1–5). In this instrument, the block is sequentially swept over the blade of a knife that cuts the block into a series of thin sections that adhere to one another end-to-end and thereby form a *ribbon*. Between each stroke, the distance between the block and knife edge is shortened. For light microscopy, the microtome knives are usually constructed of polished steel and can provide sections several micrometers thick. The

FIGURE 1–5   An ultramicrotome. (Courtesy of DuPont Instruments.)

sections for electron microscopy must be much thinner (typically 100 to 500 Å) and require more elaborate microtomes (called ultramicrotomes). Moreover, either diamond knives or knives prepared by fracturing plate glass are used in place of polished steel. Figure 1–6 illustrates the preparation of a ribbon of sections during ultramicrotomy.

Mounting Sections prepared for light microscopy are mounted on glass slides and may be stained with dyes of various colors that specifically attach to different molecular constituents of the cells. Sections to be examined with the electron microscope are generally not stained (no colors are seen with the electron microscope), although contrast may be improved by "poststaining" with electron-dense materials such as uranyl acetate, uranyl nitrate, and lead citrate. The sections are mounted on copper "grids" (small disks perforated with numerous openings) that have been coated with a thin (sometimes monomolecular) film of carbon (Fig. 1–7). The grid supports the film, which in turn supports the thin section. Thus the beam of electrons must pass through the spaces of the grid, the supporting film, and the section before striking the fluorescent screen. A comparison of photomicrographs obtained with the light microscope and the TEM is given in Figure 1–8; the difference in resolution is striking.

Specialized Applications of Transmission Electron Microscopy

Shadow Casting In shadow casting, the sample (usually containing small particles such as viruses or macromolecules) is spread on a coated grid, which is then placed in an evacuated chamber. A chromium or platinum wire is heated until the metal is vaporized, and the vapor is deposited onto the sample *at a precise*

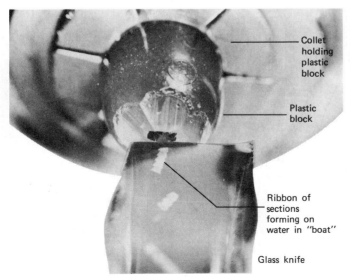

FIGURE 1–6 Sectioning of embedded tissue to form a *ribbon* of thin sections. The thin sections float off the edge of the knife onto water that fills a "boat" attached to the glass knife.

FIGURE 1–7 (*a*) Various grids used for mounting sections for electron microscopy. (*b*) Appearance of a ribbon of thin sections on the grid. (Photo courtesy of R. Chao.)

FIGURE 1–8 Comparison of photomicrographs of a white blood cell (granulocyte) obtained by (*a*) light and (*b*) transmission electron microscopy. (*a*, courtesy of Dr. M. H. Doolittle; *b*, courtesy of R. Chao.)

(a)

(b)

angle. The metal piles up in front of the sample particles but leaves clear areas behind them. If the resulting electron photomicrographs are printed in reverse, the areas containing the electron-dense metal that had piled up against the particles appear light, while the electron-transparent areas behind the particles appear as dark shadows (Fig. 1–9). Because the vaporized metal atoms tend to be projected in a straight line, the shadows are cast at precise angles, and in this manner the general shape and profile of a particle may be discerned.

Negative Staining   In the negative-staining procedure, the sample (again small particles such as viruses or macromolecules) is surrounded by an electron-dense material, such as phosphotungstic acid, that permeates the open superficial interstices of the sample. When the excess stain is carefully washed away, the sample particles appear as light (i.e., electron-transparent) areas that are highlighted by the surrounding dark background (Fig. 1–10).

Freeze-Fracturing   Freeze-fracturing is a technique in which the tissue is first *fractured* (i.e., cracked) along planes of natural weakness that run through each cell. These planes generally occur between the two layers of lipid molecules that comprise part of the limiting membrane around the cells' various vesicular organelles (see Chapter 15). Figure 1–11 depicts the basic differences between sectioning and fracturing.

The tissue to be freeze-fractured is first impregnated with glycerol and then frozen at -130°C in liquid Freon. The frozen tissue is transferred to an evacuated chamber containing a microtome and steel knife (also maintained at about -100°C using liquid nitrogen). The microtome knife is used to produce a fracture plane through the tissue (Figs. 1–12*a* and 1–12*b*). When the plane of the fracture intersects the

FIGURE 1–9   *Shadow casting.* (*a*) Shadowed bacteriophage showing polyhedral head (×55,000). (Photo by R. C. Williams; courtesy of John Wiley & Sons Photo Library.) (*b*) Shadowed SP3 viruses. From the shadowing angle and shadow contours, the size and shapes of the viral parts may be determined. (Courtesy of Dr. F. A. Eiserling.)

(a)   (b)

FIGURE 1–10 Tobacco mosaic (TMV), T4, and φX174 viruses visualized by negative staining. (Courtesy of Dr. F. A. Eiserling.)

membrane of a vesicular structure (e.g., nucleus, mitochondrion, vacuole, etc.), the membrane is split along its center, producing two "half-membranes." These are called the **E** (for "exterior") half and **P** (for "protoplasmic") half. The **E** half formerly faced the cell's external phase (see below) and the **P** half faced the internal phase (i.e., the protoplasm). One surface of each half-membrane is the original membrane surface (the **E** and **P** surfaces of Figs. 1–12c and 1–12d) and the other surface is the newly exposed fracture face (the **EF** and **PF** surfaces of Figs. 1–12c and 1–12d). The vacuum of the chamber is then used to *sublimate* water on the cut surface to a depth of several hundred Angstroms. New membrane faces exposed by sublimation are termed **E**$_s$ and **P**$_s$ (Fig. 1–12c). An electron-dense combination of metal (usually carbon

FIGURE 1–11 A comparison of sectioning and fracturing tissue.

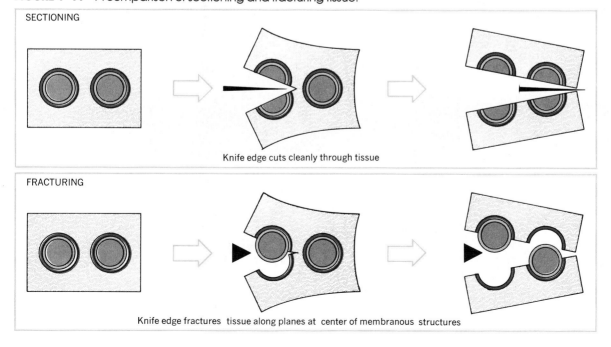

SECTIONING

Knife edge cuts cleanly through tissue

FRACTURING

Knife edge fractures tissue along planes at center of membranous structures and platinum) is then deposited on the cut surface at an angle and piles up in front of and behind projections from the surface, as well as in pits and depressions (Fig. 1-12d). Additional carbon is added to form an electron-transparent backing.

The shadowed and coated tissue is removed from the chamber and the tissue itself is either floated off or dissolved away, thereby leaving only the carbon-platinum "replica" (Fig. 1-12e). The replica is trimmed to the proper size, placed on a grid, and ex-

amined with the transmission electron microscope. Note that the replica is actually a templatelike impression of the distribution of particles in the original specimen. The electron beam readily passes through portions of the replica containing the carbon but is absorbed by the areas containing the platinum. The resulting images, which have a three-dimensional impact, are considerably different from those obtained with sectioned materials (Fig. 1-13). The *shadowing angle* used in preparing the replica is usually indicated

FIGURE 1-12 Stages in the freeze-fracturing procedure (see text for explanation).

(a)

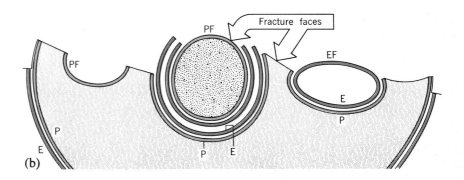

(b)

in the resulting photomicrograph (shown in the bottom-left corner of Fig. 1–13b). It is essential that the photomicrographs be viewed along the shadowing angle, otherwise depressions in the surface may be seen as projections (and vice versa). This is vividly demonstrated in the paired photographs of Figure 1–14.

The rate at which the specimen is cooled prior to fracturing markedly influences the effectiveness of the technique. Effective *cryofixation* of the specimen demands very rapid cooling such that water in the specimen passes quickly from an amorphous liquid state into an amorphous solid state (forming what is called "vitreous ice"). If the cooling rate is too slow, the water forms ice crystals, which, because of their ordered struture, deform and damage cellular components. Early efforts employing the freeze-fracture technique produced replicas containing many crystalline arrays resulting from insufficiently rapid cryofixation.

(c)

(d)

(e)

(a)

(b)

FIGURE 1–13 A comparison of electron photomicrographs of similar regions of a liver cell obtained by sectioning (*a*) and freeze-fracturing (*b*). *N*, nucleus; *P*, pores; *RER*, rough endoplasmic reticulum; *SER*, smooth endoplasmic reticulum; *Go*, Golgi body; *M*, mitochondria; *Mb*, microbody; *BC*, bile canaliculus; *Gl*, glycogen. The direction in which metal was applied to the fractured surface (i.e., the shadowing angle) in *b* is shown by the circled arrow. (Courtesy of Dr. L. Orci.)

FIGURE 1–14 Knowledge of the shadowing angle is important in interpreting photomicrographs of freeze-fractured tissue. The two photos shown here are of depressions made in wet sand viewed from two different orientations 180° apart. The left figure clearly reveals the depressions, but in the right figure they appear as elevations. (Photo courtesy of Dr. Theodore G. Rochow; copyright © 1979 by International Scientific Communications.)

The Scanning Electron Microscope

Scanning electron microscopy has become an increasingly popular technique since its introduction as a biological tool in the 1960s. With this technique, the *surface topography* of a specimen may be examined in considerable detail. At the present time, resolution is on the order of 30 Å. The organization of the scanning electron microscope (SEM) is shown in Figure 1–15 and is similar in many respects to that of the TEM. However, instead of the electron beam passing through (i.e., being "transmitted" by) the specimen, the interaction of the electrons of the beam (called "primary" electrons) with the surface of the specimen causes the emission of "secondary" electrons from the surface. The beam rapidly *scans* back and forth over the surface of the specimen, thereby producing bursts of secondary electrons. Greater numbers of secondary electrons are produced when the beam strikes projec-

tions from the specimen surface than when the beam enters a pit or depression in the surface. Hence, the number of secondary electrons produced at each point on the specimen surface, as well as the direction in which scattering occurs, depends on the surface topography. Therefore, there are quantitative and qualitative differences in the secondary electron bursts produced by the scanning electron beam. These ultimately give rise to an image in the following way.

Secondary electrons ejected at each point on the specimen surface are accelerated toward a positively charged scintillator located to one side of the specimen. Light scintillations created on impact of these electrons with the scintillator are conducted by a light guide to the photocathode of a photomultiplier tube. Electrical pulses produced in the photomultiplier tube are then amplified, and the resulting signal is relayed to a cathode-ray tube. The result is an image much like that of a television picture, consisting of light and

FIGURE 1–15　A scanning electron microscope. (Photo courtesy of International Scientific Instruments.)

dark spots. The scanning of the specimen surface by the primary electron beam is synchronized with the projection of a beam on the television screen in such a way that each portion of the specimen is reproduced in a corresponding region of the television image.

Samples to be examined are usually coated first with a metal (typically a gold–palladium alloy), forming a layer 10 nm or more thick, and then are affixed to a supporting disk that is placed in the beam path. The metal coating efficiently reflects the primary electrons of the beam and also produces large numbers of secondary electrons. It is to be noted that the thickness of the metal coating directly influences the maximum resolution attainable; for example, if the metal coating is 10 nm thick, then two particles that are less than 20 nm apart could not be resolved because their metallic coatings are contiguous. Figure 1–16 contains examples of photomicrographs obtained with the SEM.

Because the specimen being examined with the SEM can be rotated, it is possible to obtain views from different angles. This provides additional information about the size, shape, and organization of the material being studied. For example, Figure 1–17 contains two scanning electron micrographs of the same cluster of chains of the bacterium *Simonsiella* taken from different angles.

Stereo Microscopy (Stereoscopy)

True three-dimensional (i.e., stereoscopic) images of the specimen being studied can be obtained if one photomicrograph is taken as though the specimen were being viewed with the left eye *only*, and a second is obtained representing the right eye view. (The two views are obtained by a minor tilting of the sample in the horizontal plane.) When the two photomicrographs are placed side-by-side and the *stereo pair* is viewed

(b)

(a)

(c)

(b)

FIGURE 1–16 Examples of photomicrographs obtained using a scanning electron microscope. (*a*) Astrocyte (supportive but non-conducting cells of nerve tissue); (*b*) fractured pancreatic acinar cell; (*c*) mitochondria of heart muscle cell; in this view, portions of the outer membrane of some of the organelles have been fractured away revealing their internal membranous organization. (*a*, courtesy of Dr. N. B. Justesen; © 1983 by Pergamon Press, *Micron* **14**, 33. *b*, courtesy of Dr. T. Naguro, *Yonaga Acta Medica* **24**, 23 (1980); *c*, courtesy of Dr. Y. Masunaga, *J. Yonago Med. Assn.* **30**, 519 (1979).)

(a)

(b)

FIGURE 1–17 Scanning electron micrographs of chains of bacterial cells (*Simonsiella*) viewed from two different angles showing the various perspectives attainable. Magnification, 4000X. (Courtesy of Drs. D. A. Kuhn, J. Pangborn, and J. R. Woods.)

through the appropriate pair of lenses (called "stereo viewers"; see the Preface), a striking three-dimensional impression is seen, revealing details and geometric relationships that cannot be discerned from a single photomicrograph. Stereo views of the surface topography of tissues and cells are readily obtained with specimens prepared for SEM study; illustrations are presented in Figure 1–18.

The internal organization of cells is revealed in three dimensions by *high-voltage transmission electron stereoscopy*. In this procedure, cells are placed or cultured on a conventional grid and then fixed and dehydrated. The grid and cells are sandwiched between layers of carbon and are examined in a TEM in which the accelerating voltage is great enough to penetrate the entire thickness of the cell (about 1 million volts).

The cells are photographed at various tilt angles to produce the stereo pairs needed for the three-dimensional image (see Fig. 1–19).

The viewing of stereomicrographs may present some difficulties, especially for the novice. Generally, fewer problems are encountered with SEM stereoscopic views (e.g., Fig. 1–18) because the objects in the photomicrographs are opaque (i.e., certain objects are clearly in front of others). Transmission stereomicrographs are more difficult to assimilate and interpret because most of the objects are translucent. However, no other procedure provides direct images of the three-dimensional morphology of the cell's interior. A single stereo pair can reveal the entire population of mitochondria, lysosomes, or other organelles (see below) distributed through the cell.

(a)

(b)

FIGURE 1-18 Stereo electron micrographs obtained using scanning electron microscopy. To obtain the three-dimensional effect, these stereo pairs should be viewed through special stereo viewers (see the Preface). (*a*) Multiple lenses (*ommatidia*) of an insect eye (the spinelike structures are fine hairs); (*b*) a cluster of teratoma tumor cells; (*c*) a lymphocyte. (Photos courtesy of R. Chao.)

(c)

FIGURE 1–18 *continued*

FIGURE 1–19 High-voltage stereo transmission electron photomicrograph. In this view, the ruffles and microvilli that cover the surface of the cell are clearly seen. A fine, irregular lattice of microtrabeculae (filamentous structures) can be seen to occupy the space between the upper and lower cell surfaces as well as the surface projections. Magnification, 11,000X. (Photomicrograph courtesy of Drs. J. J. Wolosewick and K. R. Porter.)

CELL STRUCTURE: A PREVIEW

Free-living cells and the cells of multicellular organisms are subdivided into two major classes—**eukaryotes** (i.e., "true nucleus") and **prokaryotes** (i.e., "before nucleus"). In eukaryotes, the constituents of the cell nucleus are separated from the rest of the cell by a membranous envelope, whereas in prokaryotes these materials are not separated. Although the presence or absence of a true nucleus is the most obvious distinction between eukaryotic and prokaryotic cells, it will soon become clear that these two groups of cells also differ in many other important respects. Essentially all animal and plant cells are eukaryotic, whereas prokaryotic cells include bacteria, blue-green algae (or cyanobacteria), and the so-called pleuropneumonia-like organisms (PPLO) or mycoplasmas.

Eukaryotic Cells: The Composite Animal Cell

Animal cells vary considerably in size, shape, organelle composition, and physiological roles. Consequently, there is no "typical" cell that can serve as an example of all animal cells. There are, however, a num-

FIGURE 1–20   A composite animal cell.

ber of cell structures common to the majority of animal cells that are similar or identical in organization. These structures are depicted in the *composite* animal cell diagrammed in Figure 1–20 and described briefly in the following sections. They are dealt with in greater detail in later chapters that are individually devoted to the structure and functions of cell organelles. Figure 1–21 is an electron photomicrograph of an animal cell containing many of the structures to be discussed.

The Plasma Membrane The contents of the cell (**cytoplasm** and cytoplasmic organelles) are separated from the external surroundings by a limiting membrane, the **plasma membrane** (also called **cell membrane** or **plasmalemma**), which is composed of protein, lipid, and carbohydrate. This structure regulates the passage of materials between the cell and its surroundings and in some tissues is involved in intercellular communication (e.g., nerve tissue). These subjects are treated in depth in Chapters 15 and 24. In some tissue cells, portions of the plasma membrane are modified to form a large number of fingerlike projections called **microvilli** (Fig. 1–22*a*) because of their resemblance to the much larger *villi* of the wall of the small intestine. The microvilli greatly increase the surface area of the cell and provide for the increased passage of materials across the plasma membrane.

FIGURE 1–21
TEM photomicrograph of a pancreas cell. Most of the organelles characteristic of an animal cell are revealed in this thin section. (Courtesy of Dr. V. Imhoff; copyright © 1981, *Biol. Cellulaire* **41**, 179.)

Endoplasmic reticulum

Mitochondrion

Secretory granules

Nucleus

Plasma membrane

Cytoplasm

Golgi body

Cristae

Chromatin

Nucleolus

(a)

FIGURE 1-22 Specializations of the plasma membrane. The plasma membrane at the apical surface of secretory and absorptive tissues forms numerous tiny projections called *microvilli*. (*a*) A scanning electron photomicrograph of the surface of apocrine tissue (i.e., sweat gland cells). Below their apical surfaces, the plasma membranes of adjacent cells form a variety of different junctions such as *desmosomes* (*b*) and *gap junctions* (*c*). (*a*, courtesy of Dr. W. H. Wilborn; *b*, courtesy of Dr. D. E. Kelly; *c*, courtesy of Dr. N. B. Gilula; from Gilula, N. B., *Cell Communications* (R. P. Cox, Ed.), John Wiley & Sons, 1974; reprinted with permission of the Rockefeller University Press.)

(b)

(c)

When large numbers of cells are in close contact with one another (for example, in a tissue), it is not unusual to observe special junctions between neighboring plasma membranes. These take the form of **tight junctions**, **desmosomes**, and **gap junctions** (illustrated in Fig. 1–22 and considered in greater depth in Chapter 15).

The plasma membrane should not be thought of as a homogeneous structure that has the same chemical composition over its entire surface. Instead, the composition and organization vary in different regions of the membrane. Some areas of the plasma membrane of a liver cell, for example, face the plasma membranes of neighboring liver cells; other areas face the bile channels (*bile canaliculi*) into which substances are secreted by the liver cell. Still other portions of

(a)

(b)

(c)

the plasma membrane face the epithelial lining of *capillaries* from which substances are absorbed. Each of these regions of the plasma membrane is differently composed and differently organized and, in fact, is continually undergoing change and reorganization.

The Cytoskeleton and Microtrabecular Lattice Radiating through the cytosol of many cells are components of the **cytoskeleton** and **microtrabecular lattice**. The cytoskeleton consists of arrays of **thin filaments**, **intermediate filaments**, **thick filaments** and **microtubules** (see Figs. 1–3 and 23–2). These structures give shape and form to the cell and are also involved in cell movement. The cytoskeletal elements appear to be interconnected by a network of finer threadlike structures comprising what is called the

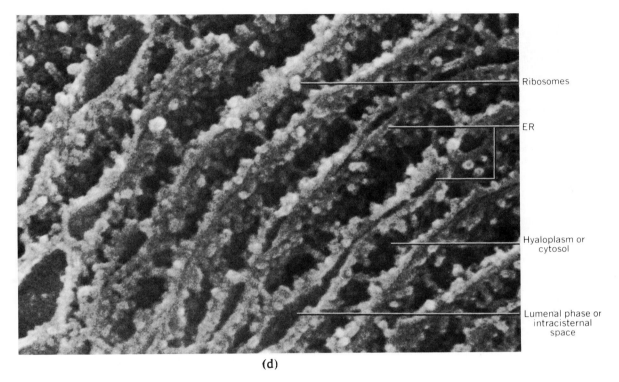

Ribosomes

ER

Hyaloplasm or cytosol

Lumenal phase or intracisternal space

(d)

Ribosomes (attached to hyaloplasmic surface)

Rough endoplasmic reticulum

Smooth endoplasmic reticulum

Intracisternal (lumenal) phase

Intracisternal phase   (e)

FIGURE 1–23 The *endoplasmic reticulum* in thin section (*a*) and freeze-fracture (*b*) through comparable regions of the cell. (*c*) High magnification view. *N*, nucleus; *M*, mitochondria; *R*, ribosomes; *C*, cisternae; *H*, hyaloplasm (cytosol); *RER*, rough endoplasmic reticulum; *SER*, smooth endoplasmic reticulum; *Go*, Golgi body; *V*, vesicle. In *b* the dashed arrows note perforations in ER membranes. Also, one of the mitochondria (m_2), has been fractured in such a way that faces of both the outer and inner mitochondrial membranes may be seen. The spaces separating the two nuclear membranes are indicated by the white arrows. The direction of shadowing is indicated by the circled arrow. (*d*) Scanning photomicrograph of rough ER; (*e*) diagram showing the continuity that exists between adjacent membranes and cisternae of both rough and smooth ER; rough ER membranes are characteristically sheetlike, whereas smooth ER membranes are characteristically tubular. (*b*, courtesy of Dr. L. Orci; *c*, courtesy of R. Chao; *d*, courtesy of Dr. K. Tanaka; copyright © 1980 by Academic Press, *Intern. Rev. Cytol.* **68**, 102.)

microtrabecular lattice. This lattice also interconnects a number of membraneous organelles and ribosomes. The cytoskeleton and microtrabecular lattice are considered in depth in Chapters 15 and 23.

The Endoplasmic Reticulum and Ribosomes   Within the cytoplasm of most animal cells is an extensive network of branching and anastomosing membrane-limited channels or cisternae collectively called the **endoplasmic reticulum** (abbreviated ER) (Fig. 1–23). The membranes of the endoplasmic reticulum divide the cytoplasm into two phases: the *lumenal* or *intracisternal* phase and the *hyaloplasmic* phase or **cytosol.** The lumenal phase consists of the material enclosed within the cisternae of the endoplasmic reticulum, whereas the cytosol surrounds the ER membranes. In the typical cell, the surface area that is represented by ER membranes is more than ten times that of the plasma membrane.

In the cytosol are large numbers of small particles called **ribosomes.** These particles are distributed either along the hyaloplasmic surface of the endoplasmic reticulum ("attached" ribosomes) or free in the hyaloplasm ("free" ribosomes). There is some evidence that the free ribosomes are interconnected by fine filaments of the microtrabecular lattice. Ribosomes carry out the synthesis of the cell's proteins, a subject covered at length in Chapter 22 (which also deals with the molecular architecture of ribosomes).

Endoplasmic reticulum with associated ribosomes is called **rough ER (RER),** the membranes of the RER typically being sheetlike (as in Fig. 1–23*e*). Endoplasmic reticulum that is devoid of attached ribosomes is called **smooth ER (SER),** the membranes usually forming a network of branching and fusing tubes. Smooth ER appears to be the site for the synthesis of lipids destined to be included in lipoproteins. Certain portions of the endoplasmic reticulum may be continuous with the plasma membrane and the nuclear envelope.

Mitochondria   Within the cytoplasm are numerous vesicular organelles called **mitochondria** (Figs. 1–13*a*, 1–13*b* and 1–24). Each mitochondrion is bordered by two membranes. The *outer* membrane is smooth, but the *inner* membrane displays numerous infoldings called **cristae** that greatly increase the surface area of the inner membrane. The space between neighboring cristae is called the mitochondrial **matrix** and often contains inclusions. Mitochondria are en-

gaged in numerous metabolic functions in the cell, including the energy-producing phases of carbohydrate and fat metabolism (called respiration), ATP synthesis, and porphyrin synthesis. The mitochondrion and its functions are considered at length in Chapter 16.

The Golgi Apparatus   The **Golgi apparatus** (also called Golgi body or Golgi complex) consists of a set of smooth, flattened cisternae that are usually stacked together in parallel rows; in this state, the organelle is sometimes referred to as a **dictyosome.** The Golgi apparatus is frequently surrounded by vesicles of various sizes, some of which are discharged from the margins of the main body of the organelle (Fig. 1–25). As related in Chapter 18, a variety of functions are ascribed to the Golgi apparatus, including the processing, packaging, and dispatchment of proteins destined for secretion, inclusion in the plasma membrane, and inclusion in lysosomes.

Lysosomes   Many cells contain vesicular structures that are generally smaller than mitochondria and are called **lysosomes** (Fig. 1–26). Lysosomes are bounded at their surface by a single membrane and contain quantities of various hydrolytic enzymes capable of digesting protein, nucleic acid, polysaccharide, and other materials. Under normal conditions, the activity of these enzymes is confined to the interior of the organelles and is therefore isolated from the surrounding cytosol. However, if the lysosomal membrane is ruptured, the released enzymes can quickly degrade the cell. Among their various roles, lysosomes take part in the intracellulr digestion of particles that are ingested by the cell during **endocytosis** and the intracellular scavenging of worn and poorly functioning organelles. Lysosomes are considered in depth in Chapter 19.

Peroxisomes and Glyoxysomes   Many cells contain small numbers of **peroxisomes** and/or **glyoxysomes.** These small organelles, which are bounded by a single membrane, contain a number of enzymes whose functions are related to the metabolism of hydrogen peroxide and glyoxylic acid. The structure and functions of peroxisomes, glyoxysomes, and other microbodies are considered further in Chapter 19.

The Nucleus   The **nucleus** is a relatively large structure frequently but not always located near the center

(a)

FIGURE 1-24 The *mitochondrion.* (Electron photomicrograph courtesy of Dr. K. Tanaka; copyright © 1980 by Academic Press, *Intern Rev. Cytol. 68*, 104.)

(b)

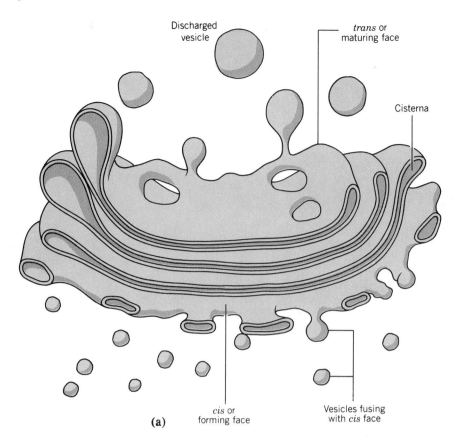

Discharged vesicle

trans or maturing face

Cisterna

cis or forming face

Vesicles fusing with *cis* face

(a)

(b)

0.5µm

(c)

FIGURE 1–25 The *Golgi apparatus*. (*a*) Diagram depicting the three-dimensional relationship between the cisternae and peripheral vesicles. (*b*) Typical appearance of Golgi bodies (G) in thin sections. (*c*) SEM view of Golgi body in a fractured cell. (*b*, courtesy of Dr. E. G. Pollock; *c*, courtesy of Dr. Y. Kinose; *J. Yonaga Med. Assn.* **30**, 527 (1979).)

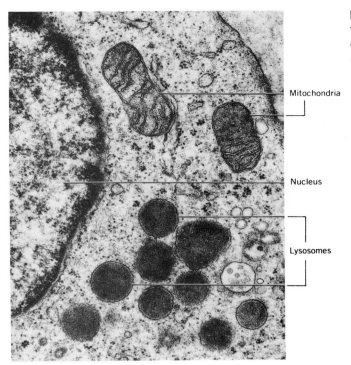

FIGURE 1–26 *Lysosomes.* In cells that have been specially treated to reveal these organelles, the lysosomes appear as electron-dense bodies somewhat smaller than mitochondria. (Courtesy of R. Chao.)

Mitochondria

Nucleus

Lysosomes

FIGURE 1–27 The cell nucleus. (Courtesy of R. Chao.)

Nuclear envelope

Pore

Pore

Outer nuclear membrane

Inner nuclear membrane

Pore

Chromatin

of the cell. The contents of the nucleus are separated from the cytosol by *two* membranes that together form the **nuclear envelope**. At various positions, the outer membrane of the envelope fuses with the inner membrane to form **pores** (Fig. 1–27). Nuclear pores provide continuity between the cytosol and the contents of the nucleus (sometimes referred to as the *nucleoplasm*). Often, the nuclear pores are plugged by a granular material (Fig. 1–28). The outer membrane of the nuclear envelope may have ribosomes attached to its hyaloplasmic side (Fig. 1–29) and may also form continuities with the membranes of the endoplasmic reticulum. Because the latter may be continuous with the plasma membrane, the **perinuclear space** (i.e., the space between the inner and outer membranes of the

nuclear envelope) corresponds to the lumenal phase of the cell and may be considered external to the cell (see Fig. 1–30).

The nuclear envelope and the pores that penetrate it are dramatically revealed in freeze-fracture preparations. Fig. 1–31 shows the cytosol-contacting half of the outer nuclear membrane fractured away, exposing the inner half of that membrane (i.e., face EF_1 of Fig. 1–30). Also fractured away are pieces of the inner nuclear membrane (the half-membrane that faced the perinuclear space), leaving only the half-membrane that faced the nucleoplasm (i.e., PF_2 of Fig. 1–30). The nuclear pores penetrate both membranes and in Figure 1–31 appear to be nonrandomly distributed in the nuclear envelope.

FIGURE 1–28 Plugged pores of the nuclear envelope. *N*, nucleus; *M*, mitochondrion. Magnification, 30,000X. (Photomicrograph courtesy of Dr. E. G. Pollock.)

FIGURE 1–29 In this electron micrograph, one can clearly discern ribosomes (*R*) attached to the outer membrane of the nuclear envelope. Also seen are nuclear pores, mitochondria (*M*), lysosomes (*L*), Golgi bodies (*G*), and nuclear chromatin (*C*). Magnification, 20,000X. (Courtesy of R. Chao.)

FIGURE 1–30 "Internal" and "external" phases of the cell. A region of the nuclear envelope is enlarged to illustrate the relationship among various faces of the inner and outer nuclear membranes. This illustration should be compared with Figure 1–12.

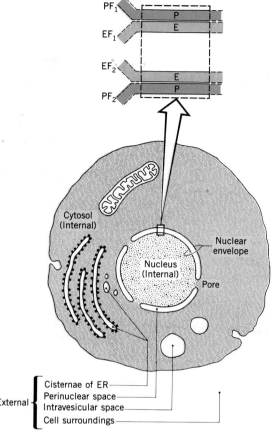

FIGURE 1–31 Freeze-fracture through the nuclear envelope exposing the EF_1 and PF_2 fracture faces. This freeze-fracture view is especially unusual because the fracture plane has passed through so large a region of the cell nucleus. Magnification, 13,000X. (Courtesy of Dr. E. G. Pollock.)

FIGURE 1-32 Dense, granular *nucleolus* in the nucleus of a plant cell. Magnification, 6000X. (Photomicrograph courtesy of Dr. E. G. Pollock.)

FIGURE 1-33 *Cilia.* The cilia have been fractured (*a*) and sectioned (*b*) in a plane parallel to the long axis of the organelles. *C*, cilium; *mv*, microvilli; *L*, lumen; *k*, kinetosome or basal body; *cn*, ciliary necklace. In *c*, the cilia are seen in cross section (note the arrangements of the microtubules). (*a* and *b*, courtesy of Dr. E. Boisvieux-Ulrich; copyright © 1977, *Biol. Cellulaire* **30**, 245; *c*, courtesy of R. Chao.)

(a)

(c)

(b)

The nucleus contains the genetic machinery of the cell (chromosomal DNA, chromosomal proteins, etc.), which is discussed in Chapters 20 and 21. Using either the light microscope or the electron microscope, the nucleus often reveals one or more dense, granular structures called **nucleoli** (Fig. 1–32). Nucleoli are not bounded by a membrane and are formed in part from localized concentrations of ribosomal precursors (see Chapters 20 and 22).

Flagella and Cilia Many free-living cells (such as protozoa and other microorganisms) possess *locomotor* organelles that project from the cell surface; these are either **flagella** or **cilia** (Fig. 1–33). The tissue cells of multicellular animals may also contain cilia, but they are employed here to move a substrate across the cell surface (such as mucus in the respiratory tract or the egg cell during its passage through the oviduct) and not for cell locomotion. The organelles are called cilia when they are short but present in large numbers and are called flagella when long but few in number. Each cilium or flagellum is covered by an extension of the plasma membrane. Internally, these organelles contain a specific array of **microtubules** that run from the **basal body** or **kinetosome** toward the tip of the structure. This array consists of two central microtubules and nine pairs of peripheral (outer) microtubules (Fig. 1–33*b*). Just below the basal body are rows of membrane particles referred to as the *ciliary necklace*.

Eukaryotic Cells: The Composite Plant Cell

All the organelles described in the preceding section as regular constituents of animal cells are also found in similar form in many plant cells. Several other organelles are unique to plant tissues and include the carbohydrate-rich *cell wall*, *plasmodesmata*, *chloroplasts*, and large *vacuoles*. A composite plant cell is depicted in Figure 1–34*a*.

The Cell Wall The **cell wall** is a thick, polysaccharide-containing structure immediately surrounding the plasma membrane (Fig. 1–34). In multicellular plants, the plasma membranes of neighboring cells are separated by these walls, and adjacent plant cells have their walls fused together by a layer called the *middle lamella*. The cell wall serves both a protective

and supportive function for the plant. The degree to which the cell wall may be involved in the regulation of the exchange of materials between the plant cell and its surroundings is difficult to assess but is most likely restricted to macromolecules of considerable size. As in animal cells, most of the regulation of exchanges between the cytoplasm and the extracellular surroundings of plant cells is a function of the plasma membrane.

Plasmodesmata At intervals the plant cell wall may be interrupted by cytoplasmic bridges between one cell and its neighbor (Fig. 1–35). These bridges are called **plasmodesmata** and represent regions in which channellike extensions of the plasma membranes of neighboring cells merge. The channels serve in intercellular circulation of materials.

Chloroplasts The ability to use light as a source of energy for sugar synthesis from water and carbon dioxide is a special feature of certain plant cells. This process, termed *photosynthesis,* is carried out in organelles called **chloroplasts** (Fig. 1–36). These organelles are commonly ellipsoidal structures bounded by two membranes. The *outer* membrane is smooth and continuous, whereas the *inner* membrane gives rise to extensive parallel infoldings called **lamellae** supported in a homogeneous matrix called the **stroma**. Most of the lamellae are arranged to form disk-shaped sacs (called **thylakoids**) that contain chlorophyll and may be stacked on top of one another, forming structures called **grana**. Lamellar membranes connecting the grana are called **stroma lamellae** (Fig. 1–36). A more detailed description of the strucure of chloroplasts and the metabolic pathways of photosynthesis is found in Chapter 17.

Vacuoles Although **vacuoles** are present in both animal and plant cells, they are particularly large and abundant in plant cells (Fig. 1–34), often occupying a major portion of the cell volume and forcing the remaining intracellular structures into a thin peripheral layer. These vacuoles are bounded by a single membrane and are formed by the coalescence of smaller vacuoles during the plant's growth and development. Vacuoles serve to expand the plant cell without diluting its cytoplasm and also function as sites for the storage of water and cell products or metabolic intermediates.

Cell wall

Central lamella

Plasma membrane

Chloroplast

Ribosomes

Mitochondrion

Endoplasmic reticulum

Large vacuole

Cytosol

Plasmodesmata

Golgi apparatus

Nucleus

Nucleolus

(a)

FIGURE 1–34   The plant cell. (*a*) A composite cell. (*b*) Photomicrograph of a tobacco leaf cell. (Photo courtesy of Dr. James Cronshaw.)

Cell wall (separating neighboring plant cells)

Plasmodesmata

FIGURE 1–35   *Plasmodesmata.* (Photomicrograph courtesy of Dr. N. Herzog.)

(b)

Stroma lamellae

Grana lamellae

Stroma

Outer membrane

Granum

FIGURE 1–36   A corn leaf cell chloroplast. (Photo by Dr. H. J. Arnott; courtesy of John Wiley & Sons Photo Library.)

Prokaryotic Cells: Bacteria

The bacteria are structurally distinct from eukaryotic microorganisms such as protozoa and contain a number of unique cellular organelles. The typical bacterium is about the size of a mitochondrion of an animal or plant cell, and in view of this small size, it is to be expected that the organelles of bacteria would be correspodingly smaller. The composite structure of a bacterium is depicted in Figure 1–37.

The Bacterial Cell Wall and Capsule The bacterial cell is enclosed within a wall that differes chemically from the cell wall of plants in that it contains protein and lipid as well as polysaccharide. The presence of a particular *peptidoglycan* (a protein–carbohydrate complex; see Chapter 5) is the basis of the histochemical classification of bacteria, being high in the so-called "gram-positive" bacteria (such as *Bacillus subtilis*) and low in the "gram-negative" bacteria (such as *Escherichia coli* and the *Simonsiella*). In some bacteria, the cell wall is surrounded by an additional structure called a **capsule**. The cell wall and capsule confer shape and form to the bacterium and also act as a physical barrier between the cell and its environment. This is important because osmotic forces usually result in a positive hydrostatic pressure inside the bacterium; in the absence of a cell wall and capsule, mechanically fragile bacteria would readily rupture.

Plasma Membrane Intrusions Infoldings of the plasma membrane of gram-positive bacteria give rise to structures called **mesosomes** (or **chondrioids**) (Fig. 1–38). Mesosomes are believed by many investigators to play a role in the division of the cell. Intrusions of the plasma membrane also form the photosynthetic organelles (**chromatophores**) of the photosynthetic bacteria.

Cytoplasmic Lamellae In some bacteria, there is a lamellar arrangement of membranes within the cytoplasm (Fig. 1–38b). However, there are no structures comparable to the endoplasmic reticulum of animal

(a)

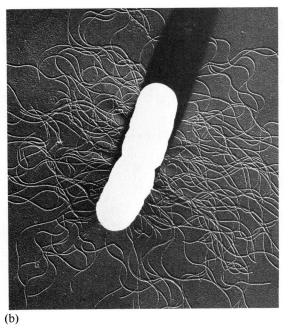

(b)

FIGURE 1–37 The *composite* bacterium (*a*) and TEM photomicrograph of shadow-cast bacterium *Caryophanum latum* (*b*). (Photo courtesy of John Wiley & Sons Photo Library.)

and plant cells. Bacteria contain large numbers of ribosomes, but most of these organelles are free in the bacterial cytosol; some ribosomes may be attached to the interior surface of the plasma membrane. Although bacterial ribosomes, like the ribosomes of eukaryotic cells, are the sites of protein synthesis, considerable differences exist between the organelles of these two groups (see Chapter 22). Cytoplasmic lamellae are particularly abundant in the *autotrophic* bacteria, which support their growth through photosynthesis or similar processes.

Nucleoids   In bacteria the nuclear material is not separated from the cytosol by membranes as it is in eukaryotic cells. However, the nuclear material is usually concentrated in a specific region of the cell, referred to as a **nucleoid**. During bacterial cell division, nucleoidal DNA becomes anchored to the plasma membrane and is distributed to the daughter cells without formation of observable chromosomes. Nucleoli are not present in the nucleoid.

Bacterial Flagella   Many bacteria contain one or more **flagella** employed for cellular locomotion. These organelles arise from the cytoplasm and penetrate the plasma membrane and cell wall (and capsule, if present). Bacterial flagella are smaller than the flagella of animal and plant cells and are simpler in organization, containing a single filament of globular proteins (called *flagellin*) surrounded by a sheath. Some bacteria are multiflagellated.

Flagellalike *axial filaments* are characteristic of some spirally shaped bacteria; these structures do not project away from the cell but are wrapped around the cell surface. Some bacteria contain nonflagellar appendages called **fimbriae** that arise from the cytosol and project a short distance above the cell surface (Fig. 1–37*a*). These structures are believed to play a role in the attachment of the bacterium to a surface or to other bacteria.

Some bacteria, such as *E. coli*, *Proteus mirabilis*, and *B. subtilis*, occur as separate, individual cells. However, in a number of groups, the daughter cells re-

(a)   (b)

FIGURE 1–38   TEM photomicrographs of bacteria. (*a*) Thin section through *Bacillus subtilis*. *cw*, cell wall; *pm*, plasma membrane; *n*, nucleoid; *m*, mesosome. Note the formation of the transverse septum, *s*, as the cell divides. Magnification, 27,000X. (*b*) Cross-fracture through *Rhodopseudomonas palustris* revealing the lamellar arrangement of photosynthetic membranes. Bar = μm. (*a*, courtesy of Dr. F. A. Eiserling; *b*, courtesy of Drs. A. R. Varga and L. A. Staehelin.)

(a)

Cell wall   Plasma membrane   Capsule

Ribosomes   Nuclear area

Capsule

Dorsal

(b)

Fibrils

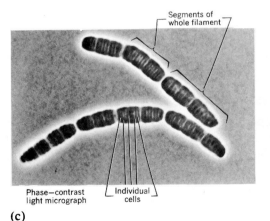

Segments of whole filament

Phase–contrast light micrograph   Individual cells

(c)

FIGURE 1–39 The filamentous bacterium *Simonsiella*. (*a*) Section through several cells of a filament revealing internal organization and dorsal–ventral differentiation. (*b*) Whole filament seen in section. (*c*) Phase-contrast photomicrograph showing that each filament is composed of several segments, each segment formed by a number of cells. Magnification: (*a*) 50,000X; (*b*) 6000X; (*c*) 4000X. Courtesy of Drs. D. A. Kuhn, J. Pangborn, and J. R. Woods.)

main attached to each other following division, so that chains (e.g., *streptococci*) or filaments are formed. An example of a filamentous genus is *Simonsiella*, which colonizes the mucosal epithelial surface of the mouth (see Fig. 1–17). The individual cells of some filamentous bacteria reveal a dorsal–ventral differentiation, that is, the ventral surface (which in *Simonsiella* attaches to and glides along the epithelium) is struc-

tured differently than the dorsal surface (which faces away from the epithelium). This is apparent not only in the scanning electron micrographs of whole filaments shown in Figure 1–17 but also in transmission electron photomicrographs of thin sections through the filaments (Fig. 1–39). Individual cells of a filament exhibit features common to single-cell (i.e., nonfilamentous) forms like *E. coli* and *B. subtilis*.

- Gelatinous layer
- Cell wall
- Plasma membrane
- Photosynthetic lamellae studded with phycobilosomes
- Septum (of dividing cell)
- Gas vacuole
- Nucleoid
- Cytosol
- Ribosome

(a)

(b)

FIGURE 1–40 (*a*) Generalized cyanobacterium. (*b*) Electron photomicrograph of dividing *Anabaena*. Note the formation of the septum. Magnification, 15,000X. (Courtesy of Drs. H. W. Beams and R. G. Kessel.)

Prokaryotic Cells: Cyanobacteria

The **cyanobacteria** (also called *blue-green algae*) are photosynthetic prokaryotes and occur as individual cells, as small clusters or colonies of cells, or as long, filamentous chains (Figs. 1–40 and 1–41). Cyanobacteria lack flagella but are able to glide over a gelatinous layer secreted through the cell surface. The photosynthetic apparatus consists of lamellae called **thylakoids**; the thylakoids are lined with pigment granules referred to as **phycobilosomes**. Cyanobacteria also contain several kinds of membrane-bound inclusions such as **gas vacuoles** (which provide

buoyancy in aquatic cyanobacteria) and **carboxysomes** (which contain enzymes involved in carbon dioxide fixation).

Prokaryotic Cells: Mycoplasmas

The **mycoplasmas** or **PPLO** (i.e., pleuropneumonia-like organisms), which cause a number of diseases in humans and other animals, are the smallest (i.e., about 0.1 μm in diameter) and simplest of all cells capable of autonomous growth. They are smaller even than some of the larger viruses. A mycoplasma is bounded at its surface by a membrane composed of

FIGURE 1–41 SEM photomicrograph of a filament of the cyanobacterium *Anabaena* (compare with the thin section of Fig. 1–40*b*). Magnification, 6000X. (Courtesy of Drs. F. A. Eiserling and S. Eipert.)

proteins and lipid, but there is no cell wall. Internally the cell's composition is more or less diffuse. The only microscopically discernible features within the cell are its genetic complement, which consists of a double-helical strand of circular DNA, and a number of ribosomes (Fig. 1–42). Mycoplasmas appear to contain the bare minimum of structural organization required for a viable, free-living cell and may represent a form intermediate between viruses and bacteria. The relative sizes of typical eukaryotic cells, bacteria, PPLOs, and viruses are compared in Figure 1–43, which dramatizes the differences that exist.

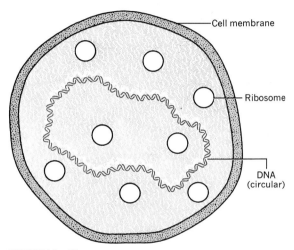

FIGURE 1–42   Structure of a mycoplasma.

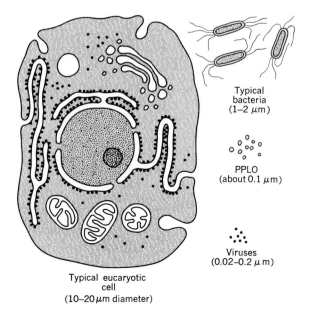

FIGURE 1–43   Relative sizes of eukaryotic cells, bacteria, mycoplasmas (PPLO), and viruses.

VIRUSES

So far, the descriptions in this chapter have been restricted to various kinds of cells. In this section, we are concerned with the organization and activities of viruses, but it should be emphasized at the outset that *viruses are not cells* and that it is a moot question whether viruses constitute living systems. Viruses are described here because of their intimate association with cells and because of their contributions to our understanding of certain cellular phenomena. It will become apparent as we deal in later chapters with the structure and interactions of nucleic acids and proteins and with gene expression that much of our present-day understanding of these subjects is based on studies initiated with viruses.

Structure of Viruses

Although all viruses or **virions** are extremely small, they are diverse in size and in organization. Generally, viruses range in diameter (or length) from about 20 to 200 nm. Thus the largest viruses are actually larger than the smallest cells. However, even the smallest of cells (e.g., bacteria and mycoplasmas) are subject to infection by viruses. Among those viruses that attack animal cells, the most notorious are the viruses that cause diseases in humans. Smallpox, chicken pox, rabies, poliomyelitis, mumps, measles, influenza, hepatitis, and the "common cold" are all produced by viruses. Even certain leukemias and cancers are of viral origin.

Most virions are either rod-shaped or quasi-spherical and contain a nucleic acid *core* surrounded by a specific geometric array of protein molecules that form a coat or **capsid** (Fig. 1–44). The proteins of the capsid are arranged to form either a *helical* pattern (when the virus is rodlike) or an *isometric* pattern (when the virus is globular). In the latter state, the virus appears much like a polyhedron. The viruses causing chicken pox, mononucleosis, fever blisters, and colds are examples of virions having polyhedral capsids. Helical capsids are more common among viruses that infect plant cells and bacteria. The **tobacco mosaic virus** (TMV), which infects the leaves of the tobacco plant, is among the most extensively studied viruses and exhibits the helical capsid pattern.

In many animal viruses and in some plant viruses, a lipoprotein **envelope** surrounds the capsid (e.g., influenza virus, herpesvirus, and smallpox virus). Among the largest and most complex virions are those that

(a)

(b)

(c)

FIGURE 1–44 Viruses. (*a*) The SPα bacteriophage of *Bacillus subtilis* and the tobacco mosaic virus. (*b*) Type D bacteriophage of *B. thuringiensis*; note the hexagonal array of the base plate (compare with Fig. 1–45). (*c, d*) T4 bacteriophage. Magnifications: (*a*) 190,000X; (*b, c, d*) 300,000X. (*a* and *c*, courtesy of Dr. F. A. Eiserling; *b*, courtesy of Dr. H. W. Ackermann; copyright © 1978, *Canadian J. Microbiol.* **24**, 821; *d*, courtesy of Dr. A. N. Broers; copyright © 1975 American Association for Advancement of Science., *Science* **189**, 638.

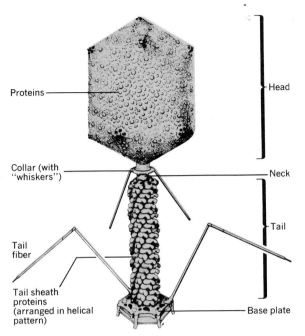

Proteins

Collar (with "whiskers")

Tail fiber

Tail sheath proteins (arranged in helical pattern)

Head

Neck

Tail

Base plate

FIGURE 1–45 Diagram of the T4 bacteriophage (compare with Fig. 1–44). Only two of the six tail fibers and collar whiskers are depicted). (Courtesy of Dr. F. A. Eiserling.)

attack bacteria (i.e., the **bacteriophages**). Most extensively studied among these are the T2, T4, and T6 (i.e., the "T-even") bacteriophages (Fig. 1–44). These bacteriophages have a taillike structure emerging from the capsid (Fig. 1–45). The tail is enclosed in a sheath of proteins arranged in a helical pattern, and the head of the virus is polyhedral. The end of the tail frequently reveals specialized structures (Fig. 1–45) involved in attachment to the surface of the host cell (see below).

Proliferative Cycle of a Virus

In the free or isolated state, viruses exhibit no metabolism and are incapable of proliferation. Proliferation of viruses requires a **host** cell and in its simplest and most direct form takes the following pattern. One or more viruses attach to specific sites on the surface of the host cell (Fig. 1–46). Following attachment or **adsorption**, the viral nucleic acid is inserted through the plasma membrane into the host's cytoplasm. Release of the core nucleic acid from a virus can be achieved experimentally; Figure 1–47 dramatically reveals the uncoiled DNA molecule released by a T4 bacteriophage. Once inside the host cell, the viral nucleic acid redirects the metabolism of the host so that new viral proteins and new viral nucleic acids are formed. These

FIGURE 1–46 SEM photomicrograph of bacteriophage infecting an *Escherichia coli* cell. (Courtesy of Dr. A. Takede; copyright © 1981 by *Biology Cellulaire*.)

0,5 μm

viral components combine in the host to form large numbers of new virions that egress from the cell by disruption of its plasma membranes (i.e., cell lysis). Some viruses exit the host by budding off, enclosed in a small piece of the host cell's plasma membrane; this does not lyse the host cell. The cycle of infection then repeats itself. The proliferative cycle of a virus is best understood for the bacteriophages and is depicted diagrammatically in Figure 1–48. The electron photomicrographs of Figure 1–49 show stages of the process.

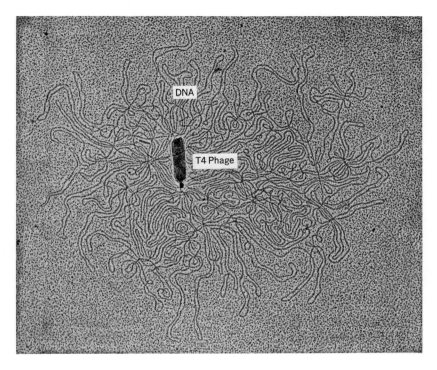

FIGURE 1–47 DNA molecule released from a "giant" T4 bacteriophage particle. The single DNA molecule is more than 150 μm long. Magnification, 20,000X. (Courtesy of Dr. F. A. Eiserling.)

FIGURE 1–48 Proliferative cycle of the bacteriophage.

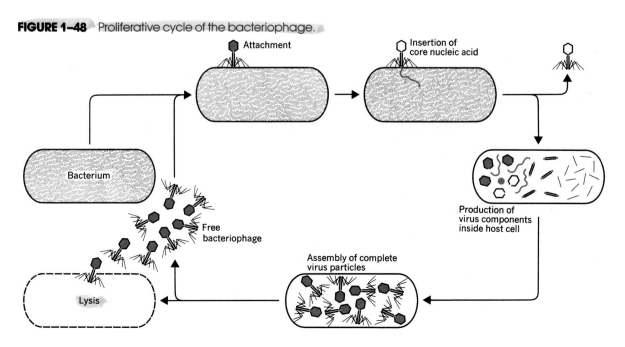

On some occasions and only for certain viruses, the injected nucleic acid does not cause proliferation and release of new virions. Instead, the injected nucleic acid is incorporated into the host's genetic material, and the host cell continues to function in its normal manner. However, duplication of the host's genetic material prior to cell division is accompanied by duplication of the incorporated viral nucleic acid. Several generations of cells may be produced, each containing a copy of the viral nucleic acid. Viruses exhibiting this phenomenon are called **temperate** viruses, because they do not cause the death of the *immediate* host. Viruses that engage only in the cycle described earlier and that kill the host cell are called **virulent** viruses. The dormant viral nucleic acid within the host is referred to as a **provirus**, and the infected cell is said to be **lysogenic**, because sooner or later, in one of the generations of host cells, the provirus nucleic acid *will* begin to direct the replication of new virions, and this in turn will lead to cell lysis and release of new infective virus particles. Virus replication is considered further in Chapters 20 and 21.

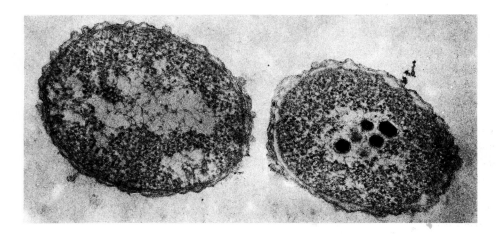

(a)

(b)

FIGURE 1–49 Stages in the proliferative cycle of a virus revealed by electron microscopy. (a) Two *Escherichia coli* cells infected by T4 phage. In one cell, the thin section reveals viral DNA condensing within the head membranes of several T4 particles. (b) In this negatively stained preparation, more than 100 T4 particles are being released from the lysing *E. coli* cell. Along the upper right edge of the cell, several empty (contracted) phages that began the infection by injecting their DNA into the host cell can be seen. Magnification: (a) 68,000X; (b) 26,000X. (a, courtesy of Dr. F. A. Eiserling; b, courtesy of Drs. L. Canedo and F. A. Eiserling.)

Classification of Viruses and the Nature of Viral Nucleic Acids

The classification of viruses poses certain problems, and several different approaches have been used. One method is to classify the virus according to the type of host cell. Hence, there are *animal viruses, plant viruses,* and *bacterial viruses* (i.e., bacteriophages). This method is not always satisfactory; for example, a few viruses infect both animals and plants. Another method employs comparisons of virus morphology (e.g., capsid shape and geometric symmetry). An interesting approach is to classify viruses according to the nature of their genetic material. Even the beginning student of biology becomes quickly aware of the functional relationships among DNA, RNA, and proteins in cells. Genetic information stored in molecules of DNA is transcribed or copied into a corresponding RNA molecule (called a messenger), and the messenger then directs the synthesis of a specific protein. The protein then acts as an enzyme or structural component of the cell. *In some viruses, however, the genetic information comprising the virion core is RNA and not DNA.* Among the RNA viruses are TMV and the virions causing polio, mumps, measles, influenza, and colds. While RNAs of cells are single-stranded molecules (Chapter 7), viral RNAs can be single stranded (e.g., TMV RNA) or double stranded (e.g., the reoviruses). Like eukaryotic and prokaryotic cells, the genetic information of many viruses is encoded in a DNA core. The DNA viruses include the T-even bacteriophages and those that cause chicken pox, herpes blisters, infectious mononucleosis, and shingles. However, the DNA may be *double-stranded linear* molecules (e.g., T5 and T7 baceriophages), *double-stranded circular* molecules (polyoma and SV 40 viruses), *single-stranded linear* molecules (the parvoviruses), or *single-stranded circular* molecules (ϕX174 bacteriophage). The chemistry of these types of nucleic acids is considered in Chapter 7. Whatever form the nucleic acid takes, it includes the genetic information for the synthesis of the variety of proteins that either become components of new viruses or are involved in the redirection of the host cell's metabolism.

It was noted at the beginning of this discussion that although viruses themselves are not cells, the study of viruses has yielded a wealth of information about cells. Research with certain viruses has provided crucial and at times astounding information about the chemistry of and interactions among nucleic acids and proteins—subjects that are dealt with at length in Chapters 7, 20, 21, and 22—and these viruses should be specifically noted.

TMV   Studies involving the tobacco mosaic virus (see Fig. 1–10) began nearly a century ago and represent the starting point in the field of **virology**. In the 1950s, TMV was at the focal point of research that verified that nucleic acids and not proteins constitute the genetic apparatus and that genetic information can be encoded in RNA as well as in DNA.

ϕX174   Studies with ϕX174 bacteriophage, which infects *E. coli*, revealed that the information for viral proliferation can be encoded in a *single* strand of DNA and does not require a double strand (i.e., double helix). Moreover, the structure of ϕX174 DNA has now been completely analyzed, that is, the entire base sequence is known. A most astounding finding yielded by these studies is that the coding sequences for several of the virus proteins are included *within* sequences for other proteins. Thus certain genes *overlap.* Similar findings are being reported for other viruses, including Simian virus 40 (SV 40), that possess double-stranded DNA and have both temperate and virulent phases.

Retroviruses   A number of RNA- and DNA-containing viruses produce tumors in animals. Among these are the retroviruses which have RNA as their core nucleic acid (e.g., Rous sarcoma virus, avian leukemia virus, and the virus causing A.I.D.S. [HTLV-3]). Unlike other RNA viruses, proliferation within the host cells requires that the inserted RNA be used for the preliminary synthesis of DNA, following which transcription and translation take the conventional pattern. The retroviruses have demonstrated that transcription can take place in the *reverse direction,* that is, from RNA to DNA.

VIROIDS

Viroids are much smaller than viruses and are also considerably simpler, for they consist of no more than a single strand of RNA. The RNA is not enclosed in any structure, and except during infection is not associated with any other chemical substances. The typical viroid is an RNA molecule about 50 nm in length. Though at the present time viroids are only suspected of being the agents of certain diseases in animal cells, they are known to be the cause of a number of plant diseases including *spindle tuber disease* in potatoes.

PRIONS

Generally, infectious agents are either viroids, viruses, prokaryotic cells (e.g., bacteria), or eukaryotic cells (e.g., certain protists). What these infectious agents have in common is that their identity is defined by the nucleic acid that each carries. Although there is still considerable debate on the matter, it now appears that there may be an exception to this rule. *Scrapie* (a disease of goats and sheep) and a disease of the nervous system in humans called Creutzfeldt–Jakob disease) appear to be caused by agents consisting *only of protein*; the agents of these diseases are called **prions**. The protein comprising a prion has a molecular weight between 50,000 and 100,000, corresponding to a particle size that is 100 times smaller than the smallest viruses.

SUMMARY

In the 300 years that followed the introduction of microscopy to biological science, the concept evolved that the cell is the fundamental unit of structure and function in all living things. This notion is referred to as the **cell doctrine**. Microscopy remains one of the cell biologist's most important and powerful research tools as new variations of light and electron microscopy have appeared, notably **freeze-fracture transmission electron microscopy**, **scanning electron microscopy**, and **stereoscopic electron microscopy**.

With these tools, the detailed structure and organization of nearly all subcellular organelles have been revealed.

Free-living cells and the cells of multicellular organisms are divided into two classes: **eukaryotes** (nearly all animal and plant cells) and **prokaryotes** (bacteria, cyanobacteria, and mycoplasmas). Eukaryotic cells are characterized by a number of discrete organelles, especially the **nucleus**, **mitochondria**, **Golgi bodies**, **lysosomes**, **peroxisomes**, **rough** and **smooth endoplasmic reticulum**, and, in plant cells, **chloroplasts** and **cell walls**. Like eukaryotic cells, many prokaryotes possess a limiting (plasma) membrane and ribosomes, but lack the true nucleus and other discrete organelles characteristic of eukaryotes.

Although **viruses** are not cells, they are intimately associated with cells, and their study has made invaluable contributions to our understanding of cell function. In the isolated state, viruses are incapable of metabolizing or proliferating. Proliferation requires preliminary infection of a host cell, which takes the form of insertion of the viral nucleic acid (DNA or RNA). Following this, the metabolism of the host is redirected to make the components of the virus. Assembly of new virus particles within the host is ultimately followed by their egress to begin a new cycle of infection. Among the smallest known agents of infection are **viroids** (naked pieces of RNA) and **prions** (almost exclusively protein).

Most of the remaining chapters are devoted to a detailed description of the biochemistry, structure, and physiological functions of cells and their organelles.

REFERENCES AND SUGGESTED READING

Articles and Reviews

Albersheim, P., The wall of growing plant cells. *Sci. Am.* **232** (4), 80 (April 1975).

Allison, A., Lysosomes and disease. *Sci. Am.* **217** (5), 62 (Nov. 1967).

Butler, P. J. G., and Klug, A., The assembly of a virus. *Sci. Am.* **239** (5), 62 (Nov. 1978).

Campbell, A. M., How viruses insert their DNA into the DNA of the host cell. *Sci. Am.* **235** (6), 102 (Dec. 1976).

Coslett, V. E., The early years of electron microscopy in biology. *Trends in Biochem Sci.* **101** (9), 361 (1985).

deDuve, C., The lysosome. *Sci. Am.* **208** (5), 64 (May 1973).

de Duve, C., Microbodies in the living cell. *Sci. Am.* **248** (5), 74 (May 1983).

Diener, T. O., Viroids. *Sci. Am.* **244** (1), 66 (Jan. 1981).

Fiddes, J. C., The nucleotide sequence of viral DNA. *Sci. Am.* **237** (6), 54 (Dec. 1977).

Fox. C. F., The structure of cell membranes. *Sci. Am.* **226** (2), 30 (Feb. 1972).

Lake, J. A., The ribosome. *Sci. Am.* **245** (2), 84 (Aug. 1981).

Marx, J. L., Organizing the cytoplasm. *Science* **222**, 1109 (1983).

Morowitz, H. J., and Tourtellotte, M. E., The smallest living cells. *Sci. Am.* **206** (3), 30 (Mar. 1962).

Neutra, M., and Leblond, C. P., The Golgi apparatus. *Sci. Am.* **220** (2), 100 (Feb. 1969).

Nomura, M., The control of ribosome synthesis. *Sci. Am.* **250** (1), 102 (Jan. 1984).

Nomura, M., Ribosomes. *Sci. Am.* **221** (4), 28 (Oct. 1969).

Orci, L., Matter, A., and Rouiller, C., A comparative study of freeze-etch replicas and thin sections of rat liver. *J. Ultr. Research* **35**, (1971).

Osborn, M., and Weber, K., The display of microtubules in transformed cells. *Cell* **12**, 561 (1977).

Pangborn, J., Kuhn, D. A,, and Woods, J. R., Dorsal–ventral differentiation in *Simonsiella* and other aspects of its morphology and ultrastructure. *Arch. Microbiol.* **113**, 197 (1977).

Porter, K. R., The cytomatrix: A short history of its study. *J. Cell Biol.* **99** (1), 3s (1984).

Prusiner, S. B., Prions, *Sci. Am.* **251** (4), 50 (Oct. 1984).

Racker, E., The membrane of the mitochondrion. *Sci. Am.* **218** (2), 32 (Feb. 1968).

Ravazzola, M., and Orci, L., Intercellular juntions in the rat parathyroid gland: A freeze-fracture study. *Rev. Cellulaire* **28**, 137 (1977).

Riesner, D., and Gross, H. J. Viroids. *Ann. Rev.* Biochem. **54**, 531 (1985).

Rothman, J. E., The Golgi apparatus: Two organelles in tandem. *Science* **213**, 1212 (1981).

Rothman, J. E., The compartmental organization of the Golgi apparatus. *Sci. Am.* **253** (3), 74 (Nov. 1985).

Satir, P., Cilia. *Sci. Am.* **204** (2), 108 (Feb. 1961).

Satir, B., The final steps in secretion. *Sci. Am.* **233** (4), 28 (Oct. 1975).

Schopf, J.W., The evolution of the earliest cells. *Sci. Am.* **239** (3), 110 (Sept. 1978).

Sharon, N., The bacterial cell wall. *Sci. Am.* **220** (5), 92 (May 1969).

Simons, K., Garof, H., and Helenius, A., How an animal virus gets into and out of its host cell. *Sci. Am.* **246** (2), 58 (Feb. 1982).

Staehelin, L. A., and Hull, B. E., Junctions between living cells. *Sci. Am.* **238** (5), 140 (May 1978).

Stolinski, C., Freeze-fracture replication in biological research: Development, current practice and future prospects. *Micron* **8**, 87 (1977).

Tanaka, K., Scanning electron microscopy of intracellular organelles. *Int. Rev. Cytol.* **68**, 97 (1980).

Taylor, D. L., and Wang, Y., Fluorescent labelled molecules as probes of the structure and function of living cells. *Nature* **284**, 405 (1970).

Woese, C. R., Archaeobacteria. *Sci. Am.* **244** (6), 98 (June 1981).

Books, Monographs, and Symposia

Bodenheimer, F. S., *The History of Biology*, Wm. Dawson and Sons, Ltd., London, 1958.

Boyd, R. F., *General Microbiology*. Times Mirror/Mosby College Publishing, St. Louis, 1984.

Brock, T. D., Smith, D. W., and Madigan, M. T., *Biology of Microorganisms* (4th ed.), Prentice–Hall, Englewood Cliffs, N.J., 1984.

Dean, R. T., *Lysosomes*, Camelot Press, Ltd., Southampton, U.K., 1977.

Dingle, J. T., *Lysosomes*, North-Holland, Amsterdam, 1977.

Echlin, P., The blue-green algae. In *Cellular and Organismal Biology* (D. Kennedy, Ed.), Freeman, San Francisco, 1974.

Fawcett, D. W., *Anatomy of Fine Structure*, Saunders, Philadelphia, 1966.

Florkin M., and Stotz, E. H. (Eds.), *Comprehensive Biochemistry*, Vol. 3, *A History of Biochemistry*. Part IV. *Early Studies on Biosynthesis*, Elsevier, Amsterdam, 1977.

Gardner, E. L., *History of Biology*, Burgess Publishing Co., Minneapolis, 1965.

Griffith, J. (Ed.), *Electron Microscopy in Biology*, Vols. 1 and 2, Wiley–Interscience, New York, 1982.

Hall, J. L., Flowers, T. J., and Roberts, R. M., *Plant Cell Structure and Metabolism* (2nd ed.), Longman Publishers, London, 1981.

Huang, A. H. C., *Peroxisomes*, Academic, New York, 1983.

Luria, S. E., Darnell, J. E., Baltimore D., and Campbell, A., *General Virology* (3rd ed.), Wiley, New York, 1978.

Malhotra, S. K., *The Plasma Membrane*, Wiley-Interscience, New York, 1983.

Nordenskiold, E., *The History of Biology*, Tudor Publishing Co., New York, 1928.

Orci, L., and Perrelet, A., *Freeze-Etch Histology*, Springer-Verlag, New York, 1977.

Reinert, J. (Ed.), *Chloroplasts*, Springer–Verlag, Berlin, 1980.

Singer, C. J., *A History of Biology*, Abelard–Schuman, London, 1959.

Tedeschi, H., *Mitochondria: Structure, Biogenesis and Transducing Functions*, Springer-Verlag, New York, 1976.

Waley, W. G., *The Golgi Apparatus*, Springer–Verlag, New York, 1975.

Watt, I. M., *The Principles and Practice of Electron Microscopy*, Cambridge University Press, Cambridge, England, 1985.

CHAPTER 2

CELL GROWTH AND PROLIFERATION

It has been estimated that the various tissues and organs of the adult human contain more than 100,000,000,000,000 cells! All these cells are derived initially from a single cell (i.e., the fertilized egg cell) through *growth* and *division*. In many tissues (e.g., the epithelial lining of the digestive organs and the blood-forming tissues), the cells continue to grow and divide for most of the person's life; however, in some tissues, such as muscle and nerve, cell division ceases some time after birth and subsequent tissue growth results from individual cell growth without division. The ongoing cell growth and division that characterizes most of the body's tissues accounts for the growth of the organism as a whole and the replacement of dying cells. With the exceptions of muscle and nerve, replacement of old tissue cells with new ones results in a complete turnover of the body's cellular composition every few years. A portion of the body's total mass is represented by noncellular materials that are secreted by cells; for example, most of the mass of bone and cartilage is represented by secreted calcium salts and proteins.

Other animals and plants also grow by the processes of cellular division followed by individual cell growth. When the term "growth" is applied to microbial organisms such as bacteria, it can refer either to an increase in the size of the individual organism (e.g., the individual bacterial cell) or to an increase in the numbers of organisms (e.g., in a culture of bacteria).

If microorganisms such as bacteria or protozoa are placed in an appropriate nutrient medium, they may grow and divide until the medium is teeming with these cells. Such a population is called a **cell culture.** Similarly, cells that are removed from certain tissues of higher animals and plants can be made to grow and divide in an artificial nutrient medium, thereby forming a **tissue culture.** In most cultures of microorganisms and in some tissue cultures (e.g., cultures of white blood cells), the cells undergo optimum growth when they are suspended in the nutrient solution; such cultures are called *suspension* or *anchorage-independent* cell cultures. In most tissue cultures, however, the cells only grow when they are attached to a surface; these are known as *anchorage-dependent* cell cultures. Occasionally, *transformed* or *neoplastic* tissue cells (e.g., tumor cells) can be adapted to grow in suspension. In theory, at least, cells from virtually any tissue can be cultured, but in practice the degree of success varies widely with the organism and tissue that is used. Nevertheless, it is fair to say that much of our present understanding of the regulation, kinetics, and mechanics of cell growth and division has been obtained from studies using cultures of microorganisms and tissue cells, and cell-culturing techniques have become increasingly important research tools of the cell biologist.

THE POPULATION GROWTH CURVE

Mathematical models can be derived that describe the manner in which the number of cells in a population increases. Consider as an example the following hypothetical situation: Suppose that we begin with a culture medium containing a single cell that grows for some period of time and then divides to produce two daughter cells; in turn, these two cells grow and after an *identical* period of time, they divide to produce four cells, and so on. In such a situation, the numbers of cells present in the population would increase *exponentially* in the following way: 1, 2, 4, 8, 16, 32, 64 . . . (Fig. 2–1a). In other words, the population of cells would double in number with each generation. Consequently, after any specific length of time (say, from t_1 to t_2), the number of cells in the population would be given by the equation.

$$N_2 = N_1 \times 2^g \qquad (2\text{--}1)$$

where N_1 is the original number of cells present at time t_1, N_2 is the number of cells present at t_2, and g is the number of generations that have occurred during the time interval $t_2 - t_1$. Figures 2–1a and 2–1b depict

graphically the exponential nature of the increase in population size during unrestrained growth.

The situation described above is ideal and does not occur in nature (except perhaps for a limited time following the fertilization of an egg cell), but it can be approached artificially under conditions of **synchronous cell growth**; this will be considered later in the chapter. If a typical expanding population of cells is examined at any instant in time, some cells would be observed to be dividing, others would just have completed division, still others would be preparing to divide, and so on. Divisions of all cells present in the population would not occur during precisely the same interval of time.

Exponential Growth

If a large number of cells are cultured together in what is called a "random" culture, the individual cells will be found in various stages of their **growth–division cycle** or **cell cycle**. At any moment, the rate at which the number of cells in the culture increases is directly proportional to the number of cells present at that time. This, of course, presumes a steady state in which the needed nutrients are always available in adequate supply and in which cellular waste products excreted into the cells' environment do not interfere with the maintenance of normal growth and division. The growth of such a random culture is described by the differential equation

$$dN/dt = kN \qquad (2\text{--}2)$$

FIGURE 2–1 (*a*) A geometric expansion of cell numbers beginning with one, plotted in linear terms as a function of time. (*b*) The same geometric expansion plotted in semilog terms. Note that the latter plot is a straight line.

where N is the number of cells present at time t, dN/dt is the change in cell number with time, and k is a **growth constant** that is specific for the population. This equation may be solved by integreation and yields the expression

$$2.3 \log_{10}(N_2/N_1) = k(t_2 - t_1) \qquad (2\text{-}3)$$

If we let N_1 equal the number of cells present in the population at time t_1 and N_2 equal the number of cells present at time t_2, then equation 2-2 is solved by integration as follows.

$$\int_{N=N_1}^{N=N_2} dN/N = k \int_{t=t_1}^{t=t_2} dt$$

Thus, $\ln N_2 - \ln N_1 = k(t_2 - t_1)$
or
$\ln (N_2/N_1) = k(t_2 - t_1)$
By converting to the more familiar base 10, the last equation takes the form of equation 2-3.

Equation 2-3 indicates that the growth of the population (i.e., the rate at which the number of cells in the population increases) is *exponential*.

Doubling Time

Although the number of cells in a population increases exponentially with time, different types of cell populations (i.e., different species of microorganisms or cells from different tissues) grow at different rates. Even populations of the same type of cell may grow at different exponential rates if the temperature, nutrients, or other growth conditions vary. Differences in growth rates are reflected by differences in the value of the growth constant, k, in equation 2-3. A convenient value that expresses the specific rate of growth of a population of cells under a specified set of conditions is the **doubling time**. The doubling time is defined as the time required for the number of cells in the population to **double** during exponential growth. Many cell biologists use the term **generation time** interchangeably with doubling time. Strictly speaking, however, the generation time is the interval of time between any point in one **cell cycle** (see later) and the same point in the *next* cell cycle.

Both the growth constant and the doubling time are specific to a particular cell culture. Although the growth *rate* (i.e., kN) increases as the population of cells gets larger, the doubling time (and, of course, the

FIGURE 2-2 A hypothetical exponential increase in cell numbers. The doubling time (see text) may be determined by the substitution of values for t and N into equation 2-3.

value of k) remains the same. An equation for the doubling time may be derived as follows.

After a time interval equal to the doubling time has elapsed, the ratio N_2/N_1 is equal to 2; therefore, from equation 2-3

$$2.3 \log 2 = kT \qquad (2\text{-}4)$$

where T is the doubling time, $t_2 - t_1$. Hence,

$$0.693 = kT \qquad (2\text{-}5)$$

and

$$T = 0.693/k \qquad (2\text{-}6)$$

The actual value of k or T may easily be determined when experimental data are used to make a semilogarithmic plot of cell number versus time (Fig. 2-2).

Sample Problem Suppose that at time t_1, the number of cells in a population (i.e., N_1) is 62,400 and at time t_2, 18.5 hours later, there are 473,000 cells. What is the doubling time for this population of cells?

From equation 2-3,

$$2.3 \log (473,000/62,400) = k (18.5)$$
$$2.3 \log 7.58 = 18.5k$$
$$k = 0.1094$$

The dimensions of k in this instance are hr^{-1}; that is, the number of cells in the population increases by 10.94% per hour. Now from equation 2-6,

$$T = 0.693/0.1094 \ hr^{-1} = 6.33 \ hr$$

Therefore, during exponential growth, the number of cells in the population doubles every 6.33 hours.

The Lag Phase of Growth

Typically, when cells are placed in a nutrient medium that favors their growth and proliferation, exponential growth of the population does not begin immediately. Instead, there is a short interval of time in which there is little or no increase in the number of cells in the population. This time interval preceding exponential growth is known as the **lag phase** (Fig. 2–3). The length of the lag phase is quite variable, even when comparing different cultures of the same type of cell. A number of factors are believed to influence the length of the lag phase of the population growth curve.

Experiments with bacteria and other microorganisms have shown that variations in the concentrations of certain constituents of the growth medium, such as carbon dioxide, and certain cations, such as H^+(i.e., pH), markedly influence the length of the lag phase. Therefore, the chemical composition of the nutrient medium influences the time interval that precedes the onset of exponential population growth.

The cells that are used to "seed" a new culture are acquired from a previous culture that was at some particular stage of its growth curve. The cells used to start a new culture are referred to as the **inoculum**. The stage of the parent culture used to provide the inoculum influences the length of the lag phase. For example, the lag phase of cultures of the bacterium *Aerobacter aerogenes* is longer when the inoculum is drawn from a previous culture that was in *early* exponential growth and shorter when drawn from a culture that was in *late* exponential growth. For the protozoan *Paramecium caudatum,* little or no lag period is observed when the inoculum consists of cells that had been growing exponentially; however, when the inoculum consists of cells from the **stationary phase** (see below), a lag period is observed.

Generally, the greater the number of cells in the inoculum, the shorter is the lag period. It has been suggested that this is due to the more rapid accumulation in the nutrient medium of a cellular secretory substance that is required during exponential growth and that reaches a "threshold" concentration earlier when there is a large number of initial cells.

The Stationary Phase

If cells are growing in a medium whose size and contents are initially fixed, then it is clear that growth cannot continue indefinitely. Nutrients in the medium will eventually be depleted, and potentially harmful metabolic waste products excreted into the medium by the cells will accumulate in high concentration. Consequently, the cell population eventually reaches some limiting size, and following this the number of cells in the population no longer increases. It should be noted that the attainment of a maximum population size does not necessarily mean that cells are no longer growing and dividing, but rather that any additional cells produced by division are balanced by the death and disruption of other cells. The interval of time in which the number of cells in the population ceases to increase and remains fairly constant is known as the **quiescent** or **stationary phase** (Fig. 2–3).

FIGURE 2–3 Various phases of the population growth curve of a typical random culture of cells.

The Death or Declining Phase

Unless the cells of stationary phase cultures are transferred to fresh medium (or fresh medium is added to the culture), stationary cultures eventually enter a **death** or **declining phase**. In this phase, the number of cells lost by death and disruption exceeds the number being produced by cell division. As a result, the number of cells in the culture rapidly diminishes (Fig. 2–3).

SENESCENCE

Although the onset of a declining phase will be triggered by the exhaustion of nutrients in the medium or the accumulation of cellular excretory products, declining phases are ultimately observed even in cultures in which the growth medium is replaced with fresh medium. Such observations have led to the notion that "programmed" into each type of cell is the capacity for a *limited* and *specific* number of divisions. For example, cultures of fibroblasts from human embryos will undergo about 50 divisions, after which the cells become *senescent*; that is, they slowly stop dividing and eventually die. If the fibroblasts are frozen and stored in liquid nitrogen for months (or even years) after having undergone 20 generations of growth, and later these cells are thawed and cultured again, they will resume division and growth for another 30 generations before becoming senescent. In other words, the sum of the population doublings before and after freezing is still 50. It is as though the cells had a built-in clock or calculator that kept track of the number of divisions. Similar finite life spans have been observed when liver cells, nerve cells, skin cells, and smooth muscle cells are repeatedly subcultured. It is widely believed that aging of the body's tissues and the death of the individual that eventually ensues may be related to the cellular senescence that characterizes cell and tissue cultures.

Although most tissue culture lines may be propagated for a limited number of generations, beyond which they become senescent, occasionally subculturing eventually produces a *continuous cell line*—that is, cells that may be indefinitely propagated. This *in vitro* **transformation** may be induced chemically (i.e., by the addition of "carcinogens"), by radiation (X rays or ultraviolet light), through viral infection of the cells, or may occur "spontaneously" within a culture (that is to say, the agent remains unidentified). Tumor-forming cancer cells are examples of cells that fail to exhibit the senescence that characterizes most cultures of tissue cells.

In vivo, all cells of a tissue usually exhibit the same ploidy, that is, the cell nuclei have the same number of complete sets of chromosomes. Typically, human tissue cells are *diploid,* having *two* complete genomes apportioned among 23 pairs of chromosomes. In *in vitro* cultures of normal (i.e., untransformed) cells, the ploidy is more variable, but the cells are *euploid,* that is, their nuclei have exact multiples of the normal genome. Some cells may be diploid and others *tetraploid* (having four complete genomes), etc. Continuous lines of transformed cells are characteristically *aneuploid*, having inexact multiples of the normal genome.

CONFLUENCE AND DENSITY-DEPENDENT INHIBITION

A number of cell types, especially those from mammalian tissues, are more readily cultured if they are allowed to grow on some sort of surface, that is, the cells are anchorage-dependent. The exponential increase in cell numbers generally halts when the population density is high enough for the cells to attain a state of *confluence,* that is, when they physically come in contact with one another along their edges. After confluence, the cells enter the stationary phase, growth is markedly reduced, and cell motility and *ruffling* of the plasma membrane (Chapter 15) diminish. Because the cells will not grow on top of one another, they form a *monolayer.*

Originally, the halt to cell proliferation that occurs after confluence is attained was referred to as **contact inhibition**, inferring that it was the contact of the surfaces of neighboring cells that heralded the onset of the stationary phase. It is now recognized that entry into the stationary phase is due not only to contact but also to the depletion of nutrients and growth factors in the medium. Thus, the term **density-dependent inhibition** (or **topoinhibition**) is preferable. "Contact inhibition" more appropriately applies to changes in the culture that are *directly* dependent on cell–cell contact, such as the inhibition of cellular motility or membrane ruffling.

If, after confluence has halted the growth of a population, a strip of cells is scratched away from the monolayer, cells at the edges of the "wound" will divide until the exposed substratum is once again covered. However, cells located some distance from the wound remain density inhibited. This observation supports the idea that the control of growth is exercised locally and is not a function of the entire culture.

GROWTH CHARACTERISTICS OF TRANSFORMED CELLS

Sometimes eukaryotic cells exhibit unusual and alternative growth characteristics both in culture and *in vivo.* Such cells are usually referred to as *transformed* or *neoplastic* cells. The transformation of some cells may be *induced* using viruses. For example, when cultures of mammalian cells are exposed to Polyoma virus or Simian virus 40 (SV 40), some of the cells in the culture are transformed into a cell line that manifests a number of unusual characteristics, including (1) the cells continue to grow and divide even after confluence is attained (with the result that they grow on top of one another instead of maintaining a monolayer), (2) the cells are no longer anchorage-dependent (thereby permitting growth in multilayers), (3) the cells grow to an indefinite population size, (4) growth continues for many more generations than in the corresponding untransformed cells (i.e., the cells appear to ignore or to have lost the built-in counting mechanism keeping track of the number of cell doublings), and (5) the cells grow irregularly in size and in shape and often reveal an abnormal chromosome content. One of the most famous (and useful) examples is the "HeLa" line of transformed cells originally derived in 1951 from a cancerous cervical tumor and that has been continuously subcultured since that time. (HeLa is an acronym for Henrietta Lacks, the woman from whose uterus the cervical cells were removed.)

The relationship that exists between cellular transformation and cancer should be apparent; indeed, when transformed cells are injected into laboratory animals, the cells form tumors or neoplasms, whereas normal cells do not. In tissue culture, transformed cells do not form conventional junctional complexes with neighboring cells. Hence, it is easy to understand why, when cellular transformation occurs *in vivo,* the transformed cells **metastasize**, that is, spread to other body tissues where they form new tumors, growing and crowding out the tissue's normal cells.

THE QUANTITATION OF CELLS

Fundamental to any quantitative study of the kinetics of cell population growth and to the accurate measurement of changes in the cell's physical or chemical properties are methods for determining either the mass or the number of cells present in the population. A variety of techniques are available to do this; among those that are easily used are estimations of cell mass based on the total fresh (i.e., wet) weight or dry weight of the sample and estimation of cell numbers from measurements of the turbidity of the cell suspen-

sion. However, these methods lack the level of accuracy that is generally considered necessary for quantitative studies. More frequently, the precise number of cells present in the sample is determined by the direct *optical* examination of an aliquot of cell suspension drawn from the culture, by *electronic gating,* or by *flow cytometry.* Although the latter two procedures are far more accurate than weight determinations, turbidity methods, or optical enumeration, they require that the cells occur separately and, therefore, cannot be applied directly to tissues.

Optical Enumeration of Cells

The direct microscopic enumeration of cells requires a glass counting chamber (often referred to as a **hemacytometer** because of its historical and widespread use for counting various kinds of *blood cells*). This device consists of a thick glass microscope slide, the surface of which has been etched with a number of squares of known dimensions. A drop of cell suspension is placed between the counting surface and an overlying cover slip, and because the distance between the undersurface of the cover slip and the surface of the counting area is known, the total number of cells present in a known volume of sample can be determined. Then, by extrapolation, the number of cells in the entire cell culture (or in any portion of it) can easily be calculated. The enumeration of cells by direct optical examination is tedious but extremely useful, especially if the suspensions contain different types of cells (as in the case of blood) that may be distinguished visually. Hemacytometric methods are still used in many laboratories.

ELECTRONIC ENUMERATION OF CELLS

Electronic Gating

Electronic gating and **flow cytometry** are rapid procedures for enumerating cells. Among the most commonly used instruments applying the "gating" approach is the Coulter Counter (after the inventor, Wallace Coulter), the main features of which are shown in Figure 2–4. A glass probe with a tiny aperture near its base is immersed in a sample of cell suspension containing electrolyte (usually a physiological salt solution); the aperture tube is filled with the same electrolyte. Electrical current is caused to flow from a platinum electrode inside the probe through the small

FIGURE 2-4 Basic components of an electronic cell-gating instrument such as the Coulter Counter (see text for detailed description of the instrument's operation).

aperture to a second electrode submerged in the cell suspension. At the same time, some of the cell suspension is drawn through the aperture and into the probe by a pump. The passage of a cell through the aperture temporarily alters the flow of electrical current through the aperture, because the cell displaces a volume of the electrolyte and is generally not as conductive as the electrolyte. The resulting sudden drop in voltage is recorded by the instrument as a pulse (which is displayed on an associated oscilloscope) and is counted by a *scaler*. A mercury manometer connected to the probe (and therefore also to the pump) operates two switches through two wire electrodes that are inserted through the manometer along its length. With the pump on and a vacuum applied to the aperture tube, not only is the cell suspension drawn through the aperture but the mercury in one arm of the manometer is also drawn downward below the "start" electrode. When the vacuum is halted (by closing the stopcock), the mercury automatically rises in the arm of the manometer, turning on the scaler as it passes the "start" electrode and turning off the scaler as it passes the "stop" electrode. With the stopcock closed, the upward movement of mercury in the arm

of the manometer draws cell suspension through the aperture, so that the cells present in a volume of suspension precisely equal to the volume of mercury between the "start" and "stop" electrodes are enumerated. Because this volume is known (i.e., the manometer is accurately calibrated), the number of cells present per unit volume of suspension is readily determined. With the electronic cell counter it is possible to count thousands of cells in just a few seconds. However, if the suspension contains different types of cells, those that are of similar size cannot be distinguished (and separately counted) by this procedure. In such an instance, it may be possible to use chemical methods to eliminate certain types of cells. For example, during a clinical measurement of the white cell count in a blood sample, it is customary to eliminate the red blood cells by adding a few drops of a specific lysing agent to the cell suspension.

The magnitude of each pulse produced by the electronic counter is directly proportional to the size (i.e., volume) of the cell passing through the aperture. The instrument's two pulse-height threshold controls (upper and lower) may be separately adjusted so that only those pulses whose peak height falls above the lower threshold and below the upper threshold will be counted. Consequently, for a cell suspension in which cell sizes might vary considerably, one can adjust the thresholds to determine only the number of those cells within a selected size range. By simultaneously increasing both the lower and upper threshold values by the same increment, it is possible to obtain a pulse-height distribution for the sample of cells. This pulse-height distribution is analogous to a cell-size distribution and is especially useful to cell biologists who are studying the growth characteristics of individual cells, for it reveals for any selected phase of a population growth curve the relative numbers of cells of all sizes (and therefore ages) present in the population. Thus, with an electronic cell-gating instrument it is possible not only to rapidly enumerate the cells in the population but also to obtain measurements of the sizes of the cells. It would be difficult to overstate the value of this instrument to the cell biologist.

Flow Cytometry

The use of flow cytometry for rapidly counting and analyzing cells was developed by M. J. Fulwyler and L. A. Herzenberg. Typically, the cells to be analyzed are first stained with a *fluorescent dye* (e.g., ethidium bromide, which combines with the nuclear DNA) or treated with a *fluorescent antibody* that binds to proteins in the cell surface. The cell suspension is then pumped through a small orifice and into the center of a narrow, horizontal channel in the "flow cell assembly" (Fig. 2–5). Pumped into the same channel through a second small opening at one end is "sheath fluid," which serves to sweep the cell suspension away from the measuring orifice and out of the flow cell assembly through a third opening. A beam of light from a laser is directed at the measuring orifice where it excites the fluorescent material in each emerging cell causing the cell to emit a flash of fluorescent light. The fluorescence is detected by a system of photomultiplier tubes that relay electrical signals to the instrument's analyzing system. The number of fluorescent light pulses detected by the photomultiplier tubes is directly related to the number of cells passing the measuring orifice. Moreover, the magnitude of each pulse varies in proportion to the quantity of fluorescent material associated with each cell, and this provides still another measurement parameter. For example, if the fluorescent pulses are produced by the cell's DNA, then not only are the cells counted, but a measurement is also made of each cell's DNA content. Thus, one can readily follow the changes that take place in the amount of DNA per cell during the cell cycle. Using flow cytometry, thousands of cells can be enumerated and also analyzed in just a few minutes.

THE CONTINUOUS CULTURE OF CELLS

When cells are cultured in a container of fixed volume, both the population density (i.e., the number of cells per unit volume of culture) and the population size

FIGURE 2–5 Basic components of a flow cytometer using fluorescence as a parameter of measurement (see text for details).

Large reservoir of
sterile culture medium

Inoculation
port

Filtered
air
inlet

Pump

Pump

Harvest/
exhaust
line

Culture chamber

pH electrode

Cells

Mixer

FIGURE 2–6 Simplified diagram of a chemostat. Sterile, cell-free culture medium enters the culture chamber at the same rate as medium containing cells is withdrawn. The population density of the culture remains constant and is determined by the rate at which a substrate or nutrient required for growth is added. Provision is made for monitoring and regulating the temperature and pH of the culture and for sampling the medium. Although the basic concept is represented by this diagram, most laboratory chemostats are considerably more complex.

(i.e., the total number of cells in the culture) increase at the same rate during the exponential phase of growth. If we consider a "biological unit" in such a culture as a cell plus some volume of the surrounding medium, then this unit undergoes continuous change as the population grows. The medium surrounding each cell becomes successively depleted of nutrients, while metabolic waste products released by the cells accumulate in increasing amounts in the medium, and the cell itself may be increasing (or decreasing) in size. In multicellular organisms like man, the cells of the various tissues and organs are bathed in a body fluid that continuously provides fresh nutrients while carrying away waste products. In other words, *in vivo* a cell's surroundings stay more or less constant. This condition differs markedly from that of a random cell culture but can be approximated experimentally by methods of **continuous culture** in which an effort is made to keep the biological unit constant.

A variety of techniques can be employed for the continuous culture of cells. One popular method involves the use of an instrument known as a **chemostat** (Fig. 2–6), introduced by A. Novick and L. Szilard. In the chemostat, fresh, sterile culture medium is contin-

uously pumped into the culture chamber, while at the same time an equal volume of old medium and some of the cells are removed. Thus the total volume of the cell chamber does not change. The rate of growth of the cell population is regulated by the rate at which a chemical substrate essential for cell growth is added to the medium, and this is adjusted so that the population size and population density remain constant. Because cellular waste products are continuously diluted by the new medium added to the chamber, these substances do not attain toxic levels. With the chemostat it is possible to maintain the cell population in the exponential phase of growth for a greatly extended period of time. In certain continuous culture devices, constant cell population density is maintained by monitoring the turbidity of the suspension and appropriately adjusting the rates at which medium flows into and out of the culture chamber. These devices are known as **turbidostats**. Another method for keeping a volume of medium surrounding each cell chemically constant involves the inoculation of small numbers of cells into very large volumes of growth medium. In this way, the growth and proliferation of the cells will have little influence on the chemical composition of the medium for many generations. The continuous culture of cells offers numerous advantages over growth in fixed volumes, because a balanced, steady-state growth rate can be maintained and physiological studies can be carried out with cells that are little influenced by fluctuations in their environment.

THE GROWTH OF INDIVIDUAL CELLS: THE CELL CYCLE

We have already seen that it is possible to determine the distribution of cell sizes (i.e., volumes) in a random population of cells. Because cell size is related to cell age, an age distribution is also obtained. If cells were produced in a manner like the commercial production of small spherical objects (such as ball bearings or marbles), then their sizes would be distributed normally about some mean size. However, in a random, exponentially expanding population of cells, the distribution of size or age is not normal. Instead, a distribution exists in which there are more young (small) cells than old (large) cells. In an ideal population (i.e., one in which a cell divides equally into two daughter cells, each of which grows to the size of the parent cell before dividing), the age or size distribution takes the form of the solid red curve seen in Figure 2–7. This

FIGURE 2–7 Theoretical distribution of cell ages in an ideal population of growing cells (solid curve). The areas under the curve represent the relative numbers of cells in each of the four phases of the cell cycle at any point in time. In a real population, some variation in cell cycle time is observed among the cells. The resulting deviation from the theoretical age distribution is shown by the broken curve.

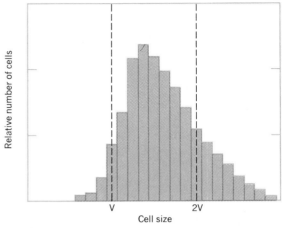

FIGURE 2–8 A typical distribution of cell sizes observed in a random culture of cells. The distribution is not "normal" but is skewed toward larger cell size. V and $2V$ are the initial and final sizes (i.e., volumes) of an ideal cell whose size doubles before equal-division into two daughter cells occurs.

curve indicates that there is an *exponential decrease* in the relative number of cells with age or size. This is so because the absolute number of cells entering division is constantly increasing in an exponential manner, so that each new generation of cells is exponentially greater in number than the one that preceded it. It should be emphasized that the curve of Figure 2–7 is derived mathematically and applies the assumption that the cells have identical doubling times.

The distribution described above probably does not exist in nature, because cells display variations in individual generation times and do not divide into progeny of exactly equal volume. Thus, the distribution of cell size ranges beyond V and $2V$ (where $2V$ is the volume of the ideal parent cell and V is the volume of each daughter cell). Furthermore, it has been shown that the small cells produced by unequal or early division do not, in turn, give rise to smaller progeny; instead, there is a random fluctuation about some inherited average size. A size distribution typical for a random population of cells is shown in Figure 2–8.

Although several possibilities exist for the rate of cell growth between divisions, most studies of cell volume distributions in either random or "synchronized" cultures (see below) indicate that cells grow either linearly or exponentially. For example, individual *Tetrahymena pyriformis* cells grow exponentially in volume between divisions, whereas *Chlorella ellipsoidea* cells grow linearly. In many cell cultures, the cell volume distribution remains constant throughout the exponential phase of the population growth curve, but in some cultures (such as cultures of the flagellate *Polytomella agilis* and the yeast *Schizosaccharomyces pombe*) the cell volume distribution shifts toward smaller size during the exponential phase. The latter observation indicates that in some cases the growth of individual cells occurs less rapidly than the growth (in numbers of cells) of the entire population.

The sizes attained by cells in culture also depend on other factors, such as the temperature at which the cells are maintained. In *T. pyriformis*, average cell size increases with increasing culture temperature between 28° and 34°C. In *P. agilis*, cells cultured at lower temperatures have a greater average size than cells cultured at higher temperatures (in the range 9° to 25°C). It is apparent that, depending on the condi-

tions of the culture, the events that result in cell division may occur either more or less rapidly than the events that result in individual cell growth.

PHASES OF THE CELL CYCLE

The various phases of the growth and reproduction of cells constitute what is called the **cell cycle** (Fig. 2–9), a complete cycle taking place in one *generation time*. The principal signposts of the cell cycle are the replication of the nuclear DNA and its distribution among the progeny cells. The replication of the nuclear DNA occurs in that portion of the cell cycle known as the **interphase**. This was definitively shown for the first time in the classic experiments of A. Howard and S. Pelc in 1953. Using the technique of **autoradiography**

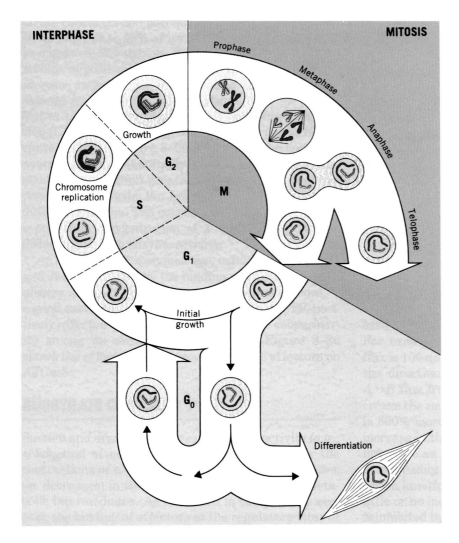

FIGURE 2–9 G_1, S, G_2, and M phases of the cell cycle in animal and plant cells and the G_0 phase of noncycling cells. In the diagram, the length of the M phase is deliberately exaggerated to show mitosis and cytokinesis.

(see Chapter 14), Howard and Pelc found that the incorporation of radioactive phosphorous into newly synthesized DNA occurred only during a short time interval within the interphase. The distribution of the replicated DNA and the physical division of the parent cell into two daughter cells occur in the "period of division" referred to as the M phase. The symbol M stands for "mitosis," the process of nuclear division familiar to even the beginning biology student and consisting of the **prophase**, **metaphase**, **anaphase**, and **telophase**. The duration of the cell cycle varies according to cell type and whether the cells are growing *in vivo* or *in vitro*. The *in vivo* cell cycle of most higher animals and plants is about 10 to 25 hours, of which only about one hour is spent in the M phase. In mammalian cell cultures, the range is somewhat broader (i.e., about 8 to 30 hours). The cell cycles of prokaryotic cells, which lack a nucleus and whose DNA is usually considerably smaller in content and is differently organized, are generally much shorter. For example, in *Benekea natriegens*, the cell cycle is completed in 10 minutes.

In eukaryotic cells, the replication of nuclear DNA occurs in a specific portion of the interphase called the S phase (S for "synthesis"); this phase typically lasts six to eight hours. Usually several hours elapse between the conclusion of the M phase of the parent cell and the onset of the S phase in the two daughter cells. This interval is known as the G_1 phase (G for "gap"). Usually, completion of the S phase is not immediately followed by the M phase. The gap between the end of the S phase and the onset of the M phase is called the G_2 phase and also lasts several hours. The G_1 phase of the cell cycle is characterized by the synthesis and accumulation of RNA and protein in the cytoplasm and is accompanied by considerable cell growth. Much of the protein synthesized during the S phase is *histone* synthesis; these histones enter the cell nucleus, where they become closely associated with the newly synthesized DNA. By the end of the S phase, the cell nucleus contains two full complements of genetic material; the cell then enters the G_2 phase, in which some cytoplasmic growth resumes and preparations are made for nuclear division.

The lengths of the G_1, S, G_2, and M phases vary among different types of cells, but variations between cells of the same type are usually very small. The greatest variation is seen in the G_1 phase. When the cell cycle is very long, most of the prolongation can be accounted for by a lengthened G_1 phase. In contrast, in egg cells, where the cell cycle is very short, the G_1

phase is very brief (or even nonexistent). The cells of animal embryos undergo a rapid sequence of divisions with very little intervening cell growth. The cell cycle may last only an hour with no G_1 phase and with some DNA synthesis occurring during mitosis or soon after mitosis is completed. Thus, the overall length of the cell cycle depends primarily on the length of the G_1 phase. The four areas under the solid curve of Figure 2–7 show for a random population the numbers of cells in each of the four phases of the cell cycle.

During the M phase, the nuclear contents undergo a series of changes and rearrangements. Prior to the onset of mitosis, the DNA is dispersed throughout the nucleus, but as the prophase of mitosis commences, the nuclear material condenses to form the clearly visible **chromosomes**. The dramatic nature of this condensation process can be appreciated when one considers that in a human cell, the DNA, which is 10–15 feet long in its dispersed state, condenses to form 46 duplicated chromosomes whose combined length is less that $1/25$ of an inch. In the remaining phases of mitosis, the duplicated chromosomes are distributed to the daughter cells, leading to cell division or **cytokinesis** itself. These events are clearly depicted in the excellent interference optics photomicrographs of A. S. Bajer, reproduced in Figure 20–21. Mitosis is considered further in Chapter 20, which deals at length with the structure, organization, and functions of the cell nucleus. Additional discussion of this subject is therefore deferred until then.

Noncycling (Arrested) Cells

The cells of some tissues (e.g., liver tissue cells) may exist in a nongrowing, nondividing state for a long period of time. Such cells are said to have been *arrested* in the G_1 phase or to have withdrawn from the normal cell cycle and entered a separate phase called G_0 (Fig. 2–9). If a separate G_0 phase exists, it is entered from the G_1 phase. Noncycling or arrested cells may ultimately be "recruited" from the G_0 phase into the G_1 phase and then continue through the cell cycle, duplicating their DNA, carrying out mitosis, and dividing. For example, the surgical removal of a portion of the liver is followed by renewed growth and division of the remaining liver tissue, quickly replacing the tissue that had been excised. Prior to removal of some of the liver tissue, the cells were in a G_0 state; after excision of some of the liver tissue, cells were recruited from the G_0 state into the G_1 phase.

It has been suggested that a *restriction point*

(called R) or control mechanism of some sort exists in the G_1 phase. If a G_1 phase cell passes through the restriction point, it continues through the cell cycle. The restriction point or control mechanism is responsive to the cell's environment, acting to arrest the cell cycle or permit continued cycling according to signals received from outside the cell. Arrested cells enter the G_0 phase for what may be an extended period of time. Some cells may not return to the cell cycle after leaving the G_1 phase. Indeed, the differentiation of certain cells such as muscle and nerve permanently removes them from the cell cycle (see Fig. 2–9).

Finally, it should be noted that in considering the various phases of the cell cycle, it is important to bear in mind that the stages are a part of a continuous process and that the cycle's subdivision into the G_1, S, and G_2 phases of interphase and the prophase, metaphase, anaphase, and telophase of mitosis is a matter of descriptive convenience only.

SYNCHRONOUS CELL CULTURES

The cells that are growing in a random culture represent a heterogeneous collection at various stages of the growth–division cycle or cell cycle. Some cells are dividing, some have just completed division, some are about to begin division, and so on. Any study of the progressive changes in either the chemical composition of the cells or cell morphology is difficult, if not impossible, when the cells are randomly distributed with respect to age. This problem could be avoided by studying the cycle of individual cells, but cells are usually too small for individual examination and analysis. To resolve the problem, some clever methods have been devised in which the cells in a culture are brought to the same stage of the cell cycle. Consequently, the entire population can be studied as though it were a single cell. This condition is known as **synchronous cell growth**.

The degree of synchrony that exists in a cell culture can be assessed by following changes in the culture's **mitotic index**, which is a measure of the fraction of cells undergoing division at any instant and is given by the formula

$$I_M = N_M / N \qquad (2\text{–}7)$$

where I_M is the mitotic index, N_M is the total number of cells visibly in mitosis, and N is the total number of cells present.

During the exponential expansion of a cell culture,

FIGURE 2–10 A comparison of the growth curves for idealized *random* and *synchronous* cell cultures.

FIGURE 2–11 A comparison of mitotic indices during the growth of *random* and *synchronous* cell cultures.

I_M remains constant; however, during synchronous growth, I_M changes from some minimal value (ideally 0) to some maximum value (ideally 1.0) during a short time interval. The efficiency of the synchronization procedure can be evaluated from a comparison of (1) the mitotic index, (2) the time required for the population to double during synchronous division, and (3) the doubling time observed in a corresponding asynchronous (i.e., random) culture of the cells. A comparison of the growth curves for a random culture and a synchronous culture is shown in Figure 2–10, and the corresponding mitotic indices are shown in Figure 2–11.

A variety of methods have been devised for inducing the synchronous growth of a cell culture; these fall into two major categories: *synchrony by induction* and *synchrony by selection.*

Synchrony by Induction

The most frequently employed methods for inducing synchrony involve *temperature cycles (temperature shocks), light cycles,* and *chemical manipulations.* When temperature is used to induce synchrony, the cells are subjected to alternating cold and warm periods. Little cell division occurs during the cold periods, but on entry into the warm periods, cell division commences. For example, cultures of the flagellate protozoan *Polytomella agilis* can be synchronized by a repetitive temperature cycle of 22 hours at 9°C followed by 2 hours at 25°C. Similar procedures have been successful with other protozoa, including *Astasia longa* and *T. pyriformis.* Synchrony may also be induced by a rapid succession of very short cold and warm periods. Following this sequence, the cell population enters synchronous division.

Certain cultures of photosynthetic cells can be induced to divide synchronously by exposing them to alternating periods of dark and light, the population doubling with each light cycle.

If a particular chemical intermediate required for cell division is withheld from the culture medium for some time, the onset of division for all cells is delayed; the division of the population then follows the addition of the substrate to the culture medium. Agents that chemically inhibit cell division can be added to the culture. Cell growth continues but cell division is halted. An example of the latter is the action of *colchicine,* which halts the division of cells at the metaphase of mitosis. If colchicine-treated cells are then transferred to a colchicine-free medium, inhibition is soon reversed and the cells all enter the G_1 phase. A general drawback to the use of colchicine and other drugs is the possibility of inducing abnormal behavior in the cells.

The "Double Thymidine Block"   A popular procedure for inducing synchrony in cells is to cause the random

population to accumulate at the beginning of the S phase by preventing the synthesis of DNA. One way to achieve this is through the addition of high concentrations of the nucleoside *thymidine* to the culture medium. The addition of the thymidine acts through a *feedback mechanism* within the cells (see Chapter 8) to prevent the production of other nucleosides that are needed for DNA synthesis. When the thymidine is first added to the culture, any cells that are in the S phase stop DNA synthesis immediately, thereby producing a culture in which some cells are halted at various stages of the S phase while the remaining cells are blocked at the entry into the S phase. After this, the block is lifted for an interval of time greater than the length of the S phase. This allows all cells to complete the S phase regardless of the point in the cell cycle where they were halted after the first application of the thymidine. Most of the cells in the population would now be at various stages of the G_2 phase. Thymidine is then applied again, and during this second exposure the cells continue through the cycle until they finish the G_1 phase. Therefore, as a result of the "double thymidine block" all the cells in the culture are arrested at the start of the S phase.

These are but a few examples of the variety of chemical procedures in which cyclic manipulaiton of the culture medium's chemistry results in cell synchrony.

Synchrony by Selection

Selection techniques involve the mechanical isolation of cells of similar age from a random culture; these cells are then inoculated into fresh medium, where they grow and divide synchronously for some time. Among the methods frequently employed for mechanically isolating cells of similar age is *filtration.* In this procedure, a cell culture is passed through a filter that separates the larger cells from the smaller ones. The isolated smaller cells (which are also the younger cells of the population) are then inoculated into fresh medium to start a new culture.

Cells of similar age may also be isolated from a random culture by *sedimentation.* Because young cells are generally smaller than older cells, they sediment more slowly and may be isolated for subculture.

Another technique that is used in tissue culture is the "grow off" method in which the cells are adsorbed onto some surface (such as filter paper); during cell division one of the two daughter cells that is produced detaches from the surface and can be collected for subculturing.

Regardless of the procedure that is employed, the ultimate effect is to alter the random age distribution of the cells so that for several subsequent generations all divisions occur over a short interval of time and all cells are at a similar stage of their cell cycle. Of the various procedures employed, synchrony by selection is to be preferred, because chemical or temperature variations are unnatural and may have undesirable effects on the cell population. The synchrony achieved by any one of these methods is usually observed to decay after several generations of synchronous growth.

CULTURE FRACTIONATION

Another way to study cells that are at the same stage of their cell cycle involves a technique called **culture fractionation**. This procedure avoids the potential problems of synchronization techniques, in which the inducing factor may itself alter or distort the normal events of the cycle. In culture fractionation, a random culture of cells (i.e., cells that are at various stages of the cell cycle) is *sorted* into subpopulations of varying cell size by sedimentation through a *liquid density gradient* (see Chapter 12) and is followed by the collection of the gradient and the entrained cells as a series of fractions. During sedimentation, the larger cells travel farther through the density gradient than the smaller cells. Cell sedimentation through the gradient may be achieved by simply using gravity (i.e., the cells fall through a stationary, cylindrical column of gradient) or may be accelerated using centrifugal techniques. The rationale behind this approach is that a relationship exists between cell size and cell age (i.e., they are directly related to one another), so that the population is sorted into a linear sequence of fractions of increasing cell age. By studying and comparing chemical, physiological, or morphological properties of the cells in each of the separated fractions, specific

effects occurring during the cell cycle can be assigned an age. Thus, changes taking place during cell growth may be followed even though a synchronous cell population is not being employed.

Culture fractionation requires the processing of very large quantities of material so that all of the collected fractions from the density gradient contain enough cells of the same age (i.e., size) for valid physical and/or chemical measurements. Sedimentation of large numbers of cells through a density gradient under the influence of the earth's gravity alone involves the use of large cylindrical chambers called "sta-put" devices; centrifugal separations of large numbers of cells usually require "zonal rotors." Both of these laboratory techniques are discussed further in Chapter 12, along with other methods of particle separation and analysis.

SUMMARY

All cells of multicellular animals and plants are derived initially from a single cell through growth and division. Mathematical models may be derived that describe the growth of cell populations, and proliferation may be studied experimentally using cultures of prokaryotic and eukaryotic cells. A number of alternative methods are used to culture cells and to measure population growth. The growth of a **random** culture of cells characteristically includes **lag, exponential, stationary,** and **declining** phases. The extrapolation of observations made using an entire population of cells to the cycle of events occurring within an individual cell requires that the cell culture be **synchronous.** The cycle of individual cells (i.e., the **cell cycle**) takes a characteristic form and is divided into the G_1, S, G_2, and M phases. The lengths of these phases vary from one type of cell to another but are characterized by the same physiological events and culminate in the division of one cell into two.

REFERENCES AND SUGGESTED READING

Articles and Reviews

Damjanovich, S., Gaspar, R., Tron, L., and Aszalos, A., Biotechnological application of flow cytometry. *Am. Biotech. Lab.* **3** (5), 11 (Sep. 1985).

Edmunds, L. N., and Adams, K. J., Clocked cell cycle clocks. *Science* **211**, 1002 (1981).

Feder, J., and Tolbert, W. R., The large-scale cultivation of mammalian cells. *Sci. Am.* **248** (1), 36 (Jan. 1983).

Mazia, D., The cell cycle. *Sci. Am.* **230** (1), 54 (Jan. 1974).

Muirhead, K. A., Horan, P. K., and Poste, G., Flow cytometry: present and future. *Bio/Technology* **3**, 337 (1985).

Pardee, A. B., Dubrow, R., Hamlin, J. L., and Kletzien, R. E., Animal cell cycle. In *Annual Reviews of Biochemistry*, Vol. 47 (E. E. Snell *et al.*, Eds.), Annual Reviews, Inc., Palo Alto, Calif., 1978.

Yanishevsky, R. M., and Stein, G. H., Regulation of the cell cycle in eukaryotic cells. *Int. Rev. Cytol.* **69**, 223 (1981).

Books, Monographs, and Symposia

Adams, R. L. P., *Cell Culture for Biochemists,* Elsevier/North-Holland, Amsterdam, 1980.

Calow, P., *Life Cycles,* Chapman & Hall, London, 1978.

Davenport, R., *An Outline of Animal Development,* Addison–Wesley, Reading, Mass., 1979.

Edmunds, L. N. (Ed.), *Cell Cycle Clocks,* Dekker, New York, 1984.

Freshney, R. I., *Culture of Animal Cells,* Alan R. Liss, Inc., New York, 1983.

Lloyd, D., Poole, R. K., and Edwards, S. W., *The Cell Division Cycle,* Academic, London, 1982.

Mitchison, J. M., *The Biology of the Cell Cycle,* Cambridge University Press, New York, 1972.

Nicolini, C. (Ed.), *Cell Growth,* Plenum, New York, 1982.

Shapiro, H. M., *Practical Flow Cytometry,* Alan R. Liss, Inc., New York, 1985.

Strehler, B. L., *Time, Cells and Aging* (2nd ed.), Academic, New York, 1977.

Zeuthen, E., *Synchrony in Cell Division and Growth,* Wiley–Interscience, New York, 1964.

CHAPTER 3

CELLULAR CHEMISTRY

All cells are composed of water, proteins, carbohydrates, lipids, nucleic acids, salts, and minute quantities of a diversity of organic compounds such as vitamins and intermediates of cellular metabolism. The proteins, carbohydrates, lipids, and nucleic acids are compounds called *biomolecules*, and the larger of these molecules such as the proteins and nucleic acids are also called *macromolecules*. The macromolecules are described separately in the following chapters. Water, ionized salts, and certain vitamins may complex with the macromolecules of the cell or they may remain free. The quantities of each of these classes of compounds vary widely from one type of cell to another and from one organism to another (Table 3–1). Water is the most common molecule, whereas carbohydrates and proteins are the most prevalent organic substances. In this chapter, some fundamentals of cellular chemistry and the properties of a number of common cellular constituents are considered.

STANDARD UNITS OF MEASUREMENT

One of the special concerns of any scientific endeavor is the effort to acquire accurate measurements in a form that can be understood and verified by others.

TABLE 3–1 CHEMICAL COMPOSITION OF SOME REPRESENTATIVE CELLS (PERCENTAGE OF TOTAL WET WEIGHT)

Constituent	Cell or Tissue				
	Rat Liver	Rat Skeletal Muscle	Sea Urchin Egg	E. coli	Corn Seed
Water	69–72	76	77	73	13
Protein	16–22	21	16	19	9
Polysaccharide	2–10	0.6	1.4	1.1	73
Lipid	5	4–11	4.8	1.1	4
Nucleic acid	1–5	1–2	1	3.5	1
Salt	1	1	0.3	2.3	1

To this end, cell biologists, like scientists in other disciplines, rely on the use of a number of *standard units of measure*. For many years this function has been fulfilled by the *metric system*. In the 1960s, in an effort to establish uniformity among all fields of science, an international conference on weights and measures recommended certain modifications of the metric system and gave it a new name. It is now called the *Système Internationale d'Unites* (i.e., *International System of Units*) and is abbreviated *S.I.*

The S.I. system recognizes seven basic units (*mass, length, time, electric current, temperature, luminous intensity,* and *quantity* of a substance). Table 3–2 lists these units and gives the symbol used for each. The metric and English equivalents of the S.I. are given in Table 3–3.

Of special note for the cell biologist and biochemist is the recommended use of the *joule* (*J*) rather than the *calorie* (*cal*) as the standard unit of energy. (One joule equals 1 kg m²/s².) In view of the fact that energy values expressed in calories are still to be found in much of the biological literature, the caloric energy values along with their joule equivalents will be cited in this book.

SPECIAL PROPERTIES AND BEHAVIOR OF WATER

Water is the essential solvent of life and the main chemical constituent of all cells. The essential role played by water can be attributed to its unique physical and chemical properties, which collectively act to protect living systems and are necessary to the structure and function of cells. One of the unusual characteristics of water is that despite its low molecular weight, it is a liquid at most environmental temperatures. Other molecules having similar molecular weights (e.g., ammonia, methane, and H_2S) are gases. In comparison with other solvents, water has high values for melting point, heat of fusion, boiling point, heat of vaporization, specific heat, and surface tension (see Table 3–4). Each of these properties serves to constrain major temperature fluctuations and to keep water in the liquid state. Water absorbs more heat energy per gram for every degree rise in temperature than other common solvents and therefore acts to

TABLE 3–2 THE SEVEN SI BASE UNITS

Physical Quantity	Name of Unit	Symbol
Mass	Kilogram	kg
Length	Meter	m
Time	Second	s[a]
Electric current	Ampere	A
Temperature	Kelvin	K
Luminous intensity	Candela	cd
Quantity of substance	Mole	mol

[a]Seconds may also be abbreviated "sec."

TABLE 3–3 THE SYSTEME INTERNATIONALÈ (S.I., INTERNATIONAL SYSTEM OF UNITS) AND ITS METRIC AND ENGLISH EQUIVALENTS

Measurement	S.I. Unit of Measure	Metric Equivalents	English System (fps) Equivalents
Mass	kilogram (kg)	1 kilogram (kg) 1000 grams (g)	2.204 pounds (lb)[a]
Length	meter (m)	1 meter (m) 100 centimeters (cm)	3.28 feet (ft)
Time	second (sec)[b]	1 second (sec)	1 second (sec)
Force	newton (N) (1 N = 1 kg m/sec²)	10⁵ dynes (D) (1 D = 1 g cm/sec²)	7.23 poundals (= 7.23 ft-lb/sec²)
Energy	joule (J)	1 joule (J) 0.239 calories (cal) 10⁷ ergs	23.73 ft-poundals
Volume	cubic meter (m³)	1000 liters (l) 10⁶ ml	35.3 cubic feet (ft³)

[a]Strictly speaking, the pound is a unit of *weight*.

[b]Seconds may also be abbreviated using the symbol s.

moderate temperature changes. (Indeed, the calorie is specifically defined as the amount of energy required to raise the temperature of one gram of water from 14.5° to 15.5°C.) Likewise, to convert one gram of water from the liquid to the vapor (gas) state at its boiling point, an additional 2259 J (540 cal) must be absorbed. This high heat of vaporization, together with the high surface tension of liquid water, tends to keep water in the liquid state. At the other temperature extreme, large amounts of energy (335 J or 80 cal per gram) must be lost for water to be converted from the liquid to the solid state.

Although water is often called the "universal solvent," not all substances dissolve in water. However, water does dissolve most salts and other ionic compounds, as well as nonionic polar compounds such as sugars, alcohols, and other molecules that contain hydroxyl, aldehyde, and ketone groups. Many substances that contain both polar and nonpolar groups (such as soaps, fatty acids, and glycerophosphatides) do not dissolve in water, but they do form **micelles**. A micellar arrangement is not a true solution but is a **suspension** or **dispersion**. The behavior of soap molecules in water is a good and also common example of micelle formation. Soap molecules, formed by the *saponification* of fatty acids (Chapter 6), consist of a long, nonpolar hydrocarbon chain terminating in a polar carboxyl group that is ionically bonded to a metal ion such as K^+ or Na^+. When dispersed in water, the soap molecules aggregate to form spherical clusters, called micelles, in which the polar carboxyl groups of the soap molecules are arranged at the surface of the sphere, where they form weak bonds with the surrounding water, and the nonpolar hydrocarbon chains project inward (Fig. 3–1).

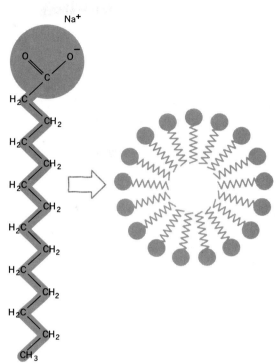

FIGURE 3–1 Soaps (fatty acids linked to metal ions) *coalesce* in aqueous suspensions to form *micelles*. The polar ends of the soap molecules (circular areas) face the surrounding water and the hydrocarbon chains (zigzag lines) are directed toward the center of the micelle.

TABLE 3–4 PHYSICAL PROPERTIES OF SEVERAL COMMON SOLVENTS

Property	Water	Methanol	Acetone	Ammonium Hydroxide
Melting point (°C)	0	−99	−94	−78
Heat of fusion				
J/g	335	92	84	351
cal/g	80	22	20	84
Boiling point (°C)	100	64.5	56.5	−20
Heat of vaporization				
J/g	2259	1100	523	1330
cal/g	540	263	125	318
Specific heat				
J/g/°C	4.2	2.4	2.1	4.2[a]
cal/g/°C	1.0	0.566	0.506	0.999[a]
Surface tension				
(dynes/cm)	72.8	22.6	23.7	73.5
(N/m)	0.0728	0.0226	0.0237	0.0735

[a]Value for a 50% solution in water.

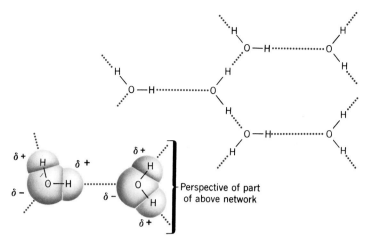

Perspective of part of above network

FIGURE 3–2 Water molecules behave like dipoles and form *hydrogen bonds* with each other. As a result, water molecules tend to arrange themselves into hexagonal lattices. (Redrawn from Holum, J. R., *Fundamentals of General Organic and Biological Chemistry*, p. 123; copyright © 1978 by John Wiley & Sons, Inc.)

The special physical properties of water are founded in its molecular structure. In water, two hydrogen atoms are each covalently bonded to a single oxygen atom. The oxygen atom, being more electrophilic than the hydrogens, attracts the shared electrons more strongly and produces an electrical asymmetry in the molecule. As a result, the oxygen atom takes on less than a full negative charge, technically called a *partial negative charge* and denoted by δ^-. The hydrogens become *partially positive* (Fig. 3–2). Because the bond angle between the hydrogen atoms is 104.5°, each water molecule acts as a **dipole** even though it has no net charge. Because of their dipole character, water molecules tend to form bonds with each other. The bonds linking neighboring water molecules are called *hydrogen bonds* and are formed by the attractions of the electronegative oxygen of one molecule for the electropositive hydrogens of two other molecules. Thus, by means of two covalent bonds and two hydrogen bonds, each oxygen atom may bond to as many as four hydrogens; the hydrogens form a *tetrahedron* about the oxygen (Fig. 3–3). The hydrogen bonding that occurs between water molecules in liquid water (Figs. 3–2 and 3–4) accounts in part for water's high heat of vaporization and surface tension.

When the temperature of water is lowered, the concomitant decrease in kinetic energy allows the water to become denser until at 4°C maximum density occurs. A continued decrease in kinetic energy allows more extensive and less transient formation of hydrogen bonds among the water molecules. This promotes the development of a lattice structure (Fig. 3–4). As the freezing point of water is approached, more space develops between the molecules and the density decreases; thus, freezing water and ice rise to the surface of the liquid.

Ionic substances are readily dissolved in water because of the water molecules' dipolar character. When salts (or other polar compounds) are dissolved in water, the orientation of the water molecules with respect to each other is disturbed. For example, sodium chloride (one of the most common salts found in cells and tissues) ionizes in water to form Na^+ and Cl^-; these ions attract the water molecules and disturb the

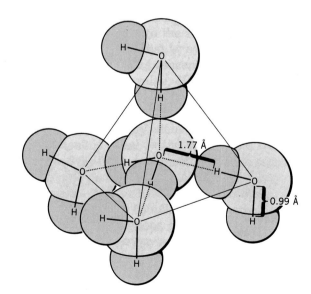

FIGURE 3–3 Tetrahedral arrangement of hydrogen atoms about oxygen in water. Two of the four hydrogen atoms are covalently bonded to the oxygen in water, and two additional hydrogens (from two other water molecules) form hydrogen bonds.

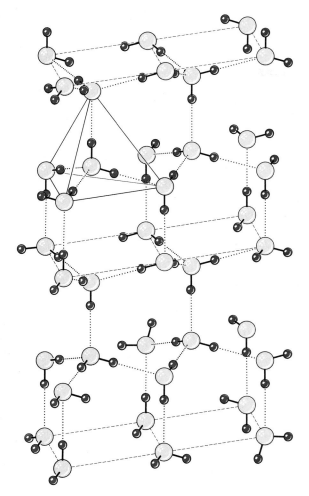

FIGURE 3–4 Three-dimensional lattice formed by water through hydrogen bonding. The introduction of ions (such as those produced when salts are dissolved in water) creates localized areas of distortion in the lattice structure.

FIGURE 3–5 *Hydration spheres* formed about cations (top) and anions (bottom).

water lattice. Dissolved cations and anions convert the water lattice into **spheres of hydration** that *encage* the ion (Fig. 3–5). Were it not for the formation of spheres of hydration about their ionic groups, many large molecules such as proteins would not be soluble in water. By encaging the ionized groups of proteins in layers of water, the groups are prevented from interacting with each other. As a general rule, ionized salts bind water more strongly than the polar groups of macromolecules, and this property is occasionally used to separate macromolecules from their solvent. For example, when salts are added to a protein solution, the salt ions form more extensive hydration spheres than the proteins. In effect, the redistribution

of the water around the salt ions diminishes the amount of solvent available for the ionized groups of the proteins. The ionized groups of the proteins interact with each other, and a protein precipitate is formed. This method for separating proteins from their solvent is called ''salting out'' and is discussed more thoroughly in Chapter 13.

Water molecules also undergo ionization; the capacity of water to form ions is another of its important biological properties. As we have already seen, in water each hydrogen atom is covalently bonded to one oxygen atom and forms a weak electrostatic bond with an oxygen atom of another water molecule (i.e., the water molecules are hydrogen bonded to one another). With measurable frequency, the covalent bond linking a hydrogen to oxygen is broken and the hydrogen establishes a closer association with the oxygen to which it was electrostatically bonded.

$$\begin{array}{c} H \\ \diagdown \\ O\text{----}H \\ \diagup \\ H \end{array} \quad \begin{array}{c} \\ \diagdown \\ O \\ \diagup \\ H \end{array} \longrightarrow \begin{array}{c} H \\ \diagdown \\ O\text{---}H + OH \\ \diagup \\ H \end{array}$$

The water molecules have thus ionized, dissociating into a *hydroxide anion* (OH^-) and a *hydronium cation* (H_3O^+). At standard temperature and pressure, one liter of water contains 1.0×10^{-7} moles of both OH^- and H_3O^+. Both the hydronium ions and hydroxide ions form hydrogen bonds with other water molecules. The notation H_3O^+ is rarely used; H^+ (called a **hydrogen ion** or **proton**) is used more frequently, even though free protons do not generally exist per se but are bonded to water to form hydronium ions.

SALTS, IONS, AND GASES

Salts are found in all cells and exist in their ionized forms. The ions may be dissolved in the cell fluid or specifically bound to other molecules such as proteins or lipids. Salts have two broad roles in the cell. One role is osmotic, in that the total concentration of dissolved salts affects the *flux of water* across the cell's membranes. The second is the more specific role that certain ions play in contributing to the structure and function of cellular particles and macromolecules. The specific cellular roles of a number of ions are given in Table 3–5.

Gases that enter the cell from the environment or are produced by the metabolism of a cell dissolve in the cytoplasmic water. The most common gases in air are nitrogen (N_2), oxygen (O_2), and carbon dioxide (CO_2). In the atmosphere at sea level, N_2 accounts for 78.03% of all gases, O_2 20.99%, and CO_2 0.03%. Both nitrogen and oxygen dissolve in water in their molecular form, but their solubility in water is low. For example, at 25°C and at 1 atmosphere pressure, only 2.83 ml of oxygen and 1.43 ml of nitrogen dissolve in 100 ml of distilled water. The behavior of carbon dioxide in

TABLE 3–5   CELLULAR FUNCTIONS OF CERTAIN IONS

Element	Ionic Form Present	Percentage Total Weight	Function
Molybdenum	MoO_4^{2-}	0.00001	Cofactor or activator of certain enzymes (e.g., nitrogen fixation, nucleic acid metabolism, aldehyde oxidation)
Cobalt	Co^{2+}	0.00001	Constituent of vitamin B_{12}
Copper	Cu^+, Cu^{2+}	0.0004	Constituent of plastocyanin and cofactor of respiratory enzymes
Iodine	I^-	0.0005	Constituent of thyroxin and other hormones
Boron	$BO_3^{3-}, B_4O_7^{2-}$	0.002	Activates arabinose isomerase
Zinc	Zn^{2+}	0.003	Cofactor or activator of certain enzymes (e.g., carbonic anhydrase, carboxypeptidase)
Manganese	Mn^{2+}	0.003	Cofactor or activator of certain enzymes (e.g., several kinases, isocitric decarboxylase)
Iron	Fe^{2+}, Fe^{3+}	0.01	Constituent of hemoglobin and the cytochromes
Magnesium	Mg^{2+}	0.1	Constituent of chlorophyll; activates ATPase
Sulfur	SO_4^{2-}	0.1	Constituent of coenzyme A, biotin, thiamine, proteins
Phosphorus	$PO_4^{3-}, H_2PO_4^-$	0.2	Constituent of lipids, proteins, nucleic acids, sugar phosphates, nucleoside phosphates
Calcium	Ca^{2+}	0.5	Constituent of plant cell walls; matrix component of bone tissue; cofactor of coagulation enzymes
Potassium	K^+	1.0	Cofactor for pyruvate kinase and K^+-stimulated ATPases

water is quite different. Some of the dissolved carbon dioxide reacts with water, forming carbonic acid, which in turn dissociates into ions:

$$CO_2 + H_2O \rightleftharpoons \underset{\substack{\text{carbonic} \\ \text{acid}}}{H_2CO_3} \rightleftharpoons H^+ + \underset{\text{bicarbonate}}{HCO_3^-}$$

Most of the carbon dioxide in cells is in the bicarbonate and carbonate forms. CO_2 gas is more soluble in water than are O_2 and N_2. For example, 75.9 ml of CO_2 dissolves in 100 ml of water at 25°C and 1 atmosphere pressure.

ACIDS, BASES, pH, AND BUFFERS

According to the Brønsted–Lowry definitions, an **acid** is a *proton* (or H^+) *donor* and a **base** is a *proton acceptor*. For example, when acetic acid is dissolved in water, a certain number of the acetic acid molecules dissociate as follows:

$$\underset{\text{(acetic acid)}}{CH_3COOH} + H_2O \rightleftharpoons \underset{\text{(acetate anion)}}{CH_3COO^-} + H_3O^+$$

In the reaction, a proton of acetic acid has been donated to water, producing an anion (acetate) and a cation (the hydronium ion). Acetic acid donated the proton (H^+) and is therefore the acid. Water accepted the proton and is a base in the reaction. If the reaction is reversed, the acetate anion acts as a proton acceptor and therefore fits the criterion of a base. Acetic acid and acetate ions are referred to as a **conjugate acid–base pair**. Protons are attracted both to the conjugate base and to water. An acid that readily donates protons to water is called a *strong acid,* while an acid that does not readily donate protons to water is termed a *weak acid.* Each acid is characterized by its tendency to dissociate; the extent of dissociation may be determined from the acid's dissociation constant, K, defined by

$$K = \frac{[H^+][A^-]}{[HA]} \qquad (3-1)$$

where for a given solution $[HA]$ is the concentration of undissociated acid, $[A^-]$ is the concentration of conjugate base, and $[H^+]$ is the hydrogen ion concentration. The dissociation constant varies somewhat according to the temperature of the solution. Customarily in physiology and biochemistry, one re-

TABLE 3–6 DISSOCIATION CONSTANTS FOR SOME COMMON ACIDS

Acid	K'	pK'
Formic acid	1.78×10^{-4}	3.75
Acetic acid	1.74×10^{-5}	4.76
Propionic acid	1.35×10^{-5}	4.87
Lactic acid	1.38×10^{-4}	3.86
Succinic acid	6.16×10^{-5}	4.21
Phosphoric acid (PO_4^{3-})	7.25×10^{-3}	2.14
Monobasic phosphate ($H_2PO_4^-$)	6.31×10^{-8}	7.20
Dibasic phosphate (HPO_4^{2-})	3.98×10^{-13}	12.4
Carbonic acid (H_2CO_3)	1.70×10^{-4}	3.77
Bicarbonate (HCO_3^-)	5.6×10^{-11}	10.26

fers to the *apparent dissociation constant, K',* of an acid. This value is based on the concentrations of reactants and products that can actually be measured following dissociation of the acid in water. Table 3–6 lists the apparent dissociation constants of some common acids.

The concentration of H^+ in cells varies widely. The range may exceed 0.1 to 10^{-10} M. Although these concentrations may not seem high, small changes in H^+ concentration have noticeable effects on cell function. It is inconvenient to express the H^+ concentration in molar terms (such as 1×10^{-7} M), and the pH scale proposed by Sørensen is used instead. Accordingly,

$$pH = -\log_{10}[H^+] \qquad (3-2)$$

As noted earlier, at 25°C, the H^+ (and OH^-) concentration of pure water is 1×10^{-7} M and the solution is at neutrality. At this concentration,

$$pH = -\log_{10}[1.0 \times 10^{-7}] = 7.0 \qquad (3-3)$$

Table 3–7 shows the range of the pH scale and lists the pH values of a number of familiar liquids. In the laboratory, the pH of a solution is determined potentiometrically using an instrument called a *pH meter.* The instrument contains two electrodes that are immersed in the liquid sample; one of these is a *reference electrode* (whose electrical potential is constant) and the other is a *glass electrode* (whose electrical potential depends on the pH of the sample). An *electrometer* measures the difference in electrical potential between the two electrodes and this value is then converted to pH units and is displayed. Most pH meters are accurate to within 0.01 pH units.

A **buffer** solution is formed by mixing together a

TABLE 3-7 THE pH SCALE

H$^+$ Concentration (M)	pH	Examples	Condition
1.0	0	0.1 M HCl	
0.1	1	Human gastric fluid	
0.01	2	Orange juice	
0.001	3	Vinegar	Acid
0.0001	4	Pineapple juice	
10^{-5}	5	Tomato juice	
10^{-6}	6	Cow's milk	
10^{-7}	7	Most body fluids; pure water	Neutral
10^{-8}	8	Seawater	
10^{-9}	9	Soil water	
10^{-10}	10	Alkaline desert ponds	
10^{-11}	11		Basic
10^{-12}	12	Limewater	
10^{-13}	13		
10^{-14}	14	1 M NaOH	

weak acid and its *conjugate base*. These solutions effectively slow the rate of pH change over a limited portion of the pH scale when a quantity of strong acid or base is added to the buffer. As shown in Figure 3-6, titration of H_3PO_4 (phosphoric acid) with up to about 8 ml of NaOH causes a gradual increase in pH to about 2.5. In contrast, the next few milliliters of NaOH that are added bring about a rapid rise in pH to about 6.0. The gradual pH change observed on addition of the first 8 ml of base is the result of the conjugate acid–base buffering capacity of the H_3PO_4 and NaH_2PO_4 (i.e., the NaOH reacted with H_3PO_4, forming the NaH_2PO_4 and H_2O), that is,

$$\underbrace{Na^+ + OH^-} + \underbrace{H^+ + H_2PO_4^-} \rightarrow \underbrace{Na^+ + H_2PO_4^-} + H_2O$$

A second buffering "plateau" is produced by the conjugate acid–base buffer system of NaH_2PO_4 and Na_2HPO_4. The third plateau is achieved through the actions of Na_2HPO_4 and Na_3PO_4.

Just as the hydrogen ion concentration can be expressed in terms of pH (or negative logarithms), the dissociation constant for a weak acid, K_a, can be similarly defined:

$$pK_a = -\log K_a \qquad (3-4)$$

The pK for each conjugate acid–base pair occurs at the *midpoint* of its buffering plateau. The most effective buffering range of an acid–base pair usually extends 0.5 pH units to each side of the pK value. The relationship between pH and pK_a is expressed in the Henderson–Hasselbach equation:

$$pH = pK_a + \log_{10}\frac{[A^-]}{[HA^-]} \qquad (3-5)$$

where the weak acid is represented by $[HA^-]$ and its conjugate base by $[A^-]$. When the concentrations of the two are equal, as at the midpoint of a plateau, the last term of equation 3-5 becomes zero (i.e., $\log_{10} 1 = 0$), and therefore pH = pK_a. The Henderson–Hasselbach equation can be used to calculate the pH if the concentrations of the acid and base are known. The equation is also useful for calculating the necessary quantities of acid and conjugate base that must be mixed together to prepare a buffer that has a par-

FIGURE 3-6 Titration curve of phosphoric acid (H_3PO_4) with sodium hydroxide (NaOH) showing the three buffering plateaus about the pK_a values. The weak acid and conjugate base for each buffering plateau are identified.

ticular pH. However, it should be pointed out that the equation is an approximation and is most accurate when almost equal amounts of the acid and base are mixed. The pK values of several buffers that are routinely used in cellular studies are listed in Table 3–8.

CHEMICAL BONDS

A molecule is a stable union of two or more atoms. The formation of the bonds that exist between the atoms comprising a molecule requires less energy than that needed to keep the atoms apart. The breakage of the bonds that join atoms consumes energy. If a large amount of energy is required to break the bond, the bond is called a *strong bond*. If small amounts of energy are required, the bond is a *weak bond*. Four major types of chemical bonds may be identified: **covalent bonds**, **ionic bonds**, **hydrogen bonds**, and **hydrophobic bonds**. The energies of these bonds are given in Table 3–9. Covalent bonds are the bonds formed by

TABLE 3–8 COMMON BUFFER SYSTEMS

Compound	pK_{a1}	pK_{a2}	pK_{a3}	pK_{a4}
Acetic acid	4.7			
Ammonium chloride	9.3			
Carbonic acid	6.4	10.3		
Citric acid	3.1	4.7	5.4	
Diethanolamine	8.9			
Ethanolamine	9.5			
Fumaric acid	3.0	4.5		
Glycine	2.3	9.6		
Glycylglycine	3.1	8.1		
Histidine	1.8	6.0	9.2	
Maleic acid	2.0	6.3		
Phosphoric acid	2.1	7.2	12.3	
Pyrophosphoric acid	0.9	2.0	6.7	9.4
Triethanolamine	7.8			
Tris-(hydroxymethyl) amino methane	8.0			
Veronal (sodium diethylbarbiturate)	8.0			
Versene (ethylenediaminotetraacetic acid)	2.0	2.7	6.2	10.3

TABLE 3–9 ENERGIES RELEASED/CONSUMED DURING BOND BREAKAGE/FORMATION

Type of Bond		Kilojoules/Mole	Kilocalories/Mole
Covalent bonds			
Single bonds	O—H	462	110.6
	H—H	436	104.2
	C—H	415	99.3
	N—H	391	93.4
	C—O	356	85.0
	C—C	347	83.1
	S—H	344	82.3
	C—N	292	69.7
	S—S	426	101.9
Double bonds	C=O	723	173
	C=N	619	148
	C=C	602	144
	P=O	623	149
Ionic bonds	NaCl	408	97.5
	KCl	424	101.3
	NaI	304	72.7
Hydrogen bonds		33–175	8–42
Hydrophobic bonds (van der Waals attractions)		33	8

the sharing of one or more electron pairs by the atoms comprising the molecule. These are generally the strongest bonds formed between atoms. Whereas ionic bonds formed between oppositely charged atoms have a high energy level in crystals (such as in crystalline NaCl, KCl, etc.; see Table 3–9), the ionic bonds that are formed in aqueous solution (the usual case for cells) have much lower energy levels and are considerably weaker than covalent bonds. As we have already seen, very weak bonds called hydrogen bonds occur when hydrogen is shared by two electronegative atoms. Hydrogen bonds most commonly occur between the oxygen atoms of water molecules, but they can also be formed between two nitrogen atoms and between nitrogen and oxygen (see also Chapters 4 and 7). Hydrophobic bonds are associations of molecules or parts of molecules that have nonpolar groups. They are not chemical bonds in the usual sense. The attraction of water molecules for one another is far greater than their attraction to hydrophobic groups and this causes the hydrophobic groups to aggregate and have minimum contact with the surrounding water. The association of the hydrophobic groups is stabilized by **van der Waals interactions**, which are weak attractions between all uncharged atoms or molecules that are in close proximity.

FIGURE 3–7 Chemical structure of cyanocobalamin (vitamin B_{12}) in which the cobalt ion is *sequestered* by the tetrapyrrole. (Redrawn from Brady, J.E., and Holum, J.R., *Fundamentals of Chemistry*, p. 704; copyright © 1981 by John Wiley & Sons, Inc.)

COORDINATION COMPOUNDS, LIGANDS, AND CHELATES

In addition to the types of bonds discussed above, oxygen and nitrogen can form a special kind of covalent association, called a **coordination bond**, with certain divalent and trivalent metal ions such as Ca^{2+}, Co^{2+}, Cu^{2+}, Fe^{2+}, Mg^{2+}, Mn^{2+}, Ni^{2+}, and Zn^{2+}. In this type of bond, all of the shared electrons that form the bond are donated by the nitrogen or oxygen atom. No electrons are donated by the metal. The energy of such bonds is about 251 kJ or 60 kcal per mole. Molecules that contain nitrogen or oxygen atoms capable of donating electron pairs to metals or metal ions are called **ligands**. If two or more electron pair donating atoms are present in the same molecule, the molecule is called a **chelate** or **chelating agent**. The metal bound by the chelating agent is said to be *sequestered*. Water may also coordinate metal ions, as in the hydrated forms of Mg, Ca, and Al salts. Figure 3–7 shows the chemical structure of cyanocobalamin (vitamin B_{12}) in which the cobalt ion forms coordination bonds (depicted as arrows) with several atoms. The oxygen-transporting protein hemoglobin (discussed in Chapter 4) is a prime example of an iron chelate. Some enzymes are activated by chelation of a metal (e.g., see the cofactors of Table 9–3).

SPECIAL COMPOUNDS

Among the compounds that have special roles in cell metabolism are the **nucleoside phosphates** (e.g., adenosine triphosphate, ATP; adenosine diphosphate, ADP; and adenosine monophosphate, AMP), the **pyridine nucleotides** (e.g., nicotinamide adenine dinucleotide, NAD^+; and nicotinamide adenine dinucleotide phosphate, $NADP^+$), and the **vitamins**.

Nucleoside Phosphates

The metabolic activities of cells are accompanied by a complex series of energy transformations, and the principal intermediaries of these energy exchanges are the nucleoside phosphates. Although ATP (Fig. 3–8) serves as the immediate energy donor in most energy-consuming (i.e., **endergonic**) reactions, other nucleoside triphosphates such as guanosine triphosphate (GTP), uridine triphosphate (UTP), and cytidine triphosphate (CTP) serve in this capacity in certain instances. Energy-producing (i.e., **exergonic**) reactions

are often accompanied by the synthesis of nucleoside triphosphates from the corresponding nucleoside diphosphate. The chemical formulas of the various nucleosides may be found in Chapter 7.

The high rate of turnover of ATP attests to its vital role in metabolism. An examination of the chemical structure of ATP (Fig. 3–8) sheds some light on why this molecule so readily serves as an energy source for energy-consuming cellular reactions. The three phosphate groups of ATP exist in an ionized state at physiological pH and the four negatively charged oxygen atoms strongly repel one another. When ATP is hydrolyzed,

$$ATP^{4-} + H_2O \rightarrow ADP^{3-} + HPO_4^-$$

both of the products are negatively charged and continue to repel each other. The amount of energy that must be consumed in order to recombine these anions to re-form ATP reflects the amount of energy that is made available during the hydrolysis. For many years, the bonds linking successive phosphate groups in ATP were represented by a "squiggle" (i.e., ~ P) and called "high-energy bonds." Actually, there is nothing special about the bonds themselves and the amount of bond energy is not that great (31 kJ or 7.3 kcal per

FIGURE 3–8 Chemical structure of ATP, ADP, and AMP. The bonds between the phosphate groups are important in energy transfer reactions.

mole). What is important is that the phosphate groups have a very high transfer potential. The study of energy transfer and transformation is a special field of biochemistry called bioenergetics; this topic is discussed in Chapter 9.

Cyclic Nucleoside Monophosphates AMP and GMP have cyclic forms called *cyclic AMP* (cAMP) (Fig. 3–9) and *cyclic GMP* (cGMP). Both cAMP and cGMP have been found to have a regulatory influence in a variety of cellular processes. cAMP is produced in the plasma membrane of cells and acts as an intracellular transmitter (sometimes called a "second messenger") by triggering certain metabolic reactions in the cell. The production of cAMP in the the plasma membrane takes place as a response to compounds (usually *hormones*) that bind to specific receptor sites on the outside surface of the membrane (see Chapter 15). A good example of the regulatory role played by the cyclic nucleotides is the action of cAMP during periods of exercise in which great demands are made by the muscle tissues for free sugars. During exercise, the hormone *epinephrine* is released into the bloodstream and on reaching the liver and muscles binds to the plasma membrane. cAMP formed in the membranes in response to epinephrine binding then diffuses into the cytosol. In the cytosol, cAMP binds to and activates an enzyme called *protein kinase*, which in turn catalytically activates a second enzyme, *phosphorylase*. Phosphorylase initiates the breakdown of *glycogen* (Chapter 5), thereby making available the large numbers of free sugar molecules whose consumption fulfills the energy demands of muscular activity.

cAMP also influences *transcription* of nuclear DNA (see Chapters 11 and 21) where it binds to the *catabolite gene activator protein* (*CAP*), and in the case of the *lac* operon (Chapter 11) promotes the transcription of genes for specific cellular enzymes. Both cAMP and cGMP appear to be involved in the control of cell proliferation. Cells that are in the log phase of growth (see Chapter 2) contain smaller amounts of cAMP and higher amounts of cGMP than cells in other phases of the growth cycle. cAMP and cGMP affect the assembly of *microtubules* in cells (Chapter 23) and this would be expected to directly influence cell growth.

Pyridine Nucleotides

The pyridine nucleotides NAD^+ and $NADP^+$ usually function as **coenzymes** for certain dehydrogenation reactions in cells. The hydrogens that are removed from compounds by various dehydrogenase enzymes are accepted by the pyridine nucleotides. Hydrogens are usually removed in pairs, with one hydrogen binding to the coenzyme as a hydride (Fig. 3–10) and the other carried as a hydrogen ion.

Vitamins

Vitamins are organic substances that are present in small amounts in all cells and are essential to certain phases of the cell's metabolism. Although vitamins are not used in large amounts by cells, the lack of a particular vitamin is usually accompanied by characteristic deficiency symptoms. Not all organisms are able to synthesize all of the vitamins that are essential to their metabolism, and they therefore rely on an external dietary source. The role of most vitamins is closely allied to enzyme function, the vitamin comprising all or part of the nonprotein portion of the enzyme (called the *coenzyme*). Table 3–10 lists the more common vitamins, the coenzymes of which they are a part, and the role of the coenzymes and/or the associated enzyme. The vitamin *niacin* is required for the synthesis of the coenzymes NAD^+ and $NADP^+$ described above (Fig. 3–10). *Riboflavin* (vitamin B_{12}) is used in the synthesis of the coenzyme *flavin adenine dinucleotide* (*FAD*), which also serves as a hydrogen acceptor in dehydrogenation reactions. The vitamin *pantothenic acid* is used in the synthesis of *coenzyme A*. The vitamin constituents of these two coenzymes are shown in Figure 3–11.

Unlike animals, which rely on an exogenous source of certain vitamins, plants are able to synthesize all their required vitamins; most of the synthesis takes place in the plant's green tissue.

FIGURE 3–9   *Chemical structure of cyclic AMP (adenosine-3, 5-monophosphate).*

FIGURE 3–10 Reduction of the coenzyme NADP+ by binding a proton (H+). The portion of the coenzyme formed by the vitamin *niacin* is shaded in color. The reduced molecule also associates with a second proton. The related coenzyme NAD+ differs by the substitution of an OH group for the phosphate group marked by an asterisk.

TABLE 3–10 VITAMINS AND COENZYMES

Vitamin	Examples of Associated Coenzymes	Function of Coenzyme and Enzyme
Vitamin A	11-*cis*-Retinal	Visual cycle reactions
Thiamine (vitamin B_1)	Thiamine pyrophosphate	Decarboxylation reactions, oxidases, transketolases
Pyridoxine (vitamin B_6)	Pyridoxal phosphate	Amino acid metabolism (decarboxylation/ transamination)
Cyanocobalamin (vitamin B_{12})	Cyanocobalamine	C—C, C—O, C—N bond cleavage
Vitamin D	1,25-Dihydroxycholecalciferol	Metabolism of calcium and phosphate
Vitamin E	—	Antioxidant activity
Vitamin K	—	Synthesis of precursor enzyme for blood coagulation (prothrombin)
Biotin	Biocytin	Carboxylation reactions
Folic acid	Tetrahydrofolic acid	Various reactions involving single carbon transfers
Lipoic acid	Lipoyllysine	Dehydrogenation, acyl group transfer
Niacin	Nicotinamide nucleotides (NAD+, NADP+)	Numerous oxidation/reduction reactions by dehydrogenases
Pantothenic acid	Coenzyme A (CoASH)	Acetyl group transfer
Riboflavin (vitamin B_2)	Flavin coenzymes (FMN, FAD)	Several oxidation/reduction reactions by dehydrogenases

FIGURE 3-11 The coenzymes flavin adenine dinucleotide (FAD) and coenzyme A. FAD contains the vitamin *riboflavin* and coenzyme A the vitamin *pantothenic acid*.

SUMMARY

Although the quantities of specific chemical constituents vary from one type of cell or organism to another, all cells contain *proteins, polysaccharides, lipids,* and *nucleic acids.* In addition to these macromolecules, all cells contain a wide variety of smaller compounds such as **water, salts, ions, metabolic intermediates,** and **vitamins.**

The physical and chemical properties of water make it a most suitable solvent for living systems. The boiling point, heat of vaporization, specific heat, and surface tension tend to keep water in the liquid state. Because of its polar character, water readily dissolves ionic and polar compounds and induces the formation of **micelles** by substances that are partly polar and partly hydrophobic. The association of water with other molecules (including other water molecules) usually takes the form of **hydrogen bonds.** These bonds have relatively low energy, yet they are vital to the molecular architecture of the cell.

Salts are usually present in ionic form and many of these are associated with macromolecules. Salt ions are also important to the osmotic behavior of cells and to buffering activities.

Some of the more important special compounds that are present in cells are **nucleoside phosphates, ligands,** and **vitamins.** Nucleoside phosphates are important sources of energy and also act to regulate specific metabolic reaction sequences in cells. Vitamins are often components of **coenzymes,** which act in concert with an enzyme to facilitate a specific chemical reaction.

REFERENCES AND SUGGESTED READING

Articles and Reviews

Bitensky, M. W., and Gorman, R. E., Cellular responses to cyclic AMP. *Prog. Biophys.* **26**, 409 (1973).

Ibers, J. A., and Holm, R. H., Modeling coordination sites in metallobiomolecules. *Science* **209**, 223 (1980).

Morton, B. A., The vitamin concept. *Vitam. Horm.* **32**, 155 (1974).

Books, Monographs, and Symposia

Brady, J. E., and Holum, J. R., *Fundamentals of Chemistry,* Wiley, New York, 1981.

Brady, J. E., and Humiston, G. E., *General Chemistry,* Wiley, New York, 1986.

Conn, E. E., and Stumpf, P. K., *Biochemistry* (4th ed.), Wiley, New York, 1976.

Greenburg, L. H., *Physics for Biology and Premed Students,* Saunders, Philadelphia, 1975.

Harrison, P. M., and Hoare, R. J., *Metals in Biochemistry,* Chapman & Hall, London, 1980.

Holum, J. R., *Elements of General and Biological Chemistry* (5th ed.), Wiley, New York, 1979.

Holum, J. R., *Fundamentals of General, Organic, and Biological Chemistry,* Wiley, New York, 1978.

Neckers, D. C., and Doyle, M. P., *Organic Chemistry,* Wiley, New York, 1977.

Solomons, T. W. G., *Organic Chemistry*, Wiley, New York, 1984.

Stryer, L., *Biochemistry* (2nd ed.), Freeman, San Francisco, 1981.

CHAPTER 4

THE CELLULAR MACROMOLECULES: PROTEINS

In most cells, over 90% of the total mass (excluding water) is represented by very large molecules called **macromolecules**. The macromolecules vary in size from several hundred to several hundred million molecular weight units. Four major classes of macromolecules may be identified; these are the **proteins**, **polysaccharides**, **lipids**, and **nucleic acids** (the latter three classes are considered in Chapters 5, 6, and 7). The relative amounts of these macromolecules in the "typical" cell are given in Table 4–1.

Of all macromolecules found in the cell, the proteins are the most chemically and physically diverse. It is to the Dutch chemist Gerard Johannes Mulder (1802–1880) that we attribute the origin of the term **protein**.

Mulder clearly recognized the primary importance of this substance to living things, for the term itself (derived from the Greek "proteios") means "of the first rank." In Mulder's time, it was generally believed that protein was a single substance, essentially the same in all cells and organisms. This, of course, is not the case, and today it is recognized that there are hundreds of thousands of chemically distinct proteins. Indeed, a single human cell may contain more than 10,000 different protein molecules.

CLASSES OF PROTEINS

Several different schemes can be used to classify proteins, no scheme being entirely satisfactory. For example, proteins may be classified according to their biological functions, being either *structural* or *dynamic*. Structural proteins include keratin, the major protein component of hair, and collagen, abundant in skin, bone, and cartilage. The dynamic proteins include the *enzymes* that serve as catalysts of metabolism, *hormonal* proteins, *respiratory pigments*, and other proteins whose principal role is not structural.

Proteins may also be classified according to their

TABLE 4–1   THE CELLULAR MACROMOLECULES

Substance	Percentage of Total Cell Weight
Water	65–85
Protein	10–25
Polysaccharide	1–5
Lipid	2–10
Nucleic acid	0.5–5

shape or *conformation*. Again, two major categories may be distinguished: **fibrous** proteins and **globular** proteins. Fibrous proteins are proteins whose overall shape is threadlike, such as *myosin* of muscle tissue and the *fibrin* threads of coagulated blood. Globular proteins are quasi-spherical, a shape characteristic of enzymes and the respiratory pigments. Generally speaking, the conformations of fibrous proteins are simpler than those of globular proteins and are more amenable to study. Although it is convenient to classify proteins according to their function or conformation, the student should recognize that all systems of classification are at best artificial.

THE AMINO ACIDS

The first intensive biochemical studies of proteins were those carried out by the German biochemist Emil Fischer around the turn of the century. In fact, Fischer received the first Nobel Prize award (in 1902) for his biochemical studies (see Table 1–1). Among Fischer's contributions to our early understanding of protein chemistry was the discovery that all proteins consist of chains of smaller units that he named "amino acids." It is now known that there are about 20 different amino acids that occur regularly as constituents of proteins and that the size (i.e., molecular weight), shape, and function of proteins are determined by the number, type, and distribution of the amino acids present in the molecule. Proteins occur in a wide spectrum of molecular sizes from small molecules such as the hormone *adrenocorticotrophic hormone* (ACTH), which consists of only 39 amino acids and has a molecular weight of 4500, to extremely large proteins such as the invertebrate blood pigment hemocyanin, which consists of 8200 amino acids and has a molecular weight greater than 900,000 (see Table 4–2 for additional examples).

All amino acids conform to the general formula shown in Figure 4–1. The central, or *alpha*, carbon atom of each amino acid is covalently bonded to four groups: (1) a hydrogen atom, (2) an amino group, (3) an acid (or carboxyl) group, and (4) a side chain called an R-group. It is the particular chemical structure of the R-group that distinguishes one amino acid from another. The structural formulae of the amino acids that regularly occur in proteins are given in Figure 4–2, where they are also subdivided into different categories according to the functional properties or chem-

TABLE 4–2 MOLECULAR WEIGHT AND AMINO ACID CONTENT OF SOME REPRESENTATIVE PROTEINS

Protein	Number of Amino Acids	Molecular Weight
Adrenocorticotrophic hormone	39	4,500
Insulin	51	5,700
Ribonuclease	124	12,000
Cytochrome-*c*	140	15,600
Horse myoglobin	150	16,000
Trypsin	180	20,000
Hemoglobin	574	64,500
Urease	4,500	473,000
Snail hemocyanin	8,200	910,000

ical nature of their R-groups. As seen in Figure 4–2, the amino acids aspartic acid and glutamic acid have *amide* configurations called asparagine and glutamine, which also occur regularly in proteins. In addition to the common amino acids, there are also a number of rare amino acids. β-alanine (found in the vitamin pantothenic acid), γ-aminobutyric acid (found in brain tissue), and ornithine, citrulline, and homoserine are amino acids that occur regularly as intermediates of cellular metabolism, but these are rarely, if ever, incorporated into proteins.

All four groups that are attached to the alpha carbon atom of an amino acid are chemically different except in the case of glycine, where R is a hyrdrogen atom; therefore, all amino acids except glycine are optically active. The four groups may be considered to lie at the four corners of a tetrahedron, with the alpha carbon atom at the center. Accordingly, these groups may be arranged in either of two ways, yielding the L and D isomers (Fig. 4–3). Only the L forms have been identified in proteins, although the D forms do occur in some antibiotics.

FIGURE 4–1 General chemical formula of an anino acid.

FIGURE 4-2 The common amino acids.

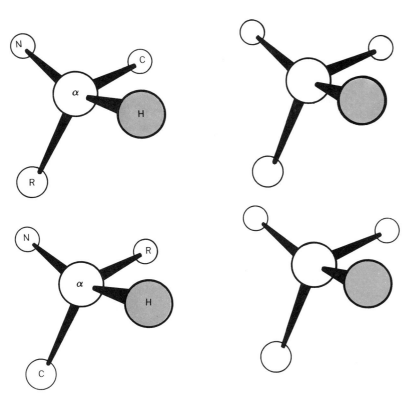

FIGURE 4–3 Stereoscopic diagram of the arrangements of groups about the alpha carbon atom of **L** (upper stereo pair) and **D** (lower stereo pair) amino acids. *R*, side chain; *N*, amino group; *C*, carboxyl group; *H*, hydrogen; α, alpha carbon.

THE PEPTIDE BOND

Proteins are composed of one or more chains of amino acids called **polypeptides**. The neighboring amino acids in the chain are linked together by **peptide bonds**; these bonds are formed, in effect, by the elimination of one molecule of water from the two amino acids linked by the bond, such that the α-carboxyl carbon atom of one amino acid is linked to the α-amino nitrogen atom of the neighboring amino acid (Fig. 4–4). This type of reaction is known as a *condensation reaction*, or, in this particular case, a *dehydration synthesis*. As will be seen in Chapter 22, which deals with the mechanism by which cells synthesize their proteins, the polymerization of the amino acids to form these chains involves a number of intermediate steps. The amino and carboxyl groups that are not directly bonded to the alpha carbon atom (such as those that occur in the side chains of lysine, glutamic acid, and aspartic acid) do not form covalent bonds with neighboring amino acids.

The two atoms that are involved in the peptide linkage and the four adjacent atoms lie in the same plane in space and form what is called a **planar amide**

group (Fig. 4–5). Primarily through the pioneering studies of L. Pauling and R. B. Corey, the specific interatomic distances and bond angles of the planar amide group are known; these are shown in Figure

FIGURE 4–4 The elimination of water between two amino acids links them together through a peptide bond. This is an example of a *dehydration synthesis*.

FIGURE 4–5 Stereoscopic diagram of the arrangement of atoms in a *planar amide group*. The alpha carbon atoms are shown in color.

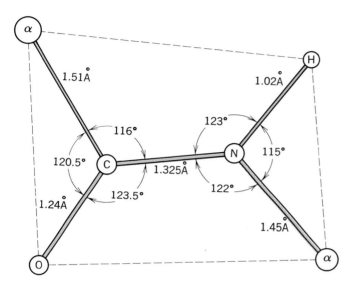

FIGURE 4–6 Dimensional properties of the planar amide group.

4-6. In 1954, Pauling received a Nobel Prize for his work on protein structure (Table 1-1). In the planar amide group, the C—N bond (i.e., the peptide bond) and the C—O bond assume a character somewhere between that of a double bond and a single bond. This prevents rotation about either bond and results in the planarity of the group.

It should be noted that a planar amide group includes parts of two neighboring amino acids, and the alpha carbon atoms at each end of a planar amide group are also included at the ends of neighboring planar amide groups. The bond angle between planar amide groups is 111°; this is the angle formed between the C—C bond of one planar amide group and the C—N bond of the next group. Note that this is an *intra*-amino acid bond angle! Figure 4-7 shows two successive planar amide groups arranged in such a manner that they are *coplanar* (i.e., both planar amide groups lie in the same plane). The alpha carbon atoms are identified simply by α. The important 111° angle formed between the C—C bond of one group (the upper group) and the C—N bond of the other (the lower group) is specifically identified. Clockwise rotation

(when viewed from the position of the alpha carbon atom) of the upper planar amide group about the C—C bond sweeps out the angle called ψ (psi). Clockwise rotation (again when viewed from the alpha carbon atom) of the lower planar amide group about the C—N bond sweeps out the angle called ϕ (phi). It should be noted that the 111° angle between the two planar amide groups remains fixed regardless of the values of ψ and ϕ. The flexibility of a chain of amino acids (i.e., a polypeptide) results from the rotations of successive planar amide groups about these bonds yielding the various values of ψ and ϕ and does not result from the linkages *between consecutive amino acids*. This point cannot be overemphasized if the conformations that can be assumed by polypeptides are to be properly understood. Figure 4-8 is a stereoscopic view showing two successive planar amide groups that are not coplanar.

The total length of two (or more) planar amide groups is maximized when ψ and ϕ are zero. Chain length is reduced as the ψ and ϕ angles are altered between 0° and 360°. If these angles are kept constant within a stretch of polypeptide, the series of planar amide groups naturally takes on the shape of a helix. In theory, all kinds of helices can be formed using various values for ψ and ϕ. However, in actuality, most combinations of ψ and ϕ are not possible, because they would result in the superimposition of atoms in space. Much of the experimental data that have been accumulated to date indicate that most polypeptides contain many regions of helical coiling and few regions of complete extension. The relative lengths of helical and extended segments of a polypeptide are determined by the specific sequence of amino acids present in the polypeptide (see below).

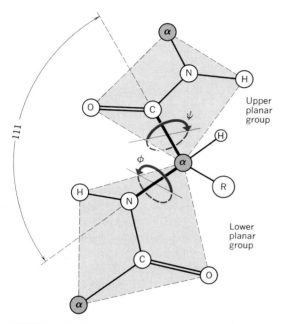

FIGURE 4–7 Two successive planar amide groups.

POLYPEPTIDES

Fundamental Properties of Helical Polypeptides

As just noted, among the diverse shapes that can be assumed by polypeptide chains is a helical shape; in fact, helically arranged polypeptides are fundamental to the shape of many fibrous and most globular proteins. To understand the properties of helical polypeptides, let us first consider the general properties of a geometric helix. Three parameters define the geometry of a helix; these helical parameters are **direction** (or *handedness*), **pitch**, and **radius** (Fig. 4–9). A helix's direction may be either *right-handed* or *left-handed*. In a right-handed helix, rotation is *clockwise* as the helix winds its way forward around its linear axis. In a left-handed helix, rotation is *counterclockwise*. Both right-handed and left-handed polypeptide helices are present in naturally occurring proteins.

The forward rotation of a helix around its axis is accompanied by a certain amount of *linear translation*. That is, every turn of the helix is accompanied by a specific amount of forward movement parallel to the helix's axis. The amount of forward movement for each 360° turn of the helix is known as the helix's pitch. The distance between a point on the backbone of a helix and the axis about which the helix winds is the helical radius; twice this value is the helical diameter. In geometric helices, the diameter is a constant. Unlike a geometric helix, the backbone of a helical polypeptide is not smooth. Rather, it is an irregular curve described by the bond angles between successive atoms of the backbone. Consequently, there is some variation in the diameter of a polypeptide helix.

For helical polypeptides, rules regarding direction

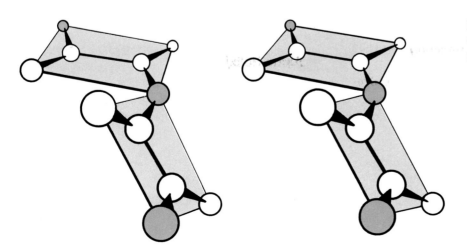

FIGURE 4–8 Stereoscopic diagram of two successive planar amide groups that are not coplanar.

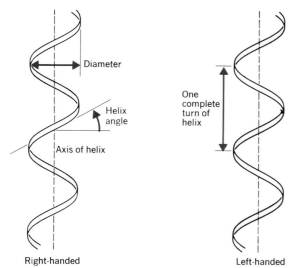

Right-handed   Left-handed

FIGURE 4–9 Fundamental properties of geometric helices.

FIGURE 4–10 Hydrogen bonds formed between carbonyl and amino groups (top) and between carbonyl and hydroxyl groups (bottom).

and diameter are the same as for geometric helices. The pitch (p) of a helical polypeptide is defined as the product of the number (n) of amino acids per complete turn of the helix and the amount of linear translation per amino acid (d), that is, $p=(n)(d)$, where n may be a whole number or a fraction. By definition, the sign of n is positive when the helix is right-handed and is negative when the helix is left-handed.

Hydrogen Bonds and Helix Stability

The helix conformation of a polypeptide is not intrinsically stable; indeed, were the peptide bonds that link the amino acids of the polypeptide chain together the only forces between amino acids, the chain would exist in any of a variety of random shapes. The stability of the highly ordered conformation of a helix is maintained by a number of additional interactions between amino acids. Particularly important among these interactions are weak electrostatic forces, called **hydrogen bonds**, that occur between one part of the helix and another.

Hydrogen bonds are a result of the differential attraction that atomic nuclei have for their orbital electrons (the attraction for an orbital electron is called **electrophilia**). For example, in a carbonyl group, the oxygen atom is more electrophilic than the carbon atom; consequently, the electron density in the vicinity of the oxygen nucleus is greater than in the vicinity of the carbon nucleus. Thus, a small negative electrostatic charge is associated with the oxygen atom and a small positive electrostatic charge is associated with the carbon atom. These charges are called *partial charges* because they do not represent full electrostatic units; they are respectively denoted δ^+ and δ^- (Fig. 4–10). Partial charges are also distributed between the nitrogen and hydrogen atoms of an amide group, because the hydrogen nucleus is far less electrophilic than the nitrogen nucleus. Weak electrostatic bonds can be formed between the partially positive hydrogen atom of an amide group and the partially negative oxygen atom of a carbonyl group (Fig. 4–10). These bonds are called hydrogen bonds because, in effect, the carbonyl oxygen and amide nitrogen are bonded together by the "sharing" of the hydrogen atom.

In helical polypeptides (and in other polypeptide conformations, see later), hydrogen bonds exist between the oxygen atoms of α-carbonyl groups and the hydrogen atoms of α-amino groups of amino acid residues that are separated by various distances along the backbone of the helix.

The Alpha Helix

One of the most common types of helix found in proteins is the **alpha helix**. The properties of this helix were determined in the 1930s and 1940s by Pauling and Corey by X-ray analysis of alpha keratin (a protein present in hair). In an alpha helix, there are 3.6 amino acids per turn (i.e., $n=3.6$), and there is a linear translation of 1.5 Å (0.15 nm) per amino acid. This results in a pitch of 5.4 Å (0.54 nm). Physical properties of the alpha helix are compared with other polypeptide conformations in Table 4–3. In the alpha helix (and in all other polypeptide helices), the R-groups of

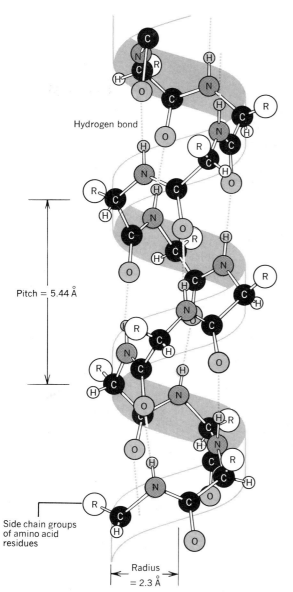

Hydrogen bond

Pitch = 5.44 Å

Side chain groups
of amino acid
residues

Radius
= 2.3 Å

FIGURE 4–11 Arrangement of atoms in, and dimensions of, the alpha helix.

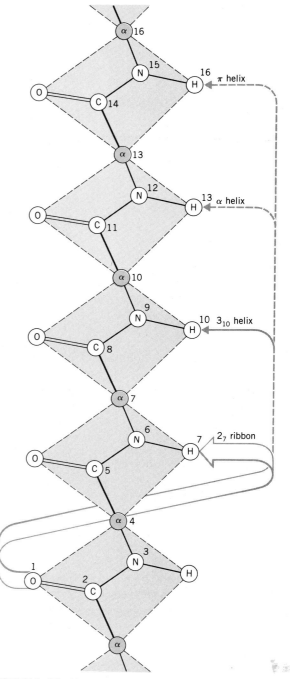

FIGURE 4–12 Hydrogen bonding patterns of the 2_7 ribbon, 3_{10}, alpha, and pi helices.

the amino acids are directed radially away from the helix's axis. The hydrogen bonds that contribute to the stability of this helix occur between the α-carbonyl oxygen atom of one residue and the α-amino hydrogen atom of another that is *three* residues farther along the backbone (Figs. 4–11 and 4–12).

Each hydrogen bond links the carbonyl oxygen and amino hydrogen atoms at the ends of a sequence of 13 covalently bonded atoms of the polypeptide (count the atoms forming this loop in Fig. 4–11). For this reason,

the alpha helix is also identified as 3.6_{13} helix (i.e., $n = 3.6$). Although the direction of an alpha helix can be right-handed or left-handed, only right-handed alpha helices are present in naturally occurring pro-

TABLE 4–3 STRUCTURAL PARAMETERS OF RIGHT-HANDED HELICES

Name of Helix	n	p	d	H-bond Atoms
Alpha helix	3.6 Å	5.4 Å	1.5 Å	13
Pi helix	4.4	3.5	0.8	16
3_{10} helix	3.0	6.0	2.0	10
2_7 ribbon	2.0	5.6	2.8	7

teins. Figure 4–13 is a stereoscopic diagram of a portion of a right-handed alpha helix.

With a single exception, any of the 20 common amino acids can occupy a position in an alpha helix. The exception is the secondary amino acid *proline*. The covalent bond that exists between the R-group of this amino acid and its α-amino group so constrains proline's spatial relationships with neighboring amino acids that it cannot satisfy the parameters of the alpha helix. For this reason, many alpha helical regions of polypeptide chains either begin or end at a position occupied by proline or are interrupted by proline, producing an abrupt bend in the helical chain.

Pi and 3_{10} Helices

Among other helix types that occur in proteins are the right-handed forms of the **pi** (or **4.4_{16}**) helix and the 3_{10} helix. The hydrogen bonding patterns that stabilize these helices are compared with that of the alpha helix in Figure 4–12. Other physical properties are compared in Table 4–3. In comparison with the pi and 3_{10} helices, the hydrogen bonding pattern of the alpha helix is strain-free, the hydrogen bonds being parallel to the helix's axis. The pi and 3_{10} helices are formed only with varying degrees of internal strain, and it is therefore not surprising that the alpha helix is so much more common in proteins than the other two right-handed forms.

Beta Conformations

Another common polypeptide conformation in proteins is the **beta conformation**. In the beta conformation, the polypeptide chain is nearly fully extended so that the backbone of the polypeptide follows a zigzag path and is not helical in the conventional sense. Stabilization of the beta conformation's ordered structure is through hydrogen bonding, but the hydrogen bonding is quite different than that in the alpha, pi, and 3_{10}

FIGURE 4–13 Stereoscopic diagram of an alpha helix. (Courtesy of C. K. Johnson, Oak Ridge National Laboratory.)

helices. In proteins possessing this conformation, stretches of polypeptide in beta conformation lie *parallel* to one another, with hydrogen bonds formed between the α carbonyl groups in one stretch and the α-amino groups in the neighboring stretch. The stretches of polypeptide may be portions of the same chain or different chains. This arrangement is depicted in Figure 4–14 and is called a **beta pleated sheet**. In the beta pleated sheet, all peptide linkages participate in the formation of the hydrogen bonds, so

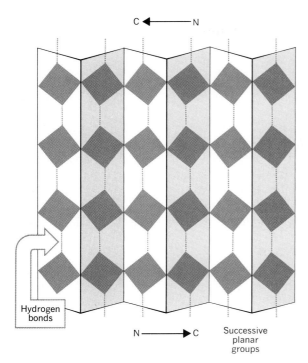

FIGURE 4–14 Structure of *silk*. Chains of nearly coplanar planar amide groups form a *pleated sheet* pattern linked together by hydrogen bonds.

that collectively they impart considerable stability to the structure.

A polypeptide chain is said to have *polarity* in that the order of covalently bonded atoms forming the backbone of the chain is not the same in both directions. One end of a chain contains an α-amino group that does not participate in a peptide linkage and the other end of the chain contains an α-carboxyl group that does not participate in a peptide linkage. These ends are respectively called the *amino* (or *N*) terminus and the *carboxyl* (or *C*) terminus of the polypeptide. If the polarities of neighboring polypeptide chains are the same, they are said to be *parallel*; if they have opposite polarities, they are said to be *antiparallel*. Parallel and antiparallel versions of the beta pleated sheet structure occur in proteins.

The polypeptide chains of *silk* fibers are arranged in the form of an antiparallel beta pleated sheet, as shown in Figure 4–14. Because neighboring chains are nearly fully extended, the protein resists stretching, although there is lateral flexibility between chains. The enzyme *superoxide dismutase* (present in red blood cells) contains regions of antiparallel beta

pleated sheet structure. Hence, the beta conformation is not restricted to fibrous proteins but is present also in globular proteins. A mixture of parallel beta pleated sheet structure and helical structure has been identified in *phosphofructokinase*, an enzyme involved in the glycolytic pathway discussed in Chapter 10.

Polyproline Helixes and the Structure of Collagen

As noted earlier, the unusual structure of the amino acid proline (see Fig. 4–2) causes this residue to be incompatible with the alpha helix conformation. However, polypeptide chains consisting only of proline residues (i.e., "polyproline") do form *left-handed* helices. The structure of polyproline is fundamental in the organization of the most abundant protein in animal tissues—*collagen*. Although species variations exist, in most collagens, proline (and hydroxyproline) accounts for more than 20% of the protein's amino acids, with the remainder predominantly represented by glycine and alanine. A. Rich, F. H. C. Crick, and G. Ramachandran have shown that the rodlike collagen molecule is formed by the braiding of three polypeptide chains. Each chain, like polyproline, is twisted to form a left-handed helix having three amino acids per turn (i.e., $n = -3$), and these are twisted around one another to form a **superhelix** that is right-handed. The organization of a portion of a collagen molecule is shown in Figure 4–15.

In collagen, every third amino acid of each left-handed helix is glycine, whose R-group is sufficiently small (i.e., the R-group is a hydrogen atom) to allow its projection into the core of the molecule. The distribution of the glycines is believed to be fundamental to the intimate intertwining of the three chains. The association between chains is stabilized by hydrogen bonds between proline and hydroxyproline residues in the neighboring chains. The arrangement of the three helices of the collagen superhelix accounts for the molecule's tensile strength and resistance to stretching.

FIGURE 4–15 Structure of *collagen*. In collagen, three left-handed polypeptide helices (color) are intertwined to form a right-handed coiled-coil or *superhelix*.

Random Coils

Most proteins contain stretches of polypeptide in which successive planar amide groups do not have identical ψ and ϕ values. Although these regions may be coiled, they do not form helices of a specific repeat pattern. Such stretches are called **random coils**. The term "random" is not to be interpreted as implying that the ψ and ϕ values between specific planar amide groups of the polypeptide undergo continuous and random change. Indeed, for a specific protein, the ψ and ϕ values of successive planar amide groups of a random coil are fixed; they just are not *identical* in successive groups. In some proteins, random coils account for a large percentage of the total structure.

LEVELS OF PROTEIN STRUCTURE

By convention, four levels of protein organization may be identified; these are called the **primary, secondary, tertiary**, and **quaternary** structures of the protein.

Primary Protein Structure

Successive amino acids forming the backbone of a polypeptide chain are linked together through peptide bonds and it is believed that these are the only covalent associations that occur between successive amino acids. The primary structure of a protein is the *order* of these amino acids in the backbone of each of the polypeptide chains comprising the molecule.

The primary structure of a polypeptide chain is de-lineated beginning with the amino acid occupying the polypeptide's N-terminus. For convenience, each amino acid is identified using its specific abbreviation (see Fig. 4–2). The first protein to have its primary structure determined was the hormone *insulin*, a relatively small protein containing only 51 amino acids. The insulin molecule consists of two polypeptide chains called the *A chain* (21 amino acids long) and the *B chain* (30 amino acids long). The structure of insulin is shown in Figure 4–16 and reveals yet another facet of the covalent associations that can exist in proteins. The A and B chains of insulin are linked together by two **disulfide bridges** and a third disulfide bridge occurs within the A chain. As shown in Figure 4–17, disulfide bridges are formed by the removal of hydrogen from the sulfhydryl groups of the side chains of two cysteine residues. When the primary structure of a polypeptide chain is determined chemically, it is customary to simultaneously determine which cysteine residues of the structure are involved in the formation of disulfide bridges.

Since the elucidation of the primary structure of insulin in 1953 by F. Sanger (for which Sanger received a Nobel Prize), several hundred proteins have been fully sequenced, many of these considerably larger than insulin. Among the fully sequenced proteins are nearly 100 forms of *hemoglobin*, the oxygen-transporting protein in the blood of vertebrates. Studies of hemoglobin have revealed some fascinating facts concerning the evolution of related proteins and the manner in which different polypeptide chains of a protein interact with one another in the molecule's biolog-

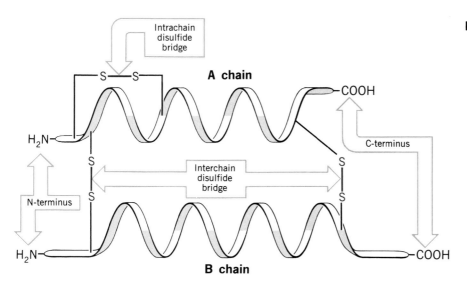

FIGURE 4–16 Structure of *insulin*.

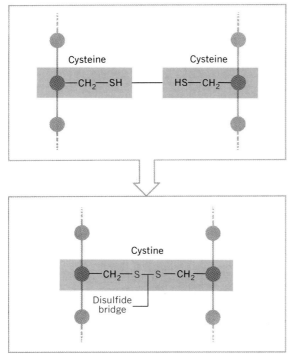

FIGURE 4-17 The *disulfide bridge*. Removal of hydrogen from the R-groups of two cysteine residues links them together through a disulfide bridge and forms *cystine*.

the resulting three-dimensional shape is considered. The three-dimensional shape is taken into account beginning with **secondary structure**. A protein's secondary structure describes any periodic spatial relationships within each of the polypeptide chains, such as (1) the locations and extent of those regions of each chain that are organized into helices and (2) the type of helices that are present.

Among the periodic structures that are common in polypeptide chains are the alpha, pi, and 3_{10} helices discussed earlier and the various beta conformations. In globular proteins, it is not uncommon for half of all the residues of each polypeptide to be organized into one or more specific secondary structures.

For convenience the various segments of a polypeptide chain can be assigned a specific nomenclature. Beginning at the N-terminus, the helical regions are denoted by the letters A, B, C, D, and so on, and the amino acids within each helix are assigned numbers (e.g., C1, C2, C3, etc.). The interhelical regions of each chain are denoted by the letters of the adjoining helices (i.e., nonhelical regions AB, BC, CD, etc.) and the amino acids within these regions are also assigned numbers (i.e., BC1, BC2, BC3, etc.). The nonhelical region at the N-terminus (if indeed the N-terminus is not part of a helix) is denoted NA and its amino acids

ical activity. We will turn to these areas later in the chapter.

Inherent Variety of Protein Primary Structures   The diversity of amino acids that may be included in proteins provides for an enormous number of different primary structures. Consider, for example, the mathematical variety that is possible in a polypeptide chain consisting of 61 amino acids (and this would be considered a relatively small protein). Each of the 61 residue positions can be occupied by any one of 20 different amino acids. Therefore, altogether there would be 20^{61} possible polypeptide molecules (i.e., 20^{61} different primary structures are possible). Now, $20^{61} = 2.3 \times 10^{79}$, and because it has been estimated that the entire universe contains 0.9×10^{79} atoms, there is greater potential variety in a polypeptide chain that is 61 amino acids long than there are atoms in the universe!

Secondary Protein Structure

When describing a protein's primary structure, the order of amino acids in each polypeptide chain *but not*

FIGURE 4-18 A hypothetical polypeptide illustrating the nomenclature used to describe amino acid positions in the chain's secondary structure.

are numbered consecutively (NA1, NA2, NA3, etc.). If there is a nonhelical segment at the C-terminus, it is identified on the basis of the last helix. For example, in a polypeptide chain containing eight helices (A through H), a nonhelical segment at the C-terminus would be identified as HC (and its amino acids numbered HC1, HC2, HC3, etc.). Using this type of nomenclature, the specific position of any amino acid can be identified (see Fig. 4–18).

Tertiary Protein Structure

Tertiary protein structure refers to the manner in which the helical and nonhelical regions of a polypeptide are *folded back on themselves* to add yet another order of shape to the molecule. In globular proteins, it is the nonhelical regions that permit the folding. The folding of a polypeptide chain is not random but occurs in a specific fashion, thereby imparting certain steric properties to the protein. Well before the three-dimensional atomic structure of the first protein was worked out, W. Kauzmann anticipated the general principles that would govern the overall shape of a protein. Kauzmann predicted in 1959 that all polar groups in the protein would either interact with each other or be solvated by the surrounding water, and that considerations of *entropy* (see Chapter 9) would draw the nonpolar parts of the protein together in the molecule's interior. This kind of specific folding is achieved and maintained by a variety of interactions between one part of the polypeptide chain and another and between the polypeptide and neighboring molecules of water. The interactions include (1) **ionic bonds** or **salt bridges**, (2) **hydrogen bonds**, (3) **hydrophobic bonds**, and (4) **disulfide bridges**.

Ionic Bonds (Salt Bridges)   In aqueous solutions, most amino acids occur in an ionized (or dissociated) state. For example, most molecules of glycine exist in the following form when glycine is dissolved in water:

$$H_3{}^+N-CH_2-C{\overset{\displaystyle O}{\underset{\displaystyle O^-}{\big\langle}}}$$

In this form, a hydrogen ion (i.e., a proton) has been dissociated from the α-carboxyl group and another has been removed from the surrounding water by the α-amino group. The resulting ion is called a **zwitterion** because it bears two different kinds of charge—positive and negative. Note that while having

both kinds of charge, the glycine molecule has no *net* charge.

The acidic amino acid aspartic acid has the following zwitterionic form:

$$\begin{array}{c} CH_2-C{\overset{\displaystyle O}{\underset{\displaystyle O^-}{\big\langle}}} \\ | \\ H_3{}^+N-CH-C{\overset{\displaystyle O}{\underset{\displaystyle O^-}{\big\langle}}} \end{array}$$

In this case, aspartic acid bears one positive charge and two negative charges and thus has a net charge (i.e., -1). Glutamic acid behaves in a similar manner.

Finally, the basic amino acid lysine yields the following zwitterion in solution:

$$\begin{array}{c} {}^+NH_3 \\ | \\ (CH_2)_4 \\ | \\ H_3{}^+N-CH-C{\overset{\displaystyle O}{\underset{\displaystyle O^-}{\big\langle}}} \end{array}$$

In this form, lysine carries two positive charges and one negative charge and has a net positive charge (i.e., $+1$).

In polypeptide chains, the α-amino and α-carboxyl groups of all of the amino acids except those that are at the N- and C-terminals are involved in peptide linkages. Therefore, except at the ends of the polypeptide chain, these groups are not ionized and contribute no charge to the polypeptide. However, the side chains of acidic and basic amino acids (as well as certain others) may contribute positive and negative charges along the length of the polypeptide if either conditions of local pH or the nature of the other side chains in the region of the tertiary structure allow dissociation or protonation. Electrostatic attraction between oppositely charged side chains of amino acids of a polypeptide may bring these regions of the chain closer together and stabilize their positions relative to one another. The bonds so formed are called **ionic bonds** or **salt bridges** (also **salt bonds**). It is also possible for ionized side chains of amino acids in the interior of the molecule to react with and bind water, and in many proteins a certain quantity of water is permanetly retained within the molecule by such interactions. Because salt ions (e.g., Na^+ and Cl^-) are also present in the surroundings of most proteins, these may also play a role in ionic bond formation between different ionized groups in the interior of the molecule. Ionic

FIGURE 4–19 The ionized R-groups and N- and C-terminals of a polypeptide chain (color) can form various ionic bonds (electrostatic interactions) with the surroundings and with each other.

FIGURE 4–20 The hydrophobic side chains of amino acids such as leu, ile, val, and phe interact with each other internally in the protein or in clefts at the protein's surface.

bonds also occur between charged side chains that project from the protein's surface and surrounding water and salt ions. The various kinds of ionic bonds are shown in Figure 4–19.

Hydrogen Bonds Hydrogen bonds formed between α-amino hydrogen atoms and α-carboxyl oxygen atoms have already been discussed in connection with the stabilization of helices and parallel chains of the beta pleated sheet structure. Hydrogen bonds can also be formed between undissociated carboxyl-containing side chains of the acidic amino acids and the amino groups of the basic amino acids lysine, tryptophan, and histidine. The hydroxyl groups of serine, theonine, and tyrosine may also participate in hydrogen bonding, as may the secondary carboxyl and amino groups of asparagine and glutamine. Although individually weak, these bonds collectively contribute to the stability of a specific tertiary structure.

Hydrophobic Bonds A third class of interactions that stabilize tertiary protein structure are **hydrophobic bonds**. These are interactions between amino acids whose side chains are hydrophobic (e.g., leucine, isoleucine, valine, and the aromatic amino acids). The side chains of these amino acids are drawn together by

their mutual hydrophobic properties, becoming organized in such a manner as to have minimal contact with the surrounding water. Placed in close proximity to one another, juxtaposed atoms of separate side chains undergo *van der Waals interactions* with each other, resulting in the formation of weak bonds. Again, it is the large numbers of these interactions that impart stability to the structure. Figure 4–20 depicts the stabilization of a fold in a polypeptide chain by the hydrophobic association between two valine side chains.

Disulfide Bridges Because they are covalent, disulfide bridges are the strongest bonds formed between one part of a polypeptide chain and another. The nature and formation of these bonds have already been discussed in connection with primary protein structure (see above). Such bonds can be formed between cysteine residues in different regions of a polypeptide (and also between cysteine residues in different polypeptide chains of a protein, see below). Where they occur, disulfide bridges contribute a considerable stabilizing influence to tertiary structure.

The four classes of bonds just discussed are depicted together in the generalized tertiary protein structure diagrammed in Figure 4–21. As you examine this diagram, it is important to note that bonds stabilizing tertiary folding may *simultaneously stabilize secondary structure*. For example, the disulfide bridge and the hydrophobic and electrostatic bonds that keep the top and middle helices of the protein depicted in Figure 4–21 parallel to each other also serve to prevent unwinding of these two helices. Thus, in a general sense, *specific interactions between one part of a protein and another can play a stabilizing role at more than one level of the protein's structure.*

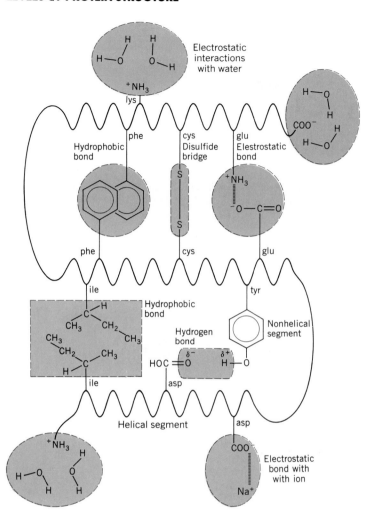

FIGURE 4–21   A hypothetical polypeptide illustrating the various types of interactions that may be involved in maintaining a particular tertiary structure.

Quaternary Protein Structure

Many proteins consist of more than one polypeptide chain. In proteins that are composed of two or more polypeptide chains, the **quaternary structure** refers to the specific orientation of these chains with respect to one another and the nature of the interactions that stabilize this orientation. The individual polypeptide chains of the protein are usually referred to as its **subunits**. Table 4–4 lists some representative proteins that are composed of subunits and gives their numbers, designations, and molecular weights. As can be seen from this sampling, proteins can contain either a small number of large subunits (e.g., *thyroglobulin*), a large number of small subunits (e.g., *apoferritin*), or any intermediate combination. Moreover, in some proteins the subunits are polypeptide chains whose primary structures are identical to each other (e.g.,

L-*arabinose isomerase*), whereas in others the subunits are different (e.g., *immunoglobulin G*).

The same classes of interactions that contribute to the stability of tertiary protein structure also serve to stabilize the quaternary association of subunits, namely, ionic bonds, hydrogen bonds, hydrophobic bonds, and disulfide bridges. As we shall see in Chapter 8, many cellular enzymes are composed of subunits, and the resulting quaternary structure is of fundamental importance in the regulation of enzyme activity. The molecular weights of proteins composed of subunits are often great enough for the molecules to be seen and studied by electron microscopy of negatively stained preparations. Electron microscopy thus provides additional information about quaternary structure, for it is often possible to discern the number and orientation of the protein's subunits.

TABLE 4–4   REPRESENTATIVE SUBUNIT COMPOSITION OF SOME PROTEINS

Protein	Molecular Weight	Subunits		
		Number	Designation	Molecular Weight
Hemoglobin A (human)	64,500	4	Alpha chains (2)	15,700
			Beta chains (2)	16,500
Lactate dehydrogenase	135,000	4	A chains (0 to 4)	33,600
			B chains (4 to 0)	33,600
Immunoglobulin G	150,000	4	Light chains (2)	25,000
			Heavy chains (2)	50,000
Tryptophan synthetase (*E. coli*)	150,000	4	Alpha chains (2)	29,500
			Beta chains (2)	45,000
Aspartate transcarbamylase	306,000	12	C chains (6)	34,000
			R chains (6)	17,000
L-arabinose isomerase (*E. coli*)	360,000	6	(identical)	60,000
Apoferritin (iron-storage protein)	456,000	24	(identical)	19,000
Thyroglobulin	670,000	2	(identical)	335,000

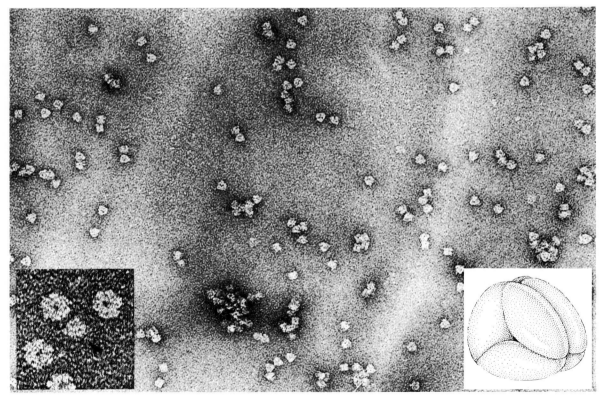

FIGURE 4–22   Molecules of the enzyme *L-arabinose isomerase* revealed by the technique of negative staining (see Chapter 1). The quaternary structure of the enzyme is formed from two stacks containing three subunits each. The arrangement is seen more clearly in the enlarged insert (the arrow identifies a second enzyme in the sample, *glutamine synthetase*) and in the diagram. (Courtesy of Drs. F. A. Eiserling and L. J. Wallace.)

The subunit organization of the enzyme L-*arabinose isomerase* is quite evident in the electron photomicrographs of Figure 4–22.

Among the groups of proteins whose quaternary structures have been extensively studied are the *hemoglobins* and *immunoglobulins*. Probably more is known about the chemistry, organization, and functions of members of these two groups than about all other proteins combined. Later in the chapter we shall consider these two groups of proteins in considerable detail.

To this point, we have been using the terms primary, secondary, tertiary, and quaternary structure exclusively in connection with proteins. However, corresponding levels of organization are recognized for *polysaccharides* and *nucleic acids* in which the order of their building blocks (i.e., *sugars* and *nucleotides*) and the coiling and folding of their chains can be delineated.

Establishment of Secondary, Tertiary, and Quaternary Structure

Although it is clear how a particular secondary, tertiary, and quaternary structure is maintained, how such a specific arrangement is initially achieved is not fully understood. Cells appear *not* to possess templates or special enzymes that function in molding a particular three-dimensional shape. Instead, most polypeptides *spontaneously* assume their biologically active shape from among the many different alternatives that exist. As we have seen, polypeptide chains are flexible. Thus, free to twist and fold and subject mainly to kinetic and steric constraints, polypeptides can progress through a variety of different shapes, progressively assuming the shape that is energetically most stable.

We shall see in Chapter 22 that a polypeptide chain is formed *in situ* by the sequential addition of individual amino acids beginning at the chain's N-terminus. Clearly, for many proteins, encoded in the polypeptide's primary structure are critical sequences of amino acids that promote the progressive assumption of a specific tertiary structure by the polypeptide as the primary structure is being laid down. Such progressive folding significantly reduces the total number of possible alternative shapes assumable by a polypeptide. In other words, once folding of a portion of the polypeptide proceeds to its thermodynamically stable state, certain alternative configurations made possible by continued elongation of the primary struc-

ture are no longer kinetically accessible. For proteins composed of two or more polypeptide chains and possessing quaternary structure, it is likely that the polypeptide subunits randomly encounter one another, automatically orienting themselves and combining in such a manner as to establish the proper (i.e., biologically active) quaternary structure. Some of the experimental evidence supporting these ideas follows.

Dissociated (or protonated) side chains of the acidic and basic amino acids create negative and positive sites along the polypeptide. These sites can be coordinated and stabilized by water molecules, which are also partially polar, or by dissolved ions. Consequently, segments of a polypeptide containing such charged residues would naturally tend to orient themselves at the outer surface of the protein where they could interact with the surrounding aqueous environment. The side chains of other amino acids such as tyrosine, glutamine, and so forth, which are electrically neutral on the whole, contain atoms of nitrogen and oxygen that are partially polar due to the differential electrophilia of these atoms. Therefore, water would also be attracted to these groups, although not to the same degree as fully charged side chains. By interacting with electrically charged surface groups, the water molecules minimize the strength of the electric fields surrounding these groups and in so doing stabilize the molecule's structure.

Amino acids such as leucine and phenylalanine are nonpolar and have hydrophobic side chains. Hydrophobic groups disturb the haphazard arrangement of water molecules, tending to impart order to the water. Any increase in order makes a system potentially less stable by reducing the system's *entropy* (which is a measure of the disorder of a system, see Chapter 9). Because entropy tends to be maximized, the hydrophobic side chains are usually turned away from the surrounding water and face each other internally.

Accordingly, among the many proteins that have been extensively studied, most of the amino acids with polar side chains are at the protein's outer surface, whereas the amino acids with hydrophobic side chains are either confined to the interior of the molecule or form crevices in its surface so as to have minimal contact with the surrounding water and ions.

Anfinsen's Studies with Ribonuclease The notion that, acting in concert, the specific primary structure of a polypeptide and the innate properties of the side chains of its amino acids cause the polypeptide to

FIGURE 4–23 Denaturation and spontaneous renaturation of the enzyme *ribonuclease*.

the four disulfide bridges re-form, and essentially all of the catalytic activity of the protein is restored (see Fig. 4–23). Similar observations have been made with other proteins, that is, they are capable of spontaneously reestablishing their biologically active tertiary structure after having undergone extensive molecular disorganization. In these proteins, the three-dimensional structure that is crucial to biological function is directed by the primary structure, and in the appropriate environment one particular configuration is overwhelmingly favored energetically over other possibilities. Polypeptides possess sufficient flexibility to explore various configurations, progressively assuming the most stable of these.

Spontaneous Formation of Quaternary Structure
Studies of a number of different proteins composed of two or more polypeptide chains support the concept of spontaneous assumption of a specific and functional quaternary structure from the separate subunits. Among those proteins with which the phenomenon is readily demonstrated is the blood protein hemoglobin. Normal adult human hemoglobin, called hemoglobin A, may be represented as

$$\alpha_2\beta_2$$

This notation indicates that the protein is a *tetramer*, that is, it consists of *four* polypeptide chains—a pair of

spontaneously assume its biologically active tertiary structure originated with the studies of the enzyme *ribonuclease* begun in the 1950s by C. B. Anfinsen; later (1972) Anfinsen received the Nobel Prize for this most definitive work. Ribonuclease is produced by cells in the pancreas and is then conveyed through ducts to the duodenal section of the small intestine, where the enzyme acts to degrade RNA in ingested food. Ribonuclease contains 124 amino acids, has a molecular weight of about 12,000, and consists of a single polypeptide chain. Anfinsen identified four disulfide bridges in the protein, suggesting that the enzyme is highly folded (Fig. 4–23). As is the case with nearly all enzymes, the catalytic activity of ribonuclease depends on the maintenance of a particular three-dimensional shape. In concentrated solutions of β-mercaptoethanol and urea, the disulfide bridges of the enzyme are broken and the resultant unfolding of the polypeptide is accompanied by a loss of enzyme activity. The enzyme is said to be *denatured*. If the β-mercaptoethanol and urea are removed by dialysis and the denatured ribonuclease reacted with oxygen,

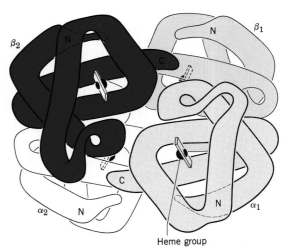

FIGURE 4–24 The hemoglobin molecule. Each molecule consists of four polypeptide chains—a pair of *alpha* globin chains and a pair of *beta* globin chains. Associated with each chain is an oxygen-binding *heme* group.

FIGURE 4–25 Stereo pair depicting the four subunits of hemoglobin. Cylindrical portions of the molecule are helices, whereas ribbon shapes represent random coils. The atoms of the heme group are shown as spheres. (Based on computer-generated protein models of Dr. A. Lesk.)

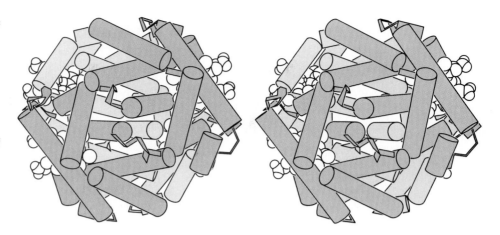

alpha chains and a pair of *beta* chains (Figs. 4–24 and 4–25). If an aqueous solution of hemoglobin is made weakly alkaline or weakly acidic, the hemoglobin tetramers undergo a stepwise dissociation into subunits, as follows:

$$\alpha_2\beta_2 \rightleftharpoons 2\,\alpha\beta \rightleftharpoons 2\,\alpha + 2\,\beta$$

(tetramer)   (hybrid or asymmetric dimer)   (monomers or subunits)

As indicated in the above reaction, the dissociation of the tetramer, first into "asymmetric dimers" and then into monomers or subunits, is reversible, that is, when normal pH is restored, fully functional hemoglobin molecules are sequentially re-formed. These and similar observations with other proteins support the view that intrinsic properties of the individual subunits of a protein are sufficient to promote the assumption of the biologically active quaternary structure. In the case of hemoglobin, it appears that the subunits seek out one another in solution and combine to form the biologically functional tetramer.

CONJUGATED PROTEINS

Conjugated proteins are proteins that contain nonprotein constituents or **prosthetic groups**. The prosthetic groups are *permanently* associated with the molecule, usually through covalent and/or noncovalent linkages with the side chains of certain amino acids. Conjugated proteins can be divided into three major classes: (1) **chromoproteins** (2) **glycoproteins**, and (3) **lipoproteins**.

Chromoproteins

The **chromoproteins** are a heterogeneous group of conjugated proteins related to each other only in that they all possess *color*. The *hemoglobins*, *myoglobins*, and other heme-containing proteins such as the *cytochromes* and *hemerythrins* belong to this group. The prosthetic groups of the chromoproteins, such as the heme groups of hemoglobin and the cytochromes, are bound to the polypeptide portion of the molecule through a combination of covalent and noncovalent bonds. We shall examine the interaction between the hemes and polypeptide chains of hemoglobin in detail later in the chapter.

Glycoproteins

Glycoproteins are proteins that contain various amounts of carbohydrate (the chemistry of carbohydrates is dealt with in Chapter 5). A number of very important proteins fall in this class, including many of the blood plasma proteins and a large number of enzymes and hormones. The surfaces (i.e., plasma membranes) of most cells also contain quantities of glycoproteins, and these molecules serve there as antigenic determinants and as receptor sites. Virtually all of the carbohydrate that is present in red blood cells occurs as membrane glycoproteins. Although more than 100 different sugars (or monosaccharides) are known, only about nine occur as regular constituents of glycoproteins (Table 4–5).

The amount of carbohydrate present in glycoproteins varies from less than 1% of the molecule's total molecular weight to more than 85% (Table 4–6). For example, in egg white *ovalbumin* (molecular weight 45,000), there is only one monosaccharide per protein

TABLE 4–5 CARBOHYDRATES OCCURRING AS REGULAR CONSTITUENTS OF GLYCOPROTEINS

Glucose
Galactose
Mannose
Fucose
Acetylglucosamine
Acetylgalactosamine
Acetylneuraminic acid
Arabinose
Xylose

TABLE 4–6 CARBOHYDRATE CONTENT OF GLYCOPROTEINS

Glycoprotein	Percentage of Carbohydrate	Function
Ovalbumin	1	Hen-egg food reserve
Follicle-stimulating hormone (FSH)	4	Hormone
Fibrinogen	5	Blood coagulation protein
Transferrin	6	Iron transport protein of blood plasma
Ceruloplasmin	7	Copper transport protein of blood plasma
Glucose oxidase	15	Enzyme
Peroxidase	18	Enzyme
Luteinizing hormone	20	Hormone
Haptoglobin	23	Hemoglobin-binding protein of blood plasma
Erythropoietin	33	Hormone
Mucin	50–60	Mucus secretion
Blood-group glycoproteins	85	Unknown

molecule, whereas in *mucin* (a secretion of the salivary glands having a molecular weight of about 1 million), about 800 monosaccharides are present. The carbohydrate moieties of glycoproteins are usually bound to the protein through covalent bonds with either asparagine, threonine, hydroxylysine, serine, or hydroxyproline (see Fig. 4–26). The carbohydrate bonded at each site of the protein may consist of a single monosaccharide unit (as in Fig. 4–26) or a linear or branched chain of several monosaccharides (called an oligosaccharide), as depicted in Figure 4–27.

Lipoproteins

Lipid-containing proteins are called **lipoproteins**. This class includes some of the blood plasma proteins and also a large number of membrane proteins. The lipid content of lipoproteins is often very high, accounting for as much as 40 to 90% of the total molecular weight of the complex. In lipoproteins, the amount of lipid present markedly affects the *density* of the molecule, and this property is often used as the basis for lipoprotein classification. Whereas uncomplexed proteins have a density in water of about 1.35, lipoproteins vary in density down to 0.9 (i.e., a lipoprotein may be less dense than water).

The interactions between the lipid and protein portions of a lipoprotein usually involve similar functional groups. For example, the hydrophobic portions of

fatty acids, glycerides, sterols, and the like (see Chapter 6) form van der Waals interactions with the hydrophobic side chains of the nonpolar amino acids. Covalent bonds are believed to occur between the phosphate moieties of certain phospholipids and the hydroxyl-containing side chains of amino acids like serine. Lipoproteins are discussed further in Chapter 15 in connection with membrane structure and properties.

Nucleoproteins

In eukaryotic cells, specific proteins called **nucleoproteins** are found intimately associated with nuclear DNA. Also, in prokaryotes as well as eukaryotes, ribonucleoprotein complexes (i.e., protein complexed with RNA) occur. These proteins are not usually classified with the conjugated proteins, because the nucleic acids involved cannot be regarded as prosthetic groups. Two types of proteins have been identified in nucleoproteins, the *histones* and *nonhistones*. Histones have a rather restricted amino acid composition (containing about 25% arginine and lysine) and are quite similar in all plant and animal cells. Their highly basic nature accounts for the close associations they form with the nucleic acids and lends credence to the notion that they are involved in the tight packing of DNA molecules during the condensation of chromatin (i.e., chromosomes) that precedes mitosis. The nonhis-

FIGURE 4–26 Covalent bonds formed between carbohydrates and the side chains of specific amino acids in *glycoproteins*.

Acetylglucosamine

Asparagine

Acetylgalactosamine

Threonine

Galactose

Hydroxylysine

Xylose

Serine

Arabinose

Hydroxyproline

○ = Amino acid

⬡ = Monosaccharide

FIGURE 4–27 The carbohydrate portion(s) of glycoproteins may be short or long chains of sugars and may be branched.

tones are considerably more heterogeneous in amino acid composition and have acidic properties. There is much evidence to suggest that by selectively combining with certain stretches of nuclear DNA, the nonhistones are involved in the regulation of gene expression. The histones and nonhistones are considered further in conjunction with the discussion of nuclear organization and function in Chapter 20.

In the remaining parts of this chapter, we shall be concerned with the structural properties of two specific proteins, the hemoglobins and the immunoglobulins. These proteins have received particularly intense study during the past few decades, studies that have revealed some rather startling facts about (1) the relationship between molecular architecture and biological activity and (2) the diversification of molecular organization in functionally related proteins within a species of organism and among different species.

VERTEBRATE HEMOGLOBINS

Our understanding of the structure of the hemoglobin tetramer and the individual globin chains has advanced rapidly during the past 15 years, primarily because information about this molecule has been acquired through a variety of independent methods of investigation. These include (1) X-ray crystallographic studies of hemoglobin crystals, providing atomic resolution on the order of just a few angstroms; (2) biochemical studies of amino acid sequences in isolated globin chains; (3) determination of the nucleotide sequences in globin chain genes (from these sequences the primary structures of the globin chains can be inferred); (4) electron spin resonance (ESR) studies of the heme groups and their manner of interaction with the globin chains; (5) radioactive tracer studies of the mechanism of globin chain biosynthesis; and (6) physiological studies of the behavior of the molecule. Most notable among those scientists who have contributed to our present understanding of the structure and function of hemoglobin is M. F. Perutz, whose work with this protein began nearly 50 years ago. In 1962, Perutz, together with his colleague J. C. Kendrew, received the Nobel Prize for his studies of hemoglobin and a closely related protein, *myoglobin.*

The following discussion of the anatomy of hemoglobin molecules will illustrate a number of general features that are characteristic of protein structure and function. There is, however, a more far-reaching lesson to be learned from these studies of hemoglobin. Hemoglobin, like many other proteins (especially enzymes), is widespread in nature and the primary structure of the molecule is now known for several dozen different species of organism. Although differences in composition do exist, there are overriding similarities among all of the hemoglobins. These similarities are not merely chemical curiosities; rather, *they provide us with an insight into the nature of evolution as it proceeds at the molecular level.*

General Structure of the Hemoglobins

The hemoglobins are conjugated globular proteins, that is, they contain some nonprotein parts. In vertebrates, the function of hemoglobin is to transport oxygen in the blood from the lungs to the tissues, and therefore hemoglobin molecules can exist in two states: oxygenated and unoxygenated. In all but the lowest vertebrates, hemoglobin is a tetramer. In lampreys, however, hemoglobin is monomeric, that is, it contains a single globin chain. In this regard, lamprey hemoglobin is similar to myoglobin (see below). In humans, the most common type of hemoglobin is hemoglobin *A* (abbreviated HbA), which consists of 574 amino acids and has a molecular weight of 64,500. Its secondary, tertiary, and quaternary structure is typical of all higher vertebrate hemoglobins. The protein portion of the molecule, called globin, is composed of four polypeptide chains, each of which is also globular in shape. The four globin chains consist of two identical pairs: two alpha chains (141 amino acids each) and two beta chains (146 amino acids each). The nonprotein portion of hemoglobin consists of four heme groups, one associated with each of the four globin chains.

Cysteine residues are not uncommon among the various hemoglobins; for example, there are six in HbA. Yet, despite the fact that disulfide bridges between cysteine residues play an important role in the maintenance of secondary, tertiary, and quaternary structure in many proteins, there are no such linkages in the hemoglobins. Instead, the tertiary structure of each chain and the quaternary structure of the whole tetramer are maintained by noncovalent linkages.

Hemoglobin molecules are highly symmetric (Fig. 4–24). The molecule can be divided into two identical halves, each consisting of an $\alpha\beta$ dimer (called a "hybrid" or "asymmetric" dimer). The complete tetramer is similar to a mildly flattened sphere (i.e., an oblate ellipsoid) having a maximum diameter of about 5.5 nm. The four polypeptides are arranged in such a manner that unlike chains have numerous stabilizing interactions, whereas like chains have few. That is, each alpha chain has many contacts with both betas, and each beta chain has many contacts with both alphas; however, alpha–alpha and beta–beta contacts are few. The like chains face each other across an axis of symmetry such that rotation of any chain 180° about this axis would make it congruent with its identical partner. A cavity about 2.5 nm long and varying in width from about 5 to 10 Å passes through the molecule along the axis. This arrangement is seen in Figure 4–24.

The interior of each subunit contains many amino acids with hydrophobic side chains; these form hydrophobic bonds with one another. Amino acids with ionized side chains predominate at the surface of the subunits and at the surface of the molecule as a whole. Amino acids with nonpolar side chains may be found along a polar segment at the surface, but in such cases the side chains are buried in a crevice formed in the molecule's surface and thereby have little contact with the surrounding water. As noted above, contacts between like subunits are rare. In the oxygenated molecule, ionic bonds (salt bridges) occur between the polar amino and carboxyl terminals of the respective chains and also between a few other residues. Contacts between unlike subunits are much more numerous and include 17 to 19 hydrogen bonds between about 35 amino acids of each hybrid dimer. Far fewer bonds link alpha and beta chains of different dimers and no covalent bonds link any of the subunits together.

Association of the Globin Chains with Heme and the Heme with Oxygen

All of the globin chains have the same fundamental secondary structure (Fig. 4–28). The helical regions are alpha helices, although there may be small excursions into pi and 3_{10} conformations. The heme group that is associated with each globin chain consists of an organic portion and an iron atom (Fig. 4–29). The organic part is called **protoporphyrin IX** and is formed from four **pyrrole rings** linked together by methene

FIGURE 4–29   Chemical formula of heme.

bridges. All of the atoms of this tetrapyrrole lie in a single plane; however, when associated with globin, the plane of the tetrapyrrole becomes slightly dome-shaped. Four *methyl*, two *vinyl*, and two *propionic acid* groups are attached to the tetrapyrrole as side chains. The iron atom is linked to each pyrrole ring, and depending on whether or not the hemoglobin is oxygenated, the iron lies either in the same plane as the tetrapyrrole or just above it. Each globin chain envelopes its heme group in a deep cleft (Figs. 4–24 and 4–25), exposing only the propionic acid side chains to the surrounding water.

Within each cleft, the heme groups interact with the R–groups of 16 amino acids from seven segments of the globin chain. Except for the iron atom, the portion of the heme group that is enclosed in the cleft is strongly hydrophobic and most of the bonds that are formed with the heme involve hydrophobic side chains of amino acids such as leucine, valine, and phenylalanine. The tertiary structure of an alpha globin chain together with its heme group is depicted stereoscopically in Figure 4–30.

Although each iron atom can form up to six bonds, only five occur in the unoxygenated form of hemoglobin. Four of these bonds are formed with the pyrrole nitrogen atoms and the fifth with the side chain of a histidine residue at position F8 (called the "proximal" histidine). In the unoxygenated state, the bond between iron and the proximal histidine "lifts" the iron atom out of the plane of the tetrapyrrole. Histidine F8 is said to be "asymmetrically tilted" in that its bond with iron is not perpendicular to the heme plane. A

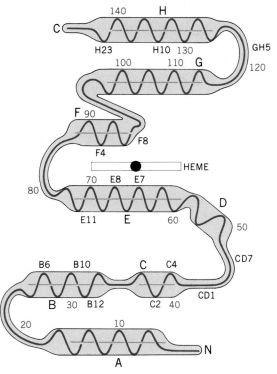

FIGURE 4–28   Secondary structure of the beta chain of human hemoglobin. The highly conserved amino acid positions discussed in the text are specifically noted.

FIGURE 4–30 Stereoscopic diagram showing the tertiary structure of the alpha globin chain of hemoglobin. (*Courtesy* of R. E. Dickerson and I. Geis.)

second histidine residue at position E7 (the "distal" histidine) interacts with the iron but does not form a bond (Fig. 4–31). During oxygenation of hemoglobin, molecular oxygen (i.e., O_2) enters the heme cleft on the distal side, where one of the two oxygen atoms then forms the sixth bond with iron. Oxygen binding causes the iron atom to be pulled back into the heme plane. This movement also draws the proximal histidine closer to the heme group, straightens the histidine–iron atom bond, and precipitates a train of additional small changes in the conformation of the polypeptide (Fig. 4–31). The changes that occur within a subunit as a result of oxygen binding lead to changes in subunit–subunit interaction and are fundamental to the "cooperativity" (discussed below) exhibited by hemoglobin.

Families of Globin Chains and Globin Chain Homologies

As noted earlier, *myoglobin* is a protein that is closely related functionally and structurally to hemoglobin. In myoglobin, however, there is a single polypeptide chain and a single heme group. Thus, three "families" of globin chains exist in nature. These are (1) the *alpha* and (2) *beta* families of globin chains that comprise all of the hemoglobins, and (3) the *myoglobin* family of globin chains. At the present time, about 60 different vertebrate alpha family globin chains, 66 beta family chains, and 60 myoglobin family chains have been fully sequenced. In addition to human hemoglobin and myoglobin, the known sequences include those of the chimpanzee, gorilla, monkey, rabbit,

mouse, horse, pig, cow, dog, sheep, camel, goat, llama, kangaroo, echidna, platypus, whale, chicken, turtle, frog, carp, shark, and lamprey. All of these globin chains share a fundamentally similar secondary and tertiary structure. Because the primary structures of these globin chains are known, it is possible to analyze the relationship between their common secondary and tertiary structures and their respective primary structures.

If two (or more) chains of the same family are found to have the same amino acid occupying a given position in the secondary structure (e.g., position B6, E7, FG2, and so on), then the chains are said to be *homologous* at that position. Such comparisons reveal an interesting relationship between the extent of homology of the globin chains of the various vertebrate species and the evolutionary relatedness of the organisms. Moreover, certain positions of the secondary structure are invariably occupied either by a specific amino acid or by a limited number of different amino acids, implying that a high degree of conservation has occurred. Such conservation is clearly related to the constraints that must be satisfied for the assumption of a specific and common tertiary structure among the various globin chains.

Highly Conserved Amino Acid Positions   Among the 60 alpha family chains that have been sequenced, identical amino acids are found at 23 of the 141 positions. The 66 beta family chains that have been sequenced have identical amino acids at 20 of their 146 positions. For the myoglobins, 27 of the 153 amino acid positions are occupied by the same amino acid. If one compares

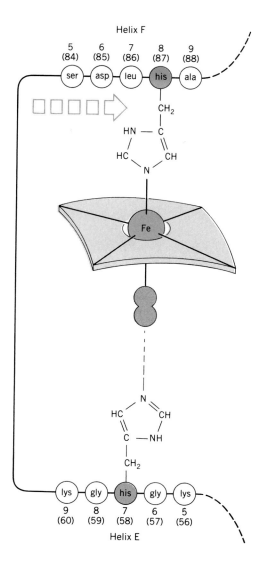

FIGURE 4–31 Tertiary structural changes occurring in an alpha subunit of hemoglobin during the oxygenation of its heme group (see text for details).

all 186 globin chains, five positions are found to be totally invariant (Table 4–7). An additional 10 positions are invariably occupied by one of two alternative amino acids whose side chains have the same properties (e.g., they are hydrophobic, positively charged, negatively charged, etc.) Of the two alternative occupants of a given position, one is overwhelmingly favored. Clearly, these 15 residues play an important part in maintaining the three-dimensional shape peculiar to the globins. For some of these positions, the role is readily explained; for example, amino acid F8 (totally invariant) is the proximal histidine and E7 (invariant in all but a few globins) is the distal histidine.

TABLE 4–7 HIGHLY CONSERVED GLOBIN CHAIN POSITIONS

Hydrophobic	Polar or Ambivalent
Invariant	
CD1 (phe)	C4 (thr)
F4 (leu)	F8 (his)
H23 (tyr)	
Almost invariant[a]	
B10 (leu, val)	B6 (gly, ala)
C2 (pro, ala)	B12 (arg, ser)
CD7 (leu, phe)	E7 (his, gln)
E11 (val, ile)	E8 (gly, ala)
GH5 (phe, leu)	H10 (lys, ala)

[a]The first amino acid listed is overwhelmingly favored.

There are 33 positions in the interior of the globin chains that do not contact the surrounding environment; among these, 30 positions are invariably occupied by amino acids with nonpolar side chains. There are also 10 positions in crevices at the surface of the molecule invariably occupied by nonpolar amino acids. Thus, more than one–fourth of the amino acid positions in globin are invariably nonpolar. These positions may be occupied by any one of several amino acids. It is generally agreed that the invariably nonpolar positions, together with the nonpolar portions of the heme group, play a major role in determining the shape of the globin chain. Internally, the hydrophobic interactions between side chains form the skeletal framework on which the shape is founded. The substitution through genetic mutation of a single polar amino acid in a position that is invariably nonpolar usually is sufficient to disrupt the normal organization of the polypeptide and render it biologically nonfunctional. Ionic and hydrogen bonds between residues of different chains (hemoglobin) and between residues and the surrounding water and ions (hemoglobin and myoglobin) also contribute to the stability and universal shape of the molecule.

Function and Action of Hemoglobin: Cooperativity in Proteins

The function of hemoglobin is to reversibly bind molecular oxygen, that is,

$$\underset{\substack{\text{unoxygenated}\\\text{hemoglobin}}}{\text{Hb}} + \underset{\text{oxygen}}{4\,O_2} \underset{\text{in tissues}}{\overset{\text{in lungs}}{\rightleftharpoons}} \underset{\text{oxyhemoglobin}}{\text{Hb}(O_2)_4}$$

The manner in which oxygen is bound by and released from hemoglobin reveals yet another characteristic of protein (especially enzyme) action, namely, that of *cooperativity*. Hemoglobin is contained within the red blood cells (erythrocytes) and is oxygenated as the blood circulates through the capillary networks of the lungs. On leaving the lungs, virtually every hemoglobin molecule has combined with four molecules of oxygen. In this state, the hemoglobin is said to be 100% *saturated*. Later, when the blood is circulated to the other body tissues, hemoglobin releases its bound oxygen. The amount of oxygen that is released is determined in part by the concentration of dissolved oxygen gas (i.e., the *partial pressure*) in the surround-

FIGURE 4–32 Oxygen association/dissociation curves of hemoglobin and myoglobin.

ing plasma and body fluid. In muscle, for example, it would not be unusual for the percentage saturation of hemoglobin to fall to 40% or lower as the released oxygen diffuses from the erythrocytes into the plasma and then into the muscle tissue. As noted earlier, the closely related and structurally similar oxygen-binding protein myoglobin, which acts to temporarily store oxygen in certain muscle tissues, consists of a single polypeptide chain and one heme group. Consequently, myoglobin can reversibly bind *only one* molecule of oxygen. Figure 4–32 compares the oxygen association/dissociation curves of hemoglobin and myoglobin.

Referring to Figure 4–32, as the oxygen partial pressure rises between 0 and 20 mm Hg, myoglobin rapidly combines with oxygen and quickly approaches complete saturation. The association/dissociation curve takes the form of a *rectangular hyperbola*. The behavior of hemoglobin is considerably different and much more complex. At low oxygen partial pressure, the affinity of hemoglobin for oxygen is considerably less than that of myoglobin. For example, at 20 mm Hg partial pressure, hemoglobin is only about 21% saturated. However, between 20 and 60 mm Hg oxygen partial pressure, the affinity of hemoglobin for oxygen is greatly increased and the hemoglobin approaches saturation. Above 60 mm Hg, hemoglobin binds only small quantities of additional oxygen. Unlike the myoglobin curve, the oxygen association/dissociation curve of hemoglobin is *sigmoid*. From a physiological standpoint, the unique oxygen-binding characteristics of hemoglobin are crucial. The partial pressure of oxygen in the blood leaving the lung capillaries is generally in excess of 100 mm Hg, but by the time the blood reaches the capillaries of the various body tissues, the partial pressure has fallen to 80 mm Hg. An examination of Figure 4–32 shows that in this interval hemoglobin will have released only a small percentage of its oxygen. However, in the tissue capillaries, where the oxygen partial pressure often falls below 40 mm Hg, hemoglobin releases much of its bound oxygen. In actively exercising muscle, where the partial pressure may drop to 20 mm Hg, still greater quantities of oxygen would be released. Thus, within the range of 60 to 20 mm Hg (a range within which most of the tissues of the body operate), a relatively small decrease in oxygen partial pressure is accompanied by a quantitative release of hemoglobin-bound oxygen.

In contrast, myoglobin retains nearly all of its bound oxygen even at a partial pressure of 20 mm Hg. Indeed, the greater affinity for oxygen displayed by myoglobin at all partial pressures provides for the efficient transfer of oxygen from the blood to myoglobin-containing musculature. If the hemoglobin oxygen association/dissociation curve was hyperbolic (like myoglobin's) instead of sigmoid, then inadequate amounts of oxygen would be released from the blood to the tissues, leading to asphyxiation even during moderate exercise. The difference between the behavior of these closely related proteins results from the *subunit interactions* that are possible in the tetrameric hemoglobin molecule but that cannot occur in the monomeric myoglobin. The subunit interactions exhibited by hemoglobin belong to a general class of interactions that are possible in multisubunit proteins; the phenomenon is called **cooperativity**.

Cooperativity in Hemoglobin   The complex behavior of hemoglobin, which is precisely what is required of an efficient oxygen-transporting system, may be attributed to its quaternary structure, for when hemoglobin is dissociated into its subunits, the separate subunits exhibit the oxygen-binding behavior of myoglobin (i.e., a hyperbolic curve). In the intact tetramer, the various subunits exhibit cooperativity. Unoxygenated hemoglobin is said to be in the "T" (i.e., *tense*) state, whereas oxygenation leads to the "R" (i.e., *relaxed*) state. In T-state hemoglobin, the iron atom of each heme group is pulled out of the plane of the tetrapyrrole and toward helix F by its interaction with the side chain of each proximal histidine. Binding of oxygen draws the iron atom into the plane,

pulling the histidine side chain toward the heme group and straightening the asymmetric tilt (see above); hence, a small change in tertiary structure occurs. Now, since each globin chain is bonded to neighboring subunits, *changes taking place in one subunit will induce changes in another* by altering the nature of the subunit associations. In particular, during oxygenation, ionic bonds linking subunits together are broken, leading to the R state. The changes in tertiary and quaternary structure that accompany the oxygenation and deoxygenation of hemoglobin have been suspected for some time, for it has long been known that crystals formed by oxyhemoglobin are quite different from those formed by deoxyhemoglobin.

The binding of the first oxygen molecule to unoxygenated hemoglobin involves one of the alpha subunits; while this occurs slowly, the conformational change in this subunit induces a change in the associated subunit (i.e., the beta member of the asymmetric dimer). The result is a more rapid binding of the second oxygen molecule by the beta subunit. The changes occurring in one subunit induce changes in the other subunit (i.e., the other asymmetric dimer), including a 15° rotation of the two dimers with respect to one another. Although this movement accelerates the binding of the third oxygen molecule, the fourth oxygen is bound less rapidly.

Effects in which conformational changes in one or more subunits of a protein alter the structure of other subunits in such a way as to modify the protein's behavior are called **cooperative effects**. In the case of hemoglobin, it may now be understood why the oxygen association/dissociation curve is sigmoid and not hyperbolic, for the individual alpha and beta subunits of the tetramer *do not operate independently of one another* in the binding of oxygen but instead influence each other. The importance of cooperative effects in proteins cannot be overstated, for we shall see in Chapter 8 that cooperativity is also exhibited by a number of enzymes whose affinities for their substrates can be modulated through subtle changes in the individual subunits of the enzyme.

EVOLUTION OF PROTEINS

Ontogeny and Phylogeny of Hemoglobin

Functionally and structurally similar proteins exist in diverse animal and plant species. Many of these are believed to have a common evolutionary origin. The picture of protein evolution is probably clearest in the case of the hemoglobins, where it is possible to make accurate comparisons of hemoglobin structure in many different animal species. Moreover, in humans (and also in other vertebrates), different kinds of hemoglobin molecules are present in the blood at different stages of development, and this ontogeny also sheds light on the phylogenetic picture.

In addition to hemoglobin A, several other normally occurring hemoglobins are present in humans. In the very early stages of embryonic development, a hemoglobin called **Gower-1** appears. This hemoglobin consists of two alpha family chains called *zeta* (ζ) chains and two beta family chains called *epsilon* (ϵ) chains; the Gower-1 tetramer is thus represented as $\zeta_2\epsilon_2$. Beginning at about the eighth week of development, the zeta and epsilon chains are gradually replaced by the adult alpha chains and also two different beta family chains that are designated *gamma*^{G} (γ^G) and *gamma*^{A} (γ^A); the two types of gamma chains are identical except that in γ^G the amino acid glycine occurs at position H13 and in γ^A this position is occupied by alanine. Alpha chain production, once begun, continues throughout embryonic, fetal, and adult life. The next hemoglobin to appear is called **Gower-2** and consists of two alpha and two epsilon chains (i.e., $\alpha_2\epsilon_2$). This is quickly followed by the appearance of hemoglobin **Portland**, which consists of two zeta and two gamma chains (i.e., $\zeta_2\gamma_2$). Both Gower-2 and Portland are soon replaced by **hemoglobin F** (designated HbF and called **fetal hemoglobin**), consisting of two alpha and two gamma chains (i.e., $\alpha_2\gamma_2$). HbF predominates in the blood through the remainder of fetal development. Beginning soon after the twelfth week of development, the adult beta chains begin to appear and there is a progressive increase in HbA (i.e., $\alpha_2\beta_2$) and an accompanying decrease in HbF. Shortly before birth, yet another beta family chain appears called the *delta (δ) chain*; together with alpha chains, the delta chains form a hemoglobin known as **A$_2$** (i.e., $\alpha_2\delta_2$). By about six months after birth, little if any HbF can be found in the blood. At this time, about 98% of the hemoglobin is HbA and the remaining 2% is HbA$_2$. In a normal individual, this ratio persists through adult life. The various human hemoglobins and their subunit compositions are listed in Table 4–8. Their differential temporal appearance is shown in Figure 4–33.

A consistent feature of the human hemoglobins is that two members of each tetramer are always alpha family chains and the other two are beta family chains. Conditions similar to this have been found in

TABLE 4–8 THE HUMAN HEMOGLOBINS

	Symbol	Chains
Embryonic hemoglobins		
Portland	P	$\zeta_2\gamma_2$
Gower-1	G$_1$	$\zeta_2\epsilon_2$
Gower-2	G$_2$	$\alpha_2\epsilon_2$
Fetal hemoglobin	F	$\alpha_2\gamma_2$
Adult hemoglobins		
Hemoglobin A	A	$\alpha_2\beta_2$
Hemoglobin A$_2$	A$_2$	$\alpha_2\delta_2$

all other mammals studied so far and also in other vertebrates. Except in the case of the lowest vertebrates (e.g., lampreys and hagfishes), hemoglobin molecules are always tetramers and are formed from one pair of alpha family chains and one pair of beta family chains. When adult organisms have two or more hemoglobins, they share a common alpha family chain. Transition from one developmental stage to another, as in the metamorphosis of amphibians, is accompanied by a change in hemoglobin in which one pair of chains (i.e., the alpha family chains) is conserved. In the lowest vertebrates, hemoglobin is a monomer (i.e., one globin chain and one heme group), and in this regard is similar to all of the vertebrate myoglobins. All of these observations play a critical role in the development of the current concept concerning the evolution of the hemoglobin molecule.

The primary structures of the alpha family and beta family chains of the human hemoglobins reveal a high degree of homology. An examination of Table 4–9 re-

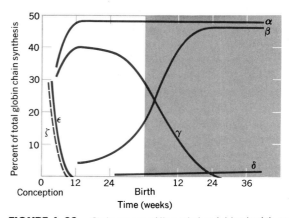

FIGURE 4–33 Ontogeny of the alpha (α) beta (β), gamma (γ), delta (δ), epsilon (ϵ), and zeta (ζ) globin chains of human hemoglobin (see text for explanation).

TABLE 4–9 NUMBERS OF AMINO ACID DIFFERENCES BETWEEN HUMAN GLOBIN CHAINS

		Alpha Family		Beta Family						Myoglobin
		Alpha	Zeta	Beta	GammaG	GammaA	Delta	Epsilon		
Alpha Family	Alpha		61	75	80	81	76	83		103
	Zeta	61		87	82	82	85	83		103
Beta Family	Beta	75	87		39	40	10	36		109
	GammaG	80	82	39		1	41	30		109
	GammaA	81	82	40	1		42	30		109
	Delta	76	87	10	41	42		40		108
	Epsilon	83	83	36	30	30	40			111
Myoglobin		103	103	109	109	109	108	111		

veals that there are 75 differences between the alpha and beta chains, implying that about one-half of all positions in each chain are occupied by the same amino acid. A common evolutionary origin for the alpha and beta chains is suggested in view of their extensive homology.

Still higher degrees of homology exist among the alpha family and among the beta family chains (Table 4–9). The alpha and zeta chains have 61 differences; the beta and gamma chains have only 39 differences; beta and epsilon chains have only 36 differences; beta and delta chains have only 10 differences; and so on. Thus, the beta, gamma, delta, and epsilon chains are more closely related to each other than any of them are to the alpha family chains; the alpha and zeta chains are more closely related to each other than either one is to the beta family chains.

Just as striking are the similarities between human alpha chains and the alpha chains of other mammals and between human beta chains and the beta chains of other mammals. Table 4–10 lists these differences for a number of mammals in order of diminishing phylogenetic relationship to man. No differences exist between the primary structures of human and chimpanzee alpha and beta globin chains. Gorilla and human globin chains differ in only two positions. Although the more distantly related mammals reveal increasingly large numbers of differences, it is consistently observed that *the alpha chains or beta chains of the different species are more closely related than are the alpha and beta chains within a species* (e.g., human alpha chains and kangaroo alpha chains have a higher degree of homology than do human alpha and

human beta chains). The latter class of observations is basic to contemporary theory on hemoglobin evolution. It is interesting (but not surprising) that a common revelation of these hemoglobin studies is that *species taxonomically more closely related have greater similarities (i.e., fewer differences) in the primary structures of their globin chains.*

The secondary and tertiary structure of myoglobin is almost identical to that of the globin chains of hemoglobin, and myoglobin's primary structure reveals an especially high degree of homology with the alpha family globin chains (see Table 4–9). In view of their similarity, myoglobin and hemoglobin are believed to have a common evolutionary origin. This notion is also supported by observations that the hemoglobin of the lowest vertebrates (e.g., lampreys and hagfishes) consists of a single polypeptide chain and not a tetramer.

On the basis of the numerous studies of vertebrate myoglobins and hemoglobins, it is generally agreed that all the globin chains have a common evolutionary

TABLE 4–10 DIFFERENCES BETWEEN HUMAN GLOBIN CHAINS AND THOSE OF OTHER MAMMALS

Mammal	Alpha Chain	Beta Chain
Chimpanzee	0	0
Gorilla	2	2
Monkey	4	8
Rabbit	18	18
Pig	18	17
Horse	18	25
Kangaroo (marsupial)	27	40
Echidna (monotreme)	36	31

origin. The proposed scheme for the evolution of human myoglobin and hemoglobin is shown in Figure 4–34. According to this scheme, the **structural genes** (see Chapters 20 and 21) for the various globin chains arose through a series of gene duplications, and subsequently these genes underwent separate and independent evolution. The degree of similarity that exists between the primary structures of the globin chains (Table 4–9) is used to establish the relative phylogenetic positions of the structural genes. Thus, the greater degree of homology that exists between the beta and delta globin chains than exists between the beta and gamma chains is presumably the consequence of the more recent duplication of the primitive beta/delta gene, and so on.

Based on the extent of homology that exists between the primary structures of polypeptide chains from different species and the presumed frequencies with which gene mutations occur, it is possible to estimate the age (and even suggest the primary structure) of the ancestral polypeptide. Such studies are encompassed in a field of biology called molecular paleogenetics. Phylogenetic relationships between species suggested on the basis of protein similarities are remarkably consistent with those that were predicted many decades ago using more conventional taxonomic criteria.

IMMUNOGLOBULINS

One of the body's most important defense mechanisms against infection is the production by the reticuloendothelial tissues of a class of proteins called **antibodies** or **immunoglobulins**. These proteins cir-

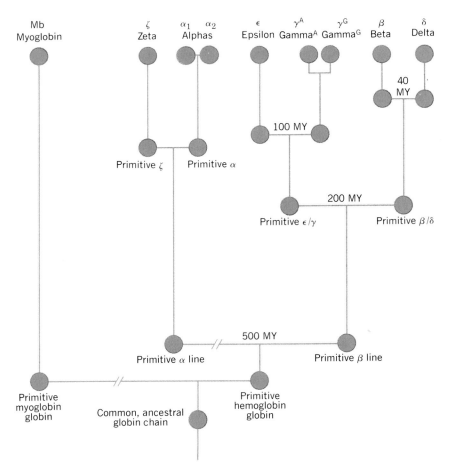

FIGURE 4–34 Evolution of the globin chains found in the human hemoglobins and in myoglobin. Beginning with a common, ancestral structural gene and a corresponding globin chain, a series of gene duplications occurred. Following each duplication, the product genes underwent individual diversification through mutation, ultimately producing the multiplicity of different normal globin chains existing today. (MY = millions of years.)

culate in the bloodstream, where they make up part of the "gamma globulin" fraction of blood plasma (see Chapter 13). The production of the immunoglobulins is stimulated by chemical substances, called *antigens*, present in or released from the infecting agent (e.g., constituents of bacterial membranes or the coats of viruses) and recognized as being foreign or alien to the body. This response is called the "immune response." Although antigens can belong to virtually any chemical category, they usually are proteins, polysaccharides, or nucleic acids. The immunoglobulin molecules that are produced during an immune response have the capacity to bind to the antigens, the reaction between the two being highly specific; that is, each type of immunoglobulin that is produced in the body reacts with a particular antigen and no other. A major source of the antibodies that circulate in the blood are the "plasma cells" derived by the differentiation of a class of white blood cells known as *B lymphocytes*.

The binding of antibodies to antigens that are in the surface of invading microorganisms (such as bacteria) triggers a series of reactions that leads to the destruction of the foreign cells. These reactions involve a host of other blood plasma proteins collectively referred to as **complement** and that are normally present in the plasma (i.e., unlike the immunoglobulins, they are not synthesized in response to the appearance of an antigen). Attachment of the immunoglobulin molecules to the surface antigens of the infecting cells is followed by the sequential binding and activation of the complement proteins. Acting in a manner similar to digestive enzymes, the complement proteins create holes in the bacterial surface. This produces a "Donnan effect" (see Chapter 15) in which ions and water enter the foreign cell, causing it to swell and eventually rupture (i.e., lyse). In effect, the immunoglobulins by binding to the surface antigens have served to identify the cells that are to be attacked and destroyed by complement. Each antibody has more than one antigen-binding site. Therefore, when free antigen molecules are encountered in the body, they can be *cross-linked* to form a large precipitable complex. When antibodies react simultaneously with antigens in the surfaces of separate cells, the cells are *agglutinated*, that is, they are clumped together into small masses. Precipitated antigens and agglutinated cells are ultimately engulfed and disposed of by *phagocytic* white blood cells. The immune response appears to be of rather recent evolutionary origin, because immunoglobulin production is characteristic only of vertebrates.

Immunoglobulin Structure

The human body is capable of synthesizing more than a million different kinds of immunoglobulin molecules, each capable of reacting with a different antigen, but all of them appear to share the same fundamental quaternary structure. During the early stages of an infection, the response to the antigen involves the production of a specific class of immunoglobulins known as **immunoglobulin M** or **IgM**, having a molecular weight of about 1,000,000. Later, the amount of IgM gradually declines as another class or "isotype" of immunoglobulin called *IgG* appears. Although other immunoglobulins are also present in blood (see Table 4–11), IgG represents the most abundant form (as much as 80%). IgG has been more extensively studied than the other immunoglobulins and much of the description that follows is based on results obtained from studies of IgG.

IgG molecules with a molecular weight of about 150,000 are composed of four polypeptide chains of two different kinds—a pair of identical high-molecular-weight chains called "heavy" or **H chains**

TABLE 4–11 IMMUNOGLOBULINS

Class	Light Chains	Heavy Chains	Tetramer	Molecular Weight	Sedimentation Coefficient
IgG	k or λ	γ^a	$k_2\gamma_2$ or $\lambda_2\gamma_2$	150,000	7 S
IgA	k or λ	α	$(k_2\alpha_2)_n$ or $(\lambda_2\alpha_2)_n$	180,000 to 500,000	7 S, 10 S, 13 S
IgM	k or λ	μ	$(k_2\mu_2)_5$ or $(\lambda_2\mu_2)_5$	950,000	18 S to 20 S
IgD	k or λ	δ	$k_2\delta_2$ or $\lambda_2\delta_2$	175,000	7 S
IgE	k or λ	ϵ	$k_2\epsilon_2$ or $\lambda_2\epsilon_2$	200,000	8 S

[a]There may be several different gamma chains.

and a pair of identical lower-molecular-weight chains called "light" or **L chains**. The L chains have a molecular weight of 20,000 to 25,000 and consist of about 214 amino acids. The H chains have a molecular weight of 50,000 to 55,000 and contain about 450 amino acids. In associating with each other, the four chains form a molecule whose quaternary structure resembles that of the letter "Y" (Fig. 4–35). Each arm of the Y contains a complete L chain and part of an H chain and the leg of the Y contains the remaining parts of the H chains. Near its C-terminus, each L chain is linked to an H chain by a disulfide bridge, and two additional disulfide bridges link the H chains together.

Each of the four polypeptide chains that form an immunoglobulin is divided into separate regions called "domains." There are *two* domains in the L chains and *four* in the H chains. Within each of the domains, folding of the polypeptide chain produces two parallel planes each containing several segments with folded beta structure. In each folded beta structure, neighboring stretches of the polypeptide have opposite polarity (i.e., they are antiparallel). The two parallel planes in each domain are held together by disulfide bridges and also by van der Waals interactions between hydrophobic side chains of amino acids in each sheet.

In each of the arms of the immunoglobulin, the H and L chains associate to establish a globular quaternary structure maintained principally by van der Waals interactions between hydrophobic side chains of amino acids in each chain; some hydrogen bonding between chains also occurs.

Antigen- and Complement-Binding Sites

Studies employing digestive enzymes such as trypsin and papain have revealed which portions of an immunoglobulin molecule combine with the antigens and which part binds complement. These digestive enzymes characteristically cleave immunoglobulins in a region (called the "hinge" region) located on the C-terminal side of the disulfide bridge that links each L chain to an H chain (see Fig. 4–35). Two types of fragments are produced by such enzymatic cleavage, one of which retains the capacity to bind the antigen (called the F_{ab} fragment). Each of the two F_{ab} fragments that is produced consists of a complete L chain and the N-terminal half of an H chain. The other type of fragment consists of the C-terminal halves of both H chains and is called the F_c fragment because it can

be *crystallized*. The antigen appears to be bound near the ends of the two F_{ab} units (the shaded area in Fig. 4–35). The angle between the F_{ab} and F_c units is variable because of the flexibility of the hinge regions of the molecule. As a result, the overall shape of an immunoglobulin can vary between that of the letter "Y" and the letter "T." This flexibility is unquestionably related to the immunoglobulin's capacity to bind to the antigens. The complement proteins associate with the F_c fragment. Immunoglobulins contain a small amount of carbohydrate (i.e., the immunoglobulins are glycoproteins). The carbohydrate portion consists of a short chain of sugars covalently bonded to an asparagine residue and sandwiched along the interface between the F_{ab} and F_c units.

ORIGIN AND EVOLUTION OF THE IMMUNOGLOBULINS

Homologous and Variable Domains

Among the immunoglobulins, certain domains of both H and L chains are homologous or *constant* (i.e., essentially the same from one immunoglobulin to another) and certain domains are *variable* (i.e., they are the basis of the differences among the immunoglobulins). Each L chain has one variable and one constant domain; each H chain has one variable and three constant domains. The variable domains occur near the N-terminals of the polypeptide chains and together create an antigen-binding site that is unique to that immunoglobulin molecule. The variable domain of each light chain is designated V_L and the constant domain C_L. The variable domain of a heavy chain is designated V_H and the constant regions are called C_H1, C_H2, and C_H3.

Not all light chains have the same constant regions; instead, there are two different classes of light chains, designated kappa (κ) and lambda (λ). About 40% of the amino acid positions of kappa and lambda chains are homologous; this degree of homology is similar to that exhibited by the alpha and beta chains of hemoglobin (see above). There are five different types of heavy chains, and these are designated γ, α, λ, δ, and ϵ. In the heavy chains, the variable region is about the same size as in the light chains but the constant region is about three times as long. Various combinations of the different light and heavy chains give rise to the five classes (or "isotypes") of immunoglobulins listed in Table 4–11.

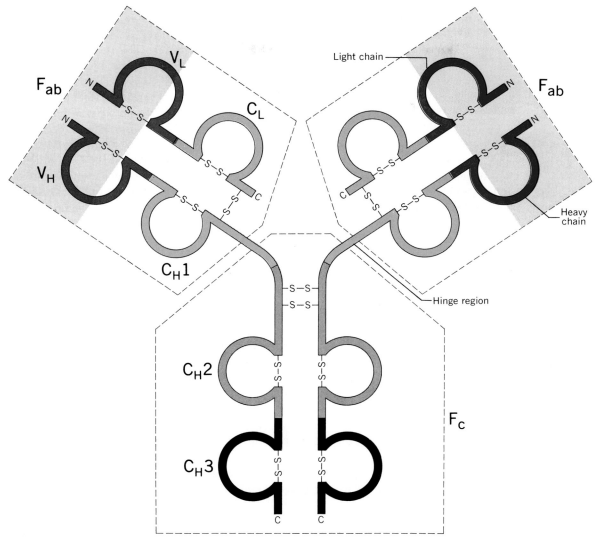

FIGURE 4–35 Fundamental architecture of an immunoglobulin. Each molecule is formed from four polypeptide chains (two "heavy" and two "light" chains) linked together by disulfide bridges. Each chain contains variable (color) and constant (gray shading) domains.

Enzymatic digestion of the immunoglobulins for subsequent chemical analysis cleaves the molecule in the "hinge" region and produces the F_{ab} and F_c fragments. The area of the immunoglobulin that binds the antigen is lightly shaded. (See text for details.)

As we have already noted, the N-terminal regions of the L and H chains have different primary structures; these differences are thought to reflect the capability of different immunoglobulins to react with different antigens. The variable regions of the L chains (V_L) are homologous with the variable regions of the H chains (V_H). Moreover, the constant regions of the H chains (C_H1, C_H2, and C_H3) are similar to one another and are also homologous with the constant region of the L chains (C_L).

The internal homologies between various parts of the immunoglobulins suggest that these polypeptides, like those that form the various hemoglobins, have a common ancestral origin. During evolution, gene duplications followed by separate diversification of the multiple genes probably produced all the immunoglobulin genes that are present today in the human genome. However, unlike the hemoglobin saga in which duplication and diversification produced the seven globin structural genes that are usually expressed, hundreds of different genes for the various regions of the L and H chains have arisen.

Basis of Immunoglobulin Diversity   For many years, the prevailing view with regard to the proteins of a cell was that each different polypeptide chain was the product of a separate structural gene. For example, the seven expressed globin chain genes of man give rise to the seven alpha family and beta family chains, which in different combinations produce all of the human hemoglobins. The immunoglobulin picture is considerably more complex. The immune system of one individual is capable of producing *more than a million* different immunoglobulins, but this is not the consequence of the existence of a corresponding number of structural genes for the H and L chains. The basis of so great an immunoglobulin diversity is the rather astounding finding that H and L polypeptide chains are the products of the combined contributions of several different genes; these are called the **C** (for **constant**), **V** (for **variable**), **J** (for **joining**), and **D** (for **diversity**) genes. While several hundred different V genes are present in the genome, there are much smaller numbers (probably less than 10) of C, J, and D genes. V, C, J, and D genes are recombined in various ways to form functional (i.e., genetically expressible) units. For example, the formation of an expressible light chain unit requires the joining of a V gene to a C gene via one of the J genes. Formation of an expressible heavy chain unit requires the joining of a V gene to a D gene and the joining of the latter complex to a J gene–C gene complex.

During embryonic development and also during an immune response, the differentiation of antibody-producing cells is accompanied by the **translocation** of chromosomal segments (see Chapter 20), thereby bringing the various V, J, D, and C genes together. The existence of a large number of different constant region genes for the H chains accounts for the greater variety of heavy chain classes seen in Table 4–11. The diversity of immunoglobulins is therefore partly explained by the very large number of possible combinations of V, J, D, and C genes. Additional diversity results from "point mutations" during the replication of the genetic material and other chromosomal phenomena that in a more general sense serve to increase genetic variation in all living things, not just in antibody-producing cells. These mechanisms are considered further in Chapters 20 and 21.

SUMMARY

Of the four major classes of macromolecules, the **proteins** are the most diverse and complex, being composed of one or more chains of **amino acids**. About 20 different amino acids occur in proteins and are covalently linked together within each chain by **peptide bonds**. The specific sequence of amino acids in each of the polypeptide chains forming a protein is called the **primary structure**. Each polypeptide chain may contain one or more regions twisted to form *helical* structures stabilized by **covalent**, **hydrophobic**, **electrostatic**, and/or **hydrogen bonds**. This is called the **secondary** level of protein structure. A more compact structure may be achieved by folding the polypeptide chain in the nonhelical regions; this establishes the polypeptide's **tertiary** structure. In proteins that have two or more polypeptides, the specific orientation of the chains with respect to one another is called **quaternary** structure.

The secondary, tertiary, and quaternary structure of proteins may be established during assembly of the primary structure or may occur spontaneously following completion of synthesis. Hydrophobic and hydrophilic amino acids are specifically distributed in relation to the final tertiary and quaternary structure of the molecule.

Proteins containing **prosthetic groups** are called **conjugated proteins** and include **chromoproteins**, **glycoproteins**, and **lipoproteins**. Chromoproteins are chemically heterogeneous and are related only in that they possess color. In glycoproteins the prosthetic groups consist of one or more branches of sugar chains covalently linked to the polypeptides. In the lipoproteins, the prosthetic groups consist of lipids linked to the protein through either covalent bonds or hydrophobic interactions.

Among all proteins studied, the structure, action, and evolution of the **hemoglobins** and **immunoglobulins** are best understood. The alpha and beta families of the globin chains of hemoglobin are encoded in separate genes descended from a common ancestor through gene duplication and mutation. In marked contrast, the heavy and light chains of the immunoglobulins are encoded by hundreds of genes belonging to four different classes. Variable recombination of members of each class of genes ultimately provides the incredible diversity of primary structures that characterizes the immunoglobulins within a species.

REFERENCES AND SUGGESTED READING

Articles and Reviews

Baldwin, R., Seeding protein folding. *Trends in Biochem. Sci.* **17** (1), 6 (Jan. 1986).

Chothia, C., and Lesk, A. M., Helix movements in proteins. *Trends in Biochem. Sci.* **10** (3), 116 (Mar. 1985).

Doolittle, R. F., Similar amino acid sequences: Chance or common ancestry? *Science* **214**, 149 (1981).

Doolittle, R. F., Proteins. *Sci. Am.* **253** (4) 88, (Oct. 1985).

Edelman, G. M., The structure and function of antibodies. *Sci. Am.* **223** (2) (Aug. 1970).

Efstratiadis, A., Paskony, J. W., Maniatis, T., Lawn, R. M., O'Connell, C. O., Spritz, R. A., DeRiel, J. K., Forget, B. G., Weissman, S. M., Slightom, J. L., Blechl, A. E., Smithies, O., Baralle, F. E., Shoulders, C. C., and Proudfoot, N. J., The structure and evolution of the β-globin gene family. *Cell* **21**, **653** (1980).

Eyre, D. R., Collagen: Molecular diversity in the body's protein scaffold. *Science* **207**, 1315 (1980).

Fraser, R. D. B., Keratins. *Sci. Am.* **221** (2), 86 (Aug. 1969).

Gale, R. W., Clegg, J. B., and Huehns, E. R., Human embryonic hemoglobins Gower-1 and Gower-2. *Nature* **280**, 162 (1979).

Goldberg, M. E., The second translation of the genetic message: Protein folding and assembly. *Trends in Biochem. Sci.* **10** (10), 388 (Oct. 1985).

Gross, J., Collagen. *Sci. Am.* **204** (5), 120 (May 1961).

Karlsson, S., and Nienhuis, A. W., Developmental regulation of human globin genes. *Ann. Rev. Biochem.* **54**, 1071 (1985).

Karplus, M., and McCammon, J. A., The dynamics of proteins. *Sci. Am.* **254** (4), 42 (April 1986).

Kendrew, J. C., The three-dimensional structure of a protein molecule. *Sci. Am.* **205** (6), 96 (Dec. 1961).

Lesk, A. M., Themes and contrasts in protein structures. *Trends in Biochem. Sci.* **9** (6), 272 (June 1984).

Perutz, M. F., The hemoglobin molecule. *Sci. Am.* **211** (5), 64 (Nov. 1964).

Perutz, M. F., Structure and mechanism of haemoglobin. *Br. Med. Bull.* **32**, 195 (1976).

Perutz, M. F., Electrostatic effects in proteins. *Science* **201**, 1187 (1978).

Perutz, M. F., Haemoglobin structure and respiratory transport. *Sci. Am.* **239** (6), 92 (Dec. 1978).

Poljak, R. J., Correlations between three-dimensional structure and function of immunoglobulins. *C. R. C. Crit. Rev. Biochem.* **5**, 435 (1978).

Sharon, N., Glycoproteins. *Sci. Am.* **230** (5), 78 (May 1974).

Sharon, N., Glycoproteins. *Trends in Biochem. Sci.* **9** (4), 198 (April 1984).

Tanford, C., The hydrophobic effect and the organization of living matter. *Science* **200**, 1012 (1978).

Uy, R., and Wold, F., Posttranslational covalent modification of proteins. *Science* **198**, 890 (1977).

Weinberg, R. A., The molecules of life. *Sci. Am.* **253** (4), 48 (Oct. 1985).

Williams, R. 'J'. P., and Moore, G. R., Protein antigenicity, organization, and mobility. *Trends in Biochem. Sci.* **10** (3), 96 (Mar. 1985).

Wood, W. G., Haemoglobin synthesis during human fetal development. *Br. Med. Bull.* **32**, 282 (1976).

Zuckerkandl, E., The evolution of hemoglobin. *Sci. Am.* **212** (5), 110 (May 1965).

Books, Monographs, and Symposia

Ayala, F. J., *Molecular Evolution,* Sinauer, Sunderland, Mass., 1976.

Bunn, H. F., Forget, B. G., and Ranney, H. M., *Human Hemoglobins,* Saunders, Philadelphia, 1977.

Dickerson, R. E., and Geis, I., *The Structure and Action of Proteins,* Harper & Row, New York, 1969.

Dickerson, R. E., and Geis, I., *Hemoglobin: Structure, Function, Evolution, and Pathology,* Benjamin/Cummings, Menlo Park, Calif., 1983.

Fox, J. L., Zdenek, D., and Blazej, A., *Protein Structure and Evolution,* Dekker, New York, 1975.

Rawn, J. D., *Biochemistry,* Harper & Row, New York, 1983.

Rocchi, R., and Giormani, V., Glycoproteins: Covalent attachment of carbohydrates or carbohydrate derivatives to amino acids, peptides, and proteins. In *Chemistry and Biochemistry of Amino Acids, Peptides, and Proteins,* (B. Weinstein, Ed.), Vol. 7, Dekker, New York, 1983.

Russel, T. R., Brew, K., Faber, H., and Schultz, J., *From Gene to Protein: Information Transfer in Normal and Abnormal Cells,* Academic, New York, 1979.

Schulz, G. E., and Schirmer, R. H., *Principles of Protein Structure,* Springer-Verlag, New York, 1979.

Stryer, L., *Biochemistry* (2nd ed.), Freeman, San Francisco, 1981.

CHAPTER 5

THE CELLULAR MACROMOLECULES: POLYSACCHARIDES

Polysaccharides are complex carbohydrates that play important and diverse roles in the physiology of cells and tissues; some (e.g., cellulose and chitin) serve as structural and supportive extracellular elements and others (e.g., starch and glycogen) are stored within cells as reserve energy sources. Carbohydrates are composed of simple sugar molecules called **monosaccharides**, many of which conform to the general formula $C_nH_{2n}O_n$. In some monosaccharides, nitrogen and sulfur are also present. The simple sugars are linked together in chains to form carbohydrates of varying size. Carbohydrates consisting of two simple sugars are called **disaccharides**; those containing three simple sugars are called **trisaccharides**. The **oligosaccharides** encompasses the disaccharides and trisaccharides and other sugars with up to 10 monosaccharide units. The **polysaccharides** contain very large numbers (often many thousands) of simple sugar units. More than 200 different naturally occurring monosaccharides have been identified but only a small number of these occur as regular constituents of the polysaccharides. The chemical structures of a number of common sugars have been known since 1900, mainly as a result of the work of Emil Fischer. For his work on carbohydrates, Fischer received the first No-bel Prize. Fischer's contributions to our early understanding of protein chemistry have already been noted (Chapter 4).

MONOSACCHARIDES

The most commonly occurring monosaccharides are either **aldoses** or **ketoses** and contain three to six carbon atoms forming an unbranched chain; those sugars containing three carbons are **trioses**, those with four carbons **tetroses**, those with five carbons **pentoses**, and those with six carbons **hexoses** (Figs. 5–1 and 5–2). Each of the carbon atoms of the chain is assigned a number beginning with the end that is closest to the carbon bearing the *double-bonded* oxygen. In the aldoses, the number 1 carbon (called the "ultimate" carbon) forms the double bond with oxygen, whereas in the ketoses, it is usually the "penultimate" carbon. The middle carbon atom (i.e., carbon number 2) of glyceraldehyde is *asymmetric*; that is, it has four chemically different groups bonded to it, and it therefore has two **optical isomers**. The two isomers of glyceraldehyde are shown in Figure 5–3 and are identified as the D and L forms. All the aldoses may

$$
\begin{array}{c}
H\!\diagdown\!_1C\!\!=\!\!O \\
\mid \\
H\!-\!_2C\!-\!OH \\
\mid \\
H\!-\!_3C\!-\!OH \\
\mid \\
H
\end{array}
$$

Glyceraldehyde
(a triose)

$$
\begin{array}{c}
H\!\diagdown\!_1C\!\!=\!\!O \\
\mid \\
H\!-\!_2C\!-\!OH \\
\mid \\
H\!-\!_3C\!-\!OH \\
\mid \\
H\!-\!_4C\!-\!OH \\
\mid \\
H
\end{array}
$$

Erythrose
(a tetrose)

$$
\begin{array}{c}
H \\
\mid \\
H\!-\!_1C\!-\!OH \\
\mid \\
_2C\!\!=\!\!O \\
\mid \\
H\!-\!_3C\!-\!OH \\
\mid \\
H
\end{array}
$$

Dihydroxyacetone
(a triose)

$$
\begin{array}{c}
H \\
\mid \\
H\!-\!_1C\!-\!OH \\
\mid \\
_2C\!\!=\!\!O \\
\mid \\
H\!-\!_3C\!-\!OH \\
\mid \\
H\!-\!_4C\!-\!OH \\
\mid \\
H
\end{array}
$$

Erythrulose
(a tetrose)

$$
\begin{array}{c}
H\!\diagdown\!_1C\!\!=\!\!O \\
\mid \\
H\!-\!_2C\!-\!OH \\
\mid \\
H\!-\!_3C\!-\!OH \\
\mid \\
H\!-\!_4C\!-\!OH \\
\mid \\
H\!-\!_5C\!-\!OH \\
\mid \\
H
\end{array}
$$

Ribose
(a pentose)

$$
\begin{array}{c}
H\!\diagdown\!_1C\!\!=\!\!O \\
\mid \\
H\!-\!_2C\!-\!OH \\
\mid \\
HO\!-\!_3C\!-\!H \\
\mid \\
H\!-\!_4C\!-\!OH \\
\mid \\
H\!-\!_5C\!-\!OH \\
\mid \\
H\!-\!_6C\!-\!OH \\
\mid \\
H
\end{array}
$$

Glucose
(a hexose)

$$
\begin{array}{c}
H \\
\mid \\
H\!-\!_1C\!-\!OH \\
\mid \\
_2C\!\!=\!\!O \\
\mid \\
H\!-\!_3C\!-\!OH \\
\mid \\
H\!-\!_4C\!-\!OH \\
\mid \\
H\!-\!_5C\!-\!OH \\
\mid \\
H
\end{array}
$$

Ribulose
(a pentose)

$$
\begin{array}{c}
H \\
\mid \\
H\!-\!_1C\!-\!OH \\
\mid \\
_2C\!\!=\!\!O \\
\mid \\
HO\!-\!_3C\!-\!H \\
\mid \\
H\!-\!_4C\!-\!OH \\
\mid \\
H\!-\!_5C\!-\!OH \\
\mid \\
H
\end{array}
$$

Xylulose
(a pentose)

$$
\begin{array}{c}
H\!\diagdown\!_1C\!\!=\!\!O \\
\mid \\
HO\!-\!_2C\!-\!H \\
\mid \\
HO\!-\!_3C\!-\!H \\
\mid \\
H\!-\!_4C\!-\!OH \\
\mid \\
H\!-\!_5C\!-\!OH \\
\mid \\
H\!-\!_6C\!-\!OH \\
\mid \\
H
\end{array}
$$

Mannose
(a hexose)

$$
\begin{array}{c}
H\!\diagdown\!_1C\!\!=\!\!O \\
\mid \\
H\!-\!_2C\!-\!OH \\
\mid \\
HO\!-\!_3C\!-\!H \\
\mid \\
HO\!-\!_4C\!-\!H \\
\mid \\
H\!-\!_5C\!-\!OH \\
\mid \\
H\!-\!_6C\!-\!OH \\
\mid \\
H
\end{array}
$$

Galactose
(a hexose)

$$
\begin{array}{c}
H \\
\mid \\
H\!-\!_1C\!-\!OH \\
\mid \\
_2C\!\!=\!\!O \\
\mid \\
HO\!-\!_3C\!-\!H \\
\mid \\
H\!-\!_4C\!-\!OH \\
\mid \\
H\!-\!_5C\!-\!OH \\
\mid \\
H\!-\!_6C\!-\!OH \\
\mid \\
H
\end{array}
$$

Fructose
(a hexose)

FIGURE 5–1 The common *aldoses*.

FIGURE 5–2 The common *ketoses*.

D-Glyceraldehyde   L-Glyceraldehyde

FIGURE 5–3   D and L forms of *glyceraldehyde*.

be considered to be derivatives of glyceraldehyde, whereas the corresponding ketoses are derivatives of dihydroxyacetone. D or L notation is determined by similarity of the second carbon atom to glyceraldehyde and is independent of whether the sugars are dextrorotatory or levorotatory. The most common aldoses and ketoses found in nature (e.g., those in Figs. 5–1 and 5–2) are the D forms. The number of possible isomers of a monosaccharide is determined by the number (n) of asymmetric carbon atoms present in the chain. Glucose ($n = 4$) has 2^4 or 16 possible isomers. In spite of the variety of possible monosaccharide isomers, only a few different forms occur naturally in any abundance. For example, there are only three naturally occurring isomers of glucose; these are D-glucose, D-mannose, and D-galactose (see Fig. 5–1).

Pyranoses and Furanoses

Some of the monosaccharides can occur in either of two structural configurations: *open-chain* and *cyclic* (or *ringed*). Cyclic forms that contain a six-member ring are called **pyranoses** and those that contain a five-member ring are called **furanoses**. Both the open-chain and cyclic configurations are in reversible equilibrium with one another in aqueous solution, although the cyclic forms are much more prevalent. As an example, the formation of the cyclic structures of D-glucose is shown in Figure 5–4. The aldehyde group of carbon atom number 1 reacts with the hydroxyl group of carbon number 5 to produce a six-member ring containing oxygen. The formation of the ring form of glucose makes carbon number 1 asymmetric and thereby increases the number of possible isomers. The two ring forms of D-glucose are called the α and β anomers. In aqueous solutions of glucose, the β anomer accounts for about 63% of the molecules present, the α form 36%, and the open-chain form about 1%.

The three-dimensional arrangement of the constituent atoms of the ring structure of monosaccharides is more readily visualized when the structure is depicted in the **Haworth projection** (after W. H. Haworth) (Fig. 5–5). In this projection, the members of the ring are arranged in a plane that is perpendicular to the plane of the page. Groups to the right of the carbon skeleton in the open-chain configuration are depicted below the plane of the ring and those to the left of the carbon skeleton extend above the plane of the ring.

α-D-Glucose   D-Glucose   β-D-Glucose
(ring form)   (open chain)   (ring form)

FIGURE 5–4   Open-chain and cyclic forms of D-glucose.

Certain bonds between members of the ring are deliberately shaded more heavily to assist the viewer's perception of the three-dimensional perspective.

The Haworth projection of glucose (and other pyranoses) in which the members of the ring are all depicted as lying in the same plane is not entirely accurate. Though carbon atoms 2, 3, and 5 and the oxygen atom lie in the same plane (to form the corners of a square), carbon atoms 1 and 4 lie either slightly above or below the plane. This gives rise to the so-called *boat* and *chair* forms of glucose (Fig. 5–6), of which the chair form is energetically favored. Other cyclic aldohexoses also exhibit boat and chair forms. For simplicity and convenience, the carbon atoms of the ring and their associated hydrogens are often omitted from the Haworth representation (Fig. 5–7). The cyclic forms of two other common monosaccharides, α-D-ribose and β-D-fructose, are shown in Figure 5–8. It should be noted that while both glucose and fructose are hexoses, glucose (an aldohexose) forms a pyranose whereas fructose (a ketohexose) forms a furanose.

FIGURE 5–5   Haworth projection of glucose.

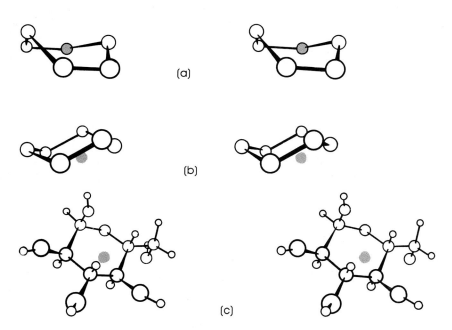

FIGURE 5–6   Stereoscopic views of the "boat" (*a*) and "chair" (*b*) forms of aldohexoses and the complete α anomer (*c*). The oxygen atom of the ring is shown in color.

FIGURE 5–7   For simplicity and convenience the carbon atoms and associated hydrogens of the ring may be omitted from the Haworth projection.

FIGURE 5–8   Cyclic forms of D-ribose and D-fructose.

FIGURE 5–9 Condensation of two molecules of glucose to form the disaccharide *maltose*.

FIGURE 5–10 *Sucrose.*

FIGURE 5–11 *Lactose.*

DISACCHARIDES

Disaccharides consist of two ringed monosaccharides. The bonds that unite neighboring monosaccharides are called **glycosidic bonds** and are formed by the condensation of a hydroxyl group of carbon atom number 1 of one monosaccharide with the hydroxyl group of either the number 2, 4, or 6 carbon atom of another. The formation of the common disaccharide **maltose** from two molecules of glucose is shown in Figure 5–9. In maltose, the oxygen bridge is formed between the number 1 carbon atom of one α-D-glucose unit and the number 4 carbon atom of the other. The bond formed is referred to as an $\alpha 1 \rightarrow 4$ glycosidic bond.

Another important disaccharide is **sucrose** (i.e., ordinary "table" sugar), which is formed by the condensation of α-D-glucose and β-D-fructose (Fig. 5–10). Milk contains the disaccharide **lactose**, which consists of the hexoses β-D-galactose and β-D-glucose (Fig. 5–11). In lactose, the glycosidic bond is of the beta variety; i.e., $\beta 1 \rightarrow 4$ (compare with maltose).

POLYSACCHARIDES

The **polysaccharides** (or **glycans**) are composed of long chains of sugars and can be divided into two main functional groups: the *structural* polysaccharides and the *nutrient* polysaccharides. The structural polysaccharides serve primarily as extracellular or intercellular supporting elements. Included in this group are **cellulose** (found in plant cell walls), **mannan** (found in yeast cell walls), **chitin** (in the shells of arthropods and the cell walls of some fungi), **hyaluronic acid**, **keratan sulfate**, and **chondroitin sulfate** (in connective tissue) and the **peptidoglycans** of bacteria. The nutrient polysaccharides serve as reserves of monosaccharides and are in continuous metabolic turnover. Included in this group are **starch** (plant cells and bacteria), **glycogen** (animal cells), and **paramylum** (in certain protozoa).

On chemical bases, the polysaccharides can be divided into two broad classes: the **homopolysaccharides** and the **heteropolysaccharides**. In the homopolysaccharides, all the constituent sugars are the same. Included in this class are cellulose, starch, glycogen, and paramylum. In the heteropolysaccharides, the constituent sugars may take different forms; included here are hyaluronic acid, keratan sulfate, and chondroitin sulfate. Some polysaccharides are unbranched (i.e., linear) chains whose structure may be ribbonlike or helical (usually a left-handed spiral). Other polysaccharides are branched and, like many proteins, assume a globular form. A description of some of the more important polysaccharides follows.

Cellulose

Cellulose is not only the most common of all polysaccharides, it is also the most abundant organic substance in the living world. Indeed, it is estimated that more that half of all organic carbon is present as cellulose. Cellulose is the major component of plant cell walls, where it plays a structural role. Cellulose is an unbranched polymer of glucose in which the neighboring monosaccharides are joined by $\beta 1 \rightarrow 4$ glycosidic bonds (Fig. 5–12). Chain lengths vary from several hundred to several thousand glucosyl units (in the algae *Valonia*, a single molecule of cellulose may contain more than 20,000 glucosyl units). Successive pyranose rings in cellulose are rotated 180° relative to one another so that the chain of sugars takes on a "flip-flop" appearance.

In plant cell walls, large numbers of cellulose molecules are organized into cross-linked, parallel **microfibrils** (Fig. 5–13), whose long axis is that of the individual chain. The cellulose microfibrils may be coated with **hemicellulose** (a smaller polysaccharide formed from xylose, mannose, or galactose). The plant cell wall is composed of a series of layers of these microfibrils, with the walls of neighboring cells cemented together by **pectin** (a polymer of **galacturonic acid**). Vast amounts of cellulose may be deposited between neighboring cells as is vividly seen in the scanning electron photomicrographs of wood tissue shown in Figure 5–13. Cellulose has also been identified in the cell walls of algae and some fungi.

Chitin

Chitin is an extracellular structural polysaccharide found in large quantities in the body covering (*cuticle*) of arthropods and in smaller amounts in sponges, mollusks, and annelids. Chitin has also been identified in the cell walls of most fungi and some green algae. The chemical structure of chitin is closely related to that of cellulose; the difference is that the hydroxyl group of each number 2 carbon atom is replaced by an acetamido group. Hence, chitin is an unbranched polymer of **N-acetylglucosamine** (Fig. 5–14) containing several thousand successive aminosugar units linked by $\beta 1 \rightarrow 4$ glycosidic bonds.

Hyaluronic Acid, Keratan Sulfate, and Chondroitin Sulfate

Cartilage tissue contains the related polysaccharides **hyaluronic acid**, **keratan sulfate**, and **chondroitin sulfate**. Hyaluronic acid is an unbranched heteropolysaccharide containing repeating disaccharides of N-acetylglucosamine and **glucuronic acid** (Fig. 5–15). Glucuronic acid is linked to N-acetylglucosamine in each disaccharide by a $1 \rightarrow 3$ glycosidic bond, but successive disaccharides are $1 \rightarrow 4$ linked. In addition to cartilage, hyaluronic acid is found in other connective tissues, in the synovial fluid of joints, in the vitreous humor of the eyes, and also in the capsules that enclose bacteria.

Keratan sulfate, like hyaluronic acid, is a repeating disaccharide forming an unbranched chain. Each disaccharide unit of the polysaccharide consists of *galactose* and *sulfated N-acetylglucosamine*. **Chondroitin sulfate** is a repeating disaccharide consisting of alternating *glucuronic acid* and *sulfated N-acetylgalactosamine* residues. Hyaluronic acid, keratan sulfate, and chondroitin sulfate, together with a number of oligosaccharides and proteins, form the

Cellulose

FIGURE 5–12 Small segment of a *cellulose* molecule.

FIGURE 5–13 (a) Transmission electron photomicrograph of cellulose microfibrils in the cell wall of cotton cells. As the wall develops the microfibrils are laid down in successive layers; within a layer, the orientation of the microfibrils is the same but the orientation may vary from one layer to the next. Magnification, 15,000X. (b) Scanning electron photomicrograph of parenchyma cells of celery tissue. Note the uniform, polyhedrol shape of the cell wall. (c) Scanning electron photomicrograph of pine wood cut perpendicular to the cells' long axis. Note the thickness of the cell walls. Magnification, 350X. (d) Longitudinally cut pine wood. Magnification, 350X. This tissue is younger than that shown in c and the cell walls are thinner; note the bordered pits through which materials can flow between adjacent cell surfaces. (a, c, and d, courtesy of Dr. J. H. M. Willison; b, courtesy of Dr. J. R. Troughton; copyright © 1973 Jacaranda Wiley Ltd.)

Chitin

FIGURE 5–14　Small segment of a *chitin* molecule.

proteoglycan (see below) that gives cartilage its unusual tensile strength and resilience.

Inulin

Inulin is an unbranched nutrient polysaccharide found in the bulbs of such plants as artichokes, dahlias, and dandelions. It consists of repeating *fructose* units in $\beta 2 \rightarrow 1$ linkage (Fig. 5–16).

Glycogen

Glycogen is a branched nutrient homopolysaccharide containing *glucose* in $\alpha 1 \rightarrow 4$ and $\alpha 1 \rightarrow 6$ linkages. It is found in nearly all animal cells and also in certain protozoa and algae. In view of its ubiquitous occurrence, it is an extremely important polysaccharide and has been the subject of numerous studies. In man and other vertebrates, glycogen is stored primarily in the liver and muscles and is the principal form of stored carbohydrate. In an unstarved animal, as much as 10% of the liver weight may be glycogen. Glycogen undergoes almost continuous biosynthesis and degradation, especially in liver tissue. Liver glycogen serves as a reservoir for glucose under starvation conditions (it may be almost completely depleted during 24 hours of fasting) and during muscular exertion. Glycogen is quickly resynthesized from newly ingested carbohydrate.

Acetylglucosamine

Glucuronic acid

Hyaluronic acid

FIGURE 5–15　Small segment of *hyaluronic acid*.

FIGURE 5–16 Small segment of an *inulin* molecule. Successive sugars are in $\beta\ 2 \rightarrow 1$ linkage.

Inulin

Glycogen molecules exist in a continuous spectrum of sizes, with the largest molecules containing many thousands of glucosyl units. A small portion of a glycogen molecule is shown in Figure 5–17. The glucosyl units that are linked by $\alpha 1 \rightarrow 4$ glycosidic bonds are organized into long chains; the chains are interconnected at branch points by $\alpha 1 \rightarrow 6$ glycosidic bonds. This yields the "bush"- or "tree"-like structure depicted in Figure 5–18. It should be noted that in a glycogen molecule there is only one glucose unit whose number 1 carbon atom bears a hydroxyl group. All of the other 1-OH groups are involved in $\alpha 1 \rightarrow 4$ and $\alpha 1 \rightarrow 6$ glycosidic bonds. The single free 1-OH group is called the "reducing end" of the molecule and is noted by the letter R in Figure 5–18. In contrast, numerous "nonreducing ends" are present (i.e., free 4-OH and

$\alpha 1 \rightarrow 4$ linkage

$\alpha 1 \rightarrow 6$ linkage

FIGURE 5–17 A portion of a glycogen molecule. Glucosyl units are linked by $\alpha 1 \rightarrow 4$ glycosidic bonds to form long chains, and these chains are interconnected by $\alpha 1 \rightarrow 6$ glycosidic bonds.

6-OH groups) at the terminals of the outermost chains.

In Figure 5–18 the individual glucose units are represented by circles and the branch points (i.e., the 1→6 linkages) by heavier connections. In this model of glycogen, a number of different kinds of chains may be distinguished. *A chains* are attached to the molecules by a single 1→6 linkage (chains of open circles in Fig. 5–18) and *B chains* (gray circles) bear one or more A chains. Each glycogen molecule contains only one *C chain* (colored circles), and this is the chain that ends in the free reducing group. *Exterior chains* are those portions of individual chains between the nonreducing end groups and the outermost branch points. Finally, those parts of individual chains between branch points are called *interior chains*. The exterior chains of glycogen are usually six to nine glucosyl units long, whereas interior chains contain only three to four glucosyl units. Approximately 8 to 10% of all glycosidic linkages are the 1→6 type.

Glycogen molecules are often sufficiently large to be studied by electron microscopy. Although the molecules can be seen in transmission electron photomicrographs of osmium tetroxide-fixed tissues (especially liver, see Fig. 1-13), morphological studies of glycogen are usually carried out with material that has been negatively stained with phosphotungstic acid and osmium tetroxide (Fig. 5–19).

Starch

Starch is a nutrient polysaccharide found in plant cells, protists, and certain bacteria and is similar in many respects to glycogen. (In fact, glycogen is often referred to as "animal starch.") Starch usually occurs in cells in the form of granules visible by both light and electron microscopy. In plant cells (such as potato or corn) these granules may be several micrometers in diameter and may account for more than half of the cell's dry weight. In microorganisms the starch granules are smaller, having diameters of only 0.5 to 2 μm. Starch granules contain a mixture of two different polysaccharides, **amylose** and **amylopectin**, and the relative amounts of these two polysaccharides vary according to the source of the starch.

The amylose component of starch is an unbranched 1→4 polymer of glucose and may be several thousand glucosyl units long. The polysaccharide chain exists in the form of a left-handed helix containing six glucosyl residues per turn (Fig. 5–20). The familiar blue color that is produced when starch is treated with iodine is believed to result from the coordination of iodine ions in the interior of the helix. The polysaccharide must contain at least six helical turns (i.e., 36 glucosyl residues) to produce the characteristic blue color when treated with iodine.

Amylopectin is a branched polysaccharide contain-

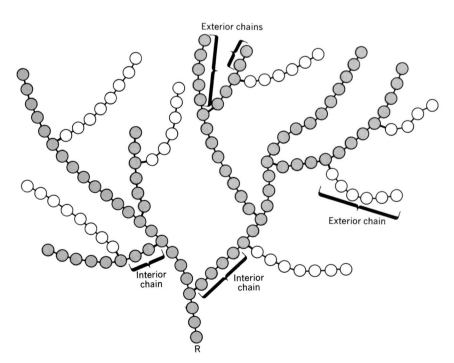

Exterior chains

Exterior chain

Interior chain

Interior chain

R

FIGURE 5–18　"Bush"- or "tree"-like structure of the glycogen molecule. A chains are shown by open circles, B chains are gray circles, and the C chain is shown in color. The reducing end of the molecule is denoted by the letter R.

FIGURE 5–19 Electron photomicrograph of negatively stained glycogen particles (rosettelike structures) from liver tissue magnified 80,000X.

FIGURE 5–20 Left-handed helix formed by the *amylose* polysaccharide.

ing 1→4- and 1→6-linked glucosyl units; in this respect it is similar to glycogen. However, amylopectin has a more open structure with fewer 1→6 linkages and longer chain lengths. Some of the characteristics of amylopectin and glycogen are compared in Table 5–1.

Other Polysaccharides

In addition to the polysaccharides already described, several others should be briefly reviewed. **Mannan**, a homopolymer of mannose, is found in the cell walls of yeast and is also stored intracellularly in some plants. In yeast, the mannan has a branched structure, whereas in plants it is a linear molecule. **Paramylum** is a nutrient homopolysaccharide stored intracellularly as large granules in certain protozoa (e.g., *Euglena*). It consists of unbranched chains of glucosyl

TABLE 5–1 SOME CHEMICAL PROPERTIES OF AMYLOPECTIN AND GLYCOGEN

	Amylopectin	Glycogen
Molecular weight	10^7	10^6–10^9
Average chain length	20–25	10–14
Percentages 1→6 linkages	4–5	8–10
Exterior chain length	12–17	6–9
Interior chain length	5–8	3–4

TABLE 5–2 POLYSACCHARIDES

Class	Name	Source	Composition	Linkages
Structural polysaccharides	Cellulose	Plant cell walls	Glucose (beta linkage)	Unbranched 1→4
	Mannan	Yeast cell walls	Mannose (beta linkage)	Branched 1→2, 1→3, and 1→6
	Chitin	Arthropod shells, fungal cell walls	Acetylglucosamine and glucuronic acid (beta linkage)	Unbranched 1→4
	Hyaluronic acid	Synovial fluid (joints), subcutaneous tissue	Acetylglucosamine and glucuronic acid (beta linkage)	Unbranched 1→3 and 1→4
	Peptidoglycans	Bacterial cell walls	Acetylglucosamine and acetylmuramic acid	Unbranched 1→4
Nutrient polysaccharides	Inulin	Artichokes, dandelions	Fructose (beta linkage)	Unbranched 2→1
	Paramylum	Certain protozoa (e.g., *Euglena*)	Glucose (beta linkage)	Unbranched 1→3
	Glycogen	Certain protozoa (e.g., *Tetrahymena*) and most animals	Glucose (alpha linkage)	Branched 1→4 and 1→6
	Starch: Amylopectin	Plant cells and some protozoa (e.g., *Polytomella*)	Glucose (alpha linkage)	Branched 1→4 and 1→6
	Amylose		Glucose (alpha linkage)	Unbranched 1→4

units in 1→3 linkage. **Dextran** is a branched homopolymer of glucose produced by certain bacteria. However, unlike glycogen and amylopectin, the glycosidic linkages vary from one species to another and may be 1→6, 1→4, 1→3, or 1→2. Some of the properties of the various polysaccharides are summarized in Table 5–2.

PROTEOGLYCANS, GLYCOPROTEINS, AND GLYCOLIPIDS

Polysaccharides also occur in covalent combination with proteins and lipids. In the **proteoglycans**, the protein portion of the molecule represents only a small portion of the total molecular weight. These macromolecules are also known as **mucopolysaccharides**. Among the proteoglycans, one of the most extensively studied is **cartilage proteoglycan**, which is found extracellularly in cartilage and bone. In cartilage proteoglycan (Fig. 5–21), strands of protein, called **core protein**, extend radially from a long, central hyaluronic acid molecule. Three carbohydrate-bearing regions of each core protein strand may be

identified. The first region contains numerous oligosaccharides, the second contains keratan sulfate chains, and the third contains chondroitin sulfate chains. This complex of protein and polysaccharide gives cartilage its special resilience and tensile properties.

The cell walls of certain bacteria contain **peptidoglycans**. These are heteropolymers of *N*-acetylglucosamine and *N*-acetylmuramic acid in which short peptides are linked to the polysaccharide backbone.

In the **glycoproteins**, the carbohydrate portion consists of much shorter chains, which are often branched. The chains may be few in number or there may be so many as to account for more than half of the total molecular weight. Glycoproteins serve diverse roles in cells and tissues and include certain enzymes, hormones, and antibodies or immunoglubins (see also Chapter 4).

Glycolipids are covalent combinations of carbohydrate and lipid. The carbohydrate portion may be a single monosaccharide or a linear or branched chain. Glycolipids are components of most cell membranes; their chemistry is considered in greater detail in Chapter 6.

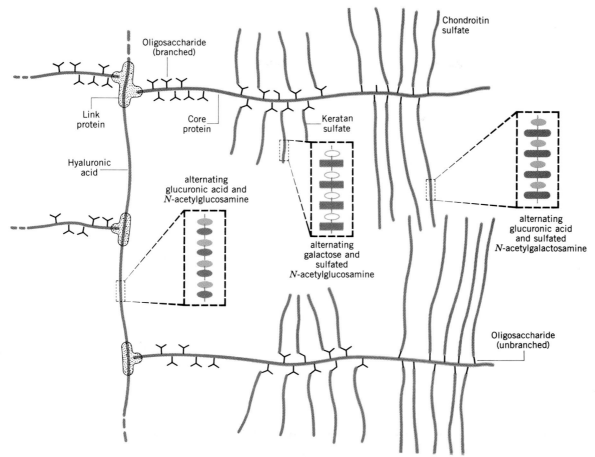

FIGURE 5–21 *Cartilage proteoglycan. Core proteins* attached to a central *hyaluronic acid* chain by *link protein* bear chains of *keratan sulfate* and *chondroitin sulfate*. This arrangement gives cartilage its resilience and tensile strength. (See text for details.)

SUMMARY

The **polysaccharides** are macromolecules composed of individual sugar units called **monosaccharides**. Among these, the most common is the hexose **glucose**. Although monosaccharides may contain three or more carbon atoms and are derivatives of **glyceraldehyde** or **dihydroxyacetone**, the six-carbon forms occur most often in polysaccharides. Individual sugars may take the open-chain configuration or may form ringed structures; only the ringed **pyranoses** and **furanoses** occur as regular constituents of polysaccharides. From a functional standpoint, polysaccharides may be divided into *nutrient* and *structural* types. Nutrient polysaccharides such as **glycogen** and starch are often stored in cells as discrete granules visible by either light or electron microscopy. The most abundant structural polysaccharide, **cellulose**, serves as the major supportive element in the cell walls of higher plants. Polysaccharides may be formed from unbranched chains of sugars (as in cellulose and the **amylose** component of starch) or from branched chains (as in glycogen and the **amylopectin** component of starch). All sugars are the same type in the **homopolysaccharides** (e.g., cellulose, glycogen, and starch) but vary in the **heteropolysaccharides**. Polysaccharides may also occur in covalent combination with proteins, short peptides, and lipids, forming **proteoglycans, glycoproteins, peptidoglycans,** and **glycolipids**.

REFERENCES AND SUGGESTED READING

Articles and Reviews

Caplan, A. I., Cartilage. *Sci. Am.* **251** (4), 84 (Oct. 1984).

Hakomori, S., Glycosphingolipids. *Sci. Am.* **254** (5), 44 (May 1986).

Manners, D. J., The molecular structure of glycogens. *Adv. Carbohyd. Chem.* **13**, 261 (1957).

Sharon, N., Carbohydrates. *Sci. Am.* **243** (5), 90 (Nov. 1980).

Books, Monographs, and Symposia

Aspinall, G. O., *Polysaccharides*, Pergamon, Oxford, 1970.

Ginsberg, V., and Robbins, P. (Eds.), *Biology of Carbohydrates*, Wiley-Interscience, New York, 1981.

Lehninger, A. L., *Principles of Biochemistry*, Worth, New York, 1982.

McGilvery, R. W., *Biochemistry, A Functional Approach* (2nd ed.), Saunders, Philadelphia, 1979.

Rawn, J. D., *Biochemistry*, Harper & Row, New York, 1983.

Rees, D. A., *Polysaccharide Shapes*, Halsted Press, New York, 1977.

Sharon, N., *Complex Carbohydrates*, Addison–Wesley, London, 1975.

Smith, E.L., Hill, R. L., Lehman, I. R., Lefkowitz, R. J., Handler, P., and White, A., *Principles of Biochemistry* (7th ed.), McGraw–Hill, New York, 1983.

Stryer, L., *Biochemistry* (2nd ed.), Freeman, San Francisco, 1981.

CHAPTER 6

THE CELLULAR MACROMOLECULES: LIPIDS

Lipids are a chemically heterogeneous collection of molecules that are only sparingly soluble in water but are highly soluble in nonpolar, organic solvents such as acetone, benzene, chloroform, and ether. It is because of their similar solubility properties that they are usually placed together in one category. Like the carbohydrates, lipids serve two major roles in cells and tissues: (1) they occur as constituents of certain structural components of cells, particularly membranous organelles; and (2) they may be stored within cells as reserve energy sources. The most common lipids include the **fatty acids**, **neutral fats**, **phospholipids** (**glycerophosphatides**, **plasmalogens**, and **sphingolipids**), **glycolipids**, **terpenes**, and **steroids**. When they occur independently, several of these lipids are not macromolecular; however, because they have similar properties and frequently combine with each other and with other kinds of molecules to form macromolecular complexes, they are considered in this section.

CELLULAR DISTRIBUTION OF LIPIDS

Table 6–1 shows the distribution of certain lipids in the four major cell fractions that are obtained when

TABLE 6–1 LIPID DISTRIBUTION IN ANIMAL CELLS

Fraction	Total Lipid (% dry weight)	Percentage of Total Lipid		
		Neutral Fats	Phospholipids[a]	Terpenes and Steroids
Nuclei	16	3	90	5
Mitochondria	21	1	90	6
Microsomes[b]	32	0	90	6
Cytosol	7	70	30	4

[a]Glycerophosphatides, plasmalogens, and sphingolipids.

[b]The microsomal fraction includes fragments of the ER, ribosomes, and other tiny cellular inclusions.

liver tissue is homogenized and fractionated by a technique called *differential centrifugation* (see Chapter 12). The distribution is believed to be fairly representative, at least of animal cells. It should be noted that the mitochondrial and microsomal fractions are richest in lipid. Most of the lipid in these fractions is phospholipid (i.e., glycerophosphatides, etc.) serving as structural constituents of the mitochondrial and microsomal membranes where the phospholipid is combined with membrane protein (see Chapter 15). Most of the lipids that are found in the nonsedimenting, soluble phase of the cell (i.e., the cytosol) are neutral fats and are in rapid metabolic turnover. The neutral fat content of the cytosol fractions of tissues that are specialized for fat storage (e.g., adipose tissue) is much higher. Prokaryotic cells contain less lipid than eukaryotic cells and lack steroids. In bacteria, more than 90% of all the cellular lipid is associated with the plasma membrane.

FATTY ACIDS

The fatty acids are an important and also abundant group of lipids. In cells, the fatty acids only sparingly occur freely; instead, they are esterified to other components and form the saponifiable lipids (see below). A fatty acid molecule may be either **saturated** or **unsaturated**. The saturated fatty acids consist of long hydrocarbon chains terminating in a carboxyl group and conform to the general formula

$$CH_3-(CH_2)_n-COOH$$

In nearly all naturally occurring fatty acids, n is an even number from 2 to 22. In the saturated fatty acids most commonly found in animal tissues, n is either 12 (i.e., *myristic acid*), 14 (i.e., *palmitic acid*), or 16 (i.e., *stearic acid*). In unsaturated fatty acids, at least two but usually no more than six of the carbon atoms of the hydrocarbon chain are linked together by double bonds. The two most common unsaturated fatty acids are *oleic acid* and *linoleic acid,* depicted in Figure 6–1 along with the saturated fatty acid stearic acid. In all the common saturated and unsaturated fatty acids, the long hydrocarbon chain is *unbranched*. Unsaturated fatty acids predominate in the *saponifiable* lipids of higher plants and in animals that live at low temperatures. Lipids in the tissues of animals inhabiting warm climates contain larger quantities of saturated fatty acids. Some animals, especially mammals, are unable to synthesize certain fatty acids and therefore require them in their diets. The essential fatty acids, like linoleic acid, must be obtained from plant material. Fatty acids found in bacterial cells are usually of short chain length and, when unsaturated, have only one double bond.

Fatty acid molecules contain both hydrophilic and hydrophobic parts. In their dissociated states (shown in the formulas of Fig. 6–1), the carboxyl ends of the molecules are mildly soluble in water, whereas the long hydrocarbon chains repel water. In neutral solu-

FIGURE 6–1 The saturated fatty acid *stearic acid* and the unsaturated fatty acids, *oleic* and *linoleic acid.*

tions, salts of the fatty acids form small spherical droplets or **micelles** in which the dissociated carboxyl groups occur at the surface, and the hydrophobic carbon chains project toward the center. (This arrangement is shown in Fig. 3–1.) As noted above, most of the fatty acids that are found in cells and tissues are esterified to other molecules. Because the ester linkages can be hydrolyzed by the addition of alkali, these lipids are called **saponifiable lipids**.

SAPONIFIABLE LIPIDS

Neutral Fats

The **neutral fats**, **triglycerides**, or **triacylglycerols** are esters of glycerol and three fatty acids and have the general formula shown in Figure 6–2. In this formula, n, n', and n'' may be the same number or different numbers. Usually, the fatty acid that is esterified to the first carbon atom of glycerol (i.e., the top carbon in Fig. 6–2) is saturated, whereas the fatty acid esterified to the middle carbon atom is unsaturated. Either saturated or unsaturated fatty acids are found at the third carbon position. Unlike the fatty acids, the neutral fats are entirely nonpolar. Neutral fats, which are often deposited in cells as potential sources of chemical energy, represent the major type of stored lipid; they occur as droplets in the cytoplasm and most of the lipid that is recovered in the soluble phase or cytosol of disrupted cells takes this form. The lipid droplets stored in abundance in muscle cells are clearly seen in Figure 6–3.

Glycerophosphatides

The major members of this group of lipids are derivatives of **phosphatidic acid**. Phosphatidic acid is similar to a triglyceride except that one of the fatty acids is replaced by a phosphate group (Fig. 6–4). Phosphatidic acid and its derivatives are present in plasma membranes, where they play an active role in membrane function (Chapter 15) in addition to serving as major structural constituents (Fig. 6–3). The most common derivatives of phosphatidic acid are **phosphatidyl choline** (also known as **lecithin**), **phosphatidyl ethanolamine** (also known as **cephalin**), **phosphatidyl serine**, and **phosphatidyl inositol** (Fig. 6–5). Free rotation about the single bonds of the glycerol backbone allow the hydrophilic phosphate and its derivatives to face away from the hydrophobic portion, as shown in the structural formulas of Figure 6–5. Consequently, glycerophosphatides (and other lipids discussed below) are *amphipathic*, that is, one end of the molecule is strongly hydrophobic (i.e., the end containing the hydrocarbon chains) while the other end is hydrophilic due to the charged nature of the dissociated phosphate group and other substituents.

In the presence of water, glycerophosphatides can behave in various ways (Fig. 6–6). For example, they can form *monolayers* (i.e., *monomolecular layers*) on the water with their polar ends projecting into the water and their hydrophobic ends directed away from the water. Also, like fatty acids, glycerophosphatides can form **micelles** with the polar ends of the molecules at the micelle's surface and the hydrocarbon chains projecting toward the center. However, of special significance is the tendency of glycerophosphatides to form *bilayers* or *bimolecular sheets* (Fig. 6–6). Bilayers formed predominantly from glycerophosphatides are believed to be fundamental to the structure of the plasma membrane, the membranes of the endoplasmic reticulum (see Chapter 15), and the membranes of vesicular organelles.

Hydrophobic interactions between the tails of the glycerophosphatides are the most important forces promoting the formation of bilayers. The assembly of a bilayer is spontaneous as long as there is adequate lipid present; the spontaneous nature of bilayer formation is reminiscent of the automatic folding of protein molecules that takes place due to the interaction of hydrophobic amino acid side chains (Chapter 4).

Liposomes When aqueous suspensions of glycerophosphatides are subjected to rapid agitation using ultrasound (i.e., *insonation*), the lipid disperses in the water and forms **liposomes** or **lipid vesicles** (Fig. 6–7). These liposomes consist of a spherical bilayer of the

$$CH_3—(CH_2)_n—\overset{\overset{\displaystyle O}{\|}}{C}—O—CH_2$$
$$CH_3—(CH_2)_{n'}—\overset{\overset{\displaystyle O}{\|}}{C}—O—CH$$
$$CH_3—(CH_2)_{n''}—\overset{\overset{\displaystyle O}{\|}}{C}—O—CH_2$$

Glycerol backbone

FIGURE 6–2 Generalized formula for a *neutral fat (triglyceride)*.

FIGURE 6–3   Lipids in cells. (Top.) Fat droplets in striatedmuscle tissue. Freeze-fracture (left) and transmission (right) electron photomicrographs. (Courtesy of A. P. Bland and C. J. Roberts; copyright © 1981 by Pergamon Press Ltd.; *Micron* **12**, 163.) (Bottom.) *Cardiolipin*, an acidic phospholipid commonly found in the inner membranes of mitochondria and chloroplasts. (Left.) Freeze-fractured granule revealing cocentric layers; (right,) crystalline lipid structure and X-ray diffraction pattern (insert). (Courtesy Drs. P. Arslan, M. Beltrame, and U. Muscatello; copyright © 1980 Pergamon Press Ltd.; *Micron* **11**, 118.)

$$CH_3-(CH_2)_n-\overset{\overset{\displaystyle O}{\|}}{C}-O-CH_2$$

$$CH_3-(CH_2)_{n'}-\overset{\overset{\displaystyle O}{\|}}{C}-O-CH$$

$$H_2C-O-\overset{\overset{\displaystyle O}{\|}}{\underset{\underset{\displaystyle O^{\ominus}}{|}}{P}}-OH$$

FIGURE 6–4 *Phosphatidic acid.*

glycerophosphatide molecules enclosing a small volume of the aqueous medium. If small molecules and ions are initially dissolved in the water, some of these will be enclosed by the liposomes. The ability of different substances inside or outside the liposomes to traverse the bilayer can be studied experimentally. When this is done, it is found that liposomes exhibit many of the permeability properties of natural cellular membranes (a subject pursued in Chapter 15).

In recent years there has been considerable interest in using liposomes as vectors for the transfer of drugs, proteins, nucleic acids, ions, and other molecules into cells. The mechanism is illustrated in Figure 6–8. Liposomes are either incubated or formed *de novo* in aqueous solutions of the substance to be transferred and are then washed free of unencapsulated material. When mixed with cells *in vitro* or introduced *in vivo* (e.g., through intravascular or intraperitoneal injection), the liposomes fuse with the plasma membranes of the target cells. The contents of the liposome may enter the cell via either of several routes: (1) following fusion, the liposome's contents may be emptied into the cytosol; (2) the entire liposome may be endocytosed and degraded, thereby releasing the entrained substances in the cell; and/or (3) the liposomes may undergo degradation extracellularly, creating a localized high concentration of material that may then be incorporated by the cell. The capability to produce liposomes that differentially bind to the surfaces of specific types of cells and tissues holds great therapeutic promise, for it would make it possible to deliver specific chemical agents to specific tissues.

Plasmalogens

Plasmalogens are a special class of lipids especially abundant in the membranes of nerve and muscle cells

FIGURE 6–5 *Phosphatidyl choline, phosphatidyl ethanolamine (cephalin), phosphatidyl serine, and phosphatidyl inositol.*

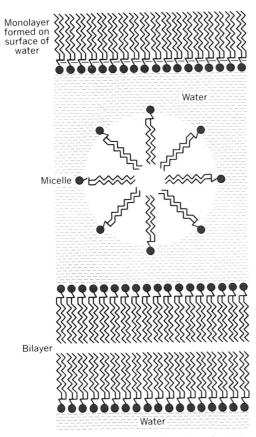

FIGURE 6–6　In the presence of water, glycero-phosphatides form monolayers on the water's surface, *micelles*, or *bimolecular* sheets (bilayers).

and also characteristic of cancer cells. A plasmalogen is similar to a glycerophosphatide except that the fatty acid at the number 1 position of the glycerol is replaced by an unsaturated ether (Fig. 6–9).

Sphingolipids

Sphingolipids are derivatives of **sphingosine**, an amino alcohol possessing a long, unsaturated hydrocarbon chain (Fig. 6–10a). The *myelin sheath* surrounding many nerve cells is particularly rich in the sphingolipid **sphingomyelin** (Fig. 6–10b). In sphingomyelin, the amino group of the sphingosine skeleton is linked to a fatty acid and the hydroxyl group is esterified to phosphorylcholine.

Glycolipids

The outer surface of the plasma membrane of most cells is studded with short chains of sugars. These sugars are either parts of **glycoproteins** (see Chapter 4) or they are attached to membrane lipids, thereby forming **glycolipids**. Glycolipids in the plasma membrane play vital roles in *immunity, blood group specificity,* and *cell–cell recognition.* The carbohydrates that are found in glycolipids vary from individual monosaccharides (e.g., glucose and galactose) to short, branched or unbranched oligosaccharides. The lipid portion of a glycolipid is similar to sphingosine with the amino group of the sphingosine skeleton acylated by a fatty acid (as in sphingomyelin) and the hydroxyl group associated with the carbohydrate. The simplest glycolipids are the **cerebrosides**, which, as their name suggests, are abundant in brain tissue, where they are found in the myelin sheaths and may account for as much as 20% of the sheath's dry weight. Small amounts of cerebrosides are also found in kidney, liver, and spleen cells. Usually, the sugar of a cerebroside is either glucose or galactose (Fig. 6–11). In the **gangliosides** (also associated with nerve tissue), the carbohydrate portion of the molecule consists of a chain of sugar molecules usually including glucose, galactose, and neuraminic acid. One particular ganglioside, called G_{M2}, may accumulate in the lysosomes of brain cells because of a genetic deficiency that results in the failure of the cells to produce a lysosomal enzyme that degrades this ganglioside. The condition is known as *Tay–Sachs* disease and frequently leads to paralysis, blindness, and retarded development.

FIGURE 6-7 *Liposomes.* A liposome is a small spherical body whose surface is formed by a lipid bilayer. Liposomes show some permeability properties that are similar to those of natural membranes enclosing cells.

External aqueous phase

Liposome wall

Internal aqueous phase

Sonicator

liposomes are washed or dialyzed

liposomes containing material to be transferred

Fusion

liposomes attach to plasma membrane

Target cell

liposomes are endocytosed

Lysosomes

FIGURE 6-8 Liposomes may be used as vectors to transfer drugs, hormones, proteins, etc., into cells. The liposomes are formed by insonation of an aqueous lipid suspension containing the material to be transported. Liposomes attach to the surfaces of cells and may fuse with the plasma membrane, following which the liposome's contents are released into the cytosol. Alternatively, the liposomes may be endocytosed and degraded intracellularly.

$$CH_3-(CH_2)_n-CH=CH-O-CH_2$$

$$CH_3-(CH_2)_{n'}-\overset{\overset{\textstyle O}{\|}}{C}-O-CH$$

$$H_2C-O-\overset{\overset{\textstyle O}{\|}}{\underset{\underset{\textstyle O^{\ominus}}{|}}{P}}-\boxed{\begin{array}{l}\text{Choline}\\\text{Serine}\\\text{Ethanolamine}\\\text{Inositol}\end{array}}$$

FIGURE 6–9 Generalized formula of a *plasmalogen*.

$$CH_3-(CH_2)_{12}-CH=CH-CH-OH$$

$$NH_2-CH$$

$$H_2C-OH$$

(a)

$$CH_3-(CH_2)_{12}-CH=CH-CH-OH$$

$$CH_3-(CH_2)_n-\overset{\overset{\textstyle O}{\|}}{C}-NH-CH$$

$$H_2C-O-\overset{\overset{\textstyle O}{\|}}{\underset{\underset{\textstyle O^{\ominus}}{|}}{P}}-O-(CH_2)_2-\overset{\overset{\textstyle CH_3}{|}}{\underset{\underset{\textstyle CH_3}{|}}{N^{\oplus}}}-CH_3$$

(b)

FIGURE 6–10 The sphingolipids
sphingosine (a) and *sphingomyelin* (b).

FIGURE 6–11 A *glycolipid*. The molecule shown here is a *cerebroside* and contains only one sugar, in this instance glucose. Cerebrosides may also contain galactose. In *gangliosides*, the hexose is replaced by a chain of sugars.

NONSAPONIFIABLE LIPIDS

The nonsaponifiable lipids (also known as simple lipids) are lipids that do not contain fatty acids as constituents. Two major classes of nonsaponifiable lipids are the **terpenes** and the **steroids**.

Terpenes

The terpenes include certain fat-soluble vitamins (such as vitamins A, E, and K), carotenoids (pigments of plant cells that are involved in photosynthesis), and certain coenzymes (such as coenzyme Q, ubiquinone). All the terpenes are synthesized from various numbers of a five-carbon building block called an **isoprene unit** (Fig. 6–12). The isoprene units are bonded together in a head-to-tail organization. Two isoprene units form a **monoterpene**, four form a **diterpene**, six a **triterpene**, and so on. The monoterpenes are responsible for the characteristic odors and flavors of plants (e.g., *geraniol* from geraniums, *menthol* from mint, and *limoneme* from lemons).

Steroids

The steroids are an especially important class of lipids consisting of a system of fused cyclohexane and cyclopentane rings. All are derivatives of **perhydrocyclopentanophenanthrene**, which consists of three fused cyclohexane rings (in a nonlinear arrangement) and a terminal cyclopentane ring (Fig. 6–13a).

FIGURE 6–13 *Perhydrocyclopentanophenanthrene* (a) and one of its derivatives, *cholesterol* (b).

Steroids have widely different physiological properties and these are determined by the nature of the chemical groups attached to the basic molecular skeleton. Some steroids are **hormones** (e.g., *estrogen, progesterone,* and *corticosterone*) and affect cellular activities by influencing gene expression (Chapters 11

FIGURE 6–12 (a) *Isoprene,* the building block of the *terpenes;* (b) *vitamin A (retinol).* The ends of neighboring isoprene units are indicated by the dashed lines.

and 20). Some steroids are **vitamins** (e.g., vitamin D_2) and influence the activities of certain cellular enzymes. Other steroids are regular constituents of plasma membranes, where they influence membrane structure, permeability, and transport. Probably the best known steroid and an important constituent of the plasma membranes of many animal cells is **cholesterol** (Fig. 6–13b).

SUMMARY

The lipids are a heterogeneous collection of molecules soluble in nonpolar solvents that play two major roles in cells and tissues: (1) they are sources of reserve energy and (2) they are structural constituents of cellular membranes. The simplest lipids, the **saturated** and **unsaturated** fatty acids, consist of unbranched hydrocarbon chains terminating in carboxyl groups. In the **neutral fats**, the carboxyl groups of fatty acids are esterified to the three carbon atoms of **glycerol** and represent the most abundant form of stored lipid. The neutral fats are in rapid metabolic turnover. **Glycerophosphatides**, in which one of the fatty acid chains of a neutral fat is replaced by the polar phosphate group or a phosphate derivative, are **amphipathic** and, as such, play an important part in the structure and organization of cell membranes. The most physiologically diverse lipids are the **terpenes** (formed from **isoprene** units) and the **steroids** (formed from a series of fused, ringed hydrocarbons). Some terpenes function as vitamins and coenzymes. Steroids serve as hormones, as vitamins, and as constituents of the plasma membranes of many cells.

REFERENCES AND SUGGESTED READING

Articles and Reviews

Bell, R. M., and Coleman, R. A., Enzymes of glycolipid synthesis in eukaryotes. *Ann. Rev. Biochem.* **49.** 459 (1980).

Bretscher, M. S., The molecules of the cell membrane. *Sci. Am.* **253** (4), 100 (Oct. 1985).

Capaldi, R. A., A dynamic model of cell membranes. *Sci. Am.* **230** (3), 26 (Mar. 1974).

Fieser, L. F., Steroids. *Sci. Am.* **192** (1), 52 (Jan. 1955).

Fox, C. F., The structure of cell membranes. *Sci. Am.* **226** (2), 30 (Feb. 1972).

Hadley, N. F., Cuticular lipids of plants and arthropods. *Biol. Rev.* **56**, 23 (1981).

Hakomori, S., Glycosphingolipids. *Sci. Am.* **254** (5), 44 (May 1986).

Hokin, L. E., and Hokin, M. R., The chemistry of cell membranes. *Sci. Am.* **231** (4), 78 (Oct. 1965).

Kirscher, H. W., and Heftman, E., Symposium: Sterol analysis. *Lipids* **15**, 697 (1980).

Lindgren, F. T., Symposium: High density lipoproteins. *Lipids* **14**, 504 (1979).

Maierhofer, G., Liposomes: Preparation and application. *Am. Lab.* **17** (10), 129 (Oct. 1985).

Morrell, P., and Norton, W. T., Myelin. *Sci. Am.* **242** (5), 88 (May 1980).

Rosenthal, M. D., Accumulation of neutral lipids by human skin fibroblasts. *Lipids* **16**, 173 (1981).

Rothman, J. E., and Lenard, J., Membrane asymmetry. *Science* **195**, 743 (1977).

Storch, J., and Kleinfield, A. M., The lipid structure of biological membrames. *Trends in Biochem. Sci.* **10** (11), 418 (Nov. 1985).

Wilschut, J., and Hoekstra, D., Membrane fusion: From liposomes to biological membranes. *Trends in Biochem. Sci.* **9** (11), 479 (Nov. 1984).

Books, Monographs, and Symposia

Conn, E. E., and Stumpf, P. K., *Outlines of Biochemistry* (4th ed.), Wiley, New York, 1976.

Lands, W. E. M., and Crawford, C. G., *Enzymes of Biological Membranes* (A. Mantonosi, Ed.), Vol. 1, Plenum, New York, 1976.

Lehninger, A. L., *Principles of Biochemistry*, Worth, New York, 1982.

McGilvery, R. W., *Biochemistry, A Functional Approach* (2nd ed.), Saunders, Philadelphia, 1979.

Ostro, M. J. (Ed.), *Liposomes*, Dekker, New York, 1983.

Stryer, L., *Biochemistry* (2nd ed), Freeman, San Francisco, 1981.

CHAPTER 7

THE CELLULAR MACROMOLECULES: NUCLEIC ACIDS

The nucleic acids were discovered over 100 years ago, but their role in genetics and in the control of cellular activity has been elucidated only during the past 50 years. This is probably due in part to the great emphasis placed on the study of proteins during the first half of this century and to the mistaken belief by many scientists during this period that it was the proteins that were endowed with the genetic information of the cell. This is understandable because the composition and organization of proteins are so diverse and complex, whereas the chemical structure of the nucleic acids is so much more regular and restricted. Indeed, nucleic acids were believed to be *regular* polymers like starch and glycogen. Consequently, it was difficult to reconcile the great diversity of life with the fundamental chemical similarities manifested by all the nucleic acids. We begin our consideration of the nucleic acids by describing some of the major observations and discoveries that ultimately led biologists to recognize that these molecules *and not proteins* were intimately involved in the transmission of genetic information and in the determination and control of the activities of the cell.

CELLULAR ROLES OF THE NUCLEIC ACIDS

The Discovery of DNA

Friedrich Miescher (Swiss, 1844–1895) is credited with the discovery of nucleic acid in 1869. He isolated nuclei from white blood cells present in pus, using dilute solutions of hydrochloric acid to dissolve away other cell structures and then adding the protein-digesting enzyme *pepsin* to further degrade residual cell protein adhering to the nuclei. Nuclei isolated in this manner were then extracted with alkali, and the chemical composition of the extract was analyzed. The chemical composition of the extract differed markedly from that of protein, so Miescher called this material "nuclein." (The term **nucleic acid** was introduced 20 years following Miescher's discovery by another biochemist, Richard Altmann.) By comparing the chemical analyses reported by Miescher with those carried out more recently, it is clear that Miescher's nuclein was, in fact, the nucleic acid deoxyribonucleic acid (DNA).

Miescher also worked with the sperm cells of

salmon, which contain particularly large nuclei (more than 90% of the cell mass is accounted for by the nucleus). In addition to isolating nuclein from the sperm nuclei, he used acid to extract an organic material having an unusually high nitrogen content. He called this substance "protamine." The protamines are proteins containing an unusually rich content of lysine and arginine residues. We now recognize that extracts of cell nuclei also contain proteins called *histones* and that these are intimately associated with the nuclear DNA (see Chapter 20). The function of the cell nucleus was unknown in Miescher's time, and while Miescher was convinced of the fundamental importance of nuclein, especially during fertilization, it was not until 60 years after his discovery that a series of experiments was carried out that established the genetic role of the nucleic acids. Also during this period, it was shown that two types of nucleic acid occur in cells: deoxyribonucleic acid and ribonucleic acid (RNA). Most notable among the experiments that established the genetic role of the nucleic acids were those on (1) the *transformation* of bacterial types and (2) *virus reproduction*.

Transformation of Bacterial Types

Two types of pneumonia bacteria (*Diplococcus pneumoniae*) exist and are readily distinguished by the appearance of their colonies when cultured on agar plates; they are called "smooth" (S) and "rough" (R) types. The S type of pneumonia bacterium (which is the normal, *virulent* kind) is enclosed within a polysaccharide capsule and gives rise to smooth, shiny colonies. In contrast, the R type is noninfective (i.e., nonvirulent), is unable to synthesize the polysaccharide capsule, and gives rise to granular (and therefore rough-appearing) colonies. In 1928, F. Griffith showed that it was possible to transform the rough bacteria into the smooth type. He simultaneously injected small numbers of live R bacteria and large numbers of heat-killed S bacteria into mice, many of which subsequently died. When these mice were examined, they were found to contain live S bacteria. Because the S bacteria originally injected into the mice were heat-killed and thus incapable of reproducing, Griffith concluded that some of the R bacteria must have been **transformed** into the S type in the presence of the dead S cells. These observations were subsequently confirmed by a number of investigators who also ruled out the possibility of contamination of the inocu-

lum by a few live S cells or of the mutation of some of the R cells into the S type following injection.

In 1932, J. L. Alloway showed that the same transformation was possible *in vitro*, for when an extract of S cells was added to a culture of R cells, some of the latter were permanently transformed into the viable S type. Therefore, there appeared to be a substance in S cells that was capable of bringing about an inheritable change in the R cells.

The substance responsible for the transformations observed by Griffith and Alloway (called "transforming principle") was finally identified in 1944 by O. T. Avery, C. M. MacLeod, and M. McCarty. Although crude extracts of heat-killed S cells were found to contain protein, polysaccharide, lipid, and nucleic acid, the removal of the protein, polysaccharide, and lipid by a combination of chemical and enzymatic procedures, including enzymatic hydrolysis, chloroform extraction, and alcohol fractionation, resulted in a product that retained the transforming activity. However, when the product was treated with the enzyme *deoxyribonuclease* (which degrades DNA), the capacity to transform R cells was lost. This evidence, together with chemical analyses, showed that the transforming principle was DNA.

Virus Reproduction

As noted in Chapter 1, viruses are composed principally of protein and nucleic acid. Depending on the type of virus, the nucleic acid is either DNA or RNA. The protein forms a coat around the virus and encloses the core of nucleic acid. In 1952, A. D. Hershey and M. W. Chase conducted a series of experiments to determine whether it was the viral protein coat or the viral nucleic acid that was required for virus reproduction. Hershey and Chase used the bacterium *Escherichia coli* and the T2 virus (a DNA-containing virus that infects *E. coli*) in two sets of experiments. In one set of experiments, *E. coli* was cultured in a medium containing the radioactive isotope of sulfur, ^{35}S (the principles of radioactive tracer experiments are discussed in Chapter 14). During the growth of the bacterial population, ^{35}S was incorporated into the cells. The culture was then infected with the T2 bacteriophage, and during the reproduction of the bacteriophage within the host cells, bacterial ^{35}S was used in the synthesis of phage protein (i.e., it was incorporated into cysteine and methionine residues). Nucleic acids do not contain sulfur. Following lysis of the bac-

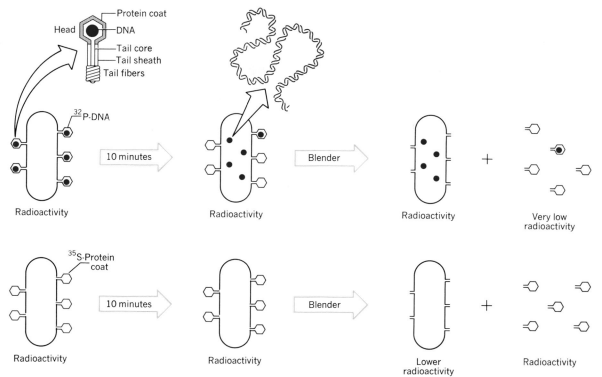

FIGURE 7–1   The experiments of A. D. Hershey and M. W. Chase. Parts shown in color represent portions of the virus that have been labeled with radioactive isotope (see text for details).

terial cells, the ^{35}S-labeled viruses were collected and used to infect *E. coli* cultured on media devoid of ^{35}S. A Waring blender was then used to separate by physical agitation what was left of the attached viruses from the surfaces of the bacterial hosts. Centrifugation was then used to separate the bacteria from unattached virus and from viruses that were detached by the action of the blender. A comparison of the radioactive sulfur content of the sedimented bacteria and unsedimented viruses revealed that nearly all the sulfur remained with the viruses and had not entered the bacterial cytoplasm.

In other experiments, *E. coli* was cultured in media containing the radioisotope ^{32}P prior to infection with the phage. The nucleic acids contain large quantities of phosphorus, and therefore ^{32}P was incorporated into newly synthesized viral DNA. When labeled viruses were used to infect additional bacteria and the blender again was employed to shake off the attached viruses, it was found that the ^{32}P radioactivity was within the infected cells. The two sets of experiments are depicted in Figure 7–1. Hershey and Chase con-

cluded that it was the DNA of the virus that entered the host cell during infection and that DNA was required for the reproduction of genetically identical virus particles by the metabolic machinery of the host cell.

The observations of Hershey and Chase also explained earlier findings by T. F. Anderson and R. M. Herriott that the T2 phage loses its capability to reproduce when distilled water is added to a suspension of the viral particles prior to their addition to a bacterial culture. Even though such viruses are still able to attach to the bacterial host, the sudden osmotic shock created by exposure to distilled water causes them to empty their nucleic acid content into the suspending medium. Therefore, it must be the nucleic acid injected into the bacterium that allows for viral replication.

Tobacco mosaic virus (TMV), the virus that infects tobacco leaves, contains RNA instead of DNA. In the early 1950s, H. Fraenkel-Conrat separated the RNA and protein components of TMV and found that the RNA separately injected into the tobacco leaves could

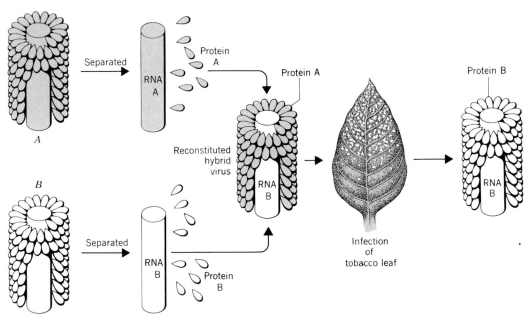

FIGURE 7–2 H. Fraenkel-Conrat's reconstitution experiments demonstrating that RNA and not protein is the genetic material in tobacco mosaic virus (see text for details).

infect the plant and produce new virus particles containing both RNA and protein. Fraenkel-Conrat also found that when protein isolated from one strain of TMV was mixed with RNA isolated from another strain, a reconstituted "hybrid" virus was produced that retained the capability to infect the tobacco leaves. When progeny viruses produced following infection by the hybrid were isolated and their RNA and protein components separated and analyzed, it was found that the type of protein in the virus coat was identical to that of the strain used as the source of RNA (Fig. 7–2). It was *not* the same as the protein of the hybrid viruses causing the infection. These observations support earlier conclusions concerning the genetic role of the nucleic acids and also prove that these molecules alone contain information that determines the specific nature of newly synthesized protein.

Although the most direct evidence supporting the notion that DNA (or in the case of some viruses, RNA) is the genetic material was obtained using microbial systems, observations supporting this idea were also made with higher organisms. Based on numerous studies of the mechanism of fertilization, including those by O. Hertwig (1865), H. Fol (1877), E. Strasburger (1884), A. Weismann (1892), and E. B. Wilson (1895), it was generally acknowledged by the turn of the century that the chromosomes contained within

the cell nucleus were concerned with the transmission of heredity. Therefore, the development of staining reactions (such as the Feulgen reaction) that are specific for DNA and the subsequent microscopic localization of the DNA in the chromosomes implicated DNA as the genetic material. Further evidence was provided during the 1940s from quantitative chemical analyses of DNA present in measured quantities of cells (i.e., numbers, dry weight, etc.). These analyses revealed that the amount of DNA per cell was more or less constant within the various tissues of an organism. Moreover, it was also found that the total quantity of DNA present in the cell nucleus was related to its **ploidy** (i.e., the number of complete sets of chromosomes). The gametes (sperm and egg cells) of an organism were shown to have one-half as much DNA as the diploid somatic cells. This is precisely what would be expected if DNA served as the genetic material.

Although RNA appears to be the genetic material of some virus particles, there is no evidence that it plays a similar role in cells. Instead, RNA serves as an intermediary between the genetic information of DNA and the expression of this information as the synthesis of the cell's enzymes and other proteins (see later and also Chapter 22). In eukaryotic cells, most of the DNA is present in the nucleus, although certain organelles (mitochondria and chloroplasts) also contain DNA (see Chapters 16 and 17). Before we explore just how the relationship between DNA and RNA can be effected and the mechanism by which DNA serves as the genetic material, it is first necessary to consider the chemistry of these two macromolecules.

COMPOSITION AND STRUCTURE OF THE NUCLEIC ACIDS

As we have already noted, there are two major classes of nucleic acids: DNA and RNA. Both are composed of unbranched chains of units called **nucleotides**, each of which contains (1) a nitrogenous base (either a **purine** or **pyrimidine**), (2) a **pentose**, and (3) **phosphoric acid**. In RNA, the pentose is **ribose**, whereas in DNA it is **2-deoxyribose**. Both DNA and RNA contain the purine nitrogenous bases **adenine** (abbreviated A) and **guanine** (G) and the pyrimidine **cytosine** (C), but in DNA a second pyrimidine is **thymine** (T), whereas in RNA it is **uracil** (U). A number of other nitrogenous bases have been identified in DNA and RNA, but these occur much less frequently. The phosphoric acid component of each nucleotide is, of course, chemically identical in both nucleic acids. These relationships are summarized in Table 7–1, and the corresponding chemical formulas are shown in Figure 7–3.

The pentose of each nucleotide unit is simultaneously bonded through its number 1 carbon atom to the nitrogenous base (forming a **nucleoside**) and through its number 5 carbon atom to phosphoric acid. The structures of the four deoxyribonucleotides of DNA together with the specific numbering system used to identify each constituent atom are shown in Figure 7–4. Successive nucleotides of DNA and RNA are joined together by phosphodiester linkages involving the 5'-phosphate group of one nucleotide unit and the 3'-hydroxyl group of the neighboring unit (Figure 7–5).

The "backbone" of a nucleic acid molecule is formed by the repeating sequence of pentose and phosphate groups, and this is the same in all molecules. What distinguishes one DNA (or RNA) molecule from another is the specific *sequence* of purine and pyrimidine bases present in the chain of nucleotides and the *total* number of nucleotides (i.e., the size of the molecule). Each chain of nucleotides is called a **polynucleotide**.

Structure of DNA

At one time, it was believed that the four purines and pyrimidines of DNA occurred in approximately equal amounts in the molecule. However, the studies of E. Chargaff and others in the late 1940s showed that this was not the case. Instead, they found that the relative amounts of the nitrogenous bases varied between species but were constant within a species. The constancy noted within a species was maintained regardless of the tissue or organ from which the DNA was isolated. Furthermore, the relative amounts of the nitrogenous bases were similar in closely related species and quite different in unrelated species. Chargaff also made the following extremely important finding. *Regardless of the species used as the source of DNA, the molar ratios of adenine and thymine were always very close to unity, and the same was true for guanine and cytosine.* No such constant relationship could be demonstrated for any other combination of nitrogenous base pairs. This implied that for some reason, every molecule of DNA contained equal amounts of adenine and thymine and also equal amounts of guanine and cytosine.

Using chemical information of this sort, together with the results of X-ray crystallographic studies of DNA, J. D. Watson, F. H. C. Crick, M. H. F. Wilkins, and R. Franklin proposed a model for the structure of DNA in the early 1950s. They suggested that a molecule of DNA consists of two helical polynucleotide chains wound around a common axis to form a right-handed "double helix." In contrast to the arrangement of amino acid side chains in helical polypeptides (where the side chains are directed radially away from the helix axis), the purine and pyrimidine bases of each polynucleotide chain were directed inward toward the center of the double helix so that they faced each other.

On the basis of stereochemical studies, Watson and Crick further suggested that the only possible arrangement of the nitrogenous bases within the double helix that was consistent with its predicted dimensions was that in which a purine always faced a

TABLE 7–1   CHEMICAL CONSTITUENTS OF DNA AND RNA

	DNA	RNA
Purines	Adenine	Adenine
	Guanine	Guanine
Pyrimidines	Cytosine	Cytosine
	Thymine	Uracil
Pentose	2-Deoxyribose	Ribose
	Phosphoric acid	Phosphoric acid

Nucleic acids are composed of repeating subunits called nucleotides.
Each nucleotide is composed of three units.

(1) A phosphate group:	
(2) A five-carbon sugar or pentose:	In RNA: Ribose / In DNA: 2-deoxyribose
(3) A Cyclic, nitrogen-containing base:	

Uracil
(2,4-oxypyrimidine)

Cytosine
(4-amino-2-oxypyrimidine)

Thymine
(2,4-oxy-5-methyl pyrimidine)

In RNA only
(with rare exceptions):

In both RNA
and DNA:

In DNA only
(with rare exceptions):

Adenine
(6-aminopurine)

Guanine
(2-amino-6-oxypurine)

FIGURE 7–3 Structural formulas of the constituents of nucleic acids. When the pentoses are present in nucelosides, nucleotides, or nucleic acids, the five carbon atoms are numbered 1', 2', 3', 4', and 5', respectively, so that they may be distinguished from the carbons of the nitrogen bases. (Modified from Gardner, E. J., and Snustad, D.P., *Principles of Genetics*, 7th ed., p. 88. Copyright © 1984 by John Wiley & Sons, Inc.)

FIGURE 7–4 The four deoxy ribonucleotides of DNA. RNA contains similar *ribo*nucleotides, which contain the pyrimidines uracil and cytosine and the purines adenine and guanine. (Modified from Gardner, E.J., and Snustad, D.P., *Principles of Genetics*, 7th ed., p. 89. Copyright © 1984 by John Wiley & Sons, Inc.)

FIGURE 7–5 Portion of one strand of a DNA molecule showing the phosphodiester linkage between successive nucleotides.

FIGURE 7–6 Formation of hydrogen bonds between thymine and adenine and between cytosine and guanine. Note that three bonds are formed between C and G and only two between T and A.

structure. Accordingly, Watson and Crick concluded that *adenine must be matched with thymine and guanine with cytosine.* This conclusion was, of course, in agreement with the chemical findings of Chargaff (see above)—in fact, Chargaff's data may have been critical to the development of Watson and Crick's proposals. The manner in which hydrogen bonds are formed between adenine and thymine and between guanine and cytosine is shown in Figure 7–6. Although individually weak, the great number of these bonds contributes appreciably to the stability of the double-helical structure. In addition, the double helix is stabilized by hydrophobic bonds between neighboring nitrogenous bases of each polynucleotide chain.

Certain other features of the structure of DNA should be noted. The two polynucleotide chains that make up the molecule are **antiparallel**. That is, beginning at one end of the molecule and progressing toward the other, successive nucleotides of one chain are joined together by 3'→5' phosphodiester linkages, whereas the complementary nucleotides of the other chain are joined by 5'→3' phosphodiester linkages. This antiparallel arrangement is depicted diagrammatically in Figure 7–7.

pyrimidine, for the diametric distance between the two polynucleotide chains is too small to accommodate two juxtaposed purines. Which purine was matched with which pyrimidine became clear from a consideration of which pairs would be able to form the hydrogen bonds necessary to stabilize the double-helical

A-DNA and B-DNA Right-handed, double-helical DNA can exist in either of two principal forms: these are called *A-DNA* and *B-DNA*. The two forms differ primarily in the positioning of the nitrogenous bases around the axis of the double helix and in the numbers of bases per helical turn. In B-DNA, there are 10 base

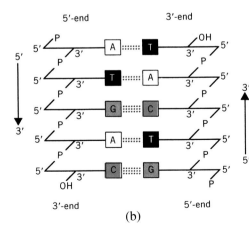

FIGURE 7–7 Molecular structure of DNA showing the pentose sugar (S)–phosphate (P) backbones of the polynucleotide chains and their antiparallel nature. (*a*) Molecular structure; (*b*) schematic representation, emphasizing the opposite polarity of the complementary strands.

(a)   (b)

pairs per turn of the helix, each turn sweeping out 3.4 nm (34 Å) of linear translation (see Fig. 7–8); in A-DNA, there are 11 bases per helical turn, each turn sweeping out 2.8 nm (28 Å) of linear translation. Thus in the A form, the double helix has a greater diameter. In B-DNA the complementary bases lie in a plane that is perpendicular to the axis of the helix, whereas in A-DNA, the planes of successive base pairs are tilted relative to the helical axis. It is the B form of the DNA that is believed to predominate in cells, although the A form may exist in DNA–RNA hybrids (see Chapter 21).

The two polynucleotides are twisted around one another in such a way as to produce two helical grooves in the surface of the molecule; these are called the *major* and *minor* grooves (Fig. 7–8). The floor of the major groove is lined with oxygen and nitrogen atoms that could form hydrogen bonds with the amino acid side chains of proteins. Indeed, it is the association of specific proteins with DNA that is believed to be involved in regulating gene expression (Chapter 21). Interaction of water molecules with the atoms lining the floor of the minor groove is believed to contribute to the stability of the B form.

FIGURE 7–8 Diagram of the DNA double helix showing complementary base pairing, major and minor grooves, and certain molecular dimensions. Two forms of the right-handed, double helix exist: the A form and the B form. In B-DNA, which is the form shown here, the planes of the complementary base pairs are perpendicular to the axis of the double helix. B-DNA has a diameter of 20 Å. Each complete turn of a helix accounts for 34 Å of linear translation and each nucleotide 3.4 Å. (Hence, there are 20 nucleotides per helical turn.) The usual symbols for the nitrogen bases are used, with S for sugar and P for phosphate.

Replication of DNA

One of the intrinsic properties of the genetic material is its capacity for replication. The manner in which DNA satisfies this requirement is apparent from the nitrogenous base pairing required in the model. Because the sequence of bases in one polynucleotide chain automatically determines the sequence of bases in the other, it is clear that one-half of a molecule (i.e., one of the two helices) contains all the information necessary for constructing a whole molecule. For example, if we know that the sequence of bases along one polynucleotide chain of DNA is **A T G A C**, and so on, then the complementary sequence in the other chain must be **T A C T G**, and so on. Therefore, if the double helix were unwound, each separate polynucleotide chain could act as a template for the production of a new, complementary chain. The result would be two identical double helices where there was only one before. Of course, one-half of each new double helix would be represented by one of the original polynucleotide chains. The basic features of this process are shown in Figures 7–9 and 7–10. A detailed description of the mechanism by which the replication of DNA occurs, together with its experimental basis, is given in Chapter 21.

Denaturation and Renaturation of DNA

DNA is readily denatured by extremes of temperature and pH. The denaturation takes the form of an unwinding of the double helix as hydrogen bonds between complementary bases are disrupted. This form of DNA denaturation is referred to as *melting* and produces separate DNA strands. Solutions of DNA absorb ultraviolet light (UVL) having a wavelength of 260 nm. When the temperature of a native solution of DNA is elevated, the resulting melting is accompanied by an increase in UVL absorption. This **hyperchromic effect** occurs because the purines and pyrimidines of separated strands can absorb more light energy than when they are part of a double helix. Some viruses contain a single-stranded form of DNA (see later), and because this form does not exhibit the hyperchromic effect, it is readily distinguished from double-stranded forms.

When DNA is melted thermally, denaturation begins in regions of the double helix that are rich in A-T base pairs and progressively shifts to regions of greater and greater G-C content. This is because the two hydrogen bonds holding each A-T pair together

FIGURE 7–9 Fundamental model for the replication of DNA. The replication of the DNA molecule could be effected if the double helix unwound and the separated chains acted as templates for the production of two new complementary chains. In this figure, the original double helix has partially unwound, and two complementary chains are being formed. The process is described in detail in Chapter 21.

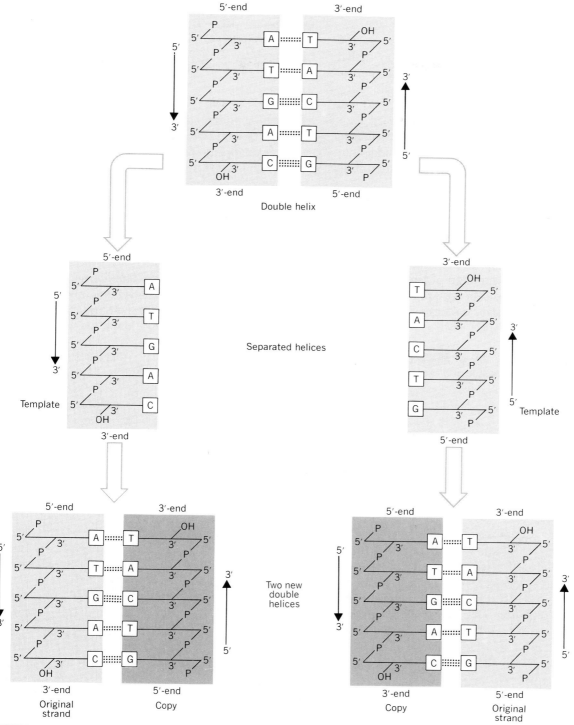

FIGURE 7–10 Replication of the base sequence of DNA. The complementary base sequence of a 5-base-pair segment of a DNA molecule is shown at the top of the figure. The complementary bases in this region are separated as the double helix unwinds and serve as templates for the production of two new polynucleotide chains (middle). Finally, two complete DNA molecules are produced with identical base sequences (bottom). One-half of each of these DNA molecules is represented by one newly synthesized chain (color) and by one original chain.

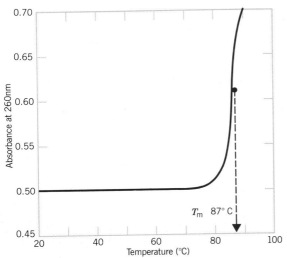

FIGURE 7–11   A typical thermal denaturation curve for DNA showing the hyperchromic effect and the evaluation of the T_m of the nucleic acid. (Redrawn from Conn, E. E., and Stumpf, P. K., *Outlines of Biochemistry*, 4th ed., p. 121. Copyright © by John Wiley & Sons, Inc.)

lowering the temperature of the solution, whereby separated strands recombine to form double helices as hydrogen bonds between complementary bases are reformed. This *reannealing* can be monitored as a decrease in UVL absorption by the DNA solution. The capacity for denatured DNA to reanneal can be used to assess the size of an organism's *genome* (the total number of genes, see Chapters 20 and 21) and the *complexity* of the DNA that is present.

When reannealing studies are to be performed, the isolated DNA is first broken by shearing force into lengths of several hundred to several thousand nucleotide pairs. The double-helical DNA is then thermally denatured yielding single strands. A known concentration of the single-stranded DNA is then incubated at the reannealing temperature (usually about 25° below the T_m) and the reannealing rate is determined from the rate of change in UVL absorbance. A large genome reanneals more slowly than a small genome because there is a greater number and variety of DNA fragments. Thus, each fragment takes a longer time to "seek out" and anneal with its complementary partner. The kinetics of DNA renaturation (Fig. 7–12) is represented by a curve relating the percentage of reassociated fragments to the "C_0t number" (i.e., the concentration of DNA in moles of nucleotides per liter (C_o) times the reaction time (t) in seconds). As seen in Figure 7–12, viral DNA (curve c) reanneals more rapidly than prokaryote DNA (curve d), and the latter reanneals more quickly than eukaryote DNA (curve e). The discovery of a eukaryotic DNA fraction in mammalian cells that reanneals unexpectedly rapidly (curve b) revealed for the first time the existence in the genomes of higher organisms of repetitive DNA sequences (see Chapter 22). In Figure 7–12, curve a shows the reannealing rate of a solution containing a

can be broken more easily (hence, at a lower temperature) than the three hydrogen bonds holding each G-C pair together. A quantitative measure of the change in UVL absorbance that takes place as the temperature of a DNA solution is slowly elevated is called a **melting curve** (Fig. 7–11). The point in the melting curve at which the change from double-stranded to single-stranded DNA is half complete is called the T_m value and is characteristic of a particular source of DNA. The species specificity that is characteristic of DNA melting curves reflects differences in the G-C and A-T compositions of different kinds of DNA (Fig. 7–12).

Thermally denatured DNA can be renatured by

FIGURE 7–12 C_0t curves for different types of DNA. The kinetics of reannealing is a function of the size and complexity of the genome. (*a*) Mixure of poly U and poly A fragments; (*b*) repetitive DNA fraction from mice; (*c*) DNA from T 4 phage of *Eschericlia coli*; (*d*) *E. coli* DNA; and (*e*) nonrepetitive DNA fraction from a calf.

mixture of synthetically produced polyuridylic acid (a nucleotide chain with only uracil bases) and polyadenylic acid strands. Even though uracil is not usually found in DNA, like thymine it can form hydrogen bonds with adenine and does so in many RNA molecules (see below).

Z-DNA

For about 25 years following the original establishment of the Watson–Crick model of DNA structure, it was presumed that all naturally occurring doublehelical DNA was right-handed. However, in 1979 A. Rich confirmed earlier observations reported by F. M. Pohl and T. Jovin that a left-handed form of DNA also exists. As in right-handed DNA, the two helices are held together by complementary base pairing and the strands are antiparallel. Because the sugar–phosphate backbones of the two polynucleotides trace a zigzag course around the axis of the helices (Fig. 7–13), this left-handed DNA has been called **Z-DNA**.

Though the structure of Z-DNA originally proposed by Rich was based on studies of DNA crystals produced in the laboratory, Z-DNA has since been identified in the chromosomes of a number of eukaryotic cells. A variety of indirect evidence also implies that Z-DNA is a normal constituent of animal cells, plant cells, and bacteria. Unlike B-DNA, Z-DNA is highly immunogenic, making it possible to readily produce antibodies (Chapter 4) against Z-DNA. The reaction of these antibodies with DNA isolated from a variety of cell types implies the presence of the Z form. Naturally occurring *Z-DNA-binding proteins* have also been isolated from a number of cells.

The Z form of DNA appears to coexist with the B form in the same DNA molecules. Indeed, it is believed that DNA can "flip" between the B and Z forms in those regions of a double helix that are rich in sequences having *alternating* purines and pyrimidines. These sequences appear to occur in selective regions of a cell's DNA, lending credence to the idea that switches in DNA helicity between right-handed and left-handed forms may be involved in selective gene expression (see Chapters 11 and 21).

"Single-Stranded" DNA

Although in nearly every case so far studied, DNA consists of two polynucleotide chains twisted about one another to form a double helix, it is now apparent that in a few bacterial viruses (i.e., the φX174 and S13

FIGURE 7–13 Double-helical DNA can exist in both *left-handed* (Z-DNA) annd *right-handed* (A-DNA and B-DNA) forms. Shown here are Z-DNA (on the left) and B-DNA (on the right). In Z-DNA, the phosphate–sugar backbone of each nucleotide (shown by heavy black and red lines) follows a zig-zag pattern around the axis of the helices.

E. coli phages) DNA exists as a single polynucleotide chain. This was initially suspected when chemical analyses of the nitrogenous base contents of these viral DNAs revealed that the amounts of adenine and thymine, as well as guanine and cytosine, were not equal. During reproduction of these viruses, the single-stranded DNA (referred to as the "+ strand") is injected into the host bacterial cell, where it acts as a template for the reproduction of a complementary polynucleotide chain (called the "– strand"); these two polynucleotides combine to form a conventional double helix, which then serves as the template for the production of additional + strands. The newly produced + strands are then enclosed in the viral protein coats to form new virus particles.

Structure of RNA

As noted earlier, RNA and DNA differ chemically in two notable ways: in RNA, **ribose** is the pentose (not deoxyribose as in DNA) and the pyrimidine **uracil** occurs in place of thymine (Fig. 7–4). Early chemical analyses of the nitrogenous base contents of RNAs from various sources revealed that the A:U and G:C molar ratios were quite different from the A:T and G:C molar ratios of DNA. On this basis, it was concluded that RNA occurs as a single polynucleotide chain. This contention is also supported by physicochemical studies, but it should be noted that there are some viral RNAs that are double stranded. Although only one polynucleotide chain is usually present, RNA does possess regions of double-helical coiling where the single chain loops back upon itself. These regions are stabilized by the formation of hydrophobic bonds between neighboring bases (as in DNA) and also by the formation of hydrogen bonds between guanine and cytosine and between adenine and uracil. In RNA, A and U can form two hydrogen bonds similar to the two bonds formed between A and T in DNA. As in DNA, the portions of the RNA strand that are twisted around each other to form a double helix are antiparallel. In double-helical RNA, the helices and the complementary base pairs are arranged in much the same manner as in A-DNA.

Synthesis of RNA

Except perhaps in the case of the reproduction of certain RNA viruses (see below), the synthesis of RNA appears to be directed by DNA and is called **transcription**. The formation of the RNA polynucleotide takes place using the base sequence along only one of the two deoxyribonucleotide helices of DNA (producing a temporary RNA–DNA hybrid) as a template and results in the release of a single, complementary polyribonucleotide chain in which the base uracil occurs in place of thymine (Figure 7–14). Transcription is considered in greater detail in Chapter 21.

Replication of DNA and RNA Viruses

The viruses may be divided into two classes: viruses whose genetic complement consists of DNA and viruses whose genetic complement consists of RNA. In cells infected with DNA viruses, the infecting viral DNA is *replicated*, forming new viral DNA that is then *transcribed* into RNA; this RNA is then *translated* into viral protein (Fig. 7–15a). The newly produced viral DNA and viral proteins combine in the assembly of new, complete virus particles that are released upon lysis of the host cell. A **latent state** can also be established in which the viral DNA is incorporated into the host cell's genome, being replicated and distributed along with the host cell's native DNA, until it is transcribed once again into additional viral RNA and then into viral proteins (see Chapter 1).

For most RNA viruses (e.g., poliomyelitis, influenza, common cold, etc.), DNA involvement is essentially bypassed. For example, during the infection of a cell with polio virus, the single-stranded RNA (called a + strand) enters the host cell, where it acts as a template for the synthesis of complementary – strands. The latter are then employed in the proliferation of new + strands, and these are translated into viral proteins (Fig. 7–15b).

The mechanism described above is varied in several other viruses in which the RNA is either double stranded (e.g., reoviruses, in which only one of the two RNA strands produced during replication is transcribed) or the infecting single RNA strand is complementary (rather than identical) to the newly produced viral RNAs that are to be translated into viral proteins (e.g., Sendai virus, Newcastle disease virus, etc.)

Yet another mechanism exists in the case of the RNA tumor viruses (e.g., Rous sarcoma virus). These viruses do not transfer information from RNA to RNA, but rather from RNA to DNA and then to new RNA. The viral RNA is employed as a template for the synthesis of DNA by the infected cell (a phenomenon that is called "reverse transcription"). Some of the resulting "viral DNA" may be incorporated into the genome of the host cell, establishing what is called the **provirus** state.

The provirus state has been suggested as the basis of a number of different RNA virus-induced and DNA virus-induced cancers. According to this view, one or more of the provirus genes—which are normally repressed by the host cell—may become derepressed and cause the production of an *oncogenic* (i.e., cancer-causing) substance that alters the cell's normal properties or behavior. Such a change may be delayed for a number of generations, depending on the period of latency.

During the 1960s and early 1970s, the so-called "central dogma" of molecular biology was the orderly and *unidirectional* flow of information encoded in the

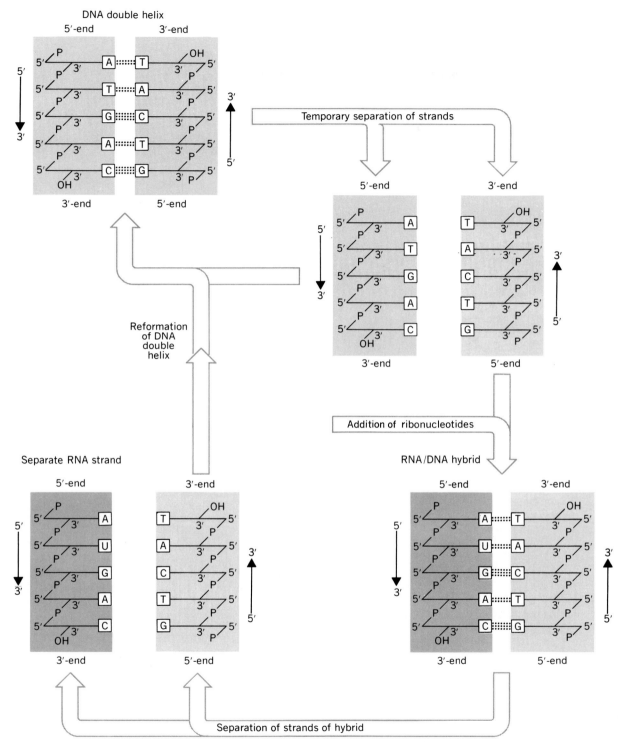

FIGURE 7–14 Transcription of the base sequence of one of the helices of DNA into the complementary base sequence of RNA. This process is considered in detail in Chapter 21.

FIGURE 7-15   Replication of DNA and RNA viruses.

base sequences of a cell's DNA to RNA and then to protein, that is,

$$\text{DNA} \rightarrow \text{RNA} \rightarrow \text{protein}$$

The discovery of reverse transcription by certain RNA viruses in which the information of RNA is passed on to DNA has necessitated a reexamination of that dogma and raised the question of whether a similar interaction between RNA and DNA might normally occur in cells (i.e., cells not infected by viruses) under specific conditions (e.g., during cellular differentiation). The central dogma might more appropriately be represented as

$$\text{DNA} \rightarrow \text{RNA} \rightarrow \text{protein}$$

Types of Cellular RNA

Cells contain three major functional types of RNA: **ribosomal RNA** (abbreviated **rRNA**), **messenger RNA (mRNA)**, and **transfer RNA (tRNA)**. All these are transcribed from DNA and are engaged in mediating the expression of the genetic message of DNA by participating in the synthesis of the cell's proteins, as discussed in Chapter 22. It has already been noted that RNA occasionally serves as the genetic material of viruses.

Of the cellular RNAs, rRNA is the most abundant, accounting for up to 85% of the total RNA of the cell. Only three or four different kinds of rRNA are present in cells, and these are confined for the most part to the cell's ribosomes. mRNA accounts for about 5 to 10% of the cell's RNA and is much more heterogeneous with respect to size and nitrogenous base content than the rRNAs. This results from the relationship (see below) between the chain lengths and base sequences of mRNAs and the variable sizes and primary structures of polypeptides synthesized in a cell. Most mRNA occurs in the cytoplasm, where it transiently combines with ribosomes during protein synthesis.

About 10 to 20% of the cell's RNA is tRNA. All tRNA molecules are similar in size and typically contain 75 nucleotide units. In spite of these similarities, a single cell may contain about 60 species of tRNA differing in their base sequences. Because most of the tRNA is recovered in the cytoplasmic (i.e., soluble) phase of disrupted cells following centrifugation, tRNA is also called *soluble* RNA (i.e., sRNA). In view of its small size and relative ease of isolation, tRNA has been more extensively studied than the other two ribonucleic acids, and the specific primary, secondary, and tertiary structures of many tRNAs have already been determined. tRNAs contain moderate amounts of unusual nucleotides such as *ribothymidine, dihydrouridine, pseudouridine*, and *methylguanosine*. These are formed by modification of the four common

RNA bases after the tRNA has been synthesized from the four unmodified ribonucleotides. The modified bases play a crucial role in establishing the unique spatial organization of these molecules.

Although the mechanisms of DNA replication and protein synthesis are considered in depth in subsequent chapters, it is appropriate that a brief accounting of the functional relationships among the nucleic acids and between nucleic acids and proteins be made at this time. Inheritable information is encoded in the various nitrogenous base sequences possessed by the cell's DNA, and by the process of *transcription* and *translation* these base sequences are employed to specify the primary structures of all proteins produced by the cell. Most important among these proteins are the enzymes that catalyze and regulate the myriad of chemical reactions characterizing the cell's metabolism. Therefore, the information of DNA confined essentially to the cell nucleus manifests itself primarily in the cytoplasm as the synthesis of a unique assemblage of proteins. The replication of DNA that precedes mitotic cell division and the equal distribution of the duplicated DNA among the progeny cells provides for the passage of complete sets of information from one generation of cells to another. In addition to serving as templates for their own replication, the nucleotide sequences of DNA are used during transcription to produce complementary base sequences of RNA. The resulting RNAs then serve as intermediaries in translating the original message into protein. Of paramount importance in this process are the mRNA molecules whose base sequences directly determine the *primary structures* of the polypeptides. These mRNA molecules leave the nucleus of the cell following their synthesis and attach in the cytoplasm to one (or several) ribosome. The rRNA of each ribosome is believed to play a role in this attachment. tRNA molecules also produced in the nucleus enter the cytoplasm, where they combine with specially activated amino acids (distinct tRNAs exist for each species of amino acid). The resulting complexes, directed by the base sequence of mRNA attached to the ribosomes, sequentially deposit their amino acids in the growing polypeptide chains. The details of this process are presented in Chapters 21 and 22.

SUMMARY

Although **nucleic acids** were discovered and chemically characterized more than 100 years ago, their genetic role was not appreciated until the phenomenon of bacterial transformation and the mechanism of virus reproduction were understood. The two major nucleic acids, DNA and RNA, are composed of unbranched chains of subunits called **nucleotides**, each nucleotide containing **phosphate**, a **pentose**, and a **nitrogenous base**. Specific differences exist in the nucleotides found in DNA and in RNA. In DNA, two **antiparallel** polynucleotide chains are twisted around one another to form a **double helix**, the helices being held together in part through hydrogen bonds between juxtaposed nitrogenous bases. Constraints imposed by the known shape and organization of the double helix and the nitrogenous bases capable of forming such stabilizing bonds suggest that the base sequence in one polynucleotide necessarily determines the sequence of bases in the other. Therein lies the capacity for replication and DNA serves as the genetic material in all cells and in many viruses.

Most RNAs are formed by a single polynucleotide twisted about itself in certain regions, thereby forming a periodic double-helical structure. In certain viruses, RNA serves as the genetic material, but in prokaryotic and eukaryotic cells, RNA serves three primary functions. The **ribosomal** RNAs are the most abundant and serve as functional components of the cell's ribosomes during protein synthesis. The **messenger** RNAs contain the sequences of nucleotides that specify the primary structures of the cell's proteins. **Transfer** RNAs function in the transport of amino acids to the messenger RNA–ribosome complex during protein synthesis or **translation**. A cell's RNAs are produced by transcription of its DNA.

REFERENCES AND SUGGESTED READING

Articles and Reviews

Bauer, W. R., Crick, F. H. C., and White, J. H., Supercoiled DNA. *Sci. Am.* **243** (1), 118 (July 1980).

Britten, R. J., and Kohne, D. E., Repeated segments of DNA. *Sci. Am.* **222** (4), 24 (April 1970).

Brown, D. D., The isolation of genes. *Sci. Am* **229** (2), 20 (Aug. 1973).

Campbell, A. M., How viruses insert their DNA into the DNA of the host cell. *Sci. Am.* **235** (6), 102 (Dec. 1966).

Crick, F. H. C., The structure of the hereditary material. *Sci. Am.* **194** (4), 54 (Oct. 1954).

Darnell, J. E., RNA. *Sci. Am.* **253** (4), 68 (Oct. 1985).

Dickerson, R. E., The DNA helix and how it is read. *Sci. Am.* **249** (6), 94 (Dec. 1983).

Felsenfeld, G., DNA. *Sci. Am.* **253** (4), 58 (Oct. 1985).

Fiddes, J. C. The nucleotide sequence of a viral DNA. *Sci. Am.* **237** (6), 54 (Dec. 1977).

Grobstein, D., The recombinant-DNA debate. *Sci. Am.* **237** (1), 22 (July 1977).

Holley, R. W., The nucleotide sequence of a nucleic acid. *Sci. Am.* **214** (2), 30 (Feb. 1966).

Kolata, G., Z-DNA moves toward "real biology." *Science* **222**, 495 (1983).

Mirsky, A. E., The discovery of DNA. *Sci. Am.* **218** (6), 78 (June 1968).

Reddy, V. B., Thimmappaya, B., Dhar, R., *et al.*, The genome of simian virus 40. *Science* **200**, 494 (1978).

Sinsheimer, R. L., Single-stranded DNA. *Sci. Am.* **207** (1), 109 (July 1962).

Temin, H. M., RNA-directed DAN synthesis. *Sci. Am.* **226** (1), 24 (Jan. 1972).

Drlica, K., *Understanding DNA and Gene Cloning*, Wiley, New York, 1984.

Hunt, T., Prentis, S., and Tooze, J. (Eds.), *DNA Makes RNA Makes Protein*, Elsevier, Amsterdam, 1983.

Kornberg, A., *DNA Replication*, Freeman, San Francisco, 1980.

Kornberg, A., *1982 Supplement to DNA Replication*, Freeman, San Francisco, 1982.

Lehninger, A. L., *Principles of Biochemistry*, Worth, New York, 1982.

Lewin, B., *Genes*, Wiley, New York, 1983.

Luria, S. F., Darnell, J. E., Baltimore, D., and Campbell, A., *General Virology* (3rd ed.), Wiley, New York, 1978.

Rawn, J. D., *Biochemistry*, Harper & Row, New York, 1983.

Saenger, W., *Prinicples of Nucleic Acid Structure*, Springer-Verlag, New York, 1984.

Stryer, L., *Biochemistry* (2nd ed.), Freeman, San Francisco, 1981.

Taylor, J. H., *Selected Papers on Molecular Genetics*, Academic, New York, 1965.

Watson, J. D., *Molecular Biology of the Gene* (3rd ed.), W. A. Benjamin, Menlo Park, Calif., 1976.

Weinberg, R. A., The molecules of life. *Sci. Am.* **253** (4), 48 (Oct. 1985).

Woodward, D. O., and Woodward, V. W., *Concepts of Molecular Genetics*, McGraw–Hill, New York, 1977.

Books, Monographs, and Symposia

Beljanski, M., *The Requlation of DNA Replication and Transcription*, Karger Publishing Co., Basel, Switzerland, 1983.

PART THREE

CELL
METABOLISM

CHAPTER 8

ENZYMES

The metabolism of a cell is characterized by a myriad of simultaneously occurring chemical reactions. Nearly all these reactions are catalyzed by a special class of proteins called **enzymes**. It has been estimated that the typical eukaryotic cell contains about 3000 different enzymes. In the absence of these enzymes, most of the cellular reactions would proceed at much slower rates. Enzymes integrate the cellular chemical reactions and provide the order without which the complex processes of life would not be possible. Most enzymes are characterized by a high degree of specificity; that is, they will catalyze one particular chemical reaction but not another. Enzyme molecules themselves are not consumed in the chemical reactions that they catalyze; instead, they become available over and over again to repeatedly catalyze their reactions.

More than a thousand different enzymes have been specifically identified, and of these, several hundred have already been isolated. Most of the enzymes responsible for such diverse functions as alimentary digestion, blood coagulation, muscle contraction, carbohydrate and lipid metabolism, and nucleic acid and protein biosynthesis are known. Before considering the properties of enzymes and the mechanisms of enzymatic catalysis, some general features of chemical reactions will be reviewed.

MOLECULARITY OF CHEMICAL REACTIONS

Chemical reactions may be classified according to their *level* of molecularity; for example, *monomolecular* reactions, *bimolecular* reactions, *trimolecular* reactions, and so on. A monomolecular reaction may be written

$$A \rightleftharpoons P \qquad (8–1)$$

and involves the conversion of one molecular species (i.e., A) to another (i.e., P) without the addition or removal of atoms. Instead, an intramolecular reorganization of the molecule occurs. The interconversion of the α and β isomers of glucose via the open-chain intermediate is an example of this type of reaction (see Chapter 5). In the case just cited, the intramolecular changes are spontaneous and do not require catalysis by an enzyme. In contrast, the conversion of glucose-6-phosphate to fructose-6-phosphate during glycolysis (Fig. 8–1) is an example of an enzyme-catalyzed monomolecular reaction. Enzymes catalyzing intramolecular reorganizations are called **isomerases**.

169

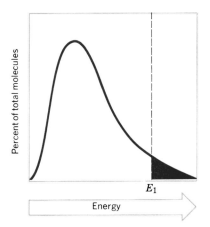

Glucose–6–phosphate   Fructose–6–phosphate

FIGURE 8–1   Reaction catalyzed by an *isomerase*. During monomolecular reactions, the atoms are rearranged, but none are added or removed.

Monomolecular reactions are relatively uncommon in cells. Much more numerous are bimolecular reactions and reactions of higher order, involving two or more reactants and/or two or more products; for example,

$$A + B \rightleftharpoons P \qquad (8\text{--}2)$$

$$A \rightleftharpoons P_1 + P_2 \qquad (8\text{--}3)$$

$$A + B \rightleftharpoons P_1 + P_2 \qquad (8\text{--}4)$$

$$A + B + C \rightleftharpoons P \qquad (8\text{--}5)$$

Among the major classes of enzymes catalyzing bimolecular and higher-order reactions are the **hydrolases**, **dehydrogenases**, **decarboxylases**, and **transferases**. Hydrolases catalyze reactions in which water is either added to or removed from the reactant(s). Dehydrogenases oxidize compounds by catalyzing the removal of hydrogen atoms. Decarboxylases remove carbon dioxide from carboxylic acids, and transferases remove reactive groups from one compound and transfer them to another. Numerous examples of these and other enzyme-catalyzed cellular reactions will be encountered in subsequent reading.

REACTION KINETICS

To begin our consideration of the *kinetics* of chemical reactions, consider the reaction described by equation 8–2 in which A and B are the reactants and P is the single product of the reaction. One prerequisite for the reaction to occur is the "collision" of molecules A and B. However, a collision alone is not sufficient to guarantee the formation of molecule P. In addition, molecules A and B must possess some minimal

amount of kinetic energy at the instant of collision and must be oriented with respect to one another at that instant such that the chemical bonds strained by the collision allow the shifts of orbital electrons necessary for the formation of the product.

The probability of a reaction occurring between A and B is determined in part by the probability of finding molecules A and B in the same region of space at the same time (i.e., this would constitute a collision). The probability of finding molecule A in a specific region of space is directly proportional to its concentration, and the same is true for molecule B. Therefore, the probability of finding *both* molecules A and B in the same region of space at the same instant in time is equal to the product of the probabilities of each being there, that is,

$$\text{Probability of collision} \sim [A][B] \qquad (8\text{--}6)$$

where $[A]$ is the concentration of molecule A and $[B]$ is the concentration of molecule B in the solution. The same line of reasoning leads us to conclude that the rate of the chemical reaction, that is, the change in the concentration of A and B with time as they are converted to P (i.e., $d[A]/dt$ or $d[B]/dt$), would be directly proportional to the product of $[A]$ and $[B]$.

In a solution of any molecular species, not all the molecules will have the same kinetic energy. Instead,

FIGURE 8–2   Maxwell–Boltzmann distribution of molecular kinetic energies. Not all molecules in solution have the same kinetic energy; instead, these energies are distributed as shown with those molecules with greater amounts of energy to the right. Only those molecules having some minimal kinetic energy (shown beyond E_1) may participate in a particular chemical reaction.

the energies are distributed as shown in Figure 8–2. This distribution, known as a *Maxwell–Boltzmann* distribution, means that different molecules in the solution have different kinetic energies. This variation results from random collisions of the molecules with one another during which some molecules gain kinetic energy while others lose kinetic energy. For each chemical reaction, there is a minimum kinetic energy that the participating molecules must have for the chemical reaction to occur. As Figure 8–2 shows, only a small percentage of the molecules in solution have this energy at any instant in time (those to the right of E_1).

Even though a collision might involve two molecules with the required kinetic energies, it is still necessary that they be oriented with respect to one another in a specific manner at the instant of collision for the reaction to proceed. Thus, only a small percentage of the collisions involving molecules with the required kinetic energy will result in the appropriate chemical reaction. Consequently, the actual rate at which the reaction takes place is equal to some constant, k_1, times the probability of a collision. This constant, called a **rate constant**, accounts for both molecular orientation and kinetic energy considerations. Thus,

$$\text{Rate of reaction} = \frac{d[A]}{dt} = \frac{d[B]}{dt} = k_1[A][B] \quad (8\text{–}7)$$

The rate of the reverse reaction is

$$\frac{d[P]}{dt} = k_2[P] \quad (8\text{–}8)$$

Once a reversible reaction has reached equilibrium, no *net* change in the concentrations of the reactants or the product(s) occurs. Therefore, at this time

$$\frac{d[A]}{dt} = \frac{d[B]}{dt} = \frac{d[P]}{dt} \quad (8\text{–}9)$$

and

$$k_1[A][B] = k_2[P] \quad (8\text{–}10)$$

Consequently,

$$\frac{k_1}{k_2} = \frac{[P]}{[A][B]} = K \quad (8\text{–}11)$$

where K is the **equilibrium constant** for the reaction and describes the relative concentrations of all reacting molecular species at equilibrium.

EFFECT OF ENZYME ON REACTION RATE

Two general mechanisms are involved in enzyme catalysis. First, the presence of the enzyme increases the likelihood that the potentially reacting molecular species will encounter each other with the required orientations in space. This occurs because the enzyme has a high affinity for the reactants (more appropriately referred to as the **substrates**) and temporarily binds them in close proximity. The association of substrates with an enzyme is not arbitrary; the substrates are bound to the enzyme in such a manner that each substrate is oriented with respect to the other in precisely the manner required for the reaction to occur. Second, the formation of temporary bonds between the enzyme and substrate forces a redistribution of electrons within the substrate molecules, and this redistribution imposes a *strain* on specific covalent bonds within the substrates—a strain that culminates in bond breakage. Biochemists refer to the introduction of bond strains in a substrate by association with an enzyme as "substrate activation." The net effect of this is to greatly increase the percentage of molecules in the population that at any instant are sufficiently reactive toward each other. This rather brief and general accounting of enzyme function is expanded later in the chapter as the specific structures and actions of certain representative enzymes are considered.

For reversible enzyme-catalyzed reactions such as

$$A + B \underset{}{\overset{E}{\rightleftharpoons}} P \quad (8\text{–}12)$$

where E is the enzyme (and can catalyze the reaction in both directions),

$$\frac{d[A]}{dt} = \frac{d[B]}{dt} = k_1[A][B]e \quad (8\text{–}13)$$

In this equation, e is the "enzyme factor"—a factor that accounts for the increase in reaction rate through catalysis—and is proportional to the concentration of the enzyme in solution. That is, if the reaction proceeds at some rate at a given enzyme concentration, then the reaction would proceed twice as rapidly at twice the enzyme concentration (Fig. 8–3).

For the reverse reaction,

$$\frac{d[P]}{dt} = k_2[P]e \quad (8\text{–}14)$$

It can be shown using equations 8–13 and 8–14 that the equilibrium constant for the reaction (i.e., k_1/k_2) is

FIGURE 8–3 Relationship between enzyme concentration and the rate of an enzyme-catalyzed reaction.

not altered by the presence of the enzyme. Thus, *the enzyme greatly affects (i.e., increases) the rate at which equilibrium is achieved, but it does not alter the respective equilibrium concentrations of the various molecular species.*

The marked effect of enzymes on reaction rates can be clearly appreciated by considering the **turnover number** of an enzyme. The turnover number is the number of moles of substrate converted to product per minute per mole of enzyme (when the enzyme is fully saturated with substrate). For example, the hydrolytic enzyme *urease*, which catalyzes the conversion of urea to ammonia, has a turnover number of about 10^6. In the absence of urease, the hydrolysis of urea is several orders of magnitude slower. *Amylase*, an enzyme involved in the breakdown of starch and other polysaccharides, can hydrolyze about 10 glycosidic bonds per minute, but *no* detectable hydrolysis occurs in the absence of amylase. The turnover numbers of most enzymes fall between 10^2 and 10^6 (Table 8–1).

TABLE 8–1 TURNOVER NUMBERS OF SOME ENZYMES

Enzyme	Turnover Number[a]
Carbonic anhydrase	3.6×10^6
Acetylcholinesterase	1.5×10^6
Urease	1.0×10^6
Amylase	1.0×10^5
Lactic dehydrogenase	6.0×10^4
Chymotrypsin	6.0×10^3
Lysozyme	3.0×10^1

[a]Moles of substrate converted to product per minute per mole of enzyme.

THE KINETICS OF ENZYME ACTION

In most instances, the association of the enzyme with the substrate is so fleeting that the complex is extremely difficult to detect. Yet, as early as 1913, L. Michaelis and M. L. Menten postulated the existence of this transient complex. On the basis of their observations with the enzyme *invertase*, which catalyzes the hydrolysis of sucrose to glucose and fructose, they proposed that enzyme-catalyzed reactions were characterized by a sequence of phases that involves (1) the formation of a complex (ES) between the enzyme (E) and the substrate (S); (2) the modification of the substrate to form the product (P) or products, which briefly remain associated with the enzyme (EP); and (3) the release of the product or products from the enzyme; that is,

$$E + S \rightleftharpoons ES \rightleftharpoons EP \rightarrow E + P \qquad (8\text{–}15)$$

These events are more conventionally described by the equation

$$E + S \underset{k_2}{\overset{k_1}{\rightleftharpoons}} ES \overset{k_3}{\longrightarrow} E + P \qquad (8\text{–}16)$$

In this equation, it is assumed that the combination of enzyme and substrate is reversible; k_1 is the rate constant for the formation of ES (the dimensions of the rate constant are *seconds^{-1}*), and k_2 is the rate constant for the dissociation of ES. After the ES complex is formed, S is converted to P with the rate constant k_3. As long as the concentration of P remains negligible (as it would be at the outset of catalysis) or if P is in some manner quickly removed from the system, then it is not necessary to consider the negligible reverse flux from $E + P$ to ES.

Figure 8–4 depicts the relationship that exists between substrate concentration and the rate at which reaction products appear for enzyme-catalyzed reactions of this type. The curve describes the *initial rate* of product formation at a fixed enzyme concentration when the substrate concentration is varied on successive trials. At low concentrations of substrate, the initial velocity of the reaction (i.e., v_0) is directly proportional to the substrate concentration (i.e., follows first-order kinetics). However, as the substrate concentration is increased, the reaction velocity levels off, approaching a maximum value. At this high substrate concentration, the reaction velocity is limited by the amount of time required to convert bound substrate

FIGURE 8–4 Relationship between substrate concentration and rate of enzyme-catalyzed reaction. Curve describes *initial* rate of product formation (P) when the enzyme concentration is held constant and the substrate concentration [S] is increased on successive trials. At low substrate concentration, the reaction flux (v_0) follows first-order kinetics. As the substrate concentration is increased, reaction rate approaches a maximum value at which time zero-order kinetics is followed.

(i.e., *ES*) to product (*P*) and free enzyme (*E*). The curve now follows zero-order kinetics. The curve of Figure 8–4 is called a Michaelis–Menten curve, and for an idealized enzyme-catalyzed reaction is a *rectangular hyperbola*.

Michaelis and Menten are also credited with the first mathematical study of the relationship between substrate concentration and reaction rates. They introduced two particularly useful mathematical expressions that for any enzyme relate [S] to V (velocity) and permit quick comparisons of various enzyme-catalyzed reactions; the two expressions are now called the **Michaelis–Menten constant** and the **Michaelis–Menten equation**. They are derived as follows.

Let

$[S]$ = the concentration of free substrate, *S*; (We may assume that the amount of available substrate is so great that the amount combined with the enzyme may be ignored in comparison. Hence, total substrate concentration and the free substrate concentration are the same);

$[E]_T$ = the total concentration of enzyme, *E*;

$[E]$ = the concentration of free enzyme (that not complexed with substrate);

$[ES]$ = the concentration of enzyme–substrate complex (Because the total amount of enzyme present is assumed to be very small in comparison with the total amount of substrate, a significant proportion of the total enzyme may be involved in the *ES* complex. Hence, separate terms for the total and free enzyme concentrations are warranted);

$[P]$ = the concentration of product, *P*.

The total enzyme concentration is given by

$$[E]_T = [E] + [ES] \qquad (8\text{–}17)$$

For equation 8–16, the initial rate of product formation, v_0, is given by

$$v_0 = \frac{d[P]}{dt} = k_3[ES] \qquad (8\text{–}18)$$

The change in the concentration of *ES* with time is equal to the rate at which *ES* is formed minus the rate at which it is being eliminated, that is,

$$\frac{d[ES]}{dt} = k_1[E][S] - (k_2[ES] + k_3[ES]) \qquad (8\text{–}19)$$

Therefore, by simplifying,

$$\frac{d[ES]}{dt} = k_1[E][S] - (k_2 + k_3)[ES] \qquad (8\text{–}20)$$

If sufficient substrate is available, then $d[ES]/dt$ would be zero, because for every *P* leaving the complex, a new *S* would enter it. Under such conditions,

$$k_1[E][S] = (k_2 + k_3)[ES] \qquad (8\text{–}21)$$

and by rearrangement

$$\frac{k_2 + k_3}{k_1} = \frac{[E][S]}{[ES]} \qquad (8\text{–}22)$$

The ratio of constants given in equation 8–22 may be set equal to a new constant, K_M, which is the Michaelis–Menten constant. Thus, the Michaelis–Menten constant is a constant that relates the steady-state concentrations of enzyme, enzyme–substrate complex, and substrate.

Each enzyme-catalyzed reaction reveals a characteristic K_M value, and this value is a measure of the tendency of the enzyme and the substrate to combine with each other. In this sense, the K_M value is an index of the affinity of the enzyme for its particular substrate. Some representative K_M values are given in

TABLE 8-2 SOME REPRESENTATIVE K_M Values

Enzyme	Source	Substrate	K_M
Sucrase	Intestine	Sucrose	$2 \times 10^{-2} M$
Urease	Soybeans	Urea	$2.5 \times 10^{-2} M$
Catalase	Liver	Hydrogen peroxide	$2.5 \times 10^{-2} M$
Carbonic anhydrase	Blood	CO_2	$9 \times 10^{-3} M$
Chymotrypsin	Pancreas	Peptides	$5 \times 10^{-3} M$
Phosphatase	Bone	Glycerophosphate	$3 \times 10^{-3} M$
Hexokinase	Liver	Glucose	$1.5 \times 10^{-4} M$
Lysozyme	Egg white	Hexa-N-acetylglucosamine	$6 \times 10^{-6} M$

Table 8–2. It is to be stressed that *the greater the affinity of an enzyme for its substrate, the lower the K_M value.* This is because the K_M value is numerically equal to the substrate concentration at which half of the enzyme molecules are associated with substrate (i.e., in the *ES* form).

The relationships derived above are based on reactions in which a single substrate molecule is bound to the enzyme. However, they also apply in situations where more than one substrate is bound to the enzyme, as long as the concentrations of all but one substrate species are held constant or are not rate-limiting (i.e., present in large excess). For example, even the pioneering studies of Michaelis and Menten involved a bimolecular enzyme-catalyzed reaction. They employed the enzyme *invertase*, which forms a complex with one molecule of sucrose and one molecule of water and releases glucose and fructose as products. However, in this reaction (as in nearly all hydrolyses), the concentration of water remains virtually unaltered during the course of the reaction.

K_M values are seldom determined using the relationship given in equation 8–22, because the enzyme-substrate concentration [ES] cannot easily be measured. Some additional mathematical manipulations may be carried out to convert equation 8–22 into a more useful form.

Because $[E] = [E]^T - [ES]$, then

$$K_M = \frac{[E][S]}{[ES]} = \frac{([E]_T - [ES])[S]}{[ES]} \qquad (8\text{–}23)$$

$$K_M[ES] + [S][ES] = [E]_T[S] \qquad (8\text{–}24)$$

$$[ES](K_M + [S]) = [E]_T[S] \qquad (8\text{–}25)$$

and

$$[ES] = \frac{[E]_T[S]}{K_M + [S]} \qquad (8\text{–}26)$$

Because $v_0 = d[P]/dt = k_3[ES]$, then

$$v_0 = \frac{k_3[E]_T[S]}{K_M + [S]} \qquad (8\text{–}27)$$

As seen in Figure 8–4, the initial reaction velocity increases with substrate concentration until the enzyme concentration becomes limiting, and at that time, the initial reaction velocity approaches a maximum value (i.e., V_{max}). This occurs because nearly all the enzyme is maintained in the *ES* form. Thus, [ES] will equal $[E]_T$, and $k_3[E]_T$ corresponds to V_{max}.

By substituting V_{max} for $k_3[E]_T$ in equation 8–27, we obtain

$$v_0 = \frac{V_{max}[S]}{K_M + [S]} \qquad (8\text{–}28)$$

Equation 8–28 is the Michaelis–Menten equation. From this equation, it may be seen that when the substrate concentration is numerically equal to the K_M value of the enzyme, then the reaction velocity is equal to one-half the maximum value. For example, if [S] and K_M are both equal to 3, then equation 8–27 simplifies to

$$v_0 = \frac{V_{max}(3)}{3 + 3} = \frac{V_{max}}{2} \qquad (8\text{–}29)$$

Consequently, the Michaelis–Menten constant for an enzyme may be determined from the substrate concentration at which the reaction velocity proceeds at one-half its maximum value (Fig. 8–5).

From the above, we may conclude that the maximum velocity of an enzyme-catalyzed reaction depends on the affinity between the enzyme and its substrate (i.e., the K_M value). The higher the K_M value is, the lower the affinity of the enzyme for the substrate; and the smaller the K_M value, the greater the affinity. In many instances, the same substrate may enter either of several different enzyme-catalyzed reactions occurring in cells. Which of the alternative reactions predominates in the cell depends in part on the K_M val-

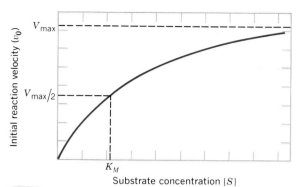

FIGURE 8-5 Relationship between substrate concentration [S] and reaction rate (v_0). When the rate of reaction is one-half its maximum initial value (V_{max}), the substrate concentration is numerically equal to the Michaelis–Menten constant (K_M). All such curves are rectangular hyperbolas. Consequently, the lower the K_M value, the lower the substrate concentration at which the reaction proceeds at a maximum rate.

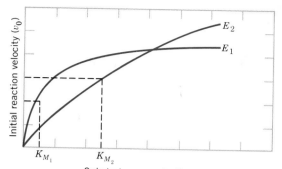

FIGURE 8-6 Relationship between substrate concentration [S] and reaction rate (v_0) for two enzymes having the same substrate. Enzyme 1 (E_1) has a lower K_M and attains maximum rate of reaction at lower substrate concentration. Enzyme 2 (E_2) has a higher K_M value and produces a maximum reaction rate at higher concentration of substrate. As a result, the reaction catalyzed by enzyme 1 is favored at low substrate levels, and the reaction catalyzed by enzyme 2 is favored at high substrate concentrations.

ues of the respective enzymes and the concentration of available substrate. At very low substrate concentrations, the specific reaction catalyzed by the enzyme with the lowest K_M will dominate, whereas at higher substrate concentrations, the reaction catalyzed by the enzyme having the greatest K_M value can dominate (if enough of the enzyme is present). Therefore, which of several different metabolic pathways is actually followed by the initial substrate in the course of a series of enzyme-catalyzed reactions may be regulated by controlling the amount of available substrate. This and other metabolic regulatory mechanisms are discussed in Chapter 11. The relationship between reaction velocity and substrate concentration for two enzymes that act on the same substrate is depicted in Figure 8–6.

To obtain the K_M value of an enzyme experimentally, it is necessary to determine v_0 for a series of substrate concentrations. In practice the evaluation of K_M from a plot similar to that in Figure 8–6 is difficult, because the precise value of V_{max} cannot be determined. This is because the curve is a rectangular hyperbola that approaches V_{max} asymptotically. In 1934, H. Lineweaver and D. Burk introduced a different form of the Michaelis–Menten equation (8–28) that simplifies the determination of K_M from experimentally obtained values of [S] and v_0. In this form, the acquired data are used to construct a straight line rather than a hyperbola. By taking the inverse of equation 8–28 we obtain

$$\frac{1}{v_0} = \frac{K_M}{V_{max}\,[S]} + \frac{[S]}{V_{max}\,[S]} \qquad (8\text{–}30)$$

or

$$\frac{1}{v_0} = \frac{K_M}{V_{max}}\frac{1}{[S]} + \frac{1}{V_{max}} \qquad (8\text{–}31)$$

The terms are now arranged in the form of the general equation of a straight line, $y = ax + b$, in which a is the slope of the line and b is the intercept on the y axis. Thus, for equation 8–31, $1/v_0$ serves as the y axis and $1/[S]$ as the x axis. Consequently, the y intercept will be $1/V_{max}$, the slope will be K_M/V_{max}, and the x intercept will be $-1/K_M$. The graph showing this relationship (known as a **double reciprocal plot**) is presented in Figure 8–7.

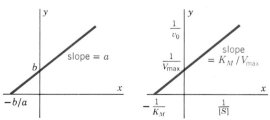

FIGURE 8-7 Lineweaver–Burk *double reciprocal plot* of $1/v_0$ and $1/[S]$ can be used to construct a line that may easily be extrapolated backward to find the K_M value for the enzyme.

EFFECTS OF INHIBITORS ON ENZYME ACTIVITY

The interaction between the substrate and the enzyme takes place in a particular region of the enzyme molecule called the *active site* (discussed later). In many instances, compounds other than the normal substrate for a particular enzyme-catalyzed reaction may bind to the enzyme's active site, and this has a significant effect on the kinetics of the normal reaction. One possible consequence of this phenomenon is the *inhibition* of normal enzyme activity, and such compounds are therefore called **enzyme inhibitors**. (Usually, the inhibitor is unaltered by its interaction with the enzyme.) In some instances, the normal substrate (S) and the inhibitor (I) *compete* with each other for the active site of the enzyme; the manner in which this affects the normal kinetics of the reaction is shown in Figure 8-8. V_{max} is not altered by the presence of a competitive inhibitor, but the K_M is elevated. As can be seen in Figure 8-8, the effect of the inhibitor is maximal at low substrate concentration (i.e., when 1/[S] is large) and minimal at high substrate concentration (i.e., when 1/[S] approaches zero).

A classic example of this form of inhibition is the competition between succinic acid and malonic acid for the enzyme *succinic acid dehydrogenase*. In this instance, competition between these two compounds for the active site of the enzyme is understandable in view of their marked chemical similarity (Fig. 8–9). Succinic acid is the normal substrate for the enzyme and,

FIGURE 8–9 (*a*) Chemical formulas of succinic acid and malonic acid. (*b*) Malonic acid is a competitive inhibitor of *succinic acid dehydrogenase*. Note the chemical similarity between the true substrate (succinic acid) and the inhibitor.

in the absence of the inhibitor, is converted to fumaric acid.

Enzyme inhibition can also be *noncompetitive* in that the binding of the inhibitor to the enzyme cannot be reversed by increasing the concentration of the normal substrate. A common example of negative inhibition is the action of heavy metals such as mercury on

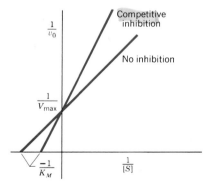

FIGURE 8–8 Effect of a competitive inhibitor on normal enzyme kinetics. V_{max} is not altered, but the K_M value is increased. At low substrate concentrations, the effect of inhibitor on reaction rate is marked but is completely reversed as substrate concentration is increased and 1/[S] approaches 0.

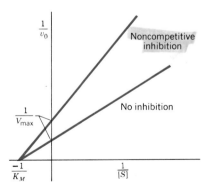

FIGURE 8–10 Effect of a noncompetitive inhibitor on normal enzyme kinetics. The K_M value is not altered, but V_{max} is lowered. Noncompetitive inhibition can be reduced but not reversed by increasing the substrate concentration.

the active sites of enzymes containing a reactive sulf-hydryl (i.e., $-SH$) group. In effect, the presence of the inhibitor prevents some percentage of the enzyme present from participating in normal catalysis. As a result, the maximum reaction velocity is depressed, even though the K_M value remains the same (Fig. 8–10).

The inhibition of enzyme activity by competitors also occurs naturally in cells and serves in the regulation of certain metabolic pathways. A special form of feedback inhibition involves **allosterism**, a topic that will be considered at length later in the chapter.

MECHANICS OF ENZYME CATALYSIS

Enzymes are proteins, and therefore their capacity for catalysis is intimately related to a specific tertiary or quaternary molecular structure. If the tertiary or quaternary structure of an enzyme is altered, a loss of enzyme activity usually follows. Thus, environmental factors that modify protein structure also influence enzyme activity. Key environmental factors that can affect enzyme activity are pH and temperature. As discussed at some length in Chapter 4, the polar side chains of certain amino acids form electrostatic bonds with each other and with surrounding ions and water

molecules; these interactions contribute in part to the specific tertiary and quaternary structure of the protein. Whether a particular amino acid side chain bears a charge is determined in part by the pH of the protein's environment. As the pH is lowered (i.e., the concentration of H^+ is increased), groups that may be negatively charged, such as the secondary COO^- of aspartic acid and glutamic acid and the O^- of tyrosine, become protonated, thereby neutralizing these negative charges. At the same time, some secondary amino groups, such as those of lysine and arginine, may accept additional protons, thereby imparting charge to these formerly neutral side chains. In contrast, as the pH is elevated (i.e., the concentration of OH^- is increased), positively charged side chains dissociate protons and are thereby neutralized, while the loss of protons from secondary COOH and OH groups renders these groups negative. Some of these relationships are depicted in Figure 8–11.

In addition to playing important roles in the maintenance of a specific tertiary or quaternary molecular structure, the polar side chains of some of the amino acids in the enzyme (i.e., those in the active site) may be involved in binding the substrate to the enzyme and thereby introduce bond strains into the substrate molecule. Consequently, most enzymes can operate only within a narrow pH range and display a pH optimum

FIGURE 8–11 Influence of a change in pH on the side chains of aspartic acid, lysine, and tyrosine.

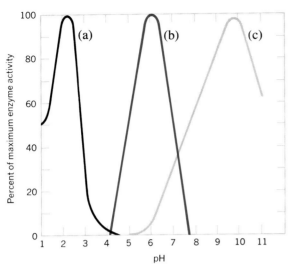

FIGURE 8–12 Effect of pH on enzyme activity. (a) *Pepsin*, (b) *glutamic acid decarboxylase*, and (c) *arginase*.

(Fig. 8–12). The pH optima of enzymes vary over a broad range of values; several examples are given in Table 8–3. On either side of the pH optimum, enzyme activity declines as the configuration of the protein is altered and/or its affinity for the substrate is correspondingly decreased.

Temperature also influences enzyme activity, and most enzymes display a temperature optimum close to the normal temperature of the cell or organism possessing that enzyme. Accordingly, the temperature optima of plant cell enzymes and enzymes of poikilothermic ("cold-blooded") animals inhabiting cold regions of the earth are usually lower than those of enzymes of homeothermic animals.

If the temperature is elevated far above the optimum, enzyme activity decreases. This is the result of an alteration of the enzyme's structure (an unraveling process called **denaturation**). Most enzymes are irreversibly denatured if maintained at temperatures above 55° to 65°C for an extended period of time.

TABLE 8–3 OPTIMUM pH VALUES FOR CERTAIN ENZYMES

Enzyme	Optimum pH
Pepsin	2.0
Glutamic acid decarboxylase	5.9
Urease	6.7
Salivary amylase	6.8
Pancreatic lipase	7.0
Trypsin	9.5
Arginase	10.0

The Active Site

The formation of the enzyme–substrate complex is not a random process. This was recognized as long ago as 1894 when Emil Fischer postulated that an enzyme allows only one or a few compounds to fit onto its surface. This is the "lock-and-key" hypothesis according to which the enzyme and its substrate have complementary shapes (see below). The specific substrate molecules (and prosthetic groups, if any) are bound to a specific region of the enzyme molecule called its **active (or catalytic) site**. The active site of an enzyme is formed by a number of amino acid residues whose side chains have two principal roles: (1) they serve to *attract* and *orient* the substrate in a specific manner within the site (such amino acids are called **contact residues** and contribute in large degree to substrate specificity) and (2) they participate in the *formation of temporary bonds* with the substrate molecule, bonds that polarize the substrate, introduce strain into certain of its bonds, and trigger the catalytic change (such amino acids are termed **catalytic residues**). The bonds formed between a substrate and the amino acid side chains forming the active site may be either covalent or noncovalent.

The contact and catalytic residues that make up the active site may be located in widely separated regions of the enzyme's primary structure, but as a result of stabilized polypeptide chain folding, they are brought into the appropriate juxtaposition. For example, the active site of the enzyme *citrate synthase* contains about 16 amino acids; eight of these form temporary bonds with the substrate (citric acid) and another eight form bonds with a coenzyme (coenzyme A). These amino acid residues are scattered through the enzyme's primary structure from amino acid position number 46 to position number 421. Even in relatively small enzymes such as *ribonuclease*, contact and catalytic residues are scattered over much of the primary structure (i.e., the shaded residues in Fig. 8–13).

"Lock-and-Key" versus "Induced-Fit" Models of Enzyme Action

Figure 8–14 depicts the interaction between enzyme and substrate according to the lock-and-key model. The substrate has polar (i.e., $\oplus$ and $\ominus$) and nonpolar ($\circledH$, hydrophobic) regions and is attracted to and associates with the active site, which is complementary in both shape and charge distribution (Figs. 8–14*a* and 8–14*b*). Positive, negative, and hydrophobic regions of the active site are created by the side chains of the

FIGURE 8–13 Structure of lysozyme showing distribution of amino acids forming the active site.

contact residues, which align the substrate for interaction with the site's catalytic residues (Ⓐ and Ⓑ). Following catalysis (Fig. 8–14c), the products are released from the active site (Fig. 8–14d), thereby freeing the enzyme for another round of catalysis. The lock-and-key model of enzyme catalysis accounts for enzyme specificity, because compounds that lack the appropriate shape or are too large or too small (Fig. 8–14e) cannot be bound to the active site.

Although the lock-and-key model accounts for much of the substrate specificity data, certain observations about enzyme behavior do not fit or are difficult to explain using this model. For example, there are a number of instances in which compounds other than the true substrate bind to the enzyme even though they fail to form reaction products. Furthermore, for many enzyme-catalyzed reactions, substrates are bound to the active site in a specific temporal order. In the 1960s, Daniel Koshland proposed the "induced-fit" theory of enzyme action according to which the active site of the enzyme does not initially exist in a shape

that is complementary to the substrate but is induced to assume the complementary shape as the substrate becomes bound. As Koshland put it, the active site is induced to assume the complementary shape "in much the same way as a hand induces a change in the shape of a glove." Thus, according to this model, the enzyme (or its active site) is *flexible*.

The induced-fit model is depicted diagrammatically in Figure 8–15. The active site and substrate initially have different shapes (Fig. 8–15a) but become complementary on substrate binding (Fig. 8–15b). The shape change places the catalytic residues in position to alter the bonds in the substrate (Fig. 8–15c), following which the products are released (Fig. 8–15d) and the active site returns to its initial state. Although molecules that are larger or smaller than the true substrate or that have different chemical properties may nonetheless be bound to the active site, none succeed in inducing the proper alignment of catalytic groups, and no catalysis occurs (Figs. 8–15e and 8–15f). The induced-fit model explains the effects of certain com-

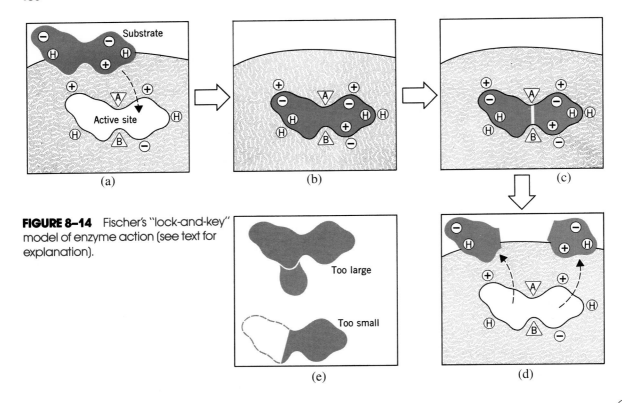

FIGURE 8–14 Fischer's "lock-and-key" model of enzyme action (see text for explanation).

petitive and noncompetitive inhibitors of enzyme action.

Before proceeding further, it should be acknowledged that some enzyme-catalyzed reactions are adequately explained by the lock-and-key model, so that a flexible active site is not a strict requirement for catalysis. By the same token, the possession of a flexible active site does not imply that any molecule may become bound to the enzyme.

A change in the shape of the active site of an enzyme can also be induced by binding of substances at sites on the enzyme's surface that are far removed from the active site. In such a case, the change is transmitted through the enzyme molecule from the site of binding to the active site. Such changes may either decrease or increase the enzyme's activity. The latter phenomenon will be discussed later in the chapter in connection with enzyme **cooperativity** and **allosterism**.

Why Enzymes Are So Large

Because nonbiological catalysts are often very small compounds, it is reasonable to wonder why biological catalysts, namely, enzymes, are so large. With the possible exception of enzymes that are secreted by cells, most of a cell's enzymes are either permanently or cyclically associated with membranes and other particulate cell components. Indeed, in a few cells (e.g., *Euglena* and *Neurospora*), there appear to be no "soluble" enzymes at all. Therefore, it is apparent that one or more parts of an enzyme's structure are involved in interactions that localize that enzyme in or on a specific cellular component and may not be involved in the catalysis itself.

Most of a cell's enzymes catalyze reactions that are part of a *metabolic pathway*; that is, each enzyme catalyzes a reaction that is one step in a series of ordered steps. It appears that in some (perhaps many) cases, the enzymes of a metabolic pathway occur in close proximity to one another and interact with each other. Thus, still another portion of an enzyme's total structure plays a role in this interaction. The localization of enzymes within a cell and the interactions between enzymes of a metabolic pathway are topics that will be pursued in detail in later chapters that deal with the organization and functions of specific cellular components.

Some parts of an enzyme may have no direct role in catalysis or in interaction with other enzymes or with

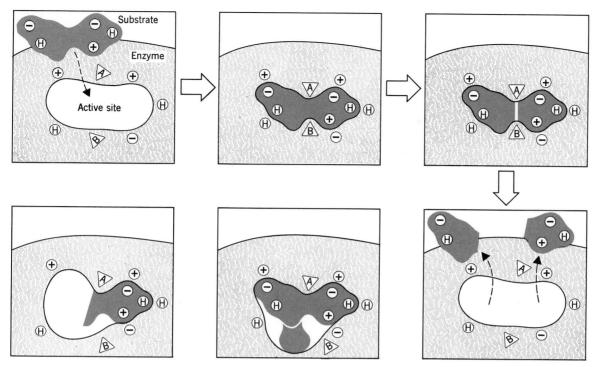

FIGURE 8–15 Koshland's "induced-fit" model of enzyme action (see text for explanation).

intracellular membranes. For example, recent studies suggest that random, nonspecific binding of a substrate to sites on an enzyme's surface other than the active site may enhance diffusion of the substrate to the active site. Physicochemical studies indicate that an enzyme would have to have a surface area in excess of 1000 Å^2 to exhibit even limited interactions with molecules other than the enzyme's substrate. For proteins, this minimum surface area corresponds to a molecular weight of about 10,000. Therefore, it is not surprising that most enzymes have molecular weights that are greater than this.

ORGANIZATION OF SPECIFIC ENZYMES

Extensive studies during the past decade have revealed the precise atomic structure of a number of important enzymes including *lysozyme, ribonuclease, chymotrypsin, trypsin, carboxypeptidase, papain, elastase,* and *thrombin.* A consideration of some of these reveals certain generalizations about the organization of enzyme molecules and provides insight into the mechanisms by which catalysis is achieved.

Lysozyme

Lysozyme is an enzyme that is produced in both plant and animal tissues and was the first to have its complete three-dimensional structure revealed (by C. C. F. Blake, D. C. Phillips, and A. C. T. North, in 1965). The enzyme cleaves polysaccharide chains found in the cell walls of certain bacteria by hydrolyzing the glycosidic bonds between neighboring hexosyl residues. Egg white lysozyme has a molecular weight of about 14,600 and consists of a single polypeptide chain of 129 amino acids. The enzyme is oval in shape with a deep cleft across its midline that divides the molecule into two parts. The shape of lysozyme and the orientation of the substrate are depicted stereoscopically in Figure 8–16. The polypeptide contains three short helical regions and a segment arranged in the form of a beta pleated sheet. All the polar residues are located on the enzyme's surface and nearly all uncharged residues are buried internally. Hydrogen bonds between one portion of the polypeptide chain and another are pronounced and appear to be vital in sustaining the active tertiary structure of the molecule.

The cleft at the center of the molecule contains the active site and is studded with hydrophobic groups

(a)

(b)

FIGURE 8–16   (*a*) Stereoscopic diagram of *lysozyme*. (Courtesy of R. E. Dickerson and I. Geis.) (*b*) Diagram illustrating how the binding of the substrate occurs in a deep cleft that forms the active site of lysozyme.

that probably form van der Waals bonds with the substrate. The bacterial polysaccharide that acts as substrate for the enzyme consists of chains of *N*-acetylglucosamine (NAG) and *N*-acetylmuramic acid (NAM) in which the two sugars alternate. In the bacterial cell wall, these chains are cross-linked by short polypeptides. Six hexosyl units of the substrate polysaccharide are simultaneously bound at the active site of lysozyme. A change in tertiary structure, which takes the form of a narrowing and deepening of the cleft in the enzyme, accompanies substrate binding and forces a modification of the conformation of one of the six hexosyl groups (i.e., NAM) bound at the active site. This conformational change involves a distortion of the chair form normally assumed by this sugar and primes this region of the substrate for catalytic alteration. It is the glycosidic bond between this group and its neighbor that is broken during catalysis. Of the 19 or more amino acids that compose the active site, only two have been identified as catalytic; these are asp (amino acid no. 52) and glu (amino acid no. 35). The remaining 17 residues align the substrate through hydrogen bonds and nonpolar interactions.

Asp 52 carries a dissociated carboxyl group (i.e., negative charge), but glu 35, being surrounded by the nonpolar side chains of other residues, is protonated. The sequence of events occurring during catalysis is depicted in Figure 8–17 in which only three of the six hexosyl units bound to the active site (i.e., numbers 3, 4, and 5) are shown. The binding of the substrate to

the enzyme and the conformational changes that occur in both as a result are followed by an attack by a hydrogen ion of the undissociated COOH group of glu 35 on the oxygen atom, forming the bridge between hexosyl units 4 and 5 (Fig. 8–17*a*). This, of course, is the specific region of the substrate placed under strain during binding. The glycosidic bond is broken, leaving carbon atom no. 1 of hexosyl unit 4 with a positive charge (i.e., a carbonium ion) (Fig. 8–17*b*). The carbonium ion is stabilized by the formation of a temporary ionic bond with the dissociated side chain of asp 52 (Fig. 8–17*b*). The dissociation of a neighboring molecule of water provides a hydroxyl group for attack on the carbonium ion and a hydrogen for glu 35 (Fig. 8–17*c*). Once the reaction is completed, the two resulting polysaccharide fragments leave the active site.

Ribonuclease

Ribonuclease is produced by the pancreas and secreted into the small intestine, where it catalyzes the hydrolytic digestion of polyribonucleotide chains. The enzyme consists of a single 124 amino acid polypeptide having a molecular weight of about 13,700. The reaction catalyzed by ribonuclease (shown in Fig. 8–18) involves the hydrolytic cleavage of the ester linkage between the 3'-phosphate group of a pyrimidine nucleotide and the 5' position of the adjacent nucleotide.

Like *lysozyme*, the active site of *ribonuclease* lies in

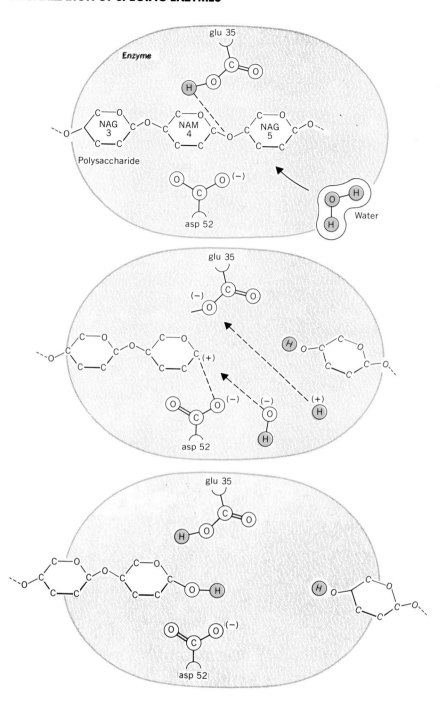

FIGURE 8-17 *Proposed mechanism for the action of lysozyme.* Only three of the five or six gluycosyl units of the polysaccharide that are bound to the active site are shown (numbered 3, 4, and 5). Hydroxyl groups, other groups, and hydrogen atoms of the pyranose rings have been omitted for clarity. Only the side chains of the two catalytic residues are shown. See text for details of the catalytic process.

a cleft in the enzyme's surface. The cleft is lined by a number of positively charged side chains of lysine and arginine residues and these probably act as contact residues by forming ionic bonds with the negatively charged phosphate groups of the ribonucleic acid backbone (Fig. 8–18). Projecting into the cleft from ei-

ther side are two histidine residues that are believed to hydrolyze the RNA substrate by general acid–base catalysis in which the P—O bond linking the ribose of one nucleotide with the ribose of a neighboring nucleotide is broken. *Ribonuclease* is depicted stereoscopically in Figure 8–19.

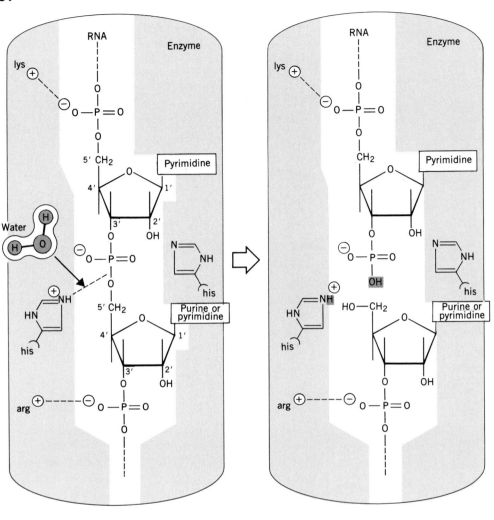

FIGURE 8–18 Proposed mechanism for the action of *ribonuclease*. The enzyme catalyzes the hydrolytic cleavage of the ester linkage between the 3′-phosphate group of a pyrimidine nucleotide and the 5′ position of the adjacent nucleotide. The positively charged R-groups of arginine and/or lysine residues in the cleft forming the active site interact with negatively charged phosphate groups of the RNA, thereby orienting the substrate. The R-groups of two histidine residues mediate the catalysis.

FIGURE 8–19 (*a*) Stereoscopic view of *ribonuclease*. Histidine and lysine residues of the active site are shown in color. (Courtesy of R. E. Dickerson and I. Geis.). (*b*) Diagram showing the general outline of the enzyme and the position of substrate in the cleft containing the active site.

(a)

(b)

Chymotrypsin

The digestive enzyme chymotrypsin is derived from the inactive precursor polypeptide *chymotrypsinogen* produced in the pancreas and containing 245 amino acids. The activation of *chymotrypsinogen* involves the preliminary cleavage of four of its peptide bonds, resulting in three separate polypeptide chains cross-linked by two disulfide bridges; two dipeptides released in the activation process do not become a part of the functional enzyme (see Fig. 22–40). *Chymotrypsin* digests alimentary protein by hydrolyzing peptide bonds on the carboxyl side of amino acids with large hydrophobic side chains (i.e., phenylalanine, tyrosine, and tryptophan residues).

Unlike *lysozyme* and *ribonuclease*, the active site of *chymotrypsin* resides in a shallow dish-shaped cavity on the surface of this roughly spherical enzyme. In the active site is a hydrophobic pocket that is believed to act as the receptor of the hydrophobic side chain of the substrate amino acid residue and thereby imparts specificity to this enzyme. In addition to these and other contact residues, the active site contains three catalytic residues: asp 102, his 57, and ser 195. Figure 8–20 depicts the series of stages that are believed to be involved in the action of *chymotrypsin*. The imidazole group of histidine attracts the hydroxyl hydrogen atom of the serine, thereby rendering the serine oxygen strongly nucleophilic and particularly reactive toward amides and esters (Fig. 8–20a). Binding of the substrate polypeptide to the enzyme (Fig. 8–20b) is followed by formation of a covalent bond between the serine oxygen and the α-carboxyl carbon atom of the substrate residue (Fig. 8–20c). This is followed by donation of the serine proton to the α-amino nitrogen atom of the neighboring substrate residue, the breakage of the peptide bond, and the release of the first product (Fig. 8–20d).

At this stage, a molecule of water enters the reaction (Fig. 8–20e), donating a hydrogen ion to serine and a hydroxyl group to the α-carboxyl carbon (Fig. 8–20f). Dissociation of this carboxyl group (Fig. 8–20g) leaves the carboxyl group negatively charged. Repulsive forces between this group and the negatively charged aspartic acid residue of the active site facilitate separation of the enzyme and final product (Fig. 8–20h).

The behavior of the catalytic serine residue is critical to the function of *chymotrypsin*. Other proteases, including *trypsin, elastase, collagenase, thrombin, fibrinolysin*, and some lysosomal *proteases*, share a similar overall tertiary structure and active site, and they cleave peptide bonds via the identical action of a catalytic serine residue. The different substrate specificities of these *serine proteases* may be attributed to the nature and location of the contact residues in their active sites. In view of their remarkable similarity, the serine proteases are believed to have a common evolutionary origin.

The accumulated information on *lysozyme, ribonuclease*, and the *serine proteases*, as well as several other enzymes, permits certain generalizations to be made concerning enzyme structure and action. The active site of the enzyme nearly always resides in a depression in its surface, and the binding of the substrate to the active site is followed by conformational change in the enzyme, the substrate, or both. The binding and/or the conformational changes that follow place a strain on certain bonds in the substrate and render them particularly susceptible to chemical attack. The catalysis itself may be facilitated by as few as one or two amino acid side chains of the enzyme, which, because of their spatial arrangement in the active site and the polar or hydrophobic nature of their surroundings, are especially reactive.

COFACTORS

Enzymes are composed of one or more polypeptide chains. However, there are a number of cases in which nonprotein constituents called **cofactors** must be bound to the enzyme (in addition to the substrate) for the enzyme to be catalytically active. In these instances, the exclusively protein portion of the enzyme is called the **apoenzyme**. Three kinds of cofactors may be identified: **prosthetic groups, coenzymes**, and **metal ions**.

Prosthetic groups are organic compounds and are distinguished from other cofactors in that they are *permanently* bound to the apoenzyme. For example, in the peroxisomal enzymes *peroxidase* and *catalase*, which catalyze the breakdown of hydrogen peroxide to water and oxygen, *heme* is the prosthetic group and is a permanent part of the enzyme's active site (see also Chapter 19).

Coenzymes are also organic compounds, but their association with the apoenzyme is transient, usually occurring only during the course of catalysis. Furthermore, the same coenzyme molecule may serve as the cofactor in a number of different enzyme-catalyzed reactions. In general, coenzymes not only assist en-

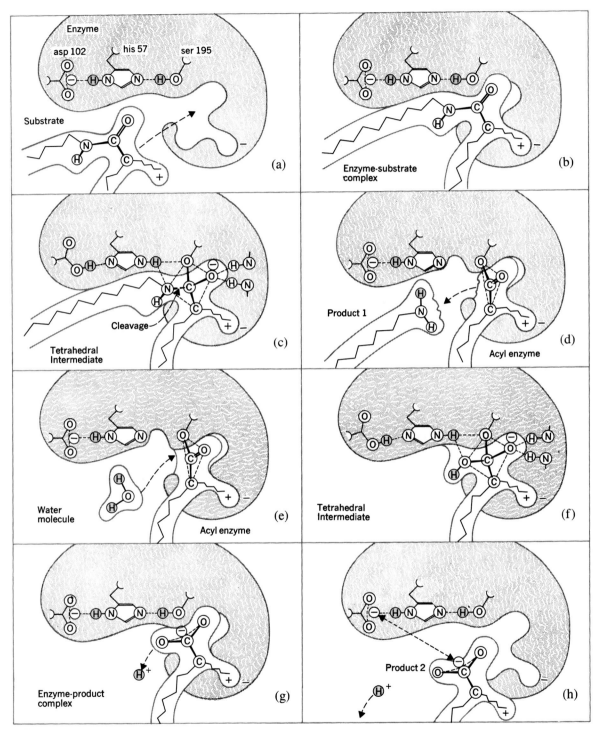

FIGURE 8-20 Stages involved in proteolysis by *chymotrypsin.*
See text for details of the catalytic process.

zymes in the cleavage of the substrate but also serve as temporary acceptors for one of the products of the reaction. The essential chemical components of many coenzymes are *vitamins*. For example, the coenzymes *nicotinamide adenine dinucleotide* (NAD) and *nicotinamide adenine dinucleotide phosphate* (NADP) contain the vitamin *niacin*; *coenzyme A* contains *pantothenic acid*; *flavin adenine dinucleotide* (FAD) contains *riboflavin* (i.e., vitamin B_2); *thiamine pyrophosphate* contains *thiamine* (i.e., vitamin B_1), and so on. The chemistry of some of these cofactors was discussed in Chapter 3.

A number of enzymes require *metal ions* for their activity. The metal ions form coordination bonds (see Chapter 3) with specific side chains at the active site and at the same time form one or more coordination bonds with the substrate. The latter assist in the polarization of the substrate bonds to be cleaved by the enzyme. For example, zinc is a cofactor for the proteolytic enzyme *carboxypeptidase* and forms coordination bonds with the side chains of two histidines and one glutamic acid residue at the active site. A fourth bond is formed between zinc and the α-carboxyl group of the substrate amino acids, and it is here that the cleavage of the peptide occurs. Table 8–4 contains a list of some of the cofactor-requiring enzymes.

The observation that catalytic activity is lost when an enzyme is stripped of its cofactor testifies to the crucial role played by these atoms or molecules. This role is diverse:

1. In some cases, the cofactor completes the active site of the enzyme or modifies it in such a manner that substrate binding can ensue.

2. The cofactor acts as a donor of electrons or atoms to the substrate and following the reaction is returned to its former state.

3. Cofactors may also serve as temporary recipients of either one of the reaction products or simply an electron or proton, again being recycled to its former state some time after the main reaction is completed.

4. Finally, the cofactor, together with the side chains of residues at the active site, may serve to polarize the substrate and prime it for catalytic alteration.

ISOENZYMES

Occasionally, several different enzyme molecules, all of which catalyze the same chemical reaction, have been isolated from a single tissue. Such families of enzymes are called **isoenzymes** or **isozymes** (see also Chapter 11). Among the various isoenzymes the *lactic dehydrogenases* have been most extensively studied, and five different forms have been identified. All are composed of four polypeptide chains of two types called M and H subunits. Thus the lactic dehydrogenase isoenzymes may take any one of the following forms: M_4, M_3H, M_2H_2, MH_3, or H_4. Similar arrange-

TABLE 8–4 SOME ENZYMES THAT REQUIRE COFACTORS

Cofactor	Enzyme	Reaction
Prosthetic groups		
Heme	Catalase	$2\,H_2O_2 \rightarrow 2\,H_2O + O_2$
Heme	Peroxidase	$2\,H_2O_2 \rightarrow 2\,H_2O + O_2$
Flavin adenine dinucleotide (FAD)	Succinic dehydrogenase	Succinic acid $\rightarrow$ fumaric acid
Coenzymes		
Flavin mononucleotide (FMN)	Some dehydrogenases	Removal of hydrogen atoms
Thiamine pyrophosphate (TPP)	Some decarboxylases	Removal of CO_2
Nicotinamide adenine dinucleotide (NAD)	Some dehydrogenases	Removal of hydrogen atoms
Lipoic acid	Some decarboxylases	Removal of CO_2
Metal ions		
Zn^{2+}	Carboxypeptidase	Hydrolysis of proteins
Zn^{2+}	Carbonic anhydrase	$CO_2 + H_2O \leftrightarrow H_2CO_3$
Cu^{2+}	Ascorbic acid oxidase	Ascorbate $\leftrightarrow$ dehydroascorbate
Cu^{2+}	Uric acid oxidase	Uric acid $\rightarrow$ allantoin $+ CO_2 + H_2O_2$
Mg^{2+}	Hexokinase	Glucose + ATP $\rightarrow$ glucose phosphate

ments are believed to exist for other groups of isoenzymes.

Isozymes should be distinguished from **allelozymes**, which are multiple forms of single polypeptide enzymes resulting from variations in a single allelic or structural gene pair. Although allelozymes act on the same substrate, and in this regard are similar to isozymes, isozymes result from combinations of polypeptide chain products of two or more separate pairs of structural genes.

ZYMOGENS

A number of enzymes arise from an inactive precursor form called a **zymogen** or **proenzyme**. A number of the alimentary digestive enzymes belong to this group, including *pepsin, trypsin,* and *chymotrypsin.* The conversion of the zymogen to the active enzyme involves the preliminary cleavage of one or more of the zymogen's peptide bonds, followed occasionally by removal of a portion of the original protein molecule. This phenomenon may best be understood by considering a few examples.

Pepsin, the major protein-digesting enzyme of the stomach, is synthesized in the form of the precursor polypeptide *pepsinogen* (molecular weight 42,000). On entering the acidic gastric juice, the pepsinogen molecule is hydrolyzed at several positions to yield a number of small peptides and the active enzyme pepsin (molecular weight 35,000). The activation of pepsinogen can also be carried out by pepsin itself.

Trypsin, another proteolytic digestive enzyme, is produced in the pancreas as the zymogen *trypsinogen* and secreted into the duodenum (the anterior portion of the small intestine). In the duodenum, another enzyme, *enterokinase,* catalyzes the removal of six amino acids from the N-*terminus* of *trypsinogen,* thereby yielding *trypsin. Trypsin* itself can convert *trypsinogen* molecules to more *trypsin.*

The enzyme *chymotrypsin* also participates in the alimentary digestion of protein and is produced in the pancreas as the inactive proenzyme *chymotrypsinogen.* The activation of this zymogen involves a series of peptide bond cleavages catalyzed by *trypsin* already present in the duodenum and also by *chymotrypsin* itself. These cleavages split the *chymotrypsinogen* molecule into three polypeptides that remain interconnected in the activated enzyme through disulfide bridges that were part of the molecule's primary structure.

The activation of the proenzyme *trypsinogen* is shown in Figure 8–21 and illustrates some of the general features of zymogen activation. The active site of the enzyme is devoid of binding and/or catalytic activity until peptide bonds of the zymogen are broken. Following this peptide cleavage, the remaining portion of the molecule is reorganized with a consequent unmasking of the active site, which can now bind and act on the substrate.

Not only proenzymes but also other proteins may be activated by a preliminary proteolysis. For example, during the coagulation of blood, the formation of the matrix of the clot is brought about through a cascade of proteolytic activations that finally convert inactive, soluble protein monomers (fibrinogen) in blood plasma to the active, polymerizable form, which then produces the insoluble protein threads (fibrin). Some protein hormones are also synthesized as inactive precursors that are activated only on peptide cleavage (e.g., the conversion of *proinsulin* to *insulin*).

THE REGULATION OF ENZYME ACTIVITY

Literally thousands of different enzyme-catalyzed reactions characterize the metabolism of a cell, and these are organized into a number of interconnected (branching and converging) pathways. Some of the more important or universal pathways are considered

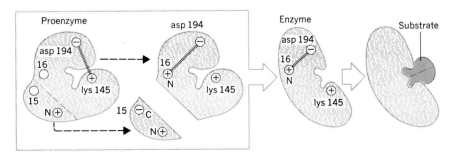

FIGURE 8–21 *Trypsinogen* activation involves the cleavage of the peptide bond linking amino acids 15 and 16, thereby freeing a small peptide. As a result, amino acid 16 of trypsinogen becomes the N-terminus of trypsin and interacts ionically with asp 194. This alters the orientation of lys 145 (which was previously neutralized by asp 194), generating the site for substrate binding and eventual cleavage (dashed line).

in Chapter 10. It is clear that the orderly functioning of a cell or organism demands that controls be placed on these reactions so that specific metabolic pathways (and, therefore, specific cell functions) are active or operative at certain times and inactive at others. One way in which such regulation can be achieved is by *altering the activity of specific enzymes under specific circumstances.*

Consider as an example the following situation. Suppose that in the course of five enzyme-catalyzed reactions, substrate A is converted to end product F, that is, where E_1, E_2, E_3, and so on are the enzymes involved in the pathway. If the end product F is able to bind to enzyme E_1 and in so doing render the enzyme inactive (i.e., unable to catalyze the conversion of A to B), then the synthesis of end product F can be regulated, for under conditions of low F concentration, the metabolic pathway leading to F would be functioning, whereas at high concentrations of F, E_1 inhibition by the end product would put a halt to the pathway leading to F. The phenomenon in which the end product of a metabolic pathway can regulate its own production by inhibition of this sort is called *feedback inhibition.*

$$A \xrightarrow{E_1} B \xrightarrow{E_2} C \xrightarrow{E_3} D \xrightarrow{E_4} E \xrightarrow{E_5} F$$

Many metabolic pathways are characterized by a number of branch points. In the example below, substrate A may be converted either to end product F or end product I. In such a pathway, the enzyme affected by feedback inhibition would occur at the branch point, that is, enzyme E_3 would be inhibited by end product F. Consequently, although the conversion of A to F would be inhibited, the formation of I would continue. The continued synthesis of I would not be possible if feedback inhibition occurred before the branch point.

$$A \xrightarrow{E_1} B \xrightarrow{E_2} C$$

That the phenomenon of feedback inhibition actually occurs in cells was definitively established in the 1950s. One of the earliest discovered examples of feedback inhibition and a classic case is the effect of cytidine triphosphate (CTP, one of the final products

FIGURE 8–22 Feedback inhibition of *aspartate transcarbamylase* by cytidine triphosphate.

formed during pyrimidine synthesis) on the enzyme *aspartate transcarbamylase* (Fig. 8-22). The biosynthesis of CTP begins with the interaction of aspartic acid and carbamyl aspartate. The reaction is catalyzed by *aspartate transcarbamylase (ATCase)* and is followed by four more reactions, culminating in the formation of CTP. When the free CTP concentration is low (i.e., when CTP is being consumed in the cell's metabolism), the formation of carbamyl asparate proceeds uninhibited. However, when the CTP level rises, carbamyl asparate synthesis is inhibited.

It is now clear that feedback inhibition is only one of several ways in which enzyme activity can be regulated. Not only end products but also other metabolites may be bound to an enzyme and in so doing alter the enzyme's activity. The binding can take place at the active site or at other sites on the enzyme's surface. Indeed, in instances where the enzyme molecule has more than one active site (see below), the binding of substrate at one site can influence subsequent substrate binding at another. The various mechanisms that regulate enzyme activity will be considered in the following section.

NUMBER OF POLYPEPTIDE CHAINS AND NUMBER OF BINDING SITES OF AN ENZYME

Many enzymes, including lysozyme and ribonuclease, are composed of a single polypeptide chain. Others consist of two or more chains. For example, the enzyme *glycogen phosphorylase* consists of two polypeptide chains; *fumarase* (a Krebs cycle enzyme) contains four polypeptide chains; and *aspartate transcarbamylase* (involved in the metabolic pathway leading to the synthesis of cytidine triphosphate) consists of 12 polypeptide chains. Each of the enzyme's constituent polypeptide chains is referred to as a **subunit**, the subunits being held together by electrostatic interactions, hydrogen bonds, van der Waals interactions, or other noncovalent forces that provide and stabilize protein quaternary structure (Chapter 4).

Enzymes may have two separated, *functionally different* binding sites. One type of site, the active site, binds the substrate of the enzyme and possesses catalytic activity, whereas the other type of site, the **allosteric** (**allo**, "other" + **steric**, "space") or **regulatory** site, lacks catalytic activity and binds an *effector* molecule. Such enzymes are usually referred to as **allosteric enzymes**. Depending on the enzyme, active and allosteric sites may be on the same polypeptide or on separate subunits. Effector molecules that inhibit enzyme activity (as in feedback inhibition) are called *negative* effectors. In some cases, binding of an effector molecule enhances enzyme activity, and these molecules are known as *positive* effectors. A single enzyme may possess regulatory sites capable of binding either negative or positive effectors; these effectors may compete with each other for the same regulatory site or be bound at separate regulatory sites.

In some enzymes, two or more subunits may each possess a catalytic site, and substrate binding to the catalytic site on one subunit may influence substrate binding at another catalytic site. Such enzymes exhibit **cooperativity**—a phenomenon already discussed in connection with the reversible oxygenation of hemoglobin (see Chapter 4). Enzymes that exhibit cooperativity and allosteric enzymes do not obey conventional Michaelis–Menten kinetics; the consequences of this are discussed later in the chapter.

Model for Allosteric Enzyme Function

A simple scheme depicting the influence of positive and negative effectors on allosteric enzyme activity is given in Figure 8–23, in which substrates and effectors and active and allosteric binding sites are represented as geometric shapes. The active and regulatory sites of the enzyme (in the absence of bound effectors) are depicted as circular areas. The binding of a positive effector (hexagon) induces a conformational change in the enzyme molecule at the regulatory site (symbolized as a change from circular to hexagonal shape). The conformational shape change is transmitted through the molecule (zigzag arrow) until it reaches and alters the active site in such a way that the substrate is bound more readily than in the absence of such activation. In Figure 8–23, the alteration of the active site is represented as a change from circular to square shape. Binding of a negative effector (diamond) induces a different type of conformational change in the regulatory site (circular to diamond shape), which is transmitted through the enzyme's structure such that the resulting change at the active site (circular to triangular shape) prevents subsequent substrate binding. The negative effector may, of course, be the end product of the pathway involving this enzyme.

The behavior of *aspartate transcarbamylase*, discussed earlier in connection with feedback inhibition, serves as a good example of the mechanism depicted in Figure 8–23. Although *aspartate transcarbamylase* activity is inhibited by the end product CTP (i.e., CTP is a negative effector), *ATCase* activity is stimulated by ATP (i.e., ATP acts as a positive effector). It appears that CTP and ATP compete for the same allosteric sites on the enzyme.

The model of Figure 8–23 deliberately fails to stipulate whether the allosteric enzyme is composed of one or more than one polypeptide chain, whether more than one active and regulatory site are present, or

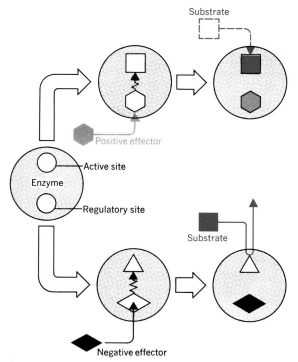

FIGURE 8–23 Simple model of allosteric enzyme function (see text for explanation).

COOPERATIVITY IN ENZYMES

When enzymes contain more than one active site, the binding of a substrate molecule to the first site may influence substrate binding to a second site. Binding of the second substrate may influence binding of a third, and so on. This phenomenon is called **cooperativity**. The influence may be positive in that binding of the first substrate molecule facilitates binding of subsequent substrate molecules (called "positive cooperativity") or the influence may be negative in that binding of a second or subsequent substrate molecule occurs less readily than binding of the first (called "negative cooperativity"). In a sense, the substrate itself is acting as either a positive or negative effector for the enzyme. These relationships are depicted in Figure 8–24. Cooperative effects are not restricted to enzymes but are observed with other proteins. Earlier (Chapter 4), we considered the cooperativity that exists among the globin chains of hemoglobin, a cooperativity that facilitates successive binding of oxygen molecules to the alpha and beta globin chains (i.e., positive cooperativity).

Both cooperative effects involving active sites on neighboring subunits of an enzyme and true allosteric effects involving regulatory sites may occur in a single enzyme molecule. A case in point is that of *cytidine triphosphate synthetase*, an enzyme involved in nucleic acid metabolism and consisting of four subunits. Two of these subunits contain active sites that bind the substrate glutamine and the other two have regulatory sites that bind GTP. When glutamine is bound to the active site of one subunit, a conformational change transmitted through the enzyme to another subunit renders the latter's active site unable to bind to glutamine (i.e., negative cooperativity). On the other hand, when GTP is bound to the regulatory site

whether the sites are on the same or on different subunits. Although all of these possibilities exist, most allosteric enzymes consist of several polypeptide subunits with active and regulatory sites on *separate* subunits. In the case of *ATCase*, there are 12 polypeptide chains—6 catalytic (C) chains (molecular weights of 34,000) and 6 regulatory (R) chains (molecular weights of 17,000). Each C chain contains one active site for aspartate and each R chain one regulatory site. Table 8–5 lists some of the allosteric enzymes, their substrates, and negative and positive effectors.

TABLE 8–5 SOME ALLOSTERIC ENZYMES

Enzyme	Substrate	Negative Effector	Positive Effector
Aspartate transcarbamylase	Aspartic acid and carbamyl phosphate	CTP	ATP
Threonine deaminase	Threonine	Isoleucine	Valine
Acetolactate synthetase	Pyruvic acid	Valine	—
Isocitric dehydrogenase	Isocitric acid	α-Ketoglutaric acid	Citric acid
Phosphofructokinase	Fructose-6-P and ATP	ATP	3'-AMP, 5'-AMP
Phosphorylase b	Glucose-1-P, glycogen, and phosphate	ATP	5'-AMP
Glycogen synthetase	Uridine diphosphoglucose	—	Glucose-6-P
Homoserine dehydrogenase	Homoserine and aspartic semialdehyde	Threonine	Isoleucine and methionine

(a)

(b)

(c)

Substrate

Substrate

Activates

Inhibits

FIGURE 8–24 *Cooperativity* in enzymes. The enzyme depicted here consists of two subunits. (*a*) Binding of the substrate to the active site of one subunit is without effect on the other subunit (i.e., *noncooperativity*). (*b*) Substrate binding to one subunit activates the site of the other subunit (i.e., *positive cooperativity*), thereby facilitating binding of the second substrate molecule (indicated by the straight path to the active site in contrast to the somewhat more devious path shown in *a*. (*c*) Binding of the first substrate molecule to the active site of one subunit produces a conformational change in the enzyme that inhibits substrate binding at the other active site (i.e., *negative cooperativity*).

of one subunit, this has a positive effect on glutamine binding but it negatively affects GTP binding at the other regulatory site. In other words, the effector GTP serves to activate catalysis but to inhibit further GTP binding.

ALLOSTERISM, COOPERATIVITY, AND MICHAELIS–MENTEN ENZYME KINETICS

As noted earlier, allosteric enzymes and enzymes that exhibit cooperative effects do not display conventional Michaelis–Menten kinetics. Figure 8–25 compares the Michaelis–Menten curves for enzymes exhibiting non-cooperativity, positive cooperativity, and negative cooperativity. The curve in Figure 8–25*a* shows the normal hyperbolic binding pattern exhibited by most

FIGURE 8–25 Michaelis–Menten curves for (*a*) an enzyme that does not exhibit cooperativity, (*b*) an enzyme exhibiting positive cooperativity, and (*c*) an enzyme exhibiting negative cooperativity.

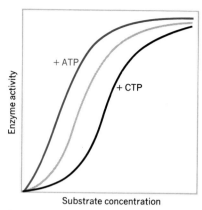

FIGURE 8–26 Effects of CTP (feedback inhibitor, negative effector) and ATP (positive effector) on the Michaelis–Menten curve of the positively cooperative enzyme *aspartate transcarbamylase.*

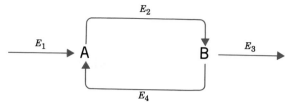

FIGURE 8–27 A *substrate cycle* (also called *futile cycle*); see text for details.

enzymes. In this example, an 81-fold increase in substrate concentration is required to elevate enzyme activity from 10 to 90% of its maximum level. The curve in Figure 8–25*b* depicts the sigmoid pattern characteristic of positive cooperativity. Here, only a ninefold increase in substrate concentration elevates enzyme activity from 10 to 90% of maximum. Note the resemblance of this curve to that for hemoglobin oxygenation (Chapter 4). Negative cooperativity produces the curve shown in Figure 8–25*c*. Although the curve appears hyperbolic, it actually is not. An increase in substrate concentration greater than 6000-fold would be required to elevate the enzyme activity from 10 to 90% of maximum. The curves have been drawn so that a substrate concentration of 1 unit corresponds to 50% of maximum enzyme activity.

As noted earlier, enzymes may exhibit cooperativity and also be affected by the binding of effectors to regulatory sites. *ATCase*, the enzyme discussed above, is a good example; it is negatively affected by CTP, positively affected by ATP, and shows positive cooperativity among its substrate binding sites. Figure 8–26 shows the effects of positive and negative effectors on *ATCase.*

SUBSTRATE CYCLES

Sudden and dramatic changes in cellular activity (e.g., conduction of nerve impulses by a nerve cell or the contractions of a muscle cell) demand rapid increases (or decreases) in the cellular level of particular metabolic intermediates. As discussed in the previous section, the binding of effectors at the regulatory sites of

allosteric enzymes acts to quickly increase (or decrease) the level of enzyme activity and in so doing quickly alter the rate at which substrate is converted to product by that enzyme. Such mechanisms for regulating the flux of a metabolic pathway through a particular reaction may be augmented by the use of **substrate cycles**.

Consider the scheme shown in Figure 8–27. Metabolic intermediate A is the *product* of enzymes E_1 and E_4 and is the *substrate* of enzyme E_2, an allosteric enzyme influenced by the binding of a positive effector. Intermediate B is the product of allosteric enzyme E_2 and is the substrate of enzymes E_3 and E_4. Thus, A and B form a substrate cycle. Substrate cycles such as the one shown in Figure 8–27 are also referred to as "futile cycles" because in the absence of any regulation the cycle would simply go around and around, dissipating energy (in the form of wasted ATP). Suppose that under "resting" cellular conditions the cycle is at idling speed and there are moderate fluxes in both the forward and reverse directions, but the *net* flux in the direction of B is very low (i.e., approaching zero) because most of B produced by E_2 is converted back to A by E_4. In such an instance, even a small increase in the activity of E_2 (through binding of a positive effector) would dramatically increase the net flux in the direction of B.

The magnitude of the effect may be illustrated by assigning some hypothetical values to these fluxes. For example, if under resting conditions the $A \rightarrow B$ flux is 100 and the $B \rightarrow A$ flux is 98, then the net flux in the direction $A \rightarrow B$ would be 2. An increase in the $A \rightarrow B$ flux from 100 to 110 (a 10% increase) could increase the net flux in the $A \rightarrow B$ direction from 2 to 12 (a 600% increase). This, of course, presumes that the increase in the rate at which B is being produced does not cause an increase in the activity of E_4. The latter could readily be avoided if (a) E_4 were already at saturation kinetics, (b) E_3 were activated so that there is little or no increase in the concentration of B, or (c) E_4 is inhibited by the same effector that stimulates E_2.

CATALYSIS BY RNA

The synthesis of RNA destined to be incorporated into a cell's ribosomes involves a series of *processing* stages in which a large polyribonucleotide is cleaved at specific points to form a smaller product (Chapter 22). Though most RNA processing involves enzymes called **endonucleases**, it has recently been discovered that one of these cleavages (at least in the protist *Tetrahymena*) is catalyzed by RNA itself; that is, the RNA molecule catalyzes the hydrolysis of one of its own phosphodiester linkages. This RNA has been called **ribozyme** to emphasize these enzymelike properties.

Ribozyme may be an example of a primordial biocatalyst; certainly, its discovery lends support to a popular notion that nucleic acid synthesis proceeded by a self-replicating process prior to the evolution of enzymes as biocatalysts.

SUMMARY

Regardless of the level of *molecularity*, nearly all reactions occurring in cells are catalyzed by the class of proteins called **enzymes**. Enzymes increase the likelihood that potentially reacting molecules (i.e., **substrates**) will encounter each other with the necessary orientation in space and "activate" their substrates by imposing strains on certain bonds. Although enzymes greatly increase reaction rates, they do not alter equilibrium concentrations. In the presence of increasing amounts of substrate, reactions approach a limiting velocity, the kinetics of which may be used to determine the enzyme's **Michaelis–Menten constant**;

this is a measure of the tendency of the enzyme and substrate to combine with each other.

The catalytic properties of enzymes are intimately related to the enzyme's primary, secondary, tertiary, and quaternary structure, and changes in structure are often accompanied by loss of catalytic activity. Consequently, most enzymes operate within narrow ranges of pH and temperature. Association of the substrate takes place in the enzyme's **active site**, which contains **contact** and **catalytic** amino acids. The active site nearly always resides in a depression or cavity in the enzyme's surface, and the binding of substrate is followed by a conformational change in the enzyme and/or the substrate. Some enzymes require a nonprotein component called a **cofactor** to be catalytically active. Cofactors may be permanently bound to the enzyme (i.e., **prosthetic groups**) or may form transient associations (e.g., **coenzymes** and **metal ions**). In some organisms, several different enzyme molecules catalyze the same chemical reaction; such families of enzymes are called **isoenzymes**. Certain enzymes possess an inactive form called a **zymogen** or **proenzyme**.

In addition to the active site, some enzymes possess an **allosteric** or **regulatory** site. The binding of a positive **effector** to the allosteric site increases the activity of the enzyme, whereas the binding of negative effector decreases enzyme activity. Changes in the levels of activity of allosteric enzymes are fundamental to the control of many metabolic pathways through feedback mechanisms. Enzymes composed of two or more polypeptide chains may contain two or more active sites, and the binding of substrate to one site often influences binding of additional substrate to other sites. Such influence, which can be positive or negative, is called **cooperativity**.

REFERENCES AND SUGGESTED READING

Articles and Reviews

Anderson, C. M., Zucker, F. H., and Steitz, T. A., Space-filling models of kinase clefts and conformation changes. *Science* **204**, 375 (1979).

Koshland, D. E., Correlation of structure and function in enzyme action. *Science* **142**, 1533 (1963).

Koshland, D. E., Protein shape and biological control. *Sci. Am.* **229** (4), 52 (Oct. 1973).

Koshland, D. E., Control of enzyme activity and metabolic pathways. *Trends in Biochem. Sci.* **9** (4), 155 (1984).

Koshland, D. E., and Neet, K. E., The catalytic and regulating properties of enzymes. *Ann. Rev. Biochem.* **37**, 359 (1968).

Naqui, A., Where are the asymptotes of Michaelis–Menten? *Trends in Biochem. Sci.* **11** (2), 16 (Feb. 1986).

Neurath, H., Protein-digesting enzymes. *Sci. Am.* **211** (6), 68 (Dec. 1964).

Newsholme, E. A., Challiss, R. A. J., and Crabtree, B., Substrate cycles: Their role in improving sensitivity in metabolic control. *Trends in Biochem. Sci.* **9** (6), 277 (1984).

Perutz, M. F., X-Ray analysis, structure and function of enzymes. *Eur. J. Biochem.* 8, 455 (1969).

Phillips, D. C., The three-dimensional structure of an enzyme molecule. *Sci. Am.* **215** (5), 78 (Nov. 1966).

Srere, P. A., Why are enzymes so big? *Trends in Biochem. Sci.* **9** (9), 387 (1984).

Stoud, R. M., A family of protein-cutting proteins. *Sci. Am.* **231** (1), 74 (July 1974).

Walsh, C. T., Enzyme mechanisms. *Trends in Biochem. Sci.* **9** (4), 159 (1984).

Books, Monographs, and Symposia

Cornish-Bowden, A., *Fundamentals of Enzyme Kinetics*, Butterworths, London, 1979.

Dickerson, R. E., and Geis, I., *The Structure and Action of Proteins*, Harper & Row, New York, 1969.

Ferdinand, W., *The Enzyme Molecule*, Wiley, New York, 1976.

Fersht, A., *Enzyme Structure and Mechanism*, Freeman, San Francisco, 1977.

Foster, R. L., *The Nature of Ezymology*, Halsted Press, New York, 1980.

Lehninger, A. L., *Principles of Biochemistry*, Worth, New York, 1982.

Rawn, J. D., *Biochemistry*, Harper & Row, New York, 1983.

Smith, E. L., Hill, R. L., Lehman, I. R., Lefkowitz, R. J., Handler, P., and White, A., *Principles of Biochemistry* (7th ed.), McGraw–Hill, New York, 1983.

Stryer, L., *Biochemistry* (2nd ed.), Freeman, San Francisco, 1981.

CHAPTER 9

BIOENERGETICS

In the study of physiology and biochemistry, cells are often thought of as tiny machines in which all events may be explained in terms of either chemical reactions, fluid dynamics, electrical fluxes across partitions, or the absorption or emission of light. In other words, cellular activities, regardless of their level of complexity, are ultimately founded on the known laws of physics and chemistry. Those laws of physics and chemistry that are fundamental to an understanding of cellular activities, especially the production and consumption of energy during cell metabolism, are considered in this chapter.

ENERGY AND METABOLISM

The **metabolism** of a cell is characterized by a myriad of chemical reactions in which energy is either *consumed, produced,* or *transduced* (i.e., converted) from one form into another. Metabolism can be subdivided into two broad categories: **catabolism** and **anabolism**. During catabolic reactions (or reaction sequences), molecules are broken down by the cell into simpler forms; whereas during anabolism, complex molecules are formed from simpler ones. The catabolic and ana-

bolic reactions that proceed in cells are accompanied by energy changes and it is the study of these changes that constitutes the field of *bioenergetics.*

Consider, for example, an anabolic process such as the synthesis of new membranes within the cell. Such biosynthesis requires (consumes) energy, and this energy must ultimately be obtained from the cell's environment in some form. Within the cell, the energy (perhaps in a new form) is consumed to "drive" the cell's membrane-synthesizing processes. Whole organisms and individual cells may be assigned to different groups according to the nature of the materials that they must acquire from their surroundings in order to support their metabolic needs. Most plant cells (i.e., those that contain chlorophyll) and many different kinds of bacteria require only CO_2, H_2O (or H_2S), simple nitrogen-containing compounds like NH_3, and trace mineral elements from their environment in order to fulfill their minimum metabolic needs. These cells or organisms are called **autotrophs**. With the exception of trace amounts of certain vitamins, they can live and grow in the complete absence of an exogenous supply of organic materials. (Indeed, most autotrophs do not even need an external source of vitamins.) When an autotroph can utilize light as a source of en-

TABLE 9–1   ENERGY AND CARBON SOURCES OF AUTOTROPHS AND HETEROTROPHS

Energy Source	Carbon Source	
	CO$_2$	Organic Substances

Light	Photoautotrophs
	  Plants
	  Most algae
	  Purple bacteria
	  Green bacteria
	  Most cyanobacteria
		Photoheterotrophs
		  A few algae
		  Some purple and
		    green bacteria
		  Some cyanobacteria

Oxidizable chemicals	Chemoautotrophs
	  Some bacteria
		Chemoheterotrophs
		  Animals
		  Most bacteria
		  Fungi
		  Protozoa

Source: Adapted from T. D. Brock, *Biology of Microorganisms* (3rd ed.), 1979, p. 207. Courtesy of Prentice–Hall, Englewood Cliffs, N.J.

ergy, it is called a **photoautotroph**. Other autotrophs can obtain their energy from the oxidation of inorganic substances such as ammonium ions (i.e., NH_4^+), ferrous iron (i.e., Fe^{2+}), or elemental sulfur (S). This kind of autotroph is called a **chemoautotroph** (see Table 9–1 for examples).

All animal cells (and certain plant cells and most bacteria) depend on an external source of organic compounds and specific vitamins for their metabolism and are therefore called **heterotrophs**. Some heterotrophs (e.g., a few algae and bacteria) can also use light as an energy source and are called **photoheterotrophs**. However, most heterotrophs require organic compounds both as a source of energy and as raw materials for the synthesis of intracellular components; such heterotrophs are called **chemoheterotrophs** (Table 9–1). Energy that is derived by the catabolism of organic materials is used to meet anabolic needs.

The primary sources of energy and raw materials for heterotrophic metabolism are polysaccharides, lipids, and proteins. Organisms that remove these macromolecules from their environment break them down in the successive catabolic stages of metabolism. As these compounds are chemically degraded, the chemical energy that is inherent in their molecular structure is both released in the form of heat and used to create the bonds that form new molecules, as in the attachment of free (inorganic) phosphate to ADP to form ATP (Fig. 9–1). The ultimate primary products of catabolism are NH_3, CO_2, and H_2O.

Although autotrophic organisms can use CO_2, H_2O, and small nitrogenous compounds from their environment, these small compounds do not by themselves

contain enough extractable chemical energy to sustain the organisms. Consequently, autotrophs also absorb energy in the form of light and, using the light energy, they synthesize simple organic acids from CO_2 and water (i.e., photosynthesis), phosphorylate ADP to form ATP, and synthesize amino acids from the organic ac-

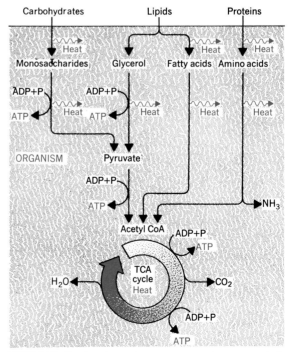

FIGURE 9–1 The flow of materials and energy during *catabolism*. Some energy is released as heat and some is used for the synthesis of ATP.

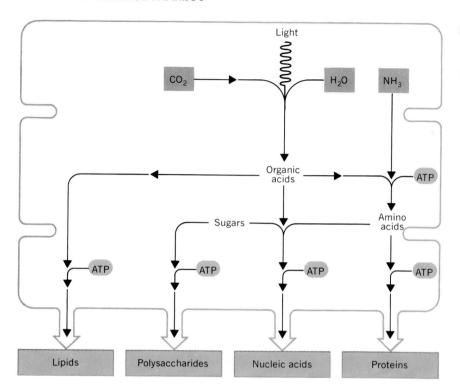

FIGURE 9–2 The flow of materials and energy during *anabolism*. ATP serves as a source of energy during many stages of anabolism.

ids using incorporated NH_3 (a process called **amination**). Drawing from the pool of ATP as a source of chemical energy, and using these simpler molecules, cells then synthesize complex molecules such as proteins, polysaccharides, nucleic acids, and lipids (Fig. 9–2). All these metabolic activities are accompanied by the loss of some unusable chemical energy as heat.

Autotrophic organisms not only possess the enzymes for the anabolic processes just described but, like heterotrophs, they also have catabolic enzyme systems. Consequently, autotrophs can produce ATP by breaking down polysaccharides, lipids, and proteins. Like autotrophs, heterotrophs have ATP-dependent anabolic enzyme systems that synthesize macromolecules, but most heterotrophs are unable to carry out photosynthesis. Compounds are cycled *within* cells and also *between* cells and *between* whole organisms as depicted in Figure 9–3. Each transition is accompanied by a specific energy change.

THE LAWS OF THERMODYNAMICS

Thermodynamics is the study of energy changes, that is, the conversion of energy from one form into an-

other. Such changes obey the first two laws of thermodynamics.

The First Law of Thermodynamics

The first law is concerned with the conversion of energy *within* a "system," where a system is defined as a body (e.g., a cell or an organism) and its surroundings. This law, which applies to both biological and nonbiological systems, states the following:

> *Energy cannot be created or destroyed but can be converted from one form into another: during such a conversion, the total amount of the energy of the system remains constant.*

This law applies to all levels of organization in the living world; it applies to organisms, cells, organelles, and to the individual chemical reactions that characterize metabolism. In practice, it is difficult to measure the energy possessed by cells (i.e., to limit the "system" to an individual cell), because energy may escape into the environment surrounding the cell during the measurement. Similarly, energy may be acquired by the cell from its environment; for example, a photosynthesizing cell absorbs energy from its environment in the form of light. A cell's *acquisition* of en-

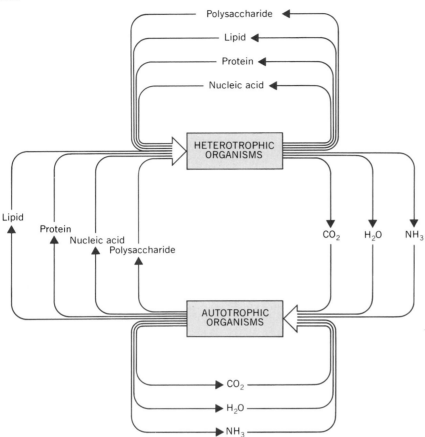

ergy from its environment (or its *loss* to the environment) should not be confused with the destruction or creation of energy, which according to the first law of thermodynamics does not occur.

From a biological viewpoint, the first law of thermodynamics indicates that at any given moment a cell possesses a specific quantity of energy. This energy takes several forms; it includes (1) **potential energy** (e.g., the energy of the bonds that link atoms together in a molecule or the pressure–volume relationships within the cell as a whole or within membrane-enclosed intracellular components); (2) **electrical energy** (e.g., the distribution of different amounts of electrical charge across cellular membranes); and (3) **thermal energy** (e.g., the temperature-dependent constant and random motions of molecules and atoms). According to the first law, these forms of energy may be interconverted; for example, some of the cell's potential energy can be converted into electrical or thermal energy, but the cell cannot create or destroy energy. When a cell breaks down a polysaccharide to ultimately form CO_2 and H_2O, some of the potential

energy present in the carbohydrate is conserved as potential energy by phosphorylating ADP, thereby forming ATP. The ATP so produced represents a new energy source (and also one that is of greater immediate utility for the cell). However, not all of the energy of the original carbohydrate is conserved as potential energy; some of it becomes thermal energy and is transferred to the surroundings as heat. It is important to recognize that none of the energy is destroyed and it should be possible to account for all of the energy originally present in the polysaccharide in other forms within the system (i.e., in the ATP that is produced and in the heat that is released).

The Second Law of Thermodynamics

The first law of thermodynamics tells us that the total energy of an isolated system consisting of a cell (or organism) and its surroundings is the *same* before and after a series of events or chemical reactions has taken place. What the first law does not tell us is the *direction* in which the reactions proceed. This problem can be illustrated using a simple example. Suppose we place a small cube of ice in a liter of hot water, seal the combination in an insulated container (e.g., a vacuum bottle), and allow the system (i.e., the ice and the water) to reach an equilibrium. In such a system, we would not be surprised to find that the ice melts and that this is accompanied by a decrease in the temperature of the water. When we later examine the system, we find that we are left only with water (no ice) and that the water is at a reduced temperature. The flow of heat, which is thermal energy, from the hot water to the ice thereby causing the ice to melt is *spontaneous* and the energy that is "lost" by the water is "gained" by the melting ice so that the total energy of the system remains the same. We certainly would not expect ice to form spontaneously in a sealed system that contains warm water, *even though such an eventuality is not prohibited by the first law.* Consequently, the important lessons to be learned from this illustration are that *energy changes have direction and may be spontaneous.*

To anticipate the spontaneity of a reaction and predict its direction, one must take into account a function called *entropy.* Entropy is a measure of the degree of *randomness* or *disorder* of a system, the entropy increasing with increasing disorder. Accordingly, the second law of thermodynamics states:

In all processes involving energy changes within a system, the entropy of the system increases until an equilibrium is attained.

In the illustration that was presented above, the highly ordered distribution of energy (i.e., large amounts of energy in the hot water and smaller amounts of energy in the ice) was lost as the ice melted to form water. In the resulting warm water, the energy was more randomly and uniformly distributed among the water molecules.

The units of entropy are J/mole (or cal/mole), indicating that entropy is measured in terms of the amount of substance present. When equal numbers of moles of a solid, liquid, and gas are compared at the same temperature, the solid has less entropy than the liquid and the liquid has less entropy than the gas (the gaseous state is the state of greatest disorder). Entropy can be thought of as the energy of a system that is of no value for performing work (i.e., it is *not* "useful" energy). For example, the catabolism of sucrose or other sugars by a cell is accompanied by the formation of energy-rich ATP. Although superficially it may appear as though useful energy has increased in the form of the ATP gained by the cell, the total amount of useful energy has actually decreased and the amount of unavailable energy increased. It is true that some of the potential energy of the sugar has been converted to potential energy in the form of ATP, but some has also been converted to thermal energy, which tends to raise the temperature of the cell and therefore its entropy. Suggestions that cells can decrease entropy by carrying out photosynthesis are misleading. Although it is true that during photosynthesis cells convert molecules with very little potential energy (CO_2 and H_2O) into larger molecules with considerably more potential energy (sugars) and that there is an accompanying decrease in the entropy of the cell, energy in the form of light was absorbed from the cell's environment. Because the light energy consumed during photosynthesis is a part of the whole system (i.e., the cell and its surroundings), it is clear that there has actually been an overall decrease in useful energy and an increase in entropy (see Fig. 9–4).

The entropy change during a reaction may be quite small. For example, when sucrose undergoes hydrolysis to form the sugars glucose and fructose, much of the potential energy of the original sucrose is present in the resulting glucose and fructose molecules. Changes in entropy are extremely difficult to calculate, but the difficulty can be circumvented by employing two other thermodynamic functions: **enthalpy** or heat content (denoted H) and **free energy** (denoted G).

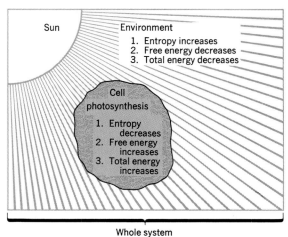

FIGURE 9–4 Relationships between the changes in entropy and free energy within a whole system and within its parts.

The change in a system's enthalpy (ΔH) is a measure of the total change in energy that has taken place, whereas the change in free energy (ΔG) is the change in the amount of energy available to do work. Changes in entropy (ΔS), enthalpy, and free energy are related by the equation

$$\Delta G = \Delta H - T\Delta S \qquad (9\text{-}1)$$

in which T is the absolute temperature of the system.

The change in free energy can also be defined as the total amount of free energy in the products of a reaction *minus* the total amount of free energy in the reactants, that is,

$$\Delta G = G_{(products)} - G_{(reactants)} \qquad (9\text{-}2)$$

A reaction that has a negative ΔG value (i.e., the sum of the free energy of the products is less than that of the reactants) will occur *spontaneously*; a reaction for which the ΔG is zero is at *equilibrium*; and a reaction that has a positive ΔG value will not occur spontaneously and proceeds only when energy is supplied from some outside source.

ΔG, ΔG⁰, and ΔG⁰′ Values

The hydrolysis of sucrose

$$\text{sucrose} + H_2O \rightarrow \text{glucose} + \text{fructose}$$

has a negative ΔG value, and therefore when sucrose is added to water, there is the spontaneous conversion of some of the sucrose molecules to glucose and fructose. However, the reverse reaction

$$\text{glucose} + \text{fructose} \rightarrow \text{sucrose} + H_2O$$

has an equal *but positive* ΔG value and therefore does not occur without an input of energy. Hence, special attention must be paid to the *direction* in which the reaction is written (i.e., the direction of the arrow) and the sign of the ΔG value. If 5 moles of sucrose are mixed with water, the formation of glucose and fructose will take place spontaneously and the ΔG may be determined; this value is, of course, greater than if 4 or 2 moles of sucrose are used. Thus, ΔG values are dependent on the amounts and concentrations of reactants and products. More uniform standards of reference that have been established by convention are the **standard free energy changes**, ΔG^0 and $\Delta G^{0\prime}$ values. ΔG^0 represents the change in free energy that takes place when the reactants and products are maintained at 1.0 molar concentrations (strictly speaking, 1.0 mo-

lal) during the course of the reaction and the reaction proceeds under standard conditions of temperature (25°C) and pressure (1 atmosphere) and at pH 0.0. The $\Delta G^{0\prime}$ value is a much more practical term for use with biological systems in which reactions take place in an aqueous environment and at a pH that usually is either equal or close to 7.0. The $\Delta G^{0\prime}$ value is defined as the standard free energy change that takes place at pH 7.0 when the reactants and products are maintained at 1.0 molar concentrations (Table 9-2).

The changes in standard free energy are independent of the route that leads from the initial reactants to the final products. For example, glucose can be converted to carbon dioxide and water either by combustion in the presence of oxygen or through the actions of cellular enzymes. Changes in standard free energy are the same, regardless of the method that is used; thus, the value of the standard free energy change provides no information about the reaction sequence by which the change has taken place. By the same token, the values obtained for changes in standard free energy tell us nothing about the *rate* at which the changes have taken place.

The $\Delta G^{0\prime}$ can be calculated from the equilibrium constant, K'_{eq}, of a reaction using the relationship

$$\Delta G^{0\prime} = -R\,T\ln K'_{eq} \qquad (9\text{-}3)$$
$$= -2.303\,R\,T\log_{10} K'_{eq} \qquad (9\text{-}4)$$

where R is the gas constant (8.314 J/mole/degree), T is the absolute temperature (in degrees Kelvin), and K'_{eq} is the equilibrium constant. Table 9-3 lists a number of $\Delta G^{0\prime}$ values for common reactions.

For the reaction

$$A + B \longrightarrow C + D$$

the equilibrium constant is defined as

$$K'_{eq} = \frac{[C]\,[D]}{[A]\,[B]} \qquad \text{at equilibrium} \qquad (9\text{-}5)$$

where $[A]$ and $[B]$ are the concentrations of the reactants and $[C]$ and $[D]$ are the concentrations of the products. If the equilibrium constant is 1.0, then the $\Delta G^{0\prime}$ value equals zero. If the equilibrium constant is *greater* than 1.0, then the $\Delta G^{0\prime}$ value is negative (e.g., -11.41 kJ/mole for a K'_{eq} value of 100), and the reaction is said to be **exergonic** (i.e., "energy releasing") because it proceeds spontaneously in the direction written when starting with unimolar concentrations of reactants and products. When the K'_{eq} value is less

TABLE 9-2   FREE ENERGY: SYMBOLS, DEFINITIONS, AND EXPLANATIONS

Symbol	Definition	Explanation
G	Free energy (also called Gibbs free energy)	The maximum energy that can be derived from a particular molecule capable of doing work under conditions of constant temperature and pressure.
ΔG	Change in free energy	The change in free energy that takes place during a chemical reaction (equal to the free energy of the products of the reaction minus the free energy of the reactants.)
ΔG^0	*Standard* free energy change[a]	An expression more frequently used in physical chemistry than in biochemistry. The change in free energy that takes place under the following standard conditions: reactants and products are maintained at 1.0 molal concentrations; the temperature is 25°C; the pressure is 1 atm; and the pH is 0.0.
$\Delta G^{0\prime}$	*Standard* free energy change[a]	Expression used in biochemistry and physiology. Change in free energy under the following standard conditions: reactants and products are maintained at 1.0 molar concentrations; the temperature is 25°C; the pressure is 1 atm; and the pH is 7.0.
$\Delta G^\prime$	*Standard* free energy change[a]	Expression rarely used anymore same as ΔG^0 but not necessarily at pH 0.0. The pH value must be specified.
$\Delta G\ddagger$	Free energy of activation	A measure of the activation energy of a chemical reaction; does not provide information about the direction or equilibrium of the reaction. (See Chapter 8.)

[a] If the reaction is independent of pH, then $\Delta G^0 = \Delta G^{0\prime} = \Delta G^\prime$.

than 1.0, the $\Delta G^{0\prime}$ value is positive (e.g., 5.71 kJ/mole for a K'_{eq} of 0.1), and the reaction is said to be **endergonic** (i.e., "energy consuming") because it does not proceed spontaneously in the direction written when starting with unimolar concentrations of reactants and products.

Calculations of $\Delta G^{0\prime}$ values are usually based on experimental measurements of isolated reactions, that is, with reactions that take place *independently* of other reactions and that are not associated with cells. ΔG^0 and $\Delta G^{0\prime}$ values do not provide information about the free energy changes of reactions as they might take place in cells or under conditions in which the concentrations of reactants and products, pH, etc., may change. This may be dramatically illustrated by considering the following example. At pH 7.0 and 25°C, the equilibrium constant for the reaction

dihydroxyacetone phosphate → glyceraldehyde-3-phosphate

is 0.0475. Therefore, using equation 9-3,

$$\begin{aligned} \Delta G^{0\prime} &= -2.303\,R\,T\log_{10} K'_{eq} \\ &= -2.303\,(8.314\ \text{J/mole/degree})\,(298)\log_{10}(0.0475) \\ &= +7.55\ \text{kJ/mole} \end{aligned}$$

The positive value indicates that this reaction does not proceed spontaneously in the direction written. However, in cells, this reaction is but one of a series of reactions in a metabolic pathway called **glycolysis** (Chapter 10). Other reactions of glycolysis that occur *prior* to this one and that have negative $\Delta G^{0\prime}$ values produce additional substrate (i.e., dihydroxyacetone phosphate) and reactions with negative $\Delta G^{0\prime}$ values that occur *after* this step remove the product glyceraldehyde-3-phosphate. As a result, the reaction proceeds in the direction written under the conditions specified above, *even though the $\Delta G^{0\prime}$ value is positive.* This example illustrates the important point that *the $\Delta G^{0\prime}$ value for a specific biological reaction cannot be used to predict reliably whether or not that particular reaction is actually taking place within the cell.*

TABLE 9–3 STANDARD FREE ENERGY CHANGES OF COMMON BIOCHEMICAL REACTIONS AT pH 7.0 AND 25°C

Reaction	$\Delta G^{0\prime}$	
	kJ/mole	kcal/mole
Hydrolysis:		
Acid anhydrides:		
Acetic anhydride + $H_2O \rightarrow$ 2 acetate	– 91.2	– 21.8
Pyrophosphate + $H_2O \rightarrow$ 2 phosphate	– 33.4	– 8.0
Esters:		
Ethyl acetate + $H_2O \rightarrow$ ethanol + acetate	– 19.7	– 4.7
Glucose-6-phosphate + $H_2O \rightarrow$ glucose + phosphate	– 13.8	– 3.3
Amides:		
Glutamine + $H_2O \rightarrow$ glutamate + $NH_4{}^+$	– 14.2	– 3.4
Glycylglycine + $H_2O \rightarrow$ 2 glycine	– 9.2	– 2.2
Glycosides:		
Sucrose + $H_2O \rightarrow$ glucose + fructose	– 29.3	– 7.0
Maltose + $H_2O \rightarrow$ 2 glucose	– 16.7	– 4.0
Esterification:		
Glucose + phosphate $\rightarrow$ glucose-6-phosphate + H_2O	+ 13.8	+ 3.3
Rearrangement:		
Glucose-1-phosphate $\rightarrow$ glucose-6-phosphate	– 7.11	– 1.7
Fructose-6-phosphate $\rightarrow$ glucose-6-phosphate	– 1.67	– 0.4
Elimination:		
Malate $\rightarrow$ fumarate + H_2O	+ 3.14	+ 0.75
Oxidation:		
Glucose + 6 $O_2 \rightarrow$ 6 CO_2 + 6 H_2O	– 2870	– 686
Palmitic acid + 23 $O_2 \rightarrow$ 16 CO_2 + 16 H_2O	– 9782	– 2338

COUPLED REACTIONS

Cells commonly employ either of two mechanisms to cause endergonic reactions to take place. One mechanism, illustrated above by the glycolysis example, is to create reactant and product concentrations that are markedly different from equilibrium values. This can be brought about either through the production of additional reactants or through the removal of products by drawing them into other cellular reactions. For example, the reaction

$$\text{glucose-6-phosphate} \xrightarrow{\text{isomerase}} \text{fructose-6-phosphate}$$

has a $\Delta G^{0\prime}$ value of 1.67 kJ/mole, and the isolated reaction would not be expected to occur spontaneously when reactants and products occur in equimolar concentrations. However, in cells, there is an enzyme that catalyzes the conversion of fructose-6-phosphate to fructose-1,6-diphosphate

$$\text{fructose-6-phosphate} \xrightarrow[\underset{\text{ATP} \quad \text{ADP}}{}]{\text{fructokinase}} \text{fructose-1,6-diphosphate}$$

and the $\Delta G^{0\prime}$ value of this reaction is – 14.22 kJ/mole. In the parlance of bioenergetics, the equilibrium of this reaction "lies far to the right." In cells, the two reactions are *coupled*; that is, by removing the product of the first reaction, its concentration is shifted away from its equilibrium value, allowing more product to be formed and therefore both reactions proceed sequentially.

The *total* free energy change for a sequence of coupled reactions is equal to the sum of the free energy changes of the individual reactions of the sequence. In the example just given, the $\Delta G^{0\prime}$ for the sequence of reactions is – 14.22 + 1.67 = – 12.55 kJ/mole.

A second type of coupling occurs when exergonic and endergonic reactions are catalyzed by the *same* enzyme. In cells, the two reactions

$$\text{ATP} + H_2O \rightarrow \text{ADP} + \text{phosphate}$$
$$(\Delta G^{0\prime} = -30.65 \text{ kJ/mole})$$

$$\text{glucose} + \text{phosphate} \rightarrow \text{glucose-6-phosphate} + H_2O$$
$$(\Delta G^{0\prime} = +13.86 \text{ kJ/mole})$$

are catalyzed (and also coupled) by the enzyme *glucokinase*; that is,

$$\text{glucose} + \text{ATP} \xrightarrow{\textit{glucokinase}} \text{glucose-6-phosphate} + \text{ADP}$$
$$(\Delta G^{0\prime} = -16.79\,\text{kJ/mole})$$

The exergonic reaction is thus coupled to the endergonic reaction by *glucokinase,* and some of the free energy yielded by the exergonic reaction is consumed to drive the exergonic reaction.

INTRACELLULAR PHOSPHATE TURNOVER

In all cells, most of the major energy changes and reaction couplings involve nucleotides. Although the cell draws on a number of different nucleotide pools,

$$\text{ATP} \rightarrow \text{ADP} + \text{phosphate}$$

exchanges are the most common. Other exchanges, such as

$$\text{ATP} \rightarrow \text{AMP} + \text{pyrophosphate}$$

occur, for example, during lipid metabolism, as do

$$\text{CTP} \rightarrow \text{CDP} + \text{phosphate}$$

reactions. Uridine triphosphate (UTP) is utilized in polysacharide synthesis and guanosine triphosphate (GTP) is consumed during protein synthesis. The deoxy derivatives of the nucleotides, dATP, dGTP, dUTP, and dCTP, are used during DNA synthesis.

Regeneration of the nucleoside triphosphates is catalyzed by a relatively nonspecific nucleoside diphosphate kinase:

$$\text{CDP} + \text{ATP} \longrightarrow \text{CTP} + \text{ADP}$$
$$\text{UDP} + \text{ATP} \longrightarrow \text{UTP} + \text{ADP}$$
$$\text{GDP} + \text{ATP} \longrightarrow \text{GTP} + \text{ADP}$$
$$\text{dCDP} + \text{ATP} \longrightarrow \text{dCTP} + \text{ADP}$$
$$\text{dADP} + \text{GTP} \longrightarrow \text{dATP} + \text{GDP}$$

The total concentration of all nucleotides in cells that are metabolizing in a steady state is relatively constant and most of the adenylate is present in the form of ATP. When cellular activity increases, the level of ATP decreases and the combined level of ADP + AMP increases. The changes that take place in the levels of these compounds cause an increase in reaction sequences such as those of glycolysis (Chapter 10), which generate ATP. Conversely, when cellular activity decreases, the transient rise in ATP acts to slow down the ATP-generating systems. The regulation of ATP synthesis (discussed in Chapter 11) is achieved through feedback control of allosteric enzymes by ATP, ADP, and AMP.

Although ATP hydrolysis is a major source of free energy in cells, cells do not form substantial reservoirs of ATP. The amount of ATP present in a cell lasts for only a short time during periods of increased cellular activity. Most cells contain additional reservoirs of compounds that can quickly be consumed to generate additional ATP. For example, striated muscle, smooth muscle, and nerve cells contain reservoirs of *phosphocreatine.* When the ATP levels in these cells undergo a marked decline, the enzyme *creatine kinase* is activated and catalyzes the reaction

$$\text{phosphocreatine} + \text{ADP} \rightarrow \text{creatine} + \text{ATP}$$
$$(\Delta G^{0\prime} = -12.6\,\text{kJ/mole})$$

Phosphoarginine and polymetaphosphate play a similar role in invertebrate and bacterial cells, respectively.

REDOX REACTIONS AND REDOX COUPLES

The movements of electrons between cellular **reductants** and **oxidants** represent another form of energy transfer in cells. A reductant (or **reducing agent**) is a substance that loses or donates electrons to another substance; the latter substance is the oxidant (or **oxidizing agent**). Conversely, an oxidant is a substance that accepts electrons from another substance, the latter being the reductant. Reactions that involve the movement of electrons between reductant and oxidant are called **redox reactions**.

Different chemical substances have different potentials for donating or accepting electrons. The tendency of hydrogen to dissociate

$$H_2 \rightleftharpoons 2\,H^+ + 2\,e^-$$

thereby releasing electrons, is used as a standard against which the tendencies of other substances to release or accept electrons is measured. The electron donor (e.g., H_2 in the above reaction) and the electron acceptor (e.g., $2\,H^+$ in the above reaction) are called a **redox couple** or **half cell**. The tendency of any chemical substance to lose or gain electrons is called the *redox potential* and is measured in volts (V). Measurements are made using an electrode that has been standardized against the H_2–$2\,H^+$ couple whose redox potential is set at 0.0 under standard conditions

TABLE 9–4 STANDARD REDOX POTENTIALS AT pH 7.0 AND 25°–37°C

Reductant	Oxidant	E_0' (volts)
Pyruvate	Acetate $+ 2\,H^+ + 2\,e-$	-0.70
Acetaldehyde	Acetate $+ 2\,H^+ + 2\,e-$	-0.58
H_2	$2\,H^+ + 2\,e-$	-0.42^a
$NADH_2$	$NAD^+ + 2\,H^+ + 2\,e-$	-0.32
Ethanol	Acetaldehyde $+ 2\,H^+ + 2\,e-$	-0.197
Lactate	Pyruvate $+ 2\,H^+ + 2\,e-$	-0.185
Succinate	Fumarate $+ 2\,H^+ + 2\,e-$	-0.031
Ubiquinol	Ubiquinone $+ 2\,H^+ + 2\,e-$	$+0.10$
2 Cytochrome $b_{(red)}$	2 Cytochrome $b_{(ox)} + 2\,e-$	$+0.030$
2 Cytochrome $c_{(red)}$	2 Cytochrome $c_{(ox)} + 2\,e-$	$+0.254$
2 Cytochrome $a_{3(red)}$	2 Cytochrome $a_{3(ox)} + 2\,e-$	$+0.385$
H_2O	$\frac{1}{2}\,O_2 + 2\,H^+ + 2\,e-$	$+0.816$

$^a E_0 = 0.0$ volts.

(pH 0.0, 1 M [H^+], 25°C, and 1 atmosphere pressure). This potential is noted by the symbol E_0. For biochemical reactions, which normally occur at pH 7.0, the redox potential of the H_2–2 H^+ couple is -0.421 V; standard redox potentials at pH 7.0 are noted by the symbol E_0'. The E_0' values of a number of biologically important redox couples are given in Table 9–4.

Any substance with a more positive E_0' value than another has the potential for oxidizing that substance (i.e., removing electrons from the substance with the more negative E_0' value). The greater the difference in redox potentials, the greater the energy changes involved. The change in standard free energy, $\Delta G^{0'}$, is related to E_0' as follows:

$$\Delta G^{0'} = n\mathscr{F}\Delta E_0'$$

where n is the number of electrons exchanged per molecule, $\mathscr{F}$ is the **Faraday** (96,406 J/V), and E_0' is the difference in redox potential between the more positive and the more negative members of the redox couple. For example, the oxidized form of cytochrome c can oxidize the reduced form of cytochrome b by removal of two electrons. The difference between the redox potentials of the two is

$$+0.254 - (+0.030) = +0.224\text{ V}$$

Therefore,

$$\begin{aligned}\Delta G^{0'} &= -2\,(96{,}406\text{ J/V})(0.224\text{ V})\\ &= -43.19\text{ kJ (per mole of each cytochrome)}\end{aligned}$$

In cells, the energy changes during redox reactions may be coupled to the synthesis of ATP; this is discussed in Chapter 16.

LIGHT AND THE TRANSDUCTION OF ENERGY

The conversion of energy from one form into another is called **transduction**. An example of transduction is the conversion of light energy (*electromagnetic radiation*) into potential chemical bond energy in such seemingly diverse processes as **photosynthesis** and **vision**. The conversion of chemical bond energy into light energy (i.e., transduction in the opposite direction) occurs in **bioluminescence** (e.g., the emission of light by fireflies and by certain microorganisms, see Fig. 9–5).

Electromagnetic radiations consist of far more than just the spectrum of colors perceived by the human eye. Each of the radiations of the **electromagnetic spectrum** (Table 9–5) can be thought of as a stream of moving packets of energy called **photons**. The stream of photons has the character of a *wave* (Fig. 9–6) in which the distance from any point on the wave to an equivalent point on the neighboring wave is the **wavelength**. Among the differences between the radiations of the electromagnetic spectrum are differences in wavelength. At one extreme are the *long* waves, such as radio and television waves, which have wavelengths up to 1000 m. At the opposite end of the spectrum are the X rays and gamma rays, which have wavelengths of 0.1×10^{-9} to 0.1×10^{-12} m. Most of the biologically important wavelengths are shorter than 10^{-6} m. In air, each photon travels at the same velocity, namely, 3×10^8 m/sec.

The energy of radiation is said to be *quantized*; that is, it has specific values or **quanta**. Quantum energy is related to wavelength by **Planck's Law:**

FIGURE 9–5   (Left) Light-emitting cells (*photocytes*) of the firefly *Photinus greeni. P,*
photocyte; *N,* nucleus; *M,* mitochondria; *V,* light-emitting vesicle (enlarged in lower left
insert); *D,* dense inclusion; *Tr,* traceheole. (Courtesy of J. F. Case and K. Linberg; copyright
© 1980 by Academic Press; *Intl. Rev. Cytol.* **68**, *173.)* (Right) Light-emitting vesicles (v) of the
bioluminescent dinoflagellate *Pyrocystis fusiformis.* (Courtesy of Beatrice Sweeney;
copyright © 1980 by Academic Press; *Intl. Rev. Cytol.* **68**, 173.)

$$q = hv = \frac{hc}{\lambda} \qquad (9\text{--}6)$$

where q is the quantum energy in joules, h is Planck's
constant (6.624×10^{-34} J sec), v is the frequency of the
radiation, c is the velocity of light, and λ is the radiation's wavelength (in meters). It may be seen from this
equation that radiations of shorter wavelength possess more energy than radiations of longer wavelength. It should be noted that there are a number of
forms of radiation that are not electromagnetic but
are particulate. Particulate radiations include the alpha rays and beta rays that are emitted from the nuclei of radioactive isotopes. These radiations are considered in Chapter 14, which deals with the uses of
radioisotopes as *tracers* of cellular metabolism.

When radiation passes through a substance, some
(perhaps all) of the energy of the radiation is absorbed
by certain electrons of the atoms and molecules comprising the substance. The energy of each photon is either *totally* transferred to an electron or *none* of the
photon energy is transferred. That is, with regard to a
specific electron, either all or none of the photon energy is absorbed. Moreover, the **Einstein–Stark Law**
of photochemical equivalence demands that the absorption of *one* quantum of light energy results in the
activation of *one* atom (or *one* molecule). The absorption of light energy by an atom or molecule often involves a shift of an electron from one orbital to another. Each electron possesses energy, the amount of
which is determined by the location of the electron or-

TABLE 9–5   WAVELENGTHS OF THE RADIATIONS OF THE ELECTROMAGNETIC SPECTRUM

Type of Radiation	Photon Energy (joules $\times 10^{-19}$)	Wavelength (nm[a])
Television waves	2.0×10^{-10}–2.0×10^{-8}	10^9–10^{11}
Radar waves	2.0×10^{-8}–2.0×10^{-5}	10^6–10^9
Radio waves	2.0×10^{-11}–2.0×10^{-5}	10^6–10^{12}
Infrared (far)	0.005–0.99	2×10^3–4×10^5
Infrared (near)	0.99–2.5	780–2×10^3
Red light	2.5–3.2	620–780
Orange light	3.2–3.4	590–620
Yellow light	3.4–3.6	545–590
Green light	3.6–4.1	490–545
Blue light	4.1–4.6	430–490
Violet light	4.6–5.1	390–430
Ultraviolet (long)	5.1–6.6	300–390
Ultraviolet (short)	6.6–9.9	200–300
Ultraviolet (very short)	9–132.4	15–200
X rays (soft)	99.3–2.0×10^4	0.1–20
X rays (hard)	2.0×10^4–3.9×10^6	0.0005–0.1
Gamma rays	1.4×10^4–19.9×10^6	0.0001–0.14

[a] 1 nm (nanometer) $= 10^{-9}$ m (meter), $1 = 1 \ \mu$m (micrometer) $= 10$ Å (angstrom).

bital in space and the speed at which the electron moves. Absorption of light energy either raises an electron to an orbital of higher energy or accelerates the electron within its orbital. Electrons may occur in pairs within an orbital, the members of the same orbital spinning in opposite directions. Most atoms at their lowest energy level (i.e., the *ground state*) have all of their electrons paired in this fashion. When the absorption of light energy raises an electron to a higher, unoccupied orbital, the electron may continue

to spin in the direction opposite to its former partner or it may spin in the same direction as its former partner. Because the energy of radiation varies with wavelength, and because the transfer of energy takes place in quantum units, it is evident that only certain wavelengths will excite specific atoms in a molecule. If the quantum transfer is great enough, the electron may altogether escape from the atom, and the atom (or molecule) would therefore be *ionized*. Radiations of very short wavelengths (e.g., gamma rays and X rays)

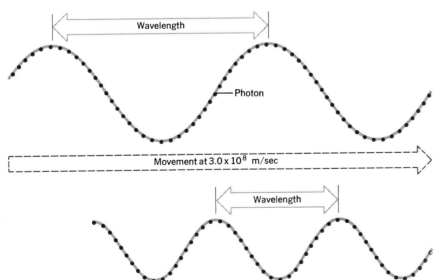

FIGURE 9–6   Wavelike character of electromagnetic radiation such as light. The energy of the light may be thought of as a stream of discrete packets of quanta called *photons*.

(a)

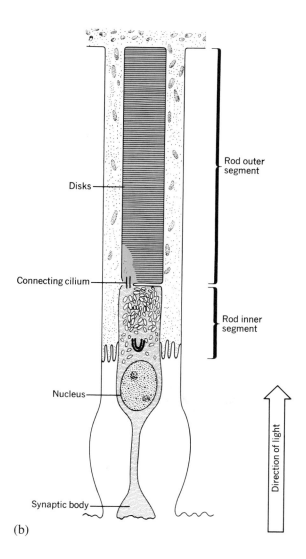

(b)

FIGURE 9–7   Rod cells of the vertebrate eye. (*a*)
Transmission electron photomicrograph of retina tissue
showing the parallel, closely packed, intracellular
membranes (disks) of the outer rod segment and the
connecting cilium that leads to the inner rod segment.
(Courtesy of Dr. S. Fisher; copyright © 1976 by Academic
Press; *J. Ultra. Res* **55**, 114.) (*b*) Diagram of a typical rod cell.

cause appreciable ionization when they pass through
cells and tissues.

   Atoms containing electrons that have been boosted
to higher orbitals are intrinsically unstable and frequently return (in less than 10^{-9} seconds) to their
ground state as the electrons move again to their former orbitals. The gain in energy that boosts an electron to a higher orbital is called a **primary reaction**,
whereas the release of energy during the return of an
electron to a lower orbital is called a **secondary reaction**. During the secondary reaction, the energy that
is released may be transferred to neighboring molecules or may take the form of light (i.e., fluorescence)
or heat. The transfer of energy to neighboring molecules during a secondary reaction underlies such biological mechanisms as vision (discussed briefly below)
and photosynthesis (considered at length in Chapter 17).

Vision

The *photoreceptor* cells of vertebrate eyes (called *rods*
and *cones*) contain a number of pigment molecules
(*carotenoids*) that are excited by light. Among these,
rhodopsin, a pigment in the rod cells, has been well
studied. The pigment consists of a protein called *opsin*
and *retinal*, a compound related to vitamin A. The
rhodopsin molecules are localized in the outer segment of the rod cells (Fig. 9–7) in disks formed by
stacks of intracellular membranes. When light energy
of the appropriate wavelength is absorbed by rhodopsin, the retinal changes from the *cis* to the *trans* iso-

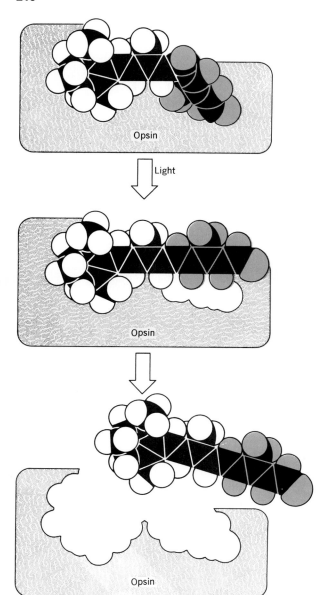

11-*cis*-Retinal

all-*trans*-Retinal

FIGURE 9–8   Chemical changes in rhodopsin induced by light.

mer (Fig. 9–8). This change or *bleaching* alters the bonding between the retinal and the opsin, causing the eventual release of the retinal from the protein. The bleaching of the rhodopsin releases cyclic GMP that in turn alters the permeability of the outer segment disk membranes to cations such as Na^+, Ca^{2+}, and K^+ and brings about a change in electrical potential across the rod cell's plasma membrane. As a result, a transmitter substance is released by the inner rod segment and this serves as a chemical stimulus for nerve cells that lead from the rods to the brain.

OTHER KINDS OF TRANSDUCTIONS

A number of additional examples of energy transductions will be encountered in succeeding chapters of the book, but a few examples will be briefly previewed here. Cells frequently utilize potential chemical energy such as that in ATP to move *permeable* solutes through their membranes in a direction that is *against* the solute's concentration gradient. These vital cellular transport mechanisms are treated more fully in Chapter 15, which deals with the structure and functions of the cellular membranes. The solute concentration gradient that is created by a transport mechanism may itself serve as a potential source of energy. Mitochondria generate an ion gradient across their membranes and this gradient is believed to serve as an energy source for the phosphorylation of ADP. The process is discussed in detail in Chapter 16.

In nerve cells, there is the selective transport of Na^+ and K^+ across the plasma membrane against the concentration gradients of these permeable ions, and this serves to create an electrical potential across the membrane. The establishment of this potential underlies the capability of nerve cells to propagate impulses (see Chapter 24). The potential chemical energy inherent in ATP is used in muscle cells to bring about the sliding of protein filaments past one another during contraction. Muscle contraction is therefore a vivid example of a transduction in which chemical energy is transformed to mechanical energy. This mechanism is also more fully described in Chapter 24.

SUMMARY

The **metabolism** of a cell can be subdivided into two categories: **catabolism** (the breakdown of molecules into simpler forms, accompanied by the release of energy) and **anabolism** (the synthesis of complex molecules from simpler ones, accompanied by the consumption of energy). **Bioenergetics** is the study of all the energy changes that take place in living systems.

Depending on the nature of the materials that must be obtained from their surroundings, cells (and organisms) are classified as **autotrophs (photoautotrophs** or **chemoautotrophs)** or **heterotrophs (photoheterotrophs** or **chemoheterotrophs)**.

Metabolic reactions that yield energy are called **exergonic** reactions and reactions that consume energy are called **endergonic** reactions. In most instances the energy of exergonic reactions is used to attach phosphate to ADP thereby forming ATP. Energy changes in cells obey the first two **laws of thermodynamics**.

> *First law:*   *Energy cannot be created or destroyed but can be converted from one form into another.*
>
> *Second law:*   *In all processes involving energy changes within a system, the entropy of the system increases until equilibrium is achieved.*

Reactions are said to be **coupled** when the energy of the exergonic reaction drives the endergonic reaction. **Redox reactions**, in which electrons are transferred from one compound (the **reductant**) to another (the **oxidant**), are also accompanied by energy exchanges.

Conversion of energy from one form to another is called **transduction**. Light energy occurs in units called **quanta**; the energy of a quantum is inversely proportional to the **wavelength** of the light. Light energy is converted to chemical energy in such biological mechanisms as photosynthesis and vision.

REFERENCES AND SUGGESTED READING

Articles and Reviews

Ingraham, L. L., and Pardee, A. B., Free energy and entropy in metabolism. In *Metabolic Pathways* (D. M. Greenberg, Ed.), Vol. 1, Academic, New York, 1967.

Lewin, R., Unexpected progress in photoreception. *Science* **227**, 500 (1985).

O'Brien, D. F., The chemistry of vision. *Science* **218**, 961 (1982).

Sweeney, B. M., Intracellular source of bioluminescence. *Int. Rev. Cytol.* **68**, 173 (1980).

Books, Monographs, and Symposia

Atkinson, D. E., *Cellular Energy Metabolism and Its Regu-*

lation, Academic, New York, 1977/Elsevier, New York, 1967.

Lehninger, A. L., *Bioenergetics,* W. A. Benjamin, Menlo Park, Calif., 1972.

Lehninger, A. L., *Principles of Biochemistry*, Worth, New York, 1982.

Rawn, J.D., *Biochemistry*, Harper & Row, 1983.

Smith, E. L., Hill, R. L., Lehman, I. R., Lefkowitz, R. J., Handler, P., and White, A., *Principles of Biochemistry* (7th ed.), McGraw–Hill, New York, 1983.

Stryer, L., *Biochemistry* (2nd ed.), Freeman, San Francisco, 1981.

Wall, F. T., *Chemical Thermodynamics*, Freeman, San Francisco, 1965.

CHAPTER 10

CELL METABOLISM

The **metabolism** of cells include (1) all the individual chemical reactions, (2) the sequences of these reactions, (3) the interrelationships that exist among the reaction sequences, and (4) the various mechanisms that regulate the reactions. As we have already seen (Chapter 9), individual reactions may be energy producing (the exergonic reactions) or energy consuming (endergonic reactions). Commonly, the *primary reactants* or *substrates* are converted into final *products* by means of a sequence of reactions, each reaction enzymatically catalyzed and requiring a product of the prior reaction as the substrate.

The overall reaction sequences may be classified as **catabolic** (i.e., degradative) if the ultimate products of the reaction sequence are considered to be subunits or parts of the initial substrate. Alternatively, if the products are a result of the combining of two or more different substrates, the sequence is considered to be **anabolic** (i.e., synthetic). A sequence of reactions is usually referred to as a **metabolic pathway**. Some metabolic pathways are common to all living organisms or cells and several of these are considered in this chapter. Some pathways are especially active, receiving as substrates the products of a variety of other, less active pathways. Alternatively, these *central pathways* may feed substrates into a number of other, less active pathways.

Figure 10–1 diagrammatically shows some of the relationships between the catabolic and anabolic pathways that are followed by the major groups of cellular compounds. Intermediates in the breakdown of carbohydrates can be diverted to lipid synthesis or to the formation of nitrogenous compounds such as nucleotides and amino acids. Lipids in microbial and plant (and to a limited extent animal) cells can be converted into carbohydrates and nitrogen compounds. Likewise, nitrogen compounds, once deaminated, can be converted into lipids or carbohydrates. All these compounds may be further degraded, their catabolism acting as sources of reducing power (e.g., NADH and NADPH) to be used in cellular anabolic reactions.

It is possible to identify specific sites within a cell where particular metabolic pathways are operative. For example, the enzymes necessary for the tricarboxylic acid (Krebs) cycle reactions are located in the mitochondria; the primary reactions of steroid synthesis are associated with the smooth endoplasmic reticulum; fatty acids are oxidized by reaction sequences in mitochondria and are synthesized in the cytosol; the reaction sequences that successively break down sug-

FIGURE 10–1   The major pathways of metabolism. By using these pathways, the cell is able to interconvert essentially all of the major compounds. Dark arrows represent degradative or *catabolic* pathways and red arrows are biosynthetic or *anabolic* pathways.

ars to form pyruvic acid or lactic acid also take place in the cytosol; proteins are synthesized by the cell's ribosomes.

The intermediates, as well as the end products, of a pathway may be drawn off and used in other pathways. For example, during the breakdown of carbohydrates, a large number of intermediate compounds are formed before the ultimate products, CO_2 and H_2O, are formed. Some of these intermediates may be diverted from the catabolic process and used in the formation of fatty acids. Other intermediates may be used in the formation of amino acids. A number of mechanisms are used by cells to regulate the activity of metabolic pathways. These *metabolic regulatory mechanisms* are discussed in the next chapter.

In this chapter, some of the major pathways common to cells are discussed. For convenience the major pathways of carbohydrate metabolism will be reviewed first, followed by those of lipid and nitrogen

metabolism. Although they are described separately, it should be remembered that in the cell many pathways are operative at the same time, and one pathway may influence both the rate and the direction of reactions of another pathway.

The terms used to identify metabolic intermediates vary somewhat in the chemical and biological literature. For example, *pyruvic* acid is sometimes identified as *pyruvate, lactic* acid as *lactate,* and so on. *In situ,* many acids dissociate and the anion (e.g., pyruvate) is often the more common form. The chemical formula may be presented as the acid or the anion.

pyruvic acid

$$\begin{array}{c} CH_3 \\ | \\ C{=}O \\ | \\ COOH \end{array}$$

pyruvate

$$\begin{array}{c} CH_3 \\ | \\ C{=}O \\ | \\ COO^- \end{array}$$

Also variable are the names assigned to phosphorylated compounds; the position of the phosphate groups in the molecules is identified at either the beginning or end of the compound's name. Hence, *3-phosphoglyceraldehyde* and *glyceraldehyde-3-phosphate* refer to the same chemical substance.

REVEALING THE INDIVIDUAL STEPS OF METABOLIC PATHWAYS

Our current understanding of cellular metabolic pathways has been assembled from a large number of individual observations and experiments. Initially the existence of a pathway is identified by observing that the consumption of certain reactants leads to the accumulation of certain products. For example, the consumption of sugar and the production of carbon dioxide and alcohol during **alcoholic fermentation** is a pathway that has been known for centuries; the overall reaction is

$$C_6H_{12}O_6 \longrightarrow 2\,CO_2 + 2\,C_2H_5OH \qquad (10\text{--}1)$$
$$\text{(glucose)} \qquad\qquad\qquad \text{(ethanol)}$$

By quantitative chemical analysis, it is possible to measure the amount of sugar consumed and the amount of CO_2 and ethanol produced. Though such an analysis reveals that all the carbon in the sugar may be accounted for in these two products (i.e., no other carbon-containing products are formed), it does not tell us anything about the number or kinds of individual steps that may characterize the overall process. Also not revealed are **coupled reactions** such as the formation of ATP during fermentation. The individual steps of a pathway may be revealed using *marker* and *tracer* techniques and through the isolation and identification of specific enzymes associated with the pathway.

Marker and Tracer Techniques

A frequently used technique for identifying the steps in a pathway is to follow the metabolism of a substrate molecule in which one of the atomic positions has been "labeled" with a radioactive isotope (Table 14–1) or in rare instances a "heavy" isotope (such as deuterium, ^2H, or heavy nitrogen, ^{15}N). Such a labeled substrate is said to be a **tracer** because specialized techniques are available to identify these labeled compounds. The tracer is then made available to the cell and metabo-

lism allowed to proceed. After a short period of time, metabolism is halted by either rapidly dropping the temperature of the cells or by adding metabolic inhibitors (Chapter 8). The cells are then broken open and potential chemical intermediates of the metabolic pathway are isolated. This procedure usually involves centrifugal fractionation of the cells followed by some combination of extraction and chromatographic procedures; these techniques are discussed at length in Chapters 12 and 13. The intermediates peculiar to the pathway being studied can be distinguished from other chemical substances present on the basis of their isotope content (e.g., by their radioactivity) and can then be chemically identified.

Because in a pathway such as

$$\text{substrate}^* \to B^* \to C^* \to D^* \to E^* \to \text{product}^* \qquad (10\text{--}2)$$

the reaction forming B occurs before the reaction forming C, which in turn occurs before the reaction forming D (and so forth), the amount of time that elapses between the addition of the tracer to the cells and the time at which metabolism is halted can be used to reveal the correct sequence. In this type of tracer experiment, the labeled substrate is made available at a designated time, and the reactions are allowed to proceed for only a short interval so that the labeled substrate can participate in no more than the first one or two reactions of the pathway. As a result, only a few of the intermediates formed will contain the label. Isolation and analysis of the labeled intermediates are then carried out. The experiment is repeated using another sample of cells and labeled substrate but is allowed to proceed for a longer period of time before being halted. As a result an additional (labeled) intermediate may be identified. The experiment is repeated again and again, each time revealing new intermediates of the pathway.

When it is suspected that after one or two reactions a substrate may be split into two or more different products,

$$*\text{substrate} \longrightarrow \overset{*}{\underset{\ddagger}{B}} \diagup\!\!\diagdown \begin{array}{c} \overset{*}{C} \\ \underset{\ddagger}{D} \end{array} \qquad (10\text{--}3)$$

then the atoms at various positions in the substrate may be labeled so that the products entering each branch may be identified and followed. In the above diagram, the labeled positions are represented by the symbols * and ‡. Because some intermediates are not formed in sufficient amounts to permit efficient extraction and identification, nonlabeled intermediates may be added to the fractionated cells. The nonlabeled intermediate mixes with the labeled intermediate and thereby provides sufficient "carrier" for extraction and analysis; this technique is known as **isotopic trapping**.

Another method used to verify the position of an intermediate in a pathway is by *overloading*. For example, suppose that the position of intermediate C in reaction sequence 10–2 is uncertain. To clarify the role of C, two parallel experiments are performed. In one experiment, the substrate is labeled and the amount of label appearing in the product is then followed as described earlier. In the second experiment, a large amount of intermediate C is added to the cells along with the labeled substrate. If C is indeed an intermediate in the pathway, then the amount of label recovered in products formed beyond the step that produces C will be greatly diminished. This is because quantities of both labeled and unlabeled C will be incorporated into the subsequent intermediates.

Enzyme Techniques

Because nearly all metabolic reactions in cells are catalyzed by enzymes, it should be possible to identify an enzyme for each metabolic reaction. In enzyme studies of this type, cells are disrupted and a cell-free *homogenate* prepared. After adding selected substrates and cofactors to the homogenate, the presence of a particular enzyme can be demonstrated by measuring the rate of disappearance of the substrate of that enzyme, or the rate of appearance of a specific product. The homogenate can be fractionated by centrifugation into its components, thereby isolating the mitochondria, plasma membranes, ribosomes, other organelles, or the cytosol. Each of these fractions can be tested for enzymatic activity so that the native location of the reactions in intact cells may be established.

Use of Enzyme Inhibitors and Enzyme-Deficient Cells

Yet another approach to determining metabolic reaction sequences is through the use of specific enzyme inhibitors or mutant cells that fail to produce a specific enzyme. Numerous enzyme inhibitors are known that block specific reactions steps. In the conversion of substrate A to product E, that is,

$$\overset{1 \quad 2 \quad 3 \quad 4}{A \rightarrow B \rightarrow C \rightarrow D \rightarrow E} \qquad (10\text{–}4)$$

an inhibitor that blocks the enzyme catalyzing reaction 2 would also prevent the formation of final product E. However, even in the presence of an inhibitor one should still be able to make the following two observations: (1) intermediate B should continue to be formed from A, because step 1 is not inhibited; in fact, B may even be found to accumulate in excess; and (2) addition of an exogenous source of C or D should be accompanied by the continued production of final product E. The use of an inhibitor for one of the reactions of a metabolic pathway thus provides a means for identifying the intermediate *before* the block. Inhibitors can also be used to test for suspected intermediates that come *after* the block.

In a similar manner, a mutant cell that lacks the genetic information for producing a specific enzyme may serve as a test organism for studying the steps of the metabolic pathway involving that enzyme.

CARBOHYDRATE METABOLISM

The individual steps in the catabolism of carbohydrates are known in great detail. For the most part, cells break down carbohydrates by similar metabolic pathways whether they are plant cells, animal cells, or bacterial cells. The central pathway, glycolysis, found in most cells is outlined in Figure 10–2; however, there are alternative pathways for oxidizing carbohydrates.

Sugars are cleaved from polysaccharides such as starch and glycogen by the action of *phosphorylase* enzymes. The *phosphorolysis* introduces inorganic phosphate and produces glucose-1-phosphate as shown in the upper half of Figure 10–3. Disaccharides, such as sucrose and maltose, are hydrolyzed by *saccharases* without the addition of phosphate, thereby releasing their constituent monosaccharides. These sugars are phosphorylated by enzymatic reaction with ATP. The function of this phosphorylation is twofold: (1) it increases the energy content of the molecule and (2) it introduces a charged moiety into the molecule, making it relatively impermeable to membranes and therefore unlikely to diffuse out of the cell or into an organelle. The phosphorylated sugars produced from either of these sources enter the enzymatic reactions outlined in Figure 10–2, producing the intermediate pyruvate. This central pathway of carbohydrate metabolism is called **glycolysis**.

GLYCOLYSIS

The major features of glycolysis are as follows:

1. The sugars are first doubly phosphorylated. In the case of monosaccharides such as glucose, fructose, mannose, and the like, 2 moles of ATP per mole of monosaccharide are utilized. Sugars that are derived from glycogen and starch require only 1 mole of ATP per mole of glucose-equivalent, because the first inorganic phosphate is acquired during the phosphorolysis of the polysaccharide.

2. The six-carbon sugar diphosphate is split by *aldolase* (10–8), producing two three-carbon units, glyceraldehyde-3-phosphate and dihydroxyacetone phosphate; the latter subsequently forms a second mole of glyceraldehyde-3-phosphate (10–9).

3. A major oxidation and phosphorylation of the substrate is catalyzed by *glyceraldehyde-3-phosphate dehydrogenase*. Two moles of hydrogen are removed per mole of substrate and reduce 2 moles of the coenzyme NAD^+. In the same reaction, inorganic phosphate is incorporated into the acid (10–10).

4. In the final steps of glycolysis, the intermediates are dephosphorylated by reaction with ADP. For each mole of monosaccharide oxidized to pyruvate, 2 moles of ATP are consumed and 4 moles of ATP are produced, resulting in a *net* production of 2 moles of ATP. Note that when the original sugar molecule is derived from glycogen or starch by phosphorolysis, there is a net production of 3 moles of ATP per mole of glucose.

Pyruvate does not accumulate in very large amounts in cells. Instead, it is converted into other products (Fig. 10–4). The enzymes that act on pyruvate vary among different kinds of organisms and with the nature of the environment. The more common fates of pyruvate are (1) its fermentation to ethanol and carbon dioxide in cells such as yeast, (2) its anaerobic conversion into lactate in cells such as muscle cells, and (3) its conversion into acetate in the mitochondria of cells and organisms living under aerobic conditions.

ANAEROBIC RESPIRATION AND FERMENTATION

The reactions of glycolysis have no specific requirement for oxygen. Oxidation reactions do occur, such as the removal of two hydrogens from glyceraldehyde-3-phosphate, and NAD^+ is reduced to NADH, but oxygen per se is not consumed. Under *anaerobic* conditions (i.e., in the absence of oxygen), pyruvate may be reduced to a variety of different compounds. Alcoholic fermentation (reactions 10–19 and 10–20, Fig. 10–4) is a common pathway in microorganisms and is of industrial importance. In these two steps, 1 mole of CO_2 is removed from each mole of pyruvate (i.e., 2 moles of CO_2 per mole of monosaccharide) and NADH is reoxidized to NAD^+, thereby producing ethanol.

The stoichiometric relationships, and the cyclic involvement of NAD are especially important in this pathway. During the oxidation of each mole of glyceraldehyde-3-phosphate (reaction 10–10, Fig. 10–2), 1 mole of NAD^+ is reduced to form NADH; and during the conversion of acetaldehyde to ethanol (reaction 10–20, Fig. 10–4), a mole of NADH is oxidized to form NAD^+. Therefore, the levels of NAD^+ and NADH are unaffected by the conversion of glyceraldehyde-3-phosphate to ethanol. The quantities of these coenzymes in cells are very small. Therefore, if the NAD^+ reduced in reaction 10–10 was not reoxidized, this central pathway would soon be blocked at the glyceraldehyde-3-phosphate step by the lack of NAD^+, and the pathway would necessarily have to cease.

Like NAD^+ and NADH, ATP and ADP are cycled between the ATP-requiring reactions in the early steps of glycolysis and the ATP-producing reactions in the later steps. Cells contain small pools of ATP, ADP, and AMP; when needed, these compounds are drawn from the pools and later returned to the pools. During the early stages of glycolysis, 2 moles of ATP are consumed per mole of glucose, whereas during the later stages 4 moles of ATP are produced. The early stages proceed by drawing ATP from the pool, while the later reactions return ATP to the pool. However, there is a net generation of ATP from glycolysis that is then available for other energy-requiring reactions within the cell.

Another common fate of pyruvate that occurs in the absence of oxygen is its conversion to lactate. This is a normal process in active muscle cells that are not receiving adequate amounts of oxygen and in many plant and bacterial cells that live under anaerobic conditions. In a single reaction, pyruvate is converted to lactate (reaction 10–21, Fig. 10–4). During the reaction, 1 mole of NADH is converted to NAD^+. This stoichiometrically balances NADH production during the earlier glyceraldehyde oxidation step.

FIGURE 10–2 The Glycolytic pathway (i.e., *glycolysis*). Although the individual reactions of glycolysis are reversible, reversal of the entire pathway does not normally occur via this sequence of steps (see *Fig. 10–9*). The $\Delta G^{o\prime}$ values given in this figure are in kJ/mole and apply to each reaction as it proceeds in the direction written. The names of the intermediates appear on the left; the names of the enzymes appear on the right. P = phosphate group.

1,3-Diphosphoglyceric acid

(10-11)

3-Phosphoglyceric acid

(10-12)

2-Phosphoglyceric acid

(10-13)

Phosphoenol pyruvic acid

(10-14)

Pyruvic acid

Phosphoglycerate kinase $\Delta G^{0'} = -18.83$

Phosphoglycerate mutase $\Delta G^{0'} = 4.44$

Enolase $\Delta G^{0'} = 1.84$

Pyruvate kinase $\Delta G^{0'} = -31.38$

FIGURE 10–3 Conversion of polysaccharides and oligosaccharides into phosphorylated monosaccharides (e.g., glucose-6-phosphate) as a preliminary to their catabolism via the glycolytic pathway. P = phosphate group.

FIGURE 10–4 The anaerobic conversion of pyruvic acid (pyruvate) to ethanol and to lactic acid (lactate).

OXIDATION OF PYRUVATE

Except in the case of the *strict* anaerobes (organisms that *cannot* live in the presence of oxygen), when oxygen is present, most cells will oxidize pyruvate rather than reduce it to lactate, ethanol, or other compounds.

Energetically, there is an advantage for the cell to oxidize the pyruvate to CO_2 and water rather than to reduce it to lactate. In the production of lactate, there is an overall change in standard free energy ($\Delta G^{0'}$) of -123.2 kJ (-29.44 kcal) per mole of glucose and a net production of 2 moles of ATP. However, in the breakdown of a mole of glucose through glycolysis and further oxidation to CO_2 and water, the $\Delta G^{0'} = -2870$ kJ (or -680 kcal) per mole of glucose, and a net of 36 (or 38, see later) moles of ATP is produced. Therefore, pyruvate still has a significant amount of stored chemical energy that the aerobic cell can extract.

All the enzymes that bring about the complete oxidation of pyruvate in eukaryotes are found in the mitochondria, either in the matrix or associated with the mitochondrial membranes. The various stages of pyruvate oxidation—(1) formation of acetyl-CoA, (2) the

tricarboxylic acid cycle (Krebs cycle), and (3) electron transport and oxidative phosphorylation—are so closely associated with the structure of the mitochondrion that a detailed description of these reactions is deferred to Chapter 16. The reactions are summarized in Figure 10–5.

OTHER PATHWAYS OF CARBOHYDRATE CATABOLISM

Pentose Phosphate Pathway

An alternative oxidative pathway for the oxidation of sugars is the **pentose phosphate pathway** (also called the **hexose monophosphate shunt** or **phosphogluconate pathway**). This pathway occurs in plants and in most animal tissues. However, its activity in comparison with glycolysis is usually lower and varies considerably from tissue to tissue. The pathway serves as the primary mechanism for (1) converting hexoses into those pentoses necessary for the synthesis of nucleotides and nucleic acids, (2) degrading pentoses so that they may be catabolized in the glycolytic pathway, and (3) generating reduced pyridine nucleotides (e.g., NADPH) to be used for synthetic reactions such as fatty acid synthesis, steroid synthesis, and amino acid synthesis. In animal tissues, the latter functions occur extensively in the liver, in the mammary glands, and in the cortex of the adrenal glands. It has been found that 20% of the sugar metabolized in the mammary glands proceeds via the pentose phosphate pathway. In heart of skeletal muscle, little synthesis of fatty acids (or other related substances) occurs, and therefore it is not surprising that the pentose phosphate pathway is not active in these tissues.

Figure 10–6 illustrates the major reaction steps of the pentose phosphate pathway. Reactions 10–22 and 10–24 describe the two oxidative steps that produce the NADPH required in the synthetic reactions. Reaction 10–24 is also the decarboxylation step that converts the hexose to a pentose. Through the action of an **isomerase** and an **epimerase**, three pentose phosphates may be formed: ribose-5-phosphate, ribulose-5-phosphate, and xylulose-5-phosphate. These compounds may be drawn off into nucleoside biosynthesis; however, if they continue through the pentose phosphate pathway, they are acted on by **transketolases** and **transaldolases** as seen in Figure 10–6. The transketolase transfers a two-carbon fragment from

FIGURE 10–5 The tricarboxylic acid or Krebs cycle.
(Photographs by K. Bendo.)

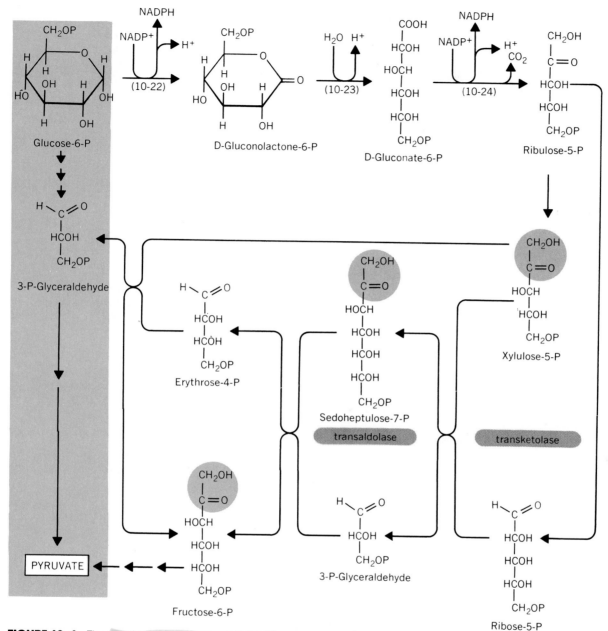

FIGURE 10–6 The *pentose phosphate pathway* (*hexose monophosphate shunt or phosphogluconate pathway*). The two-carbon fragment transferred during the *transketolase* reaction is shaded in red. The three-carbon fragment transferred during the *transaldolase* reaction includes the same two carbons transferred earlier. P = phosphate group.

xylulose-5-phosphate to ribose-5-phosphate, thereby producing sedoheptulose-7-phosphate and glyceraldehyde-3-phosphate. The transaldolase catalyzes the transfer of a three-carbon fragment from sedoheptulose-7-phosphate to glyceraldehyde-3-phosphate, thus forming erythrose-4-phosphate and fructose-6-phosphate. Because of the action of these enzymes, sugar phosphates with three to seven carbons are interconvertible, and intermediates of the pentose phosphate pathway may be channeled into glycolysis and from there to the mitochondria, where the tricarboxylic acid cycle and terminal respiration produce CO_2 and H_2O. Figure 10–7 shows the major reaction steps for the route of hexoses through the pentose phosphate pathway and into the glycolytic pathway.

In plant cells, the enzymes (the reactions) of the pentose phosphate pathway are also found in the stroma of chloroplasts and are used there to regenerate pentoses in photosynthesis (Chapter 17).

The Glyoxylate Pathway

The glyoxylate pathway is essentially a bypass of the CO_2-evolving steps of the tricarboxylic acid cycle, in fact the pathway is often called the "glyoxylate bypass." The enzymes for this bypass are commonly found in plants but are generally absent in the tissues of animals. The enzymes are localized in organelles called glyoxysomes (see Chapter 19). Figure 10–8 summarizes the key reactions of the bypass. Two moles of acetate are required for each turn of the pathway rather than the 1 mole required in the tricarboxylic acid cycle. The first mole of acetate (acetyl-CoA) condenses with oxaloacetate to form citrate, which is then converted to isocitrate. The inducible enzyme isocitric lyase then catalyzes the conversion of isocitrate to succinate and glyoxylate. The glyoxylate reacts with a second mole of acetate (acetyl-CoA) to produce malate, which is converted back to oxaloacetate.

This bypass set of reactions enables plants and microorganisms to transform acetyl-CoA derived from the breakdown of fatty acids into carbohydrate. The excess succinate produced by the glyoxylate pathway can be channeled back up the glycolytic pathway for the formation of sugar and subsequently polysaccharides. To make the synthesis of sugar energetically possible by reversing the steps of glycolysis, certain bypass reactions are employed. These are discussed below.

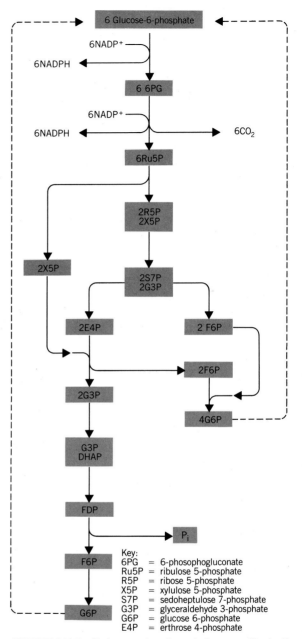

FIGURE 10–7 Balance sheet and diagram illustrating the complete oxidation of one molecule of glucose-6-phosphate to carbon dioxide via the pentose phosphate pathway.

Gluconeogenesis

Glucose can be synthesized by reversing most of the reactions of glycolysis. The initial reactions used to

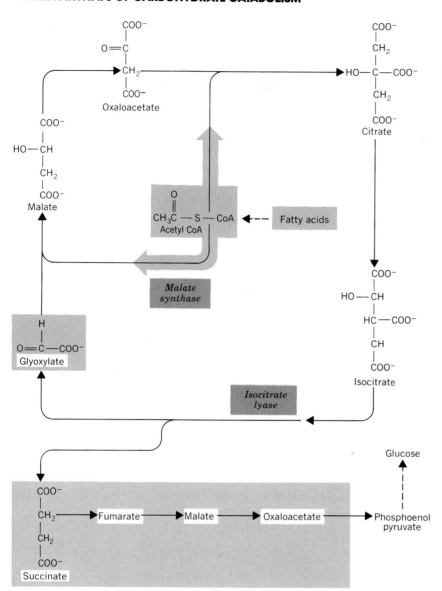

FIGURE 10–8 The *glyoxylate pathway* (*glyoxylate bypass*). Acetyl-CoA, derived from the acetate that is released from the breakdown of compounds such as fatty acids, is consumed in two steps in the cycle.

start various substrates back along the glycolytic pathway differ among different cells and tissues. Also, the carbohydrates that are finally produced vary with conditions and with cell types. Photosynthetic organisms can reduce CO_2 through a reversal of the glycolytic reactions after initially "fixing" the CO_2 using ribulose diphosphate (Chapter 17). Liver cells can regenerate glucose starting with lactate by reversing the glycolytic reactions. Almost all cells can transaminate or deaminate key amino acids, thereby producing tricarboxylic acid cycle intermediates, and by convert-

ing these into phosphoenol pyruvate, glycolysis may then be reversed to produce glucose. Plant and bacterial cells can oxidize fatty acids to acetyl-CoA and, via the glyoxylate pathway, form tricarboxylic acid intermediates; these can then be converted to phosphoenol pyruvate and thereby initiate the reversal of glycolysis. The major initial reactions and features of gluconeogenesis are shown in Figure 10–9.

In the reversal of glycolysis, there are three reactions that occur in place of those found in the catabolic sequence:

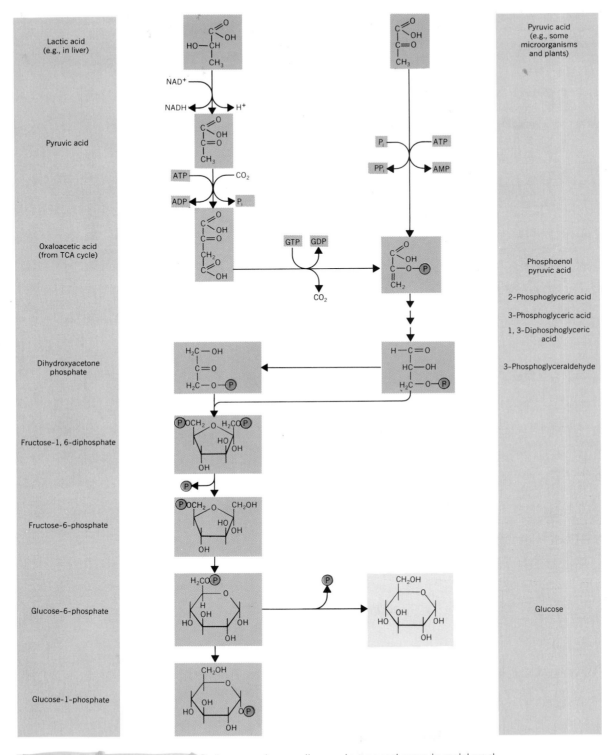

FIGURE 10–9 *Gluconeogenesis.* Cells use various pathways to convert pyruvic acid and intermediates of the tricarboxylic acid cycle into glucose-6-phosphate, glucose, and polysaccharides.

1. Pyruvate is not directly converted to phosphoenol pyruvate (the energetics of that particular reaction are not favorable; $\Delta G^{0\prime} = 31.4$ kJ/mole or 7.5 kcal/mole). Instead, the pyruvate is converted first to tricarboxylic acid cycle intermediates within the mitochondria, which then pass into the cytosol, where *phosphoenolpyruvate carboxykinase*, together with GTP, converts these to phosphoenolpyruvate ($\Delta G^{0\prime} = 4.2$ kJ/mole or 1.0 kcal/mole).

2. The dephosphorylation of fructose-1,6-diphosphate is not coupled to ATP by the catabolic sequence enzyme (*phosphofructokinase*) but instead is acted on by *fructose diphosphatase* to yield fructose-6-phosphate and inorganic phosphate.

3. In some tissues, such as liver and kidney, the glucose-6-phosphate may be converted to glucose and inorganic phosphate by *glucose-6-phosphatase* instead of *hexokinase*.

Synthesis of Glycogen and Starch

Glycogen and starch are highly branched polymers of glucose (see Chapter 5). The syntheses of these two polysaccharides proceed by similar reactions in both animals and plants, although the enzymes that catalyze the syntheses are different as are certain intermediates. In both pathways (Fig. 10–10), the first reaction is the conversion of glucose-6-phosphate to

FIGURE 10–10 Pathways employed in animals and plants for incorporating glucose into polysaccharides. UDPG, uridine diphosphoglucose; ADPG, adenine diphosphoglucose.

glucose-1-phosphate. In the case of glycogen synthesis, the glucose-1-phosphate reacts with UTP to form uridine diphosphoglucose (UDPG), whereas in plants glucose-1-phosphate reacts with ATP to form adenosine diphosphoglucose (ADPG). In both cases, the resulting nucleoside diphosphoglucose serves as a glucose donor to a preexisting polysaccharide. The glucose is added to the end of one of the outer chains of the polysaccharide via an $\alpha1\rightarrow4$ glycosidic linkage. The chain accepting the additional sugar residue is called a "primer." Repeated rounds of sugar addition continuously add to the growing glycogen or starch molecule.

Branch points are created in glycogen and the amylopectin component of starch (see Chapter 5) by separate enzymes called **branching enzymes**. These enzymes remove segments of outer chains containing a number of $\alpha1\rightarrow4$ linked glucose units and then reattach them to the polysaccharide molecule via $\alpha1\rightarrow6$ linkages.

LIPID METABOLISM

Triglycerides

The synthesis of fatty acids and their ultimate incorporation into triglycerides follows a pathway that is significantly different from that which results in the catabolic breakdown of these compounds. In effect, the linear fatty acid molecule is assembled two carbon units at a time by the consumption of acetate in the form of acetyl-CoA (Fig. 10–11). The addition of each two-carbon unit to the chain involves six steps.

1. The **priming reaction**, in which acetyl-CoA is bound first to the *nonenzymatic acyl carrier protein (ACP)* and then transferred to the enzyme *ACP-acyltransferase*.

2. The **malonyl reaction**, in which a second acetyl-CoA converted to malonyl-CoA by incorporation of CO_2 in the form of HCO_3^- is bound to ACP.

3. The **condensation reaction**, in which the malonyl group loses CO_2, becoming an acetyl group, which then attaches to the acetyl group on the enzyme.

4. A *first* **reduction step** involving NADPH.

5. A **dehydration step**.

6. A *second* **reduction step** involving NADPH.

The resulting fatty acid, which is four carbons in length, can now serve as the primer, and the sequence is repeated from step 2 onward. Each cycle adds two carbon units to the growing chain. Desaturation of the fatty acids occurs by additional enzymatic steps. The acetyl-CoA used during the synthesis is generated primarily in the mitochondria by decarboxylation of pyruvate, by oxidative degradation of some amino acids, or by oxidation of other fatty acids. The mitochondrial acetyl-CoA must be converted into other molecules in order to leave the mitochondria and enter the cytosol. In the cytosol, the acetyl-CoA is re-formed and incorporated into the fatty acids. Triglycerides are formed by the condensation of fatty acyl-CoA molecules and either dihydroxyacetone phosphate (produced by the reactions of glycolysis) or glycerol-3-phosphate (formed through the reduction of dihydroxyacetone phosphate by NADH).

Degradation of triglycerides occurs initially in the cytosol by hydrolysis to glycerol and fatty acids; the glycerol then enters the glycoytic pathway. Fatty acids must be activated and transported into the matrix of the mitochondria, where they undergo the degradative and oxidative steps outlined in Figure 10–12. The sequential repetition of steps in the mitochondria (Fig. 10–12) removes two carbon units at a time from the fatty acid until breakdown is complete and only acetyl-CoA or malonyl-CoA remains. This process is known as **β-oxidation**, because it is the bond at the second carbon atom that is cleaved. Other oxidations are known; for example, *α-oxidation* occurs in some germinating seeds producing CO_2, and *ω-oxidation* occurs in the liver of mammals, but the role of this process is not understood.

The synthesis and degradation of most other lipids such as phospholipids and sterols are also known in detail but will not be considered here. Synthesis of cholesterol occurs by the progressive buildup of two-carbon fragments (i.e., acetyl-CoA units) and involves some 25 enzymatic steps, many of which occur in the smooth ER.

NITROGEN METABOLISM

The metabolism of cells requires a variety of nitrogen-containing compounds. Central to the requirement for nitrogen is the formation of amino acids. Amino acids are necessary not only for the synthesis of proteins but also as the primary source of nitrogen in the synthesis of the nucleotide building blocks of nucleic acids. Not all cells or organisms are able to synthesize all of the amino acids necessary to their existence. For

FIGURE 10-11 Pathway for the synthesis of fatty acids (see text for explanation).

example, the cells of human tissues are able to make only 10 of the 20 amino acids required in the synthesis of human proteins. The other 10 are referred to as the *essential amino acids* and are obtained from plant or animal tissues in the diet. Nitrogen for the synthesis of the nonessential amino acids of humans and other higher animals is obtained from ammonium ions or by the transfer of an amino group from another compound (i.e., **transamination**).

An example of transamination is shown in Figure 10–13a. The amino group is transferred from the amino acid glutamic acid to the Krebs cycle intermediate oxaloacetic acid, thereby forming aspartic acid. As seen in Figure 10–13b, glutamic acid itself can be formed from another Krebs cycle intermediate, α-ketoglutaric acid, by incorporation of ammonia.

Most organisms lack the enzyme systems needed to utilize nitrate, nitrite, or atmospheric nitrogen. How-

FIGURE 10–12 Pathway for the breakdown of fatty acids via β-oxidation.

ever, many microbes, including those in symbiotic relationship with leguminous plants, and a few plants can use atmospheric nitrogen. Most higher plants can make the amino acids needed for protein synthesis by using ammonia, nitrate, and nitrite as sources of nitrogen.

Each of the 20 different amino acids required for protein synthesis is synthesized by a specific system of enzymes. Several utilize intermediates of the glycolytic pathway and the tricarboxylic acid cycle. The degradation of amino acids involves pathways that are different from the synthetic pathways but the products of deamination are usually further metabolized by the glycolytic or the tricarboxylic acid cycle reactions.

Purine and pyrimidine ribonucleotides are found in

FIGURE 10–13 Formation of amino acids by (a) transamination of organic acids (Krebs cycle acids) and (b) direct amination of organic acids.

nucleic acids, some vitamins, and certain metabolic intermediates such as ATP and NAD^+. The purine rings of purine ribonucleotides are formed by the stepwise addition of carbons (from CO_2) and nitrogens (from amino acids) to ribose-5-phosphate. In the case of pyrimidine ribonucleotides, the ribose is added *after* the assembly of the pyrimidine ring. Degradation of purines and pyrimidines occurs by a pathway in which urea and acetic acid are intermediates.

FUNCTIONS OF METABOLIC PATHWAYS

The sequences of reactions that form the metabolic pathways serve a number of functions. First, and most obvious, is the synthesis of an end product needed by the cell or organism. This product may be used as part of the structure of the cell or, in cases where it is secreted from the cell, it may be incorporated into extracellular elements. The proteoglycans in the extracellular matrix of cartilage and the pectins of the middle lamellae of plant cell walls are examples of such secretions. The end product may also function in the cell as a regulatory agent for other reactions. Hormones are such end products. End products may also be stored as reserves, such as starch, glycogen, and certain lipids.

Second, a metabolic pathway may function to provide energy-rich compounds such as ATP, GTP, and UTP and these may be used in other energy-requiring reactions. Metabolic pathways rarely yield ATP, GTP, or UTP as end products of reaction sequences, instead they are produced at one or more of the intermediate steps of a pathway. For example, in glycolysis, ATP is produced at two intermediate steps (Fig. 10–2; reactions 10–11 and 10–14). Other compounds such as NADH and NADPH are also products of intermediate reactions and may be utilized directly in other reactions of the same or a different pathway or may be oxidized to provide energy in the form of ATP.

Third, intermediates of a metabolic pathway may be drawn upon and used as substrates for other metabolic sequences. For example, the tricarboxylic acid cycle (Krebs cycle) is a sequence of reactions whose net effects are the oxidation of acetyl-CoA and the production of CO_2, NADH, and GTP. However, many of the intermediates of the Krebs cycle are drawn off and utilized for other purposes. Acetyl-CoA itself is used in lipid synthesis, and α-ketoglutarate and oxaloacetate are used in amino acid synthesis. Metabolic reactions that serve in both a catabolic (energy-producing) and an anabolic (biosynthetic) function are called **amphibolic** pathways. Amphibolic pathways, which may have their intermediates diverted to other reaction sequences, frequently include specialized reactions that replenish these intermediates. The specialized enzymatic reactions are called **anaplerotic** reactions. The "Wood–Werkman" reaction in bacteria in which carbon dioxide combines with pyruvate to form oxaloacetate is an example of such an anaplerotic mechanism. The glyoxylate pathway described earlier is also an anaplerotic mechanism, effectively forming the tricarboxylic acid cycle intermediates succinate and malate.

CALCULATIONS OF ENERGY CHANGE

For most of the major metabolic pathways of the cell the specific enzymatic reactions in which ATP, GTP, or other related compounds are formed or consumed are known. By inspecting metabolic charts, such as those of Figures 10–2, 10–3, and 10–4, one can calculate the number of moles of ATP consumed and/or produced and the *net* change for any reactant-to-product sequence. For example, when 1 mole of sucrose is converted to pyruvate, 4 moles of ATP are consumed

$$2 \text{ moles in reaction } 10\text{–}7$$
$$\underline{2 \text{ moles in reaction } 10\text{–}17}$$
$$\text{Total} = 4 \text{ moles (consumed)}$$

and 8 moles of ATP are produced

$$4 \text{ moles in reaction } 10\text{–}11$$
$$\underline{4 \text{ moles in reaction } 10\text{–}14}$$
$$\text{Total} = 8 \text{ moles (produced)}$$

Therefore, in the glycolytic oxidation of 1 mole of sucrose, there is a net production of 4 moles of ATP. These "paper calculations" reveal the theoretical amounts of ATP expected from the glycolytic sequence of reactions. In laboratory experiments, these numbers are approached but are not always obtained. This may be due to the fact that laboratory conditions are not always optimal, intermediates may be drawn into reactions that are not part of the pathway being evaluated (this is especially true in experiments using intact cells), cofactors may not be present in the proper concentration, and so on.

The *efficiency* of a metabolic pathway is usually determined by an analysis of the change in free energy. For example, in the conversion of glucose to lactate (Figs. 10–2, 10–3, and 10–4), the enzymatic sequence may be considered to be composed of the exergonic reaction

$$\underset{\text{(glucose)}}{C_6H_{12}O_6} \longrightarrow \underset{\text{(lactate)}}{2\,C_3H_6O_3}$$

and the endergonic reaction

$$2\,\text{ADP} + 2\,P_i \longrightarrow 2\,\text{ATP} + H_2O$$

The standard free energy change ($\Delta G^{0\prime}$) for the catabolic reactions is determined by summing the values for each step.

Note The $\Delta G^{0\prime}$ values given for a reaction as shown in a metabolic chart are for the reaction *as written*, including the *direction* written; for example,

$$\text{glycerate-3-P} \longrightarrow \text{glycerate-2-P}$$
$$(\Delta G^{0\prime} = 4.44 \text{ kJ or } 1.06 \text{ kcal})$$

and

$$\text{glycerate-2-P} \longrightarrow \text{glycerate-3-P}$$
$$(\Delta G^{0\prime} = -4.44 \text{ kJ or } 1.06 \text{ kcal})$$

If both an exergonic reaction and an endergonic reaction are written together, for example,

$$\text{phosphoenol pyruvate} \underset{\text{ADP}}{\overset{\text{ATP}}{\longrightarrow}} \text{pyruvate}$$
$$(\Delta G^{0\prime} = -23.9 \text{ kJ or } -5.7 \text{ kcal})$$

then the $\Delta G^{0\prime}$ given is the sum of the two separate reactions, that is,

$$\text{phosphoenol pyruvate} \longrightarrow \text{pyruvate} + P_i$$
$$(\Delta G^{0\prime} = -54.4 \text{ kJ or } 13.0 \text{ kcal})$$

and

$$\text{ADP} + P_i \longrightarrow \text{ATP}$$
$$(\Delta G^{0\prime} = 30.5 \text{ kJ or } 7.3 \text{ kcal})$$

In the case of the catabolic reactions of Figures 10–2, 10–3, and 10–4, which form lactate, the summation would be:

Reaction Number	$\Delta G^{0\prime}$ (kJ/mole)*
10–17	+ 13.80
10–6	+ 1.67
10–7	+ 16.32
10–8	+ 23.97
10–9	+ 7.66
10–10	+ 12.56 (i.e., +6.28×2)
10–11	− 98.74 (i.e., −49.37×2)
10–12	+ 8.88 (i.e., +4.44×2)
10–13	+ 3.68 (i.e., +1.84×2)
10–14	− 123.84 (i.e., −61.92×2)
10–21	− 50.20 (i.e., −25.10×2)
Total	− 184.23

*Excluding the $\Delta G^{0\prime}$ attributed to ATP→ADP + P_i or ADP + P_i→ATP

A summation of the net ADP/ATP reactions would be:

Reaction Number	$\Delta G^{0\prime}$ (kJ/mole)
10–17	–30.54
10–7	–30.54
10–11	+61.08 (i.e., 30.54×2)
10–14	+61.08 (i.e., 30.54×2)
Total	+61.08

The reactions of glycolysis and the fermentation of glucose to lactate thus produce much more free energy than they consume. The efficiency of these pathways would therefore be 61.08/184.23 = 0.332 or 33.2%.

Attempts to measure the energetics of glycolysis and fermentation in intact cells (i.e., *in vivo*) have produced striking results. In red blood cells (which are ideal in such studies because this cell derives most of its energy from glycolysis and from metabolism to lactate), the efficiency is about 53%; this is much higher than that predicted above. To determine this *in vivo* efficiency, the steady-state concentrations of all the glycolytic intermediates are measured, and from these values the actual equilibrium constants are calculated and ΔG values determined. The ΔG values (rather than the $\Delta G^{0\prime}$ values) reveal the greater efficiency of red blood cells. Skeletal muscle cells also reveal an efficiency level higher than that anticipated from calculations and summations of the type carried out above. Therefore, though $\Delta G^{0\prime}$ values provide figures for easy comparison under defined conditions (i.e., 1 molar concentrations of reactants and products, pH 7.0, and 25°C), differences in substrate concentration, pH, temperature, and other factors may produce variations in efficiency under natural conditions, that is, *in vivo*.

SUMMARY

Cells break down (i.e., **catabolize**) compounds by sequences of enzyme-catalyzed reactions to obtain usable chemical energy or to form smaller molecules that can be used to build (i.e., **anabolize**) other, larger molecules. The reaction sequences, called **metabolic pathways**, are frequently associated with specific organelles, the enzymes that catalyze the reactions being compartmentalized between or within the membranes of the organelle.

In general, carbohydrates, lipids, and proteins can be catabolized to yield energy and a pool of small molecules, and these may be used either in the synthesis of new macromolecules or excreted from the cell. The products of catabolism in one organelle may be transported to other organelles for further catabolism or anabolism. Reaction sequences and their intracellular location have been revealed using **tracer techniques**, by studies of enzyme activity and inhibition, and through the use of mutant cells.

The most common metabolic pathways in cells are:

1. **Glycolysis**—the catabolism of monosaccharides to pyruvate.

2. **Fermentation**—the catabolism of monosaccharides to products such as ethanol and CO_2 in the absence of air (anaerobic conditions).

3. **Tricarboxylic acid (Krebs) cycle**—the oxidation of pyruvate and its catabolism to CO_2 and water in the presence of oxygen.

4. **Pentose phosphate pathway** (phosphogluconate pathway or hexose monophosphate shunt)—a pathway for formation and/or catabolism of pentoses.

5. **Glyoxylate pathway**—the conversion of acetyl-CoA to carbohydrate.

6. **Gluconeogenesis**—the synthesis of glucose from simple organic acids.

7. **Glycogen** and **starch synthesis**—the incorporation of sugars into polysaccharides.

8. **Fatty acid synthesis**—the pathway leading to the formation of fatty acids and triglycerides from acetyl-CoA and glycerol.

9. **β-oxidation**—the pathway for breakdown of fatty acids.

10. **Amino acid synthesis**—the mechanism for formation of amino acids to be incorporated into proteins.

REFERENCES AND SUGGESTED READING

Articles and Reviews

Fothergill-Gilmore, L. A., The evolution of the glycolytic pathway. *Trends in Biochem. Sci.* **11** (1), 47 (Jan. 1986).

Hinkle, P. C., and McCarty, R. E., How cells make ATP. *Sci. Am.* **238** (3), 104 (Mar. 1978).

Krebs, H. A., The history of the tricarboxylic acid cycle. *Perspect. Biol. Med.* **14**, 154 (1970).

Villar-Palasi, C., and Larner, J., Glycogen metabolism and glycolytic enzymes. *Ann. Rev. Biochem.* **39**, 639 (1970).

Books, Monographs, and Symposia

Atkinson, D. E., *Cellular Energy Metabolism and Its Regulation,* Academic, New York, 1977.

Axelrod, B., Other pathways of carbohydrate metabolism. In *Metabolic Pathways* (D. M. Greenberg, Ed.), Vol. 1, Academic, New York, 1967, p. 272.

Dagley, S., and Nicholson, D. E., *Metabolic Pathways,* Wiley, New York, 1970.

Larner, J., *Intermediary Metabolism and Its Regulation,* Prentice–Hall, Englewood Cliffs, N.J., 1971.

Lehninger, A. L., *Principles of Biochemistry,* Worth, New York, 1982.

Rawn, J. D., *Biochemistry,* Harper & Row, New York, 1983.

Stanbury, J. O., Wyngarden, J. B., and Fredickson, D. S., *The Metabolic Basis of Inherited Disease,* McGraw–Hill, New York, 1972.

Stryer, L., *Biochemistry* (2nd ed.), Freeman, San Francisco, 1980.

CHAPTER 11

METABOLIC REGULATION

The diverse metabolic reactions and reaction sequences in cells have been briefly described and outlined in the preceding three chapters. From these descriptions, it is clear that a substrate can be enzymatically converted into a great variety of intermediates and products. Although there are a number of essentially unidirectional reactions, the metabolic network of reversible and cyclical reaction sequences provides for the possible conversion of almost any metabolite into any other metabolite. In a superficial comparison, one could visualize the metabolic pathways as a branching and connecting network of water pipes in which the water can be caused to flow between any two points in the network under suitable conditions. For example, simply an *excess* of water (metabolite) in one part of the system could cause flow to the other parts of the system. *In vivo* and *in vitro* experiments have indicated that the "flow" of metabolites through metabolic pathways is not as free and uncontrolled as this pipeline-network analogy suggests. Metabolic conversions are controlled or regulated by a variety of mechanisms that channel metabolites into needed compounds or into stable reserve products and prevent energetically wasteful conversions.

Cells have evolved a diverse set of regulatory mechanisms. Individual reactions may be controlled by one or more processes from simple mass action to complex hormonally controlled enzyme systems. The more common of these processes are described in this chapter.

MASS ACTION

For any reversible reaction, such as

$$A + B \rightleftharpoons C + D \tag{11-1}$$

in which A and B are reactants and C and D are products, the ratio of the concentrations of products $[C]$ and $[D]$ and reactants $[A]$ and $[B]$ at equilibrium is

$$\frac{[C][D]}{[A][B]} = K'_{eq}$$

At equilibrium concentrations, the reaction rate in one direction equals the reaction rate in the opposite direction, resulting in no net change in the concentration of reactants or products. The value of the equilibrium constant depends on temperature and pressure.

235

The **standard equilibrium constant**, K'_{eq}, is the equilibrium constant at 25°C, 1 atmosphere pressure, and pH 7.0. If the concentration of any one of the components of the reaction is altered, then the concentration of at least one other component must also change to satisfy the constancy of the K'_{eq} value. For reaction 11–1, if the K'_{eq} value is greater than 1.0, then the equilibrium of the reaction lies to the right as written; that is, when the rate of reaction forming products C and D is equal to the rate of reaction converting C and D to A and B, there is more total product than total reactant. Similarly, if the K'_{eq} value is less than 1.0, the equilibrium lies to the left as written, and there is more total reactant than product when the reactions are at equilibrium.

In a sequence of reactions, each reaction may affect the other reactions. For the reaction sequence of Figure 11–1, the equilibrium of the intermediate reaction, in which glucose-6-phosphate is converted to fructose-6-phosphate, lies to the left, because it has a K'_{eq} value that is less than 1.0. However, for the other two reactions, the equilibrium lies well to the right because their K'_{eq} values are much higher than 1.0. If the entire sequence of reactions is allowed to come to equilibrium, then the concentrations of all of the molecules present will satisfy each of the three K'_{eq} values. If additional glucose and ATP are added, the entire sequence of reactions will proceed to the right even though the K'_{eq} value of middle reactions is less than 1.0. This occurs because the production of glucose-6-phosphate by the first reaction temporarily disturbs the equilibrium that existed. The middle reaction thus proceeds to the right to a greater extent than it proceeds to the left, resulting in a net increase in fructose-6-phosphate. Some of this fructose-6-phosphate is then converted to fructose-1,6-diphosphate by the third reaction. The rate at which the net change in concentrations of reactants and products occurs and

FIGURE 11–1 *Example of a linear metabolic pathway (see text for discussion). Note that the K'_{eq} value for the summary reaction is the product of the individual K'_{eq} values of each reaction.*

FIGURE 11–2 *Example of branching in a metabolic pathway (see text for discussion).*

the direction in which the reaction proceeds is the result of *mass action*.

Which of two alternative metabolic paths is taken at a *branch point* (Fig. 11–2) is controlled to some degree by the law of mass action. Although the equilibrium lies to the right for each of the reactions of Figure 11–2, the equilibrium is "further" to the right in the reaction that produces fructose-1,6-diphosphate, and more product would be formed by reactions along that branch than along the branch that forms gluconolactone-6-phosphate. However, during glucose metabolism as it takes place *in situ*, both fructose-1,6-diphosphate and gluconolactone-6-phosphate are intermediates in a more extensive sequence of reactions than is shown in Figure 11–2 (see Figs. 10–2 and 10–3), and the K'_{eq} values of subsequent reactions may affect the direction of the reactions differently *in situ*.

REGULATION OF METABOLISM

The various interconnecting metabolic pathways and the rates of the reactions in these pathways are controlled by four groups of regulatory mechanisms that affect enzymes. The first (and most immediate) regulatory action is brought about by variations in the concentration of substrates, the influence of factors such as pH and salt concentration, and through direct modification of the enzyme's structure.

A second group of control mechanisms involves key enzymes of a metabolic pathway that are directly affected by an intermediate or end product of the pathway. By attaching to enzyme sites other than the active site, these *regulatory metabolites* or *allosteric effectors* alter the catalytic activity of the enzyme. A third group of regulatory mechanisms acts at the levels of enzyme synthesis. The synthesis of individual enzymes (or even groups of enzymes) by a cell may be initiated or halted by the sudden availability of certain metabolites. Transcription of a cell's DNA is controlled through **operons**—sets of mechanistically interrelated genes along a chromosome. The fourth and most complex group of control mechanisms occurs in complex, multicellular organisms such as ourselves, in which *hormones* and hormone-like compounds produced in one tissue are carried to other tissues where they trigger metabolic changes. Each of these regulatory mechanisms will now be considered in greater detail.

GROUP I CONTROL MECHANISMS— IMMEDIATE REGULATION

Substrate Concentration Effects

As described in Chapter 8, enzyme activity as measured by the rate of product formation increases as a hyperbolic function as the substrate concentration is raised (Fig. 11–3) until a maximum reaction velocity (V_{max}) is achieved. Excessively high substrate concentrations may actually reduce enzyme activity. Each enzyme, subject to experimental conditions, has a characteristic V_{max} as exemplified in Table 11–1 for the glycolytic enzymes in brain tissue. If the substrates present in brain tissue were to occur in excess, one would expect that the reaction catalyzed by *aldolase* (which has the lowest V_{max}) would be the rate-limiting reaction in the sequence. However, *in vivo*

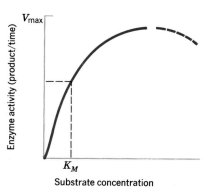

FIGURE 11–3 Effect of substrate concentration on enzyme activity.

TABLE 11–1 MAXIMUM ACTIVITY OF GLYCOLYTIC ENZYMES IN MOUSE BRAIN TISSUE UNDER ANOXIC CONDITIONS

Enzyme (in Order of Glycolysis)	V_{max} (mmoles/kg/min)
Hexokinase	15.2
Phosphoglucoisomerase	154.0
Phosphofructokinase	26.7
Aldolase	7.6
Glyceraldehyde-3-phosphate dehydrogenase	96.0
Phosphoglycerate kinase	750.0
Phosphoglycerate mutase	145.0
Enolase	36.0
Pyruvate kinase	95.0
Lactic dehydrogenase	129.0

Source: Copyright © American Society of Biological Chemists, Inc., *J. Biol. Chem.* **239**, 31 (1964).

studies indicate that enzymes are rarely saturated by substrate. Under typical conditions, *hexokinase, phosphoglucoisomerase,* and *aldolase* operate in the presence of substrate concentrations equal to or somewhat greater than the K_M value (Michaelis–Menten constant). Small changes in substrate concentration do not significantly alter the rate of metabolism in this part of the glycolytic pathway. Of greater regulatory importance for these reactions is the amount of enzyme present. The last six glycolytic enzymes in brain tissue operate at substrate levels that are significantly below the K_M value. Therefore, small changes in substrate concentration would be more likely to alter the rates of their respective reactions. Under anoxic conditions, the rate of production of lactate is not appreciably increased by supplementing the tissue with glucose. However, supplements of compounds such as glycerol, which enter the glycolytic pathway below the level of *adolase,* cause an increase in lactate production.

The *affinity* between enzymes and their substrates affects the rates of enzyme-catalyzed reactions. For example, at branch points of metabolic pathways, two enzymes compete for the same substrate. The reaction catalyzed by the enzyme with the lower K_M value is favored at low substrate concentrations whereas the reaction catalyzed by the enzyme with the higher K_M value is favored at high substrate concentrations (Fig. 11–4). Thus, when present in low concentrations, a substrate may be channeled primarily into one pathway, whereas the major direction of metabolism may shift to other pathways at higher substrate concentrations.

Effects of Salts and pH

As discussed in Chapters 4 and 8, a variety of environmental factors influence the structure of proteins and as such influence the catalytic activity of enzymes. Small changes in pH or an increase (or decrease) in the salt content of the enzyme's environment may quickly be followed by a dramatic change in the level of activity of the enzyme. A change in activity of one enzyme of a pathway will change the activity of the metabolic pathway as a whole.

Covalent Bond Modification

A number of enzymes are produced in an **inactive** or **zymogen** form and must be modified through cleavage of certain covalent bonds in order to become active. The modification may be irreversible, as is the case with hydrolytic modification of zymogens such as *pepsinogen* (to form *pepsin*) and *trypsinogen* (to form *trypsin*), which were discussed in Chapter 8. Other enzymes may be reversibly covalently activated and deactivated. The reversible activation and deactivation of *glutamine synthetase* is a well-studied example (Fig. 11–5). This enzyme catalyzes the conversion of glutamate to glutamine,

$$\text{glutamate} + NH_3 \longrightarrow \text{glutamine}$$
$$\text{ATP} \qquad \text{ADP} + \text{phosphate}$$

and the transfer of a glutamyl group to hydroxylamine,

$$\text{glutamine} + NH_2OH \rightarrow \gamma\text{-glutamyl-NHOH} + NH_3$$

Glutamine synthetase is inactivated by a reversible adenylation in which 12 molecules of AMP become covalently bonded to the hydroxyl groups of 12 tyrosine residues of the enzyme. The inactivation is catalyzed by *adenylating enzyme,* which uses ATP as the source of adenylate and releases pyrophosphate. Reactivation is brought about through the action of *deadenylating enzyme,* which, in the presence of inorganic phosphate, cleaves the bonds linking the AMP molecules to the inactive form of the enzymes, thereby producing ADP (Fig. 11–5).

FIGURE 11–4   Relationship between substrate concentration and enzyme activity for two enzymes that act on the same substrate but have different K'$_{eq}$ values.

FIGURE 11–5 Activation and deactivation of the enzyme *glutamine synthetase* by adenylation and deadenylation. The less active form of the enzyme has 12 molecules of AMP covalently bonded to 12 tyrosine residues; their removal by the deadenylating enzyme activates the enzyme.

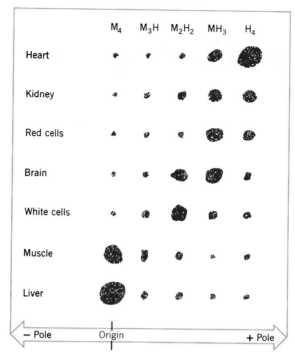

FIGURE 11–6 *Lactic dehydrogenase* isozymes. The spots show the relative amounts of each isozyme as they appear when extracts of different tissues are separated by two-dimensional electrophoresis (see Chapter 13).

Isozymes

In a number of instances, an enzyme that catalyzes a specific reaction may exist in multiple forms in a cell or organism. These multiple forms or **isozymes** have different chemical structures, and can therefore be separated from one another by electrophoresis and chromatographic procedures. Although they catalyze the same reactions, the isozymes usually have different K_M and V_{max} values and catalyze the reaction more or less effectively depending on the reaction conditions. One of the most exhaustively studied of the known isozymes is *lactic dehydrogenase*, which catalyzes one of the terminal reactions in glycolysis, that is,

$$\text{pyruvate} + \text{NADH} + \text{H}^+ \rightleftharpoons \text{lactate} + \text{NAD}^+$$

In rats and in a number of other vertebrates, this enzyme is present in five forms. Each of the five forms has a molecular weight of about 134,000, consists of four polypeptide chains of about 33,500 each, and catalyzes the same reaction. Each of the four polypeptides may be of two types, usually referred to as M and H. In rat skeletal muscle tissue, the predominant form of the isozyme contains four polypeptides of the M type; in contrast, in rat heart muscle, the predominant form contains four polypeptides of the H type. The other isozyme forms are made up of three H and one M, two H and two M, and one H and three M polypeptides (usually abbreviated H_3M, H_2M_2, and HM_3). Some of the latter forms predominate in other tissues, although each tissue has some of each isozyme (Fig. 11–6).

The M_4 isozyme is prevalent in embryonic tissue and in skeletal muscle tissue. It has a low K_M and a high V_{max}. It is well adapted to these tissues, which are frequently deprived of oxygen and must depend on the energy made available during the breakdown of glucose to lactate. Similarly, the H_4 isozyme is well suited to the metabolism of heart muscle. This isozyme has a high K_M and a low V_{max} and is inhibited by excess pyruvate. Heart muscle is primarily aerobic, converting pyruvate to CO_2 and H_2O rather than to lactate. The *lactic dehydrogenase* enzyme is most active during emergency conditions when the oxygen supply is low.

GROUP II CONTROL MECHANISMS—ALLOSTERISM

The regulatory effects exercised by substrates through mass action generally influence *all* reactions in metabolic pathways; therefore, these control mechanisms are not very specific. There are, however, a number of mechanisms by which specific reactions in a pathway can be regulated. In one such mechanism,

TABLE 11–2   COMMON ALLOSTERIC EFFECTORS

Glucose-6-phosphate
Fructose-1,6-diphosphate
1,3-Diphosphoglycerate
Citrate
Acetyl-CoA
AMP, ADP, ATP
Cyclic AMP, cyclic GMP
GTP, UTP, etc.
NAD, $NADH_2$
Fatty acids
Amino acids

enzymes catalyzing specific reactions of the pathway are influenced by the type and amount of certain **regulatory metabolites** that are present. These enzymes are called **allosteric enzymes** because their catalytic activity is modified when specific metabolites, called **allosteric effectors**, are bound to a site on the enzyme *other than the active site* (see Chapter 8). If the metabolite has an inhibitory effect, it is called a *negative* effector; if its effects are stimulatory, then it is called a *positive* effector. Some of the more common allosteric effectors are listed in Table 11–2.

Feedback Inhibition   Inhibitory allosteric effects are caused when the product of a reaction sequence binds to an allosteric site of an enzyme that catalyzes a reaction at the beginning of a linear pathway or at the branch point of a branched pathway (Figs. 11–7a and 11–7b). Binding of the effector to the enzyme causes a change in the configuration of the enzyme (called an *allosteric transition*) and this reduces the catalytic activity of the enzyme. Thus, as the end product of a reaction sequence begins to accumulate,

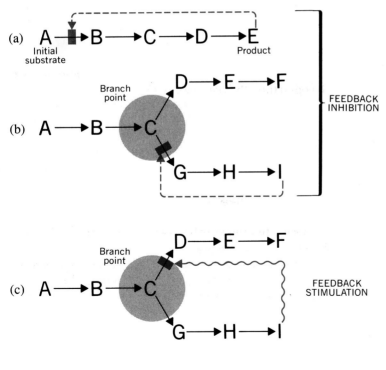

FIGURE 11–7 Feedback inhibition, feedback stimulation, and feedforward stimulation in linear and branched metabolic pathways. The small colored rectangles represent the allosteric enzymes, broken colored arrows indicate inhibition, and wavy colored arrows indicate stimulation.

more and more of the enzyme catalyzing the early stage of the reaction pathway leading to that product is rendered inactive. In this way, the rate of formation of the end product is dramatically reduced. Where branches occur in a metabolic pathway, the end product of one branch usually affects the enzyme at the branch point leading to that product (Fig. 11–7b). This type of inhibition is called **feedback inhibition** or **end product inhibition**.

Feedback Stimulation   Positive effectors generally act in one of two ways—by **feedback stimulation** or **feedforward stimulation**. In feedback stimulation, the effector may be the end product of one branch of a pathway that combines with and stimulates the enzyme catalyzing the reaction at the branch point leading into the *other* pathway (Fig. 11–7c). In other words, the reactions leading to the synthesis of the positive effector are attenuated, while the reactions that lead to the synthesis of an alternative product are enhanced.

Feedforward Stimulation   In feedforward stimulation, the initial substrate of a pathway (or an early intermediate in the pathway) stimulates the activity of an enzyme further along the metabolic pathway (Fig. 11–7d).

When an allosteric enzyme is affected by one effector it is said to be *monovalent*. However, some allosteric enzymes may bind two or more effectors and are termed *polyvalent*. A single allosteric enzyme may possess sites that combine with either negative or positive effectors, and these effectors may compete with each other for the same allosteric site or be bound at different sites on the enzyme. The kinetics of allosteric enzyme activity were discussed in Chapter 8.

METABOLIC PROCESSES CONTROLLED BY ALLOSTERIC ENZYMES

Alternate Pathways for Glycogen Synthesis and Degradation

An excellent example of allosteric enzyme regulation of metabolic processes is provided by the interrelationship in animals between the metabolic pathways that result in (1) the *synthesis* of glycogen from glucose and (2) the *oxidation* of glucose to CO_2 and water. Nearly all of the energy-consuming processes in the body proceed at the expense of ATP and much of this ATP is derived through the oxidation of glucose. During periods of elevated activity (e.g., exercise), glycogen is broken down to yield glucose, which then enters the metabolic pathway converting it to CO_2 and water, with consequent generation of ATP. In contrast, during periods of rest or low energy demand, absorbed glucose is converted to glycogen. Three of the enzymes involved in glucose metabolism are allosteric; these are *phosphofructokinase* (an enzyme required in the series of reactions that convert *glucose-6-phosphate* to CO_2 and water), *glycogen synthetase* (involved in the incorporation of glucose-1-phosphate into glycogen), and *glycogen phosphorylase* (which removes glucose as glucose-1-phosphate from glycogen during glycogen catabolism).

When ATP levels are high and no major consumption of energy is taking place in the body, glucose is diverted into glycogen (i.e., "glycogenesis" predominates). This is achieved because ATP acts as a negative effector of *phosphofructokinase* and *glycogen phosphorylase* and as a positive effector, along with glucose-6-phosphate, of *glycogen synthetase* (Fig. 11–8a).

When the ATP level falls (e.g., during exercise) and there is an increased demand for ATP, glycogen synthesis is halted as absorbed glucose is directly consumed in the production of ATP and additional glucose is made available through the catabolism of glycogen (i.e., "glycogenolysis"). This pathway is activated by the positive effects on *phosphofructokinase* and *glycogen phosphorylase* of the ATP precursor, AMP.

The hormone *epinephrine*, secreted into the bloodstream during periods of great activity, also has an effect on these metabolic pathways in muscle and in liver. When epinephrine in the bloodstream reaches the muscles, it binds to the surface of the muscle cells and promotes the synthesis of cyclic AMP (cAMP) by the enzyme *adenylcyclase*. The cAMP then allosterically activates a second enzyme (*protein kinase*), which ultimately activates *glycogen phosphorylase* but inactivates *glycogen synthetase* (Fig. 11–8b). This phenomenon is also considered later in the chapter in connection with the functions of hormones and the role of protein phosphorylation as a metabolic regulatory mechanism.

The pathways described above illustrate the mechanisms for turning allosteric enzymes on and off. In the absence of such mechanisms, both pathways would

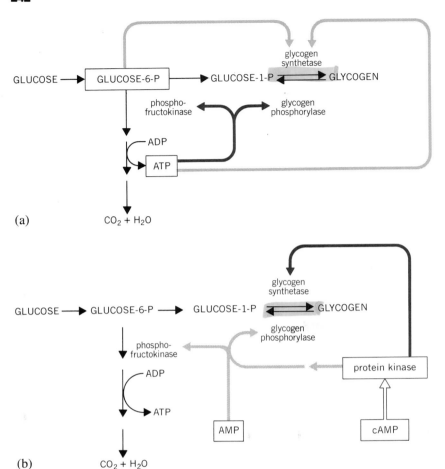

simultaneously be active so that their effects cancel one another—a most unproductive state! Allosterism thus provides a basis for regulating the levels of activity of related metabolic pathways.

The Regulation of Amino Acid Synthesis

Escherichia coli provides a clear example of control of divergent metabolic pathways by feedback inhibition. An outline of the metabolic pathways for the synthesis of three amino acids is shown in Figure 11–9. Lysine, methionine, and threonine are each synthesized from aspartate, and each may be utilized in protein synthesis. Without metabolic controls, the consumption or utilization of any one of these amino acids would stimulate the pathways and cause unneeded synthesis of the unused amino acids as well as the one utilized. Such an unregulated system would consume vital resources and energy; both factors could have survival

implications to the organism and evolutionary consequences to the species. However, in *E. coli*, the allosteric regulatory mechanisms are most effective. The accumulation of each amino acid produces a feedback inhibition of the first enzyme in the specific branch of the pathway leading to the synthesis of that amino acid. In Figure 11–9, this negative effect is shown by dashed lines. Moreover, an additional level of regulation is achieved through effects on the enzyme *aspartokinase*, which catalyzes and phosphorylation of aspartate. This enzyme exists in three forms (i.e., there are three isozymes), symbolized in Figure 11–9 by using three separate arrows to show the conversion of aspartate to aspartylphosphate. One of the isozymes is specifically and completely inhibited by threonine; the second (which is present only in small amounts) is specifically inhibited by homoserine; and the third isozyme is specifically inhibited by lysine. In addition, synthesis of the latter isozyme is *repressed*

FIGURE 11–9   Allosteric feedback control of amino acid synthesis in *Escherichia coli* cells. Small colored rectangles are allosteric enzymes and dashed arrows indicate feedback inhibition.

by lysine. (*Repression* is a regulatory mechanism that reduces the *number* of enzyme molecules in the cell and is discussed later in the chapter.)

GROUP III CONTROL MECHANISMS— REGULATION OF ENZYME SYNTHESIS IN PROKARYOTES

Another method used by cells to regulate metabolism is by altering the types and numbers of enzymes that are manufactured through the integrated actions of the cell's genetic and protein synthesizing apparatuses. This mechanism is also referred to as *controlled gene expression*. Although the topics of gene expression and protein synthesis are treated in detail in Chapters 20–22, a brief description here of the basic mechanism will set the foundation for understanding the metabolic control mechanisms that are described below.

The primary structures of all cellular enzymes are encoded in the cell's DNA or genes. To produce these enzymes, the DNA is first *transcribed* into messenger RNA (mRNA) and the mRNA is subsequently *translated* by the cell's ribosomes into polypeptides that

take on the enzymatic properties. This two-step process (transcription followed by translation) provides two additional levels at which cellular metabolism can be regulated. There can be *transcriptional* control in which there is *selective* transcription of specific genes, and there can be *translational* control operating at the level of mRNA–ribosome interaction. Most of our current understanding of metabolic regulation exercised at these levels stems from research conducted with prokaryotic organisms, for the regulation of gene expression in eukaryotic cells is much more complex.

Constitutive and Inducible Enzymes

Studies using prokaryotes indicate that enzymes fall into two categories with respect to their occurrence and numbers in cells. Those that appear to *always* be present and that occur in relatively constant concentrations are called **constitutive enzymes**. For example, the enzymes of glycolysis are usually constitutive. These enzymes are the products of genes that are continuously expressed. The other type of enzyme, called an **inducible enzyme**, is found lacking in cells or is present only in small amounts. However, on introduction of a specific metabolite, usually a substrate, the concentration of the enzyme quickly increases. The metabolite that initiates the appearance of the enzyme is called an **inducer**. Inducible enzymes are the products of genes that are selectively expressed; these genes are referred to as inducible genes.

One of the first inducible enzymes to be intensively studied was *β-galactosidase*. Wild-type *E. coli* cells metabolize glucose and will metabolize only the glucose even if lactose, another sugar, is also present. The enzymes of glucose metabolism are constitutive and are always present in the cell, whereas the enzyme needed to initiate lactose metabolism, *β-galactosidase*, is present in only minor amounts—according to one study, no more than five copies per cell. If wild-type *E. coli* cells are placed in a growth medium containing only lactose as the carbon source, they are at first unable to utilize this disaccharide (Fig. 11–10). However, within a few minutes the cells respond to the presence of lactose by synthesizing *β-galactosidase*, which hydrolyzes the lactose to glucose and galactose; these sugars are then metabolized by glycolysis. The lactose acted as an inducer of the *β-galactosidase*, and in cells grown on lactose thousands of copies of this enzyme are present.

A number of β-galactosides besides lactose can act as inducers; these include methyl B-galactoside and allolactose. Actually, the presence of these inducers initiates the synthesis of not one but three enzymes in *E. coli*: (1) *β-galactoside permease*, an enzyme formed in the plasma membrane that promotes the transfer of galactosides into the cell; (2) *β-thiogalactoside transacetylase*, the specific action of which is not understood; and (3) *β-galactosidase*, the key enzyme for initiating lactose metabolism. When, as in this case, induction can be brought about by a single agent and results in the appearance of several enzymes, the process is known as **coordinate induction**. The appearance of inducible enzymes is an example of metabolic regulation through transcriptional control.

Repressible Enzymes

The presence of a specific substance may inhibit the continued production of a specific enzyme or a sequence of enzymes in a metabolic pathway; this process is called **enzyme repression** (or in the case of the repression of a sequence of enzymes, **coordinate repression**). Like enzyme induction, enzyme repression represents regulation at the transcriptional level. *Escherichia coli* cells growing in a medium in which NH_4^+ is the only nitrogen source contain all of the enzyme systems necessary to synthesize amino acids from organic acids. However, if one of the amino acids is added exogenously, the synthesis of certain enzymes in the metabolic pathway leading to that amino acid will be repressed, and with continued growth of the culture these enzymes soon become diluted out.

In another bacterium, *Salmonella typhimurium*, a family of genes encode nine enzymes that catalyze the reactions of a pathway leading to the synthesis of the amino acid histidine. Histidine is an important constituent of proteins and continuous production of protein in this bacterium depends on the availability of histidine. The histidine must either be synthesized by the pathway that involves these nine enzymes or be supplied exogenously. When histidine is added to a culture of *S. typhimurium*, the production of the histidine biosynthetic enzymes is halted (Fig. 11–10). The genes for the histidine enzymes are said to have been *repressed*. If the bacteria are then transferred to a medium lacking histidine, the genes are *derepressed* and the histidine enzymes soon reappear. Enzyme repression is a metabolic control mechanism acting at the level of transcription and should not be confused with allosteric feedback inhibition (discussed earlier in the chapter).

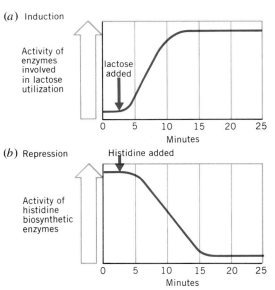

FIGURE 11–10 (*a*) *Induction* of the synthesis of enzymes required for utilization of lactose by *E. coli*; (*b*) *Repression* of the synthesis of enzymes required for the synthesis of histidine in *Salmonella typhimurium* see text for discussion). (From Gardner and Snustad, *Principles of Genetics*, 7th ed., p. 381. Copyright © 1984 by John Wiley & Sons, Inc.)

THE OPERON

In 1961, F. Jacob and J. Monod proposed the **operon model** to explain the genetic basis of enzyme induction and repression in prokaryotes. A few years later (1965), these two investigators were awarded the Nobel Prize for their most incisive work. Although Jacob and Monod's original operon model applied specifically to the regulation of the genes for lactose metabolism in *E. coli*, additional findings by these two scientists as well as the work of many others have revealed the mechanisms by which other operons function.

The components of an operon are illustrated in Figure 11–11a. The operon consists of a series of **structural genes** (denoted by SG1, SG2, SG3), a segment of DNA called an **operator** adjacent to the structural genes, and, next to the operator, an additional DNA segment called a **promoter**. Elsewhere on the chromosome, there is a segment of DNA called a **regulator gene** and next to this is a DNA segment that functions as a promoter for the regulator gene.

In the absence of an inducer (e.g., lactose in an *E. coli* culture) *RNA polymerase* binds to the promoter of the regulator gene and progresses along the DNA transcribing the regulator gene. The RNA *transcript* that is produced is then translated into a *repressor protein*, which binds to the operator of the operon. Binding of the repressor to the operator (see the top half of Fig. 11–11b) prevents RNA polymerase from attaching to this segment of the DNA. As a result, the structural genes cannot be transcribed and no mRNA or enzyme is formed. It is now known that the repressor prevents the binding of the RNA polymerase by steric hindrance.

When Jacob and Monod conducted their studies in the 1960s, the existence of the promoter and the mechanism of repressor function were not known. Today it is recognized that operons also regulate enzyme induction (Fig. 11–11b) and repression (Fig. 11–11c). In the case of the operon model for the induction of enzymes (Fig. 11–11b), the inducer (e.g., lactose) combines with the repressor to form a complex that cannot bind to the operator. Therefore, RNA polymerase can associate with the **operon promoter** and proceeds along the DNA, transcribing it into mRNA, which in turn is translated into polypeptides and enzymes.

In the operon model for enzyme repression (Fig. 11–11c), the repressor cannot bind to the operator unless it first complexes with a **corepressor**. In the example presented earlier in which histidine-synthesizing enzymes in *S. typhimurium* are repressed by histidine that is supplied exogenously, histidine serves the role of the corepressor. In the absence of the corepressor, RNA polymerase associates with the operon promoter, and transcription and translation occur. The mRNA transcripts of **multigenic operons** (i.e., operons containing two or more structural genes) contain the code for several polypeptides and are said to be *polycistronic*. Therefore, there is the **coordinate expression** of all of these genes of the operon.

The *Lac* Operon

The progressive unraveling of the molecular organization and function of the **lac operon** is a classic study in physiology and genetics. Jacob and Monod began their studies of this operon in *E. coli* in the early 1960s; since then, they and many others have continued the study so that today it is one of the best understood regulatory systems (Fig. 11–12).

In *E. coli*, the *lac* operon is inducible, consisting of three structural genes of known lengths, designated *z, y,* and *a*, which, respectively, code for β-galactosidase, β-galactoside permease, and β-galactoside transacetylase. The location and nucleotide sequences are also known for the promoter (*P*) and operator (*O*) segments of the operon. The regulator gene consists of two contiguous portions of known lengths—an *i* gene and the promoter of the *i* gene (*P(i)*) (see Fig. 11–12). The *i* gene codes for a polypeptide consisting of 360 amino acids; four of these polypeptides combine to form a biologically active tetramer that functions as a repressor of the operator gene. The repressor binds to the operator and, in so doing, prevents *RNA polymerase* from associating with the promoter of the operon.

The normal inducer of the *lac* operon is allolactose, which is produced from lactose by β-galactosidase (Fig. 11–13). A few copies of this enzyme are present in *E. coli* cells even in the uninduced state. When allolactose combines with the repressor, a steric change occurs that causes the repressor to be released from the operator. As a result, transcription of the structural genes by RNA polymerase begins and the three enzymes quickly appear in the cells.

The promoter of the *lac* operon contains two active components that regulate transcription. One is the RNA polymerase binding site and the second is the **catabolite activator protein** (*CAP*) binding site. This CAP site functions to prevent transcription of the *lac* operon when sufficient glucose is present. Today, a number of operons are known in addition to the *lac* operon. Some of these are listed in Table 11–3.

FIGURE 11–11 *Operon* model for the regulation of gene expression in prokaryotic cells. (*a*) Essential genetic components; the operon consists of one or more *structural genes* (SG1, SG2, SG3) and the adjoining *promoter* and *operator* sequences. (*b*) Mode of regulation of gene expression in an *inducible* operon; *PR*, promoter for the regulator gene; *R*, regulator gene; *PO*, promoter for the operon; *O*, operator. (*c*) Mode of regulation of gene expression for a *repressible* operon (see text for explanation). (From Gardner and Snustad, *Principles of Genetics*, 7th ed., pp. 384, 385. Copyright © 1984 by John Wiley & Sons, Inc.)

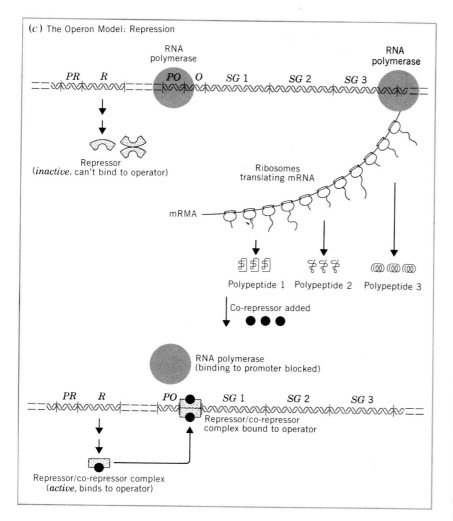

(c) The Operon Model: Repression

RNA polymerase

RNA polymerase

PR R PO O SG 1 SG 2 SG 3

Repressor
(*inactive*, can't bind to operator)

Ribosomes
translating mRNA

mRMA

Polypeptide 1 Polypeptide 2 Polypeptide 3

Co-repressor added

RNA polymerase
(binding to promoter blocked)

PR R PO SG 1 SG 2 SG 3

Repressor/co-repressor
complex bound to operator

Repressor/co-repressor complex
(*active*, binds to operator)

FIGURE 11–12 The *lac* operon is but one of several operons known to exist in the *E. coli* chromosome (upper part of figure). The *lac* operon is inducible and consists of three structural genes (*z, y,* and *a*) and the promoter (*P*) and operator (*O*) sequences adjoining the *z* gene. The regulator gene (*i*) is contiguous with the operon in this particular case. The regulator gene has its own promoter (*P(i)*). The numbers below each of the genes are their approximate lengths (in base pairs). (From Gardner and Snustad, *Principles of Genetics*, 7th ed., p. 386. Copyright © 1984 by John Wiley & Sons, Inc.)

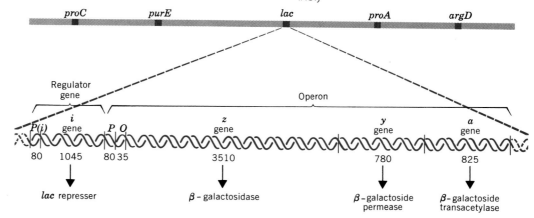

proC purE lac proA argD

Regulator gene

Operon

P(i) i gene P O z gene y gene a gene

80 1045 80 35 3510 780 825

lac represser β – galactosidase β –galactoside permease β–galactoside transacetylase

FIGURE 11-13 The two reactions that are catalyzed by the enzyme β-galactosidase. Allolactose produced in the upper reaction is an inducer of the lac operon.

TABLE 11-3 COMMON OPERONS IN PROKARYOTES

Operon	Number of Enzymes	Function
lac	3	Hydrolysis of lactose and transport of galactose
his	10	Histidine synthesis
gal	3	Galactose → UDP-glucose
leu	4	Leucine synthesis
trp	4	Tryptophan synthesis
ara	4	Metabolism and transport of arabinose
pyr	5	Aspartate → UMP

Catabolic Repression

Catabolic repression is a specific type of repression of enzyme production in which a metabolite such as glucose acts to repress the formation of enzymes that would allow the catabolism of *other*, related metabolites. For example, when *E. coli* cells are cultured in a medium that is rich in glucose, the glucose represses the formation of *β-galactosidase* even if lactose (an inducer of this enzyme) is added to the medium. Glucose is even known to repress the production of enzymes that formerly were thought to be constitutive. A common phenomenon in bacteria is the suppression of aerobic respiration and electron transport by high glucose concentrations, *even in the presence of ample oxygen*. Under these condition, the cells utilize the glycolytic and fermentative pathways.

The manner in which the catabolite brings about the effect is only partially understood. In the case of catabolite repression by glucose in bacteria, it appears that glucose affects the amount of cyclic AMP (cAMP) present in the cells. When the concentration of glucose is high, the concentration of cAMP is low; and low levels of glucose are accompanied by high concentrations of cAMP. It is possible that glucose affects the synthesis of cAMP, which is formed from ATP by *adenylcyclase* (Fig. 11-14). cAMP is necessary for the CAP to bind to the promoter site. cAMP binds to the CAP, and once the cAMP–CAP complex binds to the promoter (Fig. 11-15), the *RNA polymerase* can attach to the promoter and begin transcription.

For example, in the case of the *lac* operon, when the glucose level is high (even if lactose is also present), the cAMP is low, and therefore the cAMP—CAP complex is not available to bind to the promoter and allow transcription to start. However, in the absence of glucose and in the presence of lactose (which forms a complex with the repressor), cAMP is plentiful and is available to combine with the CAP so that transcription proceeds. The rate of *lac* operon transcription in the absence of glucose is 50 times as great as in the presence of glucose.

Translational Control

Regulatory mechanisms usually function at the beginning of a sequence of events rather than at the end. For example, allosteric feedback control is usually achieved by regulating the enzyme that catalyzes the

FIGURE 11–14 Formation of cyclic AMP from ATP by the enzyme, *adenylcyclase*.

FIGURE 11–15 Organization and nucleotide sequences of the *lac* operon. The promoter consists of two components: the site that binds the cAMP–CAP complex and the RNA polymerase binding site. The adjoining segments of the repressor (*i*) and β-galactosidase (*z*) structural genes are also shown. The horizontal arrow (labeled mRNA) shows the position at which the transcription of the operon begins (i.e., what will be the 5′ end of the mRNA). The numbers at the bottom of the figure give the distances (in base pairs) from the end of the *i* gene. (From Gardner and Snustad, *Principles of Genetics*, 7th ed., p. 389 Copyright © 1984 by John Wiley & Sons, Inc.)

first reaction in a sequence of reactions. Likewise, control of the number of enzyme molecules produced in a cell is most commonly brought about at the transcriptional level rather than at the level of translation. Regulatory mechanisms conserve cell energy and prevent the buildup of materials that will not be used. Therefore, *translational control* mechanisms would be expected to be rare and at the present time there is little evidence for very much control being exercised at this level. Indirect evidence suggests that translational control of enzyme synthesis may occur for enzymes that are part of the same operon. For example, because the three structural genes in the *lac* operon are controlled by the same operator, one would expect

that equal quantities of the three enzymes would be produced. However, induced cells contain more copies of one enzyme over another. This could be explained if the ribosome detaches from mRNA before completing the translation of the entire polycistronic message.

The economic impact of the use of control mechanisms is reflected also in the synthesis of ribosomal RNA (rRNA). When *E. coli* cells are cultured in a medium that is deficient in amino acids, they are unable to synthesize proteins. As a consequence, the cells also stop producing rRNA—a mechanism called **stringent control**. Mutants of *E. coli* do exist, however, that continue to synthesize rRNA under these conditions and are called **relaxed mutants**. Paper chromatographic analysis (see Chapter 13) of nucleotide extracts of the stringent cells and relaxed mutants reveals that there are two spots in the chromatogram of stringent cells that are absent in the chromatogram of the relaxed mutants. Initially these spots were called "magic spot I" and "magic spot II," but they have now been identified as guanosine-5'-diphosphate-2'- (or 3') diphosphate (ppGpp) and guanosine-5'-triphosphate-2'- (or 3') diphosphate (pppGpp), respectively. The studies of Cashel and Gallant and their colleagues indicate that these nucleotides trigger the events that turn off rRNA synthesis when the amino acids needed by the cells for protein synthesis are lacking.

GROUP IV CONTROL MECHANISMS— REGULATION OF ENZYME PRODUCTION IN EUKARYOTIC CELLS

As might be expected, the regulation of enzyme production in eukaryotic cells is much more complex than in prokaryotic cells. Eukaryotic cells contain a number of chromosomes instead of the single chromosome found in prokaryotes. Moreover, eukaryotic cell chromosomes are diploid (see Chapter 20) and at times polyploid. The DNA of the chromosomes is usually supercoiled and highly folded and the chromosomes themselves are physically separated from the cytoplasmic ribosomes by the nuclear envelope. Undoubtedly, these factors make control of enzyme synthesis more complex, even though transcription and translation involve what is basically the same mechanism as in prokaryotes. However, structural genes for a group of functionally related enzymes that might constitute an inducible component such as an operon are not found adjacent to one another on a chromosome and may, in fact, be distributed among different chromosomes.

Enzyme induction does occur in primitive eukaryotic organisms such as yeast and *Neurospora*, but operons are either few or nonexistent in higher eukaryotes. In all but a few cases, the mRNAs of higher eukaryotes contain the coding sequence of only one structural gene (i.e., they are **monogenic**). The induction process in the fungi is slower, and the change in concentration of enzymes is not as great as in the prokaryotes. In yeast, the induction of *β-galactosidase* takes much longer than in *E. coli*; also, the increase in enzyme activity is 10-fold and not a 1000-fold. The enzyme tryptophan-2,3-oxygenase found in the liver

cells of vertebrates is inducible but induction requires many hours.

Z-DNA

In addition to the B form of the DNA double helix, there is a Z form (see Chapter 7). In Z-DNA, the double helix is left-handed, each polynucleotide containing sequences of alternating purines and pyrimidines. A number of observations suggest that Z-DNA may play a role in the regulation of gene expression. For example, Z-DNA is formed in the transcriptionally active *macronucleus* of *Stylonychia mytilis* (a protozoan) and in the interband regions of *Drosophila* salivary chromosomes. Proteins associated with polytene chromosomes of *Drosophila* bind antibodies specific for Z-DNA. Some regulatory proteins appear to bind to Z-DNA but not B-DNA, including the catabolite activator protein described earlier.

Calcium Ions and Calmodulin

In the past few years, it has become quite clear that calcium ions play a very important role in the regulation of certain activities in eukaryotic cells. For example, such diverse phenomena as cell motility, muscle cell contraction, chromosome movement, endocytosis, exocytosis, cyclic nucleotide metabolism, phosphorylation of proteins, and glycogen metabolism are influenced by the level of Ca^{2+} in the cytosol. Some of the effects of calcium ions are mediated through a specific calcium-binding protein called **calmodulin** (Fig. 11–16) found in nearly all cells. Calmodulin is a small protein (its molecular weight is 16,720) consisting of a single polypeptide chain containing 149 amino acids. Especially interesting is the finding that its primary structure is essentially the same in all of the species that have been studied so far, indicating that the protein has undergone little change in the course of evolution.

Each molecule of calmodulin binds four calcium ions and its common occurrence in eukaryotes suggests that it has a universal regulatory function along with Ca^{2+}. When Ca^{2+} enters the cytosol, it is bound by calmodulin to form a complex that then influences cellular activity by interacting with certain other proteins. Some of the proteins affected by the Ca^{2+}–calmodulin complex are structural proteins (such as those present in the spindle apparatus formed during mitosis, see Chapter 20) but most appear to be en-

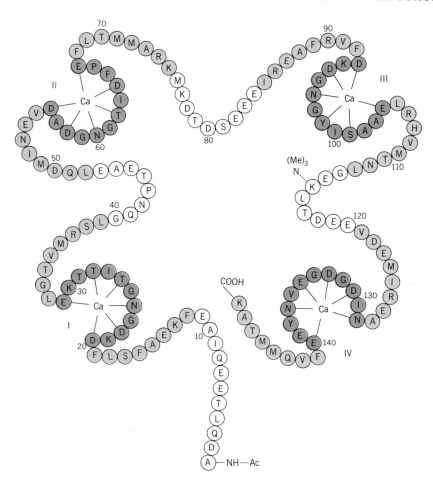

FIGURE 11-16 The *calmodulin* molecule. Calmodulin chelates Ca^{2+}, the complex having a regulatory function in many cellular processes. The primary structure of this small protein (indicated here using a one-letter code) has undergone very little change in the course of evolution. A, ala; D, asp; E, glu; F, phe; G, gly; H, his; I, ile; K, lys; L, leu; M, met; N, asn; P, pro; Q, gln; R, arg; S, ser; T, thr; V, val; and Y, tyr. (Modified from Cheung, W. Y., *J. Cylcic. Nucleo. Res.* **1**, 71; copyright © 1981 Raven Press, Courtesy of W. Y. Cheung.)

zymes. For example, the Ca^{2+}–calmodulin complex serves to directly activate a family of enzymes called **protein kinases**. Which protein kinases are activated may depend on the amount of Ca^{2+} in the cytosol. This is because calmodulin can bind up to four calcium ions and a complex containing only one Ca^{2+} may activiate a different enzyme than a complex containing two Ca^{2+} (and so on). The effect of the Ca^{2+}–calmodulin complex on cell metabolism may be *indirect*. For example, the Ca^{2+}–calmodulin complex acts to trigger *adenylate cyclase* activity in the plasma membranes of cells and this results in the production of cAMP. cAMP's role as a "second messenger" influencing cell metabolism is discussed below. Additional examples of the manner in which calmodulin and Ca^{2+} regulate cell metabolism are given in Table 11–4.

Enzyme Induction by Hormones

Enzyme induction in prokaryotic cells is usually triggered by a potential metabolite (such as the induction

TABLE 11–4 ALLOSTERIC EFFECT OF CALMODULIN–Ca^{2+} COMPLEXES ON CELL ENZYMES

Enzyme	Effect
Adenylate cyclase	Activates enzyme and thereby regulates the rate of cyclic AMP synthesis
Cyclic nucleotide diesterase	Activates enzyme and thereby regulates the rate of cyclic nucleotide degradation
Myosin kinase	Activates enzyme and thereby regulates cell movement
Phosphorylase kinase	Activates enzyme and thereby regulates glycogen breakdown
NAD kinase (plants)	Activates enzyme and thereby regulates the levels of coenzyme present
Enzymes envolved in Ca^{2+} pumps	Regulates transport of Ca^{2+} across the plasma membrane and the sarcolemma
Tubulin	Disassembles this protein, thereby leading to the disassembly of microtubules

of *β-galactosidase* by lactose). As noted above, this form of enzyme induction occurs in some eukaryotic cells as well; however, in higher animals there also is a highly developed control system in which *hormones* act as *messengers* that coordinate the biosynthetic activities of cells. This coordination may take the form of activating (or deactivating) enzymes that *already* exist in the cell or the effect may be on gene expression, resulting in the production of additional enzymes. The term "messenger" is appropriate, because the hormones are produced and secreted by one cell (or tissue) and have an effect on a different cell (or tissue). The two tissues producing and responding to the messenger may be in widely separated parts of the body, the messenger traveling from one site to the other via the bloodstream. Sometimes, the effect of the messenger is to cause the "target" cell to produce and secrete a different hormone (Fig. 11–17*a*)

From a chemical point of view, hormones are quite diverse. Some hormones (e.g., *thyroxin* and *epinephrine*) are small molecules derived from amino acids and others are proteins (e.g., *insulin* and *erythro-*

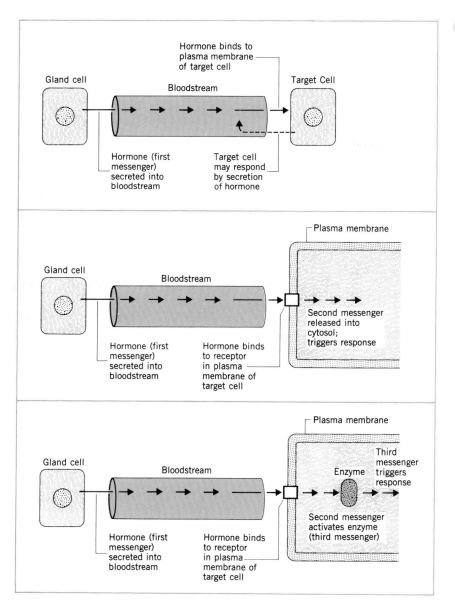

FIGURE 11–17 Certain tissues of multicellular organisms communicate with one another via chemical messengers (i.e., hormones) sent through the bloodstream. On reaching the target tissue, the first messenger may (*a*) initiate the secretion of another substance into the blood or (*b*) trigger the release of a *second messenger* from the plasma membrane into the cytosol. In some cases (*c*) the second messenger may activate a *third messenger* before an effect is produced in the cell.

poietin) or steroids (e.g., *progesterone* and *cortisol*).

Hormones regulate the production of enzymes through a variety of mechanisms. For example, the hormone may be the *first messenger* among a series of messengers that regulate a metabolic response (Figs. 11–17*b* and 11–17*c*). Such hormones attach to *receptor sites* on the plasma membrane of the "target" cell. The receptor sites consist of specific proteins having high affinities for the hormones. Hormone binding at a receptor site serves to initiate a response by the cell. Different tissues may be affected by the same hormone. The cells in these tissues may contain similar hormone-binding receptors, but the interaction between receptor and hormone is followed by a different cellular response.

Second Messengers In some cases it appears that the hormone receptor of the target cell is associated with an enzyme in the plasma membrane. Hormone binding by the receptor changes the receptor–enzyme complex and activates the enzyme. Once activated, the enzyme catalyzes a reaction that forms a **second messenger** that passes into the cytosol and triggers further responses in the target cell.

The most common second messenger is cyclic AMP. It is formed by the enzyme *adenylate cyclase*, a membrane-bound enzyme associated with hormone receptor sites. cAMP produced by *adenylate cyclase* acts as a second messenger in metabolic pathways associated with the breakdown of glycogen and triglycerides (Fig. 11–18). cAMP appears also to be involved in cellular reactions that produce and secrete hormones.

The cellular response induced when cAMP acts as a second messenger continues for only a brief period because of the presence of *phosphodiesterases* in the cytosol. These enzymes degrade cAMP, producing the inactive, noncylic mononucleotide 5'-AMP.

Calcium ions also act as second messengers. Some surface receptors are associated with channels through the plasma membrane. When the receptor binds the first messenger (i.e., the hormone), the channel is opened temporarily and Ca^{2+} enters the cell. Once inside the cell, the Ca^{2+} is bound by specific proteins such as calmodulin. The activation of enzymes by calmodulin was described earlier. It turns out that many of the enzymes activated by cAMP are also activated by Ca^{2+}–calmodulin and Ca^{2+}–calmodulin affects the activities of enzymes that form and degrade cAMP. In turn, cAMP influences the

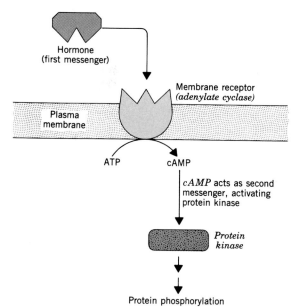

FIGURE 11–18 Hormone (i.e., first messenger) is bound at the plasma membrane of the target cell by a specific receptor associated with the enzyme *adenylate cyclase*. Binding activates the enzyme, which then produces cAMP. cAMP passes into the cytosol where it acts as a second messenger, activating *protein kinases* that regulate metabolism.

plasma membrane channels through which Ca^{2+} passes by phosphorylating the proteins that line these channels.

Cyclic GMP (cGMP), like cAMP, activates certain protein kinases. However, cGMP does not appear in response to the binding of hormones by plasma membrane receptors. Rather, the level and activity of cGMP appear to increase as the intracellular level of free Ca^{2+} increases.

"Double" Second Messengers Many hormone receptors are associated with a group of plasma membrane phospholipids called *polyphosphoinositides*. Binding of the first messenger (hormone) to the receptor initiates the breakdown of the membrane-bound polyphosphoinositides, thereby forming *diacylglycerol* and *inositol triphosphate*. The diacylglycerol enters the cytosol where it acts as a second messenger, activating a protein kinase (Fig. 11–19). (The protein kinase activated by diacylglycerol has different properties than the protein kinase activated by cAMP and Ca^{2+}–calmodulin.) The inositol triphosphate produced in the plasma membrane also enters the cytosol and

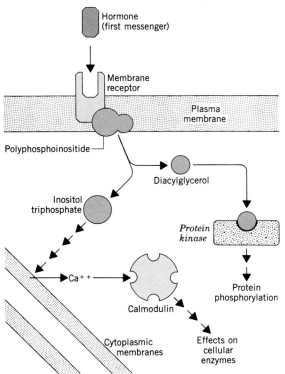

Hormone
(first messenger)

Membrane
receptor

Plasma
membrane

Polyphosphoinositide —

Diacylglycerol

Inositol
triphosphate

*Protein
kinase*

Ca++

Protein
phosphorylation

Calmodulin

Cytoplasmic
membranes

Effects on
cellular
enzymes

FIGURE 11–19 Double second messengers are produced from *polyphosphoinositide* (*PPI*). Hormone (first messenger) binding to the plasma membrane receptor activates an enzyme that degrades PPI to *inositol triphosphate* and *diacylglycerol*; both of these products then act as second messengers. Inositol triphosphate induces the release of Ca^{2+} from intracellular membranes and activates calmodulin; diacylglycerol activates protein kinases.

acts as a second messenger. When inositol triphosphate reaches the intracellular membranes, it induces the release of Ca^{2+}. Ca^{2+} released into the cytosol then activates certain proteins such as calmodulin.

Effects of Hormones on Gene Expression

The binding of certain hormones to plasma membrane receptors is followed by *internalization*. Internalization takes the form of an infolding of that portion of the plasma membrane containing the receptor–hormone complex, thereby forming a vesicle. The vesicle subsequently fuses with a Golgi body or lysosome, the enzymes of which (see Chapters 18 and 19) presumably alter the receptor–hormone complex. Ultimately, a product is formed that enters the cell nucleus where it interacts with the genome, thus altering gene expression.

Steroid hormones (e.g., estrogen and progesterone) enter the target cell by passing directly through the plasma membrane and into the cytosol. Within the cytosol, the hormones combine with receptor molecules and the complexes then migrate to the cell nucleus, where they have a direct impact on the expression of certain genes (Fig. 11–20). *In vitro* studies using chick oviduct cells have shown that the hormone estrogen enters the cytosol to form a hormone–receptor complex that migrates to the cell nucleus and induces the increased rate of transcription of the genes for lysozyme and ovalbumin.

Protein Phosphorylation and Metabolic Control

Phosphorylation by protein kinases appears to be one of the major mechanisms for activating and inactivating a great number of enzymes and thereby controlling many processes in eukaryotic cells. Generally speaking, many enzymes in degradative (catabolic) pathways are *activated* by phosphorylation, whereas the enzymes in synthetic (anabolic) pathways are *inactivated* by phosphorylation. Protein kinases activated by Ca^{2+}–calmodulin, cAMP, and diacylglycerol phosphorylate key enzymes of glycolysis and other biodegradative pathways and thereby activate these pathways. The protein kinases also phosphorylate key enzymes in glycogen metabolism, gluconeogenesis, fatty acid synthesis, cholesterol synthesis, and protein synthesis and thus inactivate these pathways. Protein kinases catalyze protein phosphorylation using ATP as the source of phosphate.

Protein phosphatases are enzymes that catalyze the removal of phosphate from phosphorylated enzymes (or other proteins). It has been proposed that protein phosphatases also function in metabolic control because they can dephosphorylate glycolytic enzymes and thereby inactivate them and can dephosphorylate glycogen-synthesizing enzymes and thereby activate them. Although the evidence is not yet available, it is likely that protein phosphatases dephosphorylate the enzymes of gluconeogenesis, fatty acid synthesis, cholesterol synthesis, and protein synthesis and therefore activate these enzymes as well.

Amplification of Signals

Signals to cells in the form of a few molecules of a hormone or other first messenger are greatly enhanced by mechanisms that use second messengers, double messengers, or even third messengers. Although only

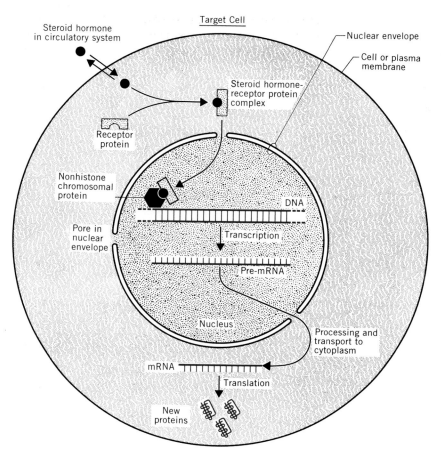

FIGURE 11-20 Metabolic regulation via steroid hormones. The small steroid hormone penetrates the cell's plasma membrane and forms a complex with a receptor protein. The hormone–receptor complex enters the cell nucleus where it affects the transcription of DNA. (From Gardner and Snustad, *Principles of Genetics*, 7th ed., p. 386. Copyright © 1984 by John Wiley & Sons, Inc.)

a few molecules of the first messenger may bind to the surface receptors, they activate enzymes each of which produces many copies of the second messenger. If additional messengers are produced by enzymes activated by second messengers, thousands of signals will be produced in the cytosol within a very short time.

Repression of the Genome in Eukaryotes

Unlike prokaryotic cells, the cells of higher animals and plants undergo extensive *differentiation*, forming tissues that have specific and limited physiological roles. Yet most differentiated cells of higher plants and animals contain *complete* genomes. This implies that large segments of the genome are repressed and go unexpressed. RNA–DNA saturation hybridization experiments (see Chapter 20) indicate that only 6–30% of the genome is expressed. In some cases, the repression is reversible. For example, when differentiated carrot root cells are isolated and grown in tissue culture, new differentiated cells are produced, includ-

ing cells that are characteristic of vascular tissue, storage tissue, and epidermal tissue. However, in many differentiated cells, such as nerve and muscle cells of higher animals, large portions of the genome are permanently repressed.

The mechanism of gene repression of eukaryotic cells is not yet clear. The chromosomes of eukaryotic cells contain large quantities of proteins and RNA in addition to DNA (Table 11-5), and over the years these have been considered potential repressors. However, their chemical properties do not adequately support such a contention. For example, Stedman and Stedman proposed as long ago as the 1940s that the

TABLE 11-5 RELATIVE CONCENTRATIONS OF CALF THYMUS CHROMATIN COMPONENTS

DNA	100
Histone	114
Nonhistone protein	33
RNA	7

histones might act as gene repressors. However, there are only five major classes of histones in eukaryotic cells and they occur in about equal amounts, with little variation between different tissues of an organism or between species. It would therefore appear that histone function is more fundamental (see Chapter 20). If the histones do have repressor activity, then the effect is nonspecific. Electrophoretic analysis (Chapter 13) of the nonhistone proteins in chromatin reveals that they occur in much greater variety than the histones. However, their diversity is insufficient to support the contention that they play a role in gene repression.

The Britten–Davidson Model of Gene Regulation in Eukaryotes

Although several models have been proposed to explain gene regulation in eukaryotes, none has been documented with evidence that would give it the degree of certainty that surrounds the Jacob–Monod operon model for prokaryotic cells. However, the model proposed by R. Britten and E. Davidson in 1969 has attracted a great deal of attention. According to

FIGURE 11–21 Model of regulation of gene expression in eukaryotic cells. Binding of an inducer to a sensor site causes transcription of adjacent integrator gene. The transcript (activator RNA) then binds to the receptor site adjacent to a structural gene on the same or different chromosome. Transcription and translation of the structural gene ensue.

this model (Fig. 11–21), the nuclear chromosomes contain DNA sequences called *sensor* genes that recognize various cellular substances such as metabolic inducers (substrates), hormone–receptor complexes, or regulatory nucleotides (e.g., ppGpp). When the inducer enters the nucleus, it binds to the sensor and promotes the transcription of an adjacent **integrator gene** whose product is a specific **activator RNA**. The activator RNA can attach to appropriate DNA sequences that constitute *receptor sites* on either the same or a different chromosome. Presumably, the function of the activator would be analogous to that of the cAMP–CAP complex described earlier in the chapter for prokaryotes. The binding of the activator to a receptor site promotes transcription of adjacent structural genes. After undergoing some modification called "processing" the mRNA transcript leaves the nucleus and is translated into protein (processing and translation of mRNA are discussed in Chapter 22).

A number of elaborate modifications of the basic model can explain the variations in gene expression and differentiation in eukaryotes. For example (Fig. 11–22a), a number of different sensor genes (S_1, S_2, and S_3) on binding different inducers (I_1, I_2, and I_3) promote the formation of different activator RNA molecules (a, b, and c) by their companion integrator genes. The presence of *multiple* receptor sites for each structural gene (i.e., L, M, and N) would imply that different combinations of structural genes would be transcribed, depending on binding of the various activators to their respective receptor sites. The binding of an activator to any one receptor would trigger transcription of the adjacent structural gene.

In an alternative model (Fig. 11–22b), transcription of structural genes in various combinations results from the binding of a specific inducer. In this variation, the sensors initiate activator synthesis in a number of adjacent integrator genes, and each activator then associates with *one* receptor. The models have a number of interesting possibilities and are supported by the observation that a large proportion of the DNA in eukaryotic cells (e.g., 40% in calf thymus gland cells) consists of repeated nucleotide sequences that are too small to be structural genes; these, perhaps, are receptor sequences.

Compartmentalization

A final regulatory mechanism in eukaryotic cells is the physical separation and isolation of groups of enzymes

(a)

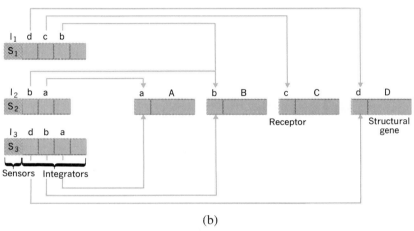

(b)

FIGURE 11-22 Britten–Davidson models for the regulation of gene expression in eukaryotic cells. In the upper model (a), there is a redundancy of receptor genes; in the lower model (b), the redundancy occurs in the integrator genes (see text for explanation).

within membranous boundaries, that is, specific groups of enzymes are *compartmentalized* within the cellular organelles. For example, in both plant and animal cells the enzymes of the tricarboxylic acid or Krebs cycle are physically separated from those of glycolysis because the former are confined within mitochondria (Chapter 16) and the latter are present in the cytosol. In plant cells, the enzymes of the dark reactions of photosynthesis are physically isolated within chloroplasts (Chapter 17). The enzymes of the glyoxylate cycle are compartmentalized in microbodies (Chapter 19), whereas many of the cell's powerful hydrolytic enzymes are restricted within the lysosomes.

Many of these compartmentalized enzymes act on substrates or employ cofactors that are produced by enzymes that are restricted to other parts of the cell.

Regulation of the transport of these compounds across cellular membranes from one cell compartment to another affords yet another level at which the control at metabolsim can be exercised.

SUMMARY

A variety of metabolic regulatory mechanisms control the direction and rate of individual enzyme-catalyzed reactions in cells as well as entire metabolic pathways. These mechanisms may be grouped into four categories:

Group I: **Immediate Regulation**   Immediate regulation of the activity of an enzyme may be achieved through such factors as substrate concentration effects, changes in pH, or changes in salt concentration.

Group II: **Allosterism**   The activities of certain enzymes of a metabolic pathway may be influenced through interaction with one or more intermediates of the pathway or the pathway's end product. Enzymes subject to such regulation are called **allosteric enzymes.** The regulatory metabolites are bound at sites other than the active site.

Group III: **Regulation of Gene Expression in Prokaryotes**   The synthesis of enzymes in prokaryotes may be initiated or halted by the sudden availability of certain metabolites. Sets of interrelated genes responding to these changes are called **operons.**

Group IV: **Regulation of Gene Expression in Eukaryotes**   Metabolic regulation is most complex in eukaroytes and is effected through diverse processes. Among these are the actions of hormones and other messengers, the activation and inactivation of enzymes by phosphorylation, and the compartmentalization of the cells activities.

REFERENCES AND SUGGESTED READING

Articles and Reviews

Carafoli, E., and Penniston, J. T., The calcium signal. *Sci. Am.* **253** (5), 70 (Nov. 1985).

Cheung, W. Y., Calmodulin. *Sci. Am.* **246** (6) 62 (June 1982).

Chock, P. B., Rhee, S. G., and Stadtman, E. R., Interconvertible enzyme cascades in cellular regulation. *Ann. Rev. Biochem.* **49**, 813 (1980).

Crawford, I. P., and Stanffer, G. V., Regulation of tryptophan biosynthesis. *Ann. Rev. Biochem.* **49**, 163 (1980).

Hokin, L. E., Receptors and phosphoinositide-generated second messengers. *Ann. Rev. Biochem.* **54**, 205 (1985).

Ingelbritsen, T. S., and Cohen, P., Protein phosphatases: Properties and role in cellular regulation. *Science* **221**, 331 (1983).

Klec, C. B., Cronch, T. H., and Richman, P. G., Calmodulin. *Ann. Rev. Biochem.* **49**, 489 (1980).

Marx, J. L., A new view of receptor action. *Science* **224**, 271 (1984).

Maniatis, T., and Ptashne, M., A DNA operator–repressor system. *Sci. Am.* **234** (1), 64 (June 1976).

O'Malley, B. W., and Schrader, W. T., The receptors of steroid hormones. *Sci. Am.* **234** (4), 32 (Feb. 1976).

Books, Monographs, and Symposia

Atkinson, D. E., *Cellular Energy Metabolism and Its Regulation*, Academic, New York, 1977.

Conn, E. E., and Stumpf, P. K., *Outline of Biochemistry* (4th ed.), Wiley, New York, 1976.

Gardner, E. J., and D. P. Snustad, *Principles of Genetics* (7th ed.), Wiley, New York, 1984.

Larner, J., *Intermediary Metabolism and Its Regulation*, Prentice–Hall, Englewood Cliffs, N.J., 1971.

Rawn, J. D., *Biochemistry*, Harper & Row, New York, 1983.

Spiro, T. G., *Calcium in Biology*, Wiley, New York, 1983.

Vogel, H. J. (Ed.), *Metabolic Pathways*, Vol. 5, *Metabolic Regulation*, Academic, New York, 1971.

Whelan, W. J., *Biochemistry of Carbohydrates*, Butterworths, London, 1975.

PART FOUR

TOOLS AND METHODS OF CELL BIOLOGY

CHAPTER 12

FRACTIONATION OF TISSUES AND CELLS

Much of our current knowledge concerning the structure, chemical composition, and function of the various cellular organelles has been obtained following the isolation of these components from cells using various **fractionation** procedures. Generally speaking, tissue fractionation involves three major phases. In the first, the tissue or cell suspension is *disrupted* in order to release the cell components. The second phase involves the *sorting* of the cell components into *fractions*, such that the constituents of any single fraction are the same but differ from the constituents that are present in any other fraction. The final phase consists of an *examination* and *analysis* of the separated cell fractions.

METHODS FOR DISRUPTING TISSUES AND CELLS

Various methods have been devised for disrupting tissues and suspended cells, but the method of choice is usually the procedure that causes *minimal damage* to the released cell constituents. Most physical procedures are based on the effects of **shearing** forces, and because the released cell parts undergo rapid deterioration at room temperature, these procedures are usually carried out at low temperature and in cold buffer solution. Among the older methods used is grinding the sample with a *mortar* and *pestle*, often with the aid of abrasive materials such as sand, alumina, or ground glass. This procedure has several disadvantages, including the loss of some cell constituents by adsorption to the abrasive and the necessity of removing the abrasive material either before or during the fractionation procedure.

Shearing forces adequate to disrupt most cells and tissues may also be obtained using a *blender* in which steel blades rapidly rotate through the cell or tissue suspension. The product of this and similar techniques is called a **homogenate** (e.g., "cell homogenate" or "tissue homogenate"). Tissues may also be homogenized by placing them, along with cold buffer solution, in a cylindrical glass tube fitted with a glass or Teflon plunger. As the plunger is driven down the tube (generally by hand), the tissue and buffer are forced upward through the narrow annular space between the wall of the tube and the plunger. The shearing forces that are generated are usually sufficient to disperse the tissue after several up-and-down strokes. This is the method of choice for the disruption of soft tissues, such as liver, brain, and kidney. One of the more rigorous methods for disrupting cells is by the use of a **pressure cell**, which consists of a steel cylinder and close-fitting steel piston. The piston is pushed into one end of the cylinder using a press or hydraulic jack, and the sample is forced out of the cylinder through a narrow opening at the other end. The size of this opening may be controlled by a needle valve. Bacteria and other microorganisms enclosed within a tough cell wall are frequently disrupted using this approach.

Cells may also be disrupted by **insonation** (ultrasound) using a sonifier. In this procedure, the probe of the sonifier is immersed in the cell suspension and caused to vibrate in the fluid (usually at about 20,000 cycles per second). These ultrasonic vibrations produce a number of effects in the fluid that act collectively to cause the disruption of the cells. Shock waves (alternate compressions and rarefactions) arising from the tip of the probe create turbulent flow of the fluid in which the cells are suspended and may disrupt the cells. The shock waves also cause **cavitation** of the fluid, that is, microscopic bubbles are formed in the fluid near the tip of the probe and these rapidly stream away from the probe along with some of the fluid. The friction created between this stream and the suspended cells also contributes to their disruption. Some of these bubbles disintegrate into still smaller bubbles that travel away from the probe like miniature projectiles; when these impinge upon the cells, the cells may be disrupted.

Cells can also be disrupted by chemical means. For example, enzymes that specifically degrade the components of the cell wall or plasma membrane may be added to the tissue or cell suspension. Alternatively, proteolytic or lipolytic agents (such as certain detergents) that dissolve the membrane may be used. Some cells are sufficiently fragile that they may be disrupted by successive freezing and thawing. Erythrocytes (red blood cells) and certain other cells can be broken by the **osmotic pressure** created within them when they are placed in a hypotonic salt solution or distilled water.

After the cells have been disrupted, the goal is to separate and isolate the structures that have been released from them. Because cellular organelles and other constituents vary in size, shape, and density, they settle through the liquid in which they are suspended at different rates. Consequently, disrupted cells are more often fractionated by some form of **centrifugation** than by any other method. Indeed,

centrifugation has become one of the most widely employed procedures in cellular research and one of the most important tools of the cell biologist.

CENTRIFUGATION

If a container is filled with a suspension of particles of varying size and density, the particles will gradually settle to the bottom of the container under the influence of gravity. The rate at which settling occurs can be greatly increased by increasing the gravitational effect on the particles. This is the rationale behind the use of centrifugation. A tube containing a suspension of particles (e.g., a tissue homogenate) is placed in the rotor of a centrifuge and then is rotated at high speed. The resulting acceleration greatly increases the force acting on the suspended particles, causing their more rapid sedimentation to the bottom of the tube along paths that are *perpendicular* to the axis of rotation (i.e., along radii of the circle being swept out by the rotating tubes; see Fig. 12–1).

Although the practical application of centrifugal force can be traced back more than a thousand years, the modern era of centrifugal methodology began in the early years of this century when the focal point of biochemistry was the study of the chemical and physi-

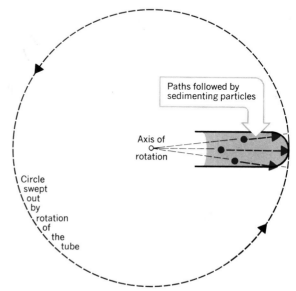

FIGURE 12–1   Paths followed by particles during centrifugation. The centrifuge tube is being viewed from a position that is above the plane in which the tube rotates.

cal properties of protein solutions and other **colloids**. The discovery by Emil Fischer in 1902 that proteins consisted of long chains of amino acids turned out to be the main impetus behind a dramatic surge in the development of centrifugal methods. Fischer and most other notable chemists of the period believed that individual protein species (myoglobin, hemoglobin, etc.) were **polydisperse**; that is, a given protein type was believed to occur in various sizes (i.e., various polypeptide chain lengths). That this is not the case was shown by Theodor Svedberg, who developed the first "analytical ultracentrifuges" (see below), in which the movements of sedimenting particles can be followed optically. Svedberg showed that a given protein type is not polydisperse but is **homogeneous (or monodisperse)**, with all molecules of that protein having the same, well-defined molecular weight. In 1926, Svedberg received the Nobel Prize in recognition of his pioneering centrifugal studies of proteins and other colloids.

Theory of Centrifugation

The force acting on a particle during centrifugation is given by the equation

$$F = m\omega^2 x \tag{12-1}$$

where m is the mass of the particle, ω (omega) is the angular velocity of the spinning rotor in radians per second (one unit of circumference contains 2π radians), and x is the distance of the particle from the axis of rotation.

Usually, the value given for the force applied to particles during centrifugation is a *relative* one; that is, it is compared with the force that the earth's gravity would have on the same particles. It is called **relative centrifugal force** or RCF and is given by

$$\text{RCF} = \frac{F_{\text{centrifugation}}}{F_{\text{gravity}}} = \frac{m\omega^2 x}{mg} = \frac{\omega^2 x}{g} \tag{12-2}$$

where g is acceleration due to gravity and equals 980 cm/sec/sec.

Because g is a constant, then

$$\frac{\omega^2}{g} = \frac{\left[\frac{(2\pi)(\text{rpm})}{60}\right]^2}{980} \tag{12-3}$$

In equation 12–3, the term $(2\pi/60)/980$ is also a constant and is equal to 1.119×10^{-5}.

Thus the equation for relative centrifugal force (i.e., equation 12–2) simplifies to

$$\text{RCF} = 1.119 \times 10^{-5}(\text{rpm})^2(x) \tag{12-4}$$

Consequently, to determine the relative centrifugal force in effect during centrifugation it is necessary to measure the revolutions per minute (rpm) of the rotor and the distance between the sample and the axis of rotation. Because RCF is the ratio of two forces, it has no units. However, it is customary to follow the numerical value of the RCF with the symbol g. This indicates that the applied RCF is a *multiple* of the earth's gravitational force.

Sample Problem What is the *maximum* relative centrifugal force applied when red blood cells are sedimented at 1000 rpm in a rotor of maximum sample radius equal to 10 cm?

$$\text{RCF}_{\text{max}} = 1.119 \times 10^{-5}(1000)^2(10)$$
$$= 1.119 \times 10^2 \text{ or } 112 \, g$$

The sedimentation of particles by centrifugation is, in effect, a method for concentrating them; therefore, one of the major physical forces opposing such concentration is **diffusion**. In the case of the sedimentation of cells or subcellular organelles (nuclei, mitochondria, etc.), the effects of diffusion are essentially nil. However, when centrifugation is used to sediment much smaller particles (such as cellular proteins, nucleic acids, or polysaccharides), the effects of diffusion become significant.

Sedimentation Rate and Sedimentation Coefficient

The RCF or "g force" applied to particles during centrifugation may readily be calculated using equation 12-4 and is independent of the physical properties of the particles being sedimented. However, a particle's **sedimentation rate** at a specified RCF depends on the properties of the particle itself. Also, because the RCF varies directly with x, it is clear that the sedimentation rate *changes* with changing distance from the axis of rotation. (For particles settling under the influence of the earth's gravity alone, the sedimentation rate is constant.) The instantaneous sedimentation rate of a particle during centrifugation is determined by three forces: (1) F_C, the **centrifugal force**, (2) F_B, the **buoyant force** of the medium, and (3) F_f, the **frictional resistance** to the particle's movement.

For a spherical particle P of volume V and density ρ_P, sedimenting through a liquid medium of density ρ_M.

$$F_C = m_P \omega^2 x \qquad (12\text{--}5)$$

The buoyant force is given by

$$F_B = (m_P/\rho_P)\rho_M \omega^2 x \qquad (12\text{--}6)$$

where (m_P/ρ_P) is the volume of the particle and $(m_P/\rho_P)(\rho_M)$ is the mass of the liquid medium displaced by the particle. The frictional force resisting the sedimentation of a sphere through a liquid at 1 cm/sec is given by **Stokes' law**, according to which

$$f = 6\pi\eta r \qquad (12\text{--}7)$$

In this equation, η is the viscosity of the medium and r is the radius of the spherical particle. For a particle sedimenting at a rate other than 1 cm/sec, the frictional resistance would be

$$F_f = f(dx/dt) = 6\pi\eta r(dx/dt) \qquad (12\text{--}8)$$

where dx/dt is the instantaneous sedimentation rate.

If the sedimentation of a particle during centrifugation is viewed as a series of small steplike sedimentation rate increases, then the sedimentation rate *between* increases results from the balance of F_C, F_B, and F_f. That is, at any instant the sedimentation rate results from the fact that

$$F_C = F_B + F_f$$

Accordingly,

$$m_P \omega^2 x = [(m_P/\rho_P)\rho_M \omega^2 x] + [6\pi\eta r(dx/dt)] \qquad (12\text{--}9)$$

Substituting the volume of a sphere (i.e., $\frac{4}{3}\pi r^3$ times its density for its mass into equation 12–9, we obtain

$$(\tfrac{4}{3}\pi r^3)(\rho_P)(\omega^2 x) = (\tfrac{4}{3}\pi r^3)(\rho_M)\omega^2 x + 6\pi\eta r(dx/dt) \qquad (12\text{--}10)$$

By factoring and transposing, we then obtain

$$\tfrac{4}{3}\pi r^3(\rho_P - \rho_M)\omega^2 x = 6\pi\eta r(dx/dt) \qquad (12\text{--}11)$$

Now, the **sedimentation coefficient**, s, is the rate at which a particle sediments under conditions of unit acceleration, that is,

$$s = (dx/dt)/\omega^2 x \qquad (12\text{--}12)$$

By transposing terms in equation 12–11 and solving for s as defined in equation 12–12, we obtain

$$s = \frac{\tfrac{4}{3}\pi r^3(\rho_P - \rho_M)}{6\pi\eta r} = \frac{2r^2(\rho_P - \rho_M)}{9\eta} \qquad (12\text{--}13)$$

Equation 12–13 is very important to the understanding of particle sedimentation and should be carefully examined, for the equation shows that those properties of a particle that determine its rate of sedimentation during centrifugation are *radius* and *effective density* (i.e., the difference between the density of the particle and the density of the liquid through which the particle is sedimenting). Because the sedimentation rate varies in proportion to the square of the particle's radius (but is a first-order function of the particle's density) it is clear that particle *size* is "more influential" than particle density in determing sedimentation rate.

From equation 12–13, it may also be noted that two particles having different sizes or different densities can have similar sedimentation coefficients. On the other hand, two particles with *either* similar size *or* similar density can have markedly different sedimentation coefficients.

The Analytical Ultracentrifuge

The sedimentation coefficient of a particle may be determined experimentally using an instrument known as an **analytical ultracentrifuge**. The rotor that spins in this ultracentrifuge typically contains two compartments. Into one compartment is placed a "reference cell" containing sample-free solvent and the other receives a cell containing the sample to be analyzed. The interior of each cell is *sector shaped* and bounded above and below by parallel quartz windows to permit light from below the rotor to pass through the reference and sample during rotation (Fig. 12–2). As the

FIGURE 12–2 Diagram of an analytical ultracentrifuge rotor (viewed from above) and its sector-shaped sample and reference cells.

particles sediment, *boundaries* are formed at the trailing edges of each particulate species. Changes in the distance between each boundary and the axis of rotation are measured as a function of time by the instrument's optical system. These measurements are then used to calculate the respective sedimentation coefficients of the particles in the suspension. The sharpness of the boundaries that are formed serves as an index of the heterogeneity of the sample. As noted earlier, the first analytical ultracentrifuges were designed and built by T. Svedberg in the 1920s.

The sedimentation coefficients of many cellular macromolecules such as proteins and nucleic acids fall in the range of 1×10^{-13} to 200×10^{-13} seconds (i.e., the dimensions of s are seconds). For convenience, a unit called the **Svedberg unit** (named in honor of T. Svedberg), abbreviated S, is used to describe sedimenta-

tion coefficients and is equal to the constant 10^{-13} seconds. Thus, most cellular proteins have sedimentation coefficients between 1 and 200 S. The sedimentation coefficients of a number of cell constituents are listed in Table 12–1.

Calculation of the sedimentation coefficient from data collected with the analytical ultracentrifuge involves the following amplification of the relationship of equation 12–12. If a boundary is x_1 centimeters from the axis of rotation at time t_1 and x_2 centimeters at time t_2, then equation 12–12 may be solved by integration as follows. First, by transposition,

$$s\,(dt) = \frac{1}{\omega^2}\,(dx/x) \qquad (12\text{–}14)$$

Integrating between the limits set above, we obtain

$$s\int_{t_1}^{t_2} dt = \frac{1}{\omega^2}\int_{x_1}^{x_2} dx/x \qquad (12\text{–}15)$$

and

$$s\,(t_2 - t_1) = \frac{1}{\omega^2}(\ln x_2 - \ln x_1) = \frac{1}{\omega^2}\left(\ln \frac{x_2}{x_1}\right) \qquad (12\text{–}16)$$

Therefore,

$$s = \frac{1}{\omega^2\,(t_2 - t_1)}\ln\frac{x_2}{x_1} \qquad (12\text{–}17)$$

Sample Problem. A solution containing a single particulate species is accelerated in the analytical ultracentrifuge at 60,000 rpm. At time t_1, the boundary between the trailing edge of the sedimenting particles and the axis of rotation is 6.0 cm, while at t_2, 60 minutes later, the boundary is 6.8 cm from the axis. What is the sedimentation coefficient of the particles?

$$\omega = \frac{60{,}000\,(2\pi)}{60}\text{radians per second}$$

$$= 6280 \text{ radians per second}$$

$$s = \frac{1}{(6280)^2(60)\,(60)}2.3\log_{10}(6.8/6.0)$$

$$= \frac{1}{1.42 \times 10^{11}}2.3\,(0.053)$$

$$= 8.6 \times 10^{-13} \text{ or } 8.6\,S$$

As its name implies, the analytical ultracentrifuge is an instrument for analysis and does not physically separate the multiple components of a mixture from

TABLE 12–1 SOME REPRESENTATIVE SEDIMENTATION COEFFICIENTS

Particle	Sedimentation Coefficient, s (in Svedberg Units)
Proteins	
Cytochrome c	1.7
Myoglobin	1.82
Insulin	1.95
Hemoglobin	4.1
Nucleic acids	
Transfer RNA	4
Ribosomal RNA	
(prokaryotes)	5, 16, and 23
Ribosomal RNA	
(eukaryotes)	5, 5.8, 18, and 28
T7 bacteriophage DNA	30
Viruses	
Poliomyelitis	154
Tobacco mosaic virus	180
Simian virus 40	240
Influenza virus	700
Subcellular components	
Nucleosomes	11
Ribosomal subunits	
Prokaryotic ribosomes	30 and 50
Eukaryotic ribosomes	40 and 60
Ribosomes	
Prokaryotic	70
Eukaryotic	80
Lysosomes	4×10^3 to 2×10^4
Peroxisomes	4×10^3
Glyoxysomes	4×10^3
Mitochondria	1×10^4 to 7×10^4
Chloroplasts	1×10^5 to 1×10^6
Nuclei	1×10^6 to 1×10^7

one another. Furthermore, the amount of material that may be studied is quite limited as a result of the small size of the rotor's sample cell.

Preparative Centrifugation

The counterpart to analytical centrifugation is **preparative centrifugation**, which provides for the isolation of cell components for further analysis. Two basic types of centrifuge rotors are regularly employed for conventional preparative centrifugation; these are the swinging-bucket rotor and the fixed-angle rotor (Fig. 12–3). The swinging-bucket rotor consists of a series of metal buckets attached to the central *harness* of the rotor. The samples to be centrifuged (previously placed in centrifuge tubes) are inserted into the buckets, which swing upward from a vertical position to a horizontal position during acceleration of the rotor (Fig. 12–3). During deceleration, the buckets return to the vertical position and the centrifuge tubes are then removed. In a fixed-angle rotor, the centrifuge tubes are maintained at a constant angle (usually 15°

FIGURE 12–3 Swinging-bucket (top) and fixed-angle (bottom) rotors. (Photos courtesy of E. I. DuPont and Co.)

to 45° from verticality) throughout centrifugation (Fig. 12–3). As a result, the radially sedimenting particles quickly strike the tube wall, where convection currents help to carry them to the bottom of the tube.

Differential Centrifugation

During centrifugation, particles sediment through the medium in which they are suspended at rates related to their size, shape, and density. Differences in the sedimentation coefficients of the various subcellular particles provide the means for their effective separation. **Differential centrifugation,** a technique introduced to cellular research in the early 1940s by the noted biologist and Nobel laureate Albert Claude, is one of the classic procedures for isolating subcellular particles and involves the *stepwise* removal of classes of particles by successive centrifugations at increasing RCF.

The material to be fractionated is subjected first to low-speed centrifugation to sediment the largest (or densest) particles present. Following this, the unsedimented material (called the **supernatant**) is transferred to another tube and centrifuged at a higher speed (and usually also for a longer period of time) to sediment particles of somewhat smaller size (and/or lower density). The sequence is repeated several times until all particles have been sedimented; each sediment is then used for further experimentation and analysis. The procedure that is routinely used for the differential fractionation of liver tissue serves as a convenient example of this method and is shown diagrammatically in Figure 12–4.

The removed liver tissue is homogenized in cold buffer and centrifuged 10 minutes at 700 *g*. This is usually sufficient to sediment all the cell nuclei to the bottom of the centrifuge tube, thereby providing the **nuclear fraction.** Depending on the effectiveness of the homogenization procedure, some unbroken cells and large cell fragments may also be recovered in this

FIGURE 12–4   Fractionation of liver tissue by differential centrifugation. Cell nuclei (pale color), mitochondria (solid color), lysosomes (white), microsomes (black), and cytosol (gray).

fraction. The overlying supernatant (called the **nuclear supernatant**) is removed and transferred to another tube for a second centrifugation at 20,000 g for 15 minutes. This sediments nearly all the mitochondria (i.e., the **mitochondrial fraction**). Again, the supernatant (i.e., *mitochondrial supernatant*) is removed and is subjected to a third centrifugation at 105,000 g for 60 minutes. This causes the sedimentation of a fraction called **microsomes**, which includes ribosomes and small fragments of intracellular membranes. The **microsomal supernatant** is referred to as the **soluble phase** of the cells, or **cytosol**, and includes soluble proteins, soluble nucleic acids, soluble polysaccharides, lipid droplets, and other minute particles. In the procedure just described the liver tissue is separated into four major fractions.

Markers   The most widely used methods for identifying the constituents of fractions produced by centrifugation involve microscopy and chemical and bioassays for specific **markers**. Although the light microscope may be used to successfully identify the larger cellular structures such as nuclei, chloroplasts, and mitochondria, the identification of smaller particles such as peroxisomes and ribosomes requires electron microscopy. Certain enzymes and other molecules may be peculiarly associated with a particular cell component and therefore readily serve as markers. For example, the enzyme *succinic dehydrogenase* is found exclusively in mitochondria; consequently, any centrifugal fraction that exhibits succinic dehydrogenase activity would be expected to contain mitochondria. Chloroplast-rich fractions exhibit *ribulose diphosphatase* activity, plasma membrane-rich fractions exhibit *adenylate cyclase* activity, and so on. Table 12–2 lists some of the more useful and reliable markers used in the identification of the components present in the tissue and cell fractions produced during centrifugation. It should be noted that not all markers are enzymes; DNA is an excellent marker of nuclei-rich fractions and chlorophyll is an excellent chloroplast marker.

Differential centrifugation has several major disadvantages. Because the homogenate is initially distributed uniformly throughout the centrifuge tube, the first particles sedimented will necessarily be contaminated with small amounts of all other constituents of the homogenate. This effect is shown in Figure 12–5. As the smaller particles are sedimented, they in turn will be contaminated by even smaller particles. In fact, the only particles obtained in relatively pure

TABLE 12–2 ENZYMES AND OTHER SUBSTANCES USED AS MARKERS DURING TISSUE FRACTIONATION

Component	Marker
Nuclei	DNA
	DNA polymerase
	RNA polymerase
	NAD pyrophosphorylase
Plasma membranes	5′-Nucleotidase
	Leucine aminopeptidase
	Aminopeptidase
	Alkaline phosphodiesterase
	Mg^{2+}-stimulated ATPase
	Leucyl-naphthylamidase
	Nucleotide triphosphatase
	Adenylate cyclase
	Cholesterol
Mitochondria	Succinate dehydrogenase
	Cytochrome oxidase
	Glutamate dehydrogenase
	Monoamine oxidase
	Cytochrome c oxidoreductase
Lysosomes	Acid phosphatase
	Aryl sulfatase c
	Phosphodiesterase-II
	β-glucuronidase
	Acridine orange
Peroxisomes	Catalase
	Uric acid oxidase
	Peroxidase
Glyoxysomes	Glycollate oxidase
	Glyoxylate reductase
Golgi bodies	UDP galactose: N-acetyl glucosamine galactosyl transferase
	Sialyl transferase
Chloroplasts	Chlorophyll
	Ribulose diphosphate carboxylase
Endoplasmic reticulum	Glucose-6-phosphatase
	NADPH-cytochrome c reductase
Cytosol	Aldolase
	Phosphoglucomutase
	Hexokinase

form will be those that sediment most slowly. In the example given above, the initial nuclear fraction would contain some mitochondria, microsomes, and cytosol; the mitochondrial fraction obtained in the second centrifugation would be contaminated with microsomes and cytosol, and so on. These disadvantages may not be serious in some instances, because the major subcellular particles of a tissue homogenate have sedimentation coefficients that differ from one another by one or more orders of magnitude. However, serious

Particles of high sedimentation coefficient
Particles of intermediate sedimentation coefficient
Particles of low sedimentation coefficient

FIGURE 12–5 Illustration of the cross-contamination that can occur during differential centrifugation. At time t_0, all particles are distributed uniformly through the centrifuge tube. During centrifugation (times t_1, t_2, and t_3), all particles sediment at rates that are determined by their sedimentation coefficients. Although the particles that have the highest sedimentation coefficient quickly reach the bottom of the tube, cross-contamination by small quantities of particles having lower sedimentation coefficients is virtually unavoidable.

difficulties arise if the particles to be separated have similar sedimentation coefficients. This problem is illustrated in Figure 12–4*f*, in which the mitochondrial fraction also contains the cells' lysosomes.

Density Gradient Centrifugation

The resolution achieved during centrifugation can be greatly improved if the mixture of particles is confined at the outset to a narrow zone at the top of the centrifuge tube and the particles then permitted to sediment from this position. Initial stability under these conditions can be obtained only if the particles are layered onto a **density gradient**, that is, a column of fluid of increasing density. This technique, known as **density gradient centrifugation**, was introduced to cellular research in the late 1940s and early 1950s by N. G. Anderson and M. K. Brakke and is an exceedingly powerful preparative tool.

Density gradients may be **stepwise (discontinuous) or continuous**. A step gradient is prepared by successively layering solutions of decreasing density in the centrifuge tube. Continuous density gradients are prepared by mixing dense and light solutions in varying proportions at a controlled rate and delivering

the mixture to a centrifuge tube in a continuous stream. A simple procedure for producing a "linear" density gradient is shown in Figure 12–6. In this procedure, the light solution (in the right cylinder) flows into and is mixed with the dense solution (in the left cylinder). The mixture is then delivered to the centrifuge tube. Using this method, a density gradient is formed that decreases linearly as a function of volume between the limiting densities originally present in the two cylinders. Gradients with other shapes (exponential, hyperbolic, etc.) may be prepared using containers of various noncylindrical shapes or gradient-generating devices specifically manufactured for this purpose. Solutes used to provide solutions for density gradients are selected on the basis of their solubility in water and their compatibility with cells and subcellular organelles. Sucrose, Ficoll (a copolymer of sucrose and epicholorhydrin), Metrizamide, Nycodenz, and Percoll (a colloidal form of silica) are the most popular, as they have minimal detrimental effects on organ-

FIGURE 12–6 Density gradients may be produced in many ways. Depicted here is the production of a "linear" density gradient using a two-chamber device. The low-density solution flows from the right chamber into the left chamber (i.e., the mixing chamber) and is rapidly mixed there with the high-density solution. The mixture is then delivered to the centrifuge tube. The density of the liquid in the mixing chamber decreases *linearly* as the volumes of the two chambers progressively diminish.

elles and provide densities up to about 1.3 g/cm Cesium chloride, cesium trifluoroacetate, and other dense salts are used when the limiting density must be considerably higher (i.e., up to 1.9 g/cm). The rate at which particles sediment through a density gradient is given by equation 12–13. Because the quantity $(\rho_P - \rho_M)$ changes as the particles sediment deeper into the gradient, s also changes. If a particle reaches a position in the gradient where the particle's density and the gradient's density are the same (i.e., $\rho_P = \rho_M$), then s becomes zero, and no further sedimentation of the particle occurs. If a particle suspension is overlaid with a gradient whose maximum density is greater than that of the particles (i.e., $\rho_M > \rho_P$), then during centrifugation the particles will *rise* through the gradient, for when ρ_M is greater than ρ_P, s becomes a negative term.

Rate Sedimentation   When the densest region of a density gradient (i.e., the liquid at the bottom of the centrifuge tube) is less dense than the particles being sedimented, all particles will eventually reach the bottom of the tube (if centrifugation is continued long enough). However, if the duration of centrifugation is carefully limited, this will not occur; instead, the particles will be distributed through the density gradient *in the order of their sedimentation coefficient*. Fractionations carried out in this manner are called **rate separations** and are based on the combined contributions of particle size and particle density. The rate fractionation of a liver tissue homogenate is depicted in Figure 12–7. The four major cell fractions (nuclear, mitochondrial, microsomal, and cytosol) are separated during a *single* centrifugation.

Isopycnic Sedimentation   When the densest region of a density gradient is denser than any of the particles to be sedimented, none can reach the bottom of the tube regardless of how long centrifugation is carried out. Instead, the particles will sediment through the gradient until they reach their **isodense** or **isopycnic** position. Fractionations carried out on the basis of particle density alone are called isopycnic separa-

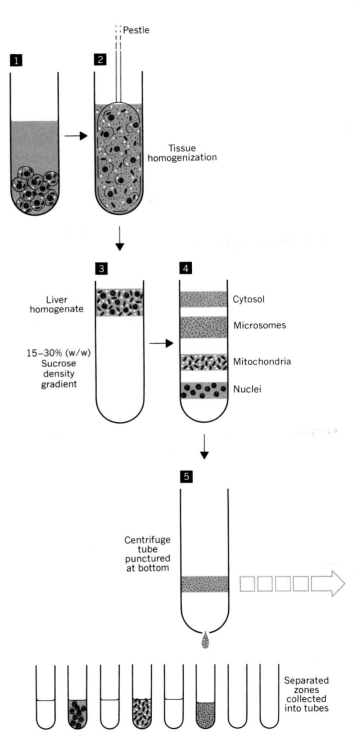

FIGURE 12–7   Homogenization and rate separation of liver tissue by density gradient centrifugation. The homogenate is carefully layered onto the surface of a 15 to 30% w/w sucrose density gradient and is centrifuged at 10,000 *g* for 20 min. The homogenate is thus separated into four major fractions: *nuclear, mitochondrial, microsomal,* and *soluble phase* (i.e., *cytosol*). Following centrifugation, the separated fractions can be collected by puncturing the bottom of the centrifuge tube and permitting the gradient to drip into a series of test tubes.

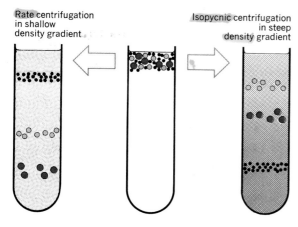

Rate centrifugation in shallow density gradient

Isopycnic centrifugation in steep density gradient

● Particles of large size and intermediate density
◎ Particles of intermediate size and low density
• Particles of small size but high density

FIGURE 12–8 Comparison of *rate* and *isopycnic* density gradient centrifugation. During rate density gradient centrifugation, the particles become distributed through the gradient in an order that is related to their respective sedimentation coefficients. Consequently, a large but not very dense particle may sediment further through the gradient in a shorter period of time than a small, very dense particle. In contrast, during isopycnic density gradient centrifugation, particles become distributed through the gradient in order of their respective densities (i.e., independent of particle size differences).

tions. It is to be noted that this implies that small, dense particles with low sedimentation coefficients sediment further through such a density gradient than large, light particles with high sedimentation coefficients (Fig. 12–8).

In rate separations, both particle size and particle density determine the final positions of particles in the gradient. However, of the two parameters, size is more influential, because s is a second-order function of particle radius but a first-order function of particle density (examine equation 12–13). In isopycnic separations, particle density alone determines final position in the density gradient. Figure 12–9 compares the sedimentation coefficients and densities of a variety of cellular components.

In certain instances, it is not necessary to layer the sample to be fractionated onto a preformed density gradient. Instead, the sample and the density gradient solutions are mixed and placed in the centrifuge tube. During centrifugation at high speed, the density gradient forms automatically within the tube, and particles migrate (upward and/or downward) to their isopycnic positions. This procedure is known as **equilibrium isopycnic centrifugation** and is regularly carried out using solutions of CsCl or other heavy salts (Fig. 12–10).

Although density gradient centrifugation is usually carried out in swinging-bucket rotors, isopycnic (but not rate) separations can also be performed in fixed-angle rotors (Fig. 12–11a). When the gradient with

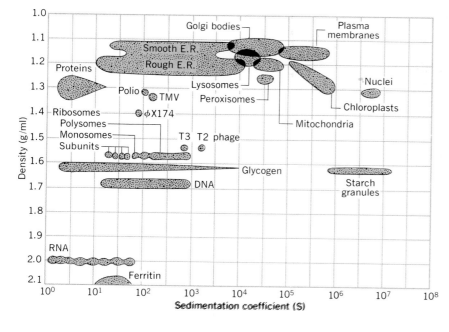

FIGURE 12–9 Densities and sedimentation coefficients of a number of biological particles and macromolecules. The isopycnic banding densities of the organelles are based on their behavior in sucrose density gradients, whereas the banding densities of the viruses and macromolecules are based on results using CsCl salt gradients.

FIGURE 12–10 Equilibrium isopycnic density gradient centrifugation. During centrifugation, the liquid in which the particles were initially (and uniformly) suspended automatically forms a density gradient. The different families of particles either sediment or float to their respective isopycnic positions as the gradient is being formed.

sample layered above is placed in the inclined position in the rotor, the sample zone and equal density planes within the gradient form ellipses. During acceleration of the rotor, each isodense elliptical layer becomes a small section of a *paraboloid of revolution* with focal point on the axis of rotation. The family of paraboloids becomes increasingly steep as rotor acceleration con-

tinues and eventually approaches verticality; this phenomenon is called *gradient reorientation*. Particles in the gradient sediment radially until they encounter the sloping tube wall; from there, they are carried down through the gradient by centrifugal force and convection until they reach the isodense region of the gradient, where they form layers. Rotor deceleration reorients the gradient again so that the separated particles form horizontal layers.

A recent modification of the fixed-angle approach to isopycnic centrifugation involves the use of **vertical tube rotors** in which the fixed angle is reduced to 0 (i.e., the centrifuge tubes are maintained in the vertical position throughout; Fig. 12–11*b*). Here, too, the gradient is reoriented during rotor acceleration and deceleration.

ZONAL CENTRIFUGATION

Density gradient centrifugation using tubes is the most widely employed technique for separating cells and cell organelles and for isolating cellular macromolecules. However, although it is one of the cell biologist's most valuable tools, it is not without disadvantages, as the amount of material that can be

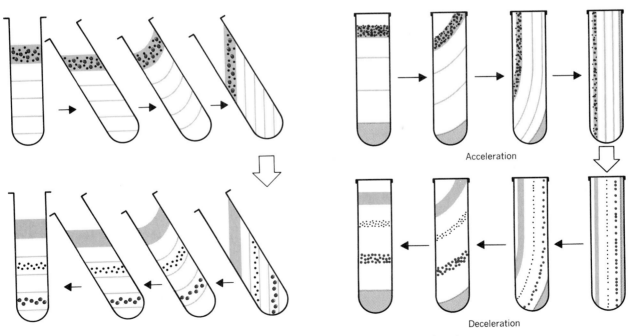

FIGURE 12–11 (*a*) Isopycnic density gradient centrifugation using a fixed-angle rotor. (*b*) Isopycnic density gradient centrifugation using a vertical tube rotor. In both cases, the axis of rotation is to the left of the tube and the density gradient is caused to reorient between the vertical and radial positions during rotor acceleration and deceleration.

fractionated in a single tube is so small. When large quantities of sample must be fractionated (to isolate sparse organelles such as lysosomes or peroxisomes), a very large number of tubes and gradients is needed. Much larger quantities of sample may be fractionated using **zonal rotors**.

A zonal rotor consists of a large cylindrical chamber subdivided into a number of sector-shaped compartments by vertical septa (or vanes) that radiate from the axial core to the rotor wall. The entire chamber is used during centrifugation and is loaded with a single density gradient, each sector-shaped compartment serving as a large centrifuge tube. The large chamber capacity of these rotors (typically 1 and 2 liters) eliminates the need for multiple runs and multiple density gradients. Two basic forms of the zonal rotor are used for tissue fractionation: (1) *dynamically unloaded (rotating-seal) rotors* and (2) *reorienting gradient ("reograd") rotors*. The similarity in appearance of these rotors (Fig. 12–12) is misleading, as the basic principles of operation are substantially different.

Dynamically Unloaded Zonal Rotors

Operation of a dynamically unloaded zonal rotor is depicted schematically in Figure 12–13. Two fluid lines connect the center and edge of the rotor chamber with a rotating-seal assembly that permits loading the density gradient (and sample) into the rotor *while the rotor is spinning*. The center fluid lines open into the rotor chamber through the core section, and the edge lines pass radially through each of the septa and open at the rotor wall.

The zonal rotor is filled with density gradient while rotating at low speed. The light end of the gradient is loaded first through the edge lines and is followed by the denser solution (Fig. 12–13a). The dense end of the gradient gradually displaces the lighter fluid toward the core of the rotor. Addition of a dense "cushion" forces some of the light end of the gradient out of the rotor (Fig. 12–13b). The sample to be fractionated is introduced through the center lines, thereby displacing some of the cushion out of the edge lines (Fig. 12–13c). Additional light fluid (called *overlay*) is then pumped into the center line to push the sample clear of the core region (Fig. 12–13d). Now the upper (stationary) portion of the seal assembly is removed and the rotor is accelerated to a higher speed for separation of the particles in the sample (Fig. 12–13e). The separated particles form a series of concentric cylindrical zones in the rotor bowl. Following particle separation, the rotor is decelerated to a lower

FIGURE 12–12 (*a*) A reorienting gradient (*reograd*) zonal rotor (TZ-28) and (*b*) a dynamically unloaded zonal rotor (B-29). Because of their large bowl size, zonal rotors greatly increase the quantities of tissue that can be fractionated during a single operation. The bowl of the TZ-28 zonal rotor is unloaded through channels in the distributor, core, and septa after the rotor has slowly decelerated to rest and the gradient has been reoriented. The B-29 rotor is unloaded while it is spinning; this is achieved by using a special rotating seal assembly that provides for continuous communication with the core and wall regions of the rotor bowl through channels in the core and septa.

FIGURE 12–13 Operation of a dynamically unloaded zonal rotor such as the B-29 shown in Figure 12–12. Lines drawn inside the rotor's sector-shaped chambers depict hypothetical planes of equal gradient density. The cushion is shown in red.

speed, the static seal is reinserted, and the entire gradient is displaced through the center lines by pumping dense fluid through the edge line (Fig. 12–13*f*). The eluting gradient may be monitored and collected in tubes for subsequent analysis.

Reograd Zonal Rotors

The operation of reograd zonal rotors is depicted in Figure 12–14. Unlike dynamically unloaded zonal rotors, the gradient and sample can be loaded into the reograd rotor either at rest through the septa lines

(a)

Static loading

Gradient in
(light end first)

(b)

(c)

(d)

(e)

(f)

Dynamic loading

(g)

(h)

Gradient in
(dense end first)

Sample in

Rotor accelerated
to speed

Particles separate

Gradient reoriented

Gradient collected
as a series of
fractions

FIGURE 12–14 Operation of a reograd zonal rotor such as the TZ-28 shown in Figure 12–12. Lines drawn inside the rotor chambers depict hypothetical planes of equal density. The cushion is shown in red.

(which communicate with the bowl floor) or while spinning using the core lines. Unloading is *always* carried out with the rotor at rest. If the rotor is loaded at rest, then the gradient is reoriented from the vertical to the radial position during acceleration (Figs. 12–14 *a* and 12–14*b*). During centrifugation, different particles in the sample sediment to form a family of concentric cylindrical zones in the radial gradient (Fig. 12–14*d*), but during rotor deceleration, they are reoriented to form horizontal layers (Figs. 12–14*e*, and 12–14*f*). The density gradient and entrained particles are then withdrawn from the stationary rotor through the septa lines and collected as fractions for further analysis.

EXAMINATION AND ANALYSIS OF SEPARATED CELL FRACTIONS

The final phase of any study involving the fractionation of cells or tissues is the *examination* and *analysis* of the separated fractions so that the morphology, chemical composition, organization, and function of the isolated components may be revealed or better understood. A host of procedures can be employed in such studies, but they will not be pursued here, as they are more properly considered individually in conjunction with closely allied topics in other chapters of the book. For example, the isolation of the plasma membrane-containing fraction (or fractions) from a tissue homogenate might be followed by morphological or biochemical analyses involving electron microscopy, chemical characterization and isolation of the membrane constituents, determination of component function in membrane transport, and so on. These topics are dealt with in Chapters 1, 13, and 15, which specifically consider the principles of electron microscopy, chemical fractionation procedures, and plasma membrane structure and function. Indeed, a considerable portion of this book is devoted to the presentation of the results of examination and analysis of isolated cell components and to the integration of those results into a functional whole.

METHODS FOR SEPARATING WHOLE CELLS

Throughout this chapter, we have been concerned with methods used to separate and isolate particles released from disrupted cells and tissues. Until quite recently, little attention was directed to a related problem—namely, how to separate *whole viable* cells from one another when the tissue or culture being studied is heterogeneously composed. For example, an organ such as the liver is composed of many different types of cells, including hepatic cells, Kupffer cells, connective tissue cells, smooth muscle cells, blood cells, and so on. Therefore, a homogenate of liver tissue contains subcellular particles from diverse kinds of cells. Even a culture of the same cell type may be heterogeneous with regard to cell ages (see Chapter 2) and therefore is representative of a broad spectrum of morphological characteristics and/or physiological activity. The development of methods for separating different types of cells present in a tissue has lagged behind technological advances in the area of subcellular fractionation. In the following section, we will examine some of the problems associated with whole-cell separations and some of the more important methods that have evolved to effect such separations.

Tissue Disaggregation

If the tissue to be fractionated consists of suspensions of individual cells (e.g., a culture of microorganisms, certain tissue cultures, blood cells, and certain tumors), the problem of whole-cell separation is far less difficult than when the cells comprise a solid tissue (such as liver, kidney, brain, etc.). It is therefore not surprising that, to date, most efforts directed toward whole-cell separations have involved natural suspensions of cells as the starting material. Some success has been obtained with solid tissues by employing chemical agents that induce tissue disaggregation—primarily *digestive enzymes* and *chelating agents*. These materials weaken the connections between neighboring cells, making it possible to mechanically disperse the tissue into individual cells without appreciable cell breakage. The tissue may be treated *after* its removal from the animal, although *perfusion* of the organ with a solution of the disaggregating agent *prior* to the organ's excision is more often preferred.

Once the tissue has been reduced to a suspension of individual cells, fractionation into subpopulations follows. If the goal of the experiment is to isolate a particular cell subpopulation for further study, then the remaining cells in the suspension may be *selectively* destroyed or removed by chemical means. For example, the leukocytes of blood may be separated from the erythrocytes by selective destruction of the erythrocytes using *osmotic* or *chemical lysis*. Purification of a particular subpopulation of cells may also be achieved by taking advantage of *differential cell agglutinability* in the mixed population. Simply *freezing* and *thawing* a suspension of cells may differentially lyse specific subpopulations. In general, chemical procedures cause some changes in *all* the cells in the mixed population, so that the method of choice is more generally one that achieves a separation by mild physical means. Among the most popular of the latter methods are **adherence** and **filtration**, conventional and zonal **centrifugation, centrifugal elutriation, unit gravity separation, countercurrent distribution, electrophoresis**, and **fluorescence-activated cell sorting**.

Adherence and Filtration

Separations of cells using differences in adherence phenomena or filtration properties are among the oldest physical procedures used. Some cells readily adhere to glass beads, nylon wool, glass wool, and so on and may be separated from nonadhering cells by passing the cell suspension through a hollow glass column packed with these materials. Success has also been obtained by coating glass or plastic beads with antibodies, antigens, or haptens so that cells will be differentially adsorbed to the beads on the basis of chemical interactions between the plasma membrane and the coating material. Sieves of varying pore diameter can also be used to separate populations of cells on the basis of differences in cell diameter.

Conventional and Zonal Centrifugation

Because of their relatively large size (i.e., in comparison with organelles and macromolecules), whole cells sediment quite rapidly. Consequently, attempts to fractionate suspensions of cells using centrifugation involve rotation at low rpm (i.e., small RCF) for short periods of time (typically less than 500 g for a few minutes). As with subcellular centrifugal fractionations, greatest resolution is obtained using density gradients in which the mixture of cells in the starting zone is separated into subpopulations on the basis of differences in average cell size and/or density. Most cells behave like miniature osmometers, so that strict attention must be paid to the selection of gradient solute. Salts are rarely used to prepare density gradients for cell separations because of their deleterious osmotic effects. Large, impermeable, and biologically inert polymers such as Percoll and Ficoll are the more fre-

quent choices. Because they offer the advantage of greatly increased sample size, reograd zonal rotors have been used with great success for separating different cell populations.

Centrifugal Elutriation (Counter-Streaming Centrifugation)

Centrifugal elutriation is an ingenious technique pioneered in the 1940s by P. E. Lindahl and brought to its present state of the art principally through the work of C. R. McEwen. In centrifugal elutriation, a suspension of cells is pumped into a specially designed rotor chamber through a marginally located entry port (Fig. 12–15), and this is followed by a continuous flow of suspending medium. Centrifugal sedimentation of the cells is opposed by the centripetal flow of the suspending medium. Both effects vary in magnitude across the radial dimension of the rotor chamber, because (1) the centrifugal force increases with distance from the rotor axis and (2) the rate of centripetal liquid flow varies according to the cross-sectional area of the chamber (the area increases exponentially as the liquid travels toward the rotor axis). Depending on its initial position in the chamber and its sedimentation coefficient, each cell will either sediment radially under the centrifugal force or be carried centripetally by the liquid flow. As a result, the cells migrate through the chamber to positions where these two forces cancel one another. Some cells (e.g., the smallest ones) may be swept from the rotor chamber entirely and others form a zone within the rotor chamber and can be collected for further study. Centrifugal elutriation has been successfully applied to separations of blood cells, algae, yeasts, and other cells in culture.

Unit Gravity Separation

The separation of particles on the basis of sedimentation rate differences may not require centrifugation if the particles are sufficiently large. For example, whole cells sediment fairly quickly even at 1 g (i.e., at *unit gravity*). Unit gravity procedures have been used effectively to separate different types of blood cells, tissue culture cells, and populations of microorganisms into subpopulations. The separation is achieved by layering the mixture of cells onto the top of a stationary density gradient and allowing the cells to settle through the gradient for some period of time. The gradient and separate cell populations are then collected as a series of fractions. Devices used to sepa-

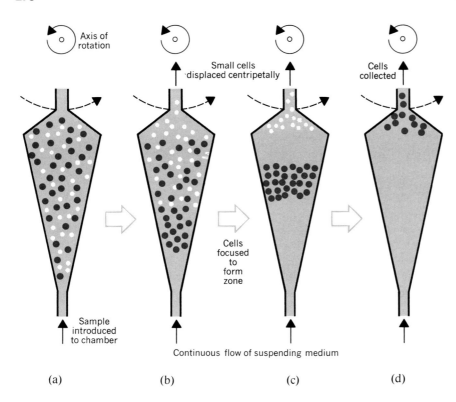

FIGURE 12–15 *Centrifugal elutriation. (a)* The sample consisting of a mixture of cells is introduced into the separation chamber through the chamber's centrifugal edge while the rotor is spinning. *(b)* The combined effects of centrifugal force and centripetal liquid flow displace the cells. *(c)* Small cells are swept from the rotor while the larger cells are focused into a narrow zone. *(d)* The focused cells may then be displaced from the rotor chamber and collected. The walls of the chamber take the shape of an exponential horn (not shown here).

rate cells in density gradients at unit gravity are called "sta-put" devices and vary from simple cylindrical chambers to more elaborate apparatuses having moving, conical end caps. The principle is illustrated in Figure 12–16.

Not only is it possible to separate heterogeneous mixtures of cells using this simple approach, but a population of a single cell type (e.g., a cell culture) can be fractionated according to cell age when age and size are related. In this way, events that occur during successive phases of the *cell cycle* (see Chapter 2) can be studied by examining the cells present in different collected fractions.

A number of methods for separating and isolating the molecular constituents of cells are considered in Chapter 13. Some of these methods have been appropriately modified and applied with varying degrees of success to separations of the different kinds of cells that make up a tissue; *countercurrent distribution* and *electrophoresis* will be mentioned briefly here.

Countercurrent Distribution

Cells (and other particles) may be separated from one another on the basis of differences in their partition between two immiscible liquids; this technique is called **countercurrent distribution** (CCD). Naturally, if the cells are to be separated without undue damage, the milieu selected must be compatible with the cells with respect to ionic composition, concentration, osmotic pressure, and so on. This demand appreciably restricts the selection of immiscible liquids usable for CCD separations of cells, especially in comparison with the range of choices available when countercurrent distribution is employed for molecular separations (see Chapter 13). Greatest success has been obtained with phases consisting of polyethylene glycol and aqueous solutions of dextran (polyglucose in which most glycosidic linkages are 1→6). The technique has been especially fruitful in separations of different microorganisms and different blood cells.

Electrophoresis

Electrophoresis is one of the most popular methods used for separating different molecular species, especially proteins (see Chapter 13). However, electrophoresis can be used to separate whole cells. Cell separations using this technique are based on the fact that the plasma membranes of cells contain ionized groups (e.g., proteins, sialic acid, and short carbohydrate chains) that impart a net electrostatic charge to the

FIGURE 12-16   The separation of cells (or other particles of comparable size) by unit gravity using a "sta-put" chamber. (a) A cushion is introduced into the bottom of the chamber and is followed (b) by introduction of the density gradient. (c) Once the gradient is loaded, the sample is carefully loaded onto its surface, followed by an overlay. (d) The cells separate from one another as they settle through the gradient under the influence of gravity and form a series of different zones (e). (f) The density gradient and separated cells are then collected from the chamber as a series of fractions.

cell surface. Different types of cells possess different net charges so that when they are placed in the appropriate conductive medium and subjected to an electrical current, they will migrate through the medium at different rates. Hence, they become separated into subpopulations that can be collected for further study. Various chemical substances may be applied to the cells to selectively alter their normal surface charge distribution and assist in their electrophoretic separation. Electrophoresis has been applied successfully to separations of microorganisms, blood cells, ascites tumor cells, HeLa cells, and other cells in culture.

Fluorescence-Activated Cell Sorting

Fluorescent dyes such as **fluorescein** can react with and bind to the surfaces of cells; the type and quantity of dye bound vary for different kinds of cells. This differential property has been used for years to *visually* distinguish different types of cells in a mixed population and very recently has been employed in an elegant instrument that physically separates the cells.

The instrument, known as a **fluorescence-activated** (or **"multiparameter"**) **cell sorter** and depicted diagrammatically in Figure 12–17, has a complex history, but its development may be credited to the combined contributions of M. J. Fulwyler, L. A. Herzenberg, R. G. Sweet, W. A. Bonner, and H. R. Hulett. The fluorescein-treated suspension of cells is mixed with electrolyte solution ("sheath fluid") and forced downward through a tiny nozzle vibrating at 40,000 cycles per second. The vibrations of the nozzle break the emerging stream into uniform droplets approximately equal in number to the frequency of vibration (i.e., 40,000 droplets per second). The population density of the original cell suspension and the flow rate are ad-

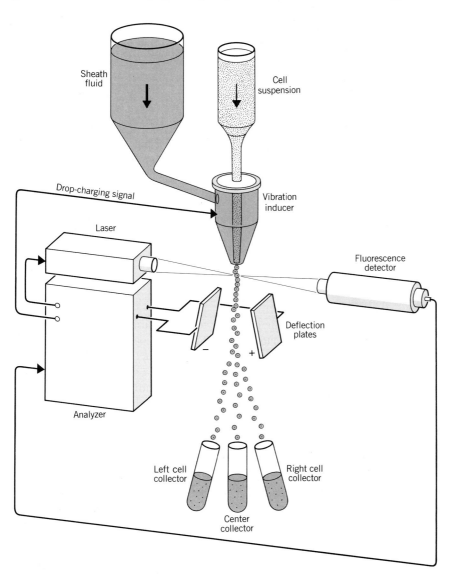

FIGURE 12–17 The major components of a fluorescence-activated (or multiparameter) cell sorter. See the text for details.

justed so that each droplet contains no more than one cell (indeed, most droplets contain no cells).

Just prior to droplet formation, the stream is illuminated with an argon-ion laser beam that excites the fluorescent material in the cell surfaces. Two detectors respectively measure the amount of fluorescent light and the volume of the cell and trigger an electrical pulse that charges each cell-containing droplet as it is formed. The amount and sign of the electrostatic charge borne by the droplet depend on the size of the entrained cell and the number of fluorescein molecules bound to its surface. These charge parameters can be selected by the operator and effectively divide the droplets into *three* classes: positively charged, negatively charged, and uncharged. The droplets then pass between two electrostatic plates; the charged droplets are appropriately deflected as they pass through the field between these plates and the uncharged droplets continue on their original course. Finally, the three droplet streams are collected in reservoirs. The left and right streams contain different population of cells and the undeflected center stream consists primarily of empty droplets, unwanted cells, and debris. The fluorescence-activated cell sorter can separate about 5000 cells per second.

HARVESTING CELLS AND SUBCELLULAR COMPONENTS: CONTINUOUS-FLOW CENTRIFUGATION

This discussion of methods for the differential isolation and separation of cells and subcellular components would not be complete without a description of the use of centrifugation for *harvesting* large quantities of particulate material from large-volume suspensions. The technique is generally known as **continuous-flow centrifugation**.

The most common application of continuous-flow centrifugation is the harvesting of bacteria, algae, protozoa, and other cells grown in multiliter cultures as a preliminary to chemical, physiological, or morphological analysis. However, the technique is also frequently employed (1) to collect cell-free culture media prior to the isolation and assay of cellular excretion products such as enzymes, vitamins, and hormones; (2) to separate blood; (3) to remove the larger subcellular components such as nuclei, chloroplasts, and mitochondria from large volumes of tissue homogenates;

and (4) to collect precipitates from large volumes of aqueous suspensions.

During continuous-flow centrifugation, the suspension of particles is introduced into the spinning centrifuge rotor as a continuous, uninterrupted stream. As the suspension passes through the rotor, particles are sedimented out of the stream and are trapped and concentrated within specific rotor chambers, while the clarified supernatant leaves the rotor and is collected separately. Continuous-flow centrifuge rotors thereby eliminate the need for a series of batch separations when very large volumes of particle suspensions must be processed. When processing is completed, the rotor is simply decelerated and opened, and the trapped cells or particles removed. Although cells or other particles present in multiliter volumes can be harvested using conventional rotors, that approach is far less efficient. Even the largest conventional rotors generally accommodate only a few liters of suspension, so that a succession of spins is necessary when larger volumes of material must be handled. Equally important, the increased size and weight of these rotors restricts their maximum operating speeds and may necessitate extended centrifugation time to ensure total particle "cleanout." Because at any instant continuous-flow rotors contain only a small fraction of the total volume of material to be centrifuged, they may be quite small. Thus, in addition to eliminating the need for successive runs, continuous-flow rotors can be operated at much higher speeds (hence greater RCF), providing more rapid and efficient particle cleanout.

The RCF experienced by the particles as they enter the collection chambers of the rotor depends on rotor speed and causes the particles to sediment at specific rates. If the rate of particle sedimentation is *greater* than the rate at which the surrounding liquid moves toward a centripetal exit port in the chamber, then the particles become trapped in the rotor. However, if the sedimentation rate is *less* than the rate of centripetal flow, the particles are carried toward the exit ports and out of the rotor. Usually, flow rate and rotor speed are selected to provide maximum cleanout of the particle suspension. However, for heterogeneous populations of particles, the flow rate and rotor speed can often be adjusted so that a *differential fractionation* of the particles is achieved, that is, depending on their sizes, shapes, and densities, some particles will be trapped in the rotor, while others are conducted out of the rotor with the supernatant.

SUMMARY

Tissue fractionation begins with the disruption of the cells and the preparation of a subcellular particle suspension or homogenate. Cells are most frequently disrupted using the shear forces generated by special **grinders, blenders, pressure cells,** or **insonators.** Chemical procedures involving lytic agents or osmotic pressure are also used with certain tissues.

Once the tissue homogenate is prepared, the method of choice for separating subcellular organelles and particles is **centrifugation.** Centrifugal fractionation may involve one or a combination of different approaches. In **differential centrifugation,** gross differences in the **sedimentation coefficients** of the subcellular particles are used to produce a series of particulate sediments at higher "*g* forces." Different families of particles may also be separated on the basis of size and/or density differences using the **density gradient** approach. A variety of centrifugal devices are used to effect the purification of subcellular particles, including conventional swinging-bucket and fixed-angle rotors and the more sophisticated vertical tube and zonal rotors.

Heterogeneous mixtures of very large particles, such as whole cells, may be fractionated by **centrifugal elutriation** or simply by **unit gravity** sedimentation. **Electronic sorting, countercurrent distribution,** and **electrophoresis** are alternative methods. When particularly large quantities of cells or particles are to be harvested, **continuous-flow** centrifugal procedures can be employed.

REFERENCES AND SUGGESTED READING

Articles and Reviews

Brakke, M. K., The origins of density gradient centrifugation. *Fractions* **1**, 1 (1979).

de Duve, C., Tissue fractionation—Past and present. *J. Cell Biol.* **50**, 20 (1971).

de Duve, C., and Beaufay, H., A short history of tissue fractionation. *J. Cell Biol.* **91**, 293 (1981).

Grabske, R. J., Separating cell populations by elutriation. *Fractions* **1**, 1 (1978).

Herzenberg, L. A., and Sweet, R. G., Fluorescence-activated cell sorting. *Sci. Am.* **234**, 108 (1976).

Horan, P. K., and Wheeless, L. L., Quantitative single cell analysis and sorting. *Science* **198**, 149 (1977).

Pretlow, T. G., Weir, E. E., and Zettergren, J. G., Problems connected with the separation of different kinds of cells. *Int. Rev. Exp. Pathol.* **14**, 91 (1975).

Rembaum, A., and Dreyer, W. J., Immunomicrospheres: Reagents for cell labelling and separation. *Science* **208**, 364 (1980).

Sharpe, P. T., Cell surface 'chromatography.' *Trends in Biochem. Sci.* **9**, 374 (Sep. 1984).

Shortman, K., Physical procedures for the separation of animal cells. *Ann. Rev. Biophys. Bioengineering* **1**, 93 (1972).

Books, Monographs, and Symposia

Anderson, N. G. (Ed.), *The Development of Zonal Centrifuges and Ancillary Systems for Tissue Fractionation and Analysis* (National Cancer Institute Monograph 21), U.S. Dept. of Health, Education, and Welfare, U.S. Govt. Printing Office, Washington, D.C., 1966.

Birnie, G. D., and Rickwood, D. (Eds.), *Centrifugal Separations in Molecular and Cell Biology*, Butterworths, London, 1978.

Cutts, J. H., *Cell Separation Methods in Hematology*, Academic, New York, 1970.

Hinton, R., and Dobrota, M., *Density Gradient Centrifugation*, North-Holland, Amsterdam, 1976.

Rickwood, D. (Ed.), *Centrifugation: A Practical Approach* (2nd ed.), IRL Press, Oxford, 1984.

Sheeler, P., *Centrifugation in Biology and Medical Science*, Wiley–Interscience, New York, 1981.

CHAPTER 13

THE ISOLATION AND CHARACTERIZATION OF CELLULAR MACROMOLECULES

Over the years, a number of sophisticated analytical and preparative techniques have been developed for separating, analyzing, and isolating the various macromolecular constituents of cells and tissues. Various forms of **electrophoresis, chromatography,** and **ultracentrifugation** are now in routine use in most laboratories engaged in molecular biological studies and have greatly increased our understanding of the chemistry and properties of the cellular macromolecules. Most of the methods that are used to separate and isolate different members of a class of macromolecules simultaneously provide information concerning their chemistry because parameters such as *molecular size, shape, density*, net and absolute *charge, differential solubility*, and so forth are used as the basis for the separation.

Nearly all the techniques that are used for separating and isolating macromolecules require that these substances initially be in a soluble state. Consequently, macromolecules present in extracellular fluids such as plasma, lymph, and hormonal and digestive secretions are most easily isolated and until recently were the subject of the most intensive studies. Now, however, a variety of procedures are available to solubilize macromolecular constituents of

membranes and other particulate components of cells. Therefore, in recent years, there has been a concerted effort to isolate and chemically characterize these macromolecular constituents as well.

In Chapter 12, a number of physical methods were described for disrupting cells. Often, more vigorous or more extensive applications of the same procedures to whole cells or to isolated organelles will also free some of the constituent macromolecules. For example, extended sonification or homogenization of mitochondrial suspensions solubilizes many of the mitochondrial enzymes and other constituents, and nucleic acids may be released from isolated nuclei under similar conditions. Chemical procedures may also be used to extract or solubilize the desired class of macromolecules. Lipids are often extracted from whole cells or isolated organelles using organic solvents such as chloroform–methanol mixtures or acetone. Proteins may be extracted from membranous elements using dissociating agents (such as urea and mercaptoethanol), chelating agents (such as ethylenediamine tetraacetic acid, EDTA) or organic detergents (such as sodium deoxycholate, sodium dodecyl sulfate, and Triton X-100). Whatever the method used, once a soluble mixture of macromolecules is obtained, these may then be separated from one another and isolated using one or a combination of methods.

In this chapter, a number of methods that are either in routine use by cell biologists for isolating and characterizing cellular macromolecules (proteins, carbohydrates, lipids, and nucleic acids) or of significant historical interest will be considered. Most of these methods (see Table 13–1) rely on differences in molecular size, shape, electrostatic charge, solubility, or biological activity to effect the separation.

SALTING IN AND SALTING OUT

Among the oldest methods for fractionating and isolating mixtures of macromolecules (especially proteins) are chemical procedures that differentially alter a molecule's solubility. Among other factors, a protein is maintained in the dissolved (i.e., solubilized) state by the interaction of its charged groups with water (which is partially polar) and with salt ions in the solution (see Chapter 4). The salt concentration (or **ionic strength**) of the protein solution significantly and differentially influences the solubility of the proteins present. For example, many proteins are insoluble in pure (i.e., salt-free) water, whereas addition of small quantities of salt renders these molecules soluble. This effect is believed to be due to an interaction between the salt ions and certain charged groups of the protein that would otherwise react with each other, resulting in insolubility. Even proteins that are soluble in distilled water may be dissolved in much greater quantities by the simultaneous addition of small amounts of salts. This phenomenon is known as **salting in**.

If the salt concentration of a protein solution is successively increased, a point is eventually reached at which some of the proteins begin to precipitate. Further addition of salt results in greater precipitation. Precipitation occurs because spheres of hydration formed around the salt ions effectively "remove" the water molecules necessary to hydrate certain surface charges of the protein (see Chapter 3). Protein–protein interactions begin to dominate over protein–solvent interactions with the resulting precipitation of the proteins. In effect, the proteins are being "squeezed out of solution." This phenomenon is called **salting out**. The effect of salt concentration on the solubility of hemoglobin is shown in Figure 13–1.

TABLE 13–1   METHODS FOR THE ISOLATION AND CHARACTERIZATION OF CELLULAR MACROMOLECULES

Method	Principal Impelling Factor(s)	Principal Retarding or Opposing Factor(s)	Separation Depends Primarily Upon
Countercurrent distribution	Mechanical	Solubility	Differential partition
Dialysis/ultrafiltration	Osmotic effects, concentration gradients, hydrodynamic force	Molecular sieve effects	Molecular size
Ultracentrifugation	Centrifugal force	Friction, bouyancy, diffusion	Molecular size, shape, effective density
Electrophoresis			
Moving-boundary	Electrostatic force	Friction, diffusion	Molecular ionic properties
Zone	Electrostatic force	Friction, diffusion, molecular sieve effects	Molecular ionic properties
Discontinuous	Electrostatic force, Kohlrausch function	Friction, diffusion, molecular sieve effects	Molecular ionic properties and molecular size
Immunoelectrophoresis	Electrostatic force	Diffusion, molecular sieve effects	Molecular ionic properties, biological activity
Isoelectric focusing	Electrostatic force	Diffusion	Molecular ionic properties
Paper chromatography	Hydrodynamic force	Association/dissociation effects, diffusion	Adsorption/partition differences
Thin-layer chromatography	Hydrodynamic force	Association/dissociation effects, diffusion	Adsorption/partition differences
Ion-exchange chromatography	Hydrodynamic force	Electrostatic forces	Molecular ionic properties
Affinity chromatography	Hydrodynamic force	Molecular affinity	Biological activity
Gel filtration	Hydrodynamic force	Molecular sieve effects	Molecular size
Gas chromatography	Gas pressure	Diffusion	Adsorption/partition differences

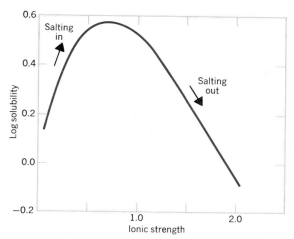

FIGURE 13-1 *Salting in and salting out of* carbonmonoxyhemoglobin using K_2SO_4.

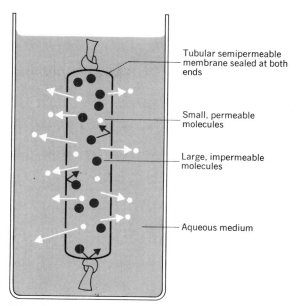

Tubular semipermeable membrane sealed at both ends

Small, permeable molecules

Large, impermeable molecules

Aqueous medium

FIGURE 13-2 Dialysis using a semipermeable membrane. Despite a concentration gradient that favors their diffusion from the dialysis bag, the large molecules are unable to pass through the bag's pores.

ISOELECTRIC PRECIPITATION

The distribution of polar groups in proteins and nucleic acids is influenced by pH. For a given protein, a pH may be identified at which there are equal amounts of positive and negative charge, and this is known as the **isoelectric point** (or **isoelectric pH**). Most proteins are **least** soluble and many proteins are *insoluble* at the isoelectric point, so that their removal from solution is most easily achieved by first adjusting the pH of the solution to the isoelectric point. Electrostatic charge distribution and the isoelectric properties of proteins and other macromolecules are considered in detail later in the chapter in connection with electrophoresis.

DIALYSIS AND ULTRAFILTRATION

Semipermeable membranes, such as those prepared from *cellophane* or *collodion* (cellulose nitrate), may be used to separate solutes on the basis of molecular weight differences. The technique is called **dialysis** and was first described in 1861 by T. Graham. In dialysis, the solute mixture is placed in a bag formed from tubular sheets of the semipermeable membrane and the bag is immersed in an aqueous medium (e.g., distilled water). Molecules larger than the pores of the membrane are confined to the tubing, whereas smaller molecules diffuse into the surrounding liquid (Fig. 13-2). Semipermeable membranes can be treated chemically or physically to alter the sizes of

the pores so that solutes of varying molecular weight are rendered permeable. Generally, dialysis is used with unmodified membranes to quickly separate low-molecular-weight solutes (i.e., molecular weight of less than 5000) such as salts, sugars, and amino acids from proteins and polysaccharides present in the solution. However, high-molecular-weight solutes do *differentially* penetrate membranes having large pore sizes (Table 13-2).

In **ultrafiltration**, force is used to drive the smaller molecules *along with solvent* through the semiperme-

TABLE 13-2 RELATIONSHIP BETWEEN MOLECULAR WEIGHT AND PERMEABILITY TO MODIFIED CELLOPHANE MEMBRANES

Solute	Molecular Weight	$T_{1/2}(min)$[a]
Tryptophan	204	4
Bacitracin	1,422	15
Cytochrome *c*	12,000	60
Ribonuclease	13,600	120
Lysozyme	14,000	138
Trypsin	20,000	240
Chymotrypsin	25,000	300
Pepsin	35,500	4,800

[a]$T_{1/2}$ is the amount of time required for one-half of the solute to permeate the membrane.

FIGURE 13–3 Ultrafiltration through a semipermeable membrane using hydrodynamic force.

able membrane. As a result, not only are permeable and impermeable molecules separated but the impermeable species is also simultaneously *concentrated* (Fig. 13–3).

ULTRACENTRIFUGATION

Centrifugation can be employed not only for the separation of cells, subcellular organelles, and other particulate constituents of cells but also for molecular separations. Since its initial development in the 1920s by T. Svedberg, the **analytical ultracentrifuge** (described in Chapter 12) has been used repeatedly to evaluate the heterogeneity or purity of molecular constituents extracted from cells and to estimate molecular sizes on the basis of sedimentation rate. Preparative separations (as opposed to analtyical studies) became possible with the development of ultracentrifuges capable of spinning rotors that could hold large amounts of sample and that could attain speeds producing an RCF in excess of 500,000 g. Using forces of this magnitude, true separations and isolations of molecular constituents of cells in rate or isopycnic gradients have become routine.

The principles of analytical and preparative centrifugation were considered at length in Chapter 12 and will not be pursued here. However, it should be noted that certain problems are encountered during the fractionation of macromolecules in density gradients that do not arise during particle or organelle fractionations. Most important among these problems is **diffusion**. Macromolecules (or, for that matter, any solute) in solution exhibit a net migration *away* from any region in which they are concentrated and *toward* regions containing a lower concentration of the solute. This movement, called diffusion, results in the *uniform* distribution of molecules throughout the space that they occupy.

The rate at which diffusion occurs is given by Ficke's law:

$$dS/dt = -D(dC/dx) \qquad (13\text{–}1)$$

in which S is the amount of substance undergoing diffusion, C is the concentration of the diffusing solute, t is time, x is the distance through which diffusion occurs, and D is a constant (called the **diffusion constant**) that characterizes the behavior of the solute molecule in a given solvent. It should be noted that the rate of diffusion is proportional to the concentration of diffusing solute, so that the tendency for diffusion increases as the concentration of the diffusing species increases. It follows that the banding of a macromolecule in a density gradient is an attempt to concentrate that molecule and is increasingly opposed by the diffusion of that molecule. For large cellular particles or organelles, the effects of diffusion are negligible because the diffusion constants of large particles are so small. However, the diffusion constants of proteins, nucleic acids, and other macromolecules are considerably greater; consequently, during centrifugation, an equilibrium between sedimentation and diffusion is eventually reached, with the result that no further concentration of a given molecular species by isopycnic banding is possible. In effect, diffusion places a limit on the resolution attainable when separating macromolecular mixtures in density gradients. Diffusion also influences the effectiveness of other methods of macromolecular fractionation, especially dialysis.

ELECTROPHORESIS

The term **electrophoresis** originally described the migration of charged particles through a liquid or semisolid medium under the influence of an electrical potential. More recently, the word has come to refer to any technique by which molecules are separated from one another in electrical potential gradients on the basis of differences in their net charges, sizes, and shapes, regardless of the conducting medium. The method is employed most often for the separation of different proteins, as the side chains of many of the constituent amino acids exist in a dissociated form and contribute some number of positive and negative charges to the macromolecule (see Chapter 4). However, other molecules such as polynucleotides are also effectively separated by electrophoresis.

The fundamental principles of electrophoresis are quite simple. If two electrodes are inserted into a solution containing an electrolyte and a suspension of macromolecules of varying net charge, the macromolecules will be accelerated toward the electrode of opposite sign with a force proportional to the magnitude of the charge on the marcomolecule (which is pH dependent) and the strength of the applied electrical potential. Because each of the migrating particles will encounter frictional resistance to its movement, each will soon attain some maximum velocity of migration. Therefore, if a mixture of these macromolecules is exposed to a constant field strength, the maximum velocities attained will be different for particles of differing net charge. In addition to net charge and field strength, several other factors influence the rate of electrophoretic migration of proteins and polynucleotides, including molecular size and shape and the na-

ture of the medium through which migration occurs. These factors will be considered later.

The rate at which a protein migrates toward one or the other electrode during electrophoresis is called its **electrophoretic mobility** (usually expressed in cm^2 sec^{-1} V^{-1}) and is dependent on the relative numbers of positively charged and negatively charged amino acid side chains. Whether or not a particular amino acid side chain carries a charge is, in turn, determined in part by the pH of the protein solution. As the pH is lowered (i.e., the concentration of H^+ is increased), negatively charged groups such as the secondary COO^- of aspartic and glutamic acid and the O^- of tyrosine become protonated, thereby neutralizing these negative charges. At the same time, some secondary amino groups such as those of lysine and arginine may accept additional protons, thereby increasing the number of positive charges associated with the protein. In contrast, as the pH is raised (i.e., the concentration of OH^- is increased), protons are dissociated from these side chains and make the protein more negative. These relationships are shown in Figure 13–4. Therefore, the numbers and types of charges associated with the amino acid side chains of a protein are determined by pH. At low pH, proteins tend to carry more positive than negative side chains and, therefore, possess a net positive charge and mi-

Aspartic acid residue

$$HC-CH_2-COO^{\ominus} \underset{OH^{\ominus}}{\overset{H^{\oplus}}{\rightleftharpoons}} HC-CH_2-COOH$$

Lysine residue

$$HC-(CH_2)_4-NH_2 \underset{OH^{\ominus}}{\overset{H^{\oplus}}{\rightleftharpoons}} HC-(CH_2)_4-NH_3^{\oplus}$$

$$HC-CH_2-\langle\bigcirc\rangle-O^{\ominus} \underset{OH^{\ominus}}{\overset{H^{\oplus}}{\rightleftharpoons}} HC-CH_2-\langle\bigcirc\rangle-OH$$

Tyrosine residue

FIGURE 13–4 Influence of pH on the charges of amino acid side chains. As the pH is lowered by addition of H^+, negative side chains of certain amino acids accept protons and are neutralized while amino groups of other side chains are protonated and contribute additional positive charges to the protein. In contrast, when the pH is raised, protons are dissociated from certain side chains, increasing the number of negative charges on the protein.

FIGURE 13–5   For each protein a characteristic curve relates electrophoretic mobility to pH. The curves for β-lactoglobulin of blood plasma and albumin of egg white are shown. The respective isoelectric points of these two proteins (indicated by arrows) are 5.1 and 4.6.

grate toward the *cathode* (the negative electrode) during electrophoresis. At high pH, negatively charged side chains predominate and the protein migrates toward the *anode.*

It follows from the above discussion that for every protein there will be a pH at which the total positive charge equals the total negative charge. If electrophoresis is carried out at this pH, no migration occurs, because the protein has no *net* charge. Above this pH, the net charge on the protein becomes increasingly negative, and its electrophoretic mobility toward the anode increases. Below this pH, electrophoretic mo-

bility toward the cathode increases. The relationships between pH and electrophoretic mobility for the proteins egg albumin and plasma β – lactoglobulin are shown in Figure 13–5. The pH at which a protein possesses no electrophoretic mobility is called the **isoelectric point** and is a characteristic of each protein (Table 13–3). Because pH markedly influences electrophoretic mobility, it is important to maintain a constant electrolyte pH during electrophoresis. This is accomplished by including buffers in the electrolyte solutions.

Moving-Boundary Electrophoresis

Electrophoresis of proteins was introduced in its original form by A. Tiselius in 1937 and called "moving-boundary" electrophoresis. For his achievements and his contributions to biochemistry through the development of electrophoretic and chromatographic procedures, Tiselius received the 1948 Nobel Prize in Chemistry. To separate the different proteins in a mixture, the sample (dissolved in a buffer solution that served as the electrolyte and maintained the desired pH) was placed in a glass tube connected to electrodes. When an electrical potential was applied across the tube, the protein molecules migrated toward one or the other electrode. Because different proteins migrate at different rates, a number of *interfaces* or *boundaries* formed between the leading (and trailing) edge of each protein type and the remaining mixture. A Schlieren optical system recorded the number of boundaries formed and their rates of migration. The moving-boundary technique was used for the analysis but not for the fractionation of complex mixtures of proteins, as the proteins in the mixture were not really separated from each other.

Zone Electrophoresis

Zone electrophoresis offers a number of important advantages over moving-boundary electrophoresis. In zone electrophoresis, the separation of the different proteins in a sample is in fact realized. For this reason, the technique can be *both* analytical and preparative. Generally, zone electrophoresis also yields greater resolution of the protein components of a mixture. In zone electrophoresis, the proteins are separated in a semisolid or porous supporting medium such as filter paper or various types of gels (polyacrylamide, cellulose acetate, hydrolyzed starch, etc.);

TABLE 13–3   ISOELECTRIC POINTS OF SOME PROTEINS

Protein	Isoelectric pH
Lysozyme	11.0
Cytochrome *c*	10.6
Ribonuclease	9.5
Normal human hemoglobin	7.1
Myoglobin	7.0
Horse hemoglobin	6.9
Transferrin	5.9
Fibrinogen	5.8
Insulin	5.4
β-Lactoglobulin	5.1
Urease	5.0
Plasma albumin	4.8
Egg albumin	4.6
Haptoglobin	4.1
Pepsin	1.0

these usually take the form of narrow sheets, slabs, or columns.

The principles of zone electrophoresis may be illustrated by considering the filter paper technique as an example. The strip of filter paper is saturated with the buffer/electrolyte solution to be used and is tautly suspended between two (inner) baths (Fig. 13–6). The inner baths communicate with two additional (outer) baths containing the electrodes; the connections may be achieved using baffles or wicks saturated with the buffer/electrolyte. The electrodes are not inserted directly into the inner baths because **electrolysis** would dramatically alter the pH of the bath and also the filter paper strip. By confining the electrolysis to the outer baths, the pH of the inner baths and the filter paper remains constant. The mixture of proteins to be separated is applied as a narrow zone perpendicular to the long axis of the filter paper.

When an electrical potential is applied, the proteins in the sample zone migrate through the filter paper toward the appropriate electrode. In so doing, they form a number of discrete zones distributed along the length of the paper (Fig. 13–6). Following separation, the individual zones or bands can be visualized using special stains. Alternatively, the paper strip may be cut into a number of sections and the proteins in each eluted and collected for further study.

Because the medium through which the proteins migrate is porous, the size of the protein influences its electrophoretic mobility—smaller proteins migrating more rapidly than larger proteins of equal net charge. Consequently, the sieving effect of the supporting medium results in a separation based on both molecular size and charge. Polyacrylamide gels are especially effective for zone electrophoretic separations, because the sieving effect may be varied by changing the concentration of acrylamide used to prepare the gel (i.e., the greater the acrylamide concentration, the smaller the pore size in the gel).

An interesting modification of zone electrophoresis is *diagonal* electrophoresis or *two-dimensional* electrophoresis. In this approach, the sample is separated first on a rectangular sheet of filter paper or polyacrylamide gel at high pH; then, after the supporting medium is rotated 90° and equilibrated at low pH, the separated zones are electrophoresed again. The result is a better resolved distribution of the components in the original sample, as it is highly unlikely that two components would have the same electrophoretic mobility at grossly different pHs.

Disk or Discontinuous Electrophoresis   Discontinuous electrophoresis is a specialized form of zone electrophoresis developed in the early 1960s by L. Ornstein and B. Davis. In view of its widespread use and importance as a research tool, it will be considered separately. The procedure provides the highest degree of resolution so far attainable by electrophoretic methods, and this is a consequence of the extreme thinness of the sample zone. Disk electrophoresis is carried out in either columns or sheets of polyacrylamide gel, the pore sizes of which may be accurately controlled by regulating the concentration of acrylamide in the gel. Consequently, the gel acts like a molecular sieve and separation is based on both the charge and the size of the macromolecule.

Pattern of protein zones produced on the filter paper during electrophoresis

Sample zone

Supporting medium

Support stand

Wick

Wick

Outer buffer/ electrolyte bath

Inner buffer/ electrolyte bath

Power supply

FIGURE 13–6 Essential components of a zone electrophoresis assembly (see text for details).

To separate two proteins by conventional methods of zone electrophoresis, it is necessary to permit migration to continue until one of the proteins has traveled at least one starting zone thickness further than the other. For example, suppose that the sample zone was 1mm thick and contained two different proteins. If one of the proteins has no electrophoretic mobility at all under the pH conditions employed, then the other would have to migrate at least 1 mm for separation of the two proteins to occur. If both proteins migrate in the same direction but at different velocities, then an even greater distance must be traversed by the protein of higher electrophoretic mobility. Because the sharpness of the zone occupied by each protein decreases with time owing to diffusion in the gel, it is most desirable to begin with the narrowest starting zone possible and to minimize the duration of electrophoresis. In disk electrophoresis, the sample zone undergoes a preliminary concentration during which its thickness is reduced from one or more centimeters to just a few hundredths of a centimeter; as a result, high resolution is achieved during very brief runs. The preliminary concentration also makes it possible to analyze samples too small or too dilute to be studied using other methods.

The ability to concentrate and narrow the starting zone during disk electrophoresis is based on a phenomenon first noted and described by F. Kohlrausch in 1897. He showed that if two solutions of ions having significantly different electrophoretic mobilities are layered over one another (e.g., slow ion above, fast ion below) and subjected to an electrical field, the boundary between the two ionic species would be sharply maintained as the ions migrate. That is, under these conditions the ions of lower electrophoretic mobility migrate at the same rates as the faster ions. This may be explained as follows: The velocity at which an ion migrates in an electrical field is determined by the product of its electrophoretic mobility and the applied voltage gradient (V/cm). Therefore, ions of low mobility can migrate as rapidly as ions of high mobility if their mobility–voltage products are equal. In the example cited above, at the instant voltage is applied, the trailing edge of the fast ions at the boundary moves away from the leading edge of the slow ions, resulting in the temporary formation of a zone of reduced conductivity and increased field strength. The increased field strength in this zone accelerates the slow ions to keep up with the fast ions, thereby creating a steady state in which the mobility–voltage products of the two ions remain constant.

Taking this a step further, consider the situation in which a region below the fast ion–slow ion boundary contains a number of ions of intermediate electrophoretic mobility (Fig. 13–7a). When an electrical potential is applied, the increased voltage gradient behind the downward-moving boundary accelerates both the ions of low mobility and those of intermediate mobility. The mobility–voltage product of the latter, however, will be somewhat greater, so that these ions will be swept up by the boundary and form a zone of continuously decreasing thickness between the fast and slow ions (Figs. 13–7b, 13–7c, and 13–7d). If instead of a single ionic species of intermediate mobility, a mixture of ions is placed in the region below the slow ion–fast ion boundary, these would be concentrated into narrow zones stacked one above the other in order of decreasing mobility. In disk electrophoresis, conditions are chosen such that the proteins in the sample have mobilities intermediate to two especially selected fast and slow ions.

Disk electrophoresis gels are cylindrical blocks or flat sheets of polyacrylamide suspended between two reservoirs of buffer/electrolyte (Fig. 13–8). Usually,

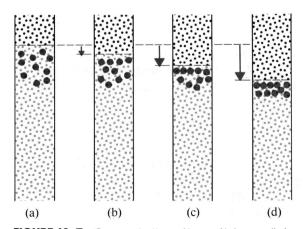

(a)   (b)   (c)   (d)

FIGURE 13–7 Concentration of ions of intermediate electrophoretic mobility (color) at the boundary between fast ions (light dots) and slow ions (dark dots). (a) Before an electrical potential is applied, the ions of intermediate mobility occupy a region below the slow ion–fast ion boundary. (b) When an electrical potential is applied, the boundary moves downward and an increased voltage gradient is formed behind. (c) and (d) The increased voltage gradient accelerates both the slow ions and the ions of intermediate mobility, causing the latter to be concentrated into a narrow zone between the trailing edge of the fast ions and the leading edge of the slow ions. This phenomenon underlies the formation of extremely thin starting zones during disk electrophoresis.

FIGURE 13–8 Essential components of discontinuous electrophoresis (see text for details).

edge of the chloride ions accelerates the slower glycine ions to keep pace. As the Cl⁻–glycine boundary moves downward, it overtakes the more slowly migrating proteins, and they too are accelerated by the increased field strength behind the boundary. As a result, by the time the trailing edge of the chloride ions reaches the end of the stacking gel, all the proteins in the original sample have been concentrated into a series of contiguous thin zones.

Once the separating gel is reached, the change in pH dramatically alters the electrophoretic mobility of the trailing ion. In the case of glycine, the degree of ionization is very low at pH 8.3 but is several times greater at pH 9.5. As a result, the glycine ions now overtake each of the protein zones and catch up with the trailing edge of the Cl⁻ (the mobility of Cl⁻ is unaf-

the polyacrylamide gel is divided into three regions called the *sample gel, stacking gel,* and *separating gel.* The sample gel contains the mixture of proteins to be separated and is prepared using low concentrations of acrylamide so that pore sizes are large and do not influence the rates of migration of different size proteins. The stacking gel is similar to the sample gel but lacks the proteins. The sample gel, stacking gel, and reservoirs have the same pH (usually 8.3). The separating gel differs from the other two regions in that it has a higher pH (usually 9.5) and is prepared with greater concentrations of acrylamide; this results in smaller pore sizes and provides the sieving effect. All three regions of the polyacrylamide gel contain the fast (leading) ion, whereas the buffer/electrolyte contains the slow (trailing) ion. In most instances, Cl⁻ serves as the fast ion and glycine ($NH_2CH_2COO^-$), as the slow ion; other amino acids or weak acids may also serve as the slow ion. At pH 8.3, nearly all proteins have an electrophoretic mobility between those of Cl⁻ and glycine.

When an electrical potential is applied across the gel, the voltage gradient causes the chloride ions to migrate down from the top of the gel. The increased field strength created immediately behind the trailing

FIGURE 13–9 Electrophoretic pattern of human plasma proteins obtained by discontinuous electrophoresis. Five samples (varying in concentration) have been separated in a single slab of polyacrylamide gel. More than 25 different proteins are quickly resolved.

It is apparent from the foregoing discussion that disk electrophoresis differs markedly from conventional zone electrophoresis and is a clever and unusual technique that provides extraordinary resolution (Fig. 13–9). The method has been widely accepted since its introduction and has been applied in diverse cell studies, especially in the analysis of the protein components of isolated cell organelles. An example of this is given in Figure 13–10, which shows more than 30 different protein components separated from the plasma membranes of liver cells.

SDS–PAGE   SDS (sodium dodecyl sulfate) is an ionic detergent that has been used for many years to solubilize membrane proteins so that they can be separated either electrophoretically or by other methods. Because membrane proteins are not usually soluble in aqueous solutions, the SDS is also incorporated into the separating medium; this prevents the proteins from forming an intractable precipitate once they enter the gel. Because the separating medium used with the SDS is polyacrylamide gel, the technique is referred to as **SDS–PAGE** (i.e., SDS–polyacrylamide gel electrophoresis).

SDS dissociates oligomeric proteins into their subunits and forms ionic bonds with the polypeptides. Therefore, the SDS alters the protein's native electrostatic charge and adds to its mass. As a result, when electrophoresis is carried out in small-pore polyacrylamide gels containing SDS, the proteins migrate through the gel at rates determined principally by molecular size (the smaller polypeptides traveling farther through the gel than the larger polypeptides). If in two neighboring *tracks* or *lanes* of the same gel, proteins of known molecular weights are electrophoresed alongside a sample containing proteins of unknown molecular weights, the latter's molecular weights can be estimated by comparing their final positions with those of the "markers." Hence, SDS–PAGE is a powerful tool for both separating and sizing macromolecules.

FIGURE 13–10   Discontinous electrophoretic separation of proteins extracted from isolated plasma membranes of liver cells.

The Maxam–Gilbert Technique   In recent years, the SDS–PAGE method has been extended to the separation of polynucleotide fragments obtained by enzymatic cleavage of DNA and RNA. Not only are the polynucleotide fragments separated on the basis of size differences, but the *order* of the nitrogenous bases in the original DNA or RNA molecule is revealed. The technique is known by the names of its developers, A. Maxam and W. Gilbert.

fected by the pH change), and the new interface then migrates rapidly through the remainder of the separating gel. The proteins, on the other hand, are physically retarded by the smaller pores in the separating gel, and because each protein zone is now in a uniform electrical field, these separate from one another strictly on the basis of net charge and size difference (just as in conventional zone electrophoresis).

To illustrate this ingenious and most valuable procedure, suppose that the base sequence of the following polynucleotide is to be determined:

5'-CTACGTAG-3'

The polynucleotides are subjected to enzymatic cleavage using *endonucleases* that introduce breaks on the 5' side of specific bases, but the amounts of the endonucleases and the lengths of the exposure times permit only *one cleavage per polynucleotide*. As a result, the following fragments are formed:

1. Cleavage at **A** yields **5'-CT-3'** and **5'-CTACGT-3'** fragments.

2. Cleavage at **T** yields **5'-C-3'** and **5'-CTACG-3'** fragments.

3. Cleavage at **G** yields **5'-CTAC-3'** and **5'-CTACGTA-3'** fragments

4. Cleavage at **C** yields **5'-CTA-3'** fragments.

Other fragments are also formed, but they are not radioactive because they do not contain the original 5' ends.

When the four families of fragments are electrophoresed in neighboring lanes of an SDS–polyacrylamide gel, the resulting pattern of radioactive fragments revealed by **autoradiography** (Chapter 14) takes the form shown in Figure 13–11. Fragment **5'-•CTACGTA-3'** produced by cleavage at **G** migrates the

Cleavage at

A   T   G   C

Starting position of each sample

Direction of electrophoresis

(a)

(b)

FIGURE 13-11 (*a*) Maxam–Gilbert procedure for determining the base sequence of polynucleotides. The colored zones show the final positions of the *radioactive* fragments (see text for details).
(*b*) Autoradiograph of a typical sequencing gel obtained using the Sanger dideoxy-chain termination method. From left to right the four lanes represent G, A, T, and C. (Courtesy of M. A. Kashdan, E. I. DuPont and Co.)

most slowly because it is the largest fragment. Therefore, lane **G** has the slowest band; migrating more rapidly is a band in the **A** lane, then the **T** lane, and so on. Reading in order beginning at the top of the pattern, we obtain **GATGCAT**, which is the base sequence of the polynucleotide beginning at its 3' end. Using the Maxam–Gilbert technique, the base sequences in fragments containing several hundred nucleotides are readily determined (Figure 13–11*b*).

Immunoelectrophoresis

Immunoelectrophoresis, a technique developed in the 1950s by C. A. Williams and P. Grabar, combines the principles of electrophoresis and immunochemistry to separate and identify antigens and antibodies. In this technique, a portion of the sample of antigens to be analyzed is first injected into an experimental animal (usually a horse or cow). After a period of time during which the animal produces antibodies against the injected antigens, the immunoglobulin-rich blood serum is collected and prepared as the **antiserum**. The remaining portion of the sample to be analyzed is then subjected to conventional zone electrophoresis in a rectangular block of agar or other supporting medium. This distributes the various proteins in the original sample into a linear series of zones (Fig. 13–12*a*). A trough cut in the agar block parallel to the direction of electrophoretic migration is then filled with the antiserum. The protein zones in the agar block diffuse radially and eventually encounter the front of inwardly diffusing antibody, where the resulting antibody–antigen reaction produces a fine, arc-shaped line of precipitation (Fig. 13–12*d*). Each line of precipitation corresponds to a *pair* of precipitants (a protein in the sample plus an antibody from the antiserum); crossed arcs indicate that all the reactants are immunochemically different proteins.

Isoelectric Focusing

Isoelectric focusing is a form of electrophoresis used to determine the isoelectric pHs of proteins and to separate proteins from one another on the basis of differences in their isoelectric points. The method was developed in the 1960s by H. Svensson and O. Vesterberg and is extremely sensitive; proteins having isoelectric pHs that differ by as little as 0.0025 units may be resolved. In one form of isoelectric focusing, a glass column is filled with a solution of synthetic

FIGURE 13–12 Immunoelectrophoresis. (*a*) Mixture of proteins separated by zone electrophoresis. (*b*) and (*c*) Trough filled with antiserum followed by diffusion of protein zones and antibodies through the gel. (*d*) Antibody–antigen reaction forms lines of precipitation (arcs) where the two diffusing fronts meet. (*e*) By drawing a perpendicular line from the widest point of the arc to the path of electrophoretic migration, the center of concentration of each protein can be determined.

charged ampholytes (aliphatic aminocarboxylic acids of various molecular weights) having a broad and continuous range of isoionic points. The ends of the column contain the electrodes and electrode solutions: a strong organic base for the cathode (usually ethanolamine) and a strong acid for the anode (usually phos-

FIGURE 13–13 Essential features of isoelectric focusing. (a) Three proteins having different isoelectric points are shown distributed through the ampholyte–sucrose density gradient. When an electrical potential is applied, the ampholytes migrate to their isoionic points, establishing a pH gradient. (b) This is followed by electrophoresis of proteins to their isoelectric positions in the gradient, forming a series of narrow zones.

phoric acid). The distribution of constituents in the column is stabilized by the presence of a sucrose density gradient that also prevents convection during electrophoresis. The mixture of proteins to be separated may initially be confined to a specific region of the column or uniformly distributed through the density gradient (Fig. 13–13).

When a potential is applied to the electrodes, the ampholytes migrate in the column to positions where they are equally attracted by both electrodes. Thus, they arrange themselves in order of their isoelectric points, the most acidic ampholytes being located near the anode and the most basic near the cathode. An examination of the pH in different regions of the column at the time would reveal that a pH gradient has been established in which pH increases in progressing from the anode to the cathode. Depending on the pH of the surrounding solution and the nature of the amino acid side chains present in the molecule, each protein assumes a characteristic net charge. Proteins close to the anode will bear a net positive charge and be re-

pelled by that electrode, whereas the same proteins near the cathode carry net negative charges and are repelled by that electrode. As a result, most proteins will migrate away from the electrodes and toward the center of the column. As this occurs, the proteins pass through the pH gradient and the charges on their amino acid side chains are altered. Proteins migrating away from the anode become less and less positive as they pass through regions of increasing pH, whereas proteins migrating away from the cathode become less negative (see Fig. 13–13). Eventually, the net charge of each protein becomes zero as that region of the pH gradient corresponding to the isoelectric pH is reached, and at that time the electrophoretic migration of the protein ceases. Consequently, the proteins become distributed through the gradient as a series of narrow zones in order of their isoelectric points (see Fig. 13–13). Once the migration of all proteins terminates, the contents of the column may be drained and the separated proteins collected as a series of fractions and further studied. The isoelectric point of each protein is determined from the pH profile of the collected gradient.

For analytical applications, the liquid density gradient is replaced by a solid medium such as polyacrylamide gel. Several different samples, including standards of reference, can be focused in neighboring lanes of the gel.

A description of all electrophoretic methods currently in use is beyond the scope of this discussion. However, it should be recognized that other forms of electrophoresis such as *density gradient* electrophoresis and *continuous-flow* electrophoresis are also regularly employed. In density gradient electrophoresis, the separation of proteins is carried out in a glass column filled with buffer and stabilized by a density gradient. Continuous-flow electrophoresis provides for the continuous application of small volumes of sample onto the supporting medium, coupled with the continuous removal of the proteins as separation is achieved. Both zone electrophoresis and isoelectric focusing have continuous-flow formats. With this technique it is possible to process very large volumes of starting material.

Although all methods of electrophoresis have proven extremely valuable for the separation and identification of proteins, certain limitations of the technique should be noted. The resolution of a protein mixture into a number of discrete zones does not guarantee that all the different proteins present in the original sample have been separated, for two or more different proteins having similar sizes may also have the same net charge under a given set of conditions and display the same electrophoretic mobilities (see Fig. 13–5). These proteins would not be resolved by electrophoresis. By the same token, two different proteins may have the same isoelectric pH and would not be separated by isoelectric focusing. Therefore, to evaluate the effectiveness of a separation and to determine the purity of the separated fractions, it is necessary to carry out electrophoresis under a variety of pH conditions or with a variety of supporting media or even to apply a combination of altogether different methods in the analysis.

COUNTERCURRENT DISTRIBUTION

A brief, preliminary description of countercurrent distribution was given in Chapter 12 in connection with the separation of whole cells in mixed populations (e.g., blood). However, the technique lends itself more directly to the separation of various molecular species in solution, and because the physicochemical principles in effect during countercurrent distribution are the foundation of many other separation procedures (i.e., paper chromatography, thin-layer chromatography, ion-exchange chromatography, affinity chromatography, gas chromatography, etc.), this method will be treated in some detail.

It has been known for a century or more that many solutes differentially distribute or *partition* themselves between the separated phases of two immiscible liquids. Solutes having gross differences in their physical properties may therefore be separated by one or a few simple extraction procedures (e.g., using separatory funnels). Reasoning that the separation of a complex mixture of solutes on the basis of their partition differences could be made considerably more effective using a *cascade* of several dozen, several hundred, or even several thousand discrete extraction steps, L. C. Craig in the late 1940s and early 1950s developed a number of special instruments to achieve this. These instruments and their more modern counterparts are generally known as Craig apparatuses and the technique as **countercurrent distribution.**

The fundamental type of Craig apparatus consists of a series or *train* of chambers or "cells" divided into upper and lower compartments that house the respective immiscible liquid phases (e.g., butanol/water, phenol/water, propanol/water, etc.). The upper phase is usually the *mobile* phase and the lower phase is the *stationary* phase. Equilibration of a dissolved solute between the phases within a chamber is achieved by agitation, so that one phase becomes finely dispersed in the other (thereby greatly increasing the interfacial surface area between the two phases). Once equilibration is achieved, the phases are allowed to separate. The mobile phase is then transferred to the next chamber of the train and mixed with a fresh volume of stationary phase. Simultaneously, a fresh volume of mobile phase is added to the first chamber. Agitation, equilibration, separation, and transfer are then serially repeated for the entire train of chambers.

A solute dissolved in a given pair of immiscible liquids will partition itself in a characteristic way, with a specific solute concentration ending up in each phase. The **partition coefficient,**

$$P = \frac{\text{concentration of solute in upper phase}}{\text{concentration of solute in lower phase}} = \frac{(S)_U}{(S)_L} \quad (13\text{–}2)$$

For example, if a solute has a partition coefficient of 1.0 in a butanol/water mixture, it will be *equally* divided between the butanol and the water phases once equilibration and separation have been achieved. Note that the solute is soluble in *both* liquids but partitions itself between the two in a characteristic way.

Figure 13–14 depicts the countercurrent distribution of 128 molecules of a hypothetical solute having P = 1.0 when carried through a train consisting of seven sets of chambers. In this example, the peak of the solute distribution in the train occurs in chamber number 3, which contains 40 molecules (or 31%) of the original solute. For a train consisting of (C + 1) chambers (numbered C_0, C_1, C_2, C_3, etc.), each containing phases of equal volume, the fraction F of solute in compartment C_n at the conclusion of the countercurrent distribution is given by

$$F = \frac{C! P^n}{n! \, (C-n)! \, (P+1)^C} \quad (13\text{–}3)$$

where n is the number of transfers needed to reach chamber C_n (and is equal to the chamber number). For the example given in Figure 13–14, the fraction of solute in chamber C_3 would be $6!(1.0)^3/3!(6-3)!(2.0)^6$, which equals 31%, as already noted. Figure 13–15 tabulates the final distributions of a mixture of five solutes having different P values through a train of seven chambers and illustrates how the peak concentrations of each solute would be found in different chambers in the train. Although it may be noted that in this example considerable overlap exists among the chambers, it should be borne in mind that the number of chambers that make up the train has deliberately been kept low to simplify the exam-

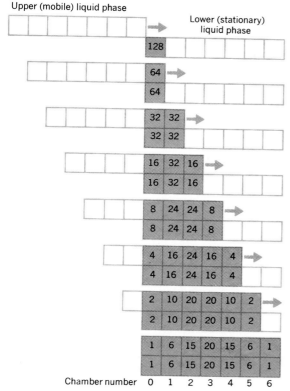

Upper (mobile) liquid phase

Lower (stationary) liquid phase

Chamber number 0 1 2 3 4 5 6

FIGURE 13–14 Countercurrent distribution of 128 molecules of a hypothetical solute having a P value of 1.0 in a train of seven chambers. At the outset, 128 molecules are dissolved in the lower half of chamber 0 and then equilibrated with an equal volume of mobile phase; this results in the equal partition of the solute in the two solvents. The upper phase of chamber 0 is then transferred and mixed with the lower phase of chamber 1, and fresh upper phase is added to chamber 0. Equilibration produces a new partition in chambers 0 and 1. The process is repeated until all seven chambers (i.e., C_0 through C_6) of the train have been used, yielding a final distribution of the solute. In practice, from several dozen to several thousand partition steps may be employed.

P \ C	0	1	2	3	4	5	6
0.2	33	40	20	5	1	0	0
0.5	9	26	33	22	8	2	0
1.0	2	9	23	31	23	9	2
2.0	0	2	8	22	33	26	9
3.0	0	0	3	13	30	36	18

Rounded off to the nearest whole percent.

FIGURE 13–15 Percentage distributions in each chamber of a seven-chamber train of five solutes having various P values. Attention is drawn (colored squares) to those chambers containing the peak concentration of each solute.

In a modified version of the original Craig approach (called *countercurrent chromatography*) a horizontal coiled length of tubing is filled with the stationary phase and the sample to be fractionated is introduced at one end. A mobile phase is then continuously pumped through the coil, progressively sweeping the sample toward the other end. In each turn of the coil, partition of the sample between the stationary and mobile phases brings about a progressive fractionation of the original mixture. In one of its most elegant forms, the coil is oriented vertically and rotated around its own axis and/or a central axis so that centrifugal and hydrodynamic forces enhance the parti-

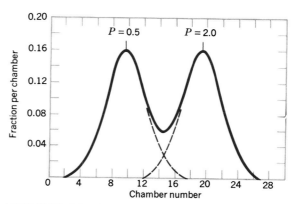

FIGURE 13–16 Countercurrent distribution of solutes having P values of 0.5 and 2.0 in a train consisting of 30 chambers.

ple. Usually, a train consists of at least several dozen chambers. Accordingly, Figure 13–16 shows the resolution attainable for solutes having P values of 0.5 and 2.0 when the train consists of 30 chambers. It can be shown that the best resolution of two solutes, A and B, is achieved when $P_A \times P_B = 1.0$. The number of chambers occupied by a particular solute as a function of the total number of chambers present in the train rapidly diminishes as n increases. Hence, it is clear why trains in which hundreds or even thousands of transfers are possible may be used to provide maximum resolution of the solutes present in the original mixture.

tion and the separation of the mobile and stationary phases.

The principles and mathematical relationships given above for a series of discrete partition transfers during countercurrent distribution apply equally well to seemingly quite distinct physical separation methods such as the various forms of chromatography described later in the chapter. In chromatography, the partition steps occur on numerous particles of tightly hydrated media such as starch, agarose, cellulose, silica, and paper fibers. This tightly bound *solvent* serves as the stationary phase and a mobile phase containing solvent and the solute mixture to be fractionated flows by. This *continuous* process may be thought of as an infinite sequence of discrete, microscopic partition steps. Each microscopic step would be analogous to the stages shown in Figure 13-14, the upper mobile phase representing a continuously flowing solvent and the lower stationary phase the stationary, hydrated supporting medium. Although equilibrium is not truly achieved at each "step," the number of steps is so great that the solute mixture is effectively fractionated.

Countercurrent distribution is particularly effective when dealing with molecules having molecular weights below 5000 (peptides, oligonucleotides, phospholipids, etc.), but the technique is also used successfully with larger molecules. The first protein isolated and purified using countercurrent distribution was the hormone insulin, which has a relatively low molecular weight (i.e., 6500). However, *ribonuclease* (MW 13,600), *lysozyme* (MW 14,000), albumin (MW 68,000), and the alpha and beta chains (MW 16,000) of human hemoglobin have also been isolated using this approach. Little success is obtained with polysaccharides because of the problem of finding an organic phase that provides useful partition coefficient differences. Ribonucleic acids are readily separated using countercurrent distribution; indeed, R. W. Holley's pioneering analysis of the primary structure of transfer RNA (Chapter 21) was carried out using alanyl-tRNA isolated and purified by countercurrent distribution. Lipids, especially steroids, are well suited for separation using this approach because of their greater solubility in organic solvents and the wide choice of solvent combinations that can be used.

PAPER CHROMATOGRAPHY

Paper chromatography is a technique in which a mixture of solutes is separated into discrete zones on a sheet of filter paper on the basis of (1) differences in solute partition between a stationary aqueous phase tightly bound to the cellulose fibers of the paper and a mobile organic liquid phase passing through the sheet by capillary action (i.e., liquid–liquid partition) and (2) differences in solute *adsorption* to the cellulose fibers and dissolution in the mobile liquid (i.e., solid–liquid partition). Although the separation is based on a combination of both phenomena, liquid–liquid partition differences are the more significant. The principles of the technique were set down in the 1940s by R. Consden, A. H. Gordon, A. J. Martin, and R. L. Synge and are not unlike those that are in effect during countercurrent distribution.

In practice, a rectangular sheet of filter paper is saturated with the aqueous phase and allowed to air dry, and the sample is applied near the end of the sheet as a narrow zone. The sheet is then suspended in a closed chamber in which the air has been saturated with the vapors of the mobile organic phase and the edge of the paper is immersed in a bath containing the mobile phase (Fig. 13-17a). Capillary action causes the mobile phase to slowly percolate through the paper from one end (the end containing the mixture to be separated) to the other. Movement of the liquid may be downward (descending chromatography) or upward (ascending chromatography). The solute mixture differentially partitions itself between the flowing solvent and the stationary phase repeatedly as the solvent front advances toward the edge of the paper. Usually, the solvent front is allowed to migrate through the paper until it has almost reached the other end, at which time the sheet is removed and dried and the solute zones located by the appropriate chemical or physical means. However, in certain instances where the solutes trail far behind the solvent front, it is desirable to allow the solvent to run off the edge of the paper sheet (descending chromatography only) so that maximum resolution of the solutes is achieved.

The rate of movement of a solute during paper chromatography is usually expressed as the dimensionless term R_f, which is the *ratio* of the distance traveled by the solute to the distance traveled by the solvent front. Naturally, the R_f can be calculated only in those instances when the solvent is not allowed to leave the end of the paper sheet.

Greater resolution of the solutes may be obtained using *two-dimensional paper chromatography* (Fig. 13-17b); after chromatography using a particular solvent system in one direction along the paper sheet,

Ascending

(a)

Descending

(b)

FIGURE 13–17 Paper chromatography. (*a*) Essential features of an ascending (left) and descending (right) chromatographic apparatus. (*b*) Stages in *two-dimensional* chromatography.

the sheet is dried and rotated 90°, and another solvent system is used to chromatograph the solutes a second time. In this manner, solutes not fully separated by partition in the first solvent may be completely separated using the second solvent. Paper chromatography can also be combined with zone electrophoresis to provide two-dimensional analysis of a solute mixture, a technique known as "fingerprinting" (see Chapter 22).

THIN-LAYER CHROMATOGRAPHY

Thin-layer chromatography (abbreviated TLC) is an especially valuable method for rapidly separating unsaturated and saturated fatty acids, triglycerides, phospholipids, steroids, peptides, nucleotides, and nu-

merous other biological substances. In effect, TLC is a modification of paper chromatography in which the sheets of filter paper are replaced by glass or plastic plates covered with a thin, uniform layer of adsorbent. The essential features of the technique may be described as follows. An aqueous slurry of the selected adsorbent is uniformly spread over a glass plate to produce a thin layer and is then dried. Following this, the sample (usually prepared in a volatile solvent) is applied near one end of the long axis of the plate as a spot or thin line. When the sample has dried, the plate is supported vertically so that the end near the sample zone is immersed in a tray containing a shallow layer of the eluting solvent (usually an organic solvent of low polarity). Capillary action causes the solvent to ascend slowly through the layer of adsorbent and, as in paper chromatography, the solutes become distrib-

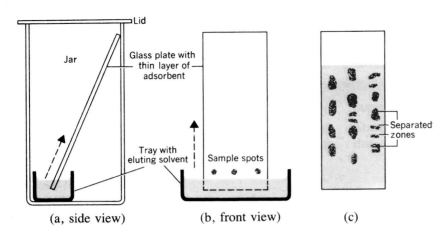

FIGURE 13–18 Thin-layer chromatography. (*a*) and (*b*) Glass plate containing thin layer of adsorbent and applied samples is immersed in tray containing shallow layer of solvent. Capillary action causes solvent to ascend through adsorbent (arrows), which separates sample spots into series of zones (*c*).

uted along the plate on the basis of differential partition between the stationary and mobile phases (Fig. 13–18). Although ascending TLC is the most common, descending and horizontal separations may also be carried out. TLC separations are very rapid, rarely exceeding 20 to 30 minutes.

After the separation has been achieved the glass plate is removed from the tray of solvent and allowed to dry. Zones containing colored substances can be detected directly, and others may be identified if they contain compounds that fluoresce when exposed to ultraviolet light. Many zones may also be rendered visible by spraying the plate with certain reagent dyes or stains. The adsorbent may also be scraped off various regions of the plate and the separated molecules eluted from the adsorbent particles. Table 13–4 lists some frequently used adsorbents and their applications.

TABLE 13–4   ADSORBENTS USED FOR THIN-LAYER CHROMATOGRAPHY

Adsorbent	Materials Separated
Silica gel	Amino acids, polypeptides, fatty acids, steroids, phospholipids, glycolipids, plasma lipids
Alumina	Amino acids, steroids, vitamins
Kieselguhr	Oligosaccharides, amino acids, fatty acids, triglycerides, steroids
Celite	Steroids
Cellulose powder	Amino acids, nucleotides
Hydroxylapatite	Polypeptides, proteins
Polyethylenimine	Nucleotides, oligonucleotides

ION-EXCHANGE CHROMATOGRAPHY

Proteins and other macromolecules may also be separated by the technique known as **ion-exchange chromatography**. The separation is usually carried out in tall glass columns packed with grains of a synthetic ion-exchange *resin* (polymers to which numerous ionizable groups have been chemically added). Resins bearing negative charges are called *cation exchangers* and positively charged resins are *anion exchangers*. As a solution of ions is passed through the column, the ions compete with each other for the charged sites on the resin. Consequently, the rate of movement of any ion through the column depends on its affinity for the resin sites, its degree of ionization, and the nature and concentration of competing ions in the solution. The differential rates of movement of ions through the column are the basis for protein and nucleic acid separations, because these molecules possess a variety of positively and negatively charged groups. Some ion exchangers used for protein and nucleic acid separations are listed in Table 13–5. Among the resins most widely used for protein separations are diethylaminoethylcellulose (DEAE-cellulose, an anion exchanger) and carboxymethylcellulose (CM-cellulose, a cation exchanger) developed by E. A. Peterson and H. A. Sober.

It is generally believed that the interaction between a resin and a protein involves the formation of electrostatic bonds between the charged sites on the resin particle and oppositely charged dissociated side chains of certain amino acids. Because a number of bonds are formed, proteins are more firmly bound than singly charged substances. For example, at the

TABLE 13-5 COMMONLY USED ION EXCHANGERS

	Polymer	Functional (Ionic) Group
Anion exchangers		
Diethylaminoethylcellulose	Cellulose	$-O-(CH_2)_2-N^+H-(CH_2H_5)_2$
QAE Sephadex	Cellulose	$-CH_2-N^+-(C_2H_5)_2CH_2\overset{\ast}{C}HOHCH_3$
Dowex-2; Bio-Rad AG2	Polystyrene-divinyl-benzene	$-N^+-(CH_3)_2C_2H_5OH$
Polylysine–kieselguhr (PLK)	Polylysine	$-(CH_2)_4-NH_3^+$
Amberlite IRA-40; Bio-Rad AG1	Polystyrene-divinyl-benzene	$-N^+(CH_3)_3$
Cation exchangers		
Carboxymethylcellulose	Cellulose	$-O-CH_2-COO^-$
Zeocarb 225; Bio-Rad AG50	Polystyrene	$-SO_3^-$
Hydroxylapatite	—	$-PO_4^{3-}$
Amberlite IRC 50	Polymethacrylate	$-COO^-$
Sephadex SE	Dextran	$-(CH_2)_2-SO_3^-$

appropriate alkaline pH, DEAE-cellulose could form bonds with the negatively charged side chains of aspartic acid, glutamic acid, and tyrosine residues present in a protein. This also implies that the affinity of a protein for the resin can be altered by a change in pH. A reduction in pH sequentially suppresses the formation of bonds between the protein and DEAE-cellulose as hydrogen ions bind to the negative side chains of the protein and displace the resin. A similar result would be effected by the addition of salt ions, which would compete with proteins for the resin sites (Fig. 13–19).

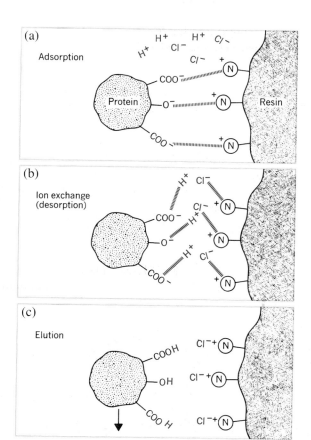

FIGURE 13–19 Desorption of protein from an anion-exchange resin by reducing the pH of the solvent. (a) A theoretical protein molecule is adsorbed to the resin by forming electrostatic bonds between certain dissociated amino acid side chains and oppositely charged sites of resin. (N^+ represents any of several quaternary nitrogen groups associated with anion exchangers; see Table 13–5.) (b) As the pH of the solvent flowing through the column is reduced, hydrogen ions compete more effectively for the dissociated amino acid side chains and break the bonds with the resin. (c) Anions in the solvent (in this case Cl⁻) bind to the resin and the released protein is carried away with the flow of solvent. A similar effect is produced by increasing the salt concentration of the solvent. Generally, altering the pH and increasing the salt concentration are carried out simultaneously so that a combination of interactions results in proten desorption.

The separation of proteins by ion-exchange chromatography is carried out as follows. The resin is suspended in a buffer or salt solution (called the starting solvent) and the resulting slurry is used to fill the chromatographic column. At this time, the charges on the resin are neutralized by ions in the solvent. The sample (also dissolved in the starting solvent) is applied at the top of the column as a narrow zone. Depending on the pH and salt concentration of the solvent, certain proteins present in the sample will form electrostatic bonds with sites on the resin (displacing solvent ions formerly bound at those sites) and the remaining proteins remain in the solvent. If a volume of the starting solvent is now passed through the column, the unbound proteins will be carried away with the flow and elute at the base of the column. To displace other proteins from the sample zone, it is necessary to change to another solvent whose pH or ionic strength will alter the degree of ionization of the amino acid side chains or compete more effectively with the proteins for the charged sites of the resin. In the case of proteins adsorbed to DEAE-cellulose or other anion exchangers, this can be accomplished by decreasing the pH and increasing the salt concentration of the solvent. If the second solvent differs only slightly from the first, only a small group of proteins will be released from the sample zone. During their descent through the column, transient bonds will be formed with the resin many times as the proteins and solvent ions compete for the resin sites. Depending on the relative affinities of the proteins for the resin and the solvent ions, different proteins may reach the bottom of the column at different times. Thus, changing the solvent not only releases an additional group of proteins but also separates them into a series of zones during their passage through the column. The sequential addition of a series of solvents of different pH and salt concentration, called *stepwise elution*, eventually displaces and separates all the proteins in the sample zone. Alternatively, the proteins may be eluted by passing a solvent of continuously changing pH and ionic strength through the column—a method known as *gradient elution*. The effluent from the column is collected as a series of fractions (Fig. 13–20).

The conditions required for protein desorption depend on the number of bonds formed between the protein and resin and on the nature of the ionized amino acid side chains. Therefore, two proteins having the same *net* surface charge density (and, consequently, similar electrophoretic mobilities) might be desorbed under quite different conditions and emerge from the

FIGURE 13–20 Diagram of components used in ion-exchange column chromatography. Gradient elution method is shown. The solvent is pumped through the optical unit of an ultraviolet light absorption monitor before entering the column. Flow out of the column is also monitored and the two absorbances are compared to identify that portion of the effluent containing proteins. The absorbances are plotted on a strip-chart recorder. Peaks in the tracing (called a chromatogram) correspond to protein zones emerging from the column.

column at different times if their *absolute* surface charges differ. For example, under a given set of conditions, one protein may bear eight negative and five positive amino acid side chains and have a net charge of − 3; another may have four negative and one positive side chain and be more weakly bound to the resin (desorbing earlier) but would have a similar electrophoretic mobility (i.e., its net charge would also be − 3). For the same reason, two proteins desorbed under the same conditions and emerging from the column together may have quite different electrophoretic mobilities. Consequently, the re-chromatography of a given protein fraction or its further examination by electrophoresis is usually recommended to evaluate the purity of isolated components.

Although the discussion of ion-exchange chromatography has centered around protein separations, it should be recalled that mixtures of different nucleic acids may also be resolved using this technique. Polylysine–kieselguhr columns have been particularly successful for chromatographic separations of both ribonucleic acids and deoxyribonucleic acids.

AFFINITY CHROMATOGRAPHY

Affinity chromatography is a novel form of column chromatography in which the molecules (principally proteins and nucleic acids) to be isolated from the sample under study are retarded in their passage through the column by their specific *biological* reaction with the column matrix. It is the biological nature of the interaction between the sample and the column matrix that distinguishes this form of chromatography from others. The specificity may take several forms, including (1) an antigen–antibody reaction, (2) the interaction of an enzyme with its substrate, allosteric effector, or coenzyme, and (3) hydrogen bonding between complementary polynucleotides.

The column is packed with porous gel particles (usually *agarose, cross-linked dextrans,* or *cellulose*) to which ligands having a high affinity for specific biological components have been covalently coupled. When the sample is applied to the column, only the constituents having a high affinity for the ligand are bound and other components are rapidly eluted. The ultimate desorption and isolation of the bound species are achieved by significantly altering the pH and/or ionic strength of the eluent (Fig. 13–21).

Affinity chromatography has been used with great success for the isolation of specific immunoglobulins from antisera. This is achieved by first coupling the gel with specific antigenic materials. When antiserum is applied to the column, specific immunoglobulins react with the antigen-coated gel particles and are retained by the column and other species are eluted. Gels to which the hormone insulin has been covalently bound have been used to isolate insulin-binding receptor sites of cell membranes. Specific cellular enzymes may be isolated by passing a tissue extract through a column in which the gel particles have previously been derivitized with the enzyme's substrate or specific cofactor or competitive inhibitor. Messenger RNA can be isolated from tissue homogenates using agarose impregnated with polyuridylic acid (polyU); mRNA's affinity for the column matrix results from the occurrence of sequences of polyA in many eukaryotic mRNAs (Chapter 21).

GEL FILTRATION

In 1959, J. Porath and P. Flodin introduced **gel filtration**, a chromatographic method for separating molecules on the basis of size (i.e., molecular weight) differences; the technique is also popularly known as **exclusion chromatography, gel permeation chromatography**, and **molecular sieving**. The procedure is carried out using glass columns packed with *nonionic,*

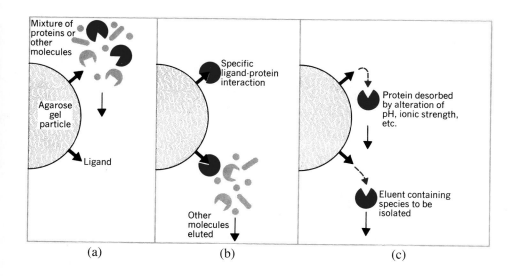

FIGURE 13–21 Affinity chromatography (see text for discussion).

porous gel particles. Because the gels do not possess ionic groups, separation of a mixture of macromolecules does not involve the temporary formation of bonds with the gel. The most commonly used gels are *cross-linked dextrans* (produced by reacting dextran with epichlorohydrin or acrylamide) in which the degree of cross-linkage determines the average pore size of the gel. Gel particles may be produced with a variety of pore sizes.

In gel filtration, the dry gel particles are first swollen by hydration in water or buffer and then packed into a glass column. The volume of the column is effectively divided into two phases: the solvent surrounding the gel particles (solvent phase) and the solvent within the gel particles (gel phase). Whether or not a given solute molecule can pass between phases depends on the molecular weight of the solute and on the pore sizes of the gel. Any solute molecules larger than the largest pore size cannot pass into the gel phase and are said to be above the *exclusion limit* of the gel. Smaller molecules are able to penetrate the gel phase to varying degrees.

If a mixture of molecules of various sizes is placed at the top of the column and is followed by the passage of solvent through the column, molecules above the gel exclusion limit will pass between the gel particles and emerge from the bottom of the column most rapidly. Molecules below the exclusion limit will pass through both the solvent and gel phases and emerge later. Therefore, molecules in the original sample are quickly separated into two different populations—molecules above and those below the gel exclusion limit (Fig. 13–22). However, if two molecules are below the gel exclusion limit but one is larger than the other, the larger molecule will pass through the column more rapidly, because it will spend less time in the gel phase than the smaller molecule (i.e., the chances that a sufficiently large gel pore will be encountered during descent through the column are less for larger molecules below the exclusion limit than for smaller molecules). Hence, even molecules below the gel exclusion limit undergo a separation and emerge from the column in order of decreasing size.

Cross-linked dextran and agarose gels may be obtained with a variety of exclusion limits and are used routinely for protein, nucleic acid, polysaccharide, and lipid separations. In most cases, a single solvent may be used to effect the separation, as molecules are not bound to the gel.

In addition to its usefulness for separating different size macromolecules, gel filtration is also used to esti-

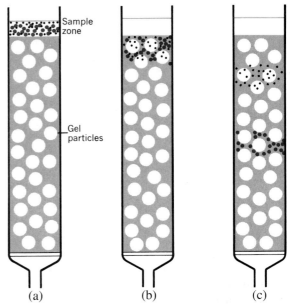

FIGURE 13–22 Separation of molecules by gel filtration. (*a*) Glass column packed with gel particles and overlaid with sample zone containing molecules of various sizes (small and large dots). (*b*) Molecules with molecular weights in excess of the gel exclusion limit percolate between the gel particles, whereas smaller molecules may enter the gel phase. (*c*) As a result, molecules reach the bottom of the column in order of decreasing size.

mate molecular weights. As in ion-exchange chromatography and electrophoresis, the determination of the purity of separated components may require additional analyses, because different molecules may have the same size and pass through the column at the same rate.

HPLC

HPLC (*high-performance liquid chromatography*) is a specialized and versatile form of chromatography that, depending on the particular combination of packing material and solvent, can incorporate the principles of partition, ion-exchange, exclusion, or affinity chromatography. Unlike the columns used in the techniques described earlier, HPLC columns are generally shorter (e.g., 10–20 cm) and much narrower (e.g., 1–5 mm) and are packed with much smaller particles of the adsorbent, resin, or gel in which the separation of the sample is to be effected. The particles comprising the column packing material rarely exceed

5–10 nm in diameter and form a tightly packed bed that offers high resistance to the flow of solvent. To withstand the very high pressures (e.g., 5000–8000 psi) that must be used to pump the solvent through the column, the columns themselves are fabricated from stainless steel. To optimize resolution, the sample must be delivered to the column as a narrow band. This is usually achieved by placing the sample in a microsyringe and quickly injecting it directly into the column through an injection port.

GAS CHROMATOGRAPHY

Gas chromatography is a special form of column chromatography in which a gas is used as the mobile phase (instead of a liquid) and either a liquid or a solid is used as the stationary phase. When a liquid is used as the stationary phase, the technique is called **gas–liquid chromatography** (GLC) and separations are based primarily on differences in the partition of the molecules in the sample between the stationary liquid and the moving gas. In **gas–solid chromatography** (GSC), separations result from the differential adsorption of sample molecules to the stationary solid phase as they are carried through the column by the gas. Of the two methods, GLC is, by far, the method most often employed.

The basic components of the gas chromatograph are the source of gas, the sample introduction chamber, the chromatographic column, the detector, and the recorder (Fig. 13–23). The gas (usually nitrogen, carbon dioxide, helium, or argon) is contained within a high-pressure cylinder connected to the column through metal tubing. A valve, pressure gauge, and flow meter are used to accurately regulate the flow of gas. The sample is introduced into the flow of gas using a microsyringe and needle inserted through a self-sealing diaphragm in the sample chamber. The chamber itself is enclosed within a heating block so that the sample (if it is not already in a gaseous form) will immediately be vaporized on introduction into the chamber and will be swept into the column by the gas. Gas chromatograph columns are made of glass, copper, or stainless steel tubing and are also enclosed in an oven; because they may be several feet long, they are often twisted to form a spiral.

The selection of packing material for the column depends on whether the separation is to be based on partition or adsorption. For GLC, the column is packed with an inert solid such as kieselguhr, which is impregnated and lightly coated with a liquid of low volatility (so that it will not be eluted from the column at the operating temperature used). For GSC, the stationary phase is an adsorbent such as charcoal or silica gel. Different molecules in the sample will be carried through the column at different rates, depending on their adsorption or partition characteristics, and will emerge from the end of the column at different times. Located near the exit of the column and also housed within an oven is the detector that monitors the composition of the emerging gas and relays electrical signals proportional to the amounts of separated components to a strip-chart recorder. The separated components are thus recorded as a series of peaks in

FIGURE 13–23 Basic components of a gas chromatograph (see text for details).

the chromatogram tracing. The most common form of detector is the *flame ionization* chamber in which the components are successively mixed with hydrogen and air and burned in a high-voltage field. Migration of the ionized fragments in this field creates a current registered by the recorder. Most chromatographic separations are analytical and the technique is so sensitive that only minute quantities of sample are required. However, separated components may also be collected as a condensate in tubes as they emerge from the heated column and are rapidly cooled.

Gas chromatography may be used to separate lipids, oligosaccharides, and amino acids after their preliminary conversion to volatile derivatives (many lipids may be chromatographed without conversion). Gas chromatography has been used with great success for the separation of different fatty acids. Depending on whether a nonpolar or a polar liquid phase is used, fatty acids may be separated according to boiling point and size or degree of saturation.

The methods described in this chapter for separating, isolating, and studying macromolecules and their constituents are widely used in cellular research. Students interested in pursuing this subject can find a more comprehensive discussion of these methods in the books and articles listed at the end of the chapter, together with descriptions of other separation methods that are used less often but are also important.

SUMMARY

Over the years, a variety of analytical and preparative techniques have been developed for separating, analyzing, and isolating the macromolecular constituents of cells and tissues. Older methods such as **salting in, salting out**, and **isoelectric precipitation** relied on differences in the solubility of the molecular species under investigation. Differences in molecular size are used to achieve separations in such diverse approaches as **dialysis, ultrafiltration, gel filtration**, and **ultracentrifugation**. Differences in the net electrostatic charges of molecules are used in various types of **electrophoretic** separations, including **moving-boundary** electrophoresis, **zone** electrophoresis, and **disk** electrophoresis. Biological activities of the molecules are used to advantage in such techniques as **immunoelectrophoresis** and **affinity chromatography**. Many **chromatographic** techniques utilize *partition* differences to achieve separations, the most popular formats being **thin-layer** chromatography, **anion-** and **cation-exchange** chromatography, **high-performance** liquid chromatography, and **gas** chromatography. Because molecular parameters such as size, charge, and solubility are used as the basis for effecting a separation, information concerning the unique physical and chemical properties of the species under investigation is simultaneously acquired.

REFERENCES AND SUGGESTED READING

Articles and Reviews

Bishop, R., Current major application areas in electrofocusing. *Sci. Tools* **26** (1), 2 (1979).

Cooper, A. R., and Atzinger, D. P., Aqueous gel permeation chromatography. *Am. Lab.*, p. 13 (Jan. 1977).

Ettre, L. S., Pioneers in chromatography, Part I. *Am. Lab.*, p. 85 (Oct. 1978).

Ettre, L. S., Pioneers in chromatography, Part II. *Am. Lab.*, p. 120 (Nov. 1978).

Gershoni, J. M., Protein blotting: Developments and perspectives. *Trends in Biochem. Sci.* **10** (3), 103 (Mar. 1985).

Gray, G. W., Electrophoresis. *Sci. Am.* **185** (6), 45 (Dec. 1951).

Laurent, T. C., and Killander, J., A theory of gel filtration and its experimental verification. *J. Chromatography* **14**, 317 (1964).

Mandava, N. B., Ito, Y., and Conway, W. D., Countercurrent chromatography: Historical development and early instrumentation. *Am. Lab.*, p. 62 (Oct. 1982).

Mandava, N. B., Ito, Y., and Conway, W. D., Countercurrent chromatography: Recent instrumentation and applications. *Am. Lab.*, p. 48 (Nov. 1982).

Maugh, T. H., A survey of separative techniques. *Science* **222**, 259 (1983).

Ornstein, D. L., and Kashdan, M. A., Sequencing DNA using ^{35}S-labeling: A troubleshooting guide. *Biotechniques* **3**, 476 (1985).

Peterson, E. A., and Sober, H. A., Chromatography of proteins. I. Cellulose ion-exchange adsorbants. *J. Am. Chem. Soc.* **78**, 751 (1956).

Sober, H. A., Gutter, F. J., Wyckoff, M. M., and Peterson, E. A., Chromatography of proteins. II. Fractionation of serum proteins on anion-exchange cellulose. *J. Am. Chem. Soc.* **78**, 756 (1956).

Vesterberg. O., Isoelectric focusing. *Am. Lab.*, p. 13 (June 1978).

Whipple, H. E. (Ed.), Gel electrophoresis. *Ann. N. Y. Acad. Sci.* **71** (1964).

Williams, C. A., Immunoelectrophoresis. *Sci. Am.* **202** (3), 130 (Mar. 1960).

Books, Monographs, and Symposia

Bobbit, J. M., *Thin-Layer Chromatography*, Reinhold Publishing Co., New York, 1963.

Determann, H., *Gel Chromatography*, Springer-Verlag, New York, 1968.

Gel Filtration: Theory and Practice, Pharmacia Fine Chemcials, Uppsala, Sweden, 1981.

Krugers, J., and Keulemans, A. I. M., *Practical Instrumental Analysis*, Elsevier, Amsterdam, 1965.

Lehninger, A. L., *Biochemistry* (2nd ed.), Worth, New York, 1975.

Lowe, C. R., *An Introduction to Affinity Chromatography*, North-Holland, Amsterdam, 1979.

Lowe, C. R., and Dean, P. D. G., *Affinity Chromatography*, Wiley, London, 1974.

Mangold, H. K., and Schmid, H. H. O., Thin-layer chromatography (TLC). In *Methods of Biochemical Analysis* (D. Glick, Ed.), Vol. XII, Wiley–Interscience, New York, 1964.

Morris, C. J. O. R., and Morris, P., *Separation Methods in Biochemsitry*, Pitmann and Sons, Ltd., London, 1964.

Randerath, K., *Thin-Layer Chromatography*, Academic, New York, 1963.

Rickwood, D., and Hames, B. D. (Eds.), *Gel Electrophoresis of Nucleic Acids*, I.R.L. Press, Oxford, 1982.

Righetti, P. G., and Drysdale, J. W., *Isoelectric Focusing*, North-Holland, Amsterdam, 1976.

Stock, R., and Rice, C. B. F., *Chromatographic Methods*, Chapman & Hall, Ltd., London, 1967.

Williams, B. L., and Wilson, K. (Eds.), *A Biologist's Guide to Principles and Techniques of Practical Biochemistry*, University Park Press, Baltimore, 1981.

Wrigglesworth, J. M. (Ed.), *Biochemical Research Techniques*. Wiley–Interscience, New York, 1983.

CHAPTER 14

RADIOACTIVE ISOTOPES AS TRACERS IN CELL BIOLOGY

Isotopes are chemical elements that have the same **atomic number** (i.e., the number of protons in the nucleus of the atom) but different **atomic masses** (i.e., the sum of the number of protons and neutrons in the nucleus). Certain isotopes are unstable and undergo spontaneous nuclear changes (called **transmutations**) accompanied by the emission of **particulate** and sometimes also **electromagnetic** radiations. These atoms are said to be radioactive and are called **radioisotopes** or **radionuclides**. Radioactive atoms may readily be detected by instruments sensitive to their radiations. Generally, an organism cannot distinguish between the stable and radioactive forms of the same element so that both are metabolized in an identical manner. It is for this reason that radioisotopes have proven extremely useful to biologists, as these elements may be conveniently employed as *tracers*. That is, the fate of a given element (or molecule) in an organism or individual cell may be studied by introducing the radioactive form of that element and following the uptake and subsequent localization of the radioactivity. Some of the radioisotopes frequently employed as tracers are listed in Table 14–1.

If the radioisotope is initially a part of a molecule, then the fate of all or part of that molecule may simi-larly be followed. The use of radioisotopically labeled compounds is particularly desirable when the compound to be administered is a normal constituent of the cell or organism and would be impossible to distinguish from stable molecules already present. Many organic and inorganic compounds of biological interest

TABLE 14–1 SOME RADIOISOTOPES FREQUENTLY USED AS TRACERS

Isotope	Used as a Tracer of
^{3}H	Virtually any organic compound
^{14}C	Virtually any organic compound
^{24}Na	Salt metabolism, exchanges across membranes
^{32}P	Nucleic acid metabolism, phospholipid metabolism, salt metabolism
^{35}S	Protein metabolism
^{36}Cl	Salt metabolism
^{42}K	Salt metabolism, exchanges across membranes
^{45}Ca	Salt metabolism, bone deposition
^{59}Fe	Heme synthesis, hemoglobin synthesis
^{125}I	Protein metabolism
^{131}I	Protein metabolism

may now be obtained that have one or more specific atomic positions occupied by radioisotopes. Because of the extremely high sensitivities of many radiation detectors, the quantities of radioisotopes employed in tracer studies can be kept small enough to preclude significant damage to cell constituents by the radiation.

ADVANTAGES OF THE RADIOISOTOPE TECHNIQUE

Results obtained from experiments in which radioisotopes are used are *quantitative*, because the amount of radioactivity present and available for detection is directly proportional to the radioisotope content of the sample being analyzed. Moreover, numerous biological studies carried out on a routine basis with radioisotopes can only be performed with great difficulty or are virtually impossible without them. Some examples are cited to illustrate the value of the radioisotope technique.

Determination of Molecular Fluxes under Conditions of Zero Net Exchange

The movement of a particular ionic or molecular species between different tissues of an organism or between a cell and its surroundings is often in *dynamic equilibrium*, that is, although a continuous exchange of a given substance between one region and another takes place, no *net* transfer of material occurs. Alternatively, the concentration of a given substance within a tissue or cell may remain fairly constant as a result of the balanced biosynthesis and degradation of that material. These situations cannot be easily detected or studied by routine chemical analyses. The continuous flux of Na^+ and K^+ across the plasma membrane of the erythrocyte is a typical example of this type of equilibrium (see Chapter 15). The concentration of K^+ within the mammalian erythrocyte is much higher than in the surrounding blood plasma, whereas the reverse is true for Na^+. These large concentration differences are sustained in spite of the continuous passage of Na^+ and K^+ across the plasma membrane in both directions. Consequently, this dynamic steady state is not revealed by chemical analysis and, prior to the utilization of radioisotopes of sodium and potassium to study this situation, the concentration differences were interpreted as the consequence of the impermeability of the erythrocyte's plasma membrane

to Na^+ and K^+. When the radioisotopes of sodium and potassium, ^{24}Na and ^{42}K, were used to label the plasma surrounding the red cells, the passage of radioactivity into the erythrocytes was observed, eventually approaching an equilibrium identical to the concentration equilibrium determined chemically. These studies unequivocally demonstrated that the plasma membrane of the red cell is readily permeable to Na^+ and K^+ and that a continuous exchange of these two ions between the cell and the surrounding plasma takes place. Observations of this kind provided great impetus to the development of concepts concerning the continuous metabolic turnover of cellular materials present in constant concentrations and the **active transport** (Chapter 15) of materials across the plasma membrane against concentration gradients.

In addition to providing qualitative information, radioisotopes may be used to accurately measure the *rates* at which the metabolic turnover of materials within a cell or tissue occurs, as well as the rates of exchange of materials across cellular membranes. Prior to the availability of radioisotopes, no satisfactory methods were available to measure these rates. In the case of the erythrocyte, it has been shown that about 2% of the cell's Na^+ and K^+ are exchanged with the plasma each hour.

Simplification of Chemical Analyses

Depending on the nature of the chemical analyses required, quantitative determinations of the distribution of a given element or compound in different tissues of the body or in different parts of an individual cell may be very difficult. This difficulty may be compounded if certain components to be analyzed contain only trace amounts of the substance to be measured. Because radioisotope-containing compounds are not distinguished from their stable (nonradioactive) counterparts by an organism, the use of labeled compounds in studies of this sort can greatly simplify and reduce the number of analyses. Two requirements must be fulfilled: (1) the administration of the labeled compound must quickly be followed by its uniform distribution among nonradioactive molecules of the same kind that are *already present* in the system and (2) once a uniform distribution is attained, the **specific activity** of the substance in question must be determined in one of the several components (i.e., tissues, cell fractions, etc.) to be analyzed. Specific activity is defined here as *the quantity of radioactive element or compound per unit mass of total element or compound* (the units of measurement are described later). If the labeled compound does become uniformly distributed in the system, then the specific activity will be equal throughout. Consequently, all subsequent measurements of the quantity of that compound may be made simply by measuring the amount of radioactivity present in the tissue or cell part to be analyzed. In nearly every case, this is far easier than performing a series of quantitative chemical measurements of the element or compound in question.

The rates at which different elements or compounds are incorporated by individual cells or by the body tissues, together with a determination of the specific loci of the deposition, can also be conveniently determined using radioisotopes. For example, cells may be incubated in a medium containing labeled material or, in the case of measurements in whole animals, the material may be introduced into the bloodstream. In these cases, the specific activity of the element or compound is determined before it is made available for incorporation and the subsequent rate of incorporation is determined either from the rate of disappearance of radioactivity from the medium or bloodstream or from the rate of appearance of radioactivity in the cells or tissue. The specific locus of deposition or utilization within the cell may be determined from separate radioactivity measurements of *fractionated* tissue (see Chapter 12) or by *autoradiographic* procedures (to be described later in this chapter) combined with light or electron microscopy. Again, it should be noted that specific quantitative chemical analyses of the material being incorporated by the cells are unnecessary when the radioisotope technique is applied.

"Isotope Dilution" Methods

A common problem confronting physiologists and one that especially lends itself to the radioisotope technique is the determination of the *total* quantity or volume of a given material in the body or in a cell when quantitative isolation of that material for analysis is not possible. Determinations of total circulating blood volume, erythrocyte mass, chloride space, exchangeable sodium, and so on fall in this category. The manner in which such unknown quantities are measured using radioisotopes may be exemplified by considering the following generalized case. Suppose that material X occurs in a system (i.e., a tissue, cell, etc.) in unknown abundance and a labeled form of this material, X^*, is either available commercially or can

be prepared experimentally with a known specific activity, SA, given in this instance by the equation

$$SA = \frac{(X^*)}{(X^* + X)} \qquad (14\text{–}1)$$

A measured quantity of the labeled material is introduced into the system and permitted to equilibrate thoroughly with the nonradioactive material that was already present. Following this, a small quantity of the material is withdrawn from the system and its specific activity is determined. Because the X^* originally introduced was uniformly mixed with unlabeled X already present, its specific activity will have been reduced in direct proportion to the total amount of X present in the system. Consequently, the total amount of X originally present can be determined from the relationship

$$X_T = X^* \left(\frac{SA_1}{SA_2} - 1 \right) \qquad (14\text{–}2)$$

where

X_T = the total amount of X originally present in the system,

X^* = the quantity of labeled material added,

SA_1 = the original specific activity of the material added, and

SA_2 = the specific activity of the sample removed for analysis.

The isotope dilution technique is a fundamental part of **radioimmunoassay** procedures (discussed later).

PRECURSOR–PRODUCT RELATIONSHIPS

Radioisotopes have been widely applied to study **precursor–product relationships**, for in many cases the introduction of a labeled compound (i.e., *precursor*) into a system is soon followed by its chemical conversion into another form (i.e., *product*). For example, soon after ^{32}P-labeled phosphate is introduced into the bloodstream, radioactivity appears in a variety of tissue phospholipids. In a similar manner, the introduction of radioactive iron is soon followed by the appearance of radioactivity in liver ferritin and also in newly synthesized hemoglobin of red blood cells maturing in the bone marrow. In other words, radioisotopically labeled compounds may be used to determine which of a variety of alternative metabolic pathways is followed by a compound by examining the radioisotope content of the alternative metabolic products. If several different pathways are open to the precursor, then the extent to which each is followed under a variety of experimental conditions may be determined from the respective radioisotope contents of the products.

The formation of an end product often involves a number of intermediate precursors and products, that is,

$$A \longrightarrow B \longrightarrow C \longrightarrow D$$

Precursor 1   Product 1   Product 2   End product
               Precursor 2 Precursor 3

Before radioisotopes were available to study such metabolic pathways, it was often difficult to determine which compound was the immediate precursor of each product. However, if each intermediate has a single precursor (as above), then the precise sequence can be determined using labeled material by following the change in specific activity of each intermediate compound with time. This may be done because *the specific activities of an intermediate and its immediate precursor are equal when the intermediate's specific activity reaches a maximum value* (Fig. 14–1). By degrading the intermediate and final products of a series of chemical reactions and determining the distribution of radioisotope among the constituent atoms, it is often possible to determine the fate of each atom in the pathway.

PROPERTIES OF RADIOACTIVE ISOTOPES

Most radiations emitted by radioisotopes are the result of changes in the unstable atomic nuclei. Whether a given atomic nucleus is stable depends in turn on the numbers of neutrons (N) and protons (Z) that it contains. The relationship between nuclear stability and the neutron: proton composition of the nucleus is shown in Figure 14–2. *All* stable (and a number of unstable) isotopes have N and Z values that fall within the shaded zone. Note that for the lighter elements, nuclear stability exists when $N \cong Z$, whereas in the stable, heavier elements, the number of neutrons exceeds the number of protons, with the allowable neutron excess increasing with atomic number. Isotopes with N and Z numbers outside of the shaded zone (as well as some that occur within the zone) shown in Figure 14–2 undergo spontaneous changes in which neutrons and protons are interconverted. The resulting change establishes a stable combination of N and Z values. These intranuclear changes are accompanied by the emission of particulate and sometimes electromagnetic radiation.

Types of Radiation Emitted by Radioisotopes

The most common types of nuclear radiations are **alpha particles**, positive and negative **beta particles**, and **gamma rays**. Alpha particles, which consist of two protons and two neutrons and are therefore identical to helium nuclei, are emitted by radioisotopes of high atomic number (e.g., uranium, polonium, thorium, and radium) and are rarely used as tracers in biological studies.

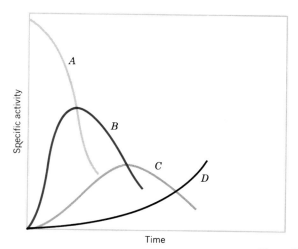

FIGURE 14–1 Relationship between the specific activities of precursors and products in the chemical pathway $A \rightarrow B \rightarrow C \rightarrow D$ following application of a pulse of radioactive tracer.

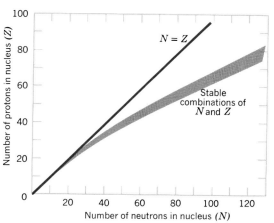

FIGURE 14–2 Relationship between stability of atomic nucleus and the number of neutrons and protons present.

Positive beta particles, also called **positrons**, are emitted from nuclei in which the *N:Z* ratio is lower than that which is stable. The nuclear transmutation involves the conversion of a proton into a neutron, positron, and neutrino, the latter two being ejected from the nucleus. The positron possesses a unit positive charge and is equal in mass to an electron, whereas the neutrino has neither mass nor charge. The isotope $^{11}_{6}C$ *decays* by positron emission (the numbers that precede the element's symbol are its atomic mass [the superscript] and its atomic number [the subscript]):

$$^{11}_{6}C \longrightarrow \ ^{11}_{5}B \ + \ _{(+1)}^{0}e \ + \ \text{neutrino}$$
$$\text{(positron)}$$

In view of the similarity between beta particles and electrons, the symbol "*e*" is often used to describe the beta particle. Very few positron-emitting radioisotopes are employed as biological tracers.

Nearly all radioisotopes used as tracers by biologists emit negative beta particles (**negatrons**). It has become customary to drop the term "negative" so that the expression "beta particle" is understood to imply negative beta particle. Although technically incorrect, this terminology is widely used and will be employed here. Beta particles are emitted from nuclei in which the *N:Z* ratio is greater than that which is stable. The nuclear change involves the conversion of a neutron to a proton with the resulting ejection of a beta particle and neutrino. $^{14}_{6}C$ and $^{32}_{15}P$ may serve as examples of this form of decay:

$$^{14}_{6}C \longrightarrow \ ^{14}_{7}N \ + \ _{(-1)}^{0}e \ + \ \text{neutrino}$$
$$\text{(beta particle)}$$

and

$$^{32}_{15}P \longrightarrow \ ^{32}_{16}S \ + \ _{(-1)}^{0}e \ + \ \text{neutrino}$$

The emission of beta particles from the nuclei of ^{3}H, ^{14}C, ^{32}P, ^{35}S, ^{36}Cl, and ^{45}Ca atoms changes the *N:Z* ratio to a stable value and lowers the energy content of the nucleus to the *ground state*. The energy lost by the nucleus in this process is divided between the beta particle and neutrino. However, for many radioisotopes (e.g., ^{24}Na, ^{59}Fe, and ^{131}I), intranuclear changes resulting in the emission of beta particles and neutrinos produce nuclei with stable *N:Z* ratios but with energy levels that are still above the ground state. In these instances, the excess energy is eliminated and the ground state is attained by the emission of one or more gamma rays; for example,

$$^{59}_{26}Fe \longrightarrow \ ^{59}_{27}Co \ + \ _{(-1)}^{0}e \ + \ \text{neutrino} \ + \ \text{gamma ray(s)}$$

Unlike alpha and beta radiation, gamma radiation is not particulate but is *electromagnetic*.

Energy of Radiation and Its Interaction with Matter

Radiation energy is measured in **electron volts** (abbreviated eV), one electron volt being the kinetic energy acquired by an electron in an electrical potential difference of one volt. The beta particles and gamma rays emitted by radioisotopes that are often used as tracers have energies ranging from about 1×10^4 to 4×10^6 eV or 0.01 to 4.0 MeV (1 MeV equals 1 million electron volts). This energy range may be compared with the energy of chemical bonds, which is on the order of a few electron volts. The kinetic energy acquired by a beta particle during transmutation of the parent atom's nucleus can vary from 0 MeV up to some maximum value, E_{max}, which is a characteristic of the particular radioisotope. As an example, the energy spectrum for ^{32}P beta particles is shown in Figure 14–3. It should be noted that the *maximum* energy of a ^{32}P beta particle is 1.71 MeV but that the *average* energy is much less (0.70 MeV). Although the value for E_{max} varies, the *shape* of the energy curve is similar for all beta particle-emitting radioisotopes.

To reduce the energy level of the radioisotope nucleus to the ground state, a specific quantity of energy, Q, must be given off as radiation during the nuclear change. For isotopes emitting only beta particles, Q equals E_{max} and the energy of the neutrino accounts for the difference between E_{max} and the actual kinetic energy acquired by the beta particle.

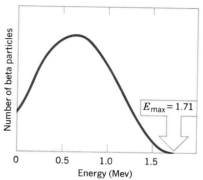

FIGURE 14–3 Energy spectrum of beta particles emitted by ^{32}P.

For example, if the transmutation of a particular ^{32}P atom results in the emission of a 1.20-MeV beta particle, then the accompanying neutrino would have an energy of 0.51 MeV (i.e., 1.20 MeV + 0.51 MeV = 1.71 MeV = E_{max}). If the beta particle has escaped the nucleus with the maximum possible energy, then no neutrino would have been emitted. Table 14–2 lists the maximum beta particle energies of some radioisotopes.

In contrast to beta particles, gamma rays are emitted at specific energy levels. Thus, all gamma rays emitted by ^{42}K have an energy of 1.5 MeV (Table 14–2). The Q value of an isotope emitting both beta and gamma radiation is equal to the sum of the energies possessed by the beta particle, neutrino, and subsequent gamma ray(s). Occasionally, as in the case of ^{24}Na, the emission of a beta particle may be followed by the emission of two (or more) successive gamma rays in order to reduce the nuclear energy level to the ground state. It may also be noted from Table 14–2 that the decay of some radioactive isotopes (e.g., ^{42}K, ^{59}Fe, and ^{131}I) proceeds along two or more alternative pathways, that is, beta particles having more than one E_{max} may be emitted, and because the Q value is constant, these are necessarily followed by gamma rays of different energies.

Beta particles and gamma rays interact with matter by *ionizing* and exciting atoms in their path. In the case of beta particles, ionization results from the repulsion of orbital electrons by the negatively charged particle. As a result beta particles tracking through matter produce a *wake* of electrons and positively charged ions (called **ion pairs**). The number of ion pairs produced per unit path length, called the **specific ionization**, is not constant but increases as the beta particle slows down. Also, the path of the beta particle is not linear but is quite erratic as a result of its repulsion by the orbital electrons of atoms with which it interacts. High-energy beta particles may traverse the linear equivalent of 1 to 2 m in air. Because gamma radiation is electromagnetic, its probability of interacting with matter is less than that for beta particles. Consequently, gamma rays have a much lower specific ionization and a much longer and also linear path length.

Half-life

The number of atoms in a sample of radioisotope that disintegrate during a given time interval decreases logarithmically with time and is unaffected by chemical and physical factors that normally alter the rates of chemical processes (i.e., temperature, concentration, pressure, etc.). Radioactive decay is therefore a classic example of a first-order reaction. A convenient term used to describe the rate of decay of a radioisotope is the **physical half-life**, T_p. This is the amount of time required to reduce the amount of radioactive material to one-half its previous value. Each radioactive

TABLE 14–2   ENERGIES OF BETA PARTICLES AND GAMMA RAYS EMITTED BY RADIOISOTOPES FREQUENTLY USED AS TRACERS

Isotope	Energy (MeV)		Q (MeV)
	Beta Particle (E_{max})	Gamma Ray	
^{3}H	0.0176	None	0.0176
^{14}C	0.154	None	0.154
^{24}Na	1.39	1.37 and 2.75	5.51
^{32}P	1.71	None	1.71
^{35}S	0.167	None	0.167
^{36}Cl	0.714	None	0.714
^{42}K	3.60	None	3.60
	or 2.10	1.50	3.60
^{45}Ca	0.254	None	0.254
^{59}Fe	0.27	1.29	1.56
	or 0.46	1.10	1.56
^{60}Co	0.31	1.17 and 1.33	2.81
^{125}I	None[a]	0.035	—
^{131}I	0.61	0.36	0.97
		or 0.28 and 0.08	
	or 0.34	0.63	0.97

[a] ^{125}I decays by electron capture.

isotope decays at a characteristic rate and therefore has a specific half-life (see Table 14–1). For example, the amount of radioactivity arising from a sample of ^{59}Fe is reduced to one-half its original value in 45.1 days, to one-fourth in 90.2 days, to one-eighth in 135.3 days, and so on. The amount of decay occurring in the course of a tracer experiment must be taken into account when radioisotopes of short physical half-life such as ^{24}Na, ^{32}P, ^{42}K, ^{59}Fe, ^{131}I, and ^{125}I are used. Of course, this is not a problem in tracer experiments employing ^{3}H and ^{14}C (Table 14–1) because the length of the experiment is insignificant in comparison with the half-life.

When radioisotopes are used in *in vivo* experiments of extended duration, the turnover rate of the element in the body (or in the cell) must also be considered, for the rate of decrease of radioactivity will be a function in both radioactive decay and metabolic turnover. In these instances, a more useful term is the **effective half-life**, T_e, which is the amount of time required to reduce the radioisotope content of the body (or cell) to one-half its original value by the *combined* effects of decay and turnover; it is determined using the relationship

$$T_e = \frac{T_b \times T_p}{T_b + T_p} \qquad (14\text{–}3)$$

where

T_e = effective half-life,

T_p = physical half-life, and

T_b = **biological half-life** and is defined as the normal amount of time required for the turnover of one-half of the body content of a given element (radioactive or nonradioactive).

The physical, biological, and effective half-lives of several elements are compared in Table 14–3. It should be noted that T_e can never be greater than T_p and that the slower the rate of turnover of an element, the closer T_e approaches T_p.

DETECTION AND MEASUREMENT OF RADIATION

The selection of instruments for the detection and measurement of radioisotopes is based primarily on the type and energy of the emitted radiation. The most commonly used detectors are: (1) **Geiger–Müller counters**, which are employed primarily with isotopes emitting beta particles of intermediate or high energy (E_{max} above 0.2 MeV) and which may also be used at low efficiency for the measurement of gamma radiation; (2) **solid scintillation counters**, which are generally employed with gamma ray-emitting isotopes; and (3) **liquid scintillation counters**, which are used with isotopes emitting low-energy beta particles (E_{max} below 0.2 MeV).

Geiger–Müller Counters

The most widely used instrument for the detection and measurement of radiation is the Geiger–Müller (or G–M) counter. The detector itself, called a G–M tube, consists of a cylinder several inches long containing two electrodes and filled with a readily ionizable inert gas such as helium or argon. The insulated metallic internal surface of the cylinder serves as the cathode and a narrow wire passing down the center of the tube serves as the anode (Fig. 14–4). One end of the G–M tube is covered by a thin material such as mylar plastic or mica and is called the **end-window**. The anode and cathode terminals at the other end of the

TABLE 14–3 PHYSICAL, BIOLOGICAL AND EFFECTIVE HALF-LIVES OF SOME RADIOISOTOPES

Element	Radio-isotope	Half-life		
		Physical	Biological	Effective
Hydrogen	^{3}H	12.3 yr	19 days	19 days
Carbon	^{14}C	5570 yr	180 days	180 days
Phosphorus	^{32}P	14.3 days	3 yr	14.1 days
Calcium	^{45}Ca	164 days	73 yr	163 days
Iron	^{59}Fe	45.1 days	3.4 yr	43.5 days
Cobalt	^{60}Co	5.3 yr	8.1 days	8.1 days
Iodine	^{125}I	60 days	156 days	43.3 days
Iodine	^{131}I	8.1 days	156 days	7.7 days
Radium	^{226}Ra	1620 yr	104 days	104 days

FIGURE 14–4 Basic components of a Geiger–Müller counter.

tube are connected to a source of high voltage and a **scaler**, a device that simply counts electrical pulses.

When a radioactive sample (usually deposited on a small metal disk called a **planchet**) is placed near the end-window, radiation enters the G–M tube, ionizing some of the gas molecules and forming a number of ion pairs (i.e., positively charged argon or helium atoms and electrons). If a sufficiently high electrical potential is applied to the electrodes, the ion pairs will migrate toward the appropriate electrode. During this migration, the ions collide with and ionize additional gas molecules, so that the passage of a single beta particle through the gas results in a large number of ions being collected at the electrodes. These events produce an electrical pulse that is recorded by the scaler as a *count*. Ideally, each ionizing ray entering the G–M tube is registered as a count and the amount of radioactivity is expressed as *counts per minute* (cpm).

Because radiation is emitted in all directions from a radioactive source, it is apparent that only a small percentage of the rays arising from the sample are directed toward the end-window. Therefore, even if all the rays entering the G–M tube are detected and counted, the cpm recorded for the sample is only a fraction of the true *rate* of disintegration (i.e., *disintegrations per minute*, dpm) of the isotope. This does not pose a serious problem when the *relative* isotope contents of a number of samples are to be determined (this is generally the case) and if constant geometric

conditions are maintained for each sample (i.e., distance of the sample from the end-window, volume of sample, etc.).

For some radioisotopes such as ^{3}H, ^{14}C, ^{35}S, and others of low E_{max}, much or all of the energy of the emitted beta particles may be dissipated before the ray enters the ionizing gas. For example, the energy may be expended within the sample itself (called **self-absorption**), in the air between the sample and the end-window, or in the material of the end-window. Even with radioisotopes emitting beta particles of high E_{max} (Table 14–2), beta particles in the low region of the energy spectrum may go undetected. Because of the low specific ionization of gamma rays and the low density of the gas in the G–M tube, gamma rays may pass through the tube without causing ionizations and therefore go undetected. For these reasons, Geiger–Müller counters are usually not suitable for the detection and measurement of radioisotopes emitting gamma rays or beta particles of low E_{max}.

Even when no radioactive sample is placed below the end-window of the G–M tube, a small count is recorded. This is known as the *background* count and results from cosmic radiation (primarily gamma rays), naturally occurring radioisotopes in laboratory materials (such as ^{40}K in glass and naturally occurring ^{14}C and ^{3}H in organic compounds), radioactive samples left in the vicinity of the detector, and electronic "noise" within the components of the counting system. Therefore, the background count must always be subtracted from the count obtained for a radioactive sample. The magnitude of the background count may be reduced by placing lead shielding around the detector so that much of the cosmic radiation and radiations from other sources are absorbed before reaching the detector.

The total amount of radioisotope present in a sample at any instant in time may be determined from its rate of disintegration, that is, its dpm; the basic unit of measurement is the **curie** (Ci) and is defined as that quantity of radioisotope undergoing 2.22×10^{12} dpm. (Note that the curie content of a radioactive sample decreases exponentially with time at a rate determined by the physical half-life of the radioisotope.) In most tracer experiments, the quantity of radioisotope used is generally at the millicurie (mCi) or microcurie (μCi) level. The curie content of a labeled compound is generally provided at the time of purchase so that the efficiency of the counting system may be determined by comparing the recorded cpm of an aliquot of the isotope with its known dpm. Generally, this value is

10% or less for G–M counters but is much higher in solid and liquid scintillation systems. Once the efficiency of the counting system is known, then the specific activity of a radioactive sample (which we may now define as the *number of curies per unit mass of element or compound*) collected during the course of a tracer experiment may be calculated from its observed counting rate and composition. G–M counters are effectively employed in tracer experiments involving ^{24}Na, ^{32}P, ^{36}Cl, and other "hard beta" emitters but generally are not used with ^{3}H and ^{14}C, which emit "soft beta" rays.

Solid Scintillation Counters

Solid scintillation counters are used to detect and measure radioisotopes emitting gamma rays. The detector (Fig. 14–5) consists of a large crystal of

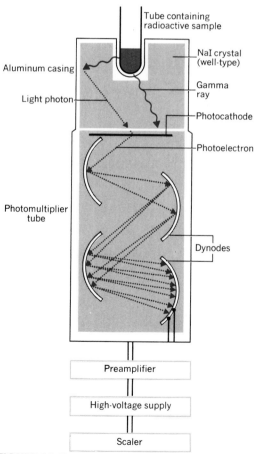

FIGURE 14–5 Basic components of a solid scintillation counter.

thallium-activated sodium iodide and a photomultiplier tube encased in an aluminum housing; the latter is interfaced with a preamplifier, a source of high voltage, and a scaler. The radioactive sample to be analyzed is placed either against the end of the detector containing the crystal or, for greatly improved counting efficiency, into a well-shaped opening drilled into the crystal's surface (Fig. 14–5). Because of its high density, the crystal absorbs much of the energy of the gamma rays, causing excitation of electrons of atoms composing the crystal and raising them to higher energy orbitals. As these electrons return again to their lower energy orbitals, flashes of light or **scintillations** are emitted and these are proportional in number to the number and energy of the gamma rays exciting the crystal. The light photons are converted by the photomultiplier tube into electrical pulses of corresponding magnitude and frequency and these are relayed to the scaler. Because the magnitude of the electrical pulses produced is proportional to the energy of the gamma rays, and because gamma rays are monoenergetic, the inclusion of the appropriate circuitry in the counting system (i.e., a pulse-height analyzer, see Chapter 2) allows different gamma ray-emitting isotopes to be distinguished.

In contrast to G–M counters, few or no problems involving self-absorption and end-window absorption are incurred when solid scintillation methods are used with gamma ray-emitting isotopes. However, the use of constant geometry and lead shielding around the detector to reduce the magnitude of the background count is important.

Radioimmunoassay At the present time, one of the most common uses of solid scintillation counting is in certain **radioimmunoassay** (RIA) procedures. The RIA technique, developed in the 1950s by S. A. Berson and R. S. Yalow, is a tracer procedure used to measure trace amounts of any substance with antigenic properties. Depending on the specific activity of the labeled compound, quantities of antigen that are as low as 10 *picograms* (i.e., 10 millionths of a microgram) may be detected. Prior to the development of RIA procedures, the assay of biological substances that did not possess enzymatic activity or some other easily measurable property was extremely difficult. However, because many different kinds of chemical substances are *antigenic* (i.e., they can elicit an antibody response), they can be reliably quantitated using the RIA procedure. The sensitivity of the RIA technique equals or surpasses all chromatographic and

TABLE 14–4   CHEMICAL SUBSTANCES MEASURED BY RADIOIMMUNOASSAY (RIA)

Steroids
    Aldosterone
    Testosterone
    Estradiol
    Digitalis
Peptides
    Antidiuretic hormone (ADH, vasopressin)
    Bradykinin
    Secretin
    Gastrin
Proteins
    Adrenocorticotrophic hormone (ACTH)
    Glucagon
    Insulin
    Calcitonin
    Somatotrophic hormone (STH, growth hormone)
    Parathyroid hormone
    Thyroglobulin
Other
    Cyclic AMP
    Morphine
    Prostaglandins

complexes can be used as a measure of the amount of unlabeled antigen present in the original sample. When the antigen is all or part protein or peptide, the labeled form is prepared by iodination with either ^{131}I or ^{125}I. The iodine is incorporated into the benzene ring of tyrosine residues. Both of these radioactive tracers emit gamma rays. In radioimmunoassays, radioactive tracers of iodine are preferred to those of hydrogen and carbon because 557 atoms of ^{3}H and 261,672 atoms of ^{14}C must be incorporated into each antigen molecule to yield the same dpm as one atom of ^{131}I or seven atoms of ^{125}I. Although ^{131}I emits more

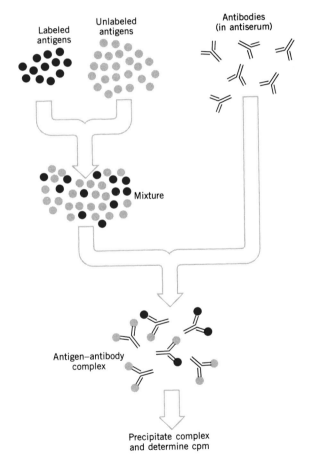

spectrophotometric methods currently in use. The technique has been especially useful for assaying hormones (Table 14–4).

The RIA procedure is depicted schematically in Figure 14–6. A sample containing the antigen to be quantified is mixed with a second sample containing radioisotopically labeled antigen of very high specific activity. To the mixture is added antiserum containing antibody previously prepared against the antigen (see Chapter 4). The amount of antibody added must be limiting so that antigen is always in excess (even when no unlabeled antigen is present). The resulting antigen–antibody reaction produces a complex in which the amount of radioisotope present is *inversely* proportional to the original quantity of unlabeled antigen in the sample being assayed. In the final stage of the radioimmunoassay, the antigen–antibody complex is separated from free antigen by precipitation.

As illustrated in Figure 14–6, the antiserum contains enough antibody to bind 12 antigen molecules. If 12 labeled and 12 unlabeled antigen molecules are present, then only one-half of the labeled antigens can be bound. If 24 unlabeled antigen molecules are present (as in the diagram), then only one-third of the labeled antigens can be bound, and so on. Thus, the radioisotope content of the isolated antigen–antibody

FIGURE 14–6   The *radioimmunoassay (RIA)* procedure. A mixture containing a known quantity of radioactively labeled antigens and an unknown quantity of unlabeled antigens is reacted with a limiting amount of antibody and the resulting antigen–antibody complexes are then isolated. The quantity of unlabeled antigen can be determined from the radioisotope content of the isolated complexes. The technique is highly sensitive.

energetic and more easily detected gamma rays, ^{125}I is usually the isotope of choice because of its longer half-life (60 days versus 8.1 days).

Liquid Scintillation Counters

Except perhaps in the case of the radioimmunoassay of proteins, ^{3}H and ^{14}C are used more extensively in tracer experiments than any other radioisotope. This is not surprising in view of the ubiquitous occurrence of hydrogen- and carbon-containing compounds in cells. The energies of the beta particles emitted by these isotopes are very low (especially tritium), so that it is only with great difficulty that they can be accurately measured using gas ionization methods. The great demand for a sensitive and accurate method for measuring ^{3}H and ^{14}C, as well as other soft beta emitters, led in the 1950s to the development of the **liquid scintillation** technique.

Before considering this important technique in detail, the basic mechanism involved in the detection process will be briefly summarized. The radioactive sample to be counted is mixed in a glass or plastic vial with a "scintillation fluid." Certain molecules dissolved in the fluid indirectly absorb the energy dissipated by the beta particles and respond by scintillating. The resulting photons of light are detected by photomultiplier tubes that relay proportional electrical pulses to the scaling units. Certain advantages of the procedure are immediately apparent, for if the sample is thoroughly mixed with the scintillation fluid, no self-absorption occurs. Moreover, because the

technique is employed almost exclusively with weak beta-emitting isotopes, the energy of *all* beta rays is dissipated within the fluid, making it possible, at least in theory, to detect most if not all of the nuclear transmutations.

The scintillation fluid contains three major components: (1) an organic **solvent** constituting the bulk of the mass, (2) a **primary solute (fluor)**, and (3) a **secondary solute (fluor)**. The most commonly used solvents and fluors are listed in Table 14–5. The energy of the beta particle is transferred to the solvent, causing ionization or excitation of solvent molecules. The latter effect is more important, because in returning to the nonexcited ground state, energy is transferred to the primary fluor molecules, causing their excitation. In returning to the ground state, the excited primary fluor emits photons of near ultraviolet light. The photocathodes of most photomultiplier tubes are not sensitive to photons having wavelengths in this region, and it is the role of the secondary fluor to absorb the light energy from the primary solute and reemit it as light of longer wavelength. (Secondary fluors are sometimes called "wave shifters.") For example, the light emitted by PPO has a wavelength maximum at 365 nm, whereas POPOP emits light having a wavelength maximum at 420 nm, which is close to the maximum sensitivity region of most photomultiplier tubes.

The size of the electrical pulse produced by the photomultiplier tube is proportional to the energy dissipated by the beta particle in the liquid scintillation fluid. Consequently, a pulse-height spectrum is pro-

TABLE 14–5   SOLVENTS AND FLUORS USED FOR LIQUID SCINTILLATION COUNTING

	Fluorescence Maximum (nm)
Solvents	
Toluene	
Methoxybenzene	
1,4-Dioxane	
Xylene	
Primary fluors	
PPO (i.e., 2,5-diphenyloxazole)	365
PBD (i.e., 2-phenyl-5-[4-biphenyl]-1,3,4-oxadiazole)	364
Naphthalene	325
Anthracene	405
Secondary fluors	
POPOP (i.e., 1,4-bis[5-phenyloxazolyl-2']benzene)	420
DMPOPOP (i.e., 1,4-bis[4-methyl-5-phenyloxazol-2']benzene)	430
BBOT (i.e., 2,5-bis[5-*tert*-butylbenzoxazolyl-2']thiophene)	438

duced having a profile almost identical to that of the beta particle energy spectrum. In the case of ^3H and ^{14}C, the sizes of many of the pulses produced are so low that spurious low-amplitude pulses arising spontaneously within the photomultiplier tube itself (i.e., "noise") result in a falsely high count. To minimize this problem, in most liquid scintillation counters the sample vial is placed between the faces of *two* photomultiplier tubes connected through a pulse-height analyzer to *coincidence* circuitry (Fig. 14–7) that permits a count to be registered only when *simultaneous* pulses are produced by the tubes. Therefore, a noise signal from one photomultiplier tube is not recorded unless another noise signal is produced by the other tube at the same time (actually, within 10^{-7} seconds of each other). Because noise signals are random, coincidental noise pulses are few in number.

As in G–M and solid scintillation counting, the background count must be subtracted from each recorded count. In liquid scintillation counting, the background count is derived from a number of sources including coincidental tube noise, naturally occurring radioisotopes in the vial and scintillation fluid (most glass contains small amounts of ^{40}K and organic solvents contain small amounts of ^3H and ^{14}C, and cosmic radiation.

Most liquid scintillation counters contain pulse-height analyzers that provide for the rejection of

FIGURE 14–8 Energy or pulse-height distributions of ^3H and ^{14}C. The energy and pulse-height distributions differ so markedly that pulses arising only from ^{14}C in a sample containing both isotopes may readily be distinguished by adjusting the settings of the pulse-height analyzer.

pulses above and below a selected size range. This feature is particularly valuable in "double-label" experiments (experiments in which two different radioisotopes are used). Because the spectrum of pulse heights produced is related to the beta particle energy spectrum of the radioisotope, the settings of the pulse-height analyzer may be adjusted to distinguish pulses arising from each of the isotopes present in the sample. This is particularly easy in the case of samples that contain both ^3H and ^{14}C, as their beta particle energy spectra differ so markedly (Fig. 14–8).

AUTORADIOGRAPHY

Any discussion of the uses of radioisotopes as tracers in cell biology would be incomplete without at least a brief description of **autoradiography**. The materials and equipment used in this technique differ significantly from those employed in the methods previously described. In autoradiography, the biological sample containing the radioisotope is placed in close contact with a sheet or film of photographic emulsion. Rays emitted by the radioisotope enter the photographic emulsion and expose it in a manner similar to visible light. After some period of time (usually several days to several weeks), the film is developed and the location of the radioisotope in the original sample is determined from the exposure spots on the film. Unlike the

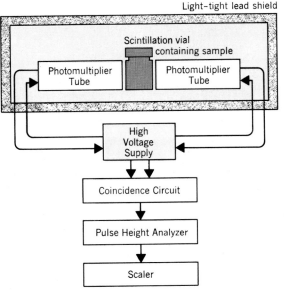

FIGURE 14–7 Basic components of a liquid scintillation counter.

methods described earlier, autoradiography is generally not employed as a quantitative technique (although it can be under certain conditions); instead, it is used to determine the specific region of *localization* of a radioactive tracer. The method is most often used with histological sections to determine the precise location of a labeled compound in a tissue or in a cell and may be applied at either the light microscope or electron microscope level. Autoradiography has also been successfully employed in conjunction with electrophoresis, chromatography (i.e., thin-layer chromatography, paper chromatography, etc.), and other molecular fractionation methods (see Chapter 13) for identifying zones containing labeled compounds (Fig. 14–9; see also Fig. 13–11*b*).

FIGURE 14–9 Autoradiogram of an electrophoresis gel. ^{14}C-labeled proteins from whole T4 phage and from virus subfractions were separated in polyacrylamide and the gel covered with a sheet of film to expose the protein bands. (A) Whole phage; (B) heads; (C) necks; (D) extended tails; (E) tails. (Courtesy of Drs. D. Coombs and F. A. Eiserling.)

Because the main goal of autoradiography is to determine the precise location of the tracer, the degree of resolution obtained in the autoradiograph is of primary importance. Resolution depends on (1) the type of photographic emulsion used, (2) the distance between the radioactive sample and the emulsion, and (3) the nature of the emitted radiation. A variety of photographic emulsions are available, varying in sensitivity and resolution according to the size and concentration of their silver halide grains, with the least sensitive films generally offering the highest degree of resolution. Emulsions can also be obtained that are particularly sensitive to specific types of radiation. Because radiation is emitted in all directions from the radioactive source, the greater the distance between the film and the source, the more diffuse is the resulting image. It is therefore important to use very thin samples and to place them in very close contact with the film. The highest degree of resolution is obtained with radioisotopes that emit rays of short path length and that have a high specific ionization. Alpha particles provide high resolution, but radioisotopes emitting this type of radiation are rarely useful in biological experiments. Autoradiographs become increasingly diffuse as the E_{max} of a beta emitter increases, but ^{3}H, ^{14}C, ^{35}S, and ^{45}Ca do yield good resolution. Gamma rays are inefficient as a result of their extremely high ranges and low specific ionizations.

Autoradiography may be employed with large but thin slices of tissues or organs (gross autoradiography) or with smaller pieces sectioned and prepared for light or electron microscopy. For example, the deposition of calcium in bone has been studied using ^{45}Ca by cutting thin, flat, longitudinal slices through bone and placing them against large sheets of photographic film. When used in conjunction with light microscopy, the paraffin-embedded tissue sections are first mounted on slides, which are then coated with photographic emulsion and stored in a light-tight, usually lead-lined box (to minimize background exposure that results from the effects of cosmic radiation and other sources of radiation). Several days or weeks later, after photochemical development, the sections are conventionally stained to better visualize the biological material and the distribution of radioisotope is determined microscopically from the location of dark exposure spots (called "grains") on the section (Fig. 14–10).

Autoradiography may also be used with thin sections of tissues prepared for electron microscopy (Fig. 14–11). This procedure involves preliminary examina-

FIGURE 14–10 Autoradiogram of chromosomes from the polytene nucleus of a *Drosophila* salivary gland cell. The dark exposure spots or "grains" reveal the incorporation of ^{3}H-labeled cytidine triphosphate into actively transcribing regions of the chromosomes. The heavily labeled regions were specifically induced to transcribe by incubating the salivary gland at 37°C in medium containing the tracer. (Courtesy of Dr. J. Lee Compton.)

FIGURE 14–11 Autoradiogram of thin section through dividing *Escherichia coli* cells. The bacteria were cultured on a medium containing ^{3}H-labeled thymidine, which was incorporated into the DNA of the nucleoids. (Courtesy of Dr. F. A. Eiserling.)

tion and photography of the thin section followed by coating the grid with a very thin layer of photographic emulsion containing silver halide crystals of particularly small size. After several days, the grid is developed and again examined and photographed with the electron microscope to identify those regions of the original section that contain clusters of metallic silver grains and therefore the radioactive tracer.

SUMMARY

Certain atomic **isotopes** are unstable and emit **particulate** (and sometimes **electromagnetic**) radiations. A number of radioactive isotopes, especially ^{3}H, ^{14}C, ^{32}P, ^{35}S, ^{59}Fe, and ^{125}I, are particularly valuable to cell biologists as **tracers** of metabolism. With radioisotopes, it is possible to follow ionic or molecular fluxes under conditions of zero net exchange, to simplify chemical analyses, to measure pool sizes, and to determine precursor–product relationships.

The most common types of radiations are **alpha particles, beta particles**, and **gamma rays**. Only negative beta particle-emitting isotopes and gamma ray-emitting isotopes are normally used by biologists. The energy acquired by a beta particle during a **nuclear transmutation** may vary over a specific range, whereas gamma rays have discrete energy values. Beta particles and gamma rays interact with matter by ionizing or exciting atoms in their path, and it is this property that is used in instruments designed to detect and measure the quantity of radioisotope present in a sample. The simplest and most common detector is the **Geiger–Müller counter** and is used principally with tracers emitting beta particles of intermediate or high energy. **Solid scintillation counters** are used with isotopes emitting gamma rays and **liquid scintillation counters** with isotopes emitting weak beta rays. ^{3}H and ^{14}C are used as tracers more often than any other isotopes, and because these emit very weak beta particles, the liquid scintillation approach is a particularly important technique in cell biology. In the typical liquid scintillation counter, the sample is mixed with a scintillation fluid in a glass vial. The energies of the beta rays emitted by the sample are ultimately converted to flashes of light, which are detected by photomultiplier tubes and counted.

Autoradiography is a special modification of the radioisotopic tracer technique and is most often used in conjunction with light and/or electron microscopy. Biological samples (usually tissue sections) containing the tracer are placed in contact with special photographic films or emulsions. The emitted rays expose the film, producing *grains*, the numbers and distribution of which within a cell or tissue provide quantitative and qualitative information about the movement and/or localization of metabolic intermediates or products.

REFERENCES AND SUGGESTED READING

Articles and Reviews

Hahn, E. J., Autoradiography: A review of basic principles. *Am. Lab.*, p. 42 (Feb. 1984).

Joftes, D. L., Radioautography, principles and procedure. *J. Nucl. Med.* **4**, 143 (1963).

Shamel, L. B., The immunoradiometric assay. *Am. Clin. Prod. Rev.*, p. 44 (Nov. 1983).

Books, Monographs, and Symposia

Chase, G. D., and Rabinowitz, J. L., *Principles of Radioisotope Methodology* (3rd ed.), Burgess Publishing Co., Minneapolis, 1967.

Cooper, T. G., *The Tools of Biochemistry*, Wiley-Interscience, New York, 1977.

Finlayson, J. S., *Basic Biochemical Calculations: Related Procedures and Principles*, Addison–Wesley, Reading, Mass., 1969.

Freifelder, D., *Physical Biochemistry*, Freeman, San Francisco, 1976.

Friedlander, G., Kennedy, J. W., and Miller, J. M., *Nuclear Radiochemistry* (2nd ed.), Wiley, New York, 1964.

Gude, W. D., *Autoradiographic Techniques*, Prentice–Hall, Englewood Cliffs, N.J., 1968.

Overman, R. T., and Clark, H. M., *Radioisotope Techniques*, McGraw–Hill, New York, 1960.

Wang, C. H., and Willis, D. L., *Radiotracer Methodology in Biological Science*, Prentice–Hall, Englewood Cliffs, N.J., 1965.

STRUCTURE AND FUNCTION OF THE MAJOR CELL ORGANELLES

CHAPTER 15

THE PLASMA MEMBRANE AND THE ENDOPLASMIC RETICULUM

One of the most striking features of cellular organization is the pervasive occurrence of membranes. Not only are cells bounded at their surface by the **plasma membrane**, but the membranous organelles compartmentalize specific cellular functions and activities. Especially in eukaryotic cells, the cytoplasmic matrix itself is pervaded by membrane-bordered channels that comprise the **endoplasmic reticulum** or **ER**. In subsequent chapters of the book, the structural organization and specific functions of the major cellular organelles are examined. In this chapter, we focus on the chemical composition of the plasma membrane and the membranes that form the endoplasmic reticulum and consider the mechanisms by which the transport of materials across these membranes can be achieved.

The *plasma membrane* (or **plasmalemma**) delimits the cell, physically separating the cytoplasm and cellular organelles from the surrounding cellular environment. This implies that all substances either entering or exiting the cell must pass *through* the plasma membrane. Although the plasma membrane may play a passive role in the exchange of very small molecules between the cell and its surroundings, the flux of many substances is actually facilitated by continuous molecular changes within the membrane. In many instances, transport through the membrane is achieved by the active participation of carrier molecules within the membrane and incurs the expenditure of large amounts of metabolic energy. The cellular ingestion (or excretion) of some materials is associated with gross movements and separations of fragments of the membrane from the main body. Stages of this activity can be seen and studied with the electron microscope (some stages also with the light microscope).

The membranes of the endoplasmic reticulum divide much of the internal cell volume into two phases. The space within the channels of the ER comprises the **intracisternal space** of the cell or the **lumenal phase** while the space *between* channels constitutes the **ground substance, cytosol**, or **cytoplasm**. The plasma membrane and the membranes of the endoplasmic reticulum have many properties in common. This is not surprising in view of the fact that portions of the plasma membrane are derived from the endoplasmic reticulum (see below). Both membranes are in a continuous state of flux as new components are added to the membranes and others are lost or degraded. Thus, the rather static appearance of these membranes in electron photomicrographs belies their dynamic nature. Moreover, periodic continuities exist between the plasma membrane and the ER with the result that the external cellular milieu is at times continuous with the intracisternal space. Much of our current knowledge of cellular membranes has been acquired through studies of the plasma membrane because plasma membranes are more easily isolated from cells in a pure state than ER membranes. To be sure, important differences between these membranes do exist; however, it is becoming increasingly apparent that their chemical composition, organization, and functions are similar. Indeed, in prokaryotes (which lack an endoplasmic reticulum), the plasma membrane plays many of the roles of the endoplasmic reticulum of eukaryotic cells.

EARLY STUDIES ON THE CHEMICAL ORGANIZATION OF THE PLASMA MEMBRANE

Among all animal and plant cells, none has been more extensively studied than the mammalian erythrocyte or red blood cell. The erythrocyte has long been the favorite of investigators studying the plasma membrane because relatively pure membrane preparations are easily obtained. The mature erythrocyte contains no nucleus, mitochondria, ribosomes, endoplasmic reticulum, or other organelles; instead, this highly specialized cell consists essentially of a concentrated (semicrystalline) solution of hemoglobin encased in a membrane. Because of the red cell's relative simplicity, its membranes are easily separated from other cytoplasmic constituents (primarily hemoglobin) by centrifugation following osmotic lysis of cells.

Results obtained using erythrocytes have frequently been extrapolated to all cells. This is unfortunate because the erythrocyte is not a typical cell, and it is therefore unlikely that the chemical composition and organization of its limiting membrane are representative. Indeed, with increased interest in studying the plasma membranes of other cells and with the advent of methods for isolating these membranes, our knowledge of the plasma membrane has expanded rapidly in recent years. During this time, it has become increasingly obvious that many of the properties of the erythrocyte membrane are unique. For historical perspective, however, we will begin by considering the early studies of the red blood cell membrane.

Existence of Lipid in the Membrane

As early as 1899, E. Overton recognized that the boundary of animal and plant cells was "impregnated" by lipid material. Overton's conclusions were based on exhaustive studies of the rates of penetration of more than 500 different chemical compounds into animal and plant cells. In general, compounds soluble in organic solvents entered the cells more rapidly than compounds soluble in water. These differences were attributed to the "selective solubility" of the membrane, that is, lipid-soluble materials would pass into the cell by dissolving in the corresponding lipid elements that made up the membrane. Overton suggested that cholesterol and lecithins might be among the lipid constituents of the plasma membrane, a suggestion that was later substantiated chemically. The pioneering studies of Overton at the turn of the century set the stage for Gorter, Grendel, Cole, Danielli, Harvey, Davson, and others who attempted to determine the specific manner in which the lipid might be organized within the membrane.

The Langmuir Trough

One of the most valuable instruments used to study the behavior of lipid films is the **Langmuir trough** (Fig. 15-1). If lipid containing hydrophilic groups (such as the carboxyl groups of fatty acids or the phosphate groups of phospholipids) is dissolved in a highly volatile solvent and several drops are then carefully applied to the surface of the water, the lipid spreads out to form a thin, monomolecular film in which the hydrophilic parts of each molecule project into the water surface and the hydrophobic parts are directed up, away from the water (Fig. 15-2). In 1917, I. Langmuir introduced a clever technique for measuring the spe-

FIGURE 15–1 The Langmuir trough.

FIGURE 15-2 Formation of a monomolecular lipid film on water. Phosphatidyl ethanolamine (a) represents a typical lipid molecule possessing polar (hydrophilic) and nonpolar (hydrophobic) regions. These regions of the phospholipid molecule are depicted diagrammatically in (b). When spread on water, the hydrophilic parts of each lipid project into the water surface and the hydrophobic parts are directed up, away from the water (c).

cific minimum surface area occupied by a monomolecular film of lipid and the force necessary to compress all the lipid molecules into this area. His device, known as the Langmuir trough or Langmuir film balance, has been used extensively over the past several decades in connection with physical measurements of membrane lipids.

The apparatus (shown in Fig. 15-1) consists of a shallow trough coated with a nonwettable material so that it can be filled with water to a level slightly higher than its edges. A bar is placed across the width of the tray and is used to sweep dirt and dust from the water. A second bar, called the "fixed barrier" (which, in fact, is slightly movable), floats on the water surface and is connected to a torsion balance. After the surface of the water is swept clean, a third bar, called the "movable barrier," is placed across the clean region of the trough and a known quantity of dissolved lipid is added to the space between it and the fixed

barrier. The lipid spreads out over the water surface, forming a layer one molecule thick (Fig. 15-2). The movable barrier is then slowly pushed toward the fixed barrier, thereby compressing the lipid until it forms a continuous and rectangular monomolecular film. The force exerted by the film on the fixed barrier can be measured using the torsion wire gauge. From the area occupied by the lipid and the amount initially added to the trough, it is possible to calculate the surface area occupied by a single lipid molecule. Langmuir himself employed this device to study the behavior of the surface films formed by a variety of organic compounds, and for this work he received the Nobel Prize in Chemistry in 1932. Others have applied the same technique in specific studies of membrane lipids.

Gorter and Grendel's Bimolecular Lipid Leaflet Model

In 1925, E. Gorter and F. Grendel published the results of their studies on the organization of lipid in the membrane of the red blood cell. Their studies were carried out using blood from a variety of mammals, including dogs, sheep, rabbits, guinea pigs, goats, and humans, and all yielded essentially the same results. The lipid present in accurately measured quantities of washed red blood cells was extracted with acetone and the acetone was then evaporated, leaving the lipid as a residue. This residue was redissolved in benzene and spread in a Langmuir trough to form a tightly packed monomolecular layer. The surface area occupied by the extracted lipid was then measured.

Gorter and Grendel determined the numbers of red cells present in each sample of blood analyzed and estimated the *total* surface area of the cells by multiplying the cell number by the average surface area per cell. (The surface area of the erythrocyte was estimated using the relationship proposed by Knoll that for red blood cells the surface area $= 2d^2$, d being the diameter of the cell determined microscopically. This estimate was subsequently shown to be in error.) By dividing the total surface area occupied by a monomolecular layer of membrane lipid extracted from these cells by the total cell surface area, the number of lipid layers present in the membrane was obtained. The value varied between 1.8 and 2.2, leading Gorter and Grendel to propose that the cell membrane was formed by a *bimolecular* lipid sheet. They further suggested that the polar ends of the lipid molecules of one layer were directed outward (from the cell) toward the

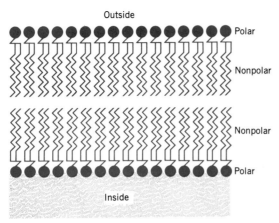

Outside

Polar

Nonpolar

Nonpolar

Polar

Inside

FIGURE 15–3 Bimolecular lipid leaflet model for the structure of the red blood cell membrane proposed by Gorter and Grendel.

surrounding plasma and the polar ends of the lipid molecules forming the other layer were directed inward toward the cell hemoglobin. Thus the nonpolar, hydrophobic ends of the lipid molecules would face one another (Fig. 15–3).

In the past 15 years, the results of Gorter and Grendel have been reexamined under improved conditions by a number of investigators. The validity of Gorter and Grendel's bimolecular lipid leaflet model depends on the assumptions that (1) all the erythrocyte lipids are in the plasma membrane, (2) all of this lipid was extracted using their acetone procedure, and (3) the average surface area of the cells was accurately estimated. The first assumption has been verified—all the erythrocyte lipid *is* in the plasma membrane. However, it is now clear that Gorter and Grendel extracted only 70 to 80% of the total lipid. This error would seriously alter the predicted ratio of *lipid film area to cell surface area* if it were not for the fact that Gorter and Grendel also *underestimated* the red blood cell surface area by a comparable amount. The two errors canceled each other out, so that the ratio of 2:1 is still obtained.

In the time since their original studies were conducted, the quantitative distributions of membrane lipids in erthrocytes have been reevaluated using newer and improved techniques that guarantee total lipid extraction from the membranes and more accurate measurements of cell surface area. Some of the new methods account also for the specific physical dimensions of various lipid molecules known to be present in the membrane (fatty acids, phospholipids, neutral lipids, etc.) in making calculations of the lipid surface area. For the most part, these studies indicate

that indeed there is enough membrane lipid present to form a layer two molecules thick.

Other sources of support for the presence of the lipid bilayer include results of X-ray diffraction analysis of dispersions of isolated cell membranes and electron spin resonance (ESR) studies that indicate that the lipid molecules are oriented with their long axes perpendicular to the plane of the membrane. Attempts have been made to reconstruct lipid bilayers using molecular models for the various membrane lipids and by considering the attractive and repulsive forces resulting from van der Waals interactions, hydrogen bonds, and electrostatic interactions. These studies indicate that a stable bimolecular lipid leaflet can be formed by a tail-to-tail interdigitation of the lipids.

The dimensions, electron-microscopic appearance, permeability, surface tension, and electrical capacitance of artificially created lipid bilayers (i.e., **liposomes**, see Chapter 6) are very similar to those of natural biological membranes. Finally, freeze-fracture techniques (see Chapter 1) used in transmission electron microscopy produce images of plasma membranes and ER membranes that suggest a natural plane of cleavage at the center of the membrane. A plane susceptible to such cleavage would be provided by the space at the center of the bimolecular lipid leaflet (see below).

THE DANIELLI–DAVSON MEMBRANE MODEL

A consistent observation made for plasma membranes that was not explained by the bimolecular lipid leaflet model was the very low surface tension of the cell membrane. In 1935, J. F. Danielli and E. N. Harvey proposed that oil droplets and other lipid inclusions in cells were bounded at their surfaces by an organized layer of the lipid and a layer of protein. It was postulated that the protein, which consisted of a monomolecular layer of hydrated molecules, faced the aqueous cytoplasm and simultaneously interacted with the polar portions of the lipid layer. The nonpolar portions of the lipid layer faced the hydrophobic oil phase of the droplet interior (Fig. 15–4). In this structure, the natural surface activity of the protein would account for the low interfacial tension of the droplet membrane. Shortly thereafter, Danielli and H. Davson suggested that the plasma membrane itself might be composed of two such lipid–protein bilayers—one facing the interior of the cell and the other facing the external milieu

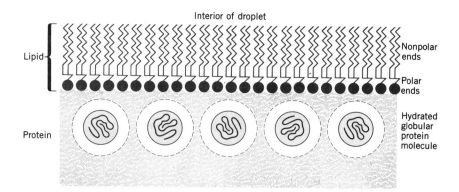

FIGURE 15–4 Danielli and Harvey's 1935 model of the protein–lipid bilayer formed at the interface between an oil droplet in a cell and the cell's aqueous cytoplasm.

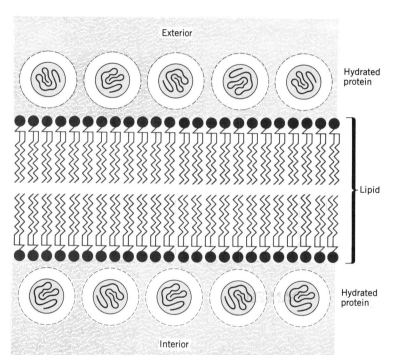

FIGURE 15–5 Danielli–Davson membrane model (1935).

(Fig. 15–5). In this arrangement, the association between the surface proteins and bimolecular lipid leaflet would be maintained primarily by electrostatic interactions between the polar ends of each lipid molecule and charged amino acid side chains of the polypeptide layers. Either electrostatic or van der Waals bonds could bind other groups to the outer protein surface. Danielli and Davson proposed that such a membrane would exhibit *selective permeability,* being capable of distinguishing between molecules of different size and solubility properties and also between ions of different charge.

By the early 1950s, several modifications were made in the Danielli–Davson membrane model. For example, it was suggested that glycoproteins might be adsorbed to the outer membrane surface, thereby accounting for the antigenic properties of cell membranes. Pores through which certain materials are exchanged between a cell and its environment were presumed to exist, the channels formed by periodic continuities (bridges) between the outer and inner protein layers.

ROBERTSON'S UNIT MEMBRANE

In the late 1950s, electron microscopy provided additional information about the structure of the plasma membrane. J. D. Robertson was a pioneer in this area, showing that membranes fixed with osmium tetroxide revealed a characteristic *trilaminar* appearance consisting of two parallel outer dark (osmiophilic) layers

FIGURE 15–6 Trilaminar appearance of the plasma membrane. (Electron photomicrograph courtesy of R. Chao.)

and a central light (osmiophobic) layer (Fig. 15–6). The osmiophilic layers typically measured 20–25 Å (2.0–2.5nm) in thickness and the osmiophobic layers measured 25–35 Å (2.5–3.5 nm), yielding a total thickness of 65–85 Å (6.5–8.5 nm). This value compared favorably with the thickness predicted on the basis of chemcial studies.

Robertson and others demonstrated that the trilaminar pattern was characteristic of many other cellular membranes, including the endoplasmic reticulum. In view of the underlying unity in the appearance of the cell membranes studied, Robertson proposed his now famous **unit membrane** model. According to Robertson, the unit membrane consisted of a bimolecular lipid leaflet sandwiched between outer and inner layers of protein organized in the pleated sheet configuration. Such an arrangement was presumed to be basically the same in all cell membranes. Though Robertson acknowledged specific chemical differences between membranes (i.e., the particular molecular species that make up each membrane differ), he proposed that the pattern of molecular organization was fundamentally the same. Although there can be no doubt about the similar electron-microscopic appearance of nearly all membranes, so strict a chemical interpretation to account for the uniformity is no longer supportable.

Robertson extended his unit membrane model to include the notion that continuity exists between the membranes of the nuclear envelope and the plasma membrane via the endoplasmic reticulum. The occurrence of such continuities has been confirmed in electron-microscopic studies of many different cells and tissues. Furthermore, Robertson suggested that vesicular organelles might arise from this continuous membrane system and that they are subsequently pinched off to form separate structures. There is some evidence in support of this notion in the case of lysosomes and microbodies (Chapter 19).

FUNCTIONAL DOMAINS IN THE PLASMA MEMBRANE

Cellular membranes vary in biological functions, in kinds and relative amounts of phospholipids, and in quantitative lipid-to-protein ratios (see later). Consider, for example, the plasma membrane of a parenchymal cell in the liver. The membrane forms junctions with at least three different neighboring structures (Fig. 15–7): (*a*) one face of the liver cell plasma membrane forms a junction with the membranes of the **capillary sinusoids**, and across this face pass various substances exchanged between the bloodstream and the liver cell cytosol (e.g., sugars, amino acids, insulin, etc.). The membrane in this region would be expected to contain specific transferases and hormone receptors. (*b*) A second face (called the **con-**

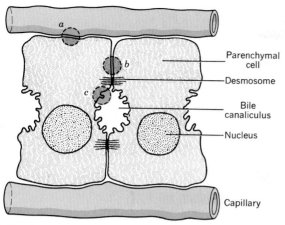

FIGURE 15-7 Three faces of the liver cell. (a) Juxtaposition of plasma membrane and sinusoid capillary membrane. (b) Juxtaposition with neighboring cell. (c) Juxtaposition with bile canaliculus. Each face contains specific membrane constituents and properties.

tiguous face) is formed with the plasma membranes of neighboring cells, and it is across this face that intercellular exchange takes place. (c) A third face of the liver cell is in contact with the **bile canaliculi**, into

which bile salts and bile pigments are continuously discharged; here too the organization and composition of the plasma membrane would be expected to be rather specific. W. H. Evans and others have shown that the different faces of the liver cell plasma membrane are clearly distinguishable in enzyme content and molecular composition. The epithelial cells that line the small intestine also reveal a differential distribution of membrane proteins. The portion of the membrane facing the intestinal lumen is rich in glycoproteins, whereas the opposite membrane face contains *sodium pumps* (discussed later). The membrane uniformity apparent during electron microscopy of cells belies the chemical heterogeneity that actually exists.

THE FLUID-MOSAIC MODEL OF MEMBRANE STRUCTURE

At the present time, the most widely accepted model of membrane structure is the **fluid-mosaic model** (an expression introduced by S. J. Singer and G. Nicolson to describe both the properties and organization of the membrane). According to this model (Fig. 15–8),

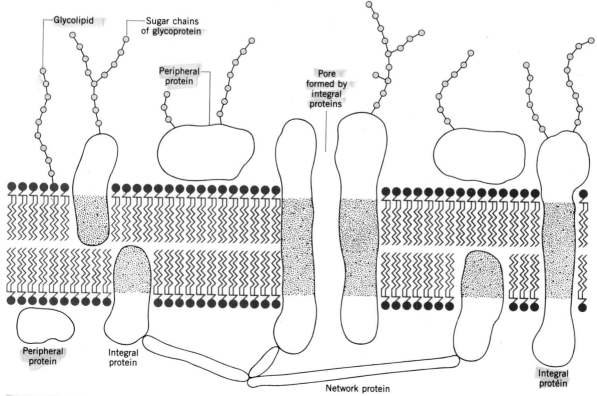

FIGURE 15-8 The *fluid-mosaic* model of membrane structure.

the membrane contains a bimolecular lipid layer, the surface of which is interrupted by proteins. Some proteins are attached at the polar surface of the lipid (i.e., the **peripheral**, or **extrinsic**, proteins), while others penetrate the bilayer or span the membrane entirely (i.e., the **integral**, or **intrinsic**, proteins). The peripheral proteins and those parts of the integral proteins that occur on the outer membrane surface frequently contain chains of sugars (i.e., they are *glycoproteins*). The sugar chains are believed to be involved in a variety of physiological phenomena including the adhesion of cells to their neighbors. Membrane lipid is primarily phospholipid, although quantities of neutral lipids may also be present. Some of the lipid at the outer surface is complexed with carbohydrate to form glycolipid.

FREEZE-FRACTURED MEMBRANES

The fluid-mosaic model of membrane structure is beautifully supported by visual evidence provided when freeze-fractured membranes are examined with the transmission electron microscope. D. Branton, who pioneered this field, showed that membranes rapidly frozen at the temperature of liquid nitrogen and cut or chipped with a microtome blade readily fracture along specific planes. When the plane of the fracture intersects the plane of the membrane, the membrane is split along the center of the lipid bilayer, producing two "half-membranes" called the **E** half and the **P** half. The **E** half is that portion of the membrane that faced the cell *exterior*, while the **P** half corresponds to the portion that faced the "protoplasm"

FIGURE 15–9   Membrane faces exposed using the freeze-fracturing technique alone and by freeze-fracturing followed by sublimation. (*a*) The fracture passes along the lipid bilayer for some distance following its intersection with the membrane. (*b*) This exposes the *PF* and *EF* fracture faces seen in the replica (*c*). Alternatively, the surface of the specimen is lowered by sublimation (*d*), exposing the outer and inner surfaces (i.e., the E_s and P_s faces) as well as the fracture faces (*e*).

(i.e., cytosol). One side of each half-membrane is the original membrane surface, called the **E** and **P** faces, whereas the other side is the newly exposed **fracture face**, called the **E** fracture face (**EF**) and **P** fracture face (**PF**). The fracture faces are extremely delicate and are not examined directly. Instead, a thin film of platinum and carbon is evaporated onto the surface of the fracture faces to produce a **replica** (see Chapter 1), which is then examined by transmission electron microscopy.

In many instances, before the replica is made, water (as well as other volatile materials) on or near the fracture surfaces is eliminated by sublimation (i.e., by carefully raising the temperature of the sample). This step, which used to be called "freeze-etching," exposes additional surface features of the fracture face (see Fig. 15–9 and especially Fig. 1–11).

Electron micrographs of freeze-fractured cells show the membranes to be covered by numerous small particles (Fig. 15–10). There is convincing evidence that the particles are membrane proteins (e.g., they disappear when the membranes are first treated with proteolytic enzymes). This suggests that the plane of fracture passes around the protein molecules rather than through them. This relationship is depicted in Figure 15–11. The relatively uniform background apparent in the fracture face in Figure 15–10 corresponds to the surface of one-half of the lipid bilayer.

FIGURE 15–10 Freeze-fractured and sublimated plasma membrane of an erythrocyte. The fracture face of the protoplasmic half of the membrane (i.e., *PF*) is covered with particles believed to be integral proteins. Sublimation reveals the outer surface of the exterior half of the membrane (i.e., E_s). The fracture face is seen to be separated by a step (St) from the outer surface of the membrane. Circled arrow indicates direction of shadowing. Magnification, 90,000x. (Photomicrograph courtesy of Dr. C. Stolinski.)

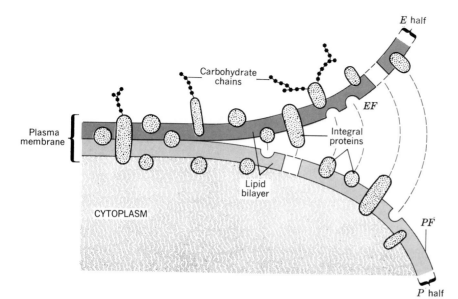

FIGURE 15–11 Freeze-fracturing of the plasma membrane. The fracture plane occurs at the center of the lipid bilayer and passes over (or under) the integral membrane proteins.

MEMBRANE PROTEINS

Peripheral (Extrinsic) Proteins

Peripheral or *extrinsic* membrane proteins are generally loosely attached to the membrane and are more readily removed than are the integral proteins. Peripheral proteins are rich in amino acids with hydrophilic side chains that permit interaction with the surrounding water and with the polar surface of the lipid bilayer. Peripheral proteins on the cell's exterior membrane surface often contain chains of sugars (i.e., they are glycoproteins).

Integral (Intrinsic) Proteins

Integral or *intrinsic* membrane proteins contain both hydrophilic and hydrophobic regions. The hydrophilic portions of the protein interact with the polar heads of the lipid molecules at each surface of the bimolecular

FIGURE 15–12   To extend into the hydrophobic lipid bilayer of the membrane, the conformations of polypeptides that are part of integral proteins are believed to be predominatly alpha helix and beta sheet. In these conformations polar alpha amino and alpha carboxyl groups are neutralized by their formation of hydrogen bonds.

leaflet. Portions of integral proteins that project beyond the surface of the lipid bilayer are also rich in hydrophilic amino acids. Amino acids in that part of the protein projecting from the outer membrane surface may be linked to chains of sugars. Parts of the protein that are buried in the hydrophobic portion of the lipid bilayer are rich in amino acids with hydrophobic side chains. These side chains are believed to form hydrophobic bonds with the hydrocarbon tails of the membrane phospholipids. It is speculated that within the hydrophobic interior of the membrane, the secondary structure of integral proteins is *alpha helix* and/or *beta sheet* (Fig. 15–12). In the alpha helix conformation, amino and carboxyl groups along a stretch of the polypeptide's backbone form hydrogen bonds with one another; in the beta sheet, hydrogen bonds are formed between amino and carboxyl groups in stretches of polypeptide that lie parallel to one another (Chapter 4). In the absence of such hydrogen bonding, these amino and carboxyl groups would have polar properties, their hydrophilic nature being incompatible with the membrane's hydrophobic interior. Alpha amino and carboxyl groups that are not "neutralized" by hydrogen bonding would be expected only in those parts of the integral protein that extend into the aqueous milieu on either side of the membrane.

Integral Proteins That Span the Membrane

M. Bretscher first demonstrated the existence of integral proteins that span the entire membrane. In a series of elegant experiments, Bretscher showed that radioactive ligands specific for membrane proteins of the erythrocyte were bound in smaller quantities to intact cells than to disrupted cells. Disruption of the cells was shown to expose portions of the membrane proteins previously facing the cell interior, thereby allowing additional radioactive ligand to associate with the protein.

T. L. Steck developed a technique for converting fragments of disrupted erythrocyte membranes into small vesicles that were either "right-side-out" (i.e., the external face of the membrane also formed the external face of the vesicle) or "inside-out" (Fig. 15–13). When proteolytic enzymes were added to separate suspensions of each type of vesicle, certain of their membrane proteins were found to be equally susceptible to digestion and could therefore be enzymatically attacked from either membrane surface. These proteins clearly spanned the membrane. Other proteins were susceptible to enzymatic digestion only when

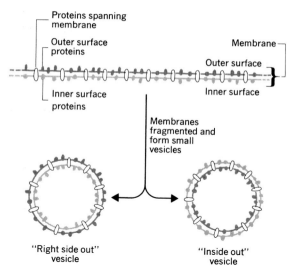

FIGURE 15–13 Experiments of T. L. Steck showing that proteins of the erythrocyte membrane may be confined to one or the other surface or may span the plasma membrane. Membranes were fragmented and allowed to form suspensions of vesicles. When the vesicles were "right-side-out," proteolytic enzymes added to the suspension destroyed those proteins normally present at the outer surface of intact membranes but had no effect on proteins normally present at the inner surface. In contrast, when "inside-out" vesicles were similarly treated, proteins normally present at the inner surface were destroyed. Proteins spanning the membrane were destroyed in both cases.

present in right-side-out or inside-out vesicles, indicating their differential distribution in the membrane's outer and inner surfaces.

Integral proteins that span the entire membrane contain *two* outer regions that are hydrophilic (i.e., one at each surface of the membrane); the *central* region is hydrophobic (Fig. 15–12). Carbohydrate associated with the hydrophilic region facing the cell's surroundings is believed to play a role in maintaining the orientation of the protein within the membrane. The hydrophilic sugars, together with the hydrophilic side chains of amino acids in the outer region of the protein, effectively prevent reorientation of the protein in the direction of the hydrocarbon core of the lipid bilayer.

Asymmetric Distribution of Membrane Proteins

The outer and inner regions of the plasma membrane do not contain either the same types or equal amounts of the various peripheral and integral proteins. For example, the outer half of the erythrocyte membrane contains far less protein than does the inner half. In addition, various membrane proteins may be present in significantly different quantities; the membranes of some cells contain a hundred times as many molecules of one protein species as another. Moreover, regardless of absolute quantity, all copies of a given membrane protein species have exactly the same orientation in the membrane. The differential distribution of proteins in the various regions of the plasma membrane within a single cell was described earlier in connection with liver parenchymal cells and intestinal epithelium. This irregular distribution of membrane proteins is known as **membrane asymmetry**. Not only are the proteins of plasma membranes asymmetrically distributed but so too are the membranes of the endoplasmic reticulum and vesicular organelles (e.g., mitochondria).

Mobility of Membrane Proteins

When cells are grown in culture, there is an occasional fusion of one cell with another to form a larger cell. The frequency of cell fusion can be greatly increased by adding *Sendai virus* to the cell culture. In the presence of this virus, even different strains of cells can be induced to fuse, producing *hybrid* cells or **heterokaryons**. D. Frye and M. Edidin utilized this phenomenon to demonstrate that membrane proteins may not maintain fixed positions in the membrane but may move about laterally through the bilayer. Frye and Edidin induced the fusion of human and mouse cells to form heterokaryons and, using fluorescent antibody labels, followed the distribution of human and mouse membrane proteins in the heterokaryon during the time interval that followed fusion. At the onset of fusion, human and mouse membrane proteins were respectively restricted to their "halves" of the hybrid cell, but in less than an hour both protein types became uniformly distributed through the membrane (Fig. 15–14). The distribution of the membrane proteins was not dependent on the availability of ATP and was not prevented by metabolic inhibitors, indicating that lateral movement of proteins in the membrane occurred by diffusion.

Although some membrane proteins are capable of lateral diffusion, many are not. G. Nicolson and others have obtained evidence suggesting that many integral proteins are restrained within the membrane by a protein *network* lying just under the membrane's inner

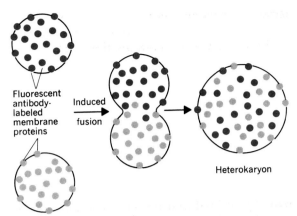

Fluorescent antibody-labeled membrane proteins

Induced fusion

Heterokaryon

FIGURE 15–14 Movement of proteins in the plasma membrane. When fluorescent antibody labeling of the membrane proteins of different cells is followed by *Sendai virus*-induced cell fusion, the proteins are soon distributed throughout the membrane of the heterokaryon.

surface (Fig. 15–9). In many cells, this network is associated with a system of cytoplasmic filaments and microtubules that radiate through the cytosol forming a **cytoskeleton** (see later).

Enzymatic Properties of Membrane Proteins

Membrane proteins have been shown to possess enzymatic activity. Table 15–1 lists some of the enzymes

TABLE 15–1 ENZYMES PRESENT IN PLASMA MEMBRANE

Acetylphosphatase
Acetylcholinesterase
Acid phosphatase
Adenosine triphosphatase
Adenylate cyclase
Alkaline phosphatase
Alkaline phosphodiesterase
Aminopeptidase (several forms)
Cellobiase
Cholesterol esterase
Guanylate cyclase
Lactase
Maltase
Monoglyceride lipase
NADH oxidizing enzymes
NAD glycohydrolase
5′-Nucleotidase
Phospholipase A
Sialidase
Sphingomyelinase
Sucrase
UDP glycosidase

that are now recognized as constituents of the plasma membrane of various cells. To this list of proteins must be added *receptor* proteins (such as the insulin-binding sites of the liver plasma membrane) and *structural* or nonenzymatic proteins.

Ectoenzymes and Endoenzymes Enzymes disposed in the plasma membrane may be characterized according to the membrane face containing the enzymatic activity. Accordingly, **ectoenzymes** are those enzymes whose catalytic activity is associated with the exterior surface of the plasma membrane; the activity of plasma membrane **endoenzymes** is associated with the interior of the cell. Many (perhaps all) plasma membrane ectoenzymes are glycoproteins.

Isolation and Characterization of Membrane Proteins

Because of the relative ease with which they may be purified, the plasma membranes of erythrocytes provided much of the early information on the chemistry of proteins (and lipids) present in membranes. Now, however, plasma membranes can be obtained from many cell types in a reasonably uncontaminated state using various forms of density gradient centrifugation. Nonetheless, the individual protein constituents of the membrane are not so easily extricated for individual study because of their high degree of insolubility. Varying degrees of success in extracting proteins from the plasma membrane have been achieved using organic detergents (especially sodium dodecyl sulfate, SDS) and concentrated solutions of urea, *n*-butanol, and ethylene diamine tetraacetic acid (EDTA). These chemicals have a disaggregating effect on membranes and cause the release of many of the membrane proteins by dissociating the bonds that link the proteins together or to other membrane constituents. Often, the removal of these agents from a preparation of solubilized membrane proteins is quickly followed by the reassociation or reaggregation of the proteins to form an intractable matrix.

Once solubilized, the membrane proteins can be separated into discrete classes using electrophoresis, chromatography, or other procedures (see Fig. 13–10). This generally demands that the dissociating agents be present in the separating medium (e.g., the electrophoresis gel, the column eluent, etc.); otherwise, application of the membrane extract to the medium is followed by membrane protein reaggregation into insoluble complexes that will not separate into distinct fractions. For example, the separation of liver plasma membrane proteins shown in Figure 13–10 is achieved only if the electrophoresis gel contains SDS. The solubility problem has been one of the greatest barriers to progress in isolating and fully characterizing the proteins of membranes.

Some of the plasma membrane enzymes listed in Table 15–1 have not actually been isolated from the membrane, as removal and isolation of the enzyme is *not* a prerequisite for establishing its presence. Instead, the enzyme activity can be measured directly in the (unsolubilized) membrane preparation.

MEMBRANE LIPIDS

Much more is known about the specific lipid composition of cell membranes, because the lipids are more readily extracted from the membranes using a variety of organic solvents. Once extracted from isolated membranes, the lipids may be separated and identified using chromatographic or other procedures. Nearly all the membranes studied so far appear to contain the same types of lipid molecules. Phospholipids such as phosphatidyl ethanolamine, phosphatidyl serine, phosphatidyl inositol, phosphatidyl choline (lecithin), and sphingomyelin are the most common constituents, but cholesterol is also present. The chemical structures of these lipids may be found in Chapter 6. Table 15–2 lists the most common lipids found in a variety of cell membranes and also shows their protein-to-lipid weight ratios; the latter vary considerably. In addition to its widespread occurrence in plasma membranes, cholesterol is also found in many intracellular membranes. The rigid (i.e., planar) nature of cholesterol (Chapter 6) imparts an ordering effect to those cellular membranes that contain this lipid.

Mobility of Membrane Lipids

Lipids exhibit a higher degree of mobility in membranes than do proteins, although lateral mobility is very much greater than transverse ("flip-flop") mobility. A single lipid molecule may move several microns laterally through the membrane in just 1 or 2 seconds!

The mobility of lipid and protein molecules in the plasma membrane attests to the membrane's fluidity. C. F. Fox and H. M. McConnell have shown that the degree of fluidity is dependent, in turn, on the fatty acid contents of side chains of phospholipids in the membrane.

Fatty acid side chains of membrane phospholipids can be either *saturated* or *unsaturated*. In saturated side chains, all the carbon–carbon bonds are single, with the remaining carbon bonds carrying hydrogen atoms; in unsaturated side chains, one or more pairs of neighboring carbon atoms are linked by double bonds (see Chapter 6). In phospholipid layers consisting exclusively of saturated fatty acids, the side chains are aligned next to one another in an ordered, crystalline array; the result is a relatively rigid structure (Fig. 15–15). In phospholipid layers consisting of a mixture of saturated and unsaturated fatty acid side chains, the packing of neighboring molecules is less orderly (and therefore more fluid). The double bonds of the unsaturated side chains produce bends in the hydrocarbon chains, and these give rise to structural

TABLE 15–2   LIPIDS PRESENT IN THE PLASMA MEMBRANE

Plasma Membrane	Major Lipids Present	Protein/Lipid (wt/wt)
Liver cell	Cholesterol, phosphatidyl choline, phosphatidyl ethanolamine, phosphatidyl serine, sphingomyelin	1.0–1.4
Intestinal epithelial cell	Cholesterol, phosphatidyl choline, phosphatidyl ethanolamine, phosphatidyl serine, sphingomyelin	4.6
Erythrocyte	Phosphatidyl inositol, cholesterol, phosphatidyl choline, phosphatidyl ethanolamine, phosphatidyl serine, sphingomyelin	1.6–1.8
Myelin	Cholesterol, cerebrosides, phosphatidyl ethanolamine, phosphatidyl choline	0.25
Gram-positive bacteria	Diphosphatidyl glycerol, phosphatidyl glycerol, phosphatidyl ethanolamine	2.0–4.0

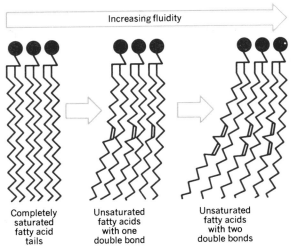

Increasing fluidity

Completely saturated fatty acid tails

Unsaturated fatty acids with one double bond

Unsaturated fatty acids with two double bonds

FIGURE 15–15 Disruption of the orderly stacking of phospholipids having saturated fatty acid tails by unsaturated fatty acids having one or more double bonds.

deformations that prevent formation of the more rigid crystalline structure. The greater the number of double bonds, the more disordered (and fluid) is the lipid bilayer (Fig. 15–15).

The rigidity of lipid layers is also affected by temperature. Almost everyone is familiar with the "melting" of fats and waxes at elevated temperatures. To maintain membrane fluidity, cells living at low temperatures have higher proportions of unsaturated fatty acids in their membranes than do cells at higher temperatures. Evidence also exists suggesting that cells can alter the balance of saturated and unsaturated fatty acids in their membranes as an adjustment to changing temperature or other factors.

In recent years, the degree of membrane fluidity has been linked to the capability of various metabolites and hormones to bind to surface receptors. An increase in membrane fluidity may be accompanied by the withdrawal of exposed receptors (i.e., they are drawn deeper into the lipid bilayer), whereas a decrease in membrane fluidity is accompanied by greater accessibility of the receptor through increased exposure above the bilayer.

Lipid Asymmetry

The various membrane lipids are not equally distributed in both monolayers, although the asymmetry is not nearly as marked as in the case of protein. The distribution of lipids in the erythrocyte membrane is shown in Table 15–3 and reveals that the choline phosphatides are primarily in the outer monolayer and the aminophosphatides are in the inner monolayer. Although lipid asymmetry is a general property of membranes, the type of asymmetry varies considerably from one membrane to another. Asymmetry, once established, is most likely maintained because of the high activation energy that would be required to move the polar groups through the hydrophobic center of the bilayer. Just as proteins are differentially distributed in the plasma membrane areas comprising the various functional faces of a tissue cell, so are the lipids. This is vividly seen in Figure 15–16, which shows the distributions of phosphatidyl choline, sphingomyelin, phosphatidyl ethanolamine, and phosphatidyl serine in the three major plasma membrane regions (faces) of the liver parenchymal cell.

MEMBRANE CARBOHYDRATE

It has already been noted that carbohydrate is present in the plasma membrane as short, sometimes branched chains of sugars attached either to exterior peripheral proteins (forming glycoproteins) or to the polar ends of phospholipid molecules in the outer lipid layer (forming glycolipids). No membrane carbohydrate is located at the interior surface.

The oligosaccharide chains of membrane glycopro-

TABLE 15–3 DISTRIBUTION OF LIPIDS IN THE ERYTHROCYTE MEMBRANE

	Interior Lipid Monolayer (%)	Exterior Lipid Monolayer (%)
Total	50	50
Sphingomyelin	6	20
Phosphatidyl choline	9	23
Phosphatidyl ethanolamine	25	6
Phosphatidyl serine	10	0
Phosphatidyl inositol	0	0

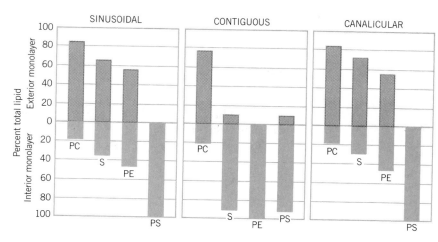

FIGURE 15–16 Relative distributions of lipids in the internal and external monolayers of the sinusoidal, contiguous, and canalicular faces of the liver cell plasma membrane. PC, phosphatidyl choline; S, sphingomyelin; PE, phosphatidyl ethanolamine; PS, phosphatidyl serine. See also Figure 15–7.

teins and glycolipids are formed by various combinations of six principal sugars: *D*-galactose, *D*-mannose, *L*-fucose, *N*-acetylneuraminic acid (also called sialic acid), *N*-acetyl-*D*-glucosamine, and *N*-acetyl-*D*-galactosamine (see Chapters 4 and 5 for chemical structures). All of these may be derived from glucose.

Possible Functions of Membrane Carbohydrate

Several roles have been suggested for the carbohydrate present on the outer surface of the plasma membrane. One possibility is that because they are highly hydrophilic, the sugars help to orient the glycoproteins (and glycolipids) in the membrane so that they are kept in contact with the external aqueous environment and are unlikely to rotate toward the membrane's hydrophobic interior. Certain plasma transport proteins, hormones, and enzymes are glycoproteins, and in these molecules, carbohydrate is important to physiological activity. It would therefore not be inappropriate to expect that in certain glycoproteins of the plasma membrane the carbohydrate moiety is basic to either enzymatic or some other activity.

Carbohydrate chains of surface glycoproteins are clearly responsible for the various human blood types (e.g., ABO types, MN types, etc.) and other tissue types. That is, the sugar sequence and the arrangement of the sugar chains in the membranes of blood cells of an individual with type A blood differ from those of an individual with type B blood, and so on. The carbohydrate is responsible for cell type specificity and is therefore fundamental to the specific antigenic properties of cell membranes. These antigenic properties are linked in some manner to the body's *immune system* and the capability of that system to distinguish between cells that should be present in the organism (native cells) and foreign cells. Foreign cells (such as bacteria or other microorganisms, transplanted tissue, or transfused blood) may be recognized as foreign because their membrane glycoproteins contain different carbohydrate markers than those present in the individual's own tissues. Such a situation triggers the immune response. In contrast, an individual's own plasma membrane carbohydrate organization is recognized as being native (referred to as "recognition of self") and does not normally trigger an immunological response. Of course, neither does blood transfusion nor tissue transplantation if the carbohydrate organization in the membranes of the "donor's" and "recipient's" cells is the same. Cell-specific membrane carbohydrate organization is considered further in connection with the actions of **lectins** and **antibodies**.

Oppenheimer, Roseman, Roth, and others have clearly implicated surface carbohydrate in the adhesion of a cell to its neighbors in a tissue; presumably, the carbohydrate acts as an adhesive maintaining the integrity of the tissue by linking neighboring cells together. *Density-dependent inhibition*, the phenomenon in which cells grown in culture stop dividing when their numbers attain a certain level, may be attributable to a mechanism triggered by interaction of carbohydrates on neighboring cells.

Carbohydrate is also present at the surface of cells as glycolipid, especially *glycosphingolipids*. Although the role of this glycolipid is uncertain, it is believed to function in cell-to-cell recognition.

LECTINS, ANTIBODIES, ANTIGENS, AND THE PLASMA MEMBRANE

Lectins

Originally, the term **lectin** (from the Latin *legeres*, which means to select) referred to a special class of carbohydrate-binding proteins present in plant seeds that could differentially agglutinate human blood cells according to a person's blood type. Since their original discovery in plant cells by H. Stilmark some 90 years ago (at which time they were referred to as **phyto-hemagglutinins**), lectins have been identified in many other biological sources, including bacteria, molds, algae, sponges, snails, crabs, and fishes. Recently, lectins have been identified in the membranes of mammalian cells. Today, the term lectin has been expanded to include all carbohydrate-binding proteins that are of *nonimmune* origin (see below).

Lectins agglutinate cells by interacting with the sugars of membrane glycoproteins. The affinity and interaction of the lectin with the surface sugar is reminiscent of the manner in which an enzyme combines with its substrate or an antibody combines with an antigen. Because the interaction of the lectin with sugar is specific (see Table 15–4), lectins can be used to map the distribution of sugars on the cell surface.

Lectin molecules contain two or more sugar-binding sites, and when large numbers of lectins bind *simultaneously* to sugars on the surfaces of separate cells (thereby cross-linking the cells), the result is agglutination (Fig. 15–17). It should be noted that binding of the lectin to the cell surface sugars can occur without ensuing agglutination *if no cross-linking takes place.*

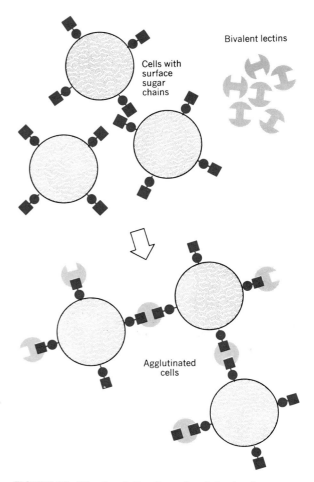

FIGURE 15–17 Agglutination of cells by lectins.

The presence and extent of cross-linking is dependent on the balance between the lectin concentration and the numbers of surface sugars. In this respect, the lectin–sugar interaction is much like that of an antibody–antigen reaction (see below). However, unlike antibodies, which chemically are very similar proteins (Chapter 4), lectins are of diverse structure, organization, and size.

Lectins will bind free sugars as well as sugars attached to cell membranes. Consequently, lectin-induced cell agglutination can be blocked by preliminary addition of the appropriate free sugar to a suspension of cells.

Lectins have been used to verify that the plasma membranes of malignant cells and normal cells differ. Malignant cells are much more readily agglutinated by lectins than the normal cells from which they are derived, that is, the malignant cells can be caused to

TABLE 15–4 SOME LECTINS AND THEIR SUGAR SPECIFICITIES

Lectin	Sugar Specificity
Concanavalin A (jack beans)	D-Mannose
Pea lectin	
Wheat germ agglutinin	N-Acetyl-D-glucosamine
Potato lectin	
Tomato lectin	
Ricinis communis agglutinin (castor beans)	D-Galactose
Peanut agglutinin	
Soybean agglutinin	N-Acetyl-D-galactosamine
Lima bean lectin	
Limulus lectin	N-Acetylneuraminic acid

agglutinate at much lower lectin concentrations than are required to agglutinate normal cells. It has been found that the increased agglutinability of the malignant cells results from increased glycoprotein mobility in the lipid bilayer of the plasma membrane. Because the malignant cell membrane is more fluid, lectins are able to cluster the glycoproteins in the membrane (i.e., draw them together) and thereby make it possible to form greater numbers of crossbridges.

How lectins can be used to determine the composition of the sugar chains of surface carbohydrate is illustrated in Figure 15–18. The lectins that are bound by unmodified cell membranes establish the nature of the terminal sugars; these may then be cleaved from the remaining carbohydrate using a specific enzyme. The newly exposed terminal sugars are now examined

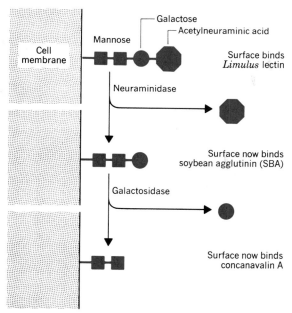

FIGURE 15–18 Structure of surface carbohydrate is established by successive treatments with lectins and terminal sugar-cleaving enzymes. In this illustration, binding of *Limulus* lectin by the cells indicates the presence of terminal acetylneruaminic acid groups. Their removal using *neuraminidase* is followed by binding of soybean agglutinin, indicating that the acetylneuraminic acid was linked to either galactose or acetylgalactosamine. Which of these two alternatives is correct is revealed by sensitivity to the specific enzyme added in the next round. In the illustration, it is *galactosidase* that now allows concanavalin A binding, thereby indicating that the terminus was galactose. Removal of galactose and binding of Con A indicates that the next sugar is mannose.

for their lectin-binding characteristics, following which another round of enzymatic sugar removal is carried out. Repetition of this sequence of treatments progressively reveals the order of sugars. Lectins may also be combined with fluorescent labels or radioactive tracers and then be used as *surface probes* to explore the types, quantities, and distributions of sugars in the plasma membranes of cells.

Recently, lectins have been used in affinity chromatography (see Chapter 13) to separate different kinds of cells from one another. A column is filled with supporting medium (e.g., cross-linked dextrans) to which specific lectins have been chemically bound. When a cell suspension is then percolated through the column, cells whose surface sugars interact with the lectin are retained by the column, whereas unbound cells are eluted.

Although a large number of lectins have been identified and their sugar-binding properties studied, the physiological role of the lectins in the cells that produce them still remains unknown. Speculations on their roles include sugar storage and transport or, perhaps, some protective function.

Antigens and Antibodies

An antigen may be loosely defined as any molecule that has the capacity to stimulate antibody production by the immune system of higher animals. Typically, antigens are glycoproteins in the membranes of cells or in other particles foreign to the animal. For example, the antigens present in the membranes of bacterial cells or in the coats of viruses act to stimulate antibody production by the immune system of the infected animal. The antibodies or immunoglobulins (see Chapter 4) produced in response to the presence of the antigen combine with the antigen to form a complex, and this is followed by a series of reactions in which the antigen-bearing agents (e.g., the bacteria) are destroyed.

Antibodies are synthesized by lymphocytes, a class of white blood cells. Two subpopulations of lymphocytes may be identified: T lymphocytes and B lymphocytes. T lymphocytes are produced by the *thymus* gland (i.e., "T" stands for "thymus") and other organs (e.g., lymph nodes and spleen) that are "seeded" by cells migrating from the thymus during development. In birds, B lymphocytes originate in an outpocket of the gut called the *bursa of Fabricius* (i.e., "B" stands for "bursa"), but in man and other mammals they are thought to originate in the hemopoietic tissues (i.e., the liver in the developing fetus and the bone marrow in children and adults). The first antibodies produced by newly formed B lymphocytes become constituents of the cell's plasma membrane, where they act as receptors for antigens; additional antibody molecules produced by B lymphocytes are secreted into the surrounding blood plasma. T lymphocytes do not secrete antibodies, instead, antibodylike proteins are incorporated into the cell's plasma membrane.

The reaction between antibody and antigen is very specific, with a particular antibody combining with only one type of antigen. An enormous variety of B and T lymphocytes are present in the body's tissues, each capable of manufacturing only a single antibody type (and therefore capable of reacting with a single type of antigen or foreign cell). The nature of this antibody-synthesizing specificity is also considered in Chapter 24 in connection with the clinical production of **monoclonal antibodies**.

In the case of B lymphocytes, binding of antigen by surface antibody leads to the proliferation of **plasma cells**, which then secrete large quantities of antibody. Following the initial reaction on the lymphocyte cell surface, subsequent antigen–antibody reactions may not involve B lymphocytes directly. Instead, the secreted antibodies react with either free antigens or more likely with antigens in the invading cells' membranes. The antibodylike proteins present in the membranes of T lymphocytes permit these cells to attach to the surfaces of foreign cells containing the appropriate antigen. This is then followed by the secretion of cytotoxic substances into the foreign cell causing its destruction. The involvement of either lymphocyte membranes or foreign cell membranes (or both) in the antigen–antibody reaction makes these molecules especially valuable tools for studying the properties of the cell surface.

It should be clear that the surface antibodies of the lymphocytes of one animal can serve as antigens if these lymphocytes are transferred to the blood stream of another animal. The transferred lymphocytes will be treated much like any other foreign cell and serve to stimulate the production of **anti-immunoglobulin antibodies** (AIA). AIA has been especially useful in probing the distribution of glycoproteins in the cell membrane.

M. C. Raff and S. dePetris prepared AIA that they then coupled with *ferritin*. Ferritin is a liver protein rich in iron and readily discernible as dark spots by transmission electron microscopy (i.e., the iron renders the ferritin electron-dense). When ferritin-coupled AIA is added to suspensions of lymphocytes, it combines with their surface immunoglobulins, thereby assisting in their identification. Raff and dePetris showed that when lymphocytes are incubated at 4°C with ferritin-coupled AIA, the electron-dense spots are distributed all over the membrane surface, whereas at 20°C, the ferritin-coupled AIA is clustered together at one pole of the cell. As in the case with lectins described earlier, the AIA is able to *cluster* the surface immunoglobulin at 20°C but not at 4°C because at the higher temperature the plasma membranes of the cell are more fluid.

ORIGIN OF THE PLASMA MEMBRANE AND ITS PROTEIN AND LIPID ASYMMETRY

All of a cell's proteins are synthesized by ribosomes, including, of course, those proteins that are destined for inclusion in the plasma membrane. Cytoplasmic ribosomes in eukaryotic cells occur in two states: (1) "attached" (ribosomes associated with the membranes of the endoplasmic reticulum) and (2) "free" (ribosomes freely dispersed in the cytosol). The differences between these ribosomes are considered in Chapter 22 together with the mechanics of protein synthesis. Both attached and free ribosomes are believed to contribute proteins to the plasma membrane.

Synthesis of Membrane Proteins and the "Signal Hypothesis"

Principally as a result of the work of G. Blobel, D. D. Sabatini, C. M. Redman, C. Milstein, J. E. Rothman, J. Lenard, and H. F. Lodish, the mechanism that routes newly synthesized proteins to their proper destinations in the cell has gradually unfolded. A major contribution to this end has been the confirmation of the **signal hypothesis** proposed in the early 1970s by Blobel and Sabatini. According to this hypothesis (Fig. 15–19), proteins that are to be either (a) secreted from the cell, (b) dispatched to lysosomes, or (c) incorporated into the plasma membrane or membranes of the endoplasmic reticulum are encoded by mRNA molecules that contain a special nucleotide sequence called a "signal." The signal segment encodes a chain of about 16 to 26 amino acids that appears at or near the beginning of the polypeptide chain. Near its N-terminus, the signal sequence contains polar, basic residues, whereas the central domain is apolar. When a ribosome attaches to the mRNA in the cytosol and begins to translate the message, the signal sequence or signal peptide emerging from the ribosome is recognized by a ribonucleoprotein complex in the cytosol called the signal recognition particle (i.e., *SRP*). SRP, which consists of a 7 S cytoplasmic RNA molecule and six polypeptides, binds to the signal sequence, bringing about a *temporary* halt to protein synthesis by that ribosome. Synthesis is resumed only if the SRP–ribosome complex attaches to the endoplasmic reticulum; ribosomes synthesizing polypeptides that lack a signal sequence do not interact with SRP and do not attach to the endoplasmic reticulum. The amino acid sequences of a number of signal peptides have now been determined. Although they contain a specific distribution of hydrophobic and charged residues, no primary sequence homologies (see Chapter 4) appear to exist. Consequently, it is believed that SRP must recognize certain features contained in the signal peptide's secondary and tertiary structure.

SRP–ribosome complexes attach to the endoplasmic reticulum at specific sites occupied by SRP receptors (also called docking proteins). Once "docking" is completed, the SRP–ribosome–docking protein interaction is replaced by a functional ribosome-membrane junction and the synthesis of the polypeptide encoded by the mRNA is resumed. SRP returns to the cytosol where it can participate in another round of signal recognition and docking (i.e., the "SRP cycle" of Fig. 15–19a). When protein synthesis is resumed by the docked ribosome, the elongating polypeptide chain passes through the ER membrane into the intracisternal space. This process, termed "translocation," is presumed to involve active participation of elements of the membrane. Translocation of the growing polypeptide into the intracisternal space is depicted in Figure 15–19 as taking place through a porelike opening in the membrane solely to indicate that the membrane's permeability barrier is transiently altered. However, it is not known with certainty whether translocation through the membrane takes place through such a proteinaceous water-filled channel or directly through the lipid bilayer. In most instances, the signal sequence is eventually cleaved from the remainder of the growing polypeptide by an extrinsic enzyme called *signal peptidase* attached to the lumenal surface of the ER. If the protein being synthesized is destined for secretion from the cell, completion of synthesis is followed by the protein's release from the ribosome into the intracisternal space. The ribosome then detaches from the membrane and the mRNA and may participate in another round of protein synthesis. At the same time, the permeability barrier of the ER is restored. Proteins discharged into the ER cisternae in this manner are ultimately conveyed to the Golgi apparatus for chemical modification prior to secretion (Chapter 18). In the case of proteins destined to be regular constituents of the endoplasmic reticulum or the plasma membrane, translocation into the intracisternal space is aborted *before synthesis is finished* so that the polypeptide is left anchored in the membrane (Fig. 15–19b). The information halting the translocation is likely contained in the polypeptide itself and is recognized by the translocation apparatus.

For peripheral membrane proteins facing the exterior of the cell or the lumenal phase of the ER, the signal sequence is followed by synthesis of the hydrophilic portion of the polypeptide. If an integral membrane protein is being synthesized, the hydrophilic portion is followed by a hydrophobic segment that remains anchored in the lipid bilayer. For proteins that span the membrane, synthesis is completed with the production of a final hydrophilic segment that faces the cytosol.

By comparing parts *a* and *b* of Figure 15–19, it is seen that the major distinction between the synthesis of secretory proteins and membrane proteins is that secretory proteins are released into the lumenal phase of the ER, whereas membrane proteins remain anchored in the ER. Addition of sugars to presumptive plasma membrane glycoproteins may occur soon after the hydrophilic portions of the molecules enter the ER cisternae. The membrane glycoprotein then migrates from the ER to the Golgi apparatus. Although the mechanism for this transfer is still uncertain, it is believed to take place either by dispatchment of small vesicles from the ER, which then migrate to and fuse with the Golgi membranes (see Chapter 18), or by lateral "flow" along the ER membranes to the Golgi. Glycosylation of the membrane proteins is completed in the Golgi apparatus. The Golgi apparatus dispatches completed plasma membrane glycoproteins as small vesicles that migrate to and fuse with the plasma membrane. The overall process is summarized in Figure 15–20, which also shows that the intracellular/extracellular orientation of the mem-

FIGURE 15–19 Synthesis of secretory proteins (*a*) and membrane proteins (*b*) by ribosomes that transiently attach to the endoplasmic reticulum. The *signal peptide* (depicted here by black squares) encoded by the signal sequence of mRNA interacts with *signal recognition protein* (SRP) in the cytosol. The complex then "docks" with the surface of the ER. After polypeptide elongation is resumed, SRP returns to the cytosol so that it can participate in another round of signal recognition and docking. This is known as the "SRP cycle." The signal peptide is removed by *signal peptidase*. In *a*, the completed polypeptide chain (red circles) is released into the cisterna of the ER. For proteins destined for the plasma membrane, addition of sugars begins in the ER (and is completed in the Golgi apparatus). Sections of the polypeptide that are rich in amino acids with hydrophobic side chains (depicted here as open circles) are embedded in the ER membrane's lipid bilayer. These side chains serve to anchor the protein in the membrane. In the example illustrated here, the protein spans the membrane and has a hydrophilic amino acid sequence at each end of the polypeptide.

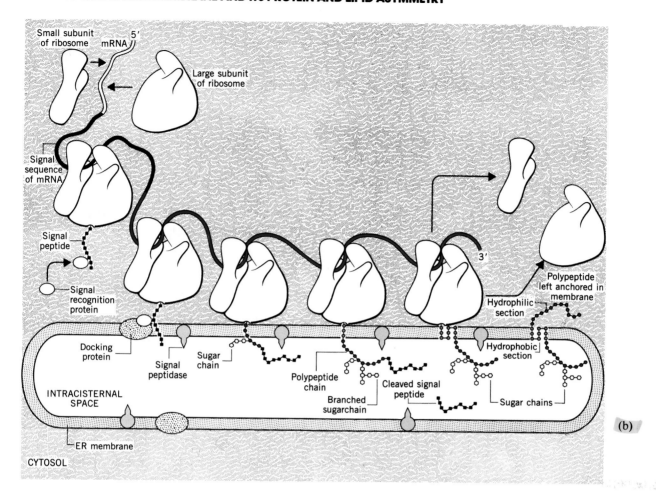

brane protein is maintained throughout its passage from the ER to the plasma membrane.

Some integral proteins have hydrophilic parts that face the interior of the cell (see Fig. 15–8) and some peripheral proteins are associated only with the membrane's cytoplasmic face. Although a similar mechanism is not precluded for the synthesis of these membrane proteins, they could be synthesized by free ribosomes. Following release of these proteins in the cytosol they may diffuse to the plasma membrane. Integral proteins would be spontaneously inserted into the membrane by a hydrophobic segment (Fig. 15–20), whereas peripheral proteins would attach to the membrane through polar interactions. Peripheral proteins reaching the plasma membrane in this manner could not pass through the membrane to the exterior surface because they could not traverse the hydrophobic membrane core.

As already noted, prokaryotic cells do not contain an endoplasmic reticulum. Secretory proteins and new plasma membrane proteins are synthesized by ribosomes that attach to the inner surface of the plasma membrane. As in eukaryotic cells, ribosome attachment follows synthesis of a signal peptide encoded in the protein's mRNA. The protein is then dispatched through the cell membrane and into the extracellular space.

Synthesis of Membrane Lipids

In eukaryotic cells, phospholipid synthesis is associated with the endoplasmic reticulum, whereas in prokaryotic cells lipid synthesis is a property of the cytoplasmic half of the plasma membrane. It is therefore likely that lipid synthesis in eukaryotic cells takes place in the cytoplasmic half of the ER membranes.

FIGURE 15–20 Mechanism accounting for the origin of plasma membrane proteins. Proteins synthesized by free ribosomes (right half of figure) may insert into the plasma membrane following the completion of the protein's synthesis and its release from the ribosome. These proteins penetrate the membrane's lipid interior via a hydrophobic amino acid sequence. Hydrophilic segments of the protein (the red portion of each polypeptide chain) project into the aqueous cytosol. Plasma membrane proteins synthesized by attached ribosomes (left half of diagram) are inserted first into the endoplasmic reticulum. They are then transferred to the Golgi apparatus, processed there, and dispatched to the plasma membrane via vesicles. See text for additional details.

Following synthesis, the phospholipid becomes at least temporarily part of the interior monolayer but may either be enzymatically translocated to the outer layer or flip-flop between the two layers. In view of the fact that outer monolayer lipids are derived from the inner monolayer, an absolute asymmetry is precluded.

Because in prokaryotic cells lipid synthesis occurs in the plasma membrane, incorporation into the membrane's structure is direct. In eukaryotic cells, however, presumptive plasma membrane lipid must make its way from the ER to the plasma membrane. This is believed to occur by one or both of two processes. Newly synthesized lipids inserted into ER membranes may make their way to the plasma membrane by the same mechanism that translocates ER membrane proteins (Fig. 15–20); that is, *both* membrane proteins and lipids pass from the ER to the Golgi apparatus and are later dispatched to the plasma membrane via small vesicles. The cytosol of eukaryotic cells contains a number of **phospholipid transport proteins** that function to transfer phospholipid molecules from one cellular membrane to another. These transport proteins might also play a major role in mediating the passage of phospholipids from the membranes of the ER to the plasma membrane.

CELL SURFACE PROPERTIES REVEALED BY ERYTHROCYTES

In view of the fact that the mature erythrocyte lacks organelles, this cell has always been a popular source of plasma membranes. Indeed, the pioneering studies of Gorter and Grendel, which were the first to indicate the existence of the lipid bilayer, were carried out using erythrocytes. Like the plasma membranes of other cells, the red blood cell membrane is asymmetric. The lipid asymmetry of the erythrocyte membrane has already been described (Table 15–3). Regarding protein asymmetry, the peripheral proteins account for about 40% of all membrane proteins *but are restricted to the membrane's interior surface.* The most abundant of these proteins and the first to be isolated is **spectrin**. Spectrin is a fibrous protein consisting of two large polypeptide chains and having a length of about 200 nm. This protein is believed to be an important component of a weblike network of proteins on the interior membrane surface. Spectrin molecules are not attached directly to the membrane's inner surface, rather, they are anchored to the membrane via other membrane proteins including **ankyrin** and **actin**. The spectrin–ankyrin–actin complexes create a weblike network or *cytoskeleton* that supports the membrane and contributes to the biconcave shape that characterizes mammalian red cells. Spectrin may not be limited to erythrocytes as spectrin or at least spectrinlike proteins have recently been isolated from the plasma membranes of other cells.

The erythrocyte membrane possesses two major integral proteins that span the lipid bilayer. One of these, called **glycophorin-A**, has been fully sequenced and reveals several very interesting properties. Glycophorin-A consists of a single chain of 131 amino acids; 16 short carbohydrate chains are linked to residues near the N-terminus of the polypeptide (primarily to serine and threonine side chains), the carbohydrate accounting for about 60% of the total mass of the glycoprotein. The N-terminal region of glycophorin-A is thought to project beyond the exterior membrane surface, the last five amino acids determining the MN blood group status of an individual. The C-terminal end of glycophorin-A is rich in acidic amino acids, especially glutamic acid, and is believed to project into the cell interior. A segment of about 20 amino acids in the middle of the polypeptide consists exclusively of nonpolar and hydrophobic amino acids and apparently is that portion of glycophorin-A that spans the lipid bilayer.

In addition to its supportive role, a cell's cytoskeleton acts to restrict the lateral movement of proteins within the membrane. In red cells, the spectrin–ankyrin–actin complexes place constraints on the mobility of glycophorin-A. Antibodies against spectrin cause aggregation of spectrin molecules and their precipitation onto the inner surface of the erythrocyte membrane, and this is accompanied by a corresponding rearrangement of glycophorin-A in the membrane.

In nucleated cells, the cytoskeleton is more extensive and includes myosin filaments and microtubules as well as actin and other proteins (Fig. 15–21). The network of cytoplasmic filaments and microtubules radiates through much of the cytosol (see Figs. 1–18 and 23–2) and provides points of attachment for many of the cellular organelles. Whereas the rearrangement of cytoskeletal components just below the cell surface manifests itself in the redistribution of integral membrane proteins, major movements of the cytoskeleton may be fundamental to such gross activities as cellular motion and endocytosis and exocytosis. The cytoskeleton is considered further in Chapter 23.

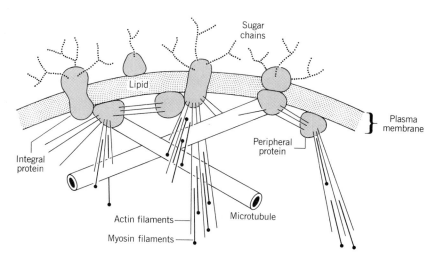

FIGURE 15–21 The *cytoskeleton* is a network of filaments and microtubules anchored at its margins to the undersurface of the plasma membrane.

The differentiation of the erythrocyte in the bone marrow is accompanied by a major reorganization of the cell in which organelles like the nucleus, mitochondria, intracellular membranes, ribosomes, and so forth are progressively lost. Despite the lack of major internal structures and its seeming simplicity, the erythrocyte retains a characteristic shape. In humans (and in most other mammals), the cell takes the form of a *biconcave disk* with a diameter of about 8 μm (Fig. 15–22), although changes in shape are readily induced by variations in osmotic pressure (see later in this chapter). The biconcave shape of the erythrocyte is important in its biological function, because such a shape maximizes oxygen diffusion from the cell into the tissues and promotes efficient stacking of cells (*rouleaux* formation) as the red cells circulate through the narrow capillary passageways. The biconcave shape of the red cell is often deformed as it circulates through the narrowest capillaries, but the shape is quickly restored in the larger passageways. As noted above, the characteristic red cell shape is maintained by the cytoskeletal protein network that lies just below the membrane surface. The importance of cytoskeletal support is documented by the observation that the shape of the cell is maintained even when nonionic detergents are used to extract the membrane's lipid bilayer and its intrinsic proteins.

In a series of rather startling experiments, B. Bull and J. D. Brailsford have shown that when an erythrocyte is attached by a portion of its undersurface to a glass slide and a laser used to make a visible mark on the membrane's surface so that membrane movement can be followed, slight lateral displacement of the cell using hydraulic force is accompanied by the membrane *rolling* in the direction of the force, much like a tank track does (Fig. 15–22). The laser mark travels over the cell surface, following the contours of the biconcave shape. In other words, though rolling laterally, the biconcave shape of the cell is maintained with the vertical biconcavity moving parallel to the glass surface. Still unanswered is whether the membrane is being displaced relative to the underlying cytoskeleton.

Erythrocytes traveling through capillaries and other blood vessels are often arranged as stacks or *rouleaux*, with their biconcave faces juxtaposed. Such an orderly procession makes it possible for so large a number of cells to pass through the body tissues in short periods of time. P. B. Canham has shown that when a cell in a rouleaux is struck by a laser beam of sufficient energy, the cell membrane is disrupted and the cell lyses (bursts). However, several cells in the rouleaux on either side of the target cell (and not directly affected by the laser) also are seen to undergo gradual lysis. This "contagious lysis" apparently results from the fact that the membranes of erythrocytes in rouleaux transiently adhere to one another. Sudden movement of the membranes of one cell (as during lysis) create sufficient shear forces at contact points with neighboring cells that their membranes are also affected. The nature of membrane interaction between neighboring erythrocytes in a rouleaux is unknown, but it is clear that much remains to be learned about the "simple" red blood cell.

CELL–CELL JUNCTIONS AND OTHER SPECIALIZATIONS OF THE PLASMA MEMBRANE

The plasma membranes of neighboring cells in a tissue frequently exhibit specialized junctional regions that play a role in cell–cell adhesion and in intercellular transport. The most common of these junctions are (1) **tight junctions (zonula occludens)**, (2) **intermediate junctions** or **belt desmosomes** (also referred to as

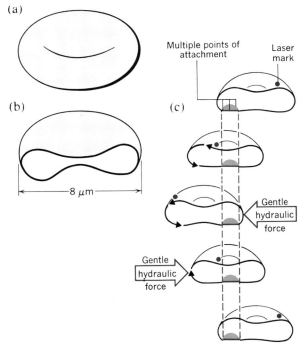

FIGURE 15–22 (*a*) Erythrocyte. (*b*) Cross section showing biconcavity. (*c*) Effect of lateral force on erythrocyte shape.

terminal bars or **zonula adherens**), (3) **spot desmosomes** (**macula adherens**), (4) **gap junctions** or **connexons** (also called **nexuses**), and (5) **plasmodesmata** (Figs. 15–23 to 15–26, and Fig. 15–31).

Tight Junctions

In **tight junctions**, the plasma membranes of neighboring cells fuse with one another at one or more points. At these points of fusion, the external half-membrane of one cell may form a continuous leaflet with the external half-membrane of the adjacent cell. Tight junctions occur at the same circumferential level of the cell, so that they give rise to *belts* of fusion points with neighboring cells. The belt obliterates the intercellular space and acts as a barrier to the flow of materials between the cell surfaces. Intracellularly, the belts of tight junctions are reinforced by a network of fine filaments that radiate into the cytoplasm. Tight junctions between cells are formed by two interdigitating rows of membrane particles (probably integral membrane proteins), one row contributed by each cell. These rows are called **sealing strands** (Fig. 15-24) and act in much the same manner as the two halves of a zipper. The number of rows of sealing strands linking neighboring cells and the extent to which they interconnect to form a network vary from one type of tissue to another. It is also thought that sealing strands act to deter the movements of other proteins within the plasma membrane. In this way, the functional specialization of different faces of the membrane can be maintained (see above).

Intermediate Junctions

Intermediate junctions, or **belt desmosomes**, are girdles of contractile filaments appressed to the cytoplasmic surfaces of juxtaposed membranes. The girdles of contractile filaments have been shown to contain actin and appear also to interweave with another web of filaments that extends into microvilli (Fig. 15–23).

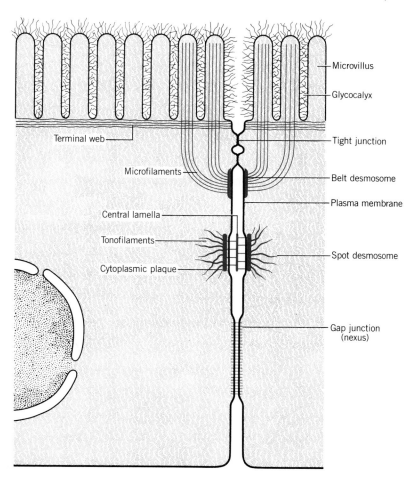

FIGURE 15–23 Specializations of the plasma membrane.

Microvillus

Glycocalyx

Terminal web

Tight junction

Microfilaments

Belt desmosome

Plasma membrane

Central lamella

Tonofilaments

Spot desmosome

Cytoplasmic plaque

Gap junction (nexus)

(a)

(c)

(b)

FIGURE 15–24 Types of cell junctions. (*a* and *b*) Typical appearances of *desmosomes* (*D*) and *gap junctions* when seen in thin section. (*c*) Freeze-fracture replica through several gap junctions (*G_j*); each junction is seen to be composed of a number of closely packed particles. (*d*) Freeze-fracture view of a *tight junction* (*T_j*) formed by a number of *sealing strands* of membrane proteins; in this view, the sealing strands (narrow ridges on right and grooves on left) create an occlusive zone that acts as a permeability barrier to vertical flow of macromolecules between the two cell surfaces. (Photomicrographs courtesy of Dr. N. B. Gilula; *a*, *b*, and *c* from Gilula, N. B., *Cell Communications* (R. P. Cox, Ed.), Wiley & Sons, Inc., 1974; *d* from Friend, D. S., and Gilula, N. B., *J. Cell Biol.* **53**, 758 (1972). Reprinted with permission from the Rockefeller University Press.)

(d)

Spot Desmosomes

Unlike tight junctions and belt desmosomes, **spot desmosomes** (or **macula adherens**) do not form a belt around the cell. Instead, they are discrete buttonlike attachment points scattered over the opposing membrane surfaces (Fig. 15–23, 15–24, and 15–25). In the region of a spot desmosome the intercellular space between the adjacent plasma membranes narrows to a distance of about 300 Å (30 nm). The intercellular space contains a central dense band (the **central lamella**). The cytoplasm adjacent to this region of the membrane is divided into two regions: a lucid region that lies immediately adjacent to the membrane and a neighboring dense band called the **cytoplasmic plaque**. Microfilaments called **tonofilaments** arise from this region and radiate into the cytoplasm, where they may also form links with other spot des-

mosomes. Tiny rodlike connectives appear to link the central lamella with the cytoplasmic plaque. Spot desmosomes are believed to be the strongest points of attachment between neighboring cells.

Epithelium is a good example of a tissue in which neighboring cells contain the various junctions just described. Beginning at the lumenal surface of the epithelium (i.e., the *apical* surface of the cells) and "descending" toward the basement membrane, the junctions occur in a characteristic order: a belt of tight junctions is followed by a belt of intermediate junctions, and spot desmosomes are more abundant near the basal ends of the cells.

Gap Junctions

The **gap junction** or **nexus** is probably the most complex modification of adjacent plasma membranes. In the gap junction, the opposing plasma membranes are separated by a distance of only 30 to 40 Å (3–4 nm) and the space is penetrated by a number of pipelike structures that extend from the cytosol of one cell into the cytosol of the neighboring cell (Fig. 15–26). Each pipe is composed of two cylindrical proteins called **connexons**, one connexon contributed by each of the two juxtaposed cells. Each connexon spans the plasma membrane, protruding about 10 Å (1 nm) into the cy-

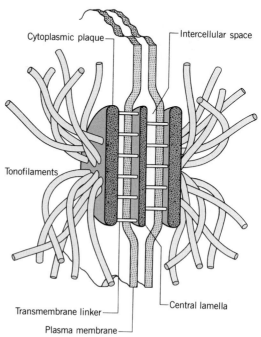

FIGURE 15–25 The *desmosome.*

tosol on one side of the membrane and about 15 to 20 Å into the extracellular space on the other. Connexons are composed of six protein subunits having molecular weights of about 30,000. Each subunit is approximately rod-shaped, with a diameter of 25 Å. Of special interest is the finding by P. N. Unwin and G. Zampighi that the long axis of each rod is *not* parallel to the long axis of the cylinder; instead, there is a slight angle of inclination (Fig. 15–26). It has been suggested that this angle is varied to open and close a channel at the center of each connexon. When opened, the connexon channel has a diameter of 15 to 20 Å. The channel is hydrophilic and allows intercellular passage of a wide variety of molecules having molecular weights up to about 800 (e.g., salt ions, amino acids, sugars, vitamins, certain hormones, and chemical messengers such as cyclic AMP). Rotation of the ends of the subunits acts to close the channel and interrupt

the passage of materials between cells. Closure of connexons seals off dying cells in a tissue, thereby preventing the loss of nutrients from adjacent healthy cells.

In freeze-fracture studies, gap junctions are revealed as patches of closely packed connexons (Fig. 15–24). In some tissues, the sealing strands characteristic of tight junctions occur at the margins of the connexon patches.

Microvilli and Surface Ruffles

In some tissues, much of the plasma membrane at the apical surface is modified to form numerous fingerlike outfoldings called **microvilli** (Fig. 15–27). The microvilli, which usually are about 1–2 μm long, greatly increase the surface area of the membrane and are especially abundant in (but not restricted to) cells having a transport activity such as absorptive epithelium or glandular epithelium. In absorptive epithelium, the microvilli are often referred to as the *brush border*. Individual cells in culture may also display microvilli.

Microvilli contain bundles of about 20 to 30 actin filaments (see Fig. 15–29) that are extensions of the cytoskeleton; at one end, the filaments are anchored to the undersurface of the plasma membrane and at the other end they enter the **terminal web** that forms a plane just below the base of the microvilli (Fig. 15–23). Filaments of the terminal web insert into the desmosomes that encircle the cell.

Depending on physiological conditions microvilli may be drawn inward toward the cell or reextended. On the basis of their extensive studies of brush borders, M. S. Mooseker and L. G. Tilney have suggested that movements of the microvilli may be brought about by an interaction between certain actin filaments of each microvillus and myosin filaments in the terminal web. The interaction is reminiscent of that which occurs between actin and myosin in contracting muscle cells (see also Chapters 23 and 24).

The plasma membranes of many cells give rise to thin, undulating folds called **surface ruffles** or **lamellipodia** (see Fig. 1–18). The intracellular **microtrabecular lattice** and the cytoskeleton that forms and supports this structure are readily discerned in high-voltage electron photomicrographs of whole cells (Fig. 15–28). Ruffles are especially characteristic of cells carrying out endocytosis (discussed later in the chapter).

FIGURE 15–26   In a *gap junction*, neighboring cells are interconnected by a number of *connexons*. Each connexon consists of two halves, each half formed by six cylinderlike protein subunits. Rotation of the subunits opens (top) and closes (bottom) the channel linking the cytosols of the neighboring cells (see also Fig. 15–24).

(a)

(b)

(c)

(d)

FIGURE 15–27 *Microvilli.* (*a*) Thin section; (*b*) freeze-fracture; G, gap junction. The terminal web at the base of the microvilli and the membrane particles that cover the *P* fracture face are seen more clearly in c and d. (*a*, courtesy of R. Chao; *b, c and d,* courtesy of Dr. A. Schiller; from *Cell and Tissue Res.* **221**, 431 (1981); copyright © Springer-Verlag.)

FIGURE 15–28 High-voltage TEM (top) and SEM (bottom) stereo views of cell surface *ruffles* and *lamellipodia*. As seen in the TEM view, the ruffles are supported by extensions of the microtrabecular lattice and cytoskeleton. Note how the supportive elements seen in the top stereo pair conform in shape and distribution to the cell surface shape seen in the bottom SEM stereo pair. The reference bar is 1 μm. (Courtesy of Drs. K. R. Porter and M. E. Stearns.)

The Glycocalyx Carbohydrate chains that are part of glycoprotein and glycolipid molecules of the plasma membrane form a networklike coat on the outer surface of the membrane. This coat is called the **glycocalyx** or "fuzzy coat." Some tissues such as the absorptive epithelium of the intestine have an especially thick glycocalyx that surrounds the microvilli and is readily discerned during electron microscopy (Figs. 15–23 and 15–29).

Plasmodesmata

In plant tissues, the cytoplasm of neighboring cells may be connected through numerous narrow channels

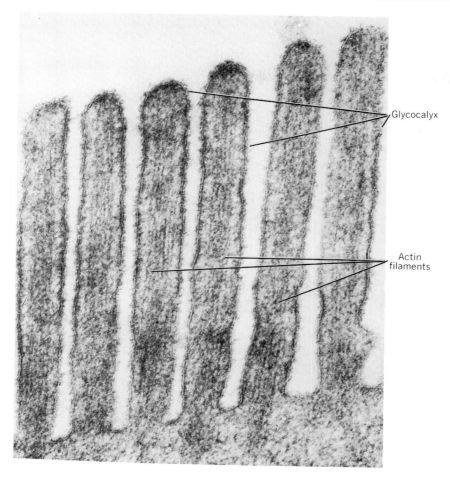

Glycocalyx

Actin
filaments

FIGURE 15–29 *Glycocalyx* covering the surface of microvilli. The network-like glycocalyx or "fuzzy coat" consists of carbohydrate chains of membrane glycoproteins and glycolipids. (Courtesy of S. Shultz.)

Cross section

Side view

Plasma membrane

Cell wall

Cytoplasm

Plasmodesmata

FIGURE 15–30 *Plasmodesmata.* (Photomicrograph courtesy of Dr. E. G. Pollock.)

that penetrate the fibrous cell wall separating the cells. The channels, called **plasmodesmata**, are formed by outward extensions of the plasma membranes of adjacent cells (Fig. 15–30) and are much larger than the channels of gap junctions. Plasmodesmata provide for the direct exchange of materials between the cytosols of neighboring cells in the plant tissue.

PASSIVE MOVEMENTS OF MATERIALS THROUGH CELL MEMBRANES

In this section, we consider some of the fundamental principles that govern the passive movements of water, ions, and various other molecules through cell membranes. The term *passive* is intended to denote that the movement of the substance through the membrane is not associated with any chemical or metabolic activities in the membrane. Passage through the membrane in such instances is regulated by such factors as the *concentration gradient* across the membrane and the chemical and physical relationships between the membrane and substances inside and outside the cell. Later, we will direct our attention to movements through the membrane that are accompanied by chemical changes, metabolic activity, or gross molecular rearrangements within the membrane itself.

Osmosis and Diffusion across Membranes

Membranes that allow substances to pass through them are said to be **permeable** to the substance. Nearly all plasma membranes are permeable to water. If water (or some other solvent) is the only substance that can pass through the membrane, the membrane is said to be **semipermeable**. Membranes that display a gradation of permeability to water and dissolved solutes (i.e., membranes that permit water to pass through more readily than salts, sugars, etc.) are said to be **selectively permeable**.

Water molecules are continuously moving into and out of cells through the plasma membrane. Such movements are generally not discernible as changes in cell size or shape because the flux in each direction is the same. When the concentrations of solutes inside and outside the cell differ, the water flux in one direction may be greater than in the other, and the cell may swell or shrink. Water moves from a region of *low* solute concentration to a region of *higher* solute concentration in order to establish a concentration equilibrium. The movement of water (or some other solvent) in response to such a solute *concentration gradient* is known as **osmosis**.

Osmosis may readily be demonstrated using artificial membranes such as cellophane, which is permeable to small molecules such as salts, sugars, and amino acids, but is impermeable to large molecules such as proteins. If a cellophane bag filled with a concentrated salt solution is connected to a vertical length of glass

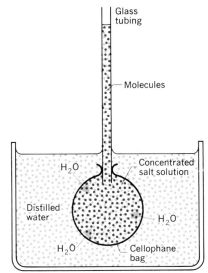

FIGURE 15–31   Osmosis of water into a cellophane bag containing a concentrated salt solution. Because the solute concentration inside the bag is greater than that outside the bag (i.e., the solute concentration outside the bag is zero), water enters through the cellophane membrane by osmosis. The influx of water causes the water level in the attached glass tubing to rise.

tubing and is then immersed in a container of distilled water, water will pass into the bag by osmosis. The entry of water into the cellophane bag will cause the level of water in the glass tubing to rise (Fig. 15–31). Salts permeate cellophane membranes more slowly than water, so that some time elapses before the salt molecules pass out of the bag into the surrounding water. The movement of solute molecules (in this case, salt molecules) from a region of *high* concentration (inside the bag) to one of *lower* concentration (outside the bag) occurs by the process of **diffusion**. In the case illustrated in Figure 15–31, the initial movement of water into the cellophane bag is followed by the outward diffusion of salt from the bag. As the solute concentration inside the bag decreases, the liquid level in the glass tubing falls as water molecules leave the bag by osmosis. The movements of salt and water molecules continue until the salt concentrations inside and outside the bag are equal.

Consider a case in which the cellophane bag is filled with a solution containing an impermeable solute. As in the previous case, water will enter the bag by osmosis, causing the liquid level in the glass tubing to rise. Because the solute is impermeable, it cannot diffuse from the bag, and a concentration equilibrium across

FIGURE 15-32 *An osmometer.*

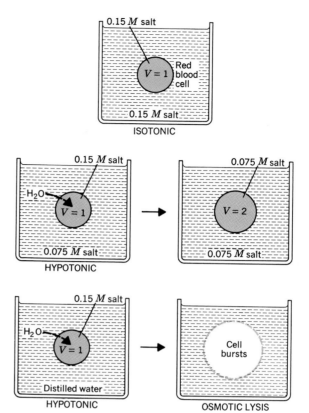

FIGURE 15-33 Behavior of red blood cells in hypotonic solutions.

the membrane cannot be achieved. Consequently, water will continue to enter the bag and rise in the glass tubing until a height is reached at which the pressure at the base of the water column is just great enough to prevent any further water influx. In theory, the water will remain at this level indefinitely. The pressure that is created inside the bag by the impermeable solute and that supports the column of water is called **osmotic pressure**. Its numerical value can be approximated by measuring the height of the water column. Usually osmotic pressure is expressed in *millimeters of mercury* (i.e., mm Hg) rather than in inches of water. Devices that are used to measure osmotic pressure are called *osmometers* (Fig. 15–32).

Osmosis and Diffusion across Cell Membranes

Cellular phenomena associated with osmosis and diffusion across the plasma membrane are readily demonstrated using red blood cells, sea urchin eggs, or certain plant cells. The plasma in which the red blood cells are normally suspended contains the same concentration of impermeable salt (0.15 M NaCl) as the erythrocyte cytoplasm (0.15 M KCl). Normal plasma is said to be **isotonic** to the red cell. If the plasma is diluted with water, its salt concentration will decrease, and the plasma will become **hypotonic** to the red blood cell. Any suspending medium containing an impermeable solute concentration that is *lower* than the corresponding solute concentration in the cells suspended in that medium is considered hypotonic. In the case of hypotonic plasma, water will enter the red cells by osmosis, causing the cells to swell. The same

effect can be produced by placing red blood cells in any hypotonic solution. Just how much water will enter the cell depends on how hypotonic the suspending medium is. For example, if the red cells are suspended in plasma containing one-half the normal salt concentration, water will enter the cells until they swell to twice their original volume. This will reduce the internal salt concentration to one-half its former value, bringing the internal and external salt concentrations into equilibrium (Fig. 15–33).

If the cells are suspended in a solution of even greater hypotonicity, then proportionately more water will have to enter the cell to reduce the internal salt concentration to that outside the cell. Obviously, cells can tolerate only a certain amount of swelling before the membrane ruptures, spilling the cell contents into the surrounding medium; this is called **osmotic lysis**. Red blood cells lyse when suspended in very dilute salt solutions or in distilled water (Fig. 15–34). In the specific case of the red blood cell, this phenomenon is called **hemolysis**, because hemoglobin from the red blood cell is released into the suspending medium.

FIGURE 15–34 When a plant cell is transferred from an isotonic environment (top) into hypotonic surroundings (bottom), water enters the cell's vacuoles, expanding them and forcing the cytoplasm into a narrow marginal band. The cell becomes *turgid*, but lysis is prevented by the cell wall, which may be distended but does not rupture.

Other animal cells behave in a similar manner in appropriately hypotonic media.

Plant cells generally do not lyse even when placed in distilled water because cell swelling is limited by the rather inflexible cellulose cell wall. In hypotonic solutions, plant cells swell as water enters the cytoplasmic vacuoles by osmosis. This forces the cytoplasm to the margins of the cell wall. Under these conditions, the plant tissue becomes **turgid** (Fig. 15–34).

Equal concentrations of impermeable salts and nondissociating (nonionizing) molecules (such as sucrose and other sugars) do *not* exhibit the same osmotic effects. For example, 0.15 M NaCl exerts twice the osmotic pressure as does 0.15 M sucrose. This is because 0.15 M NaCl undergoes dissociation in water to produce twice the number of particles (i.e., ions) per cubic

centimeter as 0.15 M sucrose. Thus, 0.15 M NaCl, 0.15 M KCl, 0.10 M CaCl$_2$, and 0.3 M sucrose would all exert the *same* osmotic pressure, because they produce the same particle concentrations in water.

Most cell membranes, including the plasma membrane of the red blood cell, are impermeable to sucrose. Therefore, both 0.15 M NaCl and 0.3 M sucrose are isotonic to red blood cells. Solutions that contain higher concentrations of impermeable solute than are found inside cells are said to be **hypertonic**. What happens to red blood cells when they are placed in a hypertonic sucrose solution is depicted in Figure 15–35. Water moves by osmosis from the cells into the medium until the concentration of solute inside and outside the cell is the same. The shrinkage associated with such water loss is called **crenation**. Other animal cells behave in a similar manner.

When plant cells are placed in hypertonic media they undergo **plasmolysis**, that is, water passes from the cytoplasmic vacuoles into the space between the cell wall and plasma membrane (Fig. 15–36).

So far, we have considered only the movement of water into and out of the cell. When the concentration of a permeable solute is greater outside a cell than inside, the solute molecules diffuse into the cell until the external and internal concentrations are in equilib-

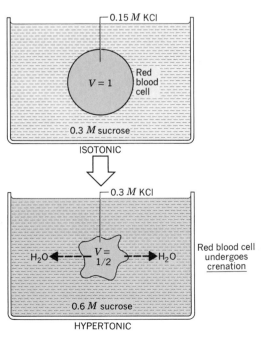

FIGURE 15–35 *Crenation* of red blood cells in hypertonic solution.

0.3 M
sucrose

ISOTONIC

0.6 M
sucrose

HYPERTONIC

FIGURE 15–36 *Plasmolysis* occurs when plant cells are transferred to a hypertonic environment. Water passes from the cell's vacuoles into the space between the plasma membrane and the cell wall.

rium. In most instances, the combined volume of all the cells present in a cell suspension is only a small fraction of the total volume of the suspending medium. Therefore, the quantity of solute entering (or leaving) the cells has a negligible effect on the solute concentration of the medium.

Consider what happens when red blood cells are placed in an isotonic saline solution (e.g., 0.15 M NaCl) containing 0.1 M urea. The red blood cell membrane is permeable to urea, and because red blood cells normally contain little or no urea, urea molecules will diffuse into the cell until the intracellular urea concentration reaches 0.1 M. Even after a concentration equilibrium is achieved, urea molecules will continue to pass through the cell membrane into and out of the cell. However, the migration in each direction will be the same, so that no *net* change in the urea concentra-

tion of the cell occurs. If the cells are transferred to an isotonic saline solution lacking urea, urea molecules will diffuse back out of the cell. Many substances enter and leave cells by diffusion through the plasma membrane.

The Permeability Constant The quantity of solute diffusing from one region to another depends on the concentration difference between the two regions, that is, the **concentration gradient**. Concentration gradients often exist between the internal and external environments of a cell. If the solute can permeate the cell membrane, diffusion in the direction of the concentration gradient into (or out of) the cell ensues.

The **diffusion rate** for a solute is given by **Ficke's equation:**

$$\frac{dS}{dt} = D\,A\frac{C_1 - C_2}{x} \qquad (15\text{--}1)$$

where

$dS/dt =$ the number of moles of solute, S, diffusing from region 1 to region 2 in the time interval dt.

$D =$ the **diffusion coefficient** of the solute. Each solute has a specific diffusion coefficient; the units of the diffusion coefficient are moles per unit cross-sectional area per unit concentration gradient per unit time. This reduces to the dimensions square centimeters per second. The diffusion coefficient of a solute is determined in part by the solute molecule's size and shape.

$A =$ the cross-sectional area through which diffusion occurs.

C_1 and $C_2 =$ the concentrations of S in regions 1 and 2, respectively.

$x =$ the distance between regions 1 and 2.

$\dfrac{C_1 - C_2}{x} =$ the concentration gradient.

In the case of diffusion into or out of the cell, we may let

$x =$ the membrane thickness.

$C_1 = C_{out} =$ the concentration of S outside of the cell.

$C_2 = C_{in} =$ the concentration of S inside of the cell.

A new term, the **permeability constant, K,** which describes the diffusion of a particular substance across a particular membrane, may be substituted for D/x in equation 15–1 to yield

$$\frac{dS}{dt} = K\,A\,(C_{out} - C_{in}) \qquad (15\text{--}2)$$

The dimensions of K are centimeters per second. Because the total amount of substance S in a cell of volume

V is VC_{in}, substitution of $V dC_{in}$ for dS in equation 15–2 followed by a transposition yields

$$\frac{dC_{in}}{dt} = K\frac{A}{V}(C_{out} - C_{in}) \qquad (15\text{--}3)$$

Thus, it can be seen that other things being equal, the rate of change of the internal solute concentration (i.e., dC_{in}/dt) depends on the surface area to volume ratio of the cell (i.e., A/V). Because the ratio A/V is at a minimum for spherical objects, the rate at which solute is exchanged by diffusion between a cell and its environment is increased by deviation from spherical shape.

If the external volume is sufficiently large in comparison with the cell volume (this generally is the case), then C_{out} will not change significantly with time as a result of diffusion. That is, during the time interval between t_0 and t_1, C_{out} will remain constant. Therefore, equation 15–3 may be solved by integration as follows:

$$\int_{C_{out}-C_{in} \text{ at } t_1}^{C_{out}-C_{in} \text{ at } t_0} \frac{dC_{in}}{C_{out} - C_{in}} = \int_{t_1}^{t_0} K\frac{A}{V}\,dt \qquad (15\text{--}4)$$

$$\ln(C_{out} - C_{in} \text{ at } t_0) - \ln(C_{out} - C_{in} \text{ at } t_1) = K\frac{A}{V}\Delta t \qquad (15\text{--}5)$$

where Δt equals the time interval between t_0 and t_1. Solving equation 15–5 for K, we obtain

$$K = \frac{V}{A\Delta t} \ln \frac{C_{out} - C_{in} \text{ at } t_0}{C_{out} - C_{in} \text{ at } t_1} \qquad (15\text{--}6)$$

C_{out}, C_{in}, A, V, and Δt may be determined experimentally, so that equation 15–6 can be employed to find the permeability constant. Evaluation of the permeability constants for a variety of substances entering a cell permits much more to be learned about the chemical nature and behavior of the cell membrane. The permeability constants of glycol, urea, glycerol, and sucrose for the membranes of the red blood cell, the bacterium *Beggiatoa*, and the alga *Chara* are compared in Table 15–5. The

marked differences that can be seen in these values reflect the variation that must exist in the composition and organization of the plasma membranes of these cells.

Factors Influencing Permeability A number of different molecular parameters influence the ability of a substance to permeate cell membranes. Important among these are the **distribution coefficient, molecular size**, and **charge**. The distribution coefficient of a substance relates its solubility in oil to its solubility in water. In general, the more *nonpolar* a substance, the more soluble it is in oil and the less soluble it is in water. In contrast, the more polar a substance is, the less soluble it is in oil and the more soluble it is in water.

One of the most extensive studies of the relationship between distribution coefficient and membrane permeability was that carried out in the 1930s by R. Collander and H. Barlund using the unicellular alga *Chara*. They found that, in general, the higher the distribution coefficient of a compound, the greater its permeability to the cell membrane. Overton had made similar but less quantitative observations many years earlier, and this led him to propose that the cell membrane was composed of lipid. The relationship between lipid solubility and membrane permeability is the basis for the proposal that lipid-soluble substances readily pass into or out of the cell by dissolving through lipid regions of the cell membrane.

For chemically related substances, permeability increases with increasing distribution coefficient *regardless of molecular size*. However, in general, when two molecules have similar distribution coefficients, the smaller molecule is more permeable than the larger molecule. The fact that water has a very high permeability constant in spite of its low lipid solubility led to the suggestion that there are *hydrophilic pores* in the cell membrane through which small polar molecules may more readily diffuse. The permeability constant for water is about 1×10^{-4} cm/sec. This value is only 0.001% of the rate at which water molecules diffuse across a water layer the thickness of the cell membrane. This implies that if they are present, hydrophilic pores must cover only a small percentage of the surface area of the cell. In the case of the red blood cell, A. K. Solomon has estimated that 0.06% of the surface area is occupied by such pores.

In addition to the distribution coefficient, the *effective size* of a molecule influences its membrane permeability. In general, small molecules are more permea-

TABLE 15–5   SOME PERMEABILITY CONSTANTS

Substance	Permeability Constant (cm/sec x 10^5)		
	Red Blood Cell	*Beggiatoa*	*Chara*
Glycol	0.21	1.39	1.2
Urea	7.8	1.58	0.11
Glycerol	0.0017	1.06	0.021
Sucrose	—[a]	0.14	0.0008

[a]Sucrose does not permeate the red blood cell membrane.

ble than large molecules. Substances of very high molecular weight (such as starch, glycogen, and many proteins) are usually unable to permeate the membrane at all. This is not to imply that cells cannot incorporate or eliminate very large molecules. Indeed, they do; however, the process involves mechanisms other than diffusion (see below).

Electrolytes enter cells more slowly than do *nonelectrolytes* of similar effective molecular size, and strong electrolytes enter more slowly than weak electrolytes. Because pH influences the degree of ionization of many electrolytes, permeability to electrolytes varies with the pH. In general, monovalent ions (e.g., Na^+, K^+, I^-, Cl^-) permeate membranes more readily than divalent ions (e.g., Ca^{2+}, Mg^{2+}, SO_4^{2-}), which in turn are more permeable than trivalent ions (e.g., Fe^{3+}). The relative permeability of the membrane to ions of the same valency number but different sign (i.e., K^+ versus Cl^-) depends on the particular cell considered.

Not all ions of the same valency and sign are equally permeable. Most differences can be explained in terms of the relative effective sizes of the ions in an aqueous medium. The charge on the ion attracts neighboring water molecules, causing them to align themselves around the ion to form **spheres of hydration**. The effective size of the ion is therefore determined by the number of these hydration spheres (see Fig. 3–5). Ions of low molecular or atomic weight attract more water molecules than do ions of higher molecular weight because there are fewer electron shells around the atomic nuclei to neutralize the ionic charge. Accordingly, the effective size of Li^+ (atomic number = 3) is greater than that of Na^+ (atomic number = 19) and so on. Therefore, membranes are usually more permeable to K^+ than to Na^+, and they are more permeable to Na^+ than to Li^+.

Ions may enter or leave a cell through pores in the plasma membrane or the endoplasmic reticulum. Cations are attracted to and pass more readily through pores lined by negative charges (e.g., negatively charged R-groups of amino acids of membrane proteins, etc.). Anions enter or leave the cell through positively charged pores.

The Gibbs–Donnan Effect

Proteins behave like ions because the R-groups of their amino acids may bear positive or negative charges. The net ionic charge of the protein molecule depends on the *relative* numbers of positive and negative side groups. Most soluble proteins behave like anions because they possess more negative than positive sites. Unlike many other ions, proteins are generally too large to permeate cell membranes. The **Gibbs–Donnan effect** (after J. W. Gibbs and F. G. Donnan) describes the effect that proteins have on the equilibrium distributions of small ions across a selectively permeable membrane. When a selectively permeable membrane separates an electrolyte solution containing proteins from one that lacks proteins, the concentrations reached at equilibrium for each permeable ionic species will *not* be the same on both sides of the membrane. Instead, the concentration of small ions having the *same* sign as the protein will be lower on the side of the membrane containing the protein (i.e., inside the cell) and the concentration of small ions of *opposite* sign will be higher. The Gibbs–Donnan effect does not influence the equilibrium distributions of nonionized substances such as glucose and urea. The Gibbs–Donnan phenomenon is responsible for the rupture and lysis of bacteria and other organisms when antibodies attach to their surface. The reason is that the binding of antibody sets off the "complement" reactions in which a succession of plasma proteins (in a cascade-type activation) form enzymes that puncture a hole in the invading bacterial cell's membrane. This is followed by an inrush of ions and water that leads to cell lysis.

The Gibbs–Donnan Effect May Be Illustrated as Follows Suppose that two chambers of *equal* and *fixed* volumes are separated by a membrane permeable to water and small ions but impermeable to proteins. Into chamber 1 is placed a sodium proteinate (NaPr) solution of concentration C_1 and into chamber 2 a NaCl solution of concentration C_2. Because the chloride ions are permeable and present at a greater concentration in chamber 2 than in chamber 1 (i.e., chamber 1 initially has no Cl^-), Cl^- will diffuse into chamber 1. To preserve the electrical neutrality of each chamber, the diffusion of Cl^- into chamber 1 must be accompanied by the diffusion of an equivalent amount of Na^+. Therefore, chamber 2 will be left with equal but reduced concentrations of Na^+ and Cl^-. According to the law of mass action, *at equilibrium the products of the diffusable ion concentrations in each chamber will be equal.* Therefore, if we let X equal the concentration of Cl^- diffusing from chamber 2 into chamber 1 (X will of course also equal the concentration of Na^+ accompanying Cl^-), then at equilibrium

$$(C_1 + X)(X) = (C_2 - X)(C_2 - X)$$

$$C_1 X + X^2 = (C_2)^2 - 2C_2 X + X^2 \qquad (15\text{-}7)$$

$$C_1 X = (C_2)^2 - 2C_2 X$$

and

$$X = \frac{(C_2)^2}{C_1 + 2C_2} \qquad (15\text{–}8)$$

Equation 15–8 is called the Gibbs–Donnan equilibrium equation and may be used to determine the equilibrium distribution of ions between the protein-containing and protein-free portions of the two-chamber system.

Let us consider a specific case in which chamber 1 initially contains $0.01\ M$ sodium proteinate (i.e., $0.01\ M\ Na^+$ and $0.01\ M\ Pr^-$) and chamber 2 contains $0.03\ M$ NaCl (i.e., $0.03\ M\ Na^+$ and $0.03\ M\ Cl^-$). The concentration of Cl^- (and Na^+) diffusing across the membrane and into chamber 1 from chamber 2 may be determined using equation 15–8 as follows:

$$X = \frac{(0.03)^2}{0.01 + 2(0.03)}$$

$$= 0.013$$

Therefore, the equilibrium distribution of ions would be (see also Fig. 15–37):

Chamber 1	Chamber 2
Na^+ $0.023\ M$	Na^+ $0.017\ M$
Cl^- $0.013\ M$	Cl^- $0.017\ M$
Pr^- $0.010\ M$	Pr^- $0.000\ M$

Note that (1) an electrical balance exists *within* each chamber, (2) the osmotic pressure is greater in chamber 1 than in chamber 2 as a result of the higher total concentration of particles (no osmosis occurs because the chamber volumes are fixed), and (3) there are more sodium ions but fewer chloride ions in chamber 1 than in chamber 2.

By extrapolating the observations made for the artificial two-chamber system just described to cells (i.e., chamber 1) and their surrounding milieu (i.e., chamber 2), it may be seen that the Gibbs–Donnan effect would result in an unequal distribution of permeable ions across the cell membrane, that is, a *stable* ionic concentration gradient would be established. Just such a phenomenon accounts in part for the resting electrochemical membrane potentials possessed by many cells (see Chapter 24).

In the cases we have considered so far, the chamber volumes have been equal and fixed. This, of course, is not necessarily the case for a cell and its surroundings. The flexibilty of the plasma membrane permits notable fluctuations in cell shape and volume. The greater osmotic pressure created inside the cell by the Gibbs–Donnan effect causes water to pass into the cell by osmosis, increasing the cell's volume and lowering its ion concentration. Because this disturbs the Gibbs–Donnan equilibrium previously established, a new equilibrium is achieved by the passage of additional permeable ions into the cell. Osmosis halts when the hydrostatic pressure created by water inside the cell balances the osmotic pressure generated by the dissolved cell solute.

FIGURE 15–37   Gibbs–Donnan effect: Initial (t_0) and equilibrium (t_1) distributions of ions across a selectively permeable membrane separating two chambers of equal and fixed volume, when one ion (Pr–) is impermeable. At equilibrium the sum of anions equals the sum of cations on each side of the membrane.

FACILITATED (MEDIATED) DIFFUSION THROUGH THE CELL MEMBRANE

A variety of compounds including sugars and amino acids pass through the plasma membrane and into the cell at a much higher rate than would be expected on the basis of their size, charge, distribution coefficient, or magnitude of the concentration gradient. The increased rate of transport through the membrane is believed to be facilitated by specific membrane *carrier* substances and is called **facilitated** (or **mediated**) **diffusion**.

During facilitated diffusion, the rate at which the solute permeates the membrane increases with increasing solute concentration *up to a limit*. Above this limiting concentration, no increase in the rate of transport across the membrane is observed. In other words, facilitated diffusion exhibits *saturation kinetics* (Fig. 15–38) and is therefore similar to the relationship between reaction rate and substrate concentration in enzyme-catalyzed reactions (see Chapter

FIGURE 15–38 Kinetics of *facilitated diffusion*. Note the similarity to the kinetics of enzyme-catalyzed reactions.

8). Other characterisitics of facilitated diffusion are also similar to enzyme catalysis. Transport is *specific*; for example, in the erythrocyte, the inward diffusion of glucose, but not fructose or lactose, is facilitated. The rate of solute permeation can also be affected by the presence of structurally similar chemical compounds, much as in *competitive enzyme inhibition*. Facilitated diffusion exhibits *pH dependency*. Although facilitated diffusion results in a more rapid attainment of a concentration equilibrium across the membrane than passive diffusion, *the normal equilibrium concentrations are not altered*. Substances are *not* transported through the membrane *against a concentration gradient*. Although facilitated diffusion is not affected by chemicals that act as metabolic inhibitors, it is affected by enzyme inhibitors such as sulfhydryl blocking agents.

Facilitated diffusion is believed to result from the interaction of solute with specific membrane molecules, presumably proteins, thereby forming, a *carrier–solute complex*. The complex undergoes a positional change within the membrane in such a way

that the solute now faces the other membrane surface and is released from the carrier (Fig. 15–39). An alternative suggestion is that the carrier may be a small molecule with the solute–carrier complex formed by an enzyme-catalyzed reaction within the membrane. Once formed, the solute–carrier complex diffuses to the other side of the membrane where the solute is released in a second reaction.

A good and well-defined example of facilitated diffusion occurs in the bacterium *Escherichia coli*. The sugar *lactose* does not readily permeate *E. coli* cells and cannot be hydrolyzed by cytoplasmic extracts of cells grown in the absence of lactose. However, when *E. coli* is in a medium containing lactose, the enzyme β-galactosidase, which hydrolyzes lactose, soon appears in the cell cytoplasm, and a specific transport protein appears in the membrane. The presence of the substrate in the growth medium is said to *induce* the formation of the enzymes by the cells (see Chapter 11). Such cells are able to transport and subsequently hydrolyze the lactose thereby forming galactose and glucose. Certain mutants of *E. coli* can be induced in this way to form β-galactosidase even though lactose remains impermeable and cannot be incorporated by the cells. Finally, other *E. coli* mutants can be found that are able to remove galactose from the medium but cannot metabolize it once it is inside the cell.

It is now clear from these studies that in wild-type cells the presence of lactose in the growth medium induces both the formation of the hydrolytic enzyme and a carrier system—called a **permease** or **translocase**. The sets of genes controlling the formation of the permease and the hydrolytic enzyme are coordinately induced (see Chapter 11) in the presence of lactose. The loss or alteration of either set of genes (as in the *E. coli* mutants) results in a corresponding inability to induce the formation of the enzyme or the permease.

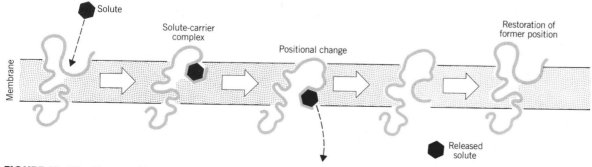

FIGURE 15–39 Stages of facilitated diffusion (see text for details).

ACTIVE TRANSPORT

During diffusion (passive or facilitated), substances pass through the plasma membrane until some sort of equilibrium is achieved. The equilibrium may be of the Gibbs–Donnan variety or may be a simple concentration equilibrium. Both involve an interplay between the concentrations of soluble solute inside and outside the cell. Cells can also accumulate solutes in quantities far in excess of that expected by any of the above mechanisms if the solute is rendered insoluble once it has entered the cell, because insoluble materials do not contribute to concentration gradients. Alternatively, once inside the cell, a solute may enter a metabolic pathway and be chemically altered, thereby reducing the concentration of that particular solute and allowing additional solute permeation. In all the cases we have so far considered, solute passage through the membrane hinges on the presence of a concentration gradient, with the solute moving in the direction of the gradient.

Substances can also move through the plasma membrane into or out of the cell *against a concentration gradient*. This requires the expenditure of energy on the part of the cell and is called **active transport**. Active transport ceases when cells are (1) cooled to very low temperatures (such as 2–4°C), (2) treated with metabolic poisons such as cyanide or iodoacetic acid, or (3) deprived of a source of energy. The best understood and most exhaustively studied cases of active transport are those that involve the movements of sodium and potassium ions across the plasma membranes of erythrocyes, nerve cells, and *Nitella* cells and that result in an ionic concentration gradient across the cell membrane. The mechanism that establishes and maintains these gradients appears to be basically similar in all of these cells and can be illustrated with the erythrocyte.

The Na$^+$/K$^+$ Exchange Pump

The cytoplasm of the erythrocyte contains 0.150 M K$^+$, whereas the surrounding blood plasma contains only 0.005 M K$^+$. In contrast, the erythrocyte contains only 0.030 M Na$^+$ and the plasma contains 0.144 M Na$^+$. Hence, marked K$^+$ and Na$^+$ concentration gradients exist across the cell membrane. In Chapter 14 we noted that tracer studies utilizing radioactive isotopes of Na and K clearly established that these ions are permeable to the erythroctye membrane and are constantly diffusing through it. Yet, in spite of this

permeability, Na$^+$ and K$^+$ concentration gradients across the membrane are maintained. The gradients are maintained because sodium ions diffusing into the cell from the plasma under the influence of the concentration gradient are transported outward again, and potassium ions diffusing out of the cell are replaced by the inward transport of K$^+$ from the plasma. That these movements involve active metabolic processes is clearly demonstrated when the temperature of a blood sample is reduced from 37°C (normal human blood temperature) to 4°C, when cyanide is added to the blood, or when plasma glucose consumed during erythrocyte metabolism is not resupplied to the blood sample. Under these conditions, cell metabolism is interrupted and is followed by the inward diffusion of Na$^+$ and the outward diffusion of K$^+$ until the ionic concentrations on both sides of the erythrocyte membrane are in a passive equilibrium.

In the case of red blood cells and nerve cells, the active transport of Na$^+$ and K$^+$ appears to be linked, that is, the mechanism responsible for the outward transport of Na$^+$ simultaneously transports K$^+$ inward. An enzyme isolated from nerve cell membranes and believed to be involved in Na$^+$ and K$^+$ transport has been shown to have two sites that bind one or more of each of these cations. The enzyme is believed to be an integral protein spanning the lipid bilayer.

Active transport of Na$^+$ and K$^+$ through the membranes of nerve cells and erythrocytes requires ATP, and ATP cannot be replaced by other nucleoside triphosphates such as GTP, UTP, and ITP. ATP is converted to ADP during active transport by a membrane-bound Na$^+$- and K$^+$-stimulated *ATPase*. This enzyme and that involved in the transport of Na$^+$ and K$^+$ may be one and the same. The membranes of cells from many other mammalian tissues seem to possess a similar ATPase activity.

Two K$^+$ and three Na$^+$ are transported through the membrane for each molecule of ATP dephosphorylated. Transport of Na$^+$ and K$^+$ through the plasma membrane is believed to occur in the following stages (see Fig. 15–40). Three sodium ions and one molecule of ATP inside the cell are bound to specific sites on the enzyme carrier, while two potassium ions are bound to a site on the same enzyme facing the exterior of the cell. Binding of the substrates results in and is followed by a change in the tertiary structure of the carrier molecule such that the bound sodium and potassium ions are "translocated" across the membrane. It is presumed that at some stage during this process, ATP is split, releasing ADP. Translocation is

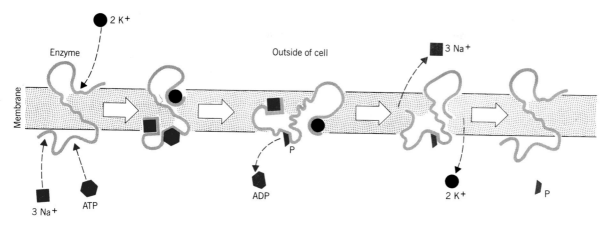

FIGURE 15–40 The Na$^+$/K$^+$ exchange pump (see text for details).

followed by an alteration of the binding sites such that the sodium ions are "released" outside the cell, while the potassium ions are released inside the cell. Once the ions are released, the carrier undergoes another change in structure, priming it for another round of the transport cycle. This stage, called "recovery," is accompanied by the release of inorganic phosphate. Although this model is widely accepted, it has also been suggested that the enzyme site that binds Na$^+$ on the inside of the cell binds K$^+$ on the outside following translocation, while the site that initially binds K$^+$ on the outside binds Na$^+$ on the inside following translocation. In this manner, the recovery phase would result in an additional movement of ions through the membrane and would be more efficient.

Cotransport

Amino acids, sugars, and other metabolites are also actively transported through the plasma membrane into the cell. In many cells, the transport of these metabolites is coupled to the movements of sodium ions, as shown in Figure 15–41. The Na$^+$/K$^+$ exchange pump creates a steep concentration gradient across the plasma membrane favoring the inward diffusion of Na$^+$. Indeed for every *two* K$^+$ pumped into the cell, *three* Na$^+$ are pumped out. Carrier proteins in the membranes bind both Na$^+$ and the metabolite, following which a change in the carrier's structure brings both substrates to the cell interior, where they are released. Release of the Na$^+$ internally is followed by its active extrusion back through the membrane. The lat-

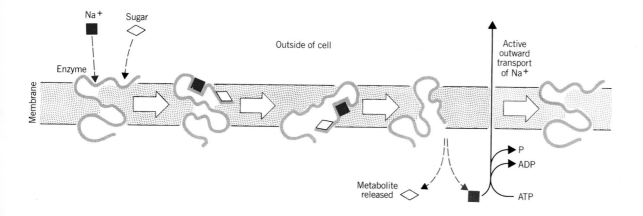

FIGURE 15–41 *Cotransport.* The coupled transport of Na$^+$ plus metabolite (in this case, sugar).

ter event is coupled to ATP hydrolysis and results in the maintenance of the steep Na$^+$ gradient. In a sense, the steep Na$^+$ gradient acts as the driving force for the inward transport of metabolites, and the simultaneous movements of Na$^+$ together with metabolites into the cell constitute **cotransport**. Because the ATP-dependent Na$^+$/K$^+$ pump pumps three Na$^+$ for every two K$^+$, an electrical gradient is created across the membrane. For this reason, the Na$^+$/K$^+$ exchange pump is called an **electrogenic pump**. As will be seen in Chapter 16, the potential energy of an electrogenic pump is coupled to ATP synthesis in mitochondria.

"Simple" Active Transport

The passage of some substances through membranes against a concentration gradient is unidirectional but not coupled to ionic movement even though ATP is consumed in the process. Such movement is called **simple active transport**. The carrier enzyme cyclically binds the solute at one membrane surface and releases it at the other. The cycle is accompanied at some point by the hydrolysis of ATP.

BULK TRANSPORT INTO AND OUT OF CELLS

Before we consider several mechanisms responsible for bulk movement of materials into and out of cells, it is important to recall what is meant by "inside the cell" and "outside the cell" (see also Chapter 1). For convenience, we have so far depicted the plasma membrane as a more or less continuous smooth sheet enclosing the cell. In reality, this is an oversimplification, for in most cells the plasma membrane exhibits numerous outfoldings and infoldings. Outfoldings of the plasma membrane cover microvilli, cilia, flagella, and other cytoplasmic extensions. Infoldings of the plasma membrane form small pockets and narrow channels that descend into the cytosol. Some of these infoldings may join the network of cisternae that form the endoplasmic reticulum and that furrows the cytoplasm. This implies that the intracisternal space of the endoplasmic reticulum may be in direct continuity with the surrounding cell environment and that the movement of materials between the lumenal phase and the cell surroundings does not require passage *across* (or *through*) any membranes. Consequently, any materials that are in the cisternae of the endo-

plasmic reticulum (i.e., in the lumenal phase) may be regarded as "outside" the cell. To get "inside" the cell, substances in the lumenal phase must pass through the membranous walls of the endoplasmic reticulum and into the cytosol.

Many intracellular vesicles appear to be derived from the endoplasmic reticulum by being "pinched off" from the latter. For example, **transitional vesicles** that merge with the forming face of the cell's Golgi bodies (Chapter 18) and **peroxisomes** and other **microbodies** (Chapter 19) are believed to be formed in this way. Technically then, the contents of these vesicles are "outside" the cell and are separated from the cytosol by membranes. Some vesicles are derived from invaginations of the plasma membrane; the contents of these vesicles are also to be considered as "outside" the cell. These relationships are depicted in Figure 15–42, in which the external (or cisternal) and internal (or cytosol) halves of the membrane are dis-

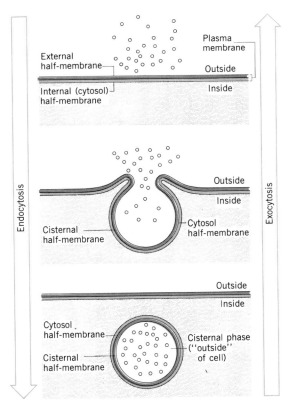

FIGURE 15–42 Relationship of external and internal (i.e., cytosol) halves of the plasma membrane to the cisternal and cytosol halves of vesicles in endocytosis and exocytosis.

tinguished. Particles within such a vesicle are in contact with the external face of the membrane.

From the preceding discussion, it should be clear that though substances may be directly exchanged between the cell surroundings and the cytosol through the plasma membrane, many exchanges also occur across membranes of cytoplasmic vesicles and the endoplasmic reticulum. These exchanges are mediated by the same mechanisms described above in connection with movements directly through the plasma membrane (i.e., diffusion, facilitated diffusion, active transport, etc.).

The formation of cytoplasmic vesicles from the plasma membrane and the consequent entrapment within these vesicles of substances formerly in the cell surroundings is called **endocytosis**. Several different kinds of endocytosis have been described, including **pinocytosis, phagocytosis**, and **receptor-mediated endocytosis**, and these are discussed below. Movements of materials from the cell into the surroundings by the fusion of cytoplasmic vesicles with the plasma membrane constitute **exocytosis**. Endocytosis continuously removes small portions of the plasma membrane, whereas exocytosis continuously adds to the membrane. From a quantitative standpoint, the turnover is quite impressive, for in some cells (e.g., cultured fibroblasts and macrophages) a membrane area equal to the total surface area of the cell is turned over every few hours. Because tracer studies indicate that many membrane constituents have a half-life of several days, it has been suggested that some of the membranous material internalized during endocytosis is returned to the plasma membrane during exocytosis. Endocytosis and exocytosis have many characteristics in common and are the subject of the next section.

ENDOCYTOSIS

Pinocytosis

Using time-lapse photography to study tissue culture cells, W. H. Lewis in 1931 described what seemed to be a curious phenomenon in which small amounts of culture medium were trapped in invaginations of the plasma membrane and then pinched off to form small cytoplasmic vesicles. Because the entire process appeared much like some form of organized cell drinking, Lewis termed the phenomenon **pinocytosis** ("pinos" means "I drink" in Greek). Lewis' observations with tissue culture cells were confirmed in 1934

by S. O. Mast and W. L. Doyle studying amoebae in which pinocytosis is readily observed with the light microscope. Using electron microscopy, it became clear in the 1950s that pinocytosis is a common phenomenon occurring at different times in many kinds of cells and tissues including leukocytes, kidney cells, intestinal epithelium, liver macrophages, and plant root cells.

Pinocytosis is induced by the presence of appropriate concentrations of proteins, amino acids, or certain ions in the medium surrounding the cell. The first step in the process (Fig. 15–43) involves the binding of the inducer substance to specific receptors in the plasma membrane. This is followed by invagination of the membrane to form either **pinocytic vesicles** or narrow channels (Fig. 15–43). Although binding of the inducer is not inhibited by cyanide or low temperature, the formation of pinocytic vesicles is, and it is therefore dependent on cell metabolism. Actin filaments are associated with the margins of pinocytic vesicles and are believed to play some role in the vesicle's invagination. Pinocytic vesicles detach from the plasma membrane and migrate toward the interior of the cell, where they may fragment into smaller vesicles or coalesce to form larger ones. Unless the vesicles are "tagged" by inducing pinocytosis in the presence of radioactive tracers, they soon become indistinguishable from other vacuoles in the cell.

Two levels of pinocytosis can be distinguished. In one of these, called **micropinocytosis**, the vesicles formed have a diameter of about 0.1 μm and are derived from small depressions in the cell surface (**caveolae**). In **macropinocytosis**, the vesicles are considerably larger, having diameters of 1–2 μm. The larger vesicles are formed either from larger invaginations of the plasma membrane or from surface ruffles (Figs. 15–29 and 15–30). Although pinocytosis is induced by the presence of specific substances in the cell surroundings, other materials are also enclosed by the pinocytic vesicles, including water, salts, and so on. In this regard, pinocytosis is partly (perhaps even predominantly) nonspecific and is to be contrasted with receptor-mediated endocytosis (discussed below), which is almost entirely specific. Molecules that induce pinocytosis, together with water and other solutes entrapped by the infolding plasma membrane, are presumed to enter the cytosol from the vesicles that are formed by diffusion, active transport, or related transport mechanisms. Pinocytosis, like other forms of endocytosis, is carried out at some energy cost on the part of the cells engaged in this activity.

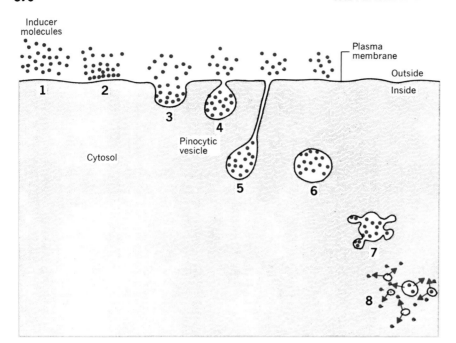

FIGURE 15–43 *Stages of pinocytosis:* 1–2, binding of inducer molecules to plasma membrane; 3–5, invagination of the membrane; 6–8, detachment from plasma membrane and fragmentation into smaller vesicles.

Receptor-Mediated Endocytosis

In **receptor-mediated endocytosis**, binding of molecules such as hormones, antibodies and other proteins, and lipids to specific receptors in the plasma membrane is followed by their *clustering* (i.e., concentration) at specific membrane sites, which are then internalized by the cell. The process is depicted diagrammatically in Figure 15–44. The receptors that mediate the endocytosis are integral proteins that span the membrane. Substances bound by these receptors, called **ligands**, are "plucked" from the extracellular milieu even when present in extremely low concentrations and surrounded by a large excess of other, unrelated solutes. Ligand–receptor complexes move laterally in the membrane toward **coated pits** in which the complexes are then concentrated. Because lateral movement through the membrane by the ligand–receptor complex is quite rapid, a coated pit is encountered within a few seconds. Once a ligand–receptor complex enters a pit, its lateral movement is halted; as a result, many such complexes are quickly accumulated in a single pit. Uncomplexed receptors, which also move laterally in the plasma membrane, may also be concentrated in a coated pit. Coated pits are present in almost all animal cells and account for about 2% of the membrane's total surface area.

The pits' "coats" consist primarily of a layer of peripheral protein called **clathrin** tightly appressed to the cytosol-facing surface of each pit. The clathrin's subunits (molecular weight about 180,000) occur in three-armed trimers called **triskelions** that interact with one another to form a cagelike net around the pit. Indeed, polymerization of the triskelions is believed to form the pit, deepening it and eventually freeing it from the membrane as a small, **coated vesicle**. (Isolated and purified triskelions have been shown to spontaneously form hemispherical and spherical cagelike structures *in vitro*.)

Coated vesicles that have just detached from the plasma membrane move deeper into the cytoplasm, progressively shed their clathrin coats, and fuse with one another to form larger, smooth-surfaced vesicles called **endosomes** (sometimes **receptosomes**) (Fig. 15–44). The clathrin released by coated vesicles is believed to be recycled to the plasma membrane, where it gives rise to new coated pits. As endosomes move even deeper into the cytoplasm, there is a progressive release of ligands from their receptors into the lumen of the vesicle. The endosome develops a tubular portion in which the membrane-bound but ligand-free receptors are concentrated and a vesicular portion containing free ligand. At this stage the endosome is

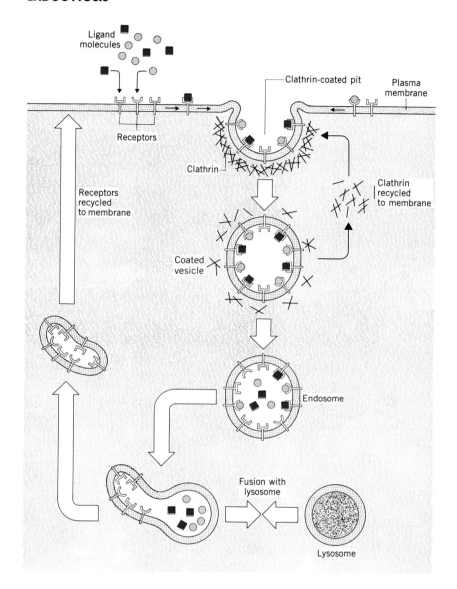

FIGURE 15–44 *Concentrative, receptor-mediated endocytosis.* Ligand molecules bind to plasma membrane receptors traveling laterally in the membrane. Receptors with and without ligand are concentrated in *clathrin-coated pits*, which invaginate to form *clathrin-coated vesicles*. As these vesicles move deeper into the cytosol, the clathrin coat is lost and recycled to the plasma membrane. Vesicles eventually form *endosomes* in which receptors and ligands are *uncoupled*. The receptors are recycled to the plasma membrane. Endosomes may fuse with lysosomes whose hydrolytic enzymes degrade the ligands.

sometimes referred to as a **CURL** ("compartment of uncoupling of receptor and ligand"). The tubular portion of the endosome is subsequently dispatched to the plasma membrane, thereby recycling its ligand receptors, and the vesicular portion continues its journey deeper into the cell. If the endosome contains proteins, other macromolecules, or particles, it may fuse with a lysosome so that enzymatic digestion of the entrained material can ensue. Figure 15–45 summarizes the overall process of receptor-mediated endocytosis, including the initial synthesis of membrane receptors by ER-bound ribosomes.

Sorting of Ligands and Receptors within a CURL

Separation of ligand and receptor within a CURL is believed to be prompted by the low pH that exists within the compartment. Still uncertain is the mechanism by which receptors and ligands are sorted. How the receptors are moved into the lipid bilayer of the tubular extensions of the CURL prior to recycling to the plasma membrane, while the ligands remain in the fluid-filled vesicular body of the organelle, is unclear. Two domains may be identified in a CURL: a membrane-rich (fluid-poor) domain consisting of the CURL's narrow, tubular extensions, and a fluid-rich

FIGURE 15–45 Synthesis and recycling of plasma membrane receptors. Receptors are synthesized by attached ribosomes and are anchored in the endoplasmic reticulum. From the margins of the endoplasmic reticulum, the receptors are dispatched to the Golgi apparatus for processing. The receptors leave the Golgi apparatus in the form of small vesicles that fuse with the plasma membrane. Receptors with and without bound ligand accumulate in clathrin-coated pits. These pits invaginate to form coated vesicles and endosomes. After uncoupling of receptors and ligands takes place, the ligand-free receptors are recycled to the plasma membrane.

(membrane-poor) vesicular domain. It has been estimated that 90% of a CURL's membrane is in the CURL's tubular extensions, whereas only 10% is in the fluid-rich domain. L. H. Rome has suggested that such an asymmetric distribution of membrane could result in an automatic localization of receptors in the membrane-rich domain (so long as the receptors are free to migrate within the lipid bilayer). In other words, the receptors accumulate in the tubular (membrane-rich) extensions of a CURL by *mass action.*

Fragments of plasma membranes, membranes of the endocytic compartment, and membranes of the Golgi apparatus can be isolated from disrupted cells using various forms of density gradient centrifugation. The isolated fractions may then be analyzed for their contents of specific proteins (e.g., clathrin, receptors, marker enzymes, etc.), lipids, and ligands. Not surprisingly, receptors, ligands, and clathrin are *differentially distributed* among these membrane fractions. However, W. H. Evans and others have shown that in fractionated liver cells the cholesterol, sphingomyelin, and marker enzyme contents of the separated membrane components are unexpectedly similar and have suggested that an *anatomical continuity* may exist between the plasma membrane and the membranes of the endocytic compartment.

Phagocytosis

Phagocytosis, which was first described by E. Metchnikoff in the late nineteenth century, involves the endocytosis of much larger quantities of material than either pinocytosis or receptor-mediated endocytosis. For example, entire ciliates, rotifers, or other microscopic organisms may be phagocytosed by an amoeba and enclosed within one or more vacuoles called **phagosomes, food vacuoles**, or **food cups** (see Fig. 15–46). During phagocytosis, the "prey" may be temporarily immobilized by secretions from the phago-

cytic cell. The phagocytosis of ciliates is characterized by the flowing of the amoeba's cytoplasm into footlike projections (**pseudopodia**) that gradually encircle and fully encapsulate the ciliate. Using a similar mechanism, certain white blood cells phagocytose hundreds of bacteria. The removal and destruction of old red blood cells in the liver, spleen, and bone marrow by reticuloendothelial cells in these organs also occurs by phagocytosis. Following phagocytosis, the phagosomes fuse with *primary lysosomes* in the cell. The hydrolytic enzymes from these lysosomes digest the en-

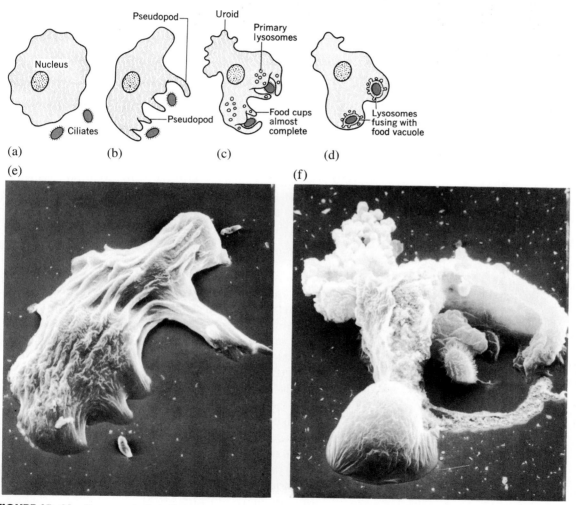

FIGURE 15–46 *Phagocytosis.* In the presence of phagocytosable material (a), the cell forms pseudopodia (b) that entrap the prey in vacuoles (c and d). The fusion of primary lysosomes with these vacuoles is followed by digestion of the prey. Scanning electron micrographs are of an amoeba prior to (e) and during (f) phagocytosis and correspond to stages b and d of the diagrams. (Photomicrographs kindly provided by Dr. K. W. Jeon. Copyright © 1976, *J. Protozool.* **23**, 83.)

gulfed material, converting it to a form that may be transported across the vacuolar membranes and into the cytosol (see Chapter 19).

EXOCYTOSIS

Exocytosis is the mechanism by which large quantities of material enclosed within a cell vacuole are transferred to the cell surroundings by fusion of the vacuole with the plasma membrane. In a sense, the process is the reverse of endocytosis, with the contents of the vacuole being emptied into the extracellular space. The best understood form of exocytosis is secretion. When the secretory vesicle touches the plasma membrane, lipids in both membranes merge to form a common bimolecular leaflet. In effect, the vesicle's membrane is incorporated into the plasma membrane, while at the same time the vesicle's contents are discharged to the exterior (Fig. 15–47). Among the intracellular structures actively involved in the processing of cytoplasmic vesicles that are destined to participate in exocytosis is the Golgi apparatus. Likewise, lysosomes are involved in processing cytoplasmic vesicles formed by endocytosis. The functions of these two organelles are considered in Chapters 18 and 19.

SUMMARY

The cell is delimited at its surface by the **plasma membrane** and the cytoplasm is furrowed by the membranes of the **endoplasmic reticulum** so that the cell volume is divided into two phases: the **cytosol** (or **cytoplasm**) and the **intracisternal phase** (or **lumenal phase**). The cell's membranes actively participate in the exchanges of material between the cytosol and the surroundings. Early studies on the chemistry of the plasma membrane focused on the content and organization of its lipids, resulting in the **bimolecular lipid leaflet** model. This concept was subsequently modified to account for the membrane proteins, culminating in the **fluid-mosaic** model.

Both the proteins and lipids of the membrane are asymmetrically distributed across the inner and outer halves. In cells, the proteins in the inner half of the membrane are anchored in position by a **cytoskeletal network** consisting of filaments and microtubules. Carbohydrate chains associated with protein (and lipid) in the outer membrane surface play roles in maintaining membrane organization, in transmembrane transport, in cell-to-cell adhesion, and in providing the membrane's antigenic properties. In some cells, the carbohydrate forms an outer covering called the **glycocalyx**.

In tissues, the plasma membranes of neighboring cells exhibit specialized junctional regions including **tight junctions, desmosomes, gap junctions**, and **plasmodesmata**. Such junctions are important in maintaining tissue integrity and in regulating the passage of a substance across a tissue and from cell to cell.

The movements of materials across cell membranes may be achieved by passive or active mechanisms. The passive movements of solutes (by **diffusion**) and water

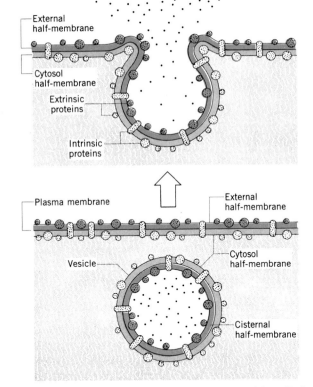

FIGURE 15–47 Fusion of an exocytic vesicle with the plasma membrane. Note that extrinsic and intrinsic proteins associated with the cisternal half of the vesicle membrane (red circles) become part of the plasma membrane's external half. In contrast, extrinsic and intrinsic proteins associated with the vesicle's cytosol half-membrane (gray circles) continue to face the cytosol after fusion with the plasma membrane is completed. The polarity of proteins that span the vesicle membrane (gray ovals) is similarly maintained.

(by **osmosis**) across the membrane occur principally as the result of concentration gradients. In some cases, diffusion through the membrane is **facilitated** by carrier molecules possessing the properties of enzymes. Substances can also move through the plasma membrane against a concentration gradient; this consumes cellular energy (i.e., ATP is hydrolyzed in the process) and is termed **active transport**. For individual ions and small molecules, active transport is effected by membrane-associated enzymes acting as pumps. However, transport into and out of cells can be in **bulk** and is then called **endocytosis** and **exocytosis**. Bulk transport involves gross movements of the plasma membrane, the three most common forms of bulk transport being **pinocytosis, receptor-mediated endocytosis**, and **phagocytosis**.

REFERENCES AND SUGGESTED READING

Articles and Reviews

Bennett, V., The membrane skeleton of human erythrocytes and its implications for more complex cells. In *Annual Review of Biochemistry* (C. C. Richardson *et al.*, Eds.), Vol. 54, Annual Reviews, Palo Alto, Calif., 1985.

Bretscher, M. S., The molecules of the cell membrane. *Sci. Am.* **253** (4), 100 (Oct. 1985.)

Bretscher, M. S., and Pearse, B. M. F., Coated pits in action. *Cell* **38**, 3 (1984).

Brown, M. S., and Goldstein, J. L., How LDL receptors influence cholesterol and atherosclerosis. *Sci. Am.* **251** (5), 598 (Nov. 1984).

Capaldi, R. A., A dynamic model of cell membranes. *Sci. Am.* **230** (3), 26 (Mar. 1974).

da Silva, P. P., and Kachar, B., On tight junctions. *Cell* **28**, 441 (1982).

Dautry-Varsat, A., and Lodish, H. F., How receptors bring proteins and particles into cells. *Sci. Am.* **250** (5), 52 (May 1984).

Davis, B. D., and Tai, P. C., The mechanism of protein secretion across membranes. *Nature* **283**, 433 (1980).

Evans, W. H., Communication between cells. *Nature* **283**, 521 (1980).

Evans, W. H., A biochemical dissection of the functional polarity of the plasma membrane of the hepatocyte. *Biochem. Biophys. Acta* **604**, 27 (1980).

Evans, W. H., and Flint, N., Subfractionation of hepatic endosomes in Nycodenz gradients and by free-flow electrophoresis. *Biochem. J.* **232**, 25 (1985).

Fox, C. F., The structure of cell membranes. *Sci. Am.* **226** (2), 30 (Feb. 1972).

Goldstein, I. J., Lectins: An overview. *Biotechs* **1**, 1 (1981).

Gorter, E., and Grendel, F., On bimolecular layers of lipoids in the chromocytes of the blood. *J. Exp. Med.* **41**, 349 (1925).

Karnovsky, M. J., Kleinfeld, A. M., Hoover, R., and Klausner, R. D., The concept of lipid domains in membranes. *J. Cell Biol.* **94**, 1 (1982).

Keynes, R. D., Ion channels in the nerve-cell membrane. *Sci. Am.* **240** (3), 126 (Mar. 1979).

Knutton, S., The mechanism of virus-induced cell fusion. *Micron* **9**, 133 (1978).

Lodish, H. F., and Rothman, J.E., The assembly of cell membranes. *Sci. Am.* **240** (1), 48 (Jan. 1979).

Lodish, H. F., Braell, W. A., Schwartz, A. L., Strous, G. J. A. M., and Zilberstein, A., Synthesis and assembly of membrane and organelle. In *International Review of Cytology* (A. L. Muggleton-Harris, Ed.), Suppl. 12, Academic, New York, 1981.

Luria, S. E., Colicins and the energetics of cell membranes. *Sci. Am.* **233** (6), 30 (Dec. 1975).

Malerhofer, G., Liposomes: preparation and application. *Am. Lab.* **17** (10), 129 (Oct. 1985).

Marchesi, V., Furthmayr, H., and Tomita, M., The red cell membrane. In *Annual Reveiw of Biochemistry* (E. E. Snell *et al.*, Eds.), Vol. 45, Annual Reviews, Palo Alto, Calif., 1976.

Marx, J. L., Newly made proteins zip through the cell. *Science* **207**, 164 (1980).

Pastan, I. H., and Willingham, M. C., Journey to the center of the cell: Role of the receptosome. *Science* **214**, 504 (1981).

Paul, S. M., and Skolnick, P., The red cell as a fluid droplet: Tank treadlike motion of the human erythrocyte membranes in shear flow. *Science* **202**, 894 (1978).

Raff, M. C., Cell surface immunology. *Sci. Am.* **234** (5), 30 (May 1976).

Ravazzola, M., and Orci, L., Intercellular junctions in the rat parathyroid gland: A freeze-fracture study. *Rev. Biol. Cellulaire* **28**, 137 (1977).

Robertson, J. D., The membrane of the living cell. *Sci. Am.* **206** (4), 65 (April, 1962).

Robertson, J. D., Membrane structure. *J. Cell Biol.* **91**, 189 (1981).

Rome, L. H., Curling receptors. *Trends in Biochem. Sci.* **10** (4), 23 (Apr. 1985).

Rothman, J. E., and Lenard, J., Membrane asymmetry. *Science* **195**, 743 (1977).

Rothman, J. E., and Lenard, J., Membrane traffic in animal cells. *Trends in Biochem. Sci.* **9** (4), 176 (1984).

Sabatini, D. D., Kreibich, G., Morimoto, T., and Adesnik, M., Mechanism for the incorporation of proteins in membranes and organelles. *J. Cell Biol.* **92**, 1 (1981).

Satir, B., The final steps in secretion. *Sci. Am.* **233** (4), 28 (Oct. 1975).

Schatz, G., Reconstituting membrane flow. *Trends in Biochem. Sci.* **10** (3), 95 (Mar. 1985).

Scott, D. E., Hall, M. N., and Silhavy, T. S., A mechanism of protein localization: The signal hypothesis and bacteria. *J. Cell Biol.* **86**, 701 (1980).

Sharon, N., Lectins. *Sci. Am.* **236** (6), 108 (June 1977).

Smith, R. L., and Oldfield, E., Dynamic structure of membranes by deuterium NMR. *Science* **225**, 280 (1984).

Staehelin, L. A., and Hull, B. E., Junctions between living cells. *Sci. Am.* **238** (5), 141 (May 1978).

Steinman, R. M., Mellman, I. S., Muller, W. A., and Cohn, Z. A., Endocytosis and the recycling of plasma membrane. *J. Cell Biol.* **96**, 1 (1983).

Stolinski, C., Freeze-fracture replication in biological research: Development, current practice and future prospects. *Micron* **8**, 87 (1977).

Storch, J., and Kleinfeld, A. M., The lipid structure of biological membranes. *Trends in Biochem Sci.* **10** (11), 418 (Nov. 1985).

Unwin, N., and Henderson, R., The structure of proteins in biological membranes. *Sci. Am.* **250** (2), 78 (Feb. 1984).

Unwin, P. N. T., and Zampighi, G. Z., Structure of the junction between communicating cells. *Nature* **283**, 545 (1980).

Walter, P., Gilmore, R., and Blobel, G., Protein translocation across the endoplasmic reticulum. *Cell* **38**, 5 (1984).

Weatherbee, J. A., Membranes and cell movement. In *International Review of Cytology* (A. L. Muggleton-Harris, Ed.), Suppl. 12, Academic, New York, (1981).

Wickner, W. T., and Lodish, H. F., Multiple mechanisms of protein insertion into and across membranes. *Science* **230**, 400 (1985).

Wilshut, J., and Hoekstra, D., Membrane fusion: From liposomes to biological membranes. *Trends in Biochem. Sci.* **9** (11), 479 (Nov. 1984).

Books, Monographs, and Symposia

Bessis, M., Weed, R. I., and Leblond, P. F. (Eds.), *Red Cell Shape*, Springer-Verlag, New York, 1973.

Bonting, S. L., and de Pont, J. J. (Eds.), *Membrane Transport*, Elsevier/North-Holland, Amsterdam, 1981.

Branton, D., and Park, R. B. (Eds.), *Papers on Biological Membrane Structure*, Little, Brown, Boston, 1968.

Evans, W. H., *Preparation and Characterization of Mammalian Plasma Membranes*, North-Holland, Amsterdam, 1979.

Gomperts, B. D., *The Plasma Membrane*, Academic, London, 1977.

Harrison, R., and Lunt, G. G., *Biological Membranes* (2nd ed.), Halsted Press, New York, 1980.

Kates, M., and Kuksis, A., *Membrane Fluidity*, Human Press, Clifton, N.J., 1980.

Lewis, J. A., Synthesis of proteins by ribosomes attached to membranes. In *Biochemistry of Cellular Regulation*, Vol. 1, *Gene Expression* (M. J. Clemens, Ed.), CRC Press, Boca Raton, Fla. 1980.

Malhotra, S. K., *The Plasma Membrane*, Wiley–Interscience, New York, 1983.

Pastan, I., and Willingham, M. C. (Eds.), *Endocytosis*. Plenum Press, New York, 1985.

Schwartz, L. M., and Azar, M. M. (Eds.), *Advanced Cell Biology*, Van Nostrand–Reinhold, New York, 1981.

Schweiger, H. G. (Ed.), *International Cell Biology 1980–1981*, Springer-Verlag, Berlin, 1981.

CHAPTER 16

THE MITOCHONDRION

Most of the energy-requiring, or *endergonic*, reactions carried out by cells either directly or indirectly consume *adenosine triphosphate* (ATP). This "energy-rich" substance is converted to adenosine diphosphate (ADP) and occasionally to adenosine monophosphate (AMP) (Fig. 16–1). Cells have evolved three major ways of producing this vital source of chemical energy: (1) ATP is produced in the cytosol during the chain of exergonic reactions called **glycolysis**, in which sugars are catabolized; (2) ATP may be produced within the **chloroplasts** of certain plant cells, utilizing the energy of sunlight; and (3) ATP may be produced within the **mitochondria** present in virtually all plant and animal cells by the **oxidation** of a variety of elementary substrates. It is for this reason that mitochondria are often referred to as the cell's "powerhouses."

ATP production in chloroplasts is considered in detail in Chapter 17, which also deals with photosynthesis, and further discussion of that subject is deferred until then. ATP synthesis during glycolysis occurs by **substrate-level phosphorylations** in which the phosphate is enzymatically transferred directly from the substrate to ADP to form ATP (see Chapters 9 and 10). Some ATP is generated in mitochondria by substrate-level phosphorylation but the greater amount is formed by special **electron transport system** (ETS) oxidation–reduction reactions. Although much of this chapter is devoted to a description of the manner in which mitochondria function in ATP production, it is to be noted that mitochondria also play several other important roles; for example, in eukaryotic cells the mitochondria are also responsible for the oxidation of fatty acids and other lipids and are one of the sites for lengthening the carbon chain of fatty acids. Some subunits of cytochromes are also synthesized in mitochondria, and final assembly occurs there. Heme is synthesized in the mitochondria of animal cells.

In recent years, it has been clearly established that mitochondria contain their own genetic apparatus, as well as the ribosomes and enzymes for the synthesis of an array of enzymatic and structural mitochondrial proteins. Mitochondria are capable of semiautonomous proliferation within the cell. These subjects are dealt with here and also in Chapters 20 and 21.

DISCOVERY OF MITOCHONDRIA

Mitochondria were first observed and isolated from cells about 130 years ago when Köllicker mechanically

377

FIGURE 16–1 Hydrolysis of ATP to form ADP and AMP.

teased these organelles from insect striated muscle tissue and studied their osmotic behavior in various salt solutions. Kölicker, whose work extended over several decades beginning around 1850, concluded that these "granules" were independent structures not directly connected to the interior structure of the cell. In 1890, Altmann identified stains specific for these granules, which he named "bioblasts"; this term was superseded when Benda introduced the expression "mitochondrion" (Greek, *mito* = "thread" + *chondrion* = "granule") because of the threadlike appearance of these granules when examined with the light microscope. In 1900, Michaelis introduced the use of the supravital dye Janus green B to specifically stain mitochondria to the exclusion of other cellular components and showed that oxidative reactions in the mitochondria caused color changes in the dye. Janus green B is still frequently employed as a histochemical marker of mitochondria.

In 1910, Warburg showed that the "large granule" fraction isolated by low-speed centrifugation of tissues disrupted by grinding contained enzymes catalyzing oxidative cellular reactions, and Kingsbury in 1912 suggested that the mitochondria were the spe-

cific loci of the oxidation. Warburg's findings were confirmed in the 1930s by Claude, Bensley, and Hoerr, who employed more sophisticated methods to isolate a mitochondrial preparation essentially free of contaminating cell structures. They disrupted liver tissue using procedures almost identical to those currently used to prepare a conventional "homogenate" (see Chapter 12) and isolated the mitochondria by repeated differential centrifugation; the final mitochondrial preparation was then examined biochemically At about the same time, Sir Hans Krebs worked out the various reactions of the tricarboxylic acid (TCA) cycle (or Krebs cycle; see below) and by 1950, Lehninger, Green, Kennedy, Hogeboom, and others had clearly shown that these reactions, as well as those of fatty acid oxidation and oxidative phosphorylation (i.e., ATP generation), were properties of mitochondria.

THE ORIGIN AND EVOLUTION OF MITOCHONDRIA

About 30 years ago, it was generally believed that mitochondria arose *de novo* from other cellular structures such as microbodies, the outer nuclear membrane, or the endoplasmic reticulum. As recently as 15 years ago, the prospect of *de novo* formation of mitochondria was still considered. However, the wealth of modern biochemical, physiological, and microscopic evidence indicates that new mitochondria arise by the growth and division of mitochondria that are already present in the cell. Using time-lapse cinematography, mitochondria are observed to divide by a fissionlike process. The mitochondrion appears to stretch out in a tubelike manner, constrict near the center assuming a dumbbell shape, and then pinch off to form two halves that once again round up (Fig. 16–2). Direct observations of mitochondrial division are also supported by biochemical studies using radioactive tracers and autoradiography (see Chapter 14). D. J. L. Luck labeled the mitochondria of *Neurospora* cells and followed the change in mitochondrial radioactivity through several generations of cells. With each cell division the amount of tracer associated with the mitochondria of the daughter cells was reduced to about one-half its previous level.

There does not appear to be a specific method for apportioning the mitochondria between the two daughter cells during cell division. In most cells the cytoplasm becomes very turbulent during cell division, and therefore a random distribution of the organelles is enhanced. In some instances, the mitochondria appear to adhere to the spindle fibers and may be drawn into the two daughter cells. During cell division in certain yeasts, the daughter cell forms as a small *bud* emerging from the parental cell. Rather than an equal division of the cytoplasm between the daughter cells, little parent cytoplasm enters the bud along with the divided nucleus. Occasionally these buds fail to obtain any mitochondria and the growth of these cells is much slower than that of normal cells, because the cells rely for the most part on glycolysis in the cytosol for ATP production. Fletcher and Sanadi in 1961 and Wilson and Dove in 1965 studied the turnover rate for mitochondria by using radioactively la-

beled compounds. The former workers labeled mitochondrial proteins and lipids and followed the disappearance of the label from the cells with time. The latter workers utilized a technique employing tritiated water, (i.e., ^3H_2O). Both studies indicated that there is a complete intracellular turnover of mitochondria every 20 days.

L. Margulis has made some interesting speculations concerning the evolutionary origin of mitochondria in eukaryotic cells. Mitochondria contain DNA and ribosomes of their own, and these have a number of functions that are independent of the nuclear and cytoplasmic ribosomal processes of the cell. (Protein synthesis and ribosome formation in mitochondria are discussed in Chapter 22.) The discovery of the presence of DNA and ribosomes in mitochondria served as the initial inspiration for the idea that mitochondria may at one time have been independent organisms, that is, mitochondria may have evolved from prokaryotelike cells. This notion is supported by the fact that mitochondrial DNA, like prokaryotic DNA, is a single, circular molecule (i.e., no free ends) and mitochondrial ribosomes are made up of subunits whose sedimentation coefficients and protein and RNA composition are more like those of prokaryotic ribosomes than eukaryotic ribosomes. It has been proposed that one type of aerobic prokaryote invaded or infected another and that through time the infective prokaryote evolved morphologically and physiologically into the mitochondrion. This fascinating theory of the evolution of mitochondria must be evaluated in the light of evidence that indicates that most mitochondrial proteins are encoded by nuclear DNA and are synthesized by ribosomes in the cytosol or on the endoplasmic reticulum. It would therefore appear that there is a greater transfer of information from nuclear DNA to the mitochondrion than arises from DNA found within the mitochondrion itself.

The genetic code of DNA isolated from the mitochondria of yeast, *Drosophila*, and human tissues has been shown to be slightly different from that of nu-

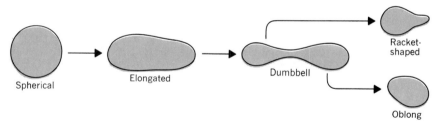

Spherical   Elongated   Dumbbell   Racket-shaped   Oblong

FIGURE 16–2 Division of mitochondria by a fissionlike process produces organelles of diverse shapes.

clear DNA. The noncoding sequences also differ. For example, DNA from human mitochondria contains 16,569 nucleotide pairs that encode 2 rRNA molecules, 22 or 23 tRNAs, and 10 to 13 proteins. Very few noncoding nucleotides are present. Eukaryotic nuclear DNA usually contains large noncoding nucleotide sequences and encodes a greater number of tRNAs (at least 32) and rRNAs (4). Frederick Sanger, who received a Nobel Prize in 1980 for his analyses of DNA nucleotide sequences, contends that mammalian mitochondrial genetic systems cannot be classified as either prokaryotic or eukaryotic.

STRUCTURE OF THE MITOCHONDRION

The size, shape, and structural organization of mitochondria, as well as the number of these organelles per cell and their intracellular location, vary considerably depending on the organism, tissue, and physiological state of the cell examined.

Some cells, usually unicellular organisms, contain a single mitochondrion. Figure 16–3 contains a photomicrograph of the single mitochondrion in the motile swarm spore of *Blastocladiella emersonia*, a fungus, and a model of the single mitochondrion of *Chlorella fusca*, an alga. At the other extreme are cells such as *Chaos chaos*, an amoeba, which contains several hundred thousand mitochondria. Cells of higher animals also contain various numbers of mitochondria. Sperm cells have fewer than 100 mitochondria. Kidney cells generally contain less that 1000 and liver cells can contain several thousand. As described in Chapter 1, prokaryotic cells such as bacteria and blue-green algae do not contain mitochondria. The functions associated with mitochondria are carried out in the cytosol or are associated with the plasma membrane.

The distribution of the mitochondria in the cell can change with time. In *Blastocladiella* (Fig. 16–3), the single mitochondrion is at the base of the flagellum. Mitochondria also appear to be concentrated in metabolically active areas of cells. In epithelial cells lining the lumen of the small intestine, the mitochondria occur in greater numbers near the cell surface that is adjacent to the lumen (where active absorption of digestive products is occurring). In general, when mitochondria are present in greater numbers in one part of the cell than another, it is usually near a site where significant ATP utilization is occurring. For example, in muscle tissue the mitochondria are aligned in rows

(a)

parallel to the contractile fibrils (Fig. 16–4). In many plant cells, cyclosis, the active streaming of the cytoplasm about the cells, tends to distribute the mitochondria uniformly.

The number and distribution of mitochondria in a cell are closely related to the activity of the cell and its organelles. Cells that are actively growing, producing especially large amounts of some product such as digestive enzymes, actively transporting materials into the cell, or undergoing movement may display increased numbers of mitochondria during periods of activity and reduced numbers during periods of quiescence. In yeast cells grown anaerobically, successive generations contain fewer and fewer mitochondria. However, cells that have been cultured in the absence of oxygen rapidly produce greater numbers of mitochondria if oxygen and appropriate nutrients are added to the culture. An increased rate of cell growth and division also occurs since greater numbers of mitochondria produce more ATP in order to facilitate absorption of nutrients and the synthesis of cell components.

FIGURE 16–3 (*a*) The single mitochondrion present in the fungus *Blastocladiella emersonia*: the mitochondrion is always located at the base of the cell's flagellum. (Courtesy of Dr. W. E. Barstow.) (*b*, *c*, and *d*) The mitochondrion of the alga *Chlorella fusca*. In *b*, the mitochondrion (*m*), cell nucleus (*n*), vacuoles (*v*), Golgi body or dictyosome (*d*), and microbody (*mb*) are depicted as three-dimensional models produced by reconstruction of serial sections through the cell. In *c*, the mitochondrion is seen alone. The torturous character of the mitochondrion is also apparent in the photomicrograph (*d*) of the cell: arrows point to DNA-containing areas of the cell's chloroplast. (Courtesy of Dr. B. E. S. Gunning: from Atkinson, A., John, P. C. L., and Gunning, B. E. S., *Protoplasma* **81**, 77 (1974), Springer-Verlag, Berlin.)

Fibrils

Mitochondria

FIGURE 16–4 The distribution of mitochondria within cells often reflects localized cellular activity. In this electron photomicrograph of muscle tissue, the mitochondria are arranged in rows between clusters of contractile fibrils. (Photomicrograph courtesy of R. Chao.)

The size and shape of mitochondria, like the number in a cell, vary from one tissue to another and with the physiological state of the cells. Most mitochondria are ovoid bodies having a diameter between 0.5 and 1.0 μm and a length up to 7 μm. Usually, the lower the number of mitochondria per cell, the larger are the individual organelles. In many electron photomicrographs, mitochondria appear to be dumbbell-shaped or racket-shaped. These odd shapes may be a reflection of the fission process by which mitochondria are believed to proliferate. Dumbbell shapes are seen just prior to separation and the "handles" (or "tails") of racket-shaped mitochondria may be the bridges that connect the separating mitochondrial halves (Fig. 16–5).

From the point of view of fine structure, mitochondria are especially interesting and very intricate organelles. Because mitochondria are so small, light microscopy reveals little about their structure. The contemporary model of mitochondrial architecture is therefore based on decades of study using the transmission electron microscope. Figure 16–6 illustrates the typical organization revealed in thin sections of these organelles. Recently, some rather spectacular and highly informative photomicrographs of mitochondrial structure have been obtained by scanning electron microscopy of organelles that have been cracked open (Fig. 16–7). In the past, SEM has not offered the degree of resolution needed to reveal organ-

elle substructure. However, using a special fracturing technique and a novel high-resolution SEM, K. Tanaka, Y. Masunaga, and T. Naguro have obtained detailed photomicrographs of mitochondria and a number of other organelles. After tissue fixation in OsO_4, the samples are frozen, cracked open, and then treated with dimethylsulfoxide (DMSO). Following a second treatment with osmium, followed by dehydration with ethanol and critical point drying, the fine structure is revealed in bold relief. The three-dimensional effect is striking and amazingly consistent with three-dimensional models formulated on the basis of the earlier TEM studies (see Fig. 1–23). Because mitochondria from diverse sources exhibit certain features in common, a "generalized" organelle may be described (Fig. 16–8). The mitochondrion is enclosed by two distinct membranes called the **outer** and **inner membranes** (Figs. 16–6 and 16–7). The inner membrane separates the organelle's volume into two phases: the **matrix**, which is a gel-like fluid enclosed by the inner membrane, and the fluid-filled **intermembrane space** between the inner and outer membranes. The matrix and intermembrane spaces as well as the outer and inner membranes themselves contain a variety of enzymes (Table 16–1). The matrix contains a number of the enzymes of the Krebs cycle (tricarboxylic acid cycle, or TCA cycle) as well as salts and water. Suspended in the matrix are strands of circular DNA (Fig. 16–9) and ribosomes. A number of

Spherical

Dumbbell

Elongate

FIGURE 16–5 Various cross-sectional shapes of mitochondria that are seen in thin sections. (Courtesy of R. Chao.)

FIGURE 16–6 Thin section through a mitochondrion, clearly revealing its outer and inner membranes. (Courtesy of Dr. K. R. Porter.)

Rough endoplasmic reticulum

Matrix

Outer membrane

Inner membrane

Cristae

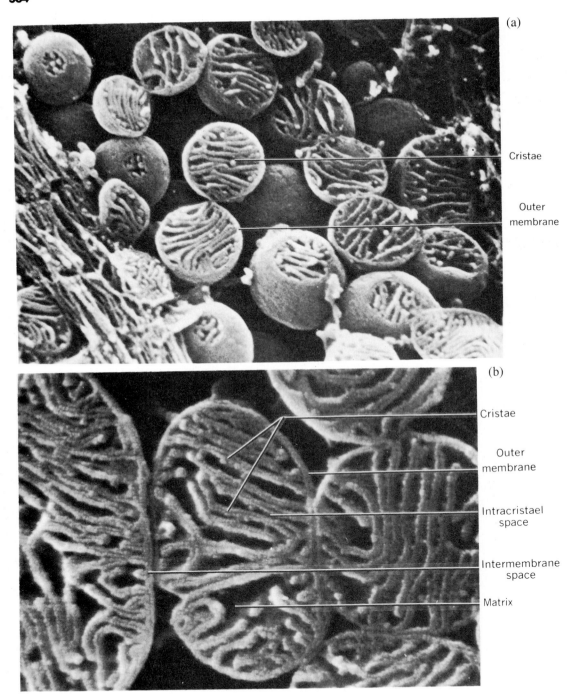

FIGURE 16–7 Scanning electron photomicrographs of mitochondria of rabbit cardiac tissue. (*a*) The outer membranes of the mitochondria have been cracked away by a freeze-fracture technique revealing the cristael membranes within the organelles. (*b*) In this higher magnification view, the intermembrane and intracristael space may be discerned; the small bumps along the matrical surface of the cristae are believed to be the *inner membrane spheres* (see text). (Photomicrographs courtesy of Dr. Y. Masunaga and Dr. K. Tanaka, *J. Yonago Med. Assn.* **30**, 519 (1979).)

FIGURE 16–8 The generalized mitochondrion.

FIGURE 16–9 DNA fibrils in mitochondria. *M*, mitochondrion; *N*, cell nucleus. (Photomicrograph courtesy of Dr. E. G. Pollock.)

TABLE 16-1 LOCATION OF SOME MITOCHONDRIAL ENZYMES

Outer membrane:
 Monoamine oxidase
 Fatty acid thiokinases
 Kynurenine hydroxylase
 Rotenone-insensitive chytochrome *c* reductase
Space between the membranes:
 Adenylate kinase
 Nucleoside diphosphokinase
Inner membrane:
 Respiratory chain enzymes
 ATP-synthesizing enzymes
 α-Keto acid dehydrogenases
 Succinate dehydrogenase
 D-β-Hydroxybutyrate dehydrogenase
 Carnitine fatty acyl transferase
Matrix:
 Pyruvate dehydrogenase complex
 Citrate synthase
 Isocitrate dehydrogenase
 Fumarase
 Malate dehydrogenase
 Aconitase
 Glutamate dehydrogenase
 Fatty acid oxidation enzymes

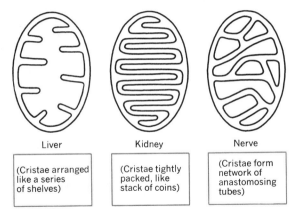

Liver
(Cristae arranged like a series of shelves)

Kidney
(Cristae tightly packed, like stack of coins)

Nerve
(Cristae form network of anastomosing tubes)

FIGURE 16-10. Various arrangements of cristae in the mitochondrion.

other inclusions have been observed in the mitochondrial matrix of diverse kinds of cells. These include filaments and tubules, what appear to be crystalline protein inclusions, and a number of small granules. The intermembrane space contains some enzymes (Table 16–1), but generally is devoid of particulate inclusions.

The Mitochondrial Membranes

The inner and outer membranes are distinctly different in structure and function. Although accurate measurements of the thicknesses of the membranes by electron microscopy are difficult because various fixatives promote different degrees of swelling, the inner membrane appears to be somewhat thicker (6.0–8.0 nm) than the outer membrane (about 6.0 nm). The inner membrane has a greater surface area because it possesses folds that extend into the matrix. These projections, called **cristae**, vary in number and shape. With distinct exceptions, the cristae of mitochondria in higher animal cells may almost bridge the matrix. Usually the cristae lie parallel to one another across the long axis of the mitochondrion but in some cells they run longitudinally or form a branching network (Fig. 16–10). In protozoa and many plants, the cristae

form a set of tubes that project into the matrix from all sides, sometimes twisting in different directions. The number of cristae may increase or decrease depending on the level of aerobic activity. Active aerobic tissue cells producing large amounts of ATP generally contain mitochondria with extensive cristae.

The organization of protein and lipid in the outer and inner mitochondrial membranes has been the subject of intense study for many years. Chemically, the two membranes are qualitatively and quantitatively distinct, differing from one another and also from other intracellular membranes. The inner membrane is much richer in protein than the outer membrane and the proteins themselves are more deeply embedded in the membrane. The outer membrane contains three or four times more phospholipid than the inner membrane and contains most of the membrane cholesterol. In contrast, the inner membrane is rich in cardiolipin. Differences in the two membranes are also apparent in freeze-fracture views of the *E* and *P* faces (Fig. 16–11). The *P* fracture face (*PF*) of the outer membrane contains more than three times as many particles as the *E* fracture face (*EF*). In contrast, the *EF* of the inner membrane contains almost as many particles as the *PF* of the outer membrane, whereas the inner membrane's *PF* contains twice as many particles as its *EF*. This is not surprising in view of the manifold enzymatic activities of the matrical surface of the inner membrane (see below). The organization of protein and lipid in the inner membrane is depicted in Figure 16–12). Many of the proteins of the mitochondrial membranes are enzymes that catalyze the reaction sequences discussed later in the chapter.

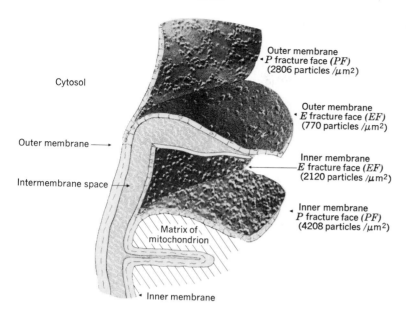

Cytosol

Outer membrane →

Intermembrane space →

Matrix of mitochondrion

← Inner membrane

Outer membrane
P fracture face (*PF*)
(2806 particles /μm²)

Outer membrane
E fracture face (*EF*)
(770 particles /μm²)

Inner membrane
E fracture face (*EF*)
(2120 particles /μm²)

Inner membrane
P fracture face (*PF*)
(4208 particles /μm²)

FIGURE 16–11   Fracture faces of and particle distributions in the outer and inner membranes of the mitochondrion. (Courtesy of Dr. L. Packer, copyright © New York Academy of Science, *Ann. N.Y. Acad. Sci.* **227**, 167 (1974).)

FIGURE 16–12   Sjöstrand and Barajas model of the structure of mitochondrial membranes. In this model, the enzymes and other proteins of the membrane are depicted as globular bodies of various shapes. Lipid molecules in areas consisting of a lipid bilayer are shown as "T"-shaped structures. (Courtesy of Dr. F. Sjöstrand, copyright © Academic Press. *J. Ultr. Research* **32**, 298 (1970).)

Much of the protein of the inner membrane can be stripped away using very dilute solutions of acetic acid, suggesting that a large proportion of the inner membrane proteins are extrinsic. Attached to the matrix side of the inner membrane are many spherical particles 8.0–9.0 nm in diameter. These *inner membrane spheres*, first described by Fernandez-Moran in 1962, are seen in the scanning electron photomicrographs of Figure 16–7 and the negatively stained preparation shown in Figure 16–13. They appear to be spaced regularly along the membrane and are borne on stalks. These inner membrane spheres have been identified as the primary sites of oxidative phosphorylation (discussed later).

Each crista is formed from two inwardly projecting leaflets of the inner mitochondrial membrane. Fine structural studies by F. Sjöstrand using a freeze-fracture technique suggest that little or no space exists between these two layers (Fig. 16–14). Instead these appear to form a single structure containing large globular proteins about 15 nm in diameter and small quantities of lipid at the matrical surfaces. Sjöstrand also contends that the inwardly projecting cris-

tael membrane leaflets are anchored to the inner membrane by one or more *stalks* or **peduncles** rather than arising as shelflike folds (Fig. 16–15). The scanning electron photomicrographs of Figure 16–7 reveal little space between the two cristael membranes and also suggest pointlike attachments of the cristae to the inner membrane.

As noted earlier, mitochondria contain some DNA, but the amount is sufficient to encode only a small number of mitochondrial proteins. Nearly all mitochondrial proteins are encoded by nuclear DNA. These proteins are synthesized by cytoplasmic ribosomes and then *imported* into the four mitochondrial compartments (i.e., outer membrane, inner membrane, intermembrane space, and matrix). Like secretory and plasma membrane proteins, proteins destined for the inner mitochondrial membrane or the matrix are synthesized with a transient *leader sequence* (see Chapter 15); their temporary anchorage to the outer membrane and their subsequent passage through it are assisted by a small protein present in the cytosol. After import, the leader sequence is enzymatically removed by a protease present in the mito-

FIGURE 16–13 Electron photomicrographs of negatively stained fragments of mitochondria showing the inner membrane and the *inner membrane spheres*. The spheres are attached by stalks to the matrix surface of the membrane. (Courtesy of Dr. B. Chance.)

(a)

FIGURE 16–14 (*a*) Cristael membranes of a mitochondrion revealed by freeze-fracturing and etching technique; the small particles in the membranes are believed to be proteins (e.g., the enzymes associated with the membrane). (*b*) Composite diagram and photomicrograph showing the angle of fracture through the cristael membranes. According to F. Sjöstrand, the intracristael space is reduced and bridged at various points by large globular proteins that are probably enzyme complexes. (Photomicrographs courtesy of Dr. F. Sjöstrand, *J. Ultr. Research* **69**, 378 (1979),)

Crista

(b)

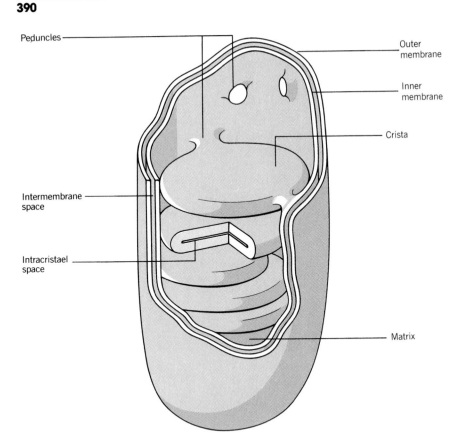

Peduncles

Outer membrane

Inner membrane

Crista

Intermembrane space

Intracristael space

Matrix

FIGURE 16–15 Model of the mitochondrion showing that the cristae are attached to the inner membrane through one or more stalks or *peduncles*.

chondrial matrix. Proteins destined for the outer mitochondrial membrane do not contain a leader sequence. Targeting and anchorage in the outer membrane is specified by the amino acid sequence in the proteins' N-terminal region. Import of proteins of the intermembrane space (e.g., cytochrome *c*) does not require a leader sequence but may involve a preliminary interaction with specific receptor proteins in the outer membrane.

In addition to their chemical and structural differences, the outer and inner membranes differ significantly in permeability. The outer membrane is permeable to a wide variety of substances of a molecular weight up to about 5000. When the fluid from the intermembrane space is isolated, it reflects the water-soluble low-molecular-weight components of the cytosol. In contrast, the inner membrane has a limited permeability especially to substances with molecular weights above 100 to 150.

The difference in permeability of the two membranes can be used to advantage to separate the outer membrane from the inner. Although several variations of the method exist, the basic procedure is to disrupt the cells and isolate intact mitochondria by differential or density gradient centrifugation. The mitochondria are then placed in a hypotonic solution, causing them to swell. Usually, a dilute phosphate buffer solution is used for this purpose. As the mitochondria swell, their outer membranes rupture and fragment; the inner membranes also swell, causing a loss of organized cristael structure, but the inner membrane does not break. The mitochondria are then transferred to a hypertonic solution, causing the matrix to shrink, pulling the inner membrane away from the outer membrane. The hypertonic solution is frequently a sucrose solution containing ATP and Mg^{2+}. Because the inner membrane may be attached to the outer membrane at several points (probably by proteinaceous connecting strands), some workers use EDTA or sonication to aid the separation of the membranes. Resuspension in an isotonic solution allows the matrix to reassume its normal size and typical morphology. Isolation of the separated membrane fractions can be achieved by centrifugation. The results of a typical isolation are shown in Figure 16–16.

(a)

(b)

(c)

FIGURE 16–16 Mitochondrial preparations isolated from rat liver tissue. (*a*) Freshly isolated whole mitochondria. (*b*) Inner membrane and matrix isolated after digitonin treatment and differential centrifugation. (*c*) Smooth vesicular outer membrane preparation isolated after digitonin treatment and differential centrifugation. (Courtesy of Dr. C. A. Schnaitman, copyright © The Rockefeller Press, *J. Cell. Biol.* **38**. 170 (1968).)

Conformational States of Mitochondria

Changes in the level or type of physiological activity occurring in mitochondria can be related to differences in the morphological appearance of the organelles. Accordingly, C. R. Hackenbrock has shown that the **orthodox conformational state** (Fig. 16–17) of mitochondria (the appearance most commonly seen in photomicrographs) is typical of inactive organelles. The **condensed conformational state** (Fig. 16–17) corresponds to periods in which phosphorylation of ADP to form ATP and electron transport occur at high rates. In the condensed conformational state, the cristae are more randomly distributed and the intermembrane space is greatly enlarged. The transition from the orthodox to the condensed conformational state is triggered by the binding of ADP to *ADP–ATP translocase* molecules in the inner membrane.

Over the years, numerous investigators have studied the relationship between the structural organization of the mitochondria and its specialized metabolic functions. As seen in Table 16–1, the location of many enzymes has been determined, and it is generally possible to assign specific functions to the outer membrane, intermembrane space, the inner membrane, and the matrix. In some cases, the enzyme activity can be assigned to a specific surface of the membrane (e.g., the ATP-synthesizing enzymes and *succinate dehydrogenase* are in the inner membrane).

Among the most thoroughly studied processes unique to mitochondria are **substrate oxidation, res-**

Orthodox conformation

Condensed conformation

0.5 μm

0.5 μm

Orthodox

Condensed

Swollen

Contracted

FIGURE 16–17 *Orthodox* and *condensed* (and swollen/contracted) conformational states of mitochondria. (Courtesy of Dr. C. R. Hackenbrock: copyright © The Rockefeller Press, *J. Cell. Biol.* **37**, 345 (1968).)

FIGURE 16–18 Relationship between the tricarboxylic acid (Krebs) cycle, respiratory chain oxidation–reductions, and oxidative phosphorylation reactions. F_p = flavoprotein: Q = ubiquinone: b, c, c_1, a, and a_3 = cytochromes.

piratory chain oxidation–reductions, and **oxidative phosphorylation** (Fig. 16–18). Products of metabolic reactions in the cytosol (such as pyruvate formation during glycolysis) enter the mitochondrion to be oxidized by the Krebs or tricarboxylic acid cycle enzymes (Fig. 16–19). The enzymes that catalyze these reactions (except *succinic dehydrogenase*; see below) are believed to be localized in the matrix or are easily removed from the surface of the inner membrane that faces the matrix.

As a result of the Krebs cycle oxidations, CO_2 and water are formed as end products and a number of special oxidation–reduction compounds are reduced. These compounds (e.g., NADH; see Chapters 3 and 10) subsequently participate in the initial step of a sequence of oxidation and reduction reactions called the **respiratory** or **electron transport chain** that is specifically associated with the inner membrane of the mitochondrion. The result of this chain of reactions is the reduction of O_2 to form H_2O. A third major process called **oxidative phosphorylation** is coupled to the respiratory chain and brings about the conversion of ADP to ATP. Oxidative phosphorylation is intimately associated with the inner membrane spheres.

In effect, pyruvate and other small molecules produced during metabolism in the cytosol cross the permeable outer mitochondrial membrane and the intermembrane space. It is on entering the inner membrane that the three major reaction sequences (Krebs or TCA cycle, respiratory chain oxidation–reductions, and oxidative phosphorylation) begin. Pyridine nucleotides that are reduced during reactions in the cytosol (e.g., NADH produced in the glycolytic pathway and NADPH produced in the pentose phosphate pathway; see Chapter 10) may also permeate the outer mitochondrial membrane, ultimately transferring their reductive capacity to the respiratory chain (see below).

TRICARBOXYLIC ACID (KREBS) CYCLE

The **tricarboxylic acid cycle** is also called the **Krebs cycle** because the major steps of the cycle were worked out in 1937 by Sir Hans Krebs. At that time, radioactively labeled compounds were not available for biological studies and the cellular site of the reactions was not known with certainty. The experiments performed by Krebs and the logic and reasoning with which his findings were interpreted in order to formulate the metabolic pathway are recognized as milestones in our progressive unraveling of cellular metabolism. In 1948 E. P. Kennedy and A. L. Lehninger showed that mitochondria isolated from liver tissue homogenates by differential centrifugation contained the enzyme activities of the Krebs cycle reactions.

GLYCOLYSIS

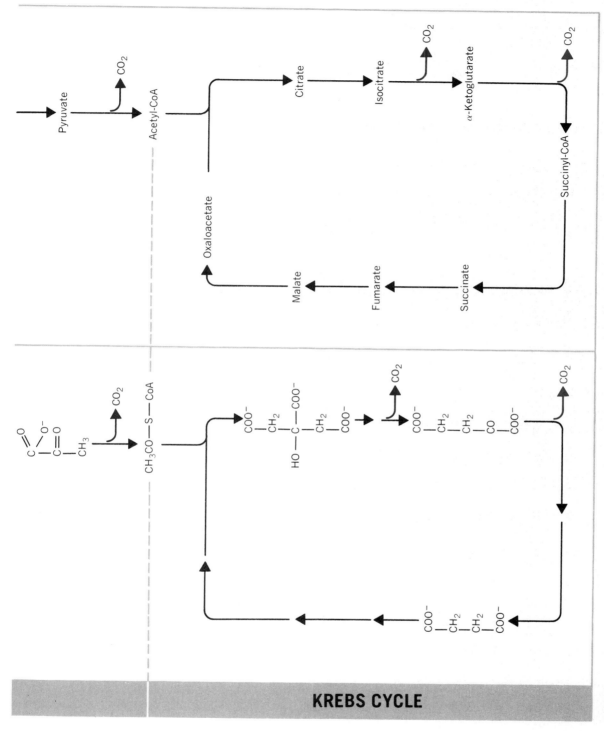

FIGURE 16–19 Summary of the reactions of glycolysis and the Krebs cycle.

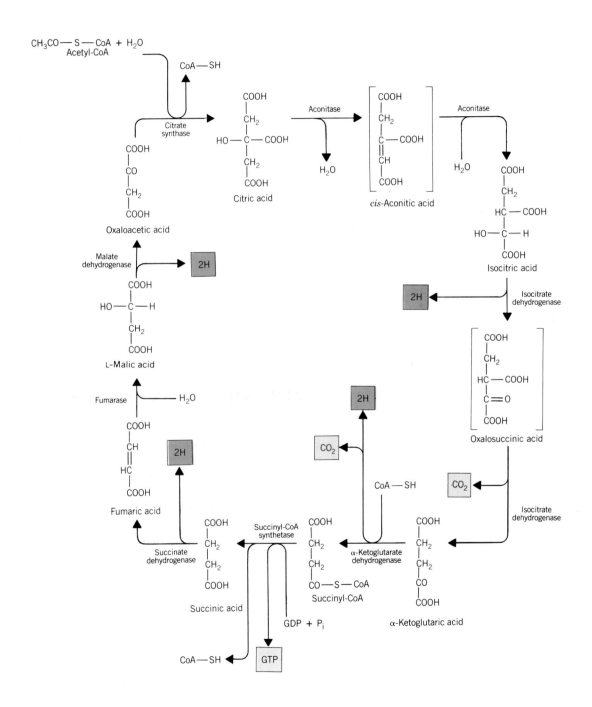

FIGURE 16–20 The Krebs (tricarboxylic acid) cycle.

Krebs received the Nobel Prize in 1953 for his elucidation of this most important metabolic pathway.

The Krebs cycle reactions are utilized by the cell to further metabolize a number of products of other reactions in the cytosol. These can be products as diverse as amino acids, fatty acids, precursors of nucleic acids, and pyruvic acid. However, the greatest metabolic merit of the cycle is its oxidation of pyruvate—the product of carbohydrate metabolism in the cytosol. The oxidation of this compound in the TCA cycle provides the reducing power to make significant amounts of ATP. The reactions are shown in Figure 16–20 and are described below.

Oxidation of Pyruvate

The oxidation of pyruvate is brought about by a set of reactions catalyzed by the **pyruvate dehydrogenase complex.** This complex contains three enzymes (E_1, *pyruvate dehydrogenase*; E_2, *dihydrolipoyl transacetylase*; and E_3, *dihydrolipoyl dehydrogenase*) and five coenzymes and is located in the matrix of the mitochondrion. The first step is the decarboxylation of the pyruvate to yield CO_2 and an α-hydroxyethyl unit attached through *thiamine pyrophosphate* (TPP, a coenzyme) to enzyme E_1, that is,

$$
\begin{array}{c}
E_1-TPP \\
+ \\
CH_3COCOOH \\
\searrow CO_2 \\
\downarrow \\
E_1-TPP-CHOH-CH_3
\end{array} \quad (16\text{–}1)
$$

The hydroxyethyl group is then dehydrogenated (oxidized) to form an acetyl group, which is transferred to one of the sulfur atoms of *lipoic acid*, a coenzyme of E_2.

$$
\begin{array}{c}
E_1-TPP-CHOH-CH_3 \\
+ \\
E_2 \overset{\displaystyle \frown}{S-S} \\
\downarrow \\
E_1-TPP \\
+ \\
E_2 \overset{\displaystyle \frown}{S \quad SH} \\
| \\
C-CH_3 \\
\| \\
O
\end{array} \quad (16\text{–}2)
$$

The acetyl group is then transferred to coenzyme A (Fig. 3–11) forming *acetyl-CoA*, which separates from the enzyme complex. One hydrogen from the lipoyl group and one from coenzyme A are transferred to flavin adenine dinucleotide (FAD) (Fig. 3–11), a coenzyme of E_3.

$$
\begin{array}{c}
E_2 \overset{\displaystyle \frown}{S \quad SH} \\
| \\
C-CH_3 \\
\| \\
O \\
+ \\
CoA-SH \\
+ \\
E_3-FAD \\
\downarrow \\
E_2 \overset{\displaystyle \frown}{S-S} \\
+ \\
E_3-FADH_2 \\
+ \\
CH_3CO-S-CoA
\end{array} \quad (16\text{–}3)
$$

The resulting $FADH_2$ is then reoxidized by the transfer of hydrogens to NAD^+.

$$FADH_2 + NAD^+ \rightarrow FAD + NADH + H^+ \quad (16\text{–}4)$$

The NADH and H^+ are then released from the enzyme complex. Thus, the overall reaction is

$$
\begin{array}{c}
\text{pyruvate} + NAD^+ + CoA \\
\downarrow \\
\text{acetyl-CoA} + NADH + H^+ + CO_2
\end{array} \quad (16\text{–}5)
$$

The $\Delta G^{0'}$ (see Chapter 9) is -33.5 kJ/mole (-8.0 kcal/mole), indicating that the reaction is highly exergonic, with the equilibrium lying well to the right. In most animal tissues, the reaction is considered irreversible.

The Krebs Cycle Reactions

The first step of the Krebs cycle is a **condensation** reaction between acetyl-CoA and oxaloacetate, a product of the cycle itself. The reaction is catalyzed by *citrate synthetase*, with citrate produced and CoA released for reutilization in another round of the previous reaction.

$$\begin{array}{c} \text{acetyl-CoA} + \text{oxaloacetate} + H_2O \\ \downarrow \\ \text{citrate} + \text{CoA} \end{array} \qquad (16\text{--}6)$$

The $\Delta G^{0\prime}$ is -32.2 kJ/mole (-7.7 kcal/mole). This reaction is the "pacemaker" or primary rate-limiting reaction of the cycle. The rate is controlled by the availability of acetyl-CoA and oxaloacetate. Succinyl-CoA, the product of the later step in the cycle, is a competitive inhibitor of this reaction, because it competes with acetyl-CoA for the active site on the enzyme.

The enzymatic conversion of citrate to isocitrate is a two-step process in which the intermediate, *cis-aconitate*, remains attached to the enzyme *aconitase* and therefore is frequently not shown in the Krebs cycle. The equilibrium of this reaction is toward citrate (i.e., $\Delta G^{0\prime}$ is 6.7 kJ/mole or 1.59 kcal/mole), but the isocitrate is rapidly oxidized in the next step, thus shifting the direction of the reaction by removal of the product.

$$\text{citrate} \xrightarrow[\;]{H_2O} \left[\begin{array}{c} \textit{cis}\text{-aconitic} \\ \text{acid} \end{array} \right] \xrightarrow[\;]{H_2O} \text{isocitrate} \qquad (16\text{--}7)$$

The oxidation of isocitrate to α-ketoglutarate is also a two-step process, with the intermediate remaining attached to the enzyme *isocitrate dehydrogenase*. The first step is an oxidation, two hydrogens being transferred to NAD^+. Actually, there are two isocitrate dehydrogenases in the mitochondrial matrix. One is linked to NAD^+ and the other to $NADP^+$. An $NADP^+$-linked isocitrate dehydrogenase is also found in the cytosol. However, it appears that the *NAD^+-isocitrate dehydrogenase* is the most active form in Krebs cycle reactions. This is an allosteric enzyme and is specifically stimulated by ADP. When large amounts of ATP are consumed in the cell and corresponding quantities of ADP produced, the rising ADP level acts as a positive effector for the allosteric enzyme. Alternatively, when the level of ATP rises, the accompanying fall in the level of ADP diminishes the positive effect on the allosteric enzyme. The overall effect of ADP-dependent *NAD^+-isocitrate dehydrogenase* in controlling the level of Krebs cycle activity is considered secondary, however, to the citrate synthetase step.

The second step of the reaction catalyzed by *isocitrate dehydrogenase* is the decarboxylation of the β-carboxyl group. The $\Delta G^{0\prime}$ for the entire reaction is -20.9 kJ/mole (-5.0 kcal/mole).

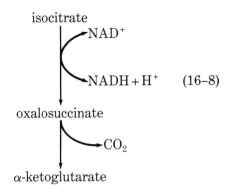

$$(16\text{--}8)$$

The oxidation of α-ketoglutarate to succinyl-CoA involves an enzyme complex called the α-**ketoglutarate dehydrogenase complex**. In the sequence of reactions catalyzed by this complex, CO_2 is removed by first complexing with thiamine pyrophosphate. During addition of CoA, two hydrogen atoms removed from lipoic acid and coenzyme A reduce NAD^+ to $NADH + H^+$. The $\Delta G^{0\prime}$ is -33.5 kJ/mole (-8.0 kcal/mole).

$$\begin{array}{c} \alpha\text{-ketoglutarate} + NAD^+ + \text{CoA} \\ \downarrow \\ \text{succinyl-CoA} + CO_2 + NADH + H^+ \end{array} \qquad (16\text{--}9)$$

The removal of CoA from succinyl-CoA is coupled to substrate-level phosphorylation of GDP and is catalyzed by the enzyme *succinyl-CoA synthetase*.

$$\begin{array}{c} \text{succinyl-CoA} + P_i + \text{GDP} \\ \downarrow \\ \text{succinate} + \text{GTP} + \text{CoA} \end{array} \qquad (16\text{--}10)$$

The $\Delta G^{0\prime}$ is -2.9 kJ/mole (-0.7 kcal/mole). Phosphate is attached to the enzyme–succinyl-CoA complex before it is transferred to GDP. In *Escherichia coli*, the phosphate is transferred to ADP; in animal and most plant tissues GTP is formed and then the GTP donates the phosphate to ADP to form ATP.

$$\text{GTP} + \text{ADP} \rightleftharpoons \text{ATP} + \text{GDP} \qquad (16\text{--}11)$$

Succinic dehydrogenase catalyzes the oxidation of succinate to fumarate. This is the one enzyme of the TCA cycle reactions that has been shown to be attached to the inner surface of the inner membrane although it is easily removed from the membrane. *Succinic dehydrogenase* employs a flavin adenine dinucleotide (FAD) as a coenzyme, and it is this coenzyme that accepts the two hydrogens removed from the succinate during oxidation.

$$succinate + FAD \rightarrow fumarate + FADH_2 \quad (16\text{–}12)$$

The $\Delta G^{0\prime}$ is 0.

Fumarate is converted to malate by *fumarase*. The reaction has a $\Delta G^{0\prime}$ of -3.7 kJ/mole (-0.88 kcal/mole).

$$fumarate + H_2O \rightarrow malate \quad (16\text{–}13)$$

Malate is oxidized by *malate dehydrogenase*, which is an NAD^+-containing enzyme. Although the $\Delta G^{0\prime}$ is 29.7 kJ/mole (7.1 kcal/mole) and indicates that the isolated reaction

$$malate + NAD^+ \rightarrow oxaloacetate + NADH + H^+ \quad (16\text{–}14)$$

is endergonic, the products of the reaction are readily removed *in vivo* and the reaction is therefore driven in the direction as written. *Malate dehydrogenase* is found not only in the matrix of mitochondria but also in the cytosol. The oxaloacetate produced by the reaction can serve to initiate the cycle again by combining with acetyl-CoA to form citrate.

Summary of the TCA Cycle

It is possible to account for all the atoms that pass through the TCA cycle. There are two carbons in acetyl-CoA, and during the cycle one is converted to CO_2 at the *isocitrate dehydrogenase* step (16–8) and the other is converted to CO_2 at the *α-ketoglutarate dehydrogenase* step (16–9). Although these are not the same carbon atoms that entered the cycle, a balance is achieved—two carbons enter the cycle and two leave the cycle.

The same accounting of carbons can be made when starting with glucose.

$$C_6H_{12}O_6 + 6\,O_2 \rightarrow 6\,CO_2 + 6\,H_2O \quad (16\text{–}15)$$

This monosaccharide is broken down during glycolysis in the cytosol (Fig. 16–19; see also Chapter 10 for a detailed description of glycolysis) to form two molecules of pyruvate. Each pyruvate molecule loses one carbon to CO_2 as it enters the Krebs cycle at the *pyruvate dehydrogenase* step (16–1 and 16–2) and another at each of the two steps described above in the TCA cycle.

$$\underset{\text{(6 carbons)}}{glucose} \longrightarrow \underset{\text{(3 carbons each)}}{2\ pyruvate} \longrightarrow 6\,CO_2 \quad (16\text{–}16)$$

An accounting can also be made for the hydrogen atoms during glycolysis and Krebs cycle reactions. One glucose ($C_6H_{12}O_6$) and 6 H_2O contribute a total of 24 hydrogens. During glycolysis, 4 hydrogens are removed at the *glyceraldehyde-3-phosphate dehydrogenase* step (10–14, Fig. 10–2) to form 2 $NADH + 2$ H^+ in the cytosol and the two pyruvates from glucose each yield two more hydrogens (2 $NADH + 2\ H^+$) at the pyruvate dehydrogenase step (16–1 and 16–2). Four more are transferred to NAD^+ at the *isocitrate dehydrogenase* step (16–8), four are transferred to NAD^+ at the *α-ketoglutarate dehydrogenase* step (16–8), four more are transferred to FAD at the *succinate dehydrogenase* step (16–12), and finally another four are transferred to NAD^+ at the *malate dehydrogenase* step (16–14). This accounts for all 24 hydrogens. Although these are not the identical hydrogen atoms that were present in the original glucose and water molecules, a numerical balance is nonetheless achieved.

In a similar way, the oxygens also balance. Glucose contributes 6 oxygens and the 6 water molecules added mean that 12 oxygens enter the system. Twelve oxygen molecules are present in the 6 CO_2 molecules produced. Taken altogether, the overall equation for the respiration of glucose through the Krebs cycle reactions is

$$C_6H_{12}O_6 + 6\,H_2O + \left\{ \begin{array}{l} 10\ NAD^+ \\ 2\ FAD \end{array} \right\}$$
$$\downarrow$$
$$6\,CO_2 + \left\{ \begin{array}{l} 10\ NADH + 10\ H^+ \\ 2\ FADH_2 \end{array} \right\} \quad (16\text{–}17)$$

The physiological importance of the Krebs cycle reactions is the production of tremendous chemical reducing power in the matrix of the mitochondrion by the accumulation of NADH, H^+, and $FADH_2$. These compounds enter the electron transport system reactions and their reducing potential drives the oxidative phosphorylation reactions that produce ATP.

ELECTRON TRANSPORT SYSTEM

The NADH, H^+, and $FADH_2$ that accumulate as the TCA cycle operates form an enormous potential energy pool. The reduced coenzymes are reoxidized and transfer their reducing capabilities through a sequence of compounds to oxygen, which accepts electrons and H^+ to form water. The sequence of carriers through which the electrons and H^+ pass is called the **electron transport system**. Each transfer of electrons (or H^+) from one compound to the next results in the oxidation (electron loss) by the donor molecule

FIGURE 16–21 Sequential reduction and oxidation of FAD linked to E_3 (*dihydrolipoyl dehydrogenase*).

and the reduction (electron gain) by the acceptor molecule. Such a transfer occurred in reaction 16–3 in the matrix during pyruvate dehydrogenase action. The enzyme-linked FAD (E_3-FAD) accepted the electrons from the substrate and was reduced, thereby forming E_3-FADH$_2$. Subsequently, the E_3-FADH$_2$ transferred hydrogens and electrons to NAD$^+$, which therefore was reduced while the FADH$_2$ was reoxidized (Fig. 16–21).

In the electron transport system the electrons are shuttled from one to another of about 12 to 17 intermediate compounds before they eventually reach oxygen to form water. Although it is possible to directly reoxidize NADH with oxygen, in the mitochondria this is usually prevented. The enzymes of the electron transport system appear to be arranged in the inner membrane in such a manner that the transfer of the electrons must proceed through the specified series of intermediates. An energy change occurs during electron transfer; a proton gradient and membrane potential are ultimately created across the inner membrane. The proton gradient and membrane potential provide the energy to phosphorylate ADP—a process termed **oxidative phosphorylation** and described in the next section.

OXIDATION–REDUCTION REACTIONS

In Chapter 3 the tendency of acids and bases to donate or accept protons was described. In acid–base systems, one compound acts as the proton (H$^+$) donor and the other as the proton acceptor, the donor being the acid and the acceptor the base. The two form a conjugate acid–base pair. In a similar manner, oxidizing and reducing agents function in pairs. In this case, they are called **redox pairs** or **redox couples**. The member of the pair that donates the electron is called the **reducing agent** or **reductant** and the electron acceptor is called the **oxidizing agent** or **oxidant**. The terms *donating* and *accepting* fail to convey the nature

of the amounts or energy change involved in the reaction. In effect, the reducing agent has a certain capability to retain electrons, as does the oxidizing agent. In a redox couple, one member attracts electrons more strongly than the other, and in effect the oxidizing agent can pull the electrons away from the reducing agent. This capability to gain (or lose) electrons can be measured and is called the **oxidation–reduction potential** or **redox potential** and is expressed in volts. (Redox potentials are described in Chapter 9.)

For convenience, the potential of most redox couples is measured against a reference standard, which usually is a hydrogen electrode. Most measurements are made under standard conditions using 1.0 M concentrations of oxidant and reductant at 25°C and at pH 7.0. The hydrogen electrode is equilibrated with H$_2$ gas at 1 atmosphere and the [H$^+$] is 1.0 M at 25°C. When the pH is adjusted to 7.0 (i.e., [H$^+$] = 10^{-7}M), the redox potential of the reference hydrogen electrode is –0.42 V. The redox potentials of the electron transport system intermediates are given in Table 16–2. The greater the E_0' value, the more strongly the oxidant accepts electrons. However, an oxidant having a lower E_0' than another compound becomes the reductant of or electron donor to that compound.

The Henderson–Hasselbach equation given in Chapter 3 (equation 3–5) describes the relationship between pH and the dissociation constant. In a similar way, the *Nernst equation* shows for a redox pair the relationship between the standard redox potential (E_0'), the observed potential at any concentration, and the concentration ratio of the oxidant and reductant.

$$E_h = E_0' + \frac{2.3RT}{n\mathscr{F}}\log\frac{[\text{oxidant}]}{[\text{reductant}]}$$

where E_0' is the standard redox potential, E_h is the observed electrode potential, R is the *gas* constant (8.31 J/deg/mole, T is the absolute temperature in degrees Kelvin, n is the number of electrons transferred, and $\mathscr{F}$

TABLE 16–2 OXIDATION–REDUCTION POTENTIALS OF THE ELECTRON TRANSPORT SYSTEM (VALUES ARE BASED ON TWO-ELECTRON TRANSFERS AT pH = 7.0 AND 25–30°C)

Electrode Equation	E'_0 (V)
$2\,H^+ + 2\,e^- \rightleftharpoons H_2$	-0.421
$NAD^+ + 2\,H^+ + 2\,e^- \rightleftharpoons NADH + H^+$	-0.320
$NADP^+ + 2\,H^+ + 2\,e^- \rightleftharpoons NADPH + H^+$	-0.324
Ubiquinone $+ 2\,H^+ + 2\,e^- \rightleftharpoons$ ubiquinol	$+0.10$
2 cytochrome $b_{(ox)} + 2\,e^- \rightleftharpoons 2$ cytochrome $b_{(red)}$	$+0.030$
2 cytochrome $c_{ox} + 2\,e^- \rightleftharpoons 2$ cytochrome c_{red}	$+0.254$
2 cytochrome $a_{3(ox)} + 2\,e^- \rightleftharpoons 2$ cytochrome $a_{3(red)}$	$+0.385$
$1/2\,O_2 + 2\,H^+ + 2e^- \rightleftharpoons H_2O$	$+0.816$

is the faraday (96,406 kJ/V). When electrons are transferred and the constants are combined, the Nernst equation becomes

$$E_h = E'_0 - \frac{0.06}{n}\log\frac{[\text{oxidant}]}{[\text{reductant}]}$$

where n = the number of electrons invovled.

CLASSES OF ELECTRON TRANSPORT SYSTEM COMPOUNDS

There are five types of compounds associated with the electron transport system. Three of these consist of enzymes whose coenzymes or prosthetic groups are known to be directly invovled in the transfer. They are (1) *pyridine-linked dehydrogenases*, which have either NAD^+ or $NADP^+$ as coenzymes; (2) *flavin-linked dehydrogenases*, which are linked to flavin adenine dinucleotides (FAD) or flavin mononucleotides (FMN); (3) *coenzyme Q* or *ubiquinone*, a lipid-soluble coenzyme functioning in electron transport; (4) the *cytochromes*, which contain iron–porphyrin prosthetic groups; and (5) the iron–sulfur proteins (which are chemically distinct from the iron-containing cytochromes).

Pyridine-linked dehydrogenases require as their coenzyme either NAD^+ or $NADP^+$ (see Fig. 3–10 for the structure of $NADP^+$ and the chemistry of its reduction by hydrogen); both NAD^+ and $NADP^+$ can accept two electrons at a time. There are about 200 dehydrogenases for which NAD^+ or $NADP^+$ serve as coenzymes. Although the NAD^+ and $NADP^+$ dehydrogenases are found in both the cytosol and the mitochondria and are known to transfer electrons between compounds in both places, it appears that only the NAD^+-linked compounds are involved in the electron transport system.

Flavin-linked dehydrogenases (often called flavoproteins, F_P) require either FAD (Fig. 3–11) or FMN. Both are prosthetic groups whose isoalloxazine ring can accept two hydrogen atoms. (Prosthetic groups are firmly bound to the protein, whereas coenzymes are not.) Flavin-linked enzymes are involved in a number of enzyme systems, the more common of which are associated with fatty acid oxidation, amino acid oxidation, and Krebs cycle activity (e.g., *pyruvate dehydrogenase* and *succinic dehydrogenase*). In general, it is not uncommon for flavin prosthetic groups and NAD^+ coenzymes to be linked to the same protein in dehydrogenases.

Ubiquinones were so named because of their occurrence in so many different organisms (ubiquitous means present everywhere) and their chemical resemblance to quinone (Fig. 16–22). They are found in several different forms including the *plastoquinones* of chloroplasts. The form present in mitochondria is of-

FIGURE 16–22 Chemical structure and sites of reduction of *ubiquinone*.

ten called *coenzyme Q* (CoQ or Q) and accepts two protons and two electrons at a time.

The *cytochromes* are proteins containing iron–porphyrin (or heme) groups (Fig. 16–23). There are a large number of cytochromes in cells; most are found in mitochondria, although some function in the endoplasmic reticulum and in chloroplasts. In mitochondria, five cytochromes appear to be associated with the inner membrane and are identified as cytochromes b, c_1, c, a, and a_3. Some occur in two or more forms, but all transfer electrons by reversible valence changes of the iron atom ($Fe^{3+} \leftrightharpoons Fe^{2+}$). In cytochromes b, c_1, c, and a, the manner of binding of iron in the porphyrin ring and its association with the protein prevent the iron ligands from forming with oxygen, and therefore these reduced cytochromes cannot be directly oxidized by molecular oxygen. Cytochrome a_3, which is the terminal carrier in the electron transport system and which together with cytochrome a forms the **cytochrome oxidase complex**, is an exception and can be directly oxidized by oxygen. In addition to cytochromes a and a_3, *cytochrome oxidase* contains copper. Electrons received from cytochrome c are picked up by cytochrome a and then transferred from a_3 to oxygen by $Cu^{2+} \rightleftharpoons Cu^+$ intermediation.

Iron–sulfur proteins of mitochondria are electron carriers containing iron and sulfur in equal amounts. The iron is reversibly oxidized during the electron transfer. Certain iron–sulfur groups occur in association with flavoproteins and others are associated with the NADH dehydrogenase complex and the cytochrome b and c complex. The term *iron–sulfur center* is used to denote any Fe–S association. Iron–sulfur proteins transfer one electron at a time.

Electron Transport Pathway

It has taken about 50 years to work out the chain of mitochondrial electron transfer shown in Figure 16–24. During the early 1900s, several investigators (notably T. Thunberg) discovered a number of dehydrogenase enzymes. In 1913, O. Warburg discovered that cyanide inhibits oxygen consumption but does not interfere with dehydrogenases. He proposed the existence of iron-containing "respiratory enzymes," now recognized as the cytochromes. The flavoproteins were identified by A. Szent-Gyorgyi as the intermediates between dehydrogenases and the respiratory enzymes. F. L. Crane discovered ubiquinone and a number of other investigators, notably Keilin, Kuhn, Green, Chance, Racker, and Lehninger, added details about the chemistry of the intermediates in electron transfer and the sequence of the reactions. Among the Nobel Prize recipients who contributed to our early understanding of electron transport are Warburg, Szent-Gyorgyi, and Kuhn.

Today we visualize the electron transport chain as an orderly sequence of compounds through which the

FIGURE 16–23 *Cytochrome c.* In cytochrome c, the heme prosthetic group is covalently bonded to the protein. In the other cytochromes (as in hemoglobin), the heme group is anchored to the protein by noncovalent bonds.

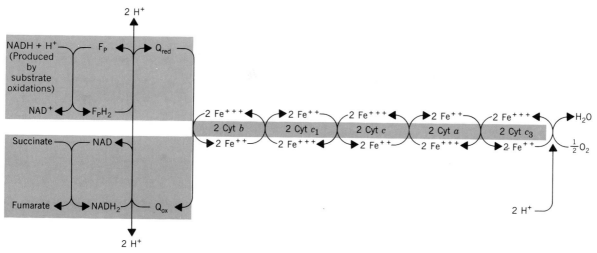

FIGURE 16-24 Oxidation–reduction reactions that transfer electrons and H^+ to O_2, thereby forming water.

hydrogens and electrons are passed. The fact that the electrons do not jump or "short-circuit" to stronger oxidizing agents is probably related to the specific physical positioning of the coenzymes in the inner membrane. However, hydrogen or electrons may enter the chain at various points.

Most electrons are removed from the substrates in the cytosol or the matrix of the mitochondria by NAD^+- (or $NADP^+$-) *linked dehydrogenases.* The NADH produced in the matrix conveys the electrons to the NAD^+-*flavoprotein-linked dehydrogenase* in the inner membrane of the mitochondrion. As shown in Figure 16–24, the NADH is then oxidized by F_P, which now being reduced is in turn reoxidized by ubiquinone (Q). Reduced Q is subsequently reoxidized by cytochrome b and the reduced iron in cytochrome b is then reoxidized by cytochrome c_1. Finally, *cytochrome oxidase* accepts the electrons from cytochrome c, passing them from cytochrome a to a_3 and subsequently to oxygen. The acceptance of the electrons by oxygen occurs at the same time as the incorporation of 2 H^+, thereby forming water. During one complete transfer, each of the coenzymes in the chain is reoxidized and is available to be reduced again. *Pyruvate dehydrogenase, malate dehydrogenase, isocitrate dehydrogenase,* and α-*ketoglutarate dehydrogenase* are examples of enzymes initiating the electron transfer chain through NAD^+. *Succinate dehydrogenase* oxidizes succinate, and the electrons removed are bound to FAD. The FAD_{suc} is reoxidized by giving up its electrons to ubiquinone (Q), thereby bypassing the NAD^+

step. Fatty acyl-CoA and glycerol phosphate are oxidized in a similar manner with entry of electrons directly from their flavoproteins to Q.

BALANCE OF ELECTRONS FROM GLYCOLYSIS AND TCA CYCLE METABOLISM

As was pointed out in the earlier section of the TCA cycle, 10 molecules of NAD^+ and 2 molecules of FAD are reduced when 1 molecule of glucose is metabolized via glycolysis and the TCA cycle. Each of these 12 compounds passes pairs of electrons through the electron transport system to oxygen. For each pair of electrons, $\frac{1}{2}$ O_2 is consumed and 1 H_2O is formed. Therefore, for 12 electron pairs, 6 O_2 are consumed and 12 H_2O are produced. You will recall that 6 of the 12 waters are consumed in the TCA cycle, so that in the overall oxidation of glucose there is a net production of only 6 H_2O.

$$C_6H_{12}O_6 + 6\ O_2 \rightarrow 6\ CO_2 + 6\ H_2O$$

THE ENERGETICS OF ELECTRON TRANSPORT

The standard free energy change, $\Delta G^{0\prime}$ (see Chapter 9), can be calculated from the redox potentials using the formula

$$\Delta G^{0\prime} = -n\mathscr{F}\Delta E_0'$$

where n = the number of electrons transferred at a time (usually 2), $\mathscr{F}$ = the number of faradays (96.4 kJ/ V), and $\Delta E_0'$ is the change in standard redox potential (the E_0' of the accepting redox couple less the E_0' of the donating couple). The $\Delta G^{0\prime}$ for the entire chain from NAD to oxygen would be -220.5 kJ/mole (-52.7 kcal/ mole). The stepwise sequence is shown in Figure 16–24. The fate of the energy released by the transfer of electrons is discussed later in the chapter under oxidative phosphorylation. It is interesting to note at this point that enough energy is produced to phosphorylate several moles of ADP at a standard free energy of formation of 30.5 kJ/mole (7.3 kcal/mole). Generally, only three molecules of ATP are formed corresponding to the three steps, producing sufficient energy for the phosphorylation (Fig. 16–25).

Transport of Protons

Transport of H^+ closely parallels electron transport. Some of the electron carriers accept and transfer elec-

trons in the form of complete hydrogen atoms. These are NAD^+, $NADP^+$, FAD, FMN, and ubiquinone. FAD and FMN carry pairs of hydrogen atoms and NAD^+ and $NADP^+$ carry a hydrogen that bears one extra electron (a hydride ion); however, a second H^+ associates closely with the NADH or NADPH. The cytochromes and nonsulfur proteins, on the other hand, only pick up and release electrons each cyctochrome carrying only one electron at a time.

Electron Transport Inhibitors

Inhibitors of the electron transport system may be assigned to three main groups. One group includes those inhibitors that act on NAD^+- (or $NADP^+$-) *dehydrogenase* and ubiquinone. Examples are rotenone (a plant substance once used by some South American Indians as a fish poison and extensively used today in insecticides), amytal, and piericidin (an antibiotic similar in structure to and possibly competitive with ubiquinone). The second group includes those that block electron transport between cytochromes b and c; an example is *antimycin A*, an antibiotic from *Strepto-*

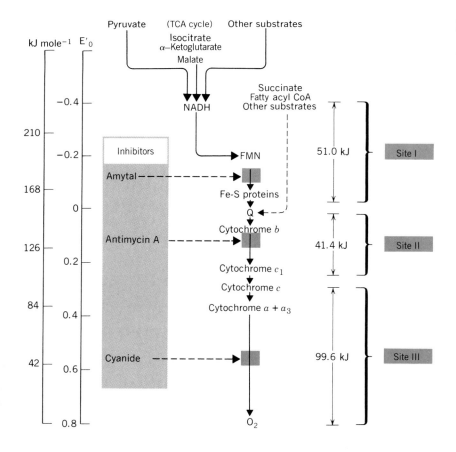

FIGURE 16–25 Changes in free energy as electrons pass through the electron transport system. Three changes in energy sufficient to drive phosphorylation occur at regions known as sites I, II, and III.

myces griseus, and a quinone analog. The third group includes those inhibitors that block transport of electrons to oxygen by *cytochrome oxidase*; included in this group are cyanide, azide, hydrogen sulfide, and carbon monoxide.

NAD[P]⁺-Transhydrogenases

The electron transport chain usually accepts NADH as the initial donor of hydrogen. In many animal tissues and in some eukaryotic microorganisms, hydrogens from NADPH are also accepted. The entry of these hydrogens into mitochondria is apparently brought about by the enzyme *NAD[P]⁺-transhydrogenase*, which is located in the outer mitochondrial membrane and which catalyzes the transfer of hydrogen to NAD⁺ located inside the mitochondrion, that is,

$$(NADPH + H^+) + NAD^+ \rightarrow (NADH + H^+) + NADP^+$$

OXIDATIVE PHOSPHORYLATION

Linkage of phosphorylation to oxidative metabolism was first proposed in the 1930s. The first evidence was the observation that phosphate was removed from the medium when TCA cycle intermediates were metabolized by cells. Subsequently, it was shown that the disappearance of inorganic phosphate was accompanied by the accumulation of organic phosphates such as glucose-6-phosphate and fructose-6-phosphate. Shortly thereafter, ATP was recognized as the primary product of oxidative phsophorylation. In 1948, Kennedy and Lehninger established that these oxidations and phosphorylations occurred exclusively in the mitochondria. By 1951, Lehninger had shown that the Krebs cycle reactions could be bypassed entirely by adding NADH and H⁺ to mitochondria treated to allow the NADH to enter the organelles. For each NADH and H⁺ added, three phosphates and one oxygen were consumed, and it was therefore concluded that the electron transport chain was important for ATP synthesis.

The potential differences of the reactions of electron transfer beginning with *NAD⁺-dehydrogenases* and ending with *cytochrome oxidase* provide the energy to drive the endergonic oxidative phosphorylation. The exergonic reactions can be summarized as

$$NADH + H^+ + \tfrac{1}{2}O_2 \rightarrow NAD^+ + H_2O$$

which has a $\Delta G^{0\prime}$ of -220.5 kJ/mole (-52.7 kcal/mole). The endergonic reactions summarized as

$$3\,ADP + 3\,P_i \rightarrow 3\,ATP$$

are associated with a standard free energy of at least 91.6 kJ (i.e., $\Delta G^{0\prime} = 30.5$ kJ/mole or 7.3 kcal/mole). The efficiency of the coupling of the two systems is at least 42% (i.e., 91.6/220.5).

The stages at which coupling occurs are known. Figure 16–25 shows the major energy changes along the electron transfer chain. The step from NADH to Q represents the reactions of the dehydrogenases that link NAD⁺ and F$_P$ to Q. This reaction sequence (Site I) provides enough energy (51.0 kJ/mole or 12.2 kcal/mole) for the formation of the first of three ATPs obtained by NADH oxidation. The step from Q to cytochrome *b* does not provide enough energy for the second phosphorylation, but the following step (from cytochrome *b* to cytochrome *c*) does yield sufficient energy (Site II). The step from cytochrome *c* to cytochrome *a* is only moderately exergonic, but the final step from cytochrome *a* to oxygen is highly exergonic (Site III). Thus, a pair of electrons transferred from one carrier to the next by these oxidation–reduction reactions at three places in the chain generates the energy for the formation of ATP. However, not all electrons from oxidized substrates enter the electron transport chain at the initial NAD⁺ position. Some electrons, like those from succinate, pass from the coenzyme of succinate dehydrogenase (i.e., FAD) directly to Q. When this happens, only two ATPs are formed per pair of electrons transferred.

Molecular Events in Oxidative Phosphorylation

The mechanism by which energy from the electron transport system is used for oxidative phosphorylation is far from clear. It would appear that there are yet to be discovered coupling factors and enzymes involved in the process. Among the more significant observations that have been made are the identification of (1) mechanisms for controlling the *rate* of electron transport and oxidative phosphorylation, (2) inhibitors that can prevent ATP formation or uncouple electron transport from oxidative phosphorylation, (3) partial oxidative phosphorylation reactions, and (4) coupling factors. Maximal electron transport in mitochondria can occur only if there is ample ADP and P$_i$ available to act as an acceptor of inorganic phosphate. When ADP is lacking or when ATP is plentiful, the rate of respiration is low and is called **state 4 respiration** (Fig. 16–26). When ADP is added, the rate of oxygen consumption rises as the ADP is phosphorylated. This elevated level of respiration is called **state 3 respiration**. Once the ADP is consumed, state 4 respiration is reestablished. The influence of ADP in respiratory rate is called *acceptor* or *respiratory control*, the "acceptor" being ADP. Interestingly, the configuration of mitochondria changes between state 4 and state 3 respiration. During state 4 respiration, the mitochondrion assumes the orthodox conformation (Fig. 16–17) but changes to the condensed conformation when added ADP induces state 3.

A number of chemical agents uncouple oxidative phosphorylation from the electron transport system (e.g., 2,4-dinitrophenol, dicumarol, and the salicylanilides). The addition of these compounds has two interesting effects. First, they speed up electron transport and oxygen consumption even in the absence of ADP, but there is no synthesis of ATP. In other words, phosphorylation is uncoupled and so is energy transfer. Second, in the presence of the uncoupling agent, the hydrolysis of ATP may occur—the *opposite* of the normal goals of mitochondrial activity.

Oligomycin is an inhibitor of oxidative phosphorylation and coupled oxygen consumption but it does not prevent electron transfer in uncoupled systems. A group of agents called ionophores also prevent oxidative phosphorylation by dissipating the energy needed for phosphorylation. Ionophores are only effective in the presence of certain monovalent cations such as K^+ and Na^+.

Some of the reactions of oxidative phosphorylation are known. An ATPase is present in the inner membrane and is stimulated by 2,4-dinitrophenol but inhibited by oligomycin. Another reaction is the exchange of inorganic phosphate with the terminal phosphate of ATP, a reaction called *phosphate–ATP exchange*. This reaction is inhibited by both 2,4-dinitrophenol and oligomycin. An exchange of atoms also occurs between the oxygens of water and those of inorganic phosphate, called *phosphate–water exchange*. The terminal phosphate can also be exchanged between ATP and ADP, a process inhibited by 2,4-dinitrophenol and oligomycin and called the *ADP–ATP exchange* reaction.

Some interesting studies that relate structure and function in mitochondria have been carried out using inverted *submitochondrial vesicles* formed from fragments of the inner membrane. In these studies, the inner membranes of mitochondria isolated from cells are subjected to sonication or detergent action. The resulting fragments of the cristael membranes then round up to form vesicles with the inner membrane spheres on the outside surface (Fig. 16–27). If *closed* vesicles are formed, they exhibit functioning electron transport system and oxidative phosphorylation reactions. There is no ATP formation with open vesicles or nonvesicular fragments, but electron transport still occurs.

If the submitochondrial vesicles are treated with urea or trypsin, the inner membrane spheres are removed and can be separated from the vesicles, which lose their capacity to carry out oxidative phosphorylation. However, if the spheres are added back to the vesicles, the capacity to perform oxidative phosphorylation is regained. The spheres contain ATPase/synthetase, which is called *coupling factor one* (F_1). This enzyme has been purified and found to have a molecular weight of about 360,000, a diameter of 9 nm, and a requirement for Mg^{2+}. In intact mitochondria the enzyme is bound to the membrane and also catalyzes the synthesis of ATP and ADP and inorganic phosphate, rather than the reverse, and as such could be termed *ATP-synthetase*. The isolated enzyme's function, however, is not inhibited by oligomycin. Another factor has been isolated that when

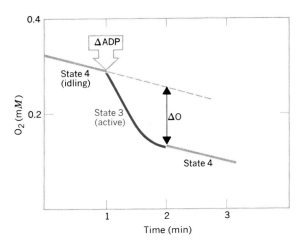

FIGURE 16–26 Changes in oxygen consumption and physiological state with the addition of ADP to mitochondria. Initially, the rate of oxygen consumption is low due to the lack of ADP (most of that available has been converted to ATP). This "resting" state is called *state 4 respiration*. Addition of ADP induces the more active *state 3 respiration* and this continues until all of the added ADP is converted to ATP, or until O_2 is depleted.

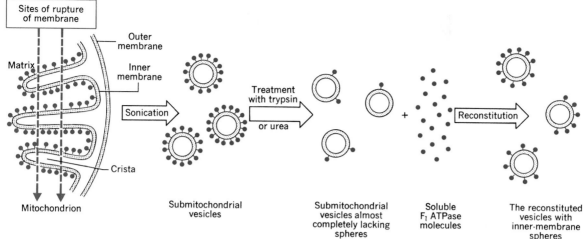

FIGURE 16–27 Under appropriate conditions the disruption of mitochondria by sonification is followed by the formation of small vesicles of inverted orientation. These vesicles function normally in electron transfer and oxidative phosphorylation. The inner membrane spheres can be detached from the outside surfaces of these vesicles using trypsin or urea. When this is done, the capacity for oxidative phosphorylation (but not respiration) is lost. Reconstitution of vesicles and spheres is accompanied by the restoration of these activities.

present with isolated F_1 renders the ATPase sensitive to oligomycin. This latter factor is called F_0 or *oligomycin-sensitivity-conferring protein (OSCP)*. F_0 is also a large protein; it is speculated that F_0 could constitute at least part of the stalk that attaches the sphere to the inner membrane.

THE CHEMIOSMOTIC-COUPLING HYPOTHESIS

The exact mechanism that couples the energy of electron transfer to the phosphorylation of ADP remains somewhat speculative. However, at the present time the *chemiosmotic-coupling hypothesis*-proposed by P. Mitchell is a widely supported model. Since its initial proposal in 1961, the hypothesis has undergone modification by Mitchell and by others. For his contributions to the development of this model, Mitchell received the Nobel Prize in 1978. According to Mitchell, a gradient of hydrogen ions is created across the inner membrane as the electron carriers pump H^+ from the matrix into the intermembrane area of intact mitochondria or into the submitochondrial vesicles. This process creates an *electrochemical gradient* (desig-

nated $\Delta\mu_{H^+}$) across the membrane consisting of a negative-inside membrane potential ($\Delta\psi$) and an acid-outside gradient of H^+ (ΔpH). An intact vesicular membrane is essential for the establishment of this electrochemical gradient. The gradient, in turn, provides the energy to drive the synthesis of ATP.

It is proposed that the electron carriers involved in electron transport are not randomly arranged in or on the inner membrane but are spatially oriented in a specific manner. As electrons are transferred through the sequence of membrane carriers, there is an accompanying expulsion of H^+ from the matrix. This causes the matrix to become especially alkaline (low H^+ and high OH^- content). In Figure 16–28, these events are depicted with the electron transport intermediates arranged in order in the membrane. The vectorial flow of protons from matrix to intermembrane space shown in Figure 16–28 creates the electrochemical gradient that is the source of energy for phosphorylation. As oxidative phosphorylation proceeds, the alkaline sink within the matrix and the external acid pool are both neutralized. Continued electron transport is necessary to maintain the pH gradient that drives phosphorylation. For many years, it was believed that only two H^+ were expelled from the matrix at each of the three electron transfer stages coupled

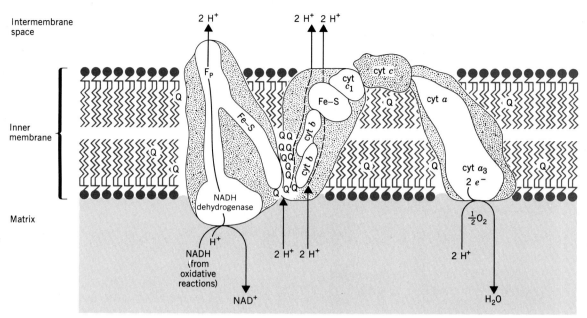

FIGURE 16–28 Transfer of H^+ through the inner mitochondrial membrane from the matrix during electron transfer. *NADH dehydrogenase* catalyzes the transfer of two protons and two electrons from NADH to H^+ to F_p (flavoprotein complex); the protons are left in the intermembrane space while the electrons are shuttled back toward the matrical surface by the iron–sulfur proteins (Fe–S). The electrons are then passed and bound to a soluble ubiquinone fraction; the passage of a pair of electrons through cytochromes *b* and *c* and an Fe–S protein effects the transfer of at least four more protons from the matrix to the intermembrane space. The electrons are then passed through cytochrome *c* and cytochrome oxidase ($a + a_3$) to oxygen on the matrical surface, where two more protons are removed from the matrix in the reduction of oxygen to form water.

to phosphorylation (i.e., sites I, II, and III in Fig. 16–25). Now there is evidence that 12 H^+ are transferred across the membrane (see Fig. 16–29).

There is good evidence, both direct and indirect, to support the chemiosmotic-coupling hypothesis. As will be seen in Chapter 17, phosphorylation of ADP also occurs in chloroplasts and can be explained by a similar chemiosmotic-coupling mechanism that creates a proton gradient across the organelle's thylakoid membranes. Indeed, A. Jagendorf has shown that a burst of ADP phosphorylation occurs when chloroplasts isolated from cells at neutral pH are placed in a buffer solution at low pH.

Whereas electron transport is accompanied by the outward transport of H^+, ATP synthesis is linked to the inward movement of H^+. Because ATP synthesis occurs on the matrix surface of the inner membrane, ADP and inorganic phosphate must be transported

from outside the mitochondrion across the membrane and into the matrix. The inward transport of P_i and ADP and the outward transport of ATP are driven by the electrochemical gradient (see Fig. 16–29). One H^+ reenters the matrix along the pH gradient during the inward movement of inorganic phosphate, and at least two more H^+ (perhaps as many as three more) reenter the matrix for each molecule of ATP synthesized (Fig. 16–30).

Although the chemiosmotic-coupling model is widely accepted, the precise mechanism resulting in ATP synthesis remains uncertain. Also unexplained is how the protons are transported chemically. The similarities in structure, properties, and function of ATP synthetases and hydrogen ion pumps that have recently been isolated from various animal, plant, and microbial cells suggest that some answers will be forthcoming.

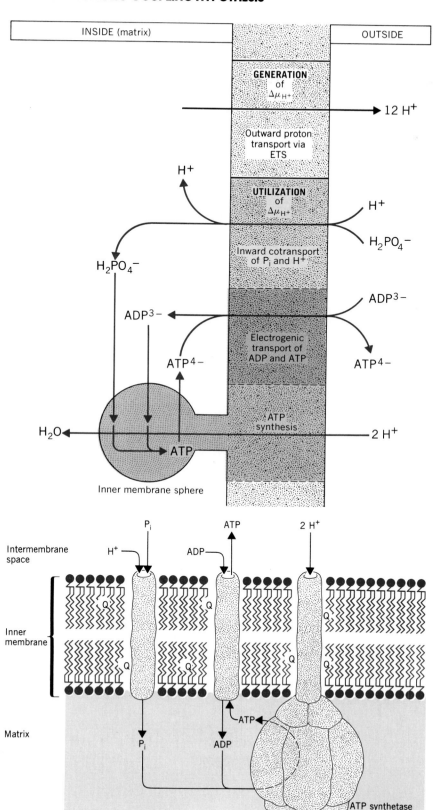

FIGURE 16–29 *Chemiosmotic coupling.* Passage of two electrons (e⁻) through the electron transport systems creates a 12 H⁺ gradient across the membrane. This gradient provides the driving force for the inward transport of phosphate and the potential for generating ATP. ADP is brought into the matrix in exchange for ATP.

FIGURE 16–30 Diagrammatic representation of the intrinsic membrane proteins associated with oxidative phosphorylation and transport of phosphate, protons, ADP, and ATP. (See also Fig. 16–29.)

OTHER FUNCTIONS OF MITOCHONDRIA

Mitochondria are generally described as the power-houses of the cell, and as such most attention is directed to the processes that evolve energy, namely, the TCA cycle, electron transport, and oxidative phosphorylation. However, a number of other processes also take place in mitochondria.

The Glyoxylate Cycle

This metabolic pathway was described in Chapter 10. The enzymes are localized in the matrix and some of them are the same as those of the TCA cycle. In effect, the glyoxylate cycle is a modified form of the Krebs cycle, but its function appears to be primarily associated with conversion of acetate produced by the catabolism of fatty acids into oxaloacetate. Oxaloacetate is also an important intermediate in the conversion of fatty acids to carbohydrates. Whereas animals lack certain enzymes of the glyoxylate cycle and are incapable of converting fatty acids to carbohydrates using this pathway, plants and microorganisms have functional glyoxylate systems. In higher plants, some of the enzymes of the system (e.g., isocitrate lyase and malate synthetase) are localized in specific organelle called *glyoxysomes* (Chapter 19) as well as in mitochondria.

Fatty Acid Oxidation

Free fatty acids are rarely found in more than trace quantities in cells because they are highly toxic. The fatty acids associated with mono-, di-, and triglycerides and in phospholipids are generally hydrolyzed from the glycerol in the cytosol and immediately activated for transport into the matrix of the mitochondrion, where they are then oxidized. These reactions were also described in Chapter 10.

Fatty Acid Chain Elongation

Fatty acids are generally synthesized by the smooth endoplasmic reticulum. However, there are a number of enzymes in mitochondria that catalyze the elongation of palmitic and other saturated fatty acids by successive additions of acetyl-CoA to the carboxyl end. In smooth ER, both unsaturated and saturated fatty acids are elongated but by the addition of malonyl-CoA rather than acetyl-CoA. Synthesis of fatty acids was discussed in Chapter 10.

Superoxide Dismutase and Catalase

During electron transport, a number of toxic reductive products of oxygen are formed. The most common are the superoxide radical (O_2^-) and hydrogen peroxide (H_2O_2). A protective enzyme has been identified in mitochondria called *superoxide dismutase*, which converts O_2^- and H^+ into hydrogen peroxide. The hydrogen peroxide, in turn, is decomposed by *catalase*, that is,

$$2\,O_2^- + 2H^+ \xrightarrow{\text{dismutase}} H_2O_2 + O_2$$

$$H_2O_2 \xrightarrow{\text{catalase}} H_2O + \tfrac{1}{2}\,O_2$$

Amphibolic and Anaplerotic Reactions

The intermediates of the TCA cycle can act as precursors of a variety of products of anabolic pathways (Table 16–3). The TCA cycle thus is *amphibolic* in that it

TABLE 16–3 TCA PRECURSORS OF ANABOLIC PATHWAYS

TCA Cycle Intermediate	Pathway or Anabolic Reaction
Citrate	Citrate + ATP + CoA → oxaloacetate + ADP + P_i + acetyl-CoA ⟶ fatty acid biosynthesis
Isocitrate	Isocitrate → glyoxylate → malate → carbohydrate synthesis
α-Ketoglutarate	α-Ketoglutarate + alanine → pyruvate + glutamate ⟶
Succinyl-CoA	Heme biosynthesis
Malate	Carbohydrate synthesis
Oxaloacetate	Oxaloacetate + alanine → pyruvate + aspartate → protein synthesis

TABLE 16–4 ANAPLEROTIC REACTIONS OF THE TCA CYCLE

TCA Cycle Intermediate Produced	Non-TCA Cycle Reaction Generating the TCA Cycle Intermediate
α-Ketoglutarate	Glutamate + pyruvate $\rightarrow$ α-ketoglutarate + alanine (from protein breakdown)
Succinate	From glyoxylate cycle
Malate	Malic enzyme: pyruvate + CO_2 + NADPH + H^+ $\rightarrow$ malate + $NADP^+$
Oxaloacetate	Pyruvate carboxylase: pyruvate + CO_2 + ATP + H_2O $\rightarrow$ oxaloacetate + ADP + P_i
Oxaloacetate	Aspartate + pyruvate $\rightarrow$ oxaloacetate + alanine (from protein breakdown)

can act in both a catabolic and anabolic manner. These anabolic reactions can drain intermediates away from the TCA cycle and thus deplete the oxaloacetate supply necessary to keep the cycle functioning. Reactions that supply the TCA cycle with intermediates that replace those lost are called **anaplerotic reactions**. Several anaplerotic reactions of the Krebs cycle are shown in Table 16–4.

PERMEABILITY OF THE INNER MEMBRANE

As noted earlier, most metabolites can readily pass through the outer mitochondrial membrane. In contrast, the inner membrane is not freely permeable. Except in special cases, sugars cannot pass through the membrane. Ions such as Na^+, K^+, and Cl^- are impermeable as are NAD^+, NADH, $NADP^+$, NADPH, AMP, CDP, GDP, CTP, GTP, CoA, and acetyl-CoA. These compounds collect in the matrix and are unable to mix or exchange with pools of these molecules in the cytosol. There are, however, special transport systems in the inner membrane that are specific for select metabolites, and via these systems, ADP, ATP, P_i, pyruvate, and a number of TCA cycle intermediates are transported. The transport systems are species-specific and in most cases are associated with membrane proteins called *carriers*, *translocases*, or *porters*. Table 16–5 lists a number of these systems and indicates their functions.

These systems can function passively, facilitating the exchange of metabolites when there is a favorable concentration gradient, or they can function in an active manner if coupled to the energy-producing electron transport system. During electron transport,

TABLE 16–5 MITOCHONDRIAL MEMBRANE TRANSPORT SYSTEMS

System	Exchange
Dicarboxylate carrier	Exchange on mole-for-mole basis of malate, succinate, fumarate, and phosphate between matrix and cytosol
Tricarboxylate carrier	Exchange on mole-for-mole basis citrate and isocitrate between matrix and cytosol
	Exchange citrate or isocitrate for dicarboxylate
Aspartate–glutamate carrier	Exchange aspartate for glutamate across membrane
α-Ketoglutarate–malate carrier	Specifically exchange α-ketoglutarate for malate across membrane
ADP–ATP carrier	Exchange of ADP for ATP

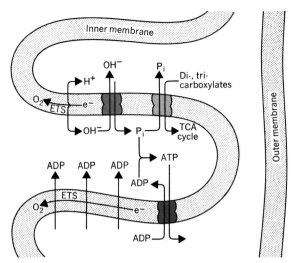

FIGURE 16–31 Carriers in the inner mitochondrial membane transfer phosphate, di- and tricarboxylic acids, and other metabolites between the matrix and intermembrane space.

there is an accumulation of hydroxyl ions in the matrix, but these hydroxyl ions can be transported externally by a *phosphate carrier*, which simultaneously carries inorganic phosphate into the matrix on an exchange basis (Fig. 16–31). The phosphate accumulated can be partly consumed in ATP synthesis, with the ADP–ATP carrier functioning to bring ADP into the matrix. The phosphate can also be exchanged indirectly with external dicarboxylate or tricarboxylate required in TCA cycle reactions using the appropriate carboxylate carrier.

One of the clearest connections between active transport across the inner mitochondrial membrane and the electron transport system involves the movement of calcium ions into the matrix. The movement of Ca^{2+} into the matrix may be accompanied by an equivalent uptake of phosphate. It has been shown that for each pair of electrons transported through the electron transport system from NADH to oxygen, six Ca^{2+} are transported across the inner membrane into the matrix. However, when the Ca^{2+} is transported, there is no phosphorylation of ADP. Studies using specific inhibitors indicate that the portions of the respiratory chain that generates the energy for phosphorylation along the electron transport chain are also the energy-providing stages for the active transport of Ca^{2+}.

CYTOSOL–MATRIX EXCHANGE OF NADH AND NADPH

Reduced coenzymes such as NADH and NADPH do not permeate the inner membrane of the mitochondrion to any significant extent. However, reduced pyridine nucleotides are known to be produced in a number of reactions in the cytosol (the reduction of NAD^+ at the glyceraldehyde step in glycolysis is an important example) and the reoxidation of NADH occurs via the mitochondrion. The mechanism involves a set of reactions called a *shuttle* (Fig. 16–32).

Glycerol Phosphate Shuttle

The **glycerol phosphate shuttle** (Fig. 16–32) involves (1) *glycerol phosphate dehydrogenase* in the cytosol, (2) *glycerol phosphate dehydrogenase* on the outer surface of the inner mitochondrial membrane, and (3) the reduction of Q in the electron transport chain. Dihydroxyacetone phosphate, NADH, and H^+ react in the cytosol to form glycerol phosphate, which diffuses through the outer mitochondrial membrane to the outer surface of the inner membrane. There the glycerol phosphate reacts with the membrane dehydrogenase to form dihydroxyacetone phosphate, which returns to the cytosol. The membrane-bound dehydrogenase employs a flavoprotein as a coenzyme, and the F_P becomes reduced during the reaction. Subsequently, the electrons from the reduced F_P are passed directly into the electron transport system at the Q step. Because the $NAD^+ \rightarrow F_P$ step of electron transport is skipped, only two ATP are generated for each pair of electrons that enters in this fashion from the cytosol.

Malate–Aspartate Shuttle

Another shuttle, the **malate–aspartate shuttle**, can also transport the hydrogens accepted during the reduction of NAD^+ in the cytosol across the inner membrane (Fig. 16–32). H^+ transported in this manner into the matrix, enters the electron transport chain via *NADH dehydrogenase*, and as a result three ATP are generated for each pair of electrons.

Total ATP Production from the Catabolism of Glucose

The complete oxidation of a molecule of glucose by glycolysis, the TCA cycle, and electron transport is ac-

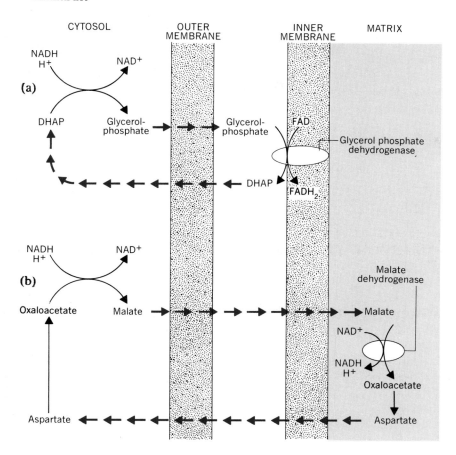

FIGURE 16–32 The glycerol phosphate and malate–aspartate *shuttles*. DHAP=dihydroxyacetone phosphate.

companied by a *gross* production of 38 or 40 ATP or a *net* production of 36 or 38 ATP. The ATPs result from substrate-level phosphorylations and electron transport coupled to oxidative phosphorylation. An accounting of ATP production is described in Figure 16–33. The different totals for ATP production (i.e., gross production of either 38 or 40) depend on which shuttle is used to transport H$^+$ from NADH into the mitochondrion. H$^+$ from the glycerol phosphate shuttle is accepted by F$_P$ and as a result one coupling step is bypassed. Hydrogen from the malate–aspartate shuttle enters the electron transport chain earlier, so that all three coupling sites are utilized.

SUMMARY

Mitochondria are the "powerhouses" of the cell and the primary sites of cell oxidations. Within these organelles, elementary substrates produced by the breakdown of carbohydrates, lipids, or nitrogenous macromolecules in other cell locations are oxidized to CO$_2$ and water. Some of the energy of these exergonic oxidations is conserved in the phosphorylation of ADP to form ATP.

Mitochondria occur in almost every type of aerobic eukaryotic cell. Although their shape may vary, all mitochondria contain two structurally and functionally different membranes—an **inner** and an **outer membrane**. Between the two membranes is the fluid-filled **intermembrane space**. The inner membrane surrounds the mitochondrial **matrix**. Projections of the inner membrane into the matrix are called **cristae**. New mitochondria form by the fission of other mitochondria.

The oxidative and phosphorylation reactions occur in the inner membrane or in the matrix. The tricarboxylic acid or *Krebs cycle* reactions constitute the first phase of the oxidation of substrates such as acetate. In these reactions, the molecules are enzy-

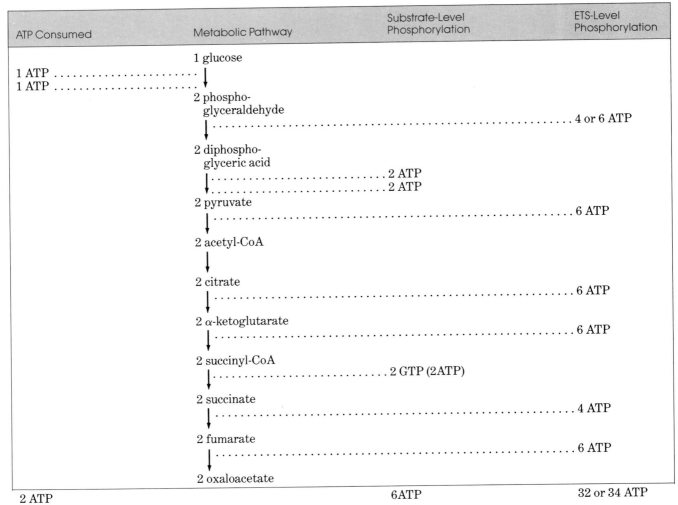

ATP Consumed	Metabolic Pathway	Substrate-Level Phosphorylation	ETS-Level Phosphorylation
1 ATP	1 glucose		
1 ATP	↓		
	2 phospho-glyceraldehyde		
	....		4 or 6 ATP
	↓		
	2 diphospho-glyceric acid		
	.... 2 ATP		
	↓ .... 2 ATP		
	2 pyruvate		
	....		6 ATP
	↓		
	2 acetyl-CoA		
	↓		
	2 citrate		
	....		6 ATP
	↓		
	2 α-ketoglutarate		
	....		6 ATP
	↓		
	2 succinyl-CoA		
	.... 2 GTP (2ATP)		
	↓		
	2 succinate		
	....		4 ATP
	↓		
	2 fumarate		
	....		6 ATP
	↓		
	2 oxaloacetate		
2 ATP		6ATP	32 or 34 ATP

FIGURE 16-33 Balance sheet for ATP production and consumption during glycolysis, the tricarboxylic acid cycle, and electron transport.

matically degraded to CO_2 and water. H^+ and electrons released from the metabolic intermediates reduce NAD^+ and FAD to NADH and $FADH_2$. Some ATP is formed at the substrate level by these reactions. Most of the ATP generated in mitochondria is via the **electron transport system** (ETS), which reoxidizes the NADH and $FADH_2$ formed in the Krebs cycle. The ETS functions as a multistep series of oxidation–reduction reactions, transferring electrons through a set of intermediates associated with the *inner membrane* and ultimately reducing molecular oxygen to water. These intermediates include *pyridine* or *pyrimidine-linked dehydrogenases, flavin-linked dehydrogenases, ubiquinones, iron-sulfur proteins,* and *cytochromes.*

Associated with the ETS is a mechanism for the formation of ATP called *oxidative phosphorylation.* For each pair of electrons transferred from NADH, three molecules of inorganic phosphate are added to three ADP to form three ATP (two ATP for each $FADH_2$ that is reoxidized). Oxidative phosphorylation is a function of the inner membrane. Mitchell's **chemiosmotic-coupling hypothesis** is the most widely accepted model explaining the source of the energy that drives phosphorylation of ADP to form ATP.

The mitochondrion is also the site of the **glyoxylate cycle** reactions, **fatty acid oxidation, fatty acid chain elongation,** *superoxide dismutase* and *catalase* reactions, and a number of **amphibolic** and **anaplerotic** reactions.

REFERENCES AND SUGGESTED READING

Articles and Reviews

Cohen, S. S., Mitochondria and chloroplasts revisited. *Am. Sci.* **61**, 437 (1973).

Cross, R. L., The mechanism and regulation of ATP synthesis by F_1-ATPases. *Ann. Rev. Biochem.* **50**, 681 (1981).

Elthon, T. E., and Stewart, C. R., A chemiosmotic model for plant mitochondria. *Bioscience* **33**, 687 (1983).

Ernster, L., and Schatz, G., Mitochondria: A historical review. *J. Cell Biol.* **91**, 227 (1981).

Fernandez-Moran, H., Cell membrane ultrastructure. *Circulation* **26**, 1039 (1962).

Fillingame, R. H. The proton-translocating pumps of oxidative phosphorylation. *Ann. Rev. Biochem.* **49**, 1079 (1980).

Fletcher, M. J., and Sanadi, D. R., Turnover of rat liver mitochondria. *Biochim. Biophys. Acta* **51**, 356 (1961).

Gray, M. W., The bacterial ancestry of plastids and mitochondria. *Bioscience* **33**, 693 (1983).

Guppy, M., and Ballantyne, J., The importance of water and oxygen in the evolution of hydrogen shuttle mechanisms. *Comp. Biochem. Physiol.* **69B**, 1 (1981).

Hateffi, Y., The mitochondrial electron transport and oxidative phosphorylation system. *Ann. Rev. Biochem.* **54**, 1015 (1985).

Hinkle, P. C., and McCarty, R. E., How cells make ATP. *Sci. Am.* **238** (3), 104 (Mar. 1978).

Mitchell, P., Keilin's respiratory chain concept and its chemiosmotic consequences. *Science* **206**, 1148 (1979).

Raven, P. H., A multiple origin for plastids and mitochondria. *Science* **169**, 641 (1970).

Schwartz, R. M., and Dayhoff, M. O., Origins of procaryotes, eucaryotes, mitochondria, and chloroplasts. *Science* **199**, 395 (1978).

Sjöstrand, F. S., The structure of mitochondrial membranes: A new concept. *J. Ultr. Research* **64**, 217 (1978).

Sjöstrand, F. S., The interpretation of pictures of freeze-fractured biological material. *J. Ultr. Research* **69**, 378 (1979).

Tanaka, K., Scanning electron microscopy of intracellular structures. *Int. Rev. Cytol.* **68**, 97 (1980).

Tedeschi, H., The mitochondrial membrane potential *Biol. Rev.* **55**, 171 (1980).

Wickner, W. T., and Lodish, H. F., Multiple mechanisms of protein insertion into and across membranes. *Science* **230**, 400 (1985).

Wilson, J. E., and Dove, J. L., Turnover of mitochondria in rat liver, kidney and heart. *J. Elisha Mitchell Sci. Soc.* **81**, 21 (1965).

Books, Monographs, and Symposia

Kozlov, I. A., Mitochondrial transhydrogenase: General principles of functioning. In *Current Topics in Membranes and Transport* (A. Kleinzeller and F. Bonner, Eds.), Academic, New York, 1982.

Lehninger, A. L., *Biochemistry* (2nd ed.), Worth, New York, 1975.

Lehninger, A. L., and Reynafarje, B., Cycles in the function of mitochondrial membrane transport systems. In *Current Topics in Cellular Regulation* (R. W. Estabrook and P. Srere, Eds.), Vol. 18, Academic, New York, 1981.

Lipmann F., The ATP-phosphate cycle, In *Current Topics in Cellular Regulation* (R. W. Estabrook and P. Srere, Eds.), Vol. 18, Academic, New York, 1981.

Margulis, L., *Symbiosis in Cell Evolution*, Freeman, San Francisco, 1981.

Meijer, A. J., and van Dam, K., Mitochondrial ion transport. In *Membrane Transport* (S. L. Bonting and J. J. de Pont, Eds.), Elsevier/North-Holland, Amsterdam, 1981.

Nicholls, D. G., *Bioenergetics: An Introduction to the Chemiosmotic Theory*, Academic, New York, 1982.

Roodyn, D. B., and Wilkie, D., *The Biosynthesis of Mitochondria*, Methuen, London, 1968.

Stryer, L., *Biochemistry* (2nd ed.), Freeman, San Francisco, 1981.

Tedeschi, H., *Mitochondria: Structure, Biogenesis and Transducing Functions*, Springer-Verlag, New York, 1976.

Wehrle, J. P., The role of electrogenic proton translocation in mitochondrial oxidative phosphorylation. In *Current Topics in Membranes and Transport* (A. Kleinzeller and F. Bonner, Eds.), Academic, New York, 1982.

CHAPTER 17

THE CHLOROPLAST

Chloroplasts are organelles found in the green tissues of plants and are responsible for the absorption of light energy, the synthesis of carbohydrates, and the evolution of molecular oxygen. The sum of these three processes is called **photosynthesis**. Light energy captured by the chloroplast is converted into potential chemical energy in the form of carbohydrates and starts the "energy chain" in nature. The oxygen evolved during the capture of light energy becomes the ultimate oxidizing agent for cellular metabolic reactions, as described in Chapters 10 and 16. Mitochondrial oxidations, as well as other oxidations in plant, animal, and many microbial cells, depend on this primary source of oxygen. It is currently believed that the entire supply of oxygen in the atmosphere today was derived from and is presently maintained by photosynthesis.

A **chloroplast** is any membrane-encased organelle containing **chlorophyll** that belongs to a group of related organelles in plants called **plastids**. The plastids have a variety of morphological forms, carry out diverse functions, and store many different compounds. For example, the **amyloplast** is the starch-storing plastid of potato tubers, and the **chromoplast** is the lycopene-containing plastid that gives the fruit of

tomatoes its red color. Each of the diverse plastids is believed to arise from a common *proplastid* precursor. All the major groups of plants, with the exception of the fungi, contain chloroplasts. There are two fundamental types of chloroplasts each associated with one of two types of photosynthesis: C_3 photosynthesis and C_4 photosynthesis. C_3 photosynthesis converts (or "fixes") carbon dioxide into three-carbon acids, whereas C_4 photosynthesis fixes CO_2 into four-carbon acids. Plants that carry out C_4 photosynthesis (called C_4 plants) are generally grasses or plants that are endemic to environments with little water. C_3 plants (those that carry out C_3 photosynthesis) are commonly broad-leafed flowering plants, cone-bearing plants, or those living in areas of adequate moisture. The characteristics of these two types of photosynthesis and their associated chloroplast types are discussed in detail later in this chapter. There may be a single chloroplast or dozens of chloroplasts in a cell. Frequently, the simpler plants, such as the algae, contain only one or a few chloroplasts per cell, whereas vascular plants, such as the cone-bearing and flowering plants, have many chloroplasts in each cell.

Photosynthesis occurs in some prokaryotic organisms such as the cyanobacteria (or blue-green algae)

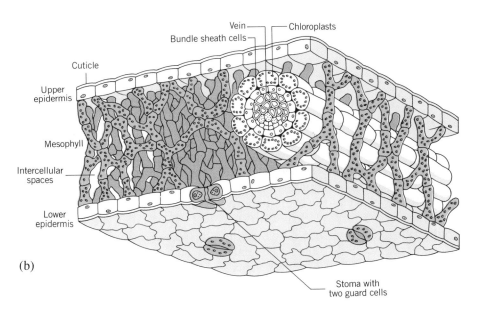

FIGURE 17–1 Arrangements of chloroplast-containing cells in leaves of plants carrying out C₃ photosynthesis (*a* and *c*) and C₄ photosynthesis (*b*). In plants employing C₃ photosynthesis (e.g., spinach, parsley, and other broad-leafed plants), chloroplasts are present in the *palisade layer* and in the *spongy mesophyll* layers that are sandwiched between the upper and lower *epidermis*. In plants that carry out C₄ photosynthesis (e.g., corn, sugar cane, and certain grasses), there is no palisade layer, but the veins of the leaf are encased by chloroplast-containing *bundle sheath cells*. (SEM photography from Troughton, J. H., and Sampson, F. B., *Plants*, John Wiley & Sons, 1973.)

and the anaerobic photosynthetic bacteria. These prokaryotes do not have true chloroplasts; instead, they have lamellated structures called **chromatophores** that carry out only the light-absorbing reactions of photosynthesis but not the carbohydrate-synthesizing reactions. For many laboratory studies of photosynthesis, the single-celled plant *Chlorella* is employed. This alga has one cup-shaped chloroplast that practically fills the cell. The organism can be easily and conveniently cultured with artificial lighting in the laboratory in solutions of inorganic salts. Because of the one-to-one relationship between cell and chloroplast and the ease with which the cells can be grown and enumerated, quantitative studies using chloroplast preparations of known content are possible.

For studies with chloroplasts of higher plants, leaves are generally used, and spinach and parsley leaves are probably the most popular source. In these leaves, the chloroplasts are found in greatest numbers in two internal tissues, the **palisade mesophyll** and the **spongy mesophyll** (Fig. 17–1a). Both tissues lie between an upper and lower epidermis and have thin cell walls that are easily broken. The number of chloroplasts in each cell varies within an organism with changing environmental conditions and varies greatly from one species to another. In spinach, there are between 20 and 40 chloroplasts in each palisade parenchyma cell. In the palisade parenchyma, the chloroplasts lie along the edges of the cell, the center of the cell being filled with large vacuoles. In the spongy parenchyma, the chloroplasts are more randomly distributed throughout the cytoplasm of the cell. In many genera, cytoplasmic streaming (i.e., **cyclosis**) moves the chloroplasts about the cell, and in a few instances an active amoeboid-type of movement by chloroplasts has been observed. Their positions in the cell and movement about the cell maximize the exposure of the chloroplasts to light.

Chloroplasts are routinely isolated from plant tissues by differential centrifugation following the disruption of the cells. Leaves are homogenized in an ice-cold buffered isotonic saline solution (e.g., 0.35 *M* NaCl) at pH 8.0. The disruption is generally carried out with bursts of a Waring blender. After filtration through a nylon gauze (20 μm pore size) to remove the larger particles of debris (cell nuclei, tissue fragments, and unbroken cells), the chloroplasts are separated by centrifugation at 200 *g* for 1 minute. The chloroplast-rich pellet is then resuspended and centrifuged again at 2000 *g* for 45 seconds to re-sediment the chloroplasts. Chloroplast preparations obtained by this procedure are generally mixtures of intact and broken organelles. Because the chemical composition, rate of photosynthetic activity, and other properties of intact chloroplasts differ significantly from those of damaged organelles, it is often desirable to separate the two populations. This may be accomplished by rate or isopycnic density gradient centrifugation of the chloroplast preparation (Chapter 12).

Chloroplast size is quite variable. Although the average diameter of a chloroplast in higher plant cells is between 4 and 6 μm, the size may fluctuate according to the amount of available illumination. In sunlight, chlorophyll is more readily synthesized by the plant, and the chloroplasts increase in size; in the shade, chlorophyll synthesis declines, and there is a corresponding reduction in chloroplast size. Changes in chloroplast shape are also observed after short-term exposure of plants to light. Short-term light exposure produces a small but measurable decrease in chloroplast volume. Presumably, this is due to a light-induced production of ATP, for the addition of ATP to chloroplasts in the dark causes a reduction in volume. Polyploid cells contain larger chloroplasts than comparable diploid cells.

The shape of most chloroplasts in higher plants is spheriod, ovoid, or discoid (Fig. 17–2). Other irregular shapes sometimes occur but are more common in lower plants. For example, in algae, cup-shaped, spiral, star-shaped, and digitate forms of chloroplasts are observed. The shape and structure of chloroplasts can also be altered by the presence of starch granules. During periods of active photosynthesis, the sugars formed in the chloroplasts are polymerized into starches that precipitate as small granules. The starch granules are usually ellipsoidal and may be up to 1.5μm long.

Fine Structure of the Chloroplast

The chloroplasts of C_3 plants are composed of two membrane layers similar to those of mitochondria. Each membrane is about 50 Å thick and the two membranes are separated by a space of about 70–100 Å. The **outer membrane**, which lacks folds or projections, serves to delimit the organelle and regulate the transport of materials between the cytoplasm and the interior of the organelle. The **inner membrane** parallels the outer membrane, but inward folds of this membrane are extensive. The inner membrane gives rise to a series of internal parallel membrane sheets

(a)

FIGURE 17–2 TEM photomicrographs of chloroplasts of sugar beet (*a*), grass (*b*), and *Chlorella* (*c*) (*chl*, chloroplast; *py*, pyrenoid; *st*, starch); (*d*) SEM photomicrograph of *Callisia* chloroplasts. (*a* and *b*, courtesy of Dr. W. Laetsch; *c*, courtesy of Drs. G. Dempsey, D. Lawrence, and V. Cassie; copyright © 1980 by Blackwell Scientific Publications, *Phycologia* **19**, 13; *d*, courtesy of Drs. R. Wise and J. Harris; copyright © 1980, *Cytologia* **45**, 113.)

(b)

(c)

(d)

FIGURE 17–3 (*a*) Cross section of a chloroplast showing the arrangement of lamellae and grana. (*b*) Grana and stroma thylakoids formed by the lamellae. (*c*) Chloroplast of a corn cell. (Courtesy of Dr. N. Rascio; copyright © 1981 by Springer-Verlag, *Protoplasma* **105**,241.)

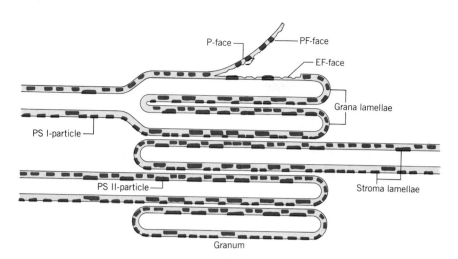

P-face — PF-face

EF-face

Grana lamellae

PS I-particle —

PS II-particle —

Stroma lamellae

Granum

FIGURE 17–4 Diagram of the stroma thylakoid lamellae and grana lamellae showing the arrangement of particles in the membranes and in its fracture faces.

called **lamellae** (Fig. 17–3). The lamellae are suspended in a granular fluid or matrix that appears somewhat electron-dense in electron photomicrographs. This matrix is referred to as the **stroma**. The lamellae form a complex series of membranes through the stroma.

Most of the lamellae in the chloroplasts of higher plants are organized to form disk-shaped sacs called **small thylakoids**. The small thylakoids are often arranged in stacks called **grana** (one stack is a *granum*) having diameters of 300 to 600 nm. Because the thylakoids are disk-shaped, the grana appear much like a stack of coins (Figs. 17–3a and 17–3b). A typical chloroplast has between 40 and 60 grana, and each granum may be composed of 2 to 100 small, flattened thylakoids. Frequently, a small portion of the thylakoid extends radially into the stroma forming a branching tube, or **large thylakoid**, that communicates with other small thylakoids and grana. Collectively, the branching and anastomosing network is called the **stroma lamellae**.

Structure of the Thylakoid

The adjacent membranes of neighboring thylakoids within each of the grana form thick layers called **grana lamellae** (Fig. 17–3). Electron photomicrographs of grana lamellae fixed with glutaraldehyde and stained with osmium reveal a five-layered arrangement consisting of three dark 40-Å-thick osmiophilic layers enclosing two 17-Å-thick osmiophobic spaces. Freeze-fracture techniques indicate that the grana membranes contain numerous particles. The

particles, which are primarily protein in composition, appear to be of two basic sizes, 105 and 140 Å in diameter (Fig. 17–4). The stroma lamellae contain mainly the smaller-size particles. The larger particles are abundant on the *EF* face of the grana membrane (see Chapter 15 for an explanation of the membrane fracture faces) and are associated with photosystem II of photosynthesis (described later in the chapter). The smaller particles are associated with photosystem I and are present in about equal numbers on the *PF* face of both the grana and stroma membranes (Fig. 17–5).

Stroma Structures

The granular stroma contains a variety of particles. The presence of starch granules was noted earlier. Electron micrographs also reveal a number of osmiophilic granules and groups of ellipsoidal structures called **stromacenters**. Strands of DNA and ribosomes are also scattered through the stroma.

The chloroplasts of C_4 plants have the same general structures as those of C_3 plants (an outer enclosing membrane, grana and stroma thylakoids, a stroma with DNA, ribosomes, and stromacenters), but there are some notable differences. The C_4 plant chloroplasts all have a **peripheral reticulum**, which is a group of anastomosing tubules about the periphery. In many C_4 plants such as corn there are two types of chloroplasts. Those found in the mesophyll cells (see Fig. 17–1b) are similar to the chloroplasts of C_3 plants but lack starch grains. The chloroplasts found in the

FIGURE 17–5 Freeze-fractured and sublimated grana membrane showing large and small membrane particles (Courtesy of Dr. K. Muhlethaler; copyright © 1965, by Springer-Verlag, *Planta* **67**, 305.)

cells surrounding the leaf veins (called *bundle sheath cells*) contain starch granules but have elongated grana.

CHEMICAL COMPOSITION OF CHLOROPLASTS

The organic constituent present in greatest quantity in the chloroplast is protein, which may represent up to 70% of the dry weight (Table 17–1). In leaf cells, 75% of the total cell nitrogen is found within the chloroplasts. Both structural and soluble proteins have been identified, but only a few of these have been extracted and purified. A peptide analysis by SDS gel electrophoresis shows compositional differences between stroma and grana lamellae, but the differences are primarily quantitative rather than qualitative. The relative compositions of stroma lamellae and grana lamellae are shown in Table 17–2.

Essentially all the pigments and cytochromes are located in the lamellae. The stroma lacks these com-

TABLE 17–1   CHEMICAL COMPOSITION OF SPINACH CHLOROPLASTS

Component	Percentage of Chloroplast Dry Weight	
	Chloroplasts Isolated in Water	Values Corrected for Loss of Soluble Protein
Total protein	50	69
Water-insoluble protein	50	31
Water-soluble protein	0	38
Total lipid	34	21
Chlorophyll	8	5
Carotenoids	1.1	0.7
Ribonucleic acids	—	1.0–7.5
Deoxyribonucleic acids	—	0.02–0.1
Carbohydrate (starch, etc.)	Variable	

Source: Modified from J. T. O. Kirk and R. A. E. Tilney-Bassett, *The Plastids,* Freeman, San Francisco, 1967.

TABLE 17–2 MAJOR COMPONENTS OF STROMA AND GRANA LAMELLAE

	Stroma Lamellae	Grana Lamellae
Total chlorophyll	278[a]	401
Chlorophyll a	238	281
Chlorophyll b	40	130
P_{700}	2.5	0.6
β-Carotene	21	17
Lutein	10	29
Violaxanthin	15	20
Neoxanthin	8	16
Phospholipid	76	66
Monogalactosyl diglyceride	231	214
Digalactosyl diglyceride	172	185
Sulfolipid	65	59
Cytochrome b (total)	1.0	3.4
Cytochrome f	0.5	0.7
Manganese	0.3	3.2

[a]Values in micromoles of component per gram of membrane protein.

pounds but contains DNA and RNA, which are not present in the lamellae. Most of the RNA is associated with the ribosomes of the stroma. The amount of DNA is low; estimates are 10^{-15} to 10^{-14} grams per chloroplast or about 0.03% of its dry weight. However, this is enough information to account for the synthesis of some chloroplast proteins, including some of the enzymes of photosynthesis. The disposition of chloroplast DNA during chloroplast division is unclear.

Lipid and lipid-soluble pigments account for about 34% of the dry weight of the spinach chloroplast. An exceedingly large number of different lipid compounds have been identified. The more common lipids are the galactosyl diglycerides, phospholipids, quinones (including vitamin K), and sterols.

The Chlorophylls

The green pigments of chloroplasts and the main sources of the color of green plants are the **chlorophylls**. Although a large number of chemically distinct chlorophylls have been identified in a variety of different plants, the structures of these chlorophylls are basically the same. (The structures of chlorophylls a and b are given in Fig. 17–6.) It is customary to identify each chlorophyll by a different letter. All photosynthetic plants have been found to contain chlorophyll a, but the presence of the secondary chlorophylls b, c, d, and e depends on the type of plant. Higher plants usually have chlorophyll b. In the photosynthetic bacteria, a chlorophyll called **bacteriochlorophyll** occurs in place of chlorophyll a. Together, chlorophylls a and b represent about 8% of the dry weight of spinach chloroplasts, with an $a{:}b$ weight ratio of 2.1 to 3.5. In most plants, the $a{:}b$ ratio varies according to the light intensity to which the plants are exposed. For example, alpine plants, which receive light of high intensity, have an average ratio of 5.5. The ratio is 2.3 in shade plants.

Chlorophyll a. $R = CH_3$

Chlorophyll b. $R = CHO$

FIGURE 17–6 Chemical structures of chlorophylls a and b.

FIGURE 17-7 *In vitro* absorption spectra of chlorophylls *a* and *b* in an ether solvent.

Each chlorophyll has a characteristic light absorption spectrum. The *in vitro* light absorption spectra of chlorophylls *a* and *b* are shown in Figure 17-7. Extracted chlorophyll *a* has absorption maxima at 430 and 670 nm, whereas the absorption maxima of chlorophyll *b* occur at 455 and 640 nm. *In vitro* absorption maxima of other plant and bacterial pigments are indicated in Table 17-3. The absorption spectrum and maxima of plant pigments vary according to the solvent used for extraction. Therefore, it is not surprising that values obtained during *in vivo* measurements differ from those yielded by extracts. For example, *in vivo* studies of chlorophyll *a* indicate that its native absorption maximum occurs at 677 nm. One very important form of chlorophyll *a* that is readily bleached by light has an absorption maximum at 700 nm. This form, which represents only about 0.1% of the total chlorophyll *a* molecules present in a sample, is called P_{700} or chlorophyll a_1. The roles of the various chlorophylls in photosynthesis are discussed later in the chapter.

The Carotenoids

The **carotenoids** are all long-chain isoprenoid compounds having an alternating series of double bonds. Although these compounds are synthesized only in plant tissue and participate in photosynthesis, they also serve as precursors of vitamin A in animal tissues. Most carotenoids are yellow, orange, or red. The formulas of α, β, and γ-carotene are shown in Figure 17-8, and their absorption maxima are given in Table 17-3. Most of these pigments are located in the chlo-

roplast lamellae and are believed to function as accessory pigments for light absorption during photosynthesis.

Location and Arrangement of the Pigments

Both the chlorophylls and the carotenoids are located almost exclusively in the chloroplast lamellae. The lamellae are about half protein and half lipid, the two pigments residing primarily in the lipid component. Some lipid is also represented by the osmiophilic granules of the stroma, but these are not believed to contain chlorophyll. Because each chlorophyll molecule has a hydrophilic portion (the tetrapyrrole) and a lipophilic portion (the phytyl chain), the chlorophyll molecules are thought to be aligned in a specific manner within the lamellae. The pyrrole groups form weak bonds with the lamellar protein. The carotenoids are dissolved in the lipid adjacent to the chlorophyll molecules.

The chloroplast stroma contains many of the enzymes associated with photosynthesis. Chloroplast protein synthesis also takes place in the stroma. Circular DNA strands about 40 μm long have been isolated from the chloroplast along with ribosomes and polyribosomes. The DNA strands have 115–200 kbp (kilobase pairs) and contain genes for about 180–280 polypeptides. Chloroplast ribosomes belong to the 70 S class and contain 23 S and 16 S RNA (see Chapter 22); thus, they are smaller than those found in the cytoplasm of plant (and animal) cells.

DEVELOPMENT OF CHLOROPLASTS

New chloroplasts are produced by the division of mature chloroplasts (Fig. 17-9) or by development from proplastids in the cell. The cells of young shoots of higher plants may contain 20 to 40 of these small ovoid bodies (Fig. 17-10). Apparently, proplastids can develop into a number of different plastid types in addition to chloroplasts. As a proplastid develops into a chloroplast, its inner membrane gives rise to infoldings that form the lamellae and thylakoids. Proplastids, as well as mature chloroplasts, increase in numbers by a form of division (Fig. 17-10b).

The control of chloroplast development is not fully understood. The differentiation of new organelles apα-Carotene

β-Carotene

γ-Carotene

FIGURE 17–8 Chemical structures of α-, β-, and γ- carotene.

TABLE 17–3 ABSORPTION MAXIMA[a] OF PLANT AND BACTERIAL PIGMENTS

Pigment	Wavelength [nm]	Occurrence
Chlorophyll a	430, 670	All green plants
Chlorophyll b	455, 640	Higher plants; green algae
Chlorophyll c	445, 625	Diatoms; brown algae
Bacteriochlorophyll	365, 605, 770	Purple and green bacteria
α-Carotene	420, 440, 470	Leaves; some algae
β-Carotene	425, 450, 480	Some plants
γ-Carotene	440, 460, 495	Some plants
Luteol	425, 445, 475	Green leaves; red and brown algae
Violaxanthol	425, 450, 475	Some leaves
Fucoxanthol	425, 450, 475	Diatoms; brown algae
Phycoerythrins	490, 546, 576	Red and blue-green algae
Phycocyanins	618	Red and blue-green algae
Allophycoxanthin	654	Red and blue-green algae

[a]Absorption maxima vary according to the solvent in which the pigment is dissolved.

Source: From E. Rabinowitch and Govindjee, *Photosynthesis*, Wiley, New York, 1969.

(a)

(b)

(c)

(d)

FIGURE 17–9 Electron photomicrographs of dividing chloroplasts. (*a*) In this SEM view of spinach cells, the chloroplasts are arranged in a single layer close to the cell wall; their constricted (i.e., dumbbell) shape is believed to represent a terminal stage in the division of the organelles by a fissionlike process. The small bodies clinging to the surface of the chloroplasts are probably mitochondria. Magnification, 6000X. (*b*) Freeze-facture TEM view of a constricted chloroplast. Magnification, 16,000X. (c) Chloroplasts revealing different degrees of constriction are seen in this light photomicrograph of a living spinach cell. Magnification, 900X. (*d*) Sequence of chloroplast division stages in *Cricosphaera*. (*a, b,* and *c,* courtesy of Dr. J. V. Possingham; copyright © 1980 by Company of Biologists, Ltd.; *J. Cell Science* **46**, 87. *d,* courtesy of Drs. T. Hori and I. Inouye; copyright © 1981 by Springer-Verlag, *Protoplasma* **106**, 121.)

(a)

FIGURE 17–10 *Proplastids.* (*a*) In this TEM view of a dividing spinach cell, several proplastids (*p*) and mitochondria (*m*) are seen on either side of the metaphase chromosomes (*c*). Magnification, 9000X. (*b*) Dividing proplastid in a spinach root tip cell. Magnification, 44,000X. (Courtesy of Dr. J. V. Possingham; copyright © 1981, Societe Francaise de Microscopie Electronique; *Biologie Cellulaire* **41**, 203.)

(b)

pears to rely on an interaction between genetic information present in the cell nucleus and information present in the chloroplast itself. Chloroplasts contain DNA that is transcribed and translated within the organelle to form some of the chloroplast proteins; however, other chloroplast proteins are derived from transcription and translation of nuclear DNA.

The intriguing but controversial hypothesis that chloroplasts are the evolutionary products of bacterialike organisms that "invaded" eukaryotic cells and established a symbiotic relationship with the host cell is similar to the proposal for the origin of mitochondria discussed in Chapter 16. In the case of the chloroplast, the host could have been a heterotrophic cell. Interestingly, chloroplasts are found in the cells of a subgroup of nudibranchs (marine snails lacking a shell) that feed on algae. As the algae cell cytoplasm passes through the intestine of the nudibranch, whole chloroplasts are absorbed into the tissue and may persist there for the life of the animal. When the animal is in light, the absorbed chloroplasts evolve oxygen.

PHOTOSYNTHESIS—HISTORICAL BACKGROUND

The overall reactions of photosynthesis may be summarized by the equation

$$6\,CO_2 + 12\,H_2O \xrightarrow[\text{chlorophyll}]{\text{light}} C_6H_{12}O_6 + 6\,O_2 + 6\,H_2O$$

Studies by many individuals over the past 300 years have led to our present understanding of this process. One of the first studies, by *Jan Baptista van Helmont* (Dutch, 1577–1640), was published in 1648 after his death. Van Helmont planted a 5-lb willow shoot in a large pot containing 200 lb of soil. The plant was regularly watered over a 5-year period and was then carefully removed and weighed. The willow tree was found to weigh almost 170 lb, but the original soil weighed only a few ounces less than the original 200 lb. Van Helmont concluded that the increase in weight of the willow tree was due primarily to the addition of water, for he was not aware of the role played by gases in the air in the growth of the plant. In 1727, *Stephen Hales* proposed that part of a plant's nourishment came from the air and that light was involved in the process.

Today, we also recognize that the small quantity of material removed from the soil included minerals, nitrogen, phosphate, and calcium salts.

Joseph Priestley (English, 1733–1804) was the first to show that plants exchange gases with the atmosphere. In particular, Priestley found that oxygen was evolved by plants during photosynthesis. Late in the eighteenth century, *Jan Ingenhousz* (Dutch, 1730–1799) showed that oxygen was produced by the green parts of plants when exposed to light. Furthermore, he observed that the amount of oxygen produced varied according to the amount of light to which the plant was exposed. In 1804, *Nicholas de Saussure* (Swiss, 1767–1845) found that the increase in the carbon content of plants resulted from the accumulation of carbon from carbon dioxide in the air. De Saussure also showed that leaves respire in darkness, that is, they take in oxygen and release carbon dioxide.

The studies of *Julius Sachs* in the mid-nineteenth century show that chlorophyll was confined to the chloroplasts and was not distributed throughout the plant cell. He also showed that sunlight caused chloroplasts to absorb carbon dioxide and that chlorophyll is formed in chloroplasts only in the presence of light. Sachs also noted that one of the products of photosynthesis was starch. It was not until 1918 that Wilstatter and Stoll isolated and characterized the green pigments chlorophylls *a* and *b*.

On the basis of these early findings, the reactions of photosynthesis were described by the equation

$$6\,CO_2 + 6\,H_2O \xrightarrow[\text{chlorophyll}]{\text{sunlight}} (C_6H_{12}O_6)_n + 6\,O_2$$

However, photosynthesis is not simply the fusing of carbon dioxide and water molecules through the use of light energy. Instead, two complex series of chemical reactions are involved. One set of reactions, called the **photochemical** or **light reactions**, occurs in the lamellae of the chloroplasts. In these reactions, light energy is used to form ATP and water molecules are split, releasing oxygen and hydrogen. The hydrogen is used in the reduction of $NADP^+$ and is not evolved as a gas. The second set of reactions, called the **synthetic** or **dark reactions**, occurs in the stroma of the chloroplast, and although these reactions do not require light, they do depend on the availability of ATP and NADPH from the photochemical reactions to reduce the carbon dioxide thereby forming sugars.

PHOTOSYNTHESIS—THE PHOTOCHEMICAL (LIGHT) REACTIONS

The Absorption of Light by Chlorophyll

The absorption of electromagnetic energy by any atom or molecule often involves a shift of electrons from one atomic orbital to another. Each electron possesses energy, and the amount is determined by the location of the electron orbital in space and the velocity at which the electron moves. When an atom absorbs light energy, an electron is either raised to an orbital of higher energy level or accelerated in its orbit. In either case, certain discrete quantities of energy are required, for when light photons have either too much or too little energy, they are not absorbed. Electrons may orbit in pairs within an orbital, the members of the same orbital spinning in *opposite* directions. Most atoms at their lowest energy level (i.e., *ground state*) have all their electrons paired in this fashion and are said to be in the *singlet* (i.e., S) state. When a photon of light is absorbed and an electron is thereby raised to a higher, unoccupied orbital, it may continue to spin in the direction opposite to that of its former partner (in which case the atom is still in the S state), or it may now spin in the same direction as its former partner (in which case the atom is said to be in the *triplet* or T state).

In a molecule, some electrons orbit exclusively about specific atomic nuclei; others may be shared be-tween two nuclei forming a bond (called localized or π^* electrons), or may orbit about several nuclei (called delocalized or π electrons). The absorption of a light photon may move a π electron to a π^* position.

Chlorophyll is normally a singlet in its ground state. Although the absorption of light causes some π electrons to be raised to a π^* orbital, chlorophyll remains in a singlet state. When red (680 nm) light is absorbed, an electron is raised to a higher, $S\pi^*$, orbital. Blue (430 nm) light possesses more energy per photon than red light, and its absorption raises an electron to a higher (but still $S\pi^*$) energy state. These transitions are summarized in Figure 17–11.

Once a molecule has absorbed light energy and is in an excited state, the ground state may be reestablished in three different ways: (1) the energy may be reemitted in the form of radiation of longer wavelength (i.e., *fluorescence*); (2) the energy may be converted to heat; and (3) the excited molecule may transfer its excess energy to another molecule. The transfer of energy from one molecule to another often involves the exchange of a high-energy electron for one of lower energy. Referring to Figure 17–11, the blue light absorbed by chlorophyll raises an electron to the $S^b\pi^*$ state. By losing energy in the form of heat, that electron could "drop" back to the $S^a\pi^*$ state. The electron could then drop back from the $S^a\pi^*$ state to the $S\pi$ state (i.e., the ground state) by the immediate loss of energy as heat or fluorescence. Another possibility also exists, for the electron could drop from the $S^a\pi^*$ state to a triplet (i.e., $T\pi^*$) state by heat loss and then to the ground state by either *phosphorescence* (delayed fluorescence) or heat loss. Concentrated solutions of extracted chlorophyll will strongly fluoresce red when placed in a beam of sunlight or ultraviolet light. Finally, it should also be noted that if the chlorophyll molecule is sufficiently excited, it may not return to the ground state by heat loss or reradiation; instead, the excited electron may be transferred to another molecule, leaving the chlorophyll in a temporary, oxidized state.

Primary Photochemical Events in Photosynthesis

As shown in Figure 17–11, chlorophyll absorbs both blue and red light, and this raises electrons to $S^b\pi^*$ or $S^a\pi^*$ states. The return of an electron from the $S^b\pi^*$ state to the $S^a\pi^*$ state is extremely fast (about 10^{-12} seconds) and does not afford an opportunity for the energy to be lost by fluorescence or by transfer to an-

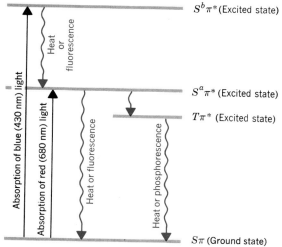

FIGURE 17–11 Main electronic states in the excitement of chlorophyll by red and blue light.

other molecule. Consequently, the energy is lost as heat. The decay of an electron from the $S^a\pi^*$ state *does* permit the transfer of energy to another molecule, and this is the event that initiates photosynthesis. Consequently, a photon of red light is just as effective in initiating photosynthesis as a photon of blue light, even though the former is much less energetic.

As we have already noted, the chloroplast contains many different pigment molecules (other chlorophylls, carotenoids, phycobilins, etc.) and the electrons of these molecules may be excited to various energy states by the absorption of light. As these excited accessory pigment molecules return to the ground state, the resulting energy is transferred to chlorophyll *a* molecules, causing their excitation. Because the chlorophyll *a* molecules present in a thylakoid vary in their absorption maxima, varying quantities of energy are required to raise their electrons to the $S^a\pi^*$ state. The molecule that requires the least amount of energy is believed to be a pigment that absorbs long, red wavelengths, namely, the P_{700} molecule. (Pigments may be identified by a letter [e.g., "P"] followed by a numerical subscript. The subscript indicates the wavelength of the light absorbed by the pigment.) It seems reasonable that light energy captured by the accessory pigments and transferred to chlorophyll *a* is in turn transferred from the latter to P_{700}. Each accessory pigment or chlorophyll *a* molecule can only pass its energy on to pigments having absorption maxima of longer wavelength because these require less energy to be activated to the $S^a\pi^*$ state. Because P_{700} has its absorption maximum at the longest wavelength, it serves as the final energy trap in a part of the photosynthetic unit called a *reaction center*. There are two types of reaction centers, each with an ultimate energy-trapping chlorophyll *a* molecule. One is P_{700} and the other is P_{680}.

Two Photosystems

When some plants are exposed to light containing only wavelengths of 690 nm or longer, photosynthetic efficiency decreases. The effect is called the **red drop** (Fig. 17–12). Because the absorbed energy is funneled to P_{700}, it would be expected that the absorption of light by the accessory pigments, chlorophyll *a*, or even P_{700} should be equally efficient. The efficiency can be increased through the addition of shorter-wavelength radiations. This *enhancement* phenomenon can increase the photosynthetic rate 30 to 40% above the rate obtained by either the short wavelength or long

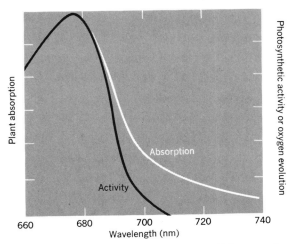

FIGURE 17–12 *The photosynthetic activity or action spectrum (i.e., oxygen evolution spectrum) and absorption spectrum of the green alga Ulva. The difference between the two curves at wavelengths above 590 nm is called the red drop.*

wavelength alone. The synergistic effect of the two different wavelengths led early investigators to conclude that two distinct photochemical reactions exist. In one of these (called **photosystem I,** PS–I), P_{700} serves as the ultimate energy trap or *reaction center*; in the other (called **photosystem II,** PS–II), the ultimate reaction center is a chlorophyll that has its absorption maximum at 680 nm and is called P_{680}. Within photosystems I and II, the organization of auxiliary pigments and chlorophylls form *antenna complexes*. These complexes act to direct light energy to the reaction center molecules P_{700} and P_{680}.

In higher plants, photosystem I is a unit containing several hundred molecules of chlorophyll (mostly chlorophyll *a*), about 50 carotenoids, one cytochrome *f*, one plastocyanin, two cytochrome b_{564} molecules, one or two ferredoxin molecules, and one molecule or a dimer of chlorophyll P_{700}. Photosystem II has about 200 molecules of chlorophylls *a* and *b* absorbing light at a wavelength that is less than 680 nm, 50 carotenoids, one P_{680} molecule or dimer (the primary electron donor), a primary electron acceptor that is a quinone, four plastoquinones, six manganese atoms, and two cytochrome b_{559} molecules.

Sequence of Energy (Electron) Flow

The absorption of light by chlorophyll P_{700} or P_{680} alters the state of the orbiting electrons, and if the

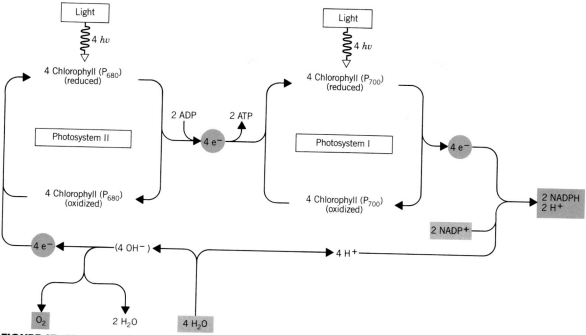

FIGURE 17–13 Absorption of light energy by photosystems I and II and the resulting movements of electrons. Electrons transferred from photosystem II to photosystem I are regained by splitting water molecules. Electrons from photosystem I are used together with H^+ from water to reduce $NADP^+$ (see text for additional details).

energy is not lost by reradiation or heat, the excited electrons can be transferred to another compound. Such an electron loss "bleaches" the chlorophyll and leaves it in an oxidized state. Oxidized P_{700} may be reduced by absorbing an electron from photosystem II, whereas the reduction of the oxidized chlorophyll of photosystem II is brought about by the oxidation of water (Fig. 17–13).

In 1938, R. Hill demonstrated that isolated chloroplasts exposed to light could evolve oxygen and reduce a variety of compounds without consuming carbon dioxide. Three years later S. Ruben and M. Kamen, using the isotope ^{18}O, were able to show that the oxygen liberated during whole plant photosynthesis was derived from water molecules, that is,

$$6\ CO_2 + 12\ H_2{}^{18}O \xrightarrow{\text{light}} C_6H_{12}O_6 + 6\ {}^{18}O_2 + 6\ H_2O$$

The method by which the water molecules are split is unknown, although four water molecules are required for the evolution of one oxygen molecule and four quanta of light are necessary:

$$4\ H_2O \xrightarrow{4\ hv} O_2 + 4\ (H^+ + e^-) + 2\ H_2O$$

Protein-bound Mn^{2+} and Cl^- may also be required. Some of these reactions are summarized in Figure 17–13.

Redox Reactions

Electrons that are released from chlorophylls P_{700} and P_{680} pass from one to another of a series of electron transfer molecules in the lamellae. Each of these molecules has a specific electrical potential called a **redox potential** (see Chapter 9). The relative values of the redox potentials of the molecules involved in the transfer of electrons determine the order in which each molecule participates in the transfer. During electron transfer, the molecules accept electrons from less positive (more negative) molecules and donate electrons to more positive (less negative) molecules. When a molecule accepts electrons, it is said to be *reduced*, and when electrons are given up, the molecule is said to be *oxidized*. The transfer of electrons following the absorption of light energy by chlorophyll therefore involves a sequence of oxidation–reduction reactions. This sequence is shown in Figure 17–14, which also identifies the intermediate electron acceptors and their redox potentials.

Photosystem I is located in the membranes of both the grana thylakoids and stroma thylakoids, whereas photosystem II is found only in the membranes of the grana thylakoids. Therefore, photosystem I functions independently in the stroma thylakoids but functions in conjunction with photosystem II in the grana thylakoids. Absorption of light energy by P_{700} of photosystem I causes the loss of electrons first to an intermediate iron–sulfur protein and then to a molecule of ferredoxin imbedded in the thylakoid membrane (i.e., "bound ferredoxin" in Fig. 17–14). Bound ferredoxin is reoxidized by the transfer of electrons either to a soluble ferredoxin bound to the FAD-containing enzyme *ferredoxin-NADP reductase* or to cytochrome b_{563} (see below). *Ferredoxin-NADP reductase* transfers the electrons to FAD, which also accepts H^+ (from water) and reduces the $NADP^+$, that is,

$$NADP^+ + 2(H^+ + e^-) \rightarrow NADPH + H^+$$

The NADPH and H^+ spill into the stroma, where the NADPH is reoxidized during the dark reactions of photosynthesis (described later).

In the alternative pathway leading to the reoxidation of bound ferredoxin, electrons are transferred to cytochrome b_{563}. Cytochrome b_{563} initiates a sequence of redox reactions, passing the electrons on to cytochrome f (also called cytochrome b_{552}), plastocyanin (PC), and ultimately to P_{700}, completing a cycle. This cyclic set of redox reactions is coupled to the phospho-

rylation of ADP (*cyclic photophosphorylation*, described below).

In the grana lamellae the P_{700} oxidized by the absorption of light can be reduced by the cyclic return of the electrons from ferredoxin or by the transfer of electrons from photosystem II. Light energy absorbed by the accessory pigments of the reaction center of photosystem II is ultimately transferred to chlorophyll P_{680} (Fig. 17–14). The ejected electrons are absorbed by an electron acceptor called "Q," whose identity is still uncertain but which is believed to be *plastoquinone* (PQ). This photoevent initiates another set of redox reactions in which the electrons pass from Q to PQ, cytochrome b_{559}, cytochrome f, plastocyanin, and ultimately to P_{700}. The electrons in photosystem II do not cycle back to chlorophyll P_{680} but their transfer through the redox reactions to P_{700} is coupled to phosphorylation of ADP (*noncyclic photophosphorylation*, described below).

The reduction of the oxidized P_{680} chlorophyll is brought about by the oxidation of water via an enzyme system that is closely associated with the structure of the thylakoid and that also produces molecular oxygen. The enzyme system presumably processes four protons simultaneously to produce O_2.

$$2 H_2O \rightarrow O_2 + 4 H^+ + 4 e^-$$

Mn^{2+} is known to be a cofactor in the system. The electrons produced act to reduce the light-oxidized

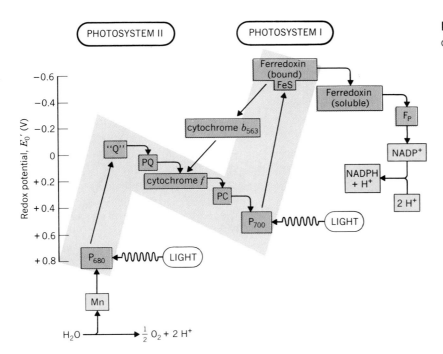

FIGURE 17–14 Movements of electrons during the light reactions of photosynthesis.

P_{680}, and the H^+ forms a pool available for the reduction of $NADP^+$.

Cyclic and Noncyclic Photophosphorylation

Phosphorylation of ADP occurs in chloroplasts during the light reactions and is called **photophosphorylation**. The photophosphorylation occurs in the lamellae of the stroma and grana thylakoids as a part of both photosystems I and II. There is a clear similarity between the mechanism of photophosphorylation in chloroplasts and electron transport system phosphorylation in mitochondria (Chapter 16). In both organelles, the mechanism is closely associated with a membrane and with a compartment separated from the rest of the organelle. In the chloroplast, the compartment is the loculus of the thylakoid; in the mitochondrion, it is the inner membrane-encapsulated matrix. In both organelles, the phosphorylation that occurs is coupled to electron transport through a sequence of redox reactions. Several of the molecules that participate in the redox reactions are similar, including the quinones and cytochromes.

Cyclic photophosphorylation occurs when electrons released from P_{700} following the absorption of light energy are cycled back to P_{700} through ferredoxin, cytochrome b_{563}, cytochrome f, and plastocyanin. The energy from these exergonic redox reactions is coupled to phosphorylation. **Noncyclic photophosphorylation** occurs when electrons released from P_{680} of photosystem II are shuttled via plastoquinone, cytochrome f, and plastocyanin to P_{700} of photosystem I. The energy from these noncyclic exergonic redox reactions is also coupled to phosphorylation.

At the present time, the mechanism of coupling is not completely understood, although much evidence supports the concept that a proton gradient generated across the thylakoid membrane supports chemiosmotic coupling much like that described in Chapter 16 for mitochondria and illustrated in Figure 17–15. Chloroplasts illuminated in an unbuffered medium quickly cause the medium to become alkaline, implying that protons are transported from the medium into the thylakoid. In darkness the system reequilibrates. If the medium is made alkaline, phosphorylation of ADP occurs in the dark. Compounds that degrade the thylakoid membrane (such as detergents) also cause leakage of protons from the thylakoid and prevent phosphorylation.

The standard free energy change that takes place when ATP is formed from ADP and P_i is about 30.5 kJ/mole (7.3 kcal/mole); this is equivalent to a redox potential of about 0.45 to 0.61V. Most of the measurements that have been made indicate that one molecule of ATP is produced for each pair of electrons moved from photosystem II to photosystem I (i.e., noncyclic photophosphorylation). Several recent studies strongly suggest that more than one molecule of ATP is produced by the reactions that begin with the splitting of water (i.e., **photolysis**) and end with the reduction of $NADP^+$. The second ADP phosphorylation is believed to be coupled to the photolysis. If the number of ATP molecules produced during the light reactions is to balance the number consumed by the dark reactions, then three ATP must be produced for each pair of electrons transported from H_2O to $NADP^+$. The third ATP may be generated by cyclic photophosphorylation.

Another light-induced phosphorylation of ADP called **pseudocyclic photophosphorylation** occurs when reduced ferredoxin transfers electrons through a series of reactions to O_2 (not to $NADP^+$ or P_{700}), with the result that water and ATP are formed.

Summary of the Light Reactions

Two photosystems function during the light reactions of photosynthesis. As each system absorbs four quanta of light energy, chlorophylls P_{680} (photosystem II) and P_{700} (photosystem I) are excited, releasing two pairs of electrons to acceptor molecules. Chlorophyll P_{680} is returned to its reduced state by the oxidation of two water molecules, and in the process one molecule of oxygen is evolved. P_{700} is returned to its reduced state by electrons that are derived from photosystem II. The transfer of two pairs of electrons from water through photosystem II to photosystem I is coupled to the photosphosphorylation of two ADP molecules, thereby producing two molecules of ATP and two molecules of water. The two pairs of electrons released from P_{700} of photosystem I reduce two $NADP^+$ forming two NADPH and two H^+. Therefore, the net result is

$$2\,H_2O + 2\,NADP^+ + (2\,ADP + 2\,P_i)$$

$$\Big\downarrow\ 8\ \text{quanta}$$

$$O_2 + 2\,NADPH + 2H^+ + (2\,ATP + 2\,H_2O)$$

The two ATP and two NADPH molecules produced by the light reactions occurring in the grana lamellae

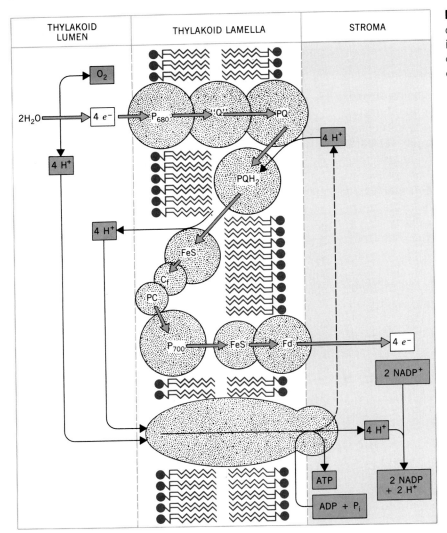

FIGURE 17-15 Model for *chemiosmotic coupling*. Locations of electron transfer intermediates in the thylakoid lemella and enzymes involved in ATP synthesis and proton (H^+) transport are tentative.

are used in the synthetic (dark) reactions that take place in the stroma. The latter reactions fix CO_2 into sugars. For each CO_2 fixed, two NADPH (and two H^+) and three ATP molecules are required (see below). Where the third ATP molecule is formed is not yet clear; it may be obtained from cyclic photophosphorylation or pseudocyclic photophosphorylation.

PHOTOSYNTHESIS—SYNTHETIC (DARK) REACTIONS

The Calvin Cycle

The elucidation of the sequence of chemical reactions that result in the incorporation of carbon dioxide into sugars and starches relied heavily on the use of radio-

active isotopes. Using ^{14}C-labeled carbon dioxide, it was possible to add [14]CO_2 at known times to an actively photosynthesizing system, halt the process a short time later, and then identify the compounds into which the labeled carbon dioxide became incorporated. Identification of the ^{14}C-containing intermediates was carried out using combined paper chromatography (Chapter 13) and autoradiography (Chapter 14).

Ruben and co-workers first showed that the active form of CO_2 in the chloroplast was carbonic acid. M. Calvin and co-workers established the sequence of reactions that follow the formation of carbonic acid and its entry into the chloroplast. They added [14]CO_2 to cultures of the alga *Chlorella* and allowed the cells to photosynthesize for given periods of time (usually between 2 and 60 seconds). The *Chlorella* cells were then killed and the soluble cell components were extracted and concentrated. The extracts containing radioactive carbon were chromatographed on paper, and the spots containing radioactivity were identified. In 1961, Calvin received the Nobel Prize for this most important series of experiments.

When photosynthesis in the presence of $^{14}CO_2$ was allowed to proceed for only 2 seconds, the major labeled compound identified was 3-phosphoglyceric acid (PGA). After 7 seconds, sugar phosphates and diphosphates were found in addition to PGA. A 60-second

exposure to $^{14}CO_2$ produced labeled phosphoenolpyruvic acid (PEP), carboxylic acids, and amino acids. Using many different time intervals, the entire sequence of reactions was uncovered, and it was found that many of the steps were the reverse of those in the glycolytic pathway (Fig. 17–16).

In the chloroplast stroma, CO_2 in the form of carbonic acid reacts with the sugar ribulose diphosphate (RuDP) to form an unstable six-carbon compound that immediately splits to form two molecules of 3-phosphoglyceric acid. The enzyme catalyzing this reaction is *ribulose-1,5-diphosphate carboxylase* and the

FIGURE 17–16 Initial steps in the fixation of CO_2 during photosynthesis. C* = ^{14}C, which traces the path of carbon through the reaction sequence. Enzymes: (1) *RDP-carboxylase* or *carboxydismutase*; (2) *phosphoglyceric acid kinase*; (3) *triose phosphate dehydrogenase*; (4) *isomerase*; (5) *fructose diphosphate aldolase*; (6) *fructose diphosphatase*.

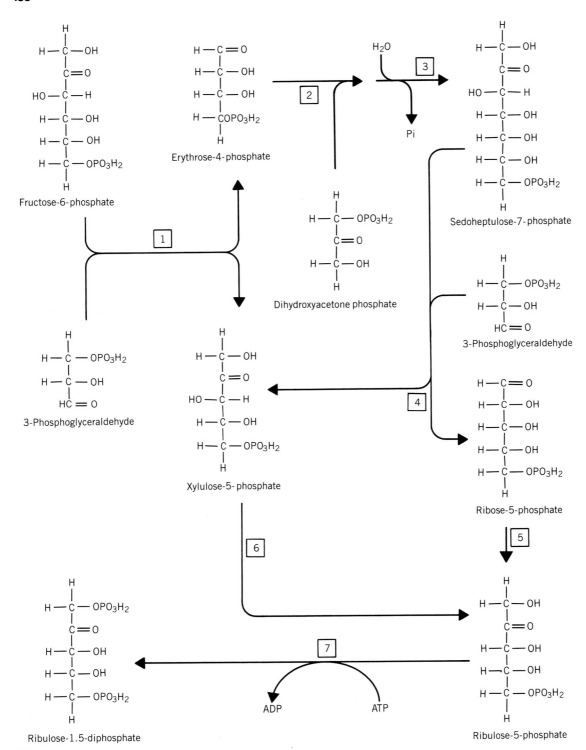

FIGURE 17–17 Resynthesis of ribulose-1,5-diphosphate. Enzymes: (1) *transketolase*;
(2) *aldolase*; (3) *sedoheptulose-1,7-diphosphatase*; (4) *transketolase*; (5) *isomerase*;
(6) *epimerase*; (7) *kinase*.

radioactive carbon of $^{14}CO_2$ is incorporated into the carboxyl group of PGA.

PGA is then reduced to 3-phosphoglyceraldehyde (PGAL) in two steps. First, each PGA is phosphorylated by ATP and then reduced by NADPH. The ATP and NADPH were produced by photochemical reactions in the grana lamellae. Thus, for each molecule of CO_2 fixed and converted to PGAL, two ATP and two NADPH molecules are required. These reactions are catalyzed by a kinase and a dehydrogenase.

Some of the PGAL is isomerized by *triose phosphate isomerase* to form dihydroxyacetone phosphate (DHAP). The enzyme *aldolase* then condenses PGAL and DHAP to produce fructose-1,6-diphosphate (FDP). *Fructose diphosphatase* splits off the phosphate group at the first carbon atom, producing fructose-6-phosphate (F6P). F6P may then be converted to fructose, glucose, or starch.

F6P and PGAL are also used for the resynthesis of RuDP (Fig. 17–17). F6P and PGAL are converted to erythrose-4-phosphate (E4P) and xylulose-5-phosphate (X5P). An aldolase then catalyzes the condensation of E4P and DHAP to form sedoheptulose-

1,7-diphosphate (SDP), which is then converted to sedoheptulose-7-phosphate (S7P). S7P and PGAL also react to form ribose-5-phosphate (R5P) and X5P. The R5P is then isomerized to form ribulose-5-phosphate (Ru5P). Ru5P is also formed from X5P. Finally, ATP is used to phosphorylate Ru5P, thereby forming RuDP.

For each CO_2 fixed, one RuDP, two ATP, and two NADPH are consumed and one phosphate sugar is produced. In actuality, all the RuDP is resynthesized. Fixation of every three molecules of CO_2 results in the formation of six PGAL. Five of the these are recycled to replenish the pool of RuDP. The extra PGAL is the photosynthetic product and is used for sugar and starch synthesis. Figure 17–18 summarizes all of the steps and the requisite numbers of molecules participating in the dark reactions; the pathway is known as the **Calvin cycle**.

Because additional ATP is consumed in the formation of RuDP, a total of nine ATP and six NADPH are required for the fixation of three CO_2 molecules. Since six CO_2 molecules are required to produce one six-carbon sugar molecule, 18 ATP molecules are consumed in the fixation (or three ATP per CO_2).

FIGURE 17–18 The path of carbon in photosynthesis. The molecule of PGAL in the box is the photosynthetic product and is employed for sugar and starch synthesis. Ru5P, ribulose-5-phosphate; RuDP, ribulose-1, 5-diphosphate; PGA, 3-phosphoglyceric acid; PGAL, 3-phosphoglyceraldehyde; DHAP, dihydroxyacetone phosphate; F6P, fructose-6-phosphate; SDP, sedoheptulose-1,7-diphosphate, S7P, sedoheptulose-7-phosphate; E4P, erythrose-4-phosphate; X5P, xylulose-5-phosphate; R5P, ribose-5-phosphate; P_i, inorganic phosphate.

It has been estimated that about 5×10^{16} grams of carbon are fixed annually by photosynthesis and this corresponds to a storage of 20.1×10^{17} kJ or 4.8×10^{17} kcal of energy. Because about 28.0×10^{21} J or 6.7×10^{21} kcal of light energy fall on the earth each year, photosynthesis traps a mere 0.0072%.

C_4 Photosynthesis (Hatch–Slack Pathway)

The C_3 or Calvin cycle is not the only metabolic pathway found in plants for fixing CO_2. M. D. Hatch and C. R. Slack showed that in some plants, CO_2 is also fixed into four-carbon compounds. This mechanism (which operates *in conjunction* with the Calvin cycle in the leaves of these plants) has since come to be known as the **Hatch–Slack** or **C_4 pathway**. Plants that employ the Hatch–Slack pathway (e.g., corn, sugar cane, and certain grasses) are characteristically found in arid environments and possess an unusual leaf anatomy (Fig. 17–1*b*). In the leaves of plants employing only the Calvin cycle, the C_3 reactions occur in both the palisade mesophyll and spongy mesophyll layers. However, in plants that can carry out the C_4 reactions, the Calvin cycle occurs only in the layer of *bundle sheath cells* that surrounds the veins of the leaf. The C_4 reactions take place in the mesophyll layers that lie immediately under the upper and lower epidermis. In addition to their unusual leaf anatomy, plants using the Hatch–Slack pathway are characterized by high photosynthetic and growth rates, low photorespiration rates, and low rates of water loss through the stomata of the leaves.

FIGURE 17–19 Electron photomicrograph of bundle sheath cells and adjacent mesophyll cells. (Courtesy of Dr. C. C. Black, *Plant Physiol.* **47**, 15 (1971). The diagram shows the C_4 photosynthetic reactions associated with each type of cell.

As in the Calvin cycle, the sequence of reactions that constitutes the C_4 pathway (and the relationship of these reactions to the Calvin cycle) was worked out using ^{14}C-labeled carbon dioxide. The reactions are summarized in Figure 17–19. Carbon dioxide passes from the air surrounding the leaves through the epidermis and into the mesophyll cells. Here the CO_2 condenses with phosphoenolpyruvate (PEP, a three-carbon compound) to form oxaloacetate (a four-carbon compound). The reaction is catalyzed by the enzyme *phosphoenolpyruvate carboxylase*. In some plant species, the oxaloacetate is then converted to malate, whereas in others it is converted to aspartate. The malate (or aspartate) then passes into the bundle sheath cells, where it is decarboxylated within the cells' chloroplasts. As seen in Figure 17–19, the CO_2 that is released contains the same carbon atom that entered the mesophyll cells. This carbon dioxide then enters the Calvin cycle reactions of the bundle sheath cells' chloroplasts, as described earlier. The three-carbon compounds formed after decarboxylating malate or aspartate are transported back into the mesophyll where they are reconverted to PEP. The sugars that accumulate from the Calvin cycle reactions are temporarily converted to starch under active photosynthetic conditions. At night or during darkness or dim light, the starches are converted back to sugars and transported out of the leaf by the vascular tissue of the veins (vascular bundles).

The presence of the C_4 pathway may seem like a needless addition to the Calvin cycle system. However, there is evidence indicating that the mesophyll is able to build up a high concentration of fixed CO_2 by this method, which could provide evolutionary advantages. In addition, the rapid and efficient fixing and storing of CO_2 as four-carbon acids decreases the leaf's need to have a large number of stomata (openings in the leaf epidermis that allow CO_2 and other gases to diffuse into the leaf). Open stomata, though allowing the passage of CO_2 into the plant, also allow water to escape from the plant—a disadvantage to plants in arid climates!

Crassulacean Acid Metabolism

Crassulacean acid metabolism (CAM) is a special form of metabolism associated with photosynthesis that is carried out by members of the plant family *Crassulaceae* (succulent herbs such as *Sedum*). Plants using this form of metabolism have closed stomata during the daylight hours and therefore cannot absorb suffi-

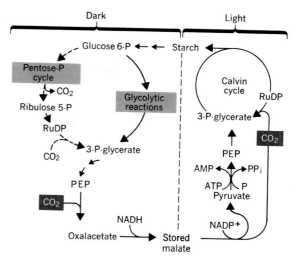

FIGURE 17–20 The major reaction steps of the light and dark phases of crassulacean acid metabolism.

cient CO_2 for photosynthesis. But during the night (dark) hours, the stomata open and the leaf cells can fix CO_2 in the dark by combining it with PEP to form oxaloacetate. The oxaloacetate is converted into malate for storage. During the following daylight hours, the malate is decarboxylated and the CO_2 is utilized in Calvin cycle reactions. The three-carbon pyruvate remaining after decarboxylation is converted first into PEP and then into phosphoglyceric acid and is also utilized in the Calvin cycle. The PEP required for the dark fixation of CO_2 is derived from some of the starch produced from the Calvin cycle products (Fig. 17–20).

BACTERIAL PHOTOSYNTHESIS

Photosynthesis in prokaryotic organisms occurs in lamellar membrane systems called **chromatophores**. The chromatophores contain the pigments for the photochemical reactions but none of the subsequent biosynthetic enzymes. The pigment system includes the chlorophylls, carotenoids, and in some cases phycobilins. However, in the purple and green bacteria, **bacteriochlorophyll** is the ultimate light-trapping molecule (not chlorophyll a). Only the cyanobacteria have chlorophyll a.

The most important distinction between plant and bacterial photosynthesis is that water is not used as the reducing agent and oxygen is not an end product. The power to reduce CO_2 may come from molecular hydrogen, H_2S, or organic compounds. Two major groups of bacteria that carry out photosynthesis are the green and purple sulfur bacteria; these organisms utilize H_2S and produce sulfur and sulfate, that is,

$$6\,CO_2 + 12\,H_2S \xrightarrow[\text{bacteriochlorophyll}]{\text{light energy}} C_6H_{12}O_6 + 6\,H_2O + 12\,S$$

During photosynthesis, sulfur accumulates as granules of elemental sulfur and may be further metabolized later.

Nonsulfur purple bacteria use organic compounds such as acetic acid as electron donors. The acetic acid is anaerobically oxidized via the Krebs cycle reactions (Chapter 10). Acetic acid can also be reduced to hydroxybutyric acid. Certain members of the sulfur and nonsulfur purple bacteria can use molecular hydrogen to reduce either CO_2 or acetic acid, that is,

$$6\,CO_2 + 12\,H_2 \rightarrow C_6H_{12}O_6 + 6H_2O$$

and

$$2\,CH_3-C{\overset{O}{\underset{OH}{}}} + H_2 \rightarrow CH_3-\overset{OH}{CH}-CH_2-C{\overset{O}{\underset{OH}{}}} + 2\,H_2O$$

hydroxybutyric
acid

OTHER PLASTIDS

The chloroplast is only one of several different plastids found in plant cells. Other plastids, such as **etioplasts, amyloplasts,** and **chromoplasts,** have different structures (Fig. 17–21) and functions. They are all called plastids because they appear to develop from a common structure or from one another.

Proplastids are small, generally colorless structures found in young or dividing cells. They have little internal structure but are delimited at their surface by a double membrane. Proplastids give rise to other types of plastids.

Etioplasts are prevalent in the leaves of plants that are grown in the dark. Their ellipsoidal and sometimes irregular structure is also delimited by a double membrane. Internally, etioplasts contain one or more paracrystalline *prolamellar bodies* (Fig. 17–21a) and a number of flattened vesicles called *primary thyla-*

koids. Etioplasts develop into chloroplasts on exposure to light.

The outer membrane of the amyloplast encloses the stroma and contains one to eight starch granules. In certain plant tissues such as the potato tuber, the starch granules within the amyloplasts may become so large that they rupture the encasing membrane. Starch granules of amyloplasts are typically composed of concentric layers of starch.

Chromoplasts contain carotenoids and are responsible for imparting color (e.g., yellow, orange, and red) to certain portions of plants such as flower petals, fruit, and some roots. The chromoplasts of carrots contain large quantities of lipid that reduce their overall density to less than 1.0; consequently, during centrifugation of root homogenates, the amyloplasts rise to the surface of the centrifuge tube. Chromoplast structure is quite diverse; they may be round, ellipsoidal, or even needle-shaped, and the carotenoids that they contain may be localized in droplets or in crystalline structures. The function of the chromoplast is not clear, but in many cases (e.g., flowers and fruit) the color that they produce probably plays a role in attracting insects and other animals for pollination or seed dispersal.

A number of other, less frequently occurring plastids have been described, such as the oil-filled *elaioplasts* including the sterol-rich *sterinochloroplasts*, and the protein-containing *proteoplasts*.

SUMMARY

In eukaryotes, the reactions of **photosynthesis**, like the mitochondrial reactions, are concerned with the formation of ATP and involve proton and electron transport in compounds like NADPH and cytochromes. The two processes differ in that photosynthesis uses light rather than chemical substrates as the source of energy, CO_2 and water are consumed rather than produced, and O_2 and carbohydrate are produced rather than consumed. The overall reaction

$$6\,CO_2 + 12\,H_2O \xrightarrow{\text{light}} (C_6H_{12}O_6)_n + 6\,O_2 + 6\,H_2O$$

can be divided into a light phase (in which **photolysis** of water occurs) and a dark phase (in which **CO_2 fixation** occurs). In the light phase, visible light is absorbed by chlorophyll or a variety of other pigments

(a)   (b)

FIGURE 17-21   (a) *Etioplast* in a cell of a bean plant grown in darkness. Note the prominent *prolamellar body* inside the organelle. Magnification, 29,000X. (b) *Amyloplast* in cotyledon cell of a spinach plant. The organelle contains prolamellar bodies at the margins of starch granules. Magnification, 16,000X. (Photomicrographs courtesy of Dr. J. V. Possingham; copyright © 1976, *Plant Physiol.* **57**, 41.)

located in the membranous **thylakoids** of the chloroplast. The light energy excites the molecules, causing them to reemit light or heat or transfer the energy to **chlorophyll** P_{700} or P_{680} molecules. The activation of chlorophyll P_{680} triggers the reactions of **photosystem II**, which generate ATP by the process called **noncyclic photophosphorylation** and terminate with the reduction of P_{700}. Absorption of light energy directly activates **photosystem I**. This photosystem can also produce ATP (by the process of **cyclic photophosphorylation**) or the reduction of $NADP^+$ to NADPH. The ATP and NADPH transported into the **stroma** of the chloroplast are consumed in the dark reactions.

During the dark reactions of C_3 plants, CO_2 is fixed by binding to ribulose diphosphate and is subsequently reduced by NADPH. ATP acts as the source of energy for these endergonic reactions. The final product is carbohydrate, usually in the form of a sugar, which may be stored as starch. In C_4 plants, CO_2 is first fixed as a four-carbon acid in the mesophyll cell chloroplasts and then transferred to the bundle sheath cells, where the **Calvin cycle** reactions occur and starch is stored.

Chloroplasts are found only in eukaryotic plant cells. They form from proplastids, which may be responsible for the formation of other plastids, such as **chromoplasts** and **leucoplasts**, or they form by a division of preexisting chloroplasts. Plastids associated with C_3 **photosynthesis** have inner membranes arranged in layers and organized into **grana**. Plastids associated with C_4 **photosynthesis** may have a similar structure but frequently lack grana.

REFERENCES AND SUGGESTED READING

Articles and Reviews

Arnon, D. I., The role of light in photosynthesis. *Sci. Am.* **203** (5), 104 (Nov. 1960).

Arnon, D. I., The discovery of photosynthetic phosphorylation. *Trends in Biochem. Sci.* **9** (6), 258 (1984).

Bassham, J. A., The path of carbon in photosynthesis. *Sci. Am.* **206** (6), 88 (June 1962).

Bogorad, L., Chloroplasts. *J. Cell Biol.* **91**, 256s (1981).

Calvin, M., The path of carbon in photosynthesis. *Science* **135**, 879 (1962).

Calvin, M., and Androes, G. M., Primary quantum conversion in photosynthesis. *Science* **138**, 867 (1962).

Cheniae, G. M., Photosystem II and O_2 evolution. *Ann. Rev. Plant Physiol.* **21**, 467 (1970).

Cramer, W. A., Widger, W. R., Herrmann, R. G., and Trebst, A., Topography of thylakoid membrane proteins. *Trends in Biochem. Sci.* **10** (3), 125 (Mar. 1985).

Ellis, R. J., Chloroplast proteins. *Ann. Rev. Plant Physiol.* **32**, 111 (1981).

Hatch, M. D., and Slack, C. R., Photosynthetic CO_2 fixation pathways. *Ann. Rev. Plant Physiol.* **21**, 141 (1970).

Heber, U., and Heldt, H. W., The chloroplast envelope. *Ann. Rev. Plant Physiol.* **32**, 139 (1981).

Jope, C. A., Atchison, B. A., and Pringle, R. C., A computer analysis of a spiral, string-of-grana model of the three-dimensional structure of chloroplasts. *Bot. Gaz.* **141**, 37 (1980).

Miller, K. R., and Lyon, M. K., Do we really know why chloroplast membranes stack? *Trends in Biochem. Sci.* **10** (6), 219 (June 1985).

Murata, N., and Miyao, M., Extrinsic membrane proteins in the photosynthetic oxygen-evolving complex. *Trends in Biochem. Sci.* **10** (3), 122 (Mar. 1985).

Oquist, G., Samuelsson, G., and Bishop, N. I., On the role of β carotene in photosystem II. *Physiol. Plant* **50**, 63 (1980).

Rabinowitch, E. I., and Govindjee, The role of chlorophyll in photosynthesis. *Sci. Am.* **213** (1), 74 (July 1965).

Shavit, N., Energy transduction in chloroplasts. *Ann. Rev. Biochem.* **49**, 111 (1980).

Stacy-French, C., Fifty years of photosynthesis. *Ann. Rev. Plant Physiol.* **30**, 1 (1979).

Staehelin, L. A., and Arntzen, C. J., Regulation of chloroplast membrane function. *J. Cell Biol.* **97**, 1327 (1983).

Stainer, R. Y., Photosynthetic mechanisms in bacteria and plants; development of a unitary concept. *Bacteriol. Rev.* **25**, 1 (1961).

Whatley, J. M., and Whatley, F. R., Chloroplast evolution. *New Phytologist* **87**, 233 (1981).

Wildman, S. G., Jope, C. A., and Atchison, B. A., Light microscopic analysis of the three dimensional structure of higher plant chloroplasts. *Bot. Gaz.* **141**, 24 (1980).

Books, Monographs, and Symposia

Bonner, J., and Varner, J. E., *Plant Biochemistry* (3rd ed.), Academic, New York, 1976).

Clayton, R. K., *Molecular Physics in Photosynthesis*, Blaisdell, New York, 1965.

Danks, D., Evans, E. H., and Whittaker, P. A., *Photosynthetic Systems*, Wiley, New York, 1983.

Foyer, C. H., *Photosynthesis*, Wiley, New York, 1984.

Hall, D. O., and Whatley, F. R., The chloroplast. In *Enzyme Cytology* (D. B. Roodyn, Ed.), Academic, London, 1967.

Hatch, M. D., Osmond, C. B., and Slayter, R. O. *Photosynthesis and Photorespiration*, Wiley–Interscience, New York, 1971.

Kok, B., Photosynthesis: The path of energy. In *Plant Biochemistry* (J. Bonner and J. E. Varner, Eds.), Academic, New York, 1965.

Mahler, H. R., and Cordes, E. H., *Biological Chemistry*, Harper & Row, New York, 1966.

Margulis, L., *Symbiosis in Cell Evolution*, Freeman, San Francisco, 1981.

Nobel, P. S., *Plant Cell Physiology*, Freeman, San Francisco, 1970.

Nobel, P. S., *Biophysical Plant Physiology*, Freeman, San Francisco, 1974.

Novikoff, A. B., and Holtzman, E., *Cells and Organelles*, Holt, Rinehart & Winston, New York, 1970.

Park, R. B., The chloroplast. In *Plant Biochemistry* (J. Bonner and J. E. Varner, Eds.), Academic, New York, 1970.

Price, C. A., *Molecular Approaches to Plant Physiology*, McGraw–Hill, New York, 1970.

Rabinowitch, E., and Govindjee, *Photosynthesis*, Wiley, New York, 1969.

Reinert, J. (Ed.), *Chloroplasts*, Springer-Verlag, Berlin, 1980.

CHAPTER 18

THE GOLGI APPARATUS

The *Golgi apparatus* or *Golgi bodies* of eukaryotic cells are organelles that play a variety of functions, including, (1) the packaging of secretory materials that are to be discharged from the cell, (2) the **processing** of proteins (e.g., **glycosylation, phosphorylation, sulfation**, and **selective proteolysis**) synthesized by ribosomes of the rough endoplasmic reticulum, (3) the synthesis of certain polysaccharides and glycolipids, (4) the sorting of proteins destined for various locations in the cell, (5) the proliferation of new membranous elements for the plasma membrane, and (6) the reprocessing of plasma membrane components that have reentered the cytosol during endocytosis.

The Golgi apparatus was first described in 1898 by C. Golgi, a pioneer of cytology and cytochemistry (Fig. 18–1). Although Golgi referred to the structure as the "internal reticular apparatus" of the cell, the organelle was later renamed in his honor. Golgi received the Nobel Prize in 1906 for his many cytological findings and his cytochemical and histochemical innovations. For many decades after its original description, the Golgi apparatus remained a controversial cell structure because it was not readily identifi-

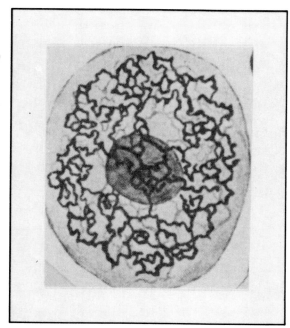

FIGURE 18–1 The "internal reticular apparatus" (i.e., Golgi apparatus) of the cell as first depicted by Camillo Golgi in 1898.

445

able in all cells. To visualize these organelles, Golgi employed stains containing silver, osmium, and other heavy metals, and it was believed by many other cytologists that the organelle was an *artifact* produced by precipitation of the metal within the cell. The controversy was not put to rest until the 1950s, when the existence of the Golgi apparatus was confirmed by electron-microscopic studies; these studies also shed light on the details of the apparatus' organization (Fig. 18–2).

STRUCTURE OF THE GOLGI APPARATUS

The contemporary model of the structure of the Golgi apparatus is based on some 25 years of transmission and scanning electron microscopic study as well as biochemical studies of isolated Golgi membranes. The photomicrographs reproduced in Figure 18–2 show the typical appearance of the organelle when studied by transmission electron microscopy. Although Golgi bodies differ somewhat in organization from one type

FIGURE 18–2 Appearance of the Golgi apparatus in transmission electron photomicrographs of thin sections. *d*, dictyosome (Golgi body); *n*, nucleus; *v*, vesicles. (Micrographs *a*, *b*, and *c* courtesy of Drs. J. DiOrio and W. F. Millington, copyright © 1978 by Springer-Verlag, *Protoplasma* **97**, 329; *d*, courtesy of Dr. R. Wetherbee, copyright © 1978 by Springer-Verlag, *Protoplasma* **97**, 347.)

of tissue or cell to another, they characteristically take the form of a stack of flattened, oval **cisternae** surrounded at their circumferencial edges and above and below by vesicles and tubular structures (Fig. 18–3). Golgi bodies are also referred to as **dictyosomes**, which means "stacklike bodies." In its strictest sense, the dictyosome does not include the vesicles that fuse with or are discharged from the margins of the cisternae. In most animal cells, the number of Golgi cisternae is five or six, the lumen of each cisterna varying in width from about 500 to 1000 nm. The Golgi bodies of plant cells often have 20 or more cisternae. The margins of each cisterna are gently curved so that the entire Golgi body takes on a bow-like appearance. The cisterna at the *convex* end of the dictyosome comprises what is called the **cis face** or **forming face**, and the cisterna at the *concave* end comprises the **trans face** or **maturing face**. The small vesicles that are adjacent to the cis face are believed to fuse with and contribute additional structure to the Golgi body. The vesicles near the trans face are larger and are believed to be

formed from the uppermost cisterna. Small vesicles may also be discharged from the margins of the cisternae that are *between* the cis and trans faces. Recent scanning electron photomicrographs of Golgi bodies (Fig. 18–4) reveal a structure that is remarkably consistent with models of the organelle based solely on transmission studies.

In many cells, particularly those in which the Golgi bodies' main function is related to secretion, the faces of the Golgi bodies are arranged in a specific manner. The forming face is located next to either the nucleus or a specialized portion of the endoplasmic reticulum that lacks bound ribosomes and is called "transitional" ER. The maturing face is usually directed toward the plasma membrane. It is believed that the nuclear membrane and endoplasmic reticulum are the source of the small vesicles that fuse with the cis face, whereas some of the larger vesicles arising from the maturing face are **secretory vesicles** and subsequently fuse with the plasma membrane. When vesicles released from the maturing face of the dictyo-

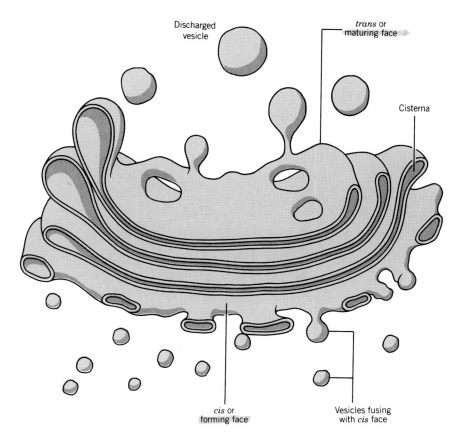

Discharged vesicle

trans or maturing face

Cisterna

cis or forming face

Vesicles fusing with *cis* face

FIGURE 18–3 Diagrammatic reconstruction of the Golgi apparatus based on electron-microscopic studies.

(a)

(b)

FIGURE 18–4 Scanning electron photomicrographs of Golgi bodies from rat epididymal tissue (Courtesy of Dr. Y. Kinose and K. Tanaka, *J. Yonago Med. Assn.* **30**, 527 (1979).)

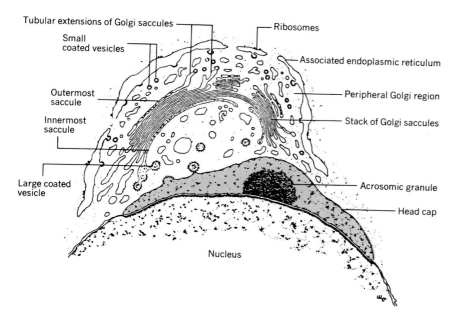

Tubular extensions of Golgi saccules

Small coated vesicles

Ribosomes

Associated endoplasmic reticulum

Outermost saccule

Peripheral Golgi region

Innermost saccule

Stack of Golgi saccules

Large coated vesicle

Acrosomic granule

Head cap

Nucleus

FIGURE 18–5 Contributions of the Golgi apparatus to the forming acrosome during spermatid development. The uppermost Golgi elements are part of the *cis* or forming face; the vesicles contributing to the acrosomal membrane are derived primarily from the *trans* or maturing face. (Copyright © Wistar Institute; F. Susi *et al.*, *Am. J. Anat.* **130**, 262 (1971).)

some form an internal cell structure, such as the developing **acrosome** of sperm cells (Fig. 18–5), the maturing face is directed toward the site of deposition. Thus, in the case of the sperm cell acrosome, the maturing face is proximal to the nucleus. Because of their intimate morphological and physiological associations with the endoplasmic reticulum and intracellular vesicles such as lysosomes, Golgi bodies are sometimes referred to as elements of the **endomembrane** or **vacuolar system** of cells.

The cisternae membranes of Golgi bodies are *smooth*, that is, they are not lined by ribosomes or other particles. At their margins, the cisternae may be perforated, creating **fenestrae**, or they may give rise to tubular extensions that branch and anastomose with each other (Fig. 18–3). The size and number of Golgi bodies vary from one type of cell to another and according to the cell's metabolic activity. The organelle is therefore to be thought of as being almost continuously in a state of flux. Some cells are reported to have only one Golgi body and others may have hundreds. Because one of the major functions of the Golgi apparatus is secretion, it is not surprising that the size and number of Golgi bodies increase during periods of active cellular secretion. In plant cells, the number of Golgi bodies increases during cell division when these organelles secrete materials that form the *cell plate*, which then develops into the cell wall that separates the two new cells. The *goblet cells* found in the intesti-

nal epithelium contain only a single, large Golgi body located in the region of the cell where *mucigen granules* are stored prior to their secretion (Fig. 18–6). In these cells, the size of the Golgi body increases dramatically during periods of digestive activity.

ORIGIN AND DEVELOPMENT OF GOLGI BODIES

The intracellular origin of Golgi bodies has been a hotly debated subject for many years. Among the proposed sources of new Golgi bodies are (1) vesicles dispatched from the endoplasmic reticulum, (2) vesicles dispatched from the outer membrane of the nuclear envelope, (3) vesicles formed by invaginations of the plasma membrane, and (4) division of Golgi bodies already present in the cell. The most widely accepted view is that Golgi bodies are formed from vesicles dispatched from the ER. These vesicles are called **transition vesicles** and are seen in Figure 18–7. Transition vesicles migrate to the forming face of the Golgi body, fuse there with existing cisterna membranes, and in so doing contribute to the organelle's growth. Presumably, a complete Golgi body can be formed in this manner.

Aggregations of transition vesicles occur in areas of the cytoplasm referred to as **zones of exclusion**, which are free of ribosomes. These zones are often

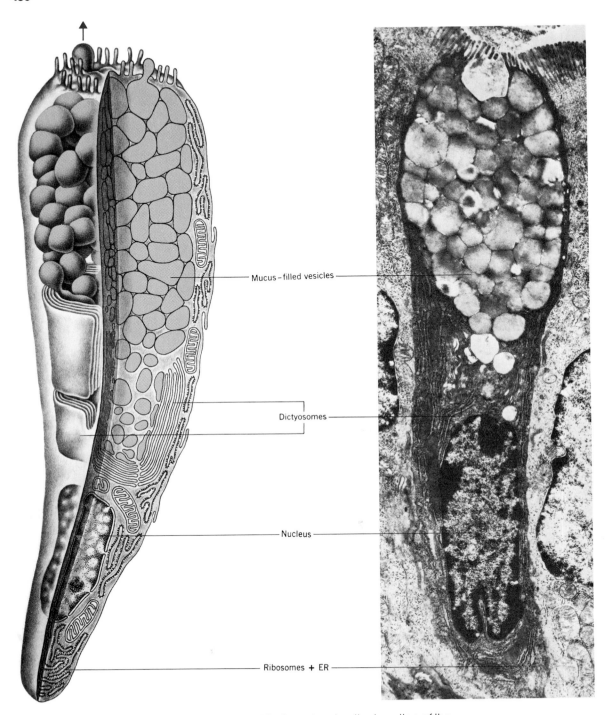

Mucus-filled vesicles

Dictyosomes

Nucleus

Ribosomes + ER

FIGURE 18–6   Goblet cell of the intestinal epithelium showing the location of the dictyosomes (Golgi bodies) and mucigen granules. (Electron photomicrograph courtesy of R. Chao.)

FIGURE 18–7 Zones of exclusion (clusters of tiny vesicles) believed to give rise to Golgi bodies. (Courtesy of Drs. H. H. Mollenhauer and J. Morre, in *Origin and Continuity of Cell Organelles*, J. Reinert and H. Ursprung, Eds.; copyright © 1971, Springer-Verlag.)

surrounded by membranes of the endoplasmic reticulum or the nuclear envelope. Small Golgi bodies, which are presumed to represent early stages in the development of the organelle, are found in the zones of exclusion. The cells of dormant seeds of higher plants generally lack Golgi bodies but they do display zones of exclusion containing aggregations of small vesicles. Photomicrographs of cells in early stages of germination suggest progressive development of Golgi bodies in these zones of exclusion (Fig. 18–7), and the development of Golgi bodies coincides with the disappearance of the aggregations of vesicles.

In frog oocytes, Golgi bodies appear to develop from clusters of vesicles in zones of exclusion (Fig. 18–8). That these early stages of Golgi body development are dependent on protein synthesis was shown by G. Werz, who found that actinomycin D (an inhibitor of protein synthesis) prevents the formation of Golgi body prestages in *Acetabularia*.

During cell division in both animals and plants, the number of Golgi bodies increases, for the number of organelles in each daughter cell just after cell division is about the same as the number in the parent cell prior to division. Photomicrographs obtained at successive stages of cell division indicate that more Golgi bodies are present in cells that are in the metaphase or anaphase of mitosis than in earlier stages. In the

multinucleate alga *Botrydium granulatum*, a single Golgi body is seen at each pole of the dividing cell just prior to the formation of the spindle. By late metaphase, two Golgi bodies are present at each end of the spindle and are separated by the centriole. Despite the seeming rapidity with which new Golgi bodies appear during cell division, there is no direct evidence for an accompanying division of Golgi bodies.

Because it is not possible to observe living Golgi bodies clearly, the developmental sequence of a single organelle cannot be followed directly. However, developmental sequences have been worked out by observing dictyosomes of cells at different stages of cell growth. For example, when cultures of the ciliate *Tetrahymena pyriformis* growing by fission are in the logarithmic phase, no typical Golgi structures are seen in the cells. However, individual, smooth, flattened *saccules* can be seen in the oral region. These saccules are of nonuniform shape and are not associated with one another. *Tetrahymena* can be induced to *conjugate* if the cells are starved for 12–48 hours. During starvation, the saccules in the oral region align to form stacks. No vesicles are seen around these stacks, suggesting that the stacks are formed from the pre-existing saccules and do not arise *de novo*. This also suggests that the stacks are not processing proteins or dispatching products. When starved cells of oppo-

FIGURE 18–8 Development of the Golgi apparatus in the frog oocyte. In advance of the appearance of a definitive Golgi apparatus, transition zones (*TZ*) containing numerous small vesicles are seen in the zone of exclusion along with the first cisternae (*CI*). (Courtesy of Drs. R. Ward and E. Ward; copyright © 1968, *J. Microscopie* **7**, 1007.)

site mating types are mixed together, conjugation occurs within 3 hours. Five minutes after mixing, and well before conjugation begins, the stacked saccules in the oral region enlarge and numerous new vesicles can be seen. The structure now has the appearance of a Golgi apparatus that is processing material. By the time the mating cells separate the Golgi bodies are well developed. If the cells are refed, the ordered arrangement of membranous sacs is lost.

The aggregation of small vesicles or saccules to form a Golgi apparatus is also seen in such diverse cells as developing zoospores and sperm and embryonic liver tissue. In embryonic liver cells, the Golgi apparatus appears to form from the coalescence of tubu-

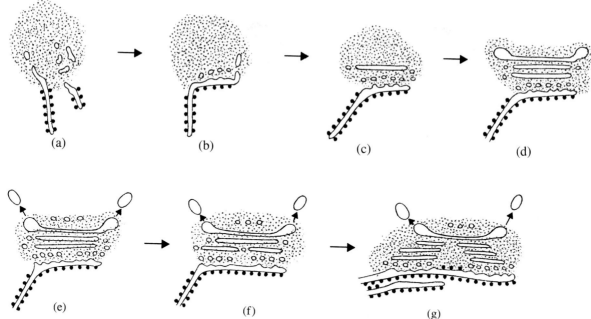

FIGURE 18–9 Proposed formation of a Golgi body from endoplasmic reticulum (*a–c*) and subsequent developmental stages: layered cisternae (*c,d*), formation of secretory vesicles (*e*), division (*f,g*). (Diagram courtesy of Dr. J. Morre *et al.*, in *Origin and Continuity of Cell Organelles*, J. Reinert and H. Ursprung, Eds.; copyright © 1971, Springer-Verlag.)

lar cisternae localized at the margins of rough endoplasmic reticulum. As the liver cells mature, the tubular cisternae become platelike, and still later secretory vesicles form at the outer edges of the cisternae. This sequence is illustrated in Figure 18–9.

FUNCTIONS OF THE GOLGI APPARATUS

A variety of proteins synthesized by rough endoplasmic reticulum pass through and are processed by the Golgi apparatus on their way to their final cellular destinations. Best understood among these are proteins destined to be secreted from the cell, plasma membrane proteins, and lysosomal proteins.

Secretion

Although the Golgi apparatus is involved in many different cellular processes, its principal role in many cells is in **secretion**. Two sets of experiments bear on the role of the Golgi apparatus in secretion. In 1964, L. Caro and G. Palade showed that Golgi bodies in the

acinar cells of pancreas tissue are involved in the packaging of enzyme precursors into **zymogen** granules prior to secretion. Caro and Palade injected radioactive amino acids into rats and followed the movements of the "label" using autoradiography (Chapter 14). This type of experiment is called a "pulse–chase" because the initial short-term application of labeled amino acids (i.e., the "pulse") is immediately followed by the more prolonged application of unlabeled forms (i.e., the "chase"). Although amino acid metabolism and protein synthesis are not interrupted, the metabolic fate of the labeled amino acids can be traced through the cell with time. As might be expected, after a 3-minute pulse the label appeared almost exclusively in the rough endoplasmic reticulum, as this is the region of protein synthesis. Following the 3-minute pulse, nonlabeled amino acids were added for 17 minutes (e.g., a total of 20 minutes from the beginning of the pulse). Although some of the radioactive label remained in the rough endoplasmic reticulum, most of the label had shifted to the Golgi apparatus. When the chase was continued for an additional 100 minutes (120 total), almost all the label had left the endoplasmic reticulum and the Golgi apparatus and was now found in the zymogen granules and in the extracellular space (as a result of the discharge of the contents of the vesicles at the plasma membrane). These experiments showed that the path of the amino acids is first into proteins in the rough ER and that these proteins are then transferred into the cisternae of the Golgi apparatus and then into the zymogen granules.

Two models have dominated thought concerning the manner in which proteins pass from the cis face of the Golgi apparatus to the trans face. According to the **cisternal progression** model, new cisternae are continuously formed as vesicles containing proteins to be processed by the Golgi are dispatched from the endoplasmic reticulum and coalesce at the cis face. Each cisterna then progresses through the stack toward the trans face. On reaching the trans face, the cisternae are broken down to form small vesicles that carry their entrained proteins to various cellular destinations.

The weight of recent scientific evidence argues against the cisternal progression model. For example, it is now known that the different cisternae of the Golgi possess different biochemical properties (e.g., different enzymes; see below), and it is difficult to ex-

plain how a cisterna having one set of properties could change into another. Consequently, the cisternal progression model is giving way to the **cisternal transfer** model, according to which the proteins proceed from one cisterna to the next via small vesicles that are released from one cisterna, move forward, and then fuse with the next cisterna.

Posttranslational Processing of Proteins

In 1966 using similar autoradiographic techniques to those employed by Caro and Palade, M. Neutra and C. P. Leblond studied the secretion of mucus by the *goblet* cells of intestinal epithelium. Mucus is a glycoprotein in which glucose and glucose derivatives are linked together to form polysaccharide side chains of the protein molecules (Fig. 18–10). Glucose labeled with tritium (^{3}H) was used to follow the assembly and fate of the glycoproteins. Fifteen minutes after injection of the radioactive sugar, the label was most concentrated in the cisternae of the Golgi apparatus. This label did not enter or associate with the rough endoplasmic reticulum first. After a 20-minute chase, the label appeared in the mucous vesicles, and after 4 hours, most had been released through the plasma

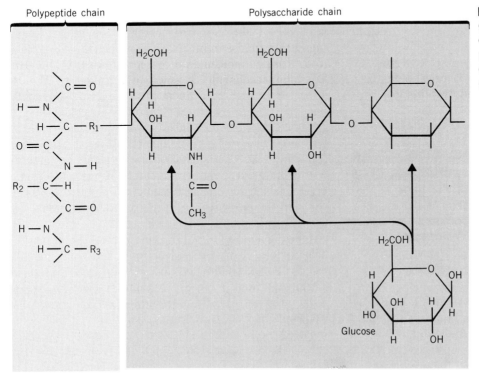

FIGURE 18–10 Mucous glycoprotein showing attachment of polysaccharide side chain to the polypeptide chain; the chain is composed of various derivatives of glucose.

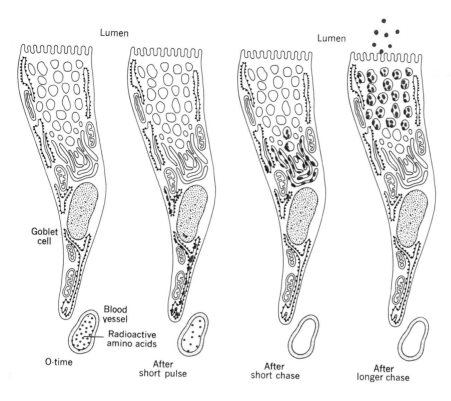

Lumen

Lumen

Goblet cell

Blood vessel

Radioactive amino acids

0-time

After short pulse

After short chase

After longer chase

FIGURE 18–11 Incorporation of amino acids into secretory proteins by the goblet cells of the intestine. Amino acids removed from the bloodstream are used in protein synthesis by rough endoplasmic reticulum. The proteins are conveyed to the Golgi apparatus for incorporation into secretory vesicles. Glycosylation of the proteins (to form glycoproteins) occurs within the Golgi apparatus. Vesicles detach from the maturing face of the Golgi apparatus and migrate to the plasma membrane, where they are discharged into the intestinal lumen.

membrane into the intestinal lumen. This experiment not only revealed the path of glucose through the cell but also demonstrated that the final stages of assembly of the glycoprotein occur *inside* the Golgi apparatus. Using the goblet cell as an example, Figure 18–11 depicts the central role of the Golgi apparatus in the packaging of newly synthesized proteins into vesicles for secretion. The assembly and processing of large molecules in the Golgi apparatus is not unique to goblet cells. Cartilage cells assemble glycoproteins in the cisternae of their Golgi bodies and sulfate groups have been shown to be added as well. Recent studies indicate that the enzymes employed in the processing reactions are selectively localized in the sequential cisternae of the Golgi apparatus *in the order of their use*. In plant cells, pectins and cellulose are assembled in the Golgi bodies prior to deposition onto the forming cell plate or cell wall.

J. E. Rothman, A. Kornfeld, H. Schachter, and others have shown that glycosylation of proteins begins in the rough endoplasmic reticulum and continues in the Golgi cisternae (Figs. 18–12 and 18–13). Preliminary glycosylation in the endoplasmic reticulum is called **core glycosylation**. In the endoplasmic reticu-

lum, a specific branched oligosaccharide is added to each glycoprotein being synthesized. The branched oligosaccharide consists of 14 sugar units: 2 molecules of *N*-acetylglucosamine (Nag), 9 molecules of mannose (man), and 3 molecules of glucose (glc) (Fig. 18–13). The oligosaccharide is covalently linked to the R-groups of asparagine (asn) residues in the proteins. Shortly after addition of the oligosaccharide, and while the protein is still in the endoplasmic reticulum, one of the mannose units and all three glucose units are removed (Fig. 18–13). Following this, the proteins are transferred to the cis face of the Golgi apparatus.

All of the glycoproteins arriving at the cis face of the Golgi apparatus from the endoplasmic reticulum have the same oligosaccharide chains. Before passing from one Golgi cisterna to the next, the proteins undergo chemical modification (i.e., processing). For example, in the cis cisterna, phosphate groups are added to the ends of the oligosaccharide chains of proteins destined for lysosomes, Secretory proteins and proteins destined for the plasma membrane experience more extensive processing. Five of the mannose units of their oligosaccharide chains are removed in the cis cisterna and are replaced in the *medial* cisternae by

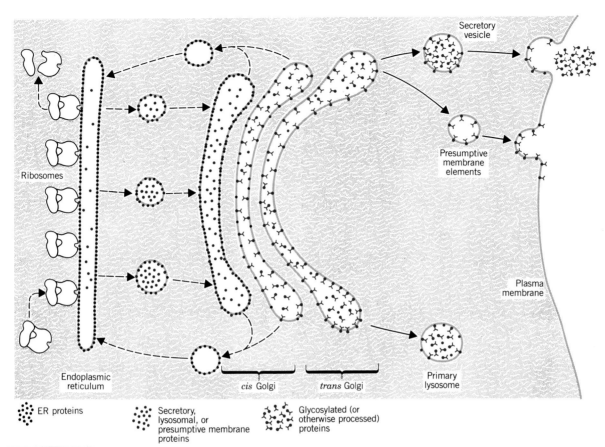

Ribosomes

Endoplasmic
reticulum

cis Golgi   *trans* Golgi

Secretory
vesicle

Presumptive
membrane
elements

Plasma
membrane

Primary
lysosome

ER proteins

Secretory,
lysosomal, or
presumptive membrane
proteins

Glycosylated (or
otherwise processed)
proteins

FIGURE 18–12 Functions of the Golgi apparatus. Proteins from the endoplasmic reticulum (black and colored circles) enter the cis face of the Golgi. As these are passed from one cisterna to the next, ER proteins (black circles) are removed from the margins of the cisterna by budding vesicles and are returned to the ER. Proteins destined for secretion or incorporation into organelles (e.g., lysosomes, plasma membrane) (colored circles) are successively purified and processed. These proteins are discharged as vesicles from the trans face (see text for additional details).

two molecules of *N*-acetylglucosamine (Fig. 18–13). After transfer to the trans cisterna, disaccharides consisting of galactose and *N*-acetylneuraminic acid (also known as sialic acid) are added to the ends of each branch of the oligosaccharide.

Proliferation of Cellular Membranes

In addition to its role in secretion, the Golgi apparatus plays a role in the preparation of membranous elements for organelles such as lysosomes and the plasma membrane (Fig. 18–12). Proteins destined to be incorporated into lysosomes or the plasma mem-

brane are synthesized by ribosomes attached to the endoplasmic reticulum (i.e., rough ER). Some of these proteins are released into the intracisternal phase of the ER and others remain anchored in the ER membranes. Within minutes of their synthesis, these proteins appear in the cis face of the Golgi apparatus. The mechanism by which intracisternal phase and membrane-associated ER proteins reach the Golgi apparatus is uncertain but is generally believed to involve one or a combination of the following processes. Transfer to the Golgi apparatus may be mediated by dispatchment of small vesicles from the ER that migrate to and fuse with the cisterna that comprises the

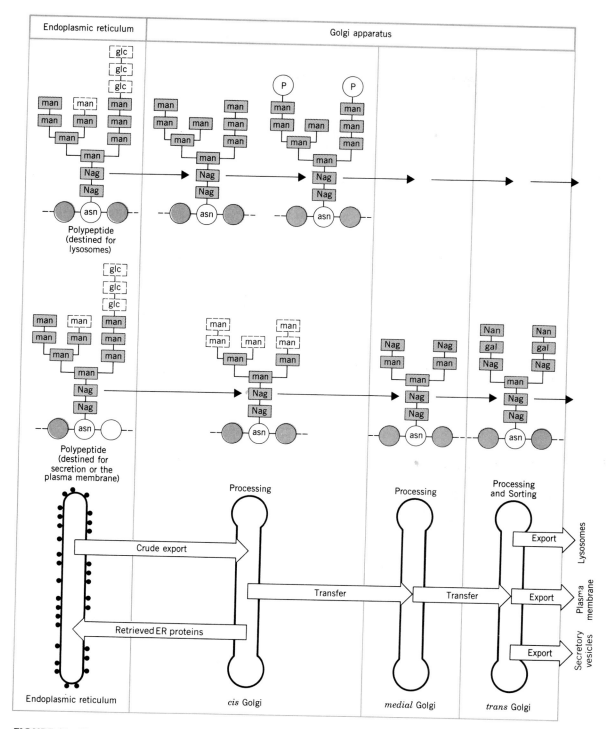

FIGURE 18-13 Division of labor among the *cis, medial,* and *trans* cisternae of the Golgi apparatus and the endoplasmic reticulum (see also Fig. 18-12). Nag, *N*-acetylglucosamine; man, mannose; glc, glucose; (P), phosphate; gal, galactose; Nan, *N*-acetylneuraminic acid (sialic acid).

cis face. Newly synthesized proteins discharged into the intracisternal phase of the ER would be enclosed within these vesicles, and proteins that were left anchored in the ER membrane would be constituents of the vesicle wall. Alternatively, proteins in the intracisternal phase of the ER reach the cisterna of the cis face by diffusion along transient connections between the channels, whereas ER membrane-anchored proteins reach the Golgi apparatus by lateral flow within the membranes that form these connections.

As glycoprotein processing progresses, the proteins are successively transferred from one Golgi cisterna to the next, ultimately reaching the trans face. The transfer may take the form of small intermediary vesicles, or diffusion (or lateral membrane flow) through transient continuities between adjacent cisternae.

Proteins that are destined to be components of lysosomal membranes or the plasma membrane and that are anchored in the ER membranes at the time of synthesis are presumed to move from the ER to the cis face and from the cis face through the medial cisternae to the trans face as membrane components. The membranes of the vesicles discharged from the trans face contain these proteins (see Fig. 18–12). Soluble lysosomal proteins (as well as proteins destined for secretion) move from the lumenal phase of the ER through the cisternal space and are enclosed by the membranes of discharged vesicles.

Vesicles containing membrane-bound and soluble lysosomal enzymes are called *primary* lysosomes and are discussed at length in connection with lysosome function in the next chapter. Vesicles containing secretory proteins fuse with the plasma membrane and empty their contents outside the cell. This process was discussed earlier in the chapter in connection with the secretory functions of the small intestine and pancreas. Vesicle membranes that are studded with presumptive plasma membrane proteins also fuse with the plasma membrane. As seen in Figure 18–12, those portions of the protein that faced the lumenal phase of the ER and that were glycosylated (or sulfated, etc.) in the Golgi cisternae face the exterior of the cell once they are incorporated into the plasma membrane.

As noted in Chapter 1, one of the characteristic features of many plant tissues is the presence of one or more large *vacuoles*. Some of these vacuoles have been shown to contain hydrolytic enzymes comparable to those present in lysosomes. For this reason, it has been suggested that the Golgi bodies of plant cells may give rise to some, if not all, of these vacuoles.

Sorting Problems Faced by the Golgi Apparatus

It is clear from the preceding discussion that proteins destined for secretion granules, lysosomes, and the plasma membrane enter the cis face of a Golgi body from a common source—the rough endoplasmic reticulum. These proteins are delivered to the cis face of the Golgi along with a vast excess of ER membrane proteins, yet the vesicles dispatched from the trans face are essentially free of ER proteins. Thus, the Golgi apparatus also acts to progressively sort these proteins. ER proteins are believed to be returned to the endoplasmic reticulum by some of the small vesicles released from the margins of the cis cisternae (Figs. 18–12 and 18–13). As noted above, there appears to be an enzymatic compartmentation of the Golgi cisternae in that certain enzymes are selectively localized in specific cisternae of the stack. Although one might have expected that processed proteins with different destinations would exit the stack at different levels, this appears not to be the case. Instead, irrespective of their final destination, secretory and lysosomal proteins pass together through the stack of cisternae and are not sorted until they reach the trans face. In 1981, J. E. Rothman suggested that a Golgi body may actually be two organelles in tandem: the **cis Golgi** and the **trans Golgi**. Beginning at the forming face, the cis Golgi consists of all but the last one or two cisternae and its role is to sort out ER proteins so that they may be returned to the ER and to process proteins having other destinations. The trans Golgi consists of the last one or two cisternae of the maturing face and it acts to receive the refined proteins and distribute them through vesicles to their specific locations throughout the cell.

More recently, Rothman and others have moderated this view and now suggest that the Golgi apparatus consists instead of *three* compartments: the *cis* compartment, the *medial* compartment, and the *trans* compartment. The cis compartment sorts out and dispatches the ER proteins and also adds phosphate to the terminal sugars of lysosomal proteins. The medial compartment (which consists of cisternae at the center of the stack) is where *N*-acetylglucosamine is added. The addition of the terminal galactose and *N*-acetylneuraminic acid units occurs in the trans compartment and this is also where the various proteins are sorted according to final destination. The addition of phosphate to the terminal sugars of lysosomal proteins in the cis compartment apparently serves to preclude the addition of *N*-acetylglucosamine in the medial compartment and galactose and *N*-acetylneuraminic acid in the trans compartment.

RECYCLING OF PLASMA MEMBRANE COMPONENTS

The fusion of secretory vesicles with the plasma membrane prior to discharging their contents outside the cell adds to the membrane's surface area. Yet, even during periods of active secretion, there is little change in the total surface area of the plasma membrane. Likewise, though endocytosis results in the internalization of portions of the plasma membrane, there appears to be no net loss of membrane surface area during periods of active endocytosis. The constancy of the plasma membrane's surface area is believed to result at least in part from the recycling of membrane components, that is, some internalized membrane components are subsequently returned to the membrane and excess membrane is removed by internalization. In addition to its role in the proliferation of new plasma membrane constituents, the Golgi

apparatus apparently is involved in the recycling or reuse of those portions of the plasma membrane that enter the cytosol during endocytosis. For example, the plasma membranes of thyroid cells can be tagged with ferritin and endocytosis induced by the addition of *thyrotropin* to the cells. Immediately following the onset of endocytosis, the ferritin appears in endosomes and lysosomes, but within about 30 minutes it can also be detected in the Golgi cisternae. Membrane components entering the Golgi apparatus following endocytosis may be reprocessed and reused in secretion, lysosome formation, or restoration of the plasma membrane itself. The various routes that may be taken by membranous elements following endocytosis are summarized in Figure 18–14.

CELL-SPECIFIC FUNCTIONS OF THE GOLGI APPARATUS

Although a discussion of each of the functions of the Golgi apparatus in different kinds of cells and tissues is beyond the scope of this book, two important examples are discussed below. A list of other well-studied cases is presented in Table 18–1.

Formation of the Cell Plate and Cell Wall in Plant Tissues

In plants, the cell plate and cell wall form during anaphase and telophase of mitosis and meiosis II (Chapter 20). During these final stages of nuclear division, the chromosomes are already separated into two masses in the cell that will become nuclei. Between these two nuclei, pectin and hemicellulose are deposited slowly, forming a plate in the center of the cell, which ultimately grows to the side walls, cutting and separating the protoplasts in two and thereby producing the two daughter cells. Prior to anaphase, the cell's Golgi bodies are found outside the spindle. During anaphase, vesicles that appear to be released from the Golgi apparatus invade the center of the spindle (Fig. 18–15) and aggregate about the spindle fibers. These vesicles are the source of the carbohydrate that forms the cell plate and eventually the cell wall. The nature of the carbohydrate secreted by the vesicles is controversial. Some investigators believe that complete cellulose fibers are secreted and others believe that the final stages of fiber assembly occur after secretion. In either case, the Golgi apparatus is clearly

FIGURE 18–14 Role of the Golgi apparatus in the recycling and reuse of plasma membrane components.

TABLE 18–1 SPECIFIC FUNCTIONS OF GOLGI STRUCTURES

Cell	Tissue or Organ	Golgi Function
Exocrine	Pancreas	Secretion of zymogen (proteases, lipases, carbohydrases, and nulceases)
Gland cell	Parotid gland	Secretion of zymogen
Goblet cell	Intestinal epithelium	Secretion of mucus and zymogens
Follicle cells	Thyroid gland	Prethyroglobulin
Plasma cells	Blood	Immunoglobulins
Myelocytes, sympathetic ganglia, Schwann cells	Nervous tissue	Sulfation reactions
Endothelial cells	Blood vessels	Sulfation reactions
Liver cells	Liver	Lipid secretion (lipid transformation?)
Alveolar epithelium	Mammary gland	Secretion of milk proteins (and lactose?)
Paneth cells	Intestines	Secretion of proteins (chitinase?)
Brunner's gland cell	Intestines	Synthesis and secretion of mucopolysaccharides, enzymes, hormones
Connective tissue	*Amblystoma* limb	Synthesis (?) and secretion of collagen
Cornea	Avian eye	Secretion of collagen
Plant cells	Most	Secretion of pectin and cellulose

FIGURE 18–15 Clusters of small vesicles released from the Golgi apparatus during anaphase in *Zea mays* root apex cell. The vesicles migrate toward the equatorial region, where they contribute polysaccharides to the forming cell plate. (Photomicrograph courtesy of Dr. G. Whaley *et al.*, copyright © Academic Press, *J. Ultr. Research* **15**, 173 (1966).)

involved in the secretion of the carbohydrate that forms the wall between the two cell halves.

The plasma membrane of plant cells does not pinch inward or grow inward during cell division as occurs in animal cells. Instead, the membrane forms on both sides of the developing cell plate and grows outward with it. Formation of the membrane results from fusion of the vesicles discharged from the Golgi apparatus.

Acrosome Development in Sperm

The development of the acrosome of sperm cells is an excellent example of the involvement of the Golgi apparatus in the formation of another cellular organelle. The acrosome is a membrane-bound structure at the anterior end of the sperm cells of most animals. A part of the acrosome membrane appears to be involved in recognition and binding of the sperm cell to the surface of the egg cell during fertilization. The acrosome contains hydrolytic enzymes of which *hyaluronidase* is the most abundant; it causes the breakdown of the protective surfaces of the egg. As shown in Figure 18–5, the singular large Golgi body of the sperm cell discharges vesicles that migrate to the forming acrosome. At the surface of the acrosome, these vesicles fuse, with the acrosome membrane contributing to the acrosome's growth.

Because the acrosome is composed of hydrolytic enzymes, it has been suggested that the acrosome is nothing more than a giant lysosome. As the acrosome grows, the Golgi body becomes reduced in size, and in many mature sperm cells disappears entirely. The outer membrane of the acrosome fuses with the plasma membrane. In mouse sperm cells, it has been shown that the area of the plasma membrane that fuses with the acrosomal membrane contains a large number of **concanavalin A** binding sites. The increased number of glycoproteins in the membrane is attributed to its origin in the Golgi apparatus.

SUMMARY

The **Golgi apparatus** or **dictyosome** consists of a parallel series of flattened vesicles or **cisternae** that form a bowl-shaped organelle. Smaller vesicles join the stack on one edge and leave the stack at the other edge.

Proteins destined for secretion, insertion in the plasma membrane, or deposition in intracellular organelles such as lysosomes and vacuoles reach the **cis face** of the Golgi apparatus from neighboring rough endoplasmic reticulum. In the cisternae comprising the **cis** and **medial Golgi**, these proteins are phosphorylated, glycosylated, sulfated, or **processed** in other ways. Glycosylation is completed in the **trans face** of the Golgi apparatus and it is here that the proteins are sorted according to their final destination. Vesicles containing Golgi products destined for other cellular compartments are dispatched from the surface of the **trans Golgi**.

Golgi bodies also play specific roles in different types of cells. For example, in plant cells Golgi bodies are associated with the formation of the **cell plate**. In sperm cells, growth of the acrosome results from the continuous influx of new materials from the Golgi apparatus.

REFERENCES AND SUGGESTED READINGS

Articles and Reviews

Brodie, D. A., Bead rings at the endoplasmic reticulum–Golgi complex boundary. *J. Cell Biol.* **90**, 92 (1981).

Farquhar, M. G., and Palade, G. E., The Golgi apparatus (complex)—(1954–1981)—From artifact to center stage. *J. Cell Biol.* **91**, 77 (1981).

Hino, Y., Minakami, S., and Muratami, H., The comparison of Golgi subfractions isolated from mitochondrial fraction with those from microsomal fraction. *Exp. Cell Res.* **133**, 171 (1981).

Lodish, H. F., and Rothman, J. E., The assembly of cell membranes. *Sci. Am.* **240** (1), 48 (Jan. 1979).

Morre, D. J., and Ovtracht, L., Dynamics of the Golgi apparatus: Membrane differentiation and membrane flow. *Int. Rev. Cytol.* (Suppl. 5), 61 (1977).

Morre, D. J., and Ovtracht, L., Structure of rat liver Golgi apparatus. *J. Ultr. Research* **74**, 28 (1981).

Neutra, M., and Leblond, C., The Golgi apparatus. *Sci. Am.* **220** (2), 100 (Feb. 1969).

Northcote, D., Chemistry of the plant cell wall. *Ann. Rev. Plant Physiol.* **23**, 113 (1972).

Rothman, J. E., The Golgi apparatus: Two organelles in tandem. *Science* **213**, 1212 (1981).

Rothman, J. E., The compartmental organization of the Golgi apparatus. *Sci. Am.* **253** (3), 74 (Sep. 1985).

Rothman, J. E., and Lenard, J., Membrane traffic in animal cells. *Trends in Biochem. Sci.* **9** (4), 176 (1984).

Schatz, G., Reconstituting membrane flow. *Trends in Biochem. Sci.* **10** (3), 95 (Mar. 1985).

Steinman, R. M., Mellman, I. S., Muller, W. A., and Cohn, Z. A., Endocytosis and the recycling of plasma membrane. *J. Cell Biol.* **96**, 1 (1983).

Susi, R., Leblond, C., and Clermont, Y., Changes in the Golgi apparatus during spermiogenesis in the rat. *Am. J. Anat.* **130**, 251 (1971).

Tartakoff, A. M., The confined function model of the Golgi complex. *Int. Rev. Cytol.* **85**, 221 (1983).

Triemer, R. E., Role of Golgi apparatus in mucilage production and cyst formation in *Euglena gracilis*. *J. Phycol.* **16**, 46 (1980).

Books, Monographs, and Symposia

Fleisher, B., and Fleischer, S., The Golgi apparatus. In *Advanced Cell Biology* (L. M. Schwartz and M. M. Azar, Eds.), Van Nostrand–Reinhold, New York, 1981.

Moore, D., Mollenhauer, H., and Bracker, C., Origin and continuity of Golgi apparatus. In *Origin and Continuity of Cell Organelles* (J. Reinhert and H. Ursprung, Eds.), Springer-Verlag, Berlin, 1971.

Waley, W. G., *The Golgi Apparatus*, Springer-Verlag, New York, 1975.

CHAPTER 19

LYSOSOMES AND MICROBODIES

Our knowledge of the structure, composition, and function of **lysosomes** and **microbodies** is considerably more recent than that of most other cell organelles. Although a variety of small bodies seen in plant and animal cells (including what are now termed lysosomes) had been variously called "microbodies" or "cytosomes" for many years, the diversity of their composition and action was not recognized until the 1950s. The "discovery" of lysosomes and microbodies during the 1950s may be attributed to the growing sophistication of electron microscopy, the application of gentler procedures for dispersing tissues and disrupting cells, and the development of improved methods for separating, fractionating, and chemically characterizing the subcellular complexes released from disrupted cells. At the present time, lysosomes are recognized as a separate category of organelles, whereas *peroxisomes, glyoxysomes*, and related vesicular organelles are collectively called microbodies.

LYSOSOMES

The existence of lysosomes was subtly suggested for the first time in the early 1950s by a series of experiments carried out by Nobel Prize Laureate Christian de Duve and his co-workers. These experiments were designed to identify the cellular locus of the two enzymes *glucose-6-phosphatase* and *acid phosphatase*. Liver tissue homogenates were separated into nuclear, mitochondrial, microsomal, and cytosol fractions by differential centrifugation (see Chapter 12) and enzyme assays were performed on each of the collected fractions. Although results with *glucose-6-phosphatase* clearly indicated that this enzyme was bound to particles sedimenting with the microsome fraction, observations on the distribution of *acid phosphatase* were at first rather confusing. The confusion centered around three seemingly peculiar but nonetheless reproducible findings: (1) the acid phosphatase activities of tissue homogenates prepared for centrifugation using a glass tube and close-fitting plunger (Dounce homogenizer) were about one-tenth the value observed when tissue was more vigorously dispersed using a Waring blender; (2) the total (i.e., combined) enzyme activity of the isolated centrifugal fractions analyzed following differential centrifugation was about twice the activity of the original homogenate; and (3) after storage for several days in a freezer, both the enzyme activity of the homogenate and the col-

TABLE 19–1 DISTRIBUTION OF ACID PHOSPHATASE ENZYME ACTIVITY IN LIVER TISSUE FRACTIONS PREPARED BY DIFFERENTIAL CENTRIFUGATION

Fraction	Acid Phosphatase Activity [μg phosphate released/20 min]	
	Before Freezing and Storage	After Freezing and Storage for 5 Days
Whole homogenate	10	89
Nuclear fraction	2	10
Mitochondrial fraction	7	46
Microsomal fraction	6	10
Soluble fraction	6	9

Source: Based on the published results of J. Berthet and C. de Duve, *Biochem. J.* **50,** 174 (1951), and C. de Duve, The lysosome in retrospect, in *Lysosomes in Biology and Pathology,* (J. T. Dingle and H. B. Fell, Eds.), North-Holland, Amsterdam, 1969, p. 6.

TABLE 19–2 SOME ENZYMES PRESENT IN LYSOSOMES

Enzyme	Substrate
Proteases and peptidases	
Cathepsin A, B, C, D, and E	Various proteins and peptides
Collagenase	Collagen
Arylamidase	Amino acid arylamides
Peptidase	Peptides
Nucleases	
Acid ribonuclease	RNA
Acid deoxyribonuclease	DNA
Phosphatases	
Acid phosphatase	Phosphate monoesters
Phosphodiesterase	Oligonucleotides, phosphodiesters
Phosphatidic acid phosphatase	Phosphatidic acids
Enzymes acting on carbohydrate chains of glycoproteins and glycolipids	
β-Galactosidase	β-Galactosides
Acetylhexosaminidase	Acetylhexosaminides, heparin sulfate
β-Glucosidase	β-Glucosides
α-Glucosidase	Glycogen
α-Mannosidase	α-Mannosides
Sialidase	Sialic acid derivatives
Enzymes acting on glycosaminoglycans	
Lysozyme	Mucopolysaccharides, bacterial cell walls
Hyaluronidase	Hyaluronic acid, chondroitin sulfates
β-Glucuronidase	Polysaccharides, mucopolysaccharides
Arylsulfatase A and B	Arylsulfates, cerebroside sulfates, chondroitin sulfate
Enzymes acting on lipids	
Phospholipase	Lecithin, phosphatidyl ethanolamine
Esterase	Fatty acid esters
Sphingomyelinase	Sphingomyelin

lected fractions, especially the mitochondrial fraction, increased dramatically (Table 19–1).

These observations were explained when de Duve and his colleagues showed that acid phosphatase activity was confined to sedimentable particles, the surrounding membranes of which limited the accessibility of the substrate (β-glycerophosphate) used in the enzyme assay. Only when these membranes were disrupted and the acid phosphatase released from the particles was the enzyme activity demonstrable. This occurred during vigorous dispersion in the Waring blender, which disrupted virtually all particles present. In contrast, only about 10% of the acid phosphatase-containing particles were disrupted by homogenization with the Dounce homogenizer, thus accounting for the low activity of the enzyme in the homogenate (Table 19–1). Some additional enzyme activity was released during and following centrifugal fractionation, but much larger quantities of enzyme were released by the membrane disruption that occurred during freezing and thawing.

At first, de Duve and his co-workers did not recognize that the acid phosphatase activity was associated with a distinct population of cellular particles. Instead, on the basis of the observed "latent" activity of the mitochondrial fraction obtained by differential centrifugation (compare the values before and after freezing in Table 19–1), de Duve believed that *acid phosphatase* resided within the mitochondria. However, continued studies during the early 1950s in which the mitochondrial fraction was further divided

centrifugally into a number of subfractions revealed that *acid phosphatase* was absent from fractions containing rapidly sedimenting mitochondria but was present in high concentrations in fractions containing slowly sedimenting mitochondria. This observation, together with a newly developed appreciation of the potential contamination of sediments occurring during differential centrifugation (see Chapter 12), led de

Duve to suspect that the *acid phosphatase* might, in fact, be associated with a special class of particles distinct from the mitochondria. Added credence was given to this idea by the finding that four other acid hydrolases, namely, *β-glucuronidase, cathepsin, acid ribonuclease*, and *acid deoxyribonuclease*, were distributed through the centrifugal fractions in an idential manner. Thus, five hydrolytic enzymes, each having an acid pH optimum and acting on completely different substrates, appeared to be present in the same subcellular particle. On the basis of the *lytic* effects of all of these enzymes, de Duve named the particles "lysosomes." A number of additional enzymes have subsequently been identified in lysosomes (Table 19-2). Most of the chemical substances present in cells, including proteins, polysaccharides, nucleic acids, and lipids, are degraded by these enzymes.

It is interesting to note that the initial postulation of the existence of lysosomes was made by de Duve purely on biochemical grounds. However, in 1955, A. Novikoff, working with de Duve, examined centrifugal fractions rich in *acid phosphatase* activity using transmission electron microscopy and provided the first morphological evidence supporting the existence of these particles. The lysosomes were identified as small, dense, membrane-enclosed particles distinct from the mitochondria.

In recent years, sophisticated centrifugal methods have been devised for obtaining preparations that are rich in lysosomes. Nearly all preparations obtained by differential centrifugation are contaminated with quantities of mitochondria. Although the *average* sedimentation coefficient of mitochondria is greater than that of lysosomes, mitochondria are polydisperse with respect to size, so that the smaller mitochondria invariably sediment with the lysosomes. Moreover, in tissues containing peroxisomes, the range of sedimentation coefficients for these organelles is almost identical to that of the lysosomes. Consequently, it is virtually impossible to obtain lysosome preparations that do not also contain peroxisomes. Somewhat greater success is obtained when *isopycnic* density gradient centrifugation is used in the last stages of the isolation procedure, for the equilibrium densities of lysosomes (1.22 g/cm³), mitochondria (1.19 g/cm³), and peroxisomes (1.23–1.25 g/cm³) in sucrose density gradients are slightly different. Most density gradient procedures used to prepare lysosomes are modifications of the technique developed by W. C. Schneider (Fig. 19–1). Using this technique, most of the mito-

chondria are banded isopycnically at a density of about 1.19 g/cm³, whereas most of the lysosomes form a separate zone at about 1.22 g/cm³ and can be recovered independently from the density gradient.

By far the greatest purity of lysosomes is obtained from tissues of animals previously treated with Triton WR-1339 (a polyethylene glycol derivative of polymerized *p-tert*-octyl phenol), Dextran (a polymer of glucose), and Thorotrast (colloidal thorium hydroxide). These compounds are rapidly incorporated in large quantities by the cell's lysosomes, significantly altering their density. For example, the incorporation of Triton WR-1339 reduces the average density of the lysosome from 1.22 to about 1.10 g/cm³. It is interesting to note that although the density of lysosomes incorporating Triton WR-1339 is significantly reduced, their size is increased; the result is that Triton WR-1339-loaded lysosomes have the same sedimentation coefficient as normal lysosomes but have a lower density.

The latent enzymatic effect originally noted by de Duve and his co-workers is still employed as a major criterion in evaluating the effectiveness of any lysosome isolation. Accordingly, the lysosome preparation is incubated under the appropriate conditions with the hydrolase substrate before and after treatments known to disrupt the lysosome membranes. If the original preparation contains intact lysosomes, then no substrate is hydrolyzed before treatment (most substrates of the lysosomal hydrolases are unable to permeate the lysosome's membrane); however, disruption of the membrane (by sonification, repeated freezing and thawing, addition of lytic agents such as bile salts, digitonin, Triton X-100, etc.) and release of the lysosomal enzymes are quickly followed by hydrolysis of the added substrates.

Structure and Forms of Lysosomes

Lysosomes are a structurally heterogeneous group of organelles and vary dramatically in size and morphology. As a result, it is difficult to identify lysosomes strictly on the basis of morphological criteria. When lysosome-rich fractions were initially isolated centrifugally by de Duve and Novikoff and examined with the electron microscope, it was found that the suspected lysosomes were generally about the same size as small mitochondria. Typically, they varied in diameter from about 0.1 to 0.8 μm, were bounded by a single membrane, and were usually somewhat electron-dense.

FIGURE 19–1 Steps in the isolation of lysosomes by combined differential and sucrose density gradient centrifugation.

Identification of lysosomes in sections of whole cells is considerably more difficult because other small, dense organelles are also bounded by a single membrane. The application of cytochemical procedures at the level of the electron microscope in which the lysosomes are identified on the basis of their enzyme content is much more reliable. Notable among such procedures is that introduced in 1952 by G. Gomori and that is routinely employed in variously modified forms for the identification of lysosomes on the basis of their high *acid phosphatase* content. In the Gomori method, the tissue to be examined is incubated at pH 5.0 in a medium containing β-glycerophosphate (a substrate for *acid phosphatase*) and a lead salt (such as lead nitrate). Phosphate enzymatically cleaved from the substrate during incubation combines with the lead ions to form insoluble lead phosphate, which precipitates at the locus of enzyme activity. Because the lead phosphate is electron-dense, electron microscopy reveals

the lysosomes as dark, granular organelles (Fig. 19–2). For identification with the light microscope, ammonium sulfide may be used to convert the lead phosphate to lead sulfide, which appears black. The Gomori reaction may be carried out with fixed and sectioned material, as well as with fresh tissue, albeit with reduced efficiency as a result of some enzyme inactivation during and following fixation.

Several different lysosomal forms have been identified within individual cells, including (1) **primary lysosomes**, (2) **secondary lysosomes**, and (3) **residual bodies**.

Primary Lysosomes *Primary lysosomes*, or **protolysosomes**, are newly produced organelles bounded by a single membrane and believed to be derived from the trans face of the Golgi apparatus (Chapter 18). Although varying somewhat in size, primary lysosomes are typically about 100 nm in diameter. Primary lyso-

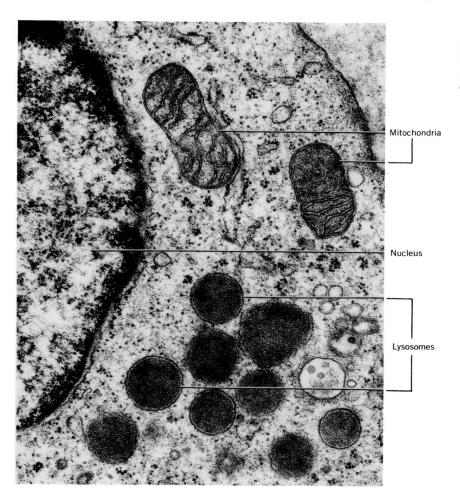

FIGURE 19–2 Electron photomicrograph of a cluster of lysosomes; these clusters are often seen next to mitochondria. (Photomicrograph courtesy of R. Chao.)

Mitochondria

Nucleus

Lysosomes

somes are virgin particles in that their digestive enzymes have not yet taken part in hydrolysis.

Secondary Lysosomes Two different kinds of *secondary lysosomes* can be identified: **heterophagic vacuoles** (also called **heterolysosomes** or **phagolysosomes**) and **autophagic vacuoles** (also called **autolysosomes**). Heterophagic vacuoles are formed by the fusion (see below) of primary lysosomes with cytoplasmic vacuoles containing *extracellular* substances brought into the cell by any of a variety of endocytic processes (see Chapter 15). Following fusion, the hydrolases of the primary lysosomes are released into the vacuole (variously called a **phagosome** or **endosome**). Autophagic vacuoles contain particles isolated from the cell's own cytoplasm, including mitochondria, microbodies, and smooth and rough fragments of the endoplasmic reticulum. The autodigestion of cellular organelles is a normal event during cell growth and repair and is especially prevalent in differentiating and dedifferentiating tissues and tissues under stress. Autophagic vacuoles containing partially degraded mitochondria are shown in Figure 19–3. The formation of heterophagic and autophagic vacuoles is soon followed by enzymatic digestion of the vacuolar contents. As digestion proceeds, it becomes increasingly difficult to identify the nature of the original secondary lysosome and the more general term **digestive vacuole** is used to describe the organelle at this stage.

Residual Bodies Endocytosed substances and parts of autophagocytosed organelles that are not digested within the secondary lysosomes and transferred to the cytoplasm are retained (usually temporarily) within the vacuoles as residues. Lysosomes containing such residues are called *residual bodies* (sometimes also called **telolysosomes** or **dense bodies**). Residual bodies are large, irregular in shape, and are usually quite electron-dense. The undigested residues often take the form of whorls of membranes, grains, amorphous masses, ferritinlike or hemosiderinlike parti-

FIGURE 19–3 Autophagic vacuoles containing partially degraded mitochondria. (Electron photomicrographs courtesy of Dr. Z. Hruban.)

FIGURE 19–4 (Left) A membrane whorl. (Electron photomicrograph courtesy of Dr. E. G. Pollock.) (Right) Residual bodies. (Electron photomicrograph courtesy of Dr. Z. Hruban.)

cles, or myelin figures (Fig. 19–4). Residual bodies often fail to display the degree of hydrolytic activity associated with the primary and secondary lysosomes.

Formation and Function of Lysosomes (the "Vacuolar System")

Lysosomal enzymes are concerned with the degradation of metabolites and not with cellular synthetic or transfer reactions. On the basis of numerous cytochemical observations, it is generally believed that primary lysosomes are formed by budding from the trans face of the Golgi apparatus. The primary lysosomes are dispatched as either smooth or coated vesicles having a diameter of about 50–100 nm. After budding, the coat (believed to be *clathrin*; see Chapter 15) is lost. The acid hydrolases that are destined for lysosomes are synthesized by ribosomes of the rough ER in the vicinity of the Golgi bodies (Chapter 18). Some of these hydrolases are discharged into the lumenal phase of the ER and others remain anchored in the ER membranes. Through either the dispatchment of tiny vesicles from the ER or via direct communication through cisternae, the hydrolases make their way to the cis face of the Golgi body. After purification and processing in successive Golgi cisternae, the hydrolases are released from the trans face of the Golgi apparatus in the form of primary lysosomes. This con-

cept, depicted diagrammatically in Figure 19–5, is supported by a large number of observations made with a variety of tissues. The process is reminiscent of the formation of zymogen granules for secretion by Golgi bodies. Because of the intimate association between Golgi bodies, ER, and primary lysosomes, regions of cells containing these organelles are sometimes called "GERL" (i.e., "Golgi–Endoplasmic Reticulum–Lysosome") complexes.

The synthesis of the lysosomal enzymes involves the preliminary assembly of a **signal sequence** that directs the large ribosomal subunit to its appropriate docking site on the endoplasmic reticulum (see Chapter 15). Following this, the elongating polypeptide is discharged through the membrane into the intracisternal space. Virtually all of the lysosomal enzymes are glycoproteins. Core glycosylation occurs during translation but final processing occurs in the cis Golgi cisterna (see Chapter 18).

Using histochemical procedures it is possible to localize lysosomal enzyme activity in the Golgi apparatus; this, of course, supports the view that lysosomal enzymes progressively transit through at least some of the Golgi cisternae. The carbohydrate portions of the lysosomal enzymes include an uncommon mannose-6-phosphate-containing oligosaccharide. Recent studies indicate that the Golgi membranes contain a specific receptor for mannose-6-phosphate and

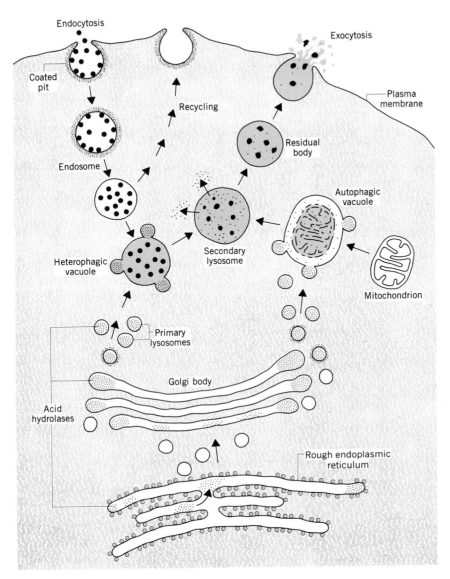

FIGURE 19–5 Formation and function of lysosomes in cellular *heterophagy* and *autophagy*. In heterophagy, endocytosed material enters cell in *coated pit*. Coat materials and ligand-binding receptors may be recycled to the plasma membrane. Fusion of *endosome* with primary lysosomes forms a secondary lysosome called a *heterophagic vacuole*. In autophagy, primary lysosomes fuse with organelles (e.g., mitochondria) to form a secondary lysosome called an *autophagic vacuole*. Some products of lysosomal digestion may be transferred to the cytosol. Lysosomes containing undegraded or unabsorbed materials are called *residual bodies*. These may migrate to and fuse with the plasma membrane during exocytosis.

it is believed that these receptors play an important role in directing the lysosomal hydrolases into newly forming primary lysosomes.

Heterophagy   Extracellular materials brought into the cell by endocytosis are enclosed within vacuoles called *endosomes*. These materials may later be rejected unaltered by exocytosis, or the endosome may fuse with one or more primary lysosomes that empty their digestive hydrolases into the newly formed particle (now called a secondary lysosome, see Fig. 19–5). Lysosomal digestion of endocytosed material is termed **heterophagy**.

The fusion of primary lysosomes with endosomes

has been demonstrated *in vivo* in a number of tissues using various exogenous markers introduced into the organism. These markers, which include horseradish peroxidase, ferritin, and hemoglobin, are engulfed by the tissue cells and are later detected within secondary lysosomes along with lysosomal hydrolases.

Z. A. Cohn and B. Benson employed ³H-labeled leucine to trace the fate of newly synthesized hydrolases in peritoneal phagocytes of mice. They found that pinocytic activity was greatly increased when these cells were incubated in blood serum. Autoradiographic analysis revealed that the labeled hydrolases appeared first in the Golgi region of the cells and later within pinocytic vesicles or endosomes. Their observations support the proposal that secondary lysosomes are formed by the fusion of endosomes and primary lysosomes. Moreover, Cohn and Benson also found that the rate at which hydrolases were produced by the cells was related to the level of pinocytic activity, suggesting that the production of primary lysosomes may somehow be regulated by endocytosis (see below).

In some cells, several small primary lysosomes may fuse with a single large endosome; in other cells, large primary lysosomes sequentially fuse with a number of small endosomes. The contents of the secondary lysosome change dramatically with time as (1) the contents of the lysosome are enzymatically degraded, (2) new materials are introduced through fusion of additional endosomes, and (3) additional hydrolases are added by the fusion of new primary lysosomes. The hydrolases in the secondary lysosome break down the endocytosed materials, producing a variety of useful substances (e.g., amino acids, sugars, etc.) as well as some useless waste products. It is generally agreed that usable materials make their way across the membrane of the secondary lysosome and enter the cytosol, where they participate in cellular metabolism. This transfer probably takes the form of passive diffusion or facilitated or active transport. Eventually, digestion and absorption are terminated, leaving only residues and denatured enzymes within the vacuole, which is now referred to as a **residual body**. In many cells, residual bodies fuse with the plasma membrane, and this is followed by exocytosis (Fig. 19–5). In some cells (especially those of higher organisms), residual bodies accumulate within the cytoplasm or continue to increase in size, eventually interfering with the normal activities of the cell and resulting in cell death. Progressive lysosome engorgement is believed to be involved in the aging process.

Autophagy   The isolation and digestion of portions of a cell's own cytoplasmic constituents by its lysosomes occurs in normal cells and is termed **autophagy** (Fig. 19–5). The phenomenon is most dramatic in the tissues of organs undergoing regression (e.g., changes in the uterus following delivery, during metamorphosis in insects, etc.). Autophagic vacuoles containing partially degraded mitochondria, smooth and rough endoplasmic reticulum, microbodies, glycogen particles, or other cytoplasmic structures are frequently observed in tissue sections examined with the electron microscope. Cellular autophagy results in a continuous turnover of mitochondria in liver tissue.

The half-life of the liver mitochondrion is about 10 days and corresponds to the destruction of one mitochondrion per liver cell every 15 minutes.

Distribution of Lysosomes

Since their initial discovery in mammalian liver, lysosomes have been identified in many different cells and tissues; some of these are listed in Table 19–3. The greatest variety of tissues found to contain lysosomes occurs in animals. Although most studies have been carried out using mammalian tissues, lysosomes have been identified in insects, marine invertebrates, fish, amphibians, reptiles, and birds. Lysosomes are particularly numerous in epithelial cells of absorptive, secretory, and excretory organs (liver, kidneys, etc.). They are also present in large numbers in the epithelial cells of the intestines, lungs, and uterus. Phagocytic cells and cells of the reticuloendothelial system (e.g., bone marrow, spleen, and liver) have also been found to contain large numbers of lysosomes. Few lysosomes occur in muscle cells or in acinar cells of the pancreas. Lysosomes are produced by certain cells in tissue culture (HeLa cells, monocytes, lymphoctyes, etc.). Although it has a number of functions not shared by lysosomes of animal cells, the large vacuole of many plant cells is a modified lysosome. Some of the various roles played by the lysosomes are summarized in Table 19–4.

TABLE 19–3   CELLS AND TISSUES CONTAINING LYSOSOMES

Protozoa	Nerve cells
Amoeba	Brain
Campanella	Intestinal epithelium
Tetrahymena	Lung epithelium
Paramecium	Uterine epithelium
Euglena	Macrophages of spleen,
Plants	bone marrow, liver,
Onion seeds	and connective tissue
Corn seedlings	Thyroid gland
Tobacco seedlings	Adrenal gland
Tissue culture cells	Bone
HeLa cells	Urinary bladder
Fibroblasts	Uterus
Monocytes	Ovaries
Macrophages	Blood
Chick cells	(leukocytes and
Lymphocytes	platelets)
Animal tissues	
Liver	
Kidney	

TABLE 19–4   SOME FUNCTIONS OF LYSOSOMES

1. Nutrition via a digestive role in protozoa and many metazoan cells
2. Nutrition via cellular autophagy during unfavorable environmental conditions
3. Lysis of organelles during cellular differentiation and metamorphosis
4. Scavenging of worn-out cell parts and denatured proteins
5. Destruction of aged red blood cells and dead cells
6. Defense against invading bacteria and viruses by circulating macrophages
7. Dissolution of blood clots and thrombi
8. Keratinization of skin
9. Secretion of hydrolases by sperm for egg penetration during fertilization
10. Yolk digestion during embryonic development
11. Bone restoration
12. Reabsorption in kidney and urinary bladder

Leukocytes, especially granulocytes, are a particularly rich source of lysosomes, and this is related to their physiological role as scavengers of microorganisms or other foreign particles in the blood. Following phagocytosis of a bacterium by a leukocyte, numerous lysosomes fuse with the endocytic vacuole containing the microorganism and initiate its digestion. The lysosomes of granular leukocytes are especially large and readily visible by light microscopy. Once the lysosome content of the leukocyte is exhausted, the white blood cell dies.

Plant Vacuoles

Many plant cells contain one or more vacuoles (see Fig. 1–33), which possess some of the properties of lysosomes. In immature and actively dividing plant cells the vacuoles are quite small. As the cells mature, the vacuoles coalesce to form larger compartments. Mature cells of higher plants usually have a larger, central vacuole that may occupy as much as (occasionally more than) 80% of the total cell volume.

The membrane enclosing the plant cell vacuole is called the **tonoplast**. Like lysosomes, plant cell vacuoles contain hydrolytic enzymes. In addition, they usually contain sugars, salts, acids, and nitrogenous compounds such as alkaloids and anthrocyanin pigments. The pH of the plant vacuole may be as high as 9 or 10 due to large quantities of alkaline substances or as low as 3 due to the accumulation of quantities of acids (e.g., citric, oxalic, and tartaric acids).

The plant vacuole is the major contributor to the *turgor* (Chapter 15) that provides support for the individual plant cell and contributes to the rigidity of the leaves and younger parts of the plant. Water accumulation in the vacuole as a result of the osmotic effects of the dissolved substances causes the vacuole to expand, pushing outward against the cytoplasm and cell wall. When there is a lack of water the turgor diminishes and this results in wilting.

Lysosome Precursors in Bacteria

Although bacterial cells do not possess lysosomes, they do contain a variety of hydrolases that are believed to be localized in the space between the plasma membrane and the cell wall. These hydrolases may be synthesized by ribosomes attached to the plasma membrane and then dispatched through it. The bacterial hydrolases play a digestive role, breaking down complex substrates in the cell's environment and providing smaller molecules required for cell growth. Bacterial hydrolases also participate in *sporulation* and *autolysis*. Although the latter process destroys the individual cells involved, it is highly beneficial to the bacterial population as a whole, for it provides for the survival of a small number of cells under unfavorable environmental conditions. Infolding of the bacterial membrane to form internalized extracellular pockets containing both hydrolases and their substrates would provide the "evolutionary link" with lysosomes of animal and plant cells.

Regulation of Lysosome Production

As noted earlier, the mechanism proposed for primary lysosome formation is strikingly similar to that proposed for zymogen granule formation in pancreatic cells and other instances of secretory protein synthesis. This similarity does not seem so unusual when one considers the following. The enzymatic contents of primary lysosomes are discharged into vacuoles (i.e., endosomes) that are derived from the plasma membrane and that contain extracellular materials. Consequently, the mechanism is similar to secretion except that the extracellular space into which the secretory products pass is internalized. In secretory cells, the production of new secretory products is regulated by a feedback mechanism in which secretion itself acts as a stimulus for the production of additional secretory materials. The experiments of Cohn and Benson described above demonstrate the relationship that exists between endocytic activity and lysosomal enzyme synthesis. It has therefore been suggested that the passage of endocytic vesicles into the Golgi regions of the cell is followed by the discharge of some primary lysosomes and that this triggers the synthesis of new acid hydrolases.

Disposition and Action of the Lysosomal Hydrolases

Many of the lysosome's enzymes are released into the surrounding environment when these organelles are physically or chemically disrupted. Those enzymes that are so readily solubilized are believed to be located in the interior of the organelle. Other lysosomal hydrolases cannot be solubilized or are extracted with great difficulty and are thought to be an integral part of the lysosome membrane together with other proteins and lipids. Some of the enzymes known to be present in lysosomes are listed in Table 19–2; it is to be noted that while this list is extensive, it is by no means complete.

All the substrates of lysosomal enzymes are either polymers or complex compounds and include proteins, DNA, RNA, polysaccharides, carbohydrate side chains of glycoproteins and glycolipids, lipids, and phosphates. The lysosomal breakdown of proteins into amino acids illustrates how these enzymes act in concert. The initial hydrolysis of protein is effected by *cathepsins D* and *E* and also by *collagenase*. These enzymes cleave peptide bonds and produce peptide fragments of varying length. The peptides, together with previously undigested proteins, are further hydrolyzed to individual amino acids by *cathepsins A* and *B. Cathepsin C, arylamidase,* and the lysosomal dipeptidases act on specific peptides, producing additional amino acids.

The breakdown of DNA and RNA is initiated by the enzymes *acid deoxyribonuclease* and *acid ribonuclease.* The resulting oligonucleotides are then degraded first by *phosphodiesterase* and then by *acid phosphatase,* producing nucleosides and inorganic phosphate. Lysosomes also possess all the enzymes necessary for hydrolysis of lipids and polysaccharides.

As noted earlier, some lysosomal enzymes are part of the membrane encasing the organelle. Among the enzymes found to be integral parts of the lysosome membrane are *acetylglucosaminidase, glucosidase,* and *sialidase. Arylsulfatase, acid phosphatase, ribonuclease,* and *glucuronidase* may also be bound to the membrane under certain conditions.

Enzymes freed from disrupted lysosomes exhibit a wide variation in stability. Some are particularly resistant to autolysis and retain their activity for months when appropriately refrigerated; others lose their activity only a few hours following tissue disruption. Of the lysosomal enzymes isolated and characterized to date, most are glycoproteins, including *cathepsin C, acid deoxyribonuclease, glucuronidase,* and *acetylglucosaminidase.*

MICROBODIES

As noted at the beginning of the chapter, the term "microbody" was used by cell biologists and cytologists for many years to describe a variety of different small cellular components. More recently, it has usually been restricted to organelles possessing *flavin oxidases* and *catalase.* Organelles possessing these activities are typically spherical or ovoid structures having a single bounding membrane, a diameter of about 0.5–1.5 μm, and containing an amorphous granular matrix, occasionally with crystalloid inclusions (Fig. 19–6). The organelles vary somewhat in structure, appearance, and function from one tissue to another and from species to species. Certain microbodies exhibit specific biochemical characteristics as well as specific distributions among animal, plant, and microbial cells. Included here are the *peroxisomes* and *glyoxysomes.*

Peroxisomes

The modern usage of the term "microbody" dates back to 1954 and the work of J. Rhodin, who described the structure and properties of these organelles in mouse kidney tissue. Since then, organelles of similar organization have been reported in many other animal tissues and also in plants (Fig. 19–6). In 1965, de Duve showed that microbodies of rat liver contained a number of oxidases that transfer hydrogen atoms to molecular oxygen, thereby forming hydrogen peroxide (Fig. 19–7). de Duve coined the term **peroxisomes** for these organelles, although a true peroxidatic activity is generally demonstrable only *in vitro. In vivo,* conditions favor the removal (or degradation) of hydrogen peroxide by *catalase* rather than by a peroxidase. However, because hydrogen peroxide is an intermediate in the reaction, the term "peroxisome" may be appropriate. The chemical and enzymatic relationships

Chloroplast

Granum

Peroxisome

Crystalloid inclusion

FIGURE 19–6 Electron photomicrograph of a peroxisome in a tobacco leaf cell; note how closely the peroxisome is appressed to a neighboring chloroplast. (Photomicrograph courtesy of Dr. E. H. Newcomb; from Frederick *et al.*, *Protoplasma* **84**, 1 (1975).)

between an oxidase, peroxidase, and catalase are shown in Figure 19–7.

A number of enzymes are characteristically present in peroxisomes, including *uric acid oxidase* (also called *uricase*), *D-amino acid oxidase, acyl-CoA oxidase, polyamine oxidase, β-hyroxyacid oxidase, NADH-glyoxylate reductase, NADP-isocitrate dehydrogenase,* and *catalase.* When *uric acid oxidase* is present in large amounts, it frequently takes the form of a paracrystalline "nucleoid" at the center of the organelle. The functions of peroxisomes in animal cells are diverse. Peroxisomal *catalase* is thought to be involved in the degradation of H_2O_2, which is extremely toxic, the source of the peroxide being other peroxiso-

mal reactions (e.g., those catalyzed by the *flavin oxidases*). *Uric acid oxidase* is important in the catabolic pathway that degrades purines. The early observation that there is an abundance of peroxisomes in cells engaged in lipid metabolism suggested that these organelles may be involved in lipid metabolism. Recently, it has been shown that liver peroxisomes contain a major system for the β-oxidation of fatty acids (see Chapter 10), however, the enzymes are different from those of mitochondria (Chapter 16) although they produce the same end product, acetyl-CoA.

Electron-microscopic studies of tissue sections often reveal a close proximity between peroxisomes and mitochondria. This is not surprising as the products of

FIGURE 19–7 *Chemical interrelationship between oxidase, peroxidase, and catalase enzymes.*

peroxisomal activity may serve as substrates for mitochondrial activity. For example, glyoxylate produced in peroxisomes may be converted there to glycine by transamination. After passage to neighboring mitochondria, the glycine may be further metabolized in a variety of ways including conversion to other amino acids or incorporation into heme.

We saw earlier that one of the characteristic features of lysosomal enzyme activity is its *latency*. No latency is exhibited by the peroxisomal enzymes, as relatively large molecules (including the peroxisomal enzyme substrates) readily permeate the peroxisome membrane.

Isolation of Peroxisomes The sedimentation coefficients and densities of peroxisomes in sucrose gradients are close to those of lysosomes and account for the fact that for some time peroxisomal enzyme activities were ascribed to lysosomes. Density gradient centrifugation of the so-called "light mitochondrial fraction" prepared by preliminary differential centrifugation is the method of choice for isolating peroxisomes. The greatest success in peroxisome purification is obtained if the lysosomes are first allowed to accumulate Triton WR-1339 (see earlier discussion). Triton-loaded lysosomes are considerably less dense than normal lysosomes and so they are easily "floated" away from the peroxisomes during sucrose density gradient centrifugation (Fig. 19–8).

Formation of Peroxisomes For many years, it was believed that peroxisomes of both plant and animal tissues arose as outgrowths of the endoplasmic reticulum and that the peroxisomal enzymes were dispatched into the cisternae of the endoplasmic retic-

ulum by attached ribosomes. The enzymes made their way into the organelle prior to its physical separation from the ER. However, the growing tide of recent evidence appears to argue in favor of an alternative model for the biogenesis of peroxisomes. In liver and in other tissues as well, many of the peroxisomal enzymes are synthesized by unattached ribosomes (i.e., ribosomes that are not bound to the ER) and are released into the cytosol. From there, the enzymes are slowly taken up by preexisting peroxisomes. Studies with peroxisomal catalase indicate that the subunits of the enzyme and the heme enter the microbody and are then assembled to form the functional enzyme. The nature of the mechanism that translocates peroxisomal proteins into peroxisomes is uncertain, but it is speculated that the proteins are first bound to a membrane receptor and are then drawn through the membrane and trapped inside as multiprotein complexes. Indeed, dense crystalline aggregates are often observed inside peroxisomes. Translocation of proteins through the peroxisome membrane is readily demonstrable *in vitro*. When newly synthesized peroxisomal proteins are tagged with a radioactive label and incubated with isolated peroxisomes, some of the protein may soon be recovered from within the organelles.

Peroxisomes often appear in clusters when examined by electron microscopy and occasional dumbbell-shaped images are observed. This suggests that the "tails" seen on peroxisomes are connections to other peroxisomes. Indeed, either permanent or transient connections between peroxisomes might form a "peroxisome reticulum." Such interconnections would explain the remarkable biochemical homogeneity of peroxisomes and the synchronous turnover of peroxisomal proteins. New peroxisomes may be formed either by the fission of preexisting peroxisomes or by budding from the peroxisome reticulum.

Glyoxysomes

In 1967, R. W. Breidenbach and H. Beevers discovered that microbodies of the fat-storing cells of germinating fatty seeds contain enzymes of the **glyoxylate cycle** (see Chapter 10) in addition to peroxisomal enzymes. They coined the specific term **glyoxysomes** for these particles, although it is to be understood that glyoxysomes are a form of peroxisome. Glyoxysomes contain not only enzymes specific to the glyoxylate cycle (i.e., *isocitrate lyase* and *malate synthetase*) but they also contain several of the essential enzymes of the Krebs cycle (Chapters 10 and 16), which therefore

FIGURE 19–8 The "light" mitochondrial fraction produced during differential centrifugation of a tissue homogenate contains lysosomes and peroxisomes in addition to small mitochondria (steps 1–4). The three organelles may then be separated by density gradient centrifugation (steps 5 and 6). Using tissues of normal animals, peroxisomes are separated from lysosomes only with great difficulty (steps 6A). However, in animals treated with Triton WR-1339, the density of lysosomes is diminished, thereby permitting more effective separation of the peroxisomes (step 6B).

function simultaneously in both mitochondria and glyoxysomes. The name "glyoxysome" takes precedence over "peroxisome" whenever glyoxylate cycle enzymes are identified in the microbody.

The relationship between the Krebs cycle and the glyoxylate cycle is shown in Figure 19–9. Both cycles employ the same reactions to produce isocitrate from acetyl-CoA and oxaloacetate, but beyond this point the pathways differ. In the Krebs cycle, isocitrate is successively decarboxylated, producing two molecules of carbon dioxide and succinate. In the glyoxylate cycle, isocitrate is converted to succinate and glyoxylate. Instead of being lost as two molecules of CO_2, the two-carbon glyoxylate condenses with another acetyl-CoA to form the four-carbon dicarboxylic acid malate. The four carbon atoms of the two acetyl-CoA molecules are thus conserved as one four-carbon compound that, after conversion to succinate and migration to the mitochondrion, may be converted to oxaloacetate. The oxaloacetate may then be utilized in gluconeogenesis.

Oxaloacetate formed in mitochondria from glyoxysomal succinate is presumed to serve as a direct precursor of phosphoenol pyruvate (PEP). The conversion of PEP to glucose and other carbohydrates occurs essentially by the reversal of the steps of glycolysis (see Chapter 10). Glyoxysome-containing tissues are thus able to convert simple two-carbon sources such as acetate into carbohydrate. In some tissues, such as fat-storing cells in seeds, the acetate is obtained through the degradation of fatty acids (Chapter 10); therefore glyoxysomes participate in the conversion of fat to carbohydrate.

Distribution and Origin of Glyoxysomes The glyoxylate cycle is especially significant for cells growing exclusively on acetate or fatty acids (e.g., a number of microorganisms), where the cycle acts as a source of four-carbon dicarboxylic acids. Certain microorganisms including *Euglena, Chlorella, Neurospora,* and *Polytomella* contain "glyoxysomelike" particles, but the term glyoxysome is normally reserved for the organelles of fat-storing endosperm or cotyledons of germinating fatty seeds.

Reports have appeared in the literature from time to time indicating the DNA is present in glyoxysomes, raising the possibility that these organelles possess some degree of autonomy. However, this notion is not generally accepted. Glyoxysomes appear to be formed by a process similar to that which gives rise to perox-

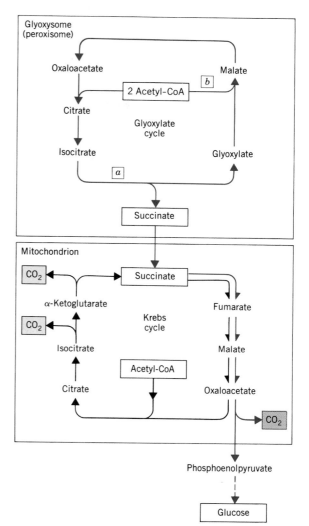

FIGURE 19–9 Comparison of the glyoxylate cycle in glyoxysomes (top) and the Krebs cycle in mitochondria (bottom). Succinate produced as a by-product of the glyoxylate cycle can be converted to glucose and other carbohydrates (red arrows). Enzymes [a] and [b] are *isocitrate lyase* and *malate synthetase,* respectively.

isomes (see above), namely, by the fission or budding of preexisting glyoxysomes. Most (if not all) glyoxysomal enzymes are synthesized by free (i.e., unattached) ribosomes in the cytosol and are then translocated into or through the membranes of glyoxysomes already present in the cell.

In concluding this discussion of microbodies, it is important to note that some organelles fitting the general microscopic description of microbodies do not clearly fit into either the peroxisome or glyoxysome category when evaluated in terms of their enzymatic activities. It is entirely possible that microbodies may be associated with varying activities depending on the specialization of the cell and that microbodies exist whose actions and cellular functions remain to be determined. It is clear that one cannot name particles solely on the basis of microscopic characteristics and expect that all function identically.

SUMMARY

The realization that certain hydrolase activities are associated with a distinct class of organelles called **lysosomes** is quite recent. Previously, these activities were believed to be localized in mitochondria. Lysosomes function in the intracellular digestion of poorly functioning or superfluous organelles, as well as endocytosed materials. Several forms of lysosomes may be identified, including **primary lysosomes, secondary lysosomes** (e.g., **heterophagic** and **autophagic** vacu-

oles), and **residual bodies**. Primary lysosomes are released from the maturing or trans face of Golgi bodies, their hydrolase content initially derived from rough endoplasmic reticulum. Fusion of primary lysosomes with **endosomes** forms the secondary lysosomes in which digestion occurs. Usable products of this digestive activity are transferred to the cytosol. Undigested or unabsorbed materials remain in the residual bodies, which may accumulate in the cell or fuse with the plasma membrane during exocytosis.

Two types of **microbodies** occur in cells; these are **peroxisomes** and **glyoxysomes**. Peroxisomes are found in both animal and plant tissues and contain a number of *flavin oxidases* that produce hydrogen peroxide during their degradative activity. The potentially harmful peroxides are further degraded by peroxisomal catalase. Glyoxysomes are found in certain plant tissues and contain enzymes of the **glyoxylate cycle** in addition to peroxisomal enzymes.

REFERENCES AND SUGGESTED READING

Articles and Reviews

Bainton, D. F., The discovery of lysosomes. *J. Cell Biol.* **91**, 66 (1981).

Beevers, H., Microbodies in higher plants. *Ann. Rev. Plant Physiol.* **30**, 159 (1979).

Dautry-Varsat, A., and Lodish, H. G., How receptors bring proteins and particles into cells. *Sci. Am.* **250** (5), 52 (May 1984).

de Duve, C., The lysosome. *Sci. Am.* **208** (5), 64 (May 1963).

de Duve, C., The peroxisome: A new cytoplasmic organelle. *Proc. R. Soc. Lond.* **173**, 71 (1969).

de Duve, C., Microbodies in the living cell. *Sci. Am.* **248** (5), 74 (May 1983).

Fahimi, H. D., and Yokota, S., Ultrastructural and cytochemical aspects of animal peroxisomes—Some recent observations. In *International Cell Biology* (H. G. Schweiger, Ed.), Springer-Verlag, Berlin, 1981.

Farquhar, M. G., and Palade, G. E., The Golgi apparatus (complex)—(1954–1981)—From artifact to center stage. *J. Cell Biol.* **91**, 77s (1981).

Lazarow, P. B., Functions and biogenesis of peroxisomes. In *International Cell Biology* (H. G. Schweiger, Ed.), Springer-Verlag, Berlin, 1981.

Leighton, F., Poole, B., Beaufay, H., Baudhuin, P., Coffey, J. W., Fowler, S., and de Duve, C., The large-scale separation of peroxisomes, mitochondria, and lysosomes from the livers of rats injected with Triton WR-1339. *J. Cell. Biol.* **37**, 482 (1968).

Lodish, H. F., Braell, W. A., Schwartz, A. L., Strous, G. J., and Zilberstein, A., Synthesis and assembly of membrane and organelle proteins. *Int. Rev. Cytol.* (Suppl. 12), 247 (1981).

Marzella, L., and Glaumann, H., Biogenesis, translocation, and function of lysosomal enzymes. *Int. Rev. Exp. Pathol.* **25**, 239 (1983).

Masters, C., and Crane, D., The role of peroxisomes in lipid metabolism. *Trends in Biochem. Sci.* **9** (7), 314 (1984).

Newcomb, E. H., and Tandon, S. R. Microbodies (peroxisomes) in soybean root nodules. In *International Cell Biology* (H. G. Schweiger, Ed.), Springer-Verlag, Berlin, 1981.

Tolbert, N. E., Metabolic pathways in peroxisomes and glyoxysomes. *Ann. Rev. Biochem.* **50**, 133 (1981).

Tolbert, N. E., and Essner, E., Microbodies: Peroxisomes and glyoxysomes. *J. Cell Biol.* **91**, 271 (1981).

Willingham, M. C., Haigler, H. T., Dickson, R. B., and Pastan, I. H., Receptor-mediated endocytosis in cultured cells: Coated pits, receptosomes, and lysosmes. In *International Cell Biology* (H. G. Schweiger, Ed.), Springer-Verlag, Berlin, 1981.

Books, Monographs, and Symposia

Breidenbach, R. W., Microbodies. In *Plant Biochemistry* (J. Bonner and J. E. Varner, Eds.), 3rd ed., Academic, New York, 1976.

Dean, R. T., *Lysosomes*, Camelot Press, Ltd., Southampton, U.K., 1977.

de Duve, C., and Wattiaux, R., Function of lysosomes. In *Annual Review of Physiology* (V. E. Hall, A. C. Giese, and R. R. Sonnenschein, Eds.), Annual Reviews, Palo Alto, Calif., 1966.

de Reuck, A. V. S., and Cameron, M. P. (Eds.), *CIBA Foundation Symposium on Lysosomes*, Little, Brown, Boston, 1963.

Dingle, J. T., *Lysosomes*, North-Holland, Amsterdam, 1977.

Dingle, J. T., and Fell, H. B. (Eds.), *Lysosomes in Biology and Pathology*, Parts I and II, North-Holland, Amsterdam, 1969.

Hruban, Z., and Rechcigl, M., *Microbodies and Related Particles*, Academic, New York, 1969.

Huang, A. H. C., Metabolism in plants peroxisomes. In *Recent Advances in Phytochemistry*, Vol. 16, *Cellular and Subcellular Localization in Plant Metabolism* (L. L. Creasy and G. Hrazdina, Eds.), Plenum, New York, 1982.

Huang, A. H. C., Trelease, R. N., and Moore, T. S., *Plant Peroxisomes*, Academic, New York, 1983.

Pastan, I., and Willingham, M. C., (Eds.), *Endocytosis*, Plenum Press, New York, 1985.

Schweiger, H. G. (Ed.), *International Cell Biology* (1980–1981), Springer-Verlag, Berlin, 1981.

Strauss, W., Lysosomes, phagosomes and related particles. In *Enzyme Cytology* (D. B. Roodyn, Ed.), Academic, London, 1967.

CHAPTER 20

THE NUCLEUS: ORGANIZATION AND DIVISION

This is the first of two consecutive chapters dealing with the biology of the cell nucleus. In this chapter, we will examine the structural and chemical organization of the major nuclear components, namely, the **chromosomes, nucleosomes, nucleolus**, and **nuclear envelope**. We will also look at the gross nuclear changes that characterize nuclear and cellular division. In Chapter 21, we will be concerned with the operation of the genetic apparatus, focusing on the molecular basis of gene expression.

Most of the functions of the nucleus are intimately related to events that occur in the cytoplasm; for example, translation of mRNA by cytoplasmic ribosomes (a mechanism described in Chapter 22) is the end result of the transcription of DNA in the nucleus and the processing of the resulting RNA. Structurally too, the nucleus and cytoplasm are interrelated. The outer membrane of the nuclear envelope is continuous with membranes of the endoplasmic reticulum. Events in the cytoplasm also affect nuclear function, initiating or terminating specific nuclear activities. Therefore, whereas the nucleus is perhaps the single dominant structural feature of most eukaryotic cells, it is not to be viewed as an isolated organelle.

STRUCTURE OF THE NUCLEUS

The nucleus of a eukaryotic cell (Fig. 20–1) is delimited by a pair of membranes called the **nuclear envelope**. The outer and inner membranes of this envelope are separated by a narrow space called the **perinuclear space** but fuse with each other at the margins of **pores**. The fluid of the cytoplasm (the cytosol) is continuous through the nuclear pores with the fluid of the nucleus, called the **nucleoplasm**. Although ribosomes may be attached to the cytoplasmic surface of the outer nuclear membrane, these ribosomes are not considered nuclear structures. The nucleoplasm contains a number of discrete structures including one or more **nucleoli, chromosomes** (as many as several hundred in some eukaryotic cells), and other structures or regions that appear at various times depending on nuclear activity.

Prokaryotic cells do not have a "true" nucleus; indeed, identification of a cell as prokaryotic ("prenucleus") or eukaryotic ("true-nucleus") is based on the presence or absence of this membrane-enclosed organelle. In prokaryotic cells, most of the genetic material is confined to an area of the cell called the

(a)

Outer membrane

Inner membrane

Chromatin

Nuclear pore

Nucleolus

Nuclear
envelope

Perinuclear
space

FIGURE 20–1 The nucleus of the
eukaryotic cell. (*a*) Transmission electron
photomicrograph of a thin section
through the nucleus. (Courtesy of R.
Chao.) (*b*) Transmission electron
photomicrograph of freeze-etched and
-fractured cell nucleus. (Courtesy of Dr. E.
G. Pollock.)

(b)

Pores

Inner
membrane
(*P* face)

Outer
membrane
(*E* face)

nucleoid (Fig. 1–37), but the nucleoid is not delimited by a membrane.

CHROMOSOMES

The first descriptions of the chromosomes of eukaryotic cells appeared between 1840 and 1880, but it was not until 1888 that Waldeyer introduced the term *chromosome* ("colored body") for these structures. Chromosomes are composed of **chromatin**, which readily binds basic stains. Because the chromatin is highly condensed during cell division, the chromosomes are easily seen and described by light microscopy. The number of chromosomes in the cell nucleus varies considerably among different animal and plant species; however, each species has a specific *chromosome number*. For example, cells of the nematode *Ascaris megalocephala* have only 2 chromosomes, human cells have 46, and certain protozoa have over 300. Unrelated organisms may have the same chromosome number. The potato plant has 48 chromosomes, but so do plum trees and chimpanzees. Within a species, each chromosome exhibits a specific and characteris-

10 μ

1 2 3 4 5

6 7 8 9 10 11 12

13 14 15 16 17 18

19 20 21 22 X Y

(a)

FIGURE 20–2 The 23 pairs of homologous human chromosomes as seen by light microscopy. Each chromosome consists of paired *chromatids* held together at the *centromere*. These chromosomes are from the cell of a male because the pair of sex chromosomes consists of one X and one Y chromosome. (Courtesy of Dr. Y. Ohnuki; copyright © 1968 by Springer-Verlag, *Chromosoma* **25**, 416.) (*b*) Scanning electron photomicrograph of metaphase chromosomes on the mitotic spindle. (Courtesy of Dr. W. K. Heneen; copyright © 1980, Societe Francaise de Microscopie Electronique, *Biol. Cellulaire* **37**, 13.)

(b)

tic shape during the metaphase (see later) of cell division. The unique appearance of the metaphase chromosomes is retained from one generation of cells and organisms to the next (Fig. 20–2).

Chromosome shape and size change during the stages of nuclear division. Most chromosomes have two *arms*, one on each side of the **primary constriction** or **centromere** (also called **kinetochore**). Metaphase chromosomes already have undergone replication so that each chromosome consists of *sister* **chromatids** and therefore appears to have *two* sets of arms (Fig. 20–3). The centromere is the site of attach-

ment of the chromosome to the microtubules of the **spindle** and acts as the focus of chromosome (or chromatid) movement during the anaphase of division. Chromosomes that lack a centromere are said to be *acentric* and fail to segregate normally during division. *Secondary* and *tertiary constrictions* may also be identified. Secondary constrictions are associated with nucleoli and are called **nucleolar organizer regions** (NOR). The significance of the tertiary constrictions is unclear, but their characteristic disposition in each chromosome helps to distinguish one chromosome from another.

FIGURE 20–3   Stereophotomicrographs of a metaphase chromosome. Coiling of the chromatin threads to form a helical structure is apparent. The arrow identifies one gyre of the helix. (Courtesy of Dr. Hans Ris.)

Fine Structure of Chromosomes

The condensed chromosomes visible during mitosis are composed of an organized array of **chromatin fibers**, but in their decondensed state give rise to a highly disperse network (Fig. 20–4). Each chromatin fiber is believed to contain one molecule of DNA. The average diameter of *Type A* chromatin fibers is about 10 nm, whereas the diameter of the DNA double helix is only 2 nm (see Fig. 7–8). The difference is due to the coiling of the DNA molecule and the presence of large quantities of proteins. *Type B* fibers are even thicker (20–30 nm), the increased thickness resulting from additional levels of coiling. Chromatin of varying thickness exists in the interphase nucleus and is presumed to consist of alternating areas of type A and B fibers.

An indication of the extent of DNA coiling that occurs in a chromosome is obtained from the *packing raio* (i.e., the length of DNA divided by the length of the chromatin fiber or whole chromosome). For decondensed interphase chromatin this ratio is about 10:1, but for metaphase chromosomes it is more than 1000:1. The extensive amount of coiling and folding that characterizes DNA permits storage of genetic information in a highly compact form.

NUCLEOSOMES

In recent years it has been shown that chromatin consists of a repeating pattern of bodies called **nucleosomes** (Fig. 20–5) formed by the association of DNA and specific nuclear proteins called **histones**. Nucleosomes are quasi-cylindrical in shape, having a diameter of about 11 nm and a length of 5.5 nm. The histones form a core around which the 2-nm-thick molecule of DNA is wound to form a left-handed **superhelix** of diameter 10.8 nm and pitch 2.8 nm (Fig. 20–6). The DNA makes two turns around the histone core, each turn consisting of about 83 base pairs. Successive nucleosomes are connected by *linker DNA* (Fig. 20–7). Nucleosomes can be released from chromosomes by partially digesting the chromatin with a nuclease. Digestion by the nuclease is progressive and initially releases nucleosomes containing 160–240 base pairs of DNA and includes the **linker DNA** that joins neighboring nucleosomes together. Continued digestion produces an intermediate product containing about 160–170 base pairs, a *core* of histones, and a separate histone termed H1. Finally, a stable unit is formed that consists of the core of histones around which is wound a length of DNA containing about 146 base pairs.

The histone core is an *octamer* consisting of two each of four histones identified as H2A, H2B, H3, and H4. The DNA forms nearly two complete turns around the core. The H3 and H4 histones are believed to form a subnucleosome particle that holds the main DNA loop and the two H2A and H2B histones are associated with the DNA nucleotide sequences at either end of the loop. The fifth histone (i.e., H1) appears to interact with the DNA at its entry and exit points on the nucleosome.

The primary structures of the four different histone molecules that make up the core of the nucleosome show a remarkable degree of conservation. Few (if any) differences in their amino acid sequences have been found, even when diverse species of organisms are studied. Histones H2A and H2B are rich in the basic amino acid lysine; this would account for their ability to bind stretches of the DNA. In contrast to the core histones, several different H1 histones may be present in the same cell, and interspecies differences are quite large. H1 histones are also rich in lysine, thereby permitting their close association with the DNA.

The core particle, the linker DNA, and histone H1 comprise a unit called a **mononucleosome** or **chromatosome**. In the complete chromatosome, the two points at which the DNA molecule enters and leaves the core particle are located near one another on the particle's surface. When H1 is digested away, the strands tend to be displaced further from one another. Thus it seems that histone H1 locks the entering and exiting stretches of DNA into position and is closely associated with the core.

The frequency of occurrence of nucleosomes along chromatin fibers varies in different tissues. Although the number of base pairs in the DNA wound around the core particle remains the same in all cells, linker DNA length varies among species and among tissues within a species. Linker DNA may contain as few as zero bases (e.g., in yeast cells) or as many as 80 bases (e.g., in sea urchin sperm cells). The significance of the variation in linker DNA length is not yet clear.

As noted earlier, two classes of chromatin fibers exist having diameters of 10 and 30 nm. It appears that the 10-nm fibers are formed from a linear array of nucleosomes (Fig. 20–7), whereas 30-nm fibers are formed by the helical coiling of 10-nm fibers. Each turn of the helix in a 30-nm fiber consists of about six nucleosomes, and successive turns form a structure

(a)

FIGURE 20–4 (*a*) Stereophotomicrographs of a *polytene* (i.e., multistranded) chromosome in a fat cell of the larva of *Drosophila melanogaster*. The chromosome consists of 24-nm chromatin fibers that are slightly stretched in the regions between successive bands. (courtesy of Dr. Hans Ris; copyright © 1976 by Intertal, International Publications Co.) (*b*) TEM photo of a histone-depleted metaphase chromosome from a human HeLa cell. The scaffold (core) of the chromosome is surrounded by many loops of DNA. (Courtesy of Drs. J. R. Paulson and U. K. Laemmli; copyright © M.I.T.)

(b)

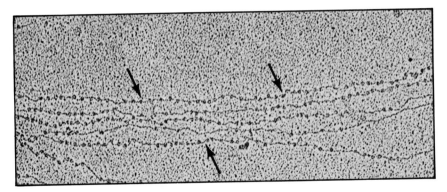

FIGURE 20–5 Electron photomicrograph of chromatin fiber revealing a succession of *nucleosomes* (arrows). (Courtesy of Dr. F. Puvion-Dutilleul.)

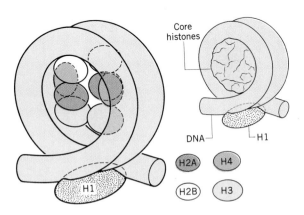

FIGURE 20–6 Structure of the nucleosome. As seen in the left half of the diagram, each nucleosome consists of an octamer of histones (two each of histones H2A, H2B, H3, and H4) around which the DNA makes nearly two full turns. A fifth histone (i.e., H1) interacts with the DNA at its entry and exit points on the nucleosome.

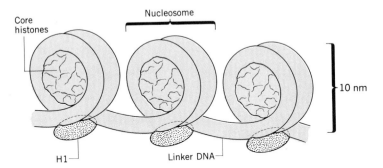

FIGURE 20–7 Successive nucleosomes are connected by *linker* DNA.

called a **solenoid** (Fig. 20–8). Whether or not chains of nucleosomes in isolated chromatin exist in the linear or solenoid configuration is influenced by the ionic strength of the medium. The linear array exists at low ionic strength, but the solenoid structure will form as the ionic strength is raised. These configurational changes may have physiological significance, as they take place during the initial stages of chromatin condensation that characterize the onset of mitosis and the decondensation that follows mitosis. Histone H1 is essential to the transition from the linear array to the solenoid structure. In the absence of H1, no ordered

structures are formed. Thus, histone's H1's roles may include promotion of close packing of successive nucleosomes.

Thirty-nanometer diameter fibers are composed of sequences of solenoids. Between solenoids, stretches of nucleosome-free DNA may be associated with *nonhistone* chromosomal proteins (Fig. 20–9). Nucleosome-free DNA is readily digested by low concentrations of the enzyme *DNAase I*, whereas nucleosomal and linker DNA are unaffected. These *DNAase-hypersensitive sites* may be important in the regulation of gene expression. Results of experiments

FIGURE 20–8 Chains of nucleosomes may be coiled to form a more compact structure called a *solenoid*. Each turn of the solenoid helix contains six nucleosomes.

with avian red blood cells and oviduct cells bear on this probability.

In avian red cells, the genes encoding the globin chains of hemoglobin are expressed, whereas the genes encoding the egg white protein ovalbumin are not. In contrast, in oviduct cells globin chain genes are not expressed and ovalbumin genes are. Treatment of the chromatin of these cells with low concentrations of *DNAase I* yields the following results. In red cells, treatment with the nuclease rapidly degrades globin chain DNA but leaves ovalbumin DNA unaffected; in contrast, treatment of oviduct cells with *DNAase I* degrades ovalbumin DNA but leaves globin chain DNA unaffected. Thus, sensitivity to *DNAase* is common to both nucleosome-free stretches of the DNA and actively expressed genes. In the SV-40 virus, a nucleosome-free stretch of the DNA about 300 base pairs long occurs at a point on the chromosome where replication and RNA synthesis begin. Association of nucleosome-free regions with nonhistone proteins suggests that the latter may play some role in "punctuating" the chromosome by marking those regions to which the transcription enzymes bind.

FIGURE 20–9 Between successive solenoids, nucleosome-free DNA may be associated with *nonhistone* chromosomal proteins.

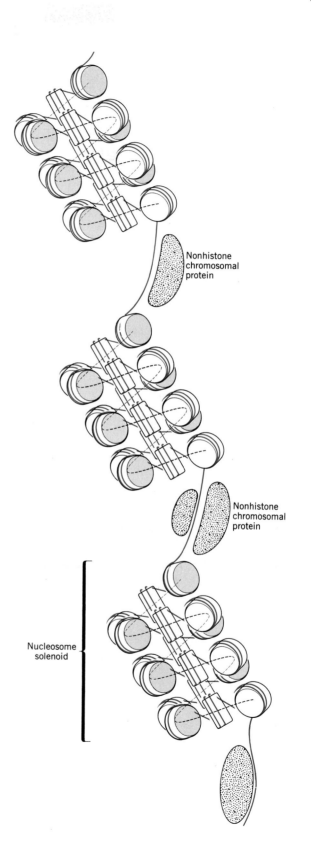

Bacterial and Viral Chromosomes

The chromosomes of bacteria and viruses are single nucleic acid molecules. In most cases, they consist of double-helical DNA, but in certain viruses they consist of only a single strand of DNA or RNA. The DNA may be linear or circular, depending on the cell or virus.

Bacteria   The chromosomes of *Escherichia coli* and *Bacillus subtilis* have been studied more extensively than those of any other bacteria. In both species the

chromosome contains a single, circular DNA molecule more than a millimeter long. The chromosome of *E. coli* is about 1100 μm long, has a diameter of 2–4 nm, and a molecular weight of 2.5×10^9. In the cell, the chromosome is condensed into an area less than 1 μm². When mild procedures are used to extract *E. coli* DNA into solutions rich in cations, the chromosomes remain in a highly condensed state (Fig. 20–10). In this form, the chromosome consists of several dozen loops. In each loop, the DNA is highly twisted and

FIGURE 20–10   Chromosome released from an *E. coli* cell. The chromosome contains many supercoiled domains. The reference bar is 1 μm long. (Courtesy of Dr. A. Worcel, copyright © 1974 by Academic Press, *J. Mol. Biol.* **82**, 108.)

forms a *supercoil* (see below). Treatment of the chromosome with nucleases that degrade RNA frees some of the loops so that the chromosome takes on a more open appearance. The loops are believed to be stabilized by RNA and protein.

Viruses   The nucleic acid of a virus may be either DNA or RNA. Among the *RNA viruses*, the nucleic acid occurs in four different structural states: (1) single-stranded, linear forms (e.g., tobacco mosaic virus); (2) single-stranded, circular forms (e.g., encephalomyocarditis and potato spindle tuber viruses); (3) double-stranded, linear forms (e.g., mammalian reo-

viruses); and (4) double-stranded, circular forms (e.g., mycoviruses). One of the best studied RNA viruses and the first to be prepared in crystallized form is the tobacco mosaic virus (TMV). This virus has the shape of a cylinder 300 nm long and 18 nm in diameter (Fig. 20–11). Within the protein sheath is the viral chromosome, which consists of a single 3300-nm-long RNA molecule composed of 6400 nucleotides. To fit within the virus, the RNA molecule is twisted to form a right-handed helix containing 49 nucleotides per turn and a pitch of 2.3 nm. The diameter of the helix is about 8 nm. Running through the center of the helix is an aqueous core about 3 nm in diameter. The protein

FIGURE 20–11   The tobacco mosaic virus. (Photomicrograph courtesy of Carl Zeiss, Inc.)

sheath or *capsid* associated with the outer surface of the RNA helix consists of 2130 identical protein subunits.

The nucleic acids of the *DNA viruses* also occur in four different structural states: (1) single-stranded, linear forms (e.g., the parvoviruses); (2) single-stranded, circular forms (e.g., ϕX174 and M13); (3) double-stranded, linear forms (e.g., Herpes virus, the T-even bacteriophages, and lambda phage); and (4) double-stranded, circular forms (e.g., SV-40 and Polyoma viruses). The double-stranded DNAs of the T4 and lambda bacteriophages are shown in Figure 20–12.

Supercoiling

The organization of eukaryotic DNA into nucleosomes demonstrates that DNA is capable of several different

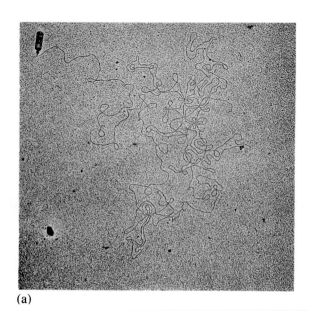

(a)

FIGURE 20–12   (*a*) DNA molecule released from a T4 bacteriophage (top left corner of the photomicrograph). (*b*) DNA from the *lambda* phage; arrows note positions of replication. The reference bar is 1 μm long. (*a*, courtesy of Dr. Hans Ris; copyright © 1963 by Cold Spring Harbor Laboratory, *C. S. H. Symp. Quant. Biol.* **28**, 2; *b*, courtesy of Drs. R. B. Imman and M. Schnos.)

(b)

levels of coiling and folding. The first level is the coiling of the two polynucleotide chains around one another to form the conventional double-helix structure (i.e., that shown in Fig. 7–8). In the second level of coiling, the double helix is wound around the histone core of a nucleosome. Finally, nucleosomes may be arranged helically to form the so-called solenoid structure. In bacteria, the length of the chromosome is many times greater than the diameter of the nucleoid area in which the DNA is concentrated; thus, it is clear that a great deal of folding of prokaryotic DNA does occur. DNA folding in prokaryotic cells and the

formation of solenoids in eukaryotic cells is called **supercoiling**. Viral nucleic acid is also supercoiled.

If one of the two strands that form a circular, double-helical DNA molecule is broken to produce two free ends and one of the free ends is then rotated around the unbroken strand while the other free end remains fixed, the resulting structure is a *supercoil* (Fig. 20–13). If the rotation of the free end tightens the double helix, the DNA is said to be *positively* supercoiled (or *overwound*); if rotation of the free end loosens the double helix, the DNA is said to be *negatively* supercoiled (or *underwound*).

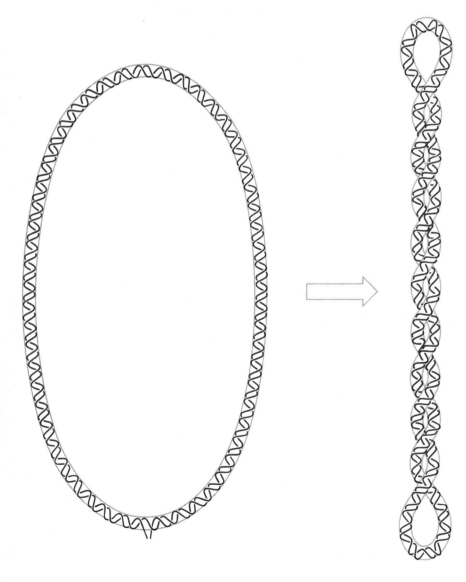

FIGURE 20–13 Supercoiling of DNA. When one of the two strands of a circular DNA molecule is nicked (left) and one free end rotated around in a clockwise direction around the unnicked strand, a positive supercoil is formed (right).

Plasmids and Episomes

Plasmids are *extrachromosomal* genetic elements found in bacteria and yeast that take the form of relatively small (i.e., they contain only a small number of genes), circular, double-stranded DNA molecules (Fig. 20-14). There may be several dozen plasmids in one cell, and although they are capable of autonomous replication, the number of plasmids remains fairly constant from one generation of cells to the next. By definition, plasmids are extrachromosomal; however, under certain conditions, plasmid DNA can be incorporated into a chromosome. Such plasmids are called **episomes**. Plasmids can be transferred from one cell to another by viruses.

In *E. coli* cells, there are three classes of plasmids: (1) F (i.e., fertility) factors, (2) R (i.e., resistance) factors, and (3) col (colicinergic) factors. The F factors promote sexual *conjugation* in *E. coli*, the R factors offer the bacteria resistance to certain drugs, and the col factors give the bacteria the capacity to secrete *colicins*, which have an antibiotic function (i.e., they kill bacteria that lack the col factor).

The F Factor and Conjugation   The circular DNA that forms the F factor plasmids of *E. coli* cells has a circumference of about 20 μm and includes several families of genes. Not all *E. coli* cells contain this plasmid; those that do are called F^+ cells, whereas those that lack the F factor are called F^-. Among the genes of this plasmid are those that control the development of *pili* or tubelike extensions from the surface of the *E. coli* cells (see Fig. 1-37). When these pili contact the surface of a neighboring F^- *E. coli* cell, they form a channel through which plasmid DNA can pass from one cell to another. Before transfer between two conjugating *E. coli* cells begins, a nick is produced in one of the two strands of the circular DNA molecule and it is at this point that replication of the plasmid is initiated. The unbroken strand serves as a template for the production of a new complementary strand and as this synthesis proceeds, the nicked strand is displaced through the pili into the neighboring cell. The transferred strand then acts as a template in the neighboring cell for the synthesis of a complementary DNA strand, and on completion of synthesis, the ends of both strands are closed to form a circular molecule. Thus, the F^+ cell and the F^- cell with which conjugation occurred end up with complete copies of the double-stranded plasmid DNA.

FIGURE 20-14   Electron photomicrograph of plasmids isolated from *E. coli* cells. (Courtesy of Dr. E. Namork.)

Mitochondrial and Chloroplast DNA

Not all of a cell's DNA is nuclear; indeed, mitochondria and chloroplasts contain DNA. Whereas nuclear DNA is complexed with protein and is apportioned among a number of chromosomes, mitochondrial DNA is "naked" and is incorporated (except in certain protozoa) into a single, circular chromosome. In multicellular animals, mitochondrial DNA (called *mtDNA*) is about 4.4 to 5.6 μm in length and has a molecular weight of about 9 to 11×10^6. In yeast, mtDNA ranges in length from 16 to 25 μm. Plant mtDNA is the longest with lengths up to 62 μm. mtDNA is found in the matrix of mitochondria (see Fig. 16–9) and at a specific point along its length is attached to the inner mitochondrial membrane.

Chloroplast DNA (*chlDNA*) is also circular and typically has a length of 35 to 45 μm, although some chlDNA may be as long as 200 μm.

Mitochondrial and chloroplast DNAs contain only a small percentage of the genes necessary to produce the organelles' RNAs and proteins. For example, mtDNA encodes the RNAs of the mitochondrial ribosomes, the organelle's transfer RNAs, and a few of the subunits of the respiratory enzymes present in the inner mitochondrial membrane. All of the remaining inner membrane proteins as well as those of the outer membrane and matrix (including the Krebs cycle enzymes) are encoded by nuclear DNA and are synthesized by cytoplasmic ribosomes. Following translation, these proteins are imported into the mitochondrial compartments (see Chapter 16).

Promiscuous DNA

From the preceding discussion, it should be clear that essentially all DNA in eukaryotic cells is confined to three membrane-bound compartments: the nucleus, mitochondria, and chloroplasts. The nuclear compartment contains all but a small number of the cell's genes and effectively controls biogenesis of the mitochondrial and chloroplast compartments. It is now apparent that DNA sequences of some genes occur in more than one cell compartment. Such DNA is called **promiscuous DNA**. For example, in maize plants a large section of chloroplast DNA encoding ribosomal RNA is also found in the DNA of the cell's mitochondria. In spinach, portions of the entire chloroplast genome are also present in the nucleus; and in sea urchins, sequence homologies have been found between mitochondrial DNA encoding 16 S ribosomal RNA and nuclear DNA.

Speculations concerning the origin of promiscuous DNA center around the notion that mitochondria and chloroplasts of eukaryotic cells orginated from autonomous, free-living prokaryotelike cells. Duplications and rearrangements of the genes of these once autonomous particles led to their establishment in the nu-

cleus. Though most of the organelle genes were lost once duplicates were established in the nucleus, some persist in more than one cellular compartment.

FORMS OF CHROMATIN

For convenience, the cell cycle (see Chapter 2) may be divided into two major phases—the **interphase** (in which the cell is engaged in its cell-specific or tissue-specific activities) and the **mitotic phase** (in which the nucleus undergoes division). During *mitosis* (discussed later in the chapter), chromosomes of eukaryotic cells *condense*, and when stained with basic dyes, the chromatin is easily studied by microscopy. During the interphase of the cell cycle, most of the chromatin does not exist in the condensed state, and it is difficult, if not impossible, to distinguish individual chromosomes. However, there are areas of the nucleus called *chromocenters* that do stain deeply even during interphase. In 1928, E. Heitz identified these chromocenters as portions of chromosomes that remain in the condensed state throughout the cell cycle.

Depending on their staining properties, two different types of chromatin may be distinguished in the interphase nucleus (Figs. 20–15 and 20–16). Portions of chromosomes that stain lightly are only partially condensed; this chromatin is termed **euchromatin** and usually represents most of the chromatin that disperses after mitosis is completed. In the dark-staining regions, the chromatin remains in the condensed state and is called **heterochromatin**. Usually there is some condensed chromatin around the nucleolus, called **perinucleolar chromatin,** and some inside the nucleolus, called **intranucleolar chromatin**. The perinucleolar and intranucleolar chromatin appear to be connected; together they are referred to as **nucleolar chromatin.**

Dense clumps of deeply staining chromatin often occur in close contact with the inner membrane of the nuclear envelope and are referred to as **condensed peripheral chromatin** (Fig. 20–15). Between the peripheral chromatin and the nucleolar chromatin are regions of lightly staining chromatin called **dispersed chromatin**.

Heterochromatin can be further divided into two types: **constitutive heterochromatin** and **facultative heterochromatin**. In constitutive heterochromatin, the DNA is permanently inactive and remains in the condensed state throughout the cell cycle. Facultative

FIGURE 20–15   Types and distribution of chromatin in the interphase nucleus (see also Fig. 20–16). (Photomicrograph courtesy of R. Chao.)

FIGURE 20–16   The interphase nucleus (compare with Fig. 20–15).

heterochromatin is not permanently maintained in the condensed state; instead, it undergoes periodic dispersal and during these times is transcriptionally active.

Heterochromatin is characterized by its especially high content of repetitive DNA sequences and contains very few, if any, structural genes (i.e., genes that encode proteins). However, heterochromatin may be involved in the regulation of gene expression, for when euchromatic genes of known functions are relocated adjacent to heterochromatin, their expression is modified. Euchromatin is believed to contain the structural genes and is expressed when decondensed in the interphase cell.

Chemically, chromatin consists for the most part of DNA and protein. Small quantities of RNA may also be present but the RNA rarely accounts for more than about 5% of the total chromatin present. Most of the protein of chromatin is histone, but "nonhistone" proteins are also present. The protein:DNA weight ratio varies from 0.8 to 1.3 and averages about 1.1. The ratio varies not only with the species but also among the tissues of a single organism. Histones are constituents of the chromatin of all eukaryotic organisms except fungi, which therefore resemble prokaryotic organisms in this respect.

Chromosome–Chromosome Associations

It has been known for some time that certain associations occur among the chromosomes arranged on the metaphase plate during nuclear division. There may be size assortment on the spindle such that long chromosomes are on the outside and short chromosomes are on the inside. Certain chromosomes are known to group together or have arms directed toward one another. G. Hoskins has shown that human chromosomes are linked by interchromosomal "connectives." The connectives are composed of DNA and protein and hold the chromosomes together even when teased from a cell by microsurgical methods. Fibers linking separate chromosomes have been visualized using electron microscopy.

Polytene Chromosomes

In the salivary glands of dipteran flies, of which the genus *Drosophila* has been most extensively studied, and in certain other tissues as well, the interphase nucleus is characterized by extremely large chromosomes. These so-called giant chromosomes are actually about 1000 parallel and tightly packed copies of the *same* chromosome (Fig. 20–4a). Because of their multiple structure, they are called **polytene** chromosomes. Along the length of the polytene chromosome, disks or bands of variably staining intensities can be distinguished. The identification of the genetic content of these bands (and interband regions) has played a major part in the genetic analysis of dipteran organisms.

Some of the bands of giant chromosomes appear to be swollen or "puffed," the specific bands exhibiting puffing varying from one tissue to another even within the same organism. It is now well established that the puffed bands of these interphase chromosomes correspond to regions in which the DNA is being actively transcribed into messenger RNA (Chapter 21).

THE NUCLEOLUS

The nucleolus is the dominant internal feature of the cell nucleus and was among the first subcellular organelles described by microscopy (by Fontona in 1774). The nucleolus is not separated from the remaining nucleoplasm by a membrane, but in many cells its margins are associated with chromatin (i.e., the perinucleolar chromatin, see Figs. 20–15 and 20–16).

The nucleolus is composed of RNA and proteins and is the site of formation of ribosomal constituents. In cells that are actively engaged in protein synthesis and which therefore have greater demands for ribosomes, the nucleoli are larger and more numerous. When the cell undergoes nuclear division, protein synthesis is decreased and ribosome production is halted. These changes are accompanied by a reduction in the size and number of nucleoli. In fact, at the metaphase stage of mitosis (see below), no nucleoli can be distinguished.

Parts of one or more chromosomes are looped into a nucleolus forming what is known as a **nucleolar organizing region** (*NOR*). These regions contain the genes that are transcribed into rRNA. Dense fibrillar regions surrounding the NOR are associated with small ribonucleoprotein particles and frequently reveal mature ribosomes. The functions of the nucleolus are described in detail in Chapter 22, which deals with ribosome structure and protein synthesis.

The Nuclear Envelope

The presence of a membrane separating the contents of the nucleus from the surrounding cytosol is one of the principal characteristics distinguishing eukaryotic organisms from prokaryotic organisms. The existence of a membrane delimiting the nucleus was first demonstrated by O. Hertwig in 1893. However, little interest was shown in studying this membrane until electron microscopy revealed that it was not simply a single membrane, but rather a double membrane in which the outer membrane had features that clearly distinguished it from the inner membrane. The two membranes are concentric and separated by a small space (called the **perinuclear space**) varying in size from about 10 to 50 nm; collectively, the two membranes form the **nuclear envelope**. A fibrous material called the **lamina densa** lies just under the inner nuclear membrane and is believed to give structural support to the envelope and also provide anchorage for the condensed peripheral heterochromatin. The inner and outer nuclear membranes fuse together at the margins of the **nuclear pores** (Figs. 20–15 and 20–16).

The functions of the nuclear envelope are diverse. Clearly, it serves to compartmentalize the nucleoplasm, acting as a barrier to the movement of large subcellular components between the nucleus and the cytosol. However, ribosome precursors from the nucleolus and mRNA and tRNA molecules do exit the nucleus through the nuclear pores. Molecules also enter the nucleus from the cytoplasm, triggering a number of nucleocytoplasmic interactions.

(a)

(b)

FIGURE 20–18 The nuclear envelope. (*a*) Thin section through the envelope. *C*, cytoplasm; *N*, nucleus; arrows point to pores that are plugged with granular material; *M*, mitochondrion; *ER*, endoplasmic reticulum. (Photomicrograph courtesy of R. Chao.) (*b*) Freeze-fractured nuclear envelope and pores. (Courtesy of Dr. E. G. Pollock.)

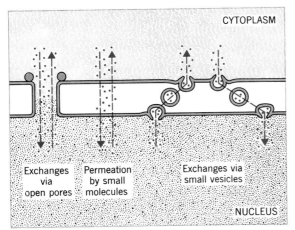

FIGURE 20–17 Various avenues for transport of material from the nucleus to the extranuclear cytoplasm.

The nuclear envelope is not just a physical barrier, but functions also in connection with cellular activities taking place on either side. Ribosomes may be attached to the cytosol face of the outer nuclear membrane, which thereby takes on a structural appearance like rough endoplasmic reticulum (Chapter 22); indeed, the outer membrane may have periodic continuities with the endoplasmic reticulum. The intimate relationship between the nuclear membranes and the endoplasmic reticulum is not surprising because a new nuclear envelope forms at the end of mitosis by the coalescence of membranes of the endoplasmic reticulum.

The inner membrane of the nuclear envelope may have special functions associated with the chromosomes. As noted earlier, in the interphase nuclei of many cells, heterochromatin is closely appressed to the inner membrane. This association does not occur at the nuclear pores; instead, the chromosomes seem to be firmly attached to the interpore areas. The role of the association between chromatin and the nuclear envelope is unclear but may be related to chromosome replication and orientation during interphase and the early prophase of mitosis and meiosis.

Some materials such as ribonucleoprotein particles

FIGURE 20–19 Models for the organization of the nuclear pores. Eight granular subunits are symmetrically arranged on the outer cytoplasmic margin of the pore. Annular projections or *spokes* are directed into the orifice and toward the central plug.

are exchanged between the nucleoplasm and the cytosol through open nuclear pores. However, the nuclear pores are not the only avenues for nucleocytoplasmic exchanges (Fig. 20–17). For example, small molecules and ions readily permeate both nuclear membranes. Larger molecules and particles may pass through the membrane by formation of small pockets and vesicles that traverse and envelope and empty on the other side.

Although many cellular membranes contain pores, the pores of the nuclear envelope are more complex and possess a special character because they are openings formed by the fusion of two membranes (Fig. 20–18). The number of nuclear pores varies widely among different cells but, in general, ranges from 1 to about 60 pores per square micrometer of membrane. The *orifice* is essentially circular, with occasional evidence of a polygonal shape. Seen in freeze-fractured preparations, the pores appear as rings and therefore are called *annuli*. Although the inside diameter of an annulus varies over a narrow range among different cells (i.e., 60 to 100 nm), it is markedly constant for a given cell type.

The rim of the pore contains eight granular *subunits* symmetrically placed about the circumference (Fig. 20–19). The center of the pore often contains distinct substructures such as conical projections from the sides into the center. More often than not the lumen of the pore is filled with "annular material" that projects into both the nucleoplasm and cytoplasm, often flowing radially beyond the pore rim.

MITOSIS

During nuclear division or **mitosis**, there is a progressive change in the structure and appearance of the chromosomes. Although mitosis is a continuous process (Figs. 20–20 and 20–21), for convenience it is usually divided into four major stages: **prophase, metaphase, anaphase**, and **telophase**.

Prophase, the first stage of mitosis, is characterized by the condensation of the chromosomes, the disappearance of nucleoli and the nuclear envelope, and the formation of the microtubules of the spindle. If, prior to prophase, the cell contained a centriole, then a second centriole is formed; the two centrioles move apart as the spindle forms. The chromosomes become distinguishable by light microscopy as a result of their pro-

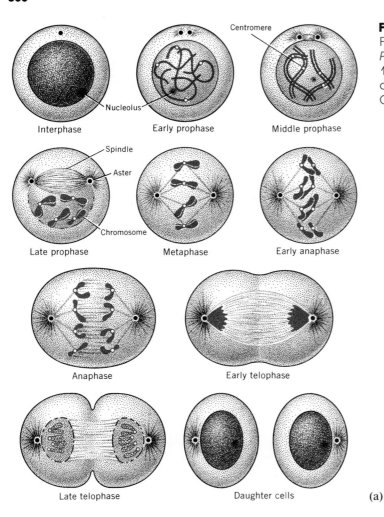

FIGURE 20–20   (*a*) The stages of mitosis (see also Fig. 20–21). (From E. J. Gardner and D. P. Snustad, *Principles of Genetics*, John Wiley & Sons, New York, 1984, p. 42.) (*b*) SEM photomicrographs of HeLa cells during mitosis. (Courtesy of Drs. R. Kessel and C. Y. Shih.)

(a)

gressive shortening and thickening and eventually are seen to be composed of two *sister chromatids* held together at the centromere or kinetochore. The sister chromatids are the products of the replication of chromosomal DNA during the interphase of the cell cycle.

Toward the end of prophase (sometimes called **prometaphase**), the spindle extends between two **poles** positioned diametrically opposite one another in the cell and the chromosomes migrate toward the center of the spindle. In metaphase, the centromeres of each chromsome are aligned midway across the spindle on a plane called the **equatorial plate**. At this time, the centromeres are linked to the spindle fibers. Some of the spindle fibers do not form associations with any chromosomes and extend directly from one pole to the other. The centromeres are duplicated so that each

chromatid becomes an independent chromosome and is attached to a spindle fiber connected to one of the two poles.

The onset of anaphase is characterized by the movement of the chromosomes toward opposite poles of the spindle. During anaphase, a process called **cytokinesis** begins and divides the cell into two halves, thereby physically separating the two complements of chromosomes. Cytokinesis is distinct from but frequently synchronized with nuclear division, occurring during the later stages of mitosis. In the final phase of mitosis, called the telophase, the chromosomes reach the poles of the spindle and begin to undergo decondensation. During the telophase, nucleoli reappear, as does a new nuclear envelope enclosing the chromosomes.

(b)

Movement of Chromosomes during Anaphase

During the anaphase of mitosis, the centromere of each chromosome advances toward one of the two poles of the spindle, with the arms of the chromosomes lagging behind (Fig. 20–21). This arrangement suggests that the chromosomes are being *pulled* toward the poles of the spindle. For a number of years many investigators have attempted to determine the nature of the mechanism that is responsible for this movement. Among the different models that have been proposed are (1) the *microtubule* model, (2) the *sliding cytoplasmic filament* model, and (3) the *dynamic equilibrium* model.

The microtubule model was first proposed in the late 1960s by J. R. McIntosh. According to McIntosh, the microtubules extend inward from the poles of the spindle and overlap at the center of the cell. During anaphase, the microtubules slide past one another in the region of overlap, thereby extending the cell and pushing the poles further apart. At the same time that the overlapping ends of the microtubules slide past each other, other microtubules connecting the centromeres to the poles are disassembled at the poles. As a result the overall length of these microtubules is decreased and the chromosomes move closer to the poles. Because microtubule sliding is known to be involved in other forms of cell movement, such as in the beating of cilia and flagella (Chapter 23), this model is entirely plausible. However, it appears not to be the complete story.

In 1974, A. Forer showed that actin microfilaments are present in the spindle, and later myosin filaments were also shown to be present. This prompted the suggestion that it is the sliding of the actin and myosin filaments past one another that accounts for chromosome movement. The role of the microtubules was

(a) Interphase

(b) Early prophase

(c) Late prophase

(g) Late anaphase

(h) Early telophase

(i) Telophase and beginning of cell plate formation

(d) Metaphase

(e) Anaphase

(f) Late anaphase

(j) Late telophase

(k) Interphase

FIGURE 20–21 Stages of mitosis as seen in a cell of the blood lily (*Haemanthus*). (Photomicrographs courtesy of Dr. A. S. Bajer; see also Fig. 20–20.)

relegated to that of serving as a structural framework on which the chromosomes are mounted. Movements of the cytoplasmic filaments are not reflected by corresponding movement of the chromosomes until anaphase begins and the microtubules begin to break down at the two poles of the spindle. In effect, the disassembly of the microtubules at the poles frees the chromosomes to move in response to the sliding of the actin and myosin.

According to the dynamic equilibrium model proposed in 1967 by S. Inoue, the microtubules that comprise the spindle fibers are in a dynamic equilibrium with a pool of microtubule subunits. The addition of new subunits to fibers that extend from one pole of the spindle to the other but do not have attached chromosomes serves to increase the interpolar distance, while at the same time the removal of subunits from either the polar ends or the centromere ends of other spindle fibers serves to shorten them and draw the chromosomes closer to the poles. In addition to biochemical evidence supporting this latter notion is the recognition that the movement of chromosomes toward the poles of the spindle during anaphase is a relativity slow process compared to the beating motions of cilia and flagella and the contraction of muscle cells. Indeed, anaphase movement takes place at a rate that is more akin to the rates at which filaments grow or shorten by either an influx or loss of subunits.

From the preceding discussion it is apparent that several different models attempting to account for chromosome movement during the anaphase of mitosis are supported by independent observations. This is not to suggest that the actual mechanism may not involve some combination of all the models. The structure of the mitotic spindle and its actions are discussed further in Chapter 23.

MEIOSIS

Meiosis is a form of nuclear division that is of fundamental importance among sexually reproducing organisms. An in-depth discussion of meiosis on a cellular as well as a genetic basis is beyond the scope of this book; such discussions are normally treated at length in textbooks of genetics (see the list of references at the end of this chapter). However, for the sake of completeness we will consider some of the major meiotic events and their implications.

Meiosis occurs in eukaryotic organisms whose cells

contain the **diploid** number ($2n$) of chromosomes. Diploid implies "double" in the sense that the genetic information present in any one chromosome can also be found in an identical (or somewhat modified) form in a second chromosome in the nucleus. The two chromosomes forming such pairs are said to be **homologous**.

As noted earlier in this chapter, human cells contain 46 chromosomes or 23 homologous pairs (i.e., in humans $n = 23$). The 46 chromosomes of the **zygote** formed at fertilization are derived equally from the sperm cell and egg cell of the male and female parents. Each of these **gametes** contributes one member of each pair of homologues. Once the zygote is formed, mitosis produces the billions of cells that ultimately make up the whole organism. Because sperm cells and egg cells contain only one member of each pair of homologues, they are said to be **haploid**. It is meiosis that produces haploid cells, the process being restricted to the reproductive tissues (i.e., ovaries and testes).

During meiosis, the replicated chromosomes of the nucleus are apportioned among four daughter nuclei, each nucleus acquiring half the number of chromosomes of a diploid cell. Although the resulting cell nuclei contain only half the diploid number of chromosomes, the chromosome set is genetically complete, because each nucleus acquires one member of each pair of homologous chromosomes. The homologous chromosomes are assorted randomly at anaphase, and this accounts in part for the genetic variation that characterizes sexually reproducing organisms. Additional genetic variation occurs during the prophase of the first nuclear division by a process called **crossing-over**. The genetic implications of random assortment and crossing-over are principal subjects of genetic courses.

The various stages of meiosis may be summarized as follows:

MEIOTIC DIVISION I

Prophase I

1. **Leptotene stage (leptonema).** The chromosomes become visible as condensation of the chromatin begins; each chromosome can be seen to consist of two chromatids.

2. **Zygotene stage (zygonema).** Homologous chromosomes are aligned side-by-side so that *allelic* genes

(a) Prophase I

(b) Metaphase I

FIGURE 20–22 The stages of meiosis as seen in the basidium of *Pholiota*. Views obtained by light (left) and electron (right) microscopy are paired. See the text for descriptions of the phases. *CH*, chromatin; *CM*, continuous microtubule; *ER*, endoplasmic reticulum; *KC*, kinetochore (centromere); *M*, membrane; *MT*, microtubule; *N*, nucleus; *NE*, nuclear envelope; *NU*, nucleolus; *PER*, perinuclear endoplasmic reticulum; *SPB*, spindle pole body. (Courtesy of Dr. K. Wells, copyright © 1978 by Springer-Verlag, *Protoplasma* **94**, 85.)

(c) Anaphase I

(d) Telophase I

(e) Interphase I

(f) Metaphase — anaphase II

(g) Anaphase II

(h) Postmeiosis

(i.e., those encoding products of similar or identical function) are situated adjacent to one another. This phenomenon is called **synapsis**. The unit consisting of two synapsed and duplicated homologous chromosomes is called a **bivalent**. As synapsis progresses, a protein framework joining adjacent, non-sister chromatids of each tetrad is formed at one or more points in the narrow space separating the homologues. It is in the region of these **synaptonemal complexes** that crossing-over occurs. Crossing-over or **chiasma** formation results from the cleavage by endonucleases of the DNA in corresponding positions of two non-sister chromatids, followed by the transposition and rejoining of the free ends of homologous strands (see Fig. 20–23 for details). As a result of crossing-over, new combinations of genes are created in the homologous chromosomes.

3. **Pachytene stage (pachynema)**. During this stage the chromatids become increasingly distinct as condensation continues.

4. **Diplotene stage (diplonema)**. The diplotene stage is characterized by the separation of the paired homologous chromosomes except at points where chiasmata are formed.

5. **Diakinesis**. Diakinesis brings prophase I to an end. During this stage chromosome condensation is completed.

Metaphase I

In this phase, the spindle apparatus forms, much as it does in mitosis, and the bivalents align on the equatorial plate. The centromeres of homologous chromosomes attach to spindle fibers arising from opposite poles of the cell.

Anaphase I

Homologous chromosomes (but *not* sister chromatids) of each tetrad separate from each other and move to opposite poles of the spindle.

Telophase I

Telophase I brings the first meiotic division to a conclusion as the separated homologues aggregate at their respective poles so that two nuclear areas are distinguishable. In most organisms, a new nuclear envelope is formed and some decondensation of the chromosomes occurs.

Interkinesis (or Interphase)

Interkinesis is the period between the end of telophase I and the onset of prophase II. This period is usually quite short. The DNA of the two nuclei produced by the first meiotic division does *not* engage in replication during interkinesis.

MEIOTIC DIVISION II

Prophase II

The events characterizing this phase are similar to mitotic prophase, although each cell nucleus has only half the number of chromosomes as a cell in prophase I, that is, the nucleus is already haploid. Each chromosome remains composed of the two sister chromatids formed prior to prophase I, except for segments that were interchanged during crossing-over.

Metaphase II

The events occurring in this phase are similar to those in mitotic metaphase. The paired chromatids migrate to the center of the spindle and are attached there to the spindle's microtubules.

Anaphase II

The events occurring in this phase are similar to those in mitotic anaphase, but differ from those of anaphase I of meiosis. In anaphase II, sister chromatids separate from one another and are drawn to opposite poles of the spindle. (Recall that sister chromatids do not separate in anaphase I.)

Telophase II

The events occurring in this phase are similar to those in mitotic telophase. The separated chromosome groups are enclosed in a newly developing nuclear envelope and begin to undergo decondensation.

Meiosis produces four cells, each with the haploid number of chromosomes. In many higher animals and some plants, meiosis in the female reproductive tissues is accompanied by an uneven division of the cytoplasm, in which case one of the two cells formed during telophase I is a nonfunctional **polar body** and may not enter prophase II (Fig. 20–24). In some organisms (such as humans) the polar body completes meiosis,

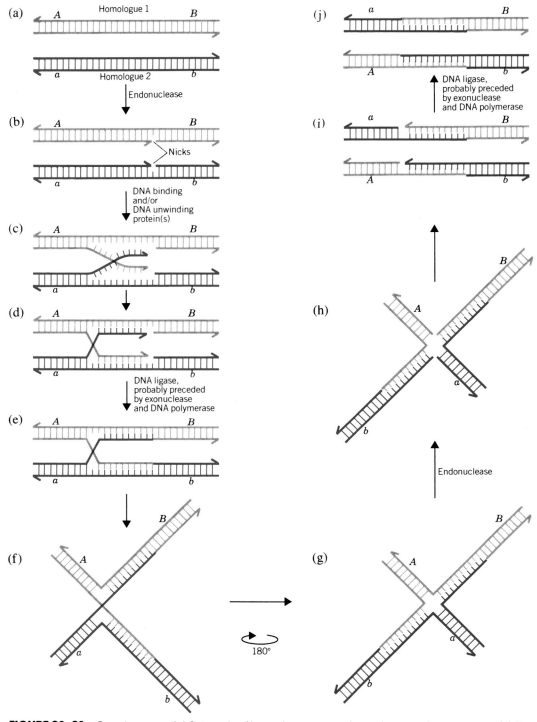

FIGURE 20–23 Crossing-over. (a) Synapsis of homologous chromosomes (gray and color) carrying markers A, B and a, b. The antiparallel nature of the DNA strands of each chromosome are shown by the arrowheads at the ends of the strands. (b) An *endonuclease* nicks one strand in each chromosome. (c) Free ends are displaced and form base pairs (d) with the unnicked complementary strand of the homologous chromosome. (e) Free ends are rejoined by a *DNA ligase*, leaving the two chromosomes joined by single-stranded bridges (f,g). Cleavage at the bridges by an endonuclease (h) frees the two chromosomes (i), following which the remaining nicks are ligated (j). (From E. J. Gardner and D. P. Snustad, *Principles of Genetics*, John Wiley & Sons, New York, 1984, p. 186.)

but the two smaller polar bodies produced during telophase II are similarly nonfunctional. The second meiotic division of the larger cell produced during telophase I is also unequal and produces an additional polar body. During the production of spermatozoa in the male reproductive tissues, division of the cytoplasm is equal, but remarkable cytoplasmic differentiation of the four spherical haploid **spermatids** pro-

duced by meiosis is required (Fig. 20–24) before functional, flagellated spermatozoa are produced.

CYTOKINESIS

In most cases, eukaryotic cells contain only one nucleus and the 1:1 nucleus:cell ratio is perpetuated by

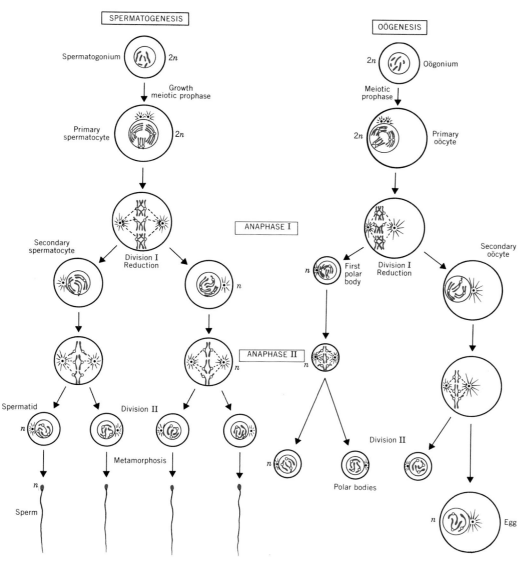

FIGURE 20–24 *Stages of meiosis in the testes (left) and ovaries (right). Spermatogenesis produces four viable sperm cells, but oogenesis produces a single viable egg with three polar bodies (see text for details). (From E. J. Gardner and D. P. Snustad, Principles of Genetics, John Wiley & Sons, New York, 1984, p. 47.)*

the synchronization of division of the cytoplasm with nuclear division. Mitotic divisions without cytoplasmic divisions occur in many species to produce multinucleate or *coenocytic* cells. Striated muscle cells, mycelia of molds, and some phloem cells of plants are examples of multinucleate cells. Few eukaryotic cells lack a nucleus. The sieve cells of plants have no nucleus but are always in close proximity to a nucleated *companion* cell. Mammalian erythrocytes (red blood cells) lose their nuclei during the final stages of differentiation in the bone marrow (Chapter 24).

In general, animal cells and many lower plant cells divide into two cells by a pinching or furrowing action in the cytoplasm that lies between the two newly formed nuclei. In plant cells, which possess a cell wall, the division of the cell occurs by the formation of a **cell plate**. The plate begins forming at the center of the spindle near the end of anaphase and it grows outward until the cell is cleaved into two halves. Though *furrowing* appears to be aided by the formation of *cytoplasmic filaments* (discussed in Chapter 23), the cell plate is formed by the progressive accumulation of carbohydrate and lipid materials that are derived from the Golgi apparatus (Chapter 18).

SUMMARY

Eukaryotic cell nuclei are delimited by a pair of membranes that form the **nuclear envelope**. These membranes fuse with each other around the margins of **nuclear pores**. A variety of substances pass between the nucleus and cytoplasm through these pores. Much of a cell's DNA is within the nucleus and is organized into a number of **chromosomes**. During **mitosis** and **meiosis**, the chromosomes condense, thereby assuming a distinctive shape and size. Metaphase chromosomes are characterized by a **primary constriction** or **centromere** and varying numbers of **secondary constrictions**. In eukaryotic cells, the double-helical DNA is associated with histones to form chains of small bodies called **nucleosomes**.

The chromosomes of bacteria and viruses are single nucleic acid molecules. In viruses, the nucleic acid may be single-stranded DNA or RNA or may be double-helical DNA. The chromosomes of prokaryotic cells are not separated from the cytosol by a nuclear envelope, but in bacteria they may be attached to the plasma membrane. **Plasmids** are small double-stranded circular pieces of DNA that exist separate from the main chromosome of a bacterium. Some plasmids are capable of being incorporated into the main chromosome and are called **episomes**.

During the interphase of the cell cycle of eukaryotic organisms, the chromatin shows varying degrees of **condensation**. The more condensed portions of interphase chromatin are called **heterochromatin** and the less condensed regions **euchromatin**. Euchromatin contains most of the cell's **structural genes** and is transcriptionally active.

REFERENCES AND SUGGESTED READING

Articles and Reviews

Bauer, W. R., Crick, F. H. C., and White, J. H., Supercoiled DNA, *Sci. Am.* **243** (1), 118 (July 1980).

Borst, P., Grivell, L. A., and Grost, G. S. P., Organelle DNA. *Trends in Biochem. Sci.* **9** (4), 128 (April 1984).

Butler, P. J. G., The folding of chromatin. *Critical Revs. Biochem.* **15** (1), 57 (1983).

Dickerson, R. E., Drew, H. R., Conner, B. N., Wing, R. M., Fratini, A. V., and Kopka, M. L., The anatomy of A-, B-, and Z-DNA. *Science* **216**, 475 (1982).

Doenecke, D., and Karlson, P., Albrecht Kossel and the discovery of histones. *Trends in Biochem. Sci.* **9** (9), 404 (Sep. 1984).

Dubochet, J., and Noll, M., Nucleosome arcs and helices. *Science* **202**, 280 (1978).

Felsenfeld, G., DNA. *Sci. Am.* **253** (4), 58 (Oct. 1985).

Gellert, M., DNA topoisomerases. *Ann. Rev. Biochem.* **50**, 879 (1981).

Gossens, G., Nucleolar structure. *Int. Rev. Cytol.* **87**, 111 (1984).

Grivell, L. A., Mitochondrial DNA. *Sci. Am.* **248** (3), 78 (Mar. 1983).

Harauz, G., and Ottensmeyer, F. P., Nucleosome reconstruction via phosphorus mapping. *Science* **226**, 936 (1984).

Khorana, H.G., Total synthesis of a gene. *Science* **203**, 614 (1979).

Kornberg, R. D., Structure of chromatin. *Ann. Rev. Biochem.* **46**, 931 (1977).

Kornberg, R. D., and Klug, A., The nucleosome. *Sci. Am.* **244** (2), 52 (Feb. 1981).

Landy, A., and Ross, W., Viral integration and excision: Structure of the lambda "att" sites. *Science* **197**, 1147 (1977).

McGhee, J. D., and Felsenfeld, G., Nucleosome structure. *Ann. Rev. Biochem.* **49**, 1115 (1980).

McIntosh, J. R., Mechanisms of mitosis. *Trends in Biochem. Sci.* **9** (4), 195 (April 1984).

Miller, O. L., The nucleolus, chromosomes, and visualization of genetic activity. *J. Cell Biol.* **91**, 15 (1981).

Takeda, Y., Ohlendorf, D. H., Anderson, W. F., and Matthews, B. W., DNA-binding proteins. *Science* **221**, 1020 (1983).

Timmis, J. N., and Scott, N. S., Promiscuous DNA: Sequence homologies between DNA of separate organelles. *Trends in Biochem. Sci.* **9** (6), 271 (1984).

Books, Monographs, and Symposia

Gardner, E. J., and Snustad, D. P., *Principles of Genetics* (7th ed.), Wiley, New York, 1984.

Lewin, B., *Genes II*, Wiley, New York, 1985.

Old, R. W., and Primrose, S. B., *Principles of Gene Manipulation*, University of California Press, Berkeley, 1980.

Rawn, J.D., *Biochemistry*, Harper & Row, New York, 1983.

Saenger, W., *Principles of Nucleic Acid Structure*, Springer-Verlag, New York, 1984.

Stryer, L., *Biochemistry* (2nd ed.), Freeman, San Francisco, 1982.

Swanson, C. P., Merz, T., and Young, W. J., *Cytogenetics*, Prentice–Hall, Englewood Cliffs, N.J., 1981.

CHAPTER 21

THE NUCLEUS AND MOLECULAR GENETICS

In Chapter 20, we were concerned primarily with the chemical composition and organization of the chromatin of the nucleus and with the behavior of the chromosomes during mitosis and meiosis. In this chapter, we will examine (1) the mechanisms by which DNA is replicated and transcribed into RNA, (2) the properties that characterize eukaryotic, prokaryotic, and viral chromosomes and plasmids, and (3) recombinant DNA technology and genetic engineering.

REPLICATION OF DNA

It has now been over 30 years since J. D. Watson, F. H. C. Crick, and M. H. F. Wilkins established the double-stranded, helical nature of the DNA molecule (Chapter 7) and suggested how DNA serves in its own replication. According to their original model for DNA replication, the two polynucleotide chains of the "parent" double helix separate and each serves as a "template" for the synthesis of a new, *complementary* polynucleotide chain. During strand separation, the nitrogen bases of each original strand are exposed and establish sites for the association of free nucleo-tides. These nucleotides are then enzymatically linked together to form a new complementary strand (see Chapter 7, Fig. 7–9). Because deoxyadenylic acid (dAMP) can form hydrogen bonds only with thymine of the template strand (and dGMP can bond only to cytosine, dCMP only to guanine, and dTMP only to adenine), the newly synthesized strand will be identical to the original complementary strand. As a result, two new double helices are formed, each consisting of one polynucleotide strand from the parent double helix and a newly synthesized polynucleotide strand. Because each of the two double helices conserves only one of the parent polynucleotide strands, the process is said to be **semiconservative**.

Replication as a "Semiconservative" Process

Although semiconservative replication of DNA was predicted by the original Watson–Crick model, it was not verified until the classic studies of M. S. Meselson and F. W. Stahl. At the time of their experiments, two other modes of replication were deemed equally feasible (Fig. 21–1): (1) **conservative replication**, in which *both* strands of the parent double helix would be con-

515

served and the new DNA molecule would consist of two newly synthesized strands; and (2) **dispersive replication**, in which replication would involve fragmentation of the parent double helix and the interspersing of pieces of the parent strands with newly synthesized pieces, thereby forming the two new double helices.

Meselson and Stahl verified the semiconservative nature of DNA replication in a series of elegant experiments using isotopically labeled DNA and a form of isopycnic density gradient centrifugation (see Chapter 12). They cultured *Escherichia coli* cells in a medium in which the nitrogen was ^{15}N (a "heavy" isotope of nitrogen, but not a radioisotope) instead of the commonly occurring and lighter ^{14}N. In time, the purines and pyrimidines of DNA in new cells contained ^{15}N (where ^{14}N normally occurs) and thus the DNA molecules were denser. DNA in which the nitrogen atoms are ^{15}N can be distinguished from DNA containing ^{14}N because during isopycnic centrifugation, the two different DNAs band at different density positions in the centrifuge tube (Fig. 21–2).

Meselson and Stahl centrifuged DNA isolated from the cells for 2–3 days at very high rotational speeds in centrifuge tubes initially containing a uniform solution of CsCl. During centrifugation, density gradients were automatically formed in the tubes as a result of the equilibrium that was established between the *sedimentation* of CsCl toward the bottom of the tube and *diffusion* of the salt toward the top of the tube. This form of centrifugation, called *equilibrium isopycnic*

centrifugation, was discussed in Chapter 12. Depending on its content of ^{15}N and ^{14}N, the DNA bands at a specific position in the density gradient. Because the DNA synthesized by cells grown in ^{15}N would be denser than ^{14}N-containing DNA it would band further down the tube (Fig. 21–2).

Cells grown for some time in the presence of ^{15}N-medium were washed free of the medium and transferred to ^{14}N-containing medium and allowed to continue to grow for specific lengths of time (i.e., for various numbers of *generation times*). DNA isolated from cells grown for one generation of time in the ^{14}N medium had a density intermediate to that of the DNA from cells grown only in ^{15}N-containing medium (identified as generation 0 in Fig. 21–3) and that of DNA from cells grown only in ^{14}N-containing medium (the controls of Fig. 21–3). Such a result immediately ruled out the possibility that DNA replication was conservative, as conservative replication would have yielded two DNA bands in the density gradient for generation 1 (i.e., F_1) cells. The single band of intermediate density (identified as "hybrid" DNA in Fig. 21–3) consisted of DNA molecules in which one strand contained ^{15}N and the other contained ^{14}N. When the incubation in ^{14}N medium was carried out for two generations of time (i.e., generation 2), *two* DNA bands were formed—one at the same density position as the DNA from cells grown exclusively in ^{14}N medium (i.e., "light controls") and one of intermediate density. Subsequent generations produced greater numbers of DNA molecules that banded at the "light" (^{14}N-

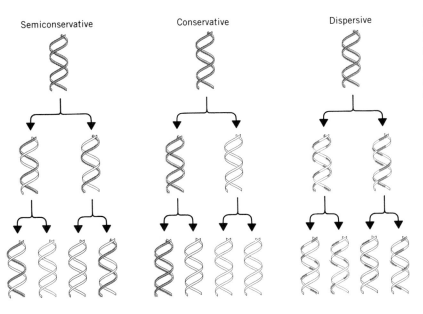

FIGURE 21–1 Three possible modes of DNA replication: *semiconservative* (left), *conservative* (center), and *dispersive* (right) (see text for explanation). (From E. J. Gardner and D. P. Snustad, *Principles of Genetics* (7th ed.), John Wiley & Sons, 1984, p. 97.)

Semiconservative   Conservative   Dispersive

containing DNA) position in the density gradient. These results are consistent only with the model of semiconservative replication. Dispersive replication would have produced a single band for each generation and the band would have been found at successively lighter density positions in the gradient. Studies using other prokaryotes as well as eukaryotes indicate that semiconservative replication of DNA is probably the universal mechanism.

Replication by Addition of Nucleotides in the 5'→3' Direction

Each nucleotide of a DNA strand is joined to the next nucleotide by a *phosphodiester* bond that links the 3' carbon of its deoxyribose to the 5' carbon of the deoxyribose of the next nucleotide (Chapter 7). At one end of the polynucleotide chain, there is a hydroxyl group attached to the 3' carbon of the last nucleotide, and at the other end there is a phosphate group attached to the 5' carbon. The two chains of a double helix have *opposite polarities* and are said to be *antiparallel*, that is, each end of the double helix contains the 5' end of one strand and the 3' end of the other. During replication, attachment of a nucleotide to a growing strand always takes place at the terminal 3' position of that strand. In other words, the forming polynucleotide chain "grows" from its 5' end toward its 3' end.

Unidirectional and Bidirectional Replication

Replication starts at a point on the chromosome where the two parental strands begin to separate; this

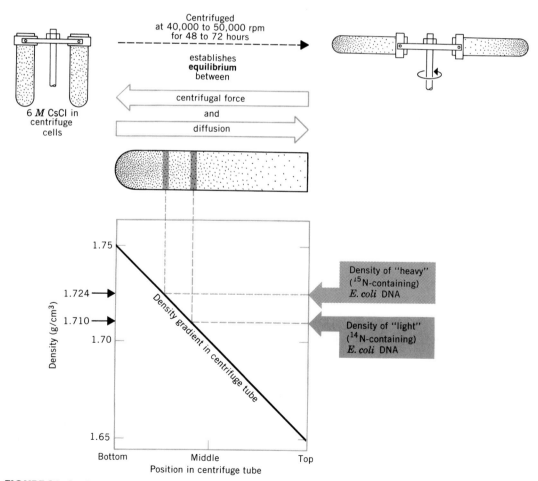

FIGURE 21–2 Separation of ^{15}N-containing DNA from ^{14}N-containing DNA by equilibrium isopycnic centrifugation (see text for details). (From E. J. Gardner and D. P. Snustad, *Principles of Genetics* (7th ed.), John Wiley & Sons, 1984, p. 98.)

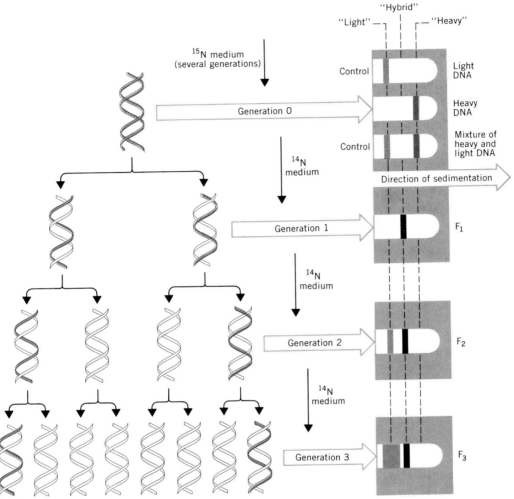

FIGURE 21–3 Results of the Meselson–Stahl experiments (right) and their interpretation (left) (see text for explanation). (From E. J. Gardner and D. P. Snustad, *Principles of Genetics* (7th ed.), John Wiley & Sons, 1984, p. 99.)

point is called the **origin**. Addition of complementary nucleotides to form two new strands takes place along both parent strand templates starting from that point (Fig. 21–4). In **unidirectional** replication, growth proceeds along both strands in the *same* direction leading from the origin. Along one of the parental template strands, synthesis of the new complementary strand takes place by the continuous addition of nucleotides to the available 3' end of the forming strand. The growing strand is called the **leading strand** or **continuous strand**. The 5' end of this strand is located at the origin and its 3' end at the moving **replication fork** (i.e., the progressing point of separation of the parental strands).

The other polynucleotide strand being formed is called the **lagging strand** or **discontinuous strand**. The elongation of this strand takes place by a somewhat modified mechanism. In contrast to the leading strand, the lagging strand has its 3' position at the origin and its 5' position at the replication fork. If nucleotides were sequentially added to the end of the lagging strand at the replication fork, then this strand's growth would proceed in a 3'→5' direction. *This does not occur*. Instead, growth takes place by the synthesis of a number of short polynucleotide chains between the replication fork and the origin. Each short chain is laid down in the direction 5' to 3' and these are later linked together and to the 5' end of the lag-

FIGURE 21–4 *Unidirectional* replication of DNA. Behind the *replication fork*, one new DNA strand grows continuously and the other grows as a series of fragments that are eventually linked together. The overall growth of both new strands is in the direction of the moving fork.

ging strand. As a result, the overall direction of growth of the lagging strand is the same as that of the leading strand. The unusual growth pattern that characterizes the synthesis of the lagging strand explains why it is also referred to as the "discontinuous" strand.

In **bidirectional replication** (Fig. 21–5), *two* replication forks are formed at the origin and these move away from the origin in *both* directions as the parental double helix is separated. The synthesis of the complementary strands also occurs in both directions. Behind each fork there is a set of leading and lagging strands. As in the case of unidirectional replication, elongation of the two leading strands is continuous, whereas elongation of the two lagging strands is discontinuous. It is to be noted that regardless of whether replication is unidirectional or bidirectional, the addition of nucleotides always occurs in the direction from 5' to 3', as new nucleotides are added to available 3' ends of either the continuous strand or the discontinuous strand.

Discontinuous synthesis of lagging strands was first demonstrated by R. Okazaki. Okazaki incubated *E. coli* cells in a medium containing ³H-thymidine for very short periods of time (a pulse of only 15 seconds) and then examined the distribution of the radioisotope in newly synthesized DNA. The radioisotope was found in a number of polynucleotides (1000–2000 nucleotides long), now referred to as **Okazaki fragments** (Figs. 21–4 and 21–5). When pulsed cells were transferred to unlabeled medium for varying lengths of time prior to analysis, the radioactive label was recovered in much longer stretches of DNA. This is because the Okazaki fragments produced during the short tritium pulse had been linked together and connected to the 5' end of the lagging strand. In eukaryotic cells, Okazaki fragments are usually smaller (about 100–200 nucleotides long).

Bidirectional replication of DNA is the mechanism employed in all eukaryotic and most prokaryotic cells. Unidirectional replication is rare and appears to occur in only a limited number of prokaryotes.

Visualization of Replication in E. coli

In 1963, J. Cairns developed a procedure employing a combination of microscopy and autoradiography that made it possible to visualize the replication of the

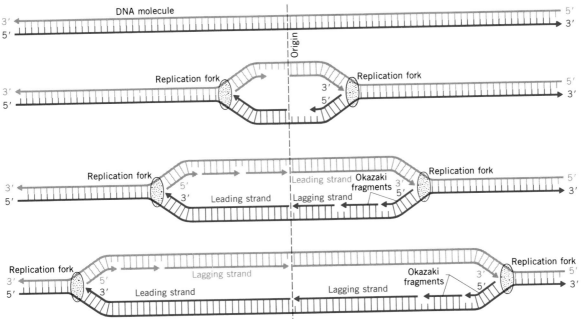

FIGURE 21–5 *Bidirectional* replication of DNA. Two replication forks move away from the origin in opposite directions. Behind each fork, one new DNA strand grows continuously and the other grows as a series of fragments that are eventually linked together. The overall growth of both sets of new strands is in the direction of the moving forks.

chromosome of *E. coli*. Cairns placed *E. coli* cells in a medium containing ³H-thymidine for various periods of time so that the radioactive thymidine was incorporated into the DNA as the chromosome was replicated in successive generations of cells. Cells were removed from the medium after various periods of incubation and *gently* lysed to release the chromosome from the cell (the shear forces created by harsh lysis break the chromosome into small pieces). The chromosomes were then transferred to glass slides and coated with a photographic emulsion sensitive to the low-energy beta particles emitted by the ³H-thymidine. After exposing the emulsion to the beta rays, the emulsion was developed and examined by light microscopy (see Chapter 14 for a discussion of combined autoradiographic–microscopic techniques). Wherever decay of labeled thymidine had occurred in a chromosome, the emulsion was exposed and created visible grains. A chromosome not engaged in replication appeared as a circular structure formed from a close succession of exposure spots. Chromosomes "caught in the act" of replication gave rise to what are called **theta** structures because they have the appearance of the Greek letter theta (i.e., Θ) (Fig. 21–6). The theta structures

reveal the positions of the replication forks in the circular chromosome.

The Replicon and the Replication Sequence

The sequence of events that takes place during DNA replication is best understood for prokaryotes and appears to be as follows (Fig. 21–7). Parental strand separation begins at a site called the **origin** which contains a special nucleotide sequence and directs the association of a number of proteins. ATP-dependent **unwinding enzymes** (also called **helicases**) promote separation of the two parental strands and establish **replication forks** that will progressively move away from the origin (Fig. 21–7a); the helicases separate the parental strand at about 1000 base pairs per second. Behind the replication fork, the single DNA strands are prevented from rewinding about one another (or forming double-stranded hairpin loops in each single strand) by the actions of a set of proteins called **helix-destabilizing proteins** or **single-strand binding proteins** (i.e., "SSBs") (Fig. 21–7b). The action of a helicase introduces a positive **supercoil** into the duplex DNA ahead of the replication fork. Enzymes called **topoisomerases** relax the supercoil by

FIGURE 21–6 (a) Autoradiogram of a replicating *E. coli* chromosome. Loops A and B have completed replication, whereas loop C remains to be replicated. X and Y are the replication forks. (b) The newly synthesized strand is shown in color and the original strand in gray. (c) Diagram illustrating successive stages during bidirectional replication. (a, courtesy of Dr. J. Cairns; copyright © 1963 by Cold Spring Harbor Laboratory. b and c modified from E. J. Gardner and D. P. Snustad, *Principles of Genetics* (7th ed.), John Wiley & Sons, 1984, p. 100.)

attaching to the transiently supercoiled duplex, nicking one of the strands, and rotating it through the unbroken strand. The nick is then resealed.

Prior to DNA synthesis beginning at the origin, short RNA polynucleotides are formed that are complementary to the DNA template. These stretches of RNA are called **primers** and are also laid down in the 5' to 3' direction. DNA nucleotides are then added one at a time to the free 3' ends of the RNA primers. Because growth of the lagging strand is discontinuous, several RNA primers and Okazaki fragments are formed. Note that an RNA primer must be formed for each Okazaki fragment to be laid down (Fig. 21–7c). The enzymes required for the synthesis of the RNA primers are a special class of RNA polymerases called **RNA primases**.

Elongation of the leading strand and synthesis of the Okazaki fragments are catalyzed by an enzyme called **DNA polymerase III**. The substrates of DNA polymerase III are the deoxynucleoside triphosphates (e.g., dATP, dGTP, dCTP, and dTTp). Addition of a nucleotide to the available 3' position of the continuously growing leading strand or an Okazaki fragment of the lagging strand involves removal of pyrophosphate to yield a deoxynucleoside monophosphate (e.g., dAMP, dGMP, dCMP, and dTMP). On completion of the Okazaki fragments, the RNA primers are excised by **DNA polymerase I**, which then fills the resulting gaps with DNA (Fig. 21–7d). After DNA polymerase I adds the final deoxyribonucleotide in the gap left by the excised primer, the enzyme **DNA ligase** forms the phosphodiester bond that links the free 3' end of the primer replacement to the 5' end of the Okazaki fragment (Fig. 21–7f).

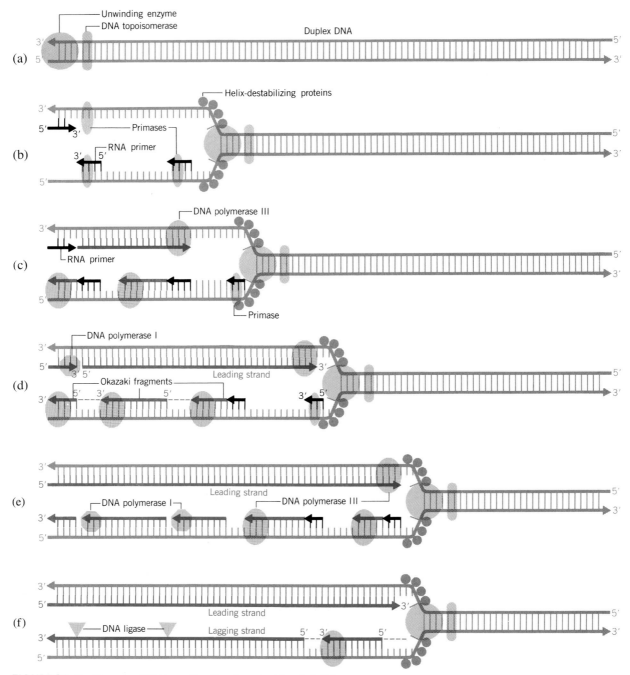

FIGURE 21-7 Stages of DNA replication (see text for details).

DNA Polymerases and "Processivity" In general, three different DNA polymerase enzymes are found in cells. In prokaryotes, these are called *DNA polymerase I, DNA polymerase II,* and *DNA polymerase III.* As noted above, *DNA polymerase I* excises the RNA primers and fills the gaps with DNA, whereas *DNA polymerase III* adds nucleotides to the growing leading strand and to the 3' ends of the RNA primers. The function of *DNA polymerase II* remains unknown. In eukaryotic cells, the DNA polymerases are *DNA polymerase α, DNA polymerase β,* and *DNA polymerase γ*; their functions are compared with the prokaryotic enzymes in Table 21–1. The rapidity and efficiency with which a DNA polymerase extends a growing chain is referred to as *processivity.* For example, the processivity of DNA polymerase III acting on the 3' end of the leading strand is very high because the enzyme remains associated with the growing end of the chain and the template strand.

During unidirectional replication, the replication fork fully circles the chromosome and the resulting DNA molecules are separated. For prokaryotes in which replication is bidirectional, the replication forks proceed around the chromosome until they meet. In eukaryotic chromosomes, where there are many replicating units or **replicons**, all replicons are linked together before the chromatids can be separated. The replicon consists of that segment of a chromosome that includes an origin and two *termination points* (i.e., points where replication ends).

The "Replisome" N. K. Sinha and A. Kornberg have suggested that the DNA polymerases, RNA primases, and helicases may be associated with each other to form a multienzyme complex, the *replisome,* that carries out the synthesis of the leading and lagging

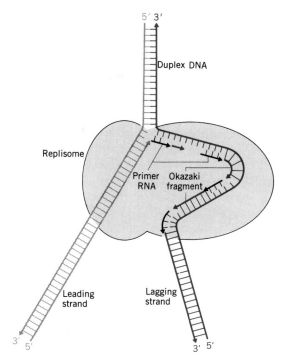

FIGURE 21–8 Model of the *replisome* (see text for explanation).

strands in a *coordinated* fashion (Fig. 21–8). Such a complex would be highly *processive* and assure rapid replication of the DNA.

High Fidelity of Replication

Despite the complexity of the process and the rapidity with which it proceeds, very few errors are made during DNA replication. For example, it is estimated that for every error that occurs, 10^9 base pairs are replicated faithfully. The high fidelity of DNA replication is attributable in part to the special properties of the DNA polymerases. The DNA polymerases will add a nucleotide to the available 3'-OH end of a growing DNA strand (or RNA primer) *only if the prior nucleotide is properly base-paired with the template nucleotide.* If a mismatched nucleotide is present, growth of the strand is transiently halted while a segment of the strand containing the error is excised. With a corrected 3' end reestablished, elongation by the DNA polymerase is resumed.

Much remains to be learned about the enzymes that catalyze the reactions of replication. So far, the major obstacle to such studies has been the difficulty of iso-

TABLE 21–1 THE DNA POLYMERASES

DNA Polymerase	Function in the Cell
Prokaryotes	
I	DNA replication and repair
II	Unknown
III	DNA replication
Eukaryotes	
α	DNA replication in the nucleus
β	DNA replication and repair in the nucleus
γ	DNA replication in mitochondria (and chloroplasts?)

lating and purifying the enzymes, either individually or in complexes. This is due in part to the fact that some of them may be associated with membranes. In prokaryotic cells, the replication forks are bound to the plasma membrane. About two dozen different proteins have been shown to be involved with the replication of DNA in *E. coli* cells. Several of the *E. coli* proteins are involved with *prepriming reactions*, that is, the reactions that occur before the formation of RNA primers.

CHROMOSOME REPLICATION IN EUKARYOTES

Replication of the chromosomes of eukaryotic cells is a much more complex process than that in prokaryotes and the DNA viruses. The added complexity is due in part to the greater length of the eukaryotic chromosome. Nonetheless, there are several fundamental similarities between DNA replication in eukaryotes, prokaryotes, and viruses. For example, (1) replication is semiconservative; (2) RNA primers are used; (3) replication occurs in both directions; (4) new DNA is synthesized along the lagging strand as a series of Okazaki fragments; and (5) the overall direction of elongation is 5' to 3'.

Semiconservative Replication in Eukaryotes

Verification of semiconservative replication of DNA in eukaryotic cells was provided in a series of experiments carried out by J. H. Taylor and P. Woods in 1957. Taylor and Woods grew *Vicia faba* bean root tip cells for one generation of time in a medium containing ^{3}H-thymidine and then transferred the root tips to ^{3}H-free medium where they were allowed to enter another round of cell division. However, *colchicine* was added to the medium to prevent anaphase separation of sister chromatids (see Chapter 20). The cells were then examined by combined light microscopy and autoradiography, and the sister chromatids of each chromosome were found to contain equal amounts of radioactive label (i.e., equal numbers of exposure grains were seen in each chromatid) (Fig. 21–9). When cells were allowed to duplicate the chromosomes and enter a second round of mitosis in unlabeled medium, the label was found in only one of the two chromatids comprising the metaphase chromosomes. The sister chromatid not distinguishable by autoradiography during this second metaphase consisted entirely of unlabeled

nucleotides. These observations are consistent with a semiconservative model of chromosome replication.

Replicons of Eukaryotic Chromosomes

Whereas the replicating prokaryotic chromosome may consist of one replicon, the average chromosome of a eukaryotic cell contains thousands of replicons. Replication of DNA does not begin in all replicons at the same time of the S phase (Fig. 21–10). Instead, some sort of control mechanism turns each replicon on at a specific time. Using the pulse-labeling technique combined with autoradiography, it is possible to show that certain regions of a chromosome always replicate early in the S phase and other regions are active at later times. Euchromatic regions are usually replicated first and heterochromatic regions later in the S phase.

Nucleosome Assembly and DNA Replication

The duplication of a eukaryotic chromosome not only requires the replication of the chromosomal DNA but also necessitates the synthesis of large quantities of new histones and the assembly of additional nucleosomes. During chromosome duplication, the histone core and DNA double helix of each nucleosome are temporarily separated, but as the growth of the leading strand proceeds these core histones rejoin the newly formed double helix, thereby quickly restoring the chromatin's nucleosomal organization (Fig. 21–11). It is believed that the temporarily detached nucleosome cores reassociate with the double-helical DNA that contains the leading strand at a position about 125 nucleotides from the leading strand's 3' end. As the lagging strand elongates, newly synthesized histones associate with this double helix, thereby creating new nucleosomes about 250 nucleotides from the 5' end of the first Okazaki fragment. In other words, the nucleosome cores in front of the replication fork are conserved and transferred to the newly formed double helix containing the leading strand, and new cores are formed for the double helix containing the lagging strand. Histone molecules required for these new nucleosomes are synthesized during the S phase of the cell cycle.

Repetitive DNA

The DNA of prokaryotic chromosomes consists essentially of *unique sequences*; that is, except for the genes

(a)

(b)

Duplication with
labeled thymidine

1st metaphase
after labeling;
(*a*) above

Duplication without
labeled thymidine

2nd metaphase
after labeling;
(*b*) above

(c)

FIGURE 21–9   Semiconservative replication of DNA in eukaryotic cells. Autoradiographs
of *Vicia faba* root tip cells at the first (*a*) and second (*b*) metaphase after an initial
replication in the presence of ^{3}H-thymidine, (c) Interpretation of the results seen in *a* and
b. ^{3}H-Thymidine-labeled DNA strands and radioactive chromatids are shown in color. (*a*
and *b* from J. H. Taylor, The replication and organization of DNA in chromosomes,
Molecular Genetics, Part I, J. H. Taylor (Ed.), Academic Press, New York, 1963; c from E. J.
Gardner and D. P. Snustad, *Principles of Genetics* (7th ed.), John Wiley & Sons, 1984,
p. 138.)

FIGURE 21–10 Electron photomicrograph of *Drosophila* DNA showing multiple sites of replication (arrows). (Courtesy of *Proc. Nat. Acad. Sci.* **71**, 135 (1974).)

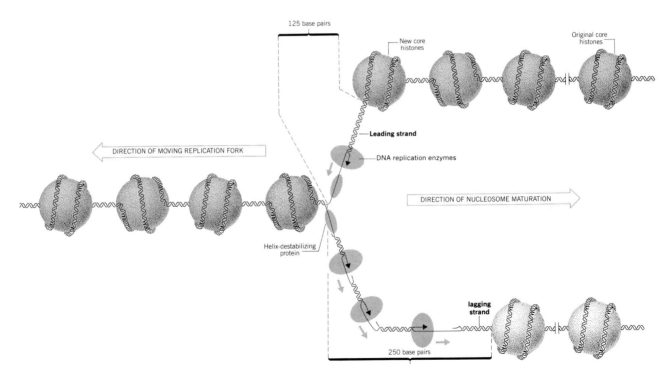

FIGURE 21–11 Disassembly and reassembly of nucleosomes during DNA replication in eukaryotic cells.

encoding ribosomal RNAs (see Chapter 22), base-pair sequences are not repeated. In contrast, the chromosomes of eukaryotic organisms contain **repetitive DNA**, that is, base sequences that are repeated many times. Indeed, it has been estimated that 20% to 50% of the genome of eukaryotes may be represented by repetitive DNA. The rest of the DNA consists of *unique sequences* (i.e., single copies). The amount of DNA represented by repeated sequences is determined experimentally by allowing fragmented and denatured DNA to reanneal by the process of complementary base pairing. The rate of reannealing is directly related to the quantity of repetitive DNA present. The method is described in some detail in Chapter 7. It is generally acknowledged that the unique sequences represent structural genes for messenger RNA and proteins. There are two broad classes of repetitive DNA: **moderately repetitive** (from as few as 2 to as many as 100,000 copies of a particular sequence) and **highly repetitive** (more than 100,000 copies). Moderately repetitive DNA includes the genes for certain proteins such as the histones and ribosomal proteins and the various ribosomal RNAs. The functions of the highly repetitive sequences are uncertain, and they are located in the inactively expressed heterochromatic regions of the nucleus (Chapter 20). Apparently, some moderately repetitive DNA can migrate from one chromosome to another; such *nomadic* sequences may be analogous to the *transposable elements* in prokaryotic cells (see later) and may have a role in the selective gene expression that characterizes an organism's development.

REPLICATION OF BACTERIOPHAGE DNA

Viruses are unable to independently replicate their nucleic acids. Instead, they infect a cell (called the *host*) and utilize the metabolic machinery of the host to replicate their own DNA. The extent to which the virus relies on the host cell's enzymatic apparatus is related to the size of the viral genome (Table 21–2). Viruses like ϕX174 whose 5375-nucleotide genome has only nine genes encoding the formation of coat proteins rely entirely on the host cell to provide the enzymes for DNA replication. In viruses with larger genomes (such as the T4 bacteriophage that attacks *E. coli* and that contains several dozen genes), the DNA encodes enzymes of the viruses' own replicative process as well as coat proteins.

Lambda (λ) Phage

Bacteriophages (viruses that infect bacteria) can be subdivided into two groups according to the manner in which they interact with their host cell. When **virulent** phages infect bacteria, they quickly multiply and then lyse the host cell. In contrast, **temperate** bacteriophages enter either a **lytic** or **lysogenic** pathway after infecting the host (Fig. 21–12). For example, when temperate viruses such as the λ phage infect a metabolically active *E. coli*, a virulent infection results and soon leads to lysis of the host and release of new phage particles. However, when λ phage infects a host cell that is metabolically inactive (i.e., the glucose supply of the cell is exhausted and its cyclic

TABLE 21–2 DNA CHARACTERISTICS OF SEVERAL VIRUSES

Virus	Related Viruses	Particle Shape	Mol. Wt. x 10^{-6}	Number of Base Pairs (kbp)	Shape	Comments
SV-40	Polyoma	Polyhedron	3.4	5.1	Duplex, circular	Animal cell host
ϕX174	S13	Polyhedron	1.8	5.4	Single strand, circular	Duplex replicative form
M13	fd, f1	Filament	1.9	5.7	Single strand, circular	Duplex replicative form
T7	T3	Head, short tail	23	35.4	Duplex, linear	Terminal redundancy
λ	ϕ80, 434 P2, 186	Head, tail	32	49	Duplex, linear	Cohesive ends form replicative circles, lysogenic
T5		Head, tail	76	115	Duplex, linear	Nicked
T4	T2, T6	Head, tail	120	180	Duplex, linear	Terminal redundancy; permuted
R17	MS2, f2 Qβ	Polyhedron	1.0	3.0	Single strand, linear	RNA, not DNA

Source: From A. Kornberg, *DNA Synthesis*, Freeman, San Francisco, copyright ©1974.

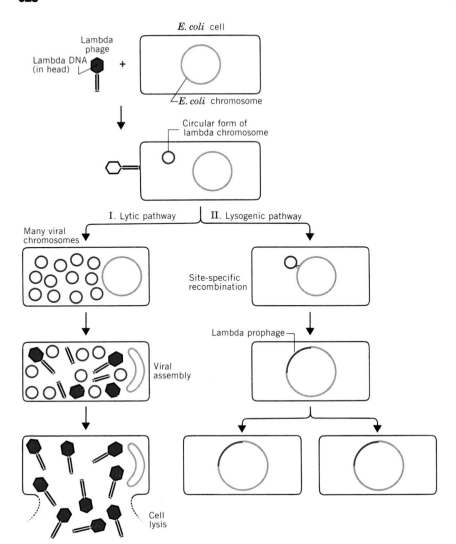

FIGURE 21–12 Alternative *lytic* and *lysogenic* pathways followed by bacteriophage after infecting a host (see text for explanation).

AMP level is elevated), a lysogenic response is initiated.

In a lysogenic response, the inserted viral DNA is integrated into the DNA of the host and is replicated along with the host's DNA during the growth and division of the host cell. The replication employs the native mechanism of the host cell. The integrated phage DNA (now called **prophage**) may be replicated in many successive generations of host cells. Eventually, the prophage DNA is expressed and produces new virus particles that are released upon lysis of the host cell.

In both the lytic and lysogenic responses, the initial stage of infection involves attachment of the virus to the surface of the host cell (in this case *E. coli*). The

specificity between the virus and the host is achieved through the interaction between specific viral adsorption proteins called **pilot proteins** and specific host cell-surface receptors. The pilot proteins not only attach the virus to the host but may also position the viral DNA (or RNA in the case of the RNA viruses) for insertion into the most appropriate part of the cell.

After attachment to the host cell, the viral nucleic acid is inserted through the cell envelope and plasma membrane into the cytosol. This may be aided by the pilot proteins and these may be carried into the cell along with the viral nucleic acid. In the case of the λ phage, the double-stranded DNA has a single-stranded tail at each end of the molecule and these tails are complementary to each other (Fig. 21–13).

FIGURE 21–13 Structure of the linear form of λ phage DNA showing the overlapping ends. The linear form can be converted to the circular form by annealing at the ends. (Redrawn from A. Kornberg, *DNA Synthesis*, W. H. Freeman and Company, copyright © 1974.)

When the λ phage DNA is inserted into *E. coli*, the complementary bases at the ends pair up (i.e., they anneal), thereby forming a circular DNA molecule. If the lytic pathway is followed, the host cell's metabolic machinery is used to produce many new copies of the circular viral nucleic acid and viral proteins and these are assembled into new virus particles. Eventually, the host cell lyses, releasing the virus particles, which may then begin a new cycle of infection. If the lysogenic pathway is followed, the circular viral DNA is integrated into the host cell's own circular chromosome and is replicated along with the host's DNA during successive cell generations, with all progeny acquiring a copy of viral DNA. Integration into the *E. coli* genome requires an enzyme (called *integrase*) encoded in one of the virus' genes (called *int*), and the viral DNA is inserted at a specific locus in the *E. coli* chromosome called *att* (Fig. 21–14). Accordingly, the *E. coli* chromosome is broken, the λ DNA inserted, and the λ prophage is formed. The prophage may be replicated along with the *E. coli* DNA for many thousands of generations.

Subsequent *excision* of the λ DNA from the *E. coli* chromosome is also an unusual phenomenon and requires the products of the *int* and *xis* genes. On rare occasions during excision, a part of the host chromosome is removed along with the λ DNA. The *E. coli* genes that are removed with the λ DNA are those for galactose utilization (i.e., *gal*) and biotin synthesis (i.e., *bio*). These genes are located on each side of the

att site. A most important implication of all this is that when these viruses are released and infect new cells, they can transfer the genes of the former host to the new host. The transfer of genes in this way is called **transduction**. The cell receiving these genes is said to be *transduced*, and the result is a form of *genetic recombination* (see below and Chapter 20).

The release of new virus particles from an infected host during a lytic response is not always accompanied by lysis and death of the host cell. Some viruses emerge from the host cell in the form of small buds and do not cause serious harm to the host cell. Occasionally, part of the host cell's plasma membrane is used to envelope the emerging virus particle.

φX174 Phage and Rolling Circle Replication

φX174 is another bacteriophage that infects *E. coli* cells. However, in contrast with the λ phage chromosome, the φX174 chromosome is circular and single stranded. The single-stranded DNA inserted into the *E. coli* cell is called the + strand (i.e., the positive strand) and it serves as the template for the synthesis of a complementary – strand (negative strand). The synthesis of the – strand begins with the synthesis of a short RNA primer at the origin, and DNA polymerase III then adds deoxyribonucleotides to the 3' end of the growing chain until the entire circular template has been used (Fig. 21–15). The RNA primer is then excised and the gap filled by the action of DNA polymerase I. DNA ligase covalently closes the free ends of the minus strand, with the result that a double-stranded, circular chromosome is formed.

The double-stranded circular chromosome (called the *parental replicative form* or simply *RF-form*) then participates in **rolling circle replication**, which produces additional double-stranded circular chromosomes (called *progeny RF*). Rolling circle replication begins when an enzyme called a *nickase* breaks the + strand at the origin—a site that is probably attached to the plasma membrane. The 5' end of the nicked + strand rolls away from the – strand and DNA synthesis ensues, using both exposed single strands as template. A new + strand is formed along the – strand beginning at its exposed 3' end, and a new – strand is formed discontinuously along the unfolded + strand beginning at a point near the 5' end of the + strand. As synthesis proceeds, the + strands continue to roll away from the – strand. When a complete – strand has been formed along the + strand template, a nickase frees the double-stranded molecule and a *ligase*

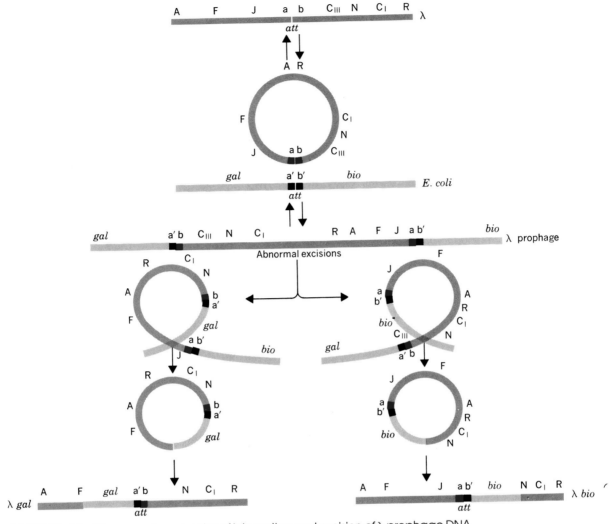

FIGURE 21–14 Diagram of mechanics of integration and excision of λ prophage DNA into chromosome of *E. coli*. (Redrawn from A. Kornberg, *DNA Synthesis*, W. H. Freeman and Company, copyright © 1974.)

joins their free ends. Thus, a complete circular DNA molecule (or progeny RF) results. Rolling circle replication by parental RF produces additional progeny.

Progeny RF are converted to single-stranded molecules by nicking the + strand and unrolling it from the double helix (Fig. 21–15); these positive strands may then be converted to circular, single-stranded molecules through the action of a ligase that links the strands' free ends. Circular single-stranded molecules may then be encased in newly synthesized viral proteins to form complete φX174 viruses. Rolling circle replication also occurs during the transfer of plasmid DNA between conjugating bacteria (see Chapter 20).

DNA REPAIR

As noted earlier in the chapter, the fidelity with which DNA is replicated is very high, so that the order of base pairs in each daughter DNA molecule is identical to the original parent duplex. However, once replication is completed, the normal structure of the DNA molecule may be altered, thereby producing a form that cannot readily be replicated. For example, bases along one of the two strands may be chemically altered or removed entirely, adjacent bases along one strand may become linked together to form a *dimer*, or breaks may be introduced into one or both chains.

Stage I: Viral positive strand → parental RF

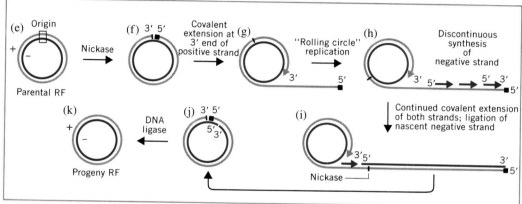

Stage II: Parental RF → progeny RF

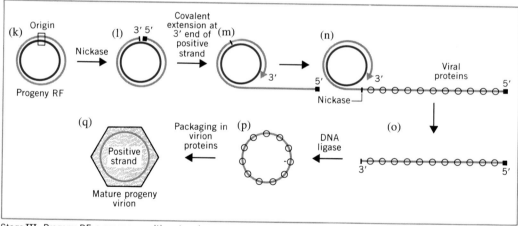

Stage III: Progeny RF → progeny positive strands

FIGURE 21–15 Rolling circle replication of the single-stranded DNA of the φX174 bacteriophage (see text for details). (Modified from E. J. Gardner and D. P. Snustad, *Principles of Genetics* (7th ed.), John Wiley & Sons, 1984, p. 114.)

Such changes in the structure of a normal DNA molecule may not only interfere with proper replication but also change normal recombination events and produce useless transcription products.

Several DNA repair mechanisms that can correct errors present in one or both strands of a molecule have been shown to exist in cells. Among these are **excision repair** and **postreplication repair**. Excision repair (Fig. 21–16), a DNA repair mechanism that has been demonstrated in both prokaryotes and eukaryotes, corrects for the presence of pyrimidine dimers within a strand. Such dimers are produced, for example, when DNA is exposed to ultraviolet radiation. P. Howard-Flanders, D. Pettijohn, P. Hanawalt, and others have shown that the nucleotides forming the dimer are enzymatically excised and the resulting gap in the strand then filled by the action of *DNA polymerase I*. The repair enzyme nicks the defective strand at the 5'

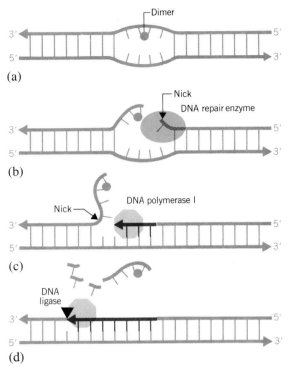

FIGURE 21–16 Excision repair of a DNA molecule in which one strand contains a pyrimidine dimer.

end of the damaged region, following which DNA polymerase I binds at the nick and adds the appropriate complementary nucleotides to the free 3' end. After adding a succession of nucleotides, the DNA polymerase produces a second nick in the strand, releasing a short polynucleotide that contains the dimer. After this, DNA ligase closes the nick, thereby completing the repair process.

If a segment of one of the two strands of a double helix contains a dimer, then during replication the complementary strand that is synthesized will lack several nucleotides in that vicinity of the chain corresponding to the position of the dimer (Fig. 21–17); thus, the new strand has a gap in it and the original (defective) strand is single stranded in the region of the dimer. Such an error cannot be corrected by excision repair, but is corrected by *postreplication* repair. In postreplication repair, the normal duplex is nicked and a repair enzyme realigns the single-stranded region with the complementary region of the sister duplex (i.e., the repair enzyme switches the free end of the nicked strand into the gap). The result is a **cross-strand exchange**. Examination of the process depicted in Figure 21–17 reveals that the upper hetero-

duplex can now be repaired by the action of *DNA polymerase I*. However, we are still left with two double helices that are cross-stranded. Another repair enzyme produces two nicks in the cross-stranded helices and recombines the free ends to produce two double helices that are no longer cross-stranded. The original dimer, which is now part of the lower double helix, can be corrected by the excision repair enzymes.

TRANSCRIPTION OF DNA

Transcription is the synthesis of RNA using DNA as template. In Chapter 7, we considered the chemistry of RNA and the basic principles of the transcription process and examined the mechanisms that regulate what portions of the genome of prokaryotic and eukaryotic cells are expressed.

Transcription of DNA produces three types of RNA: messenger RNA (mRNA), ribosomal RNA (rRNA), and transfer RNA (tRNA). The initial RNA transcripts (called **primary transcripts**) contain many more bases than the final functional ribonucleic acid molecules; the extra nucleotides are removed by a mechanism called **RNA processing**. The net effect of RNA processing (described at length in Chapter 22) is that certain nucleotides are removed, others are added, and a number are chemically modified.

Unlike replication, transcription does not progress along the entire length of a chromosome. Instead, certain parts of the chromosome are transcribed. Moreover, only one of the two strands of a DNA duplex is transcribed; this strand is called the **sense strand** (Fig. 21–18). The ribonucleotides are added to the growing 3'-OH end of the RNA transcript by **DNA-dependent RNA polymerase**. In prokaryotes, RNA polymerases are *multimeric* proteins. For example, *E. coli RNA polymerase* consists of six polypeptides and one of these (called the *sigma factor*) is responsible for initiation of transcription. The sigma factor ensures that the enzyme binds to the initiation site by recognizing the *promoter* (see Chapter 11) of the DNA. Like DNA, RNA is synthesized in the 5'→3' direction but proceeds without the need for primers. A transient *heteroduplex* consisting of the sense strand and a portion of the newly synthesized RNA is formed (Fig. 21–18b). The transcript soon separates from the sense strand, the heteroduplex structure being maintained only in the region of RNA elongation (Fig. 21–18c). As a result, a number of RNA polymerases can simulta-

(a)

(b)

(c)

(d)

(e)

FIGURE 21-17 Postreplication repair of a DNA molecule in which the presence of a dimer in one original strand results in a gap in the newly synthesized complementary strand. Filling the gap requires a cross-strand exchange. After the gap is filled, the dimer can be eliminated by excision repair, as shown in Figure 21-16.

neously transcribe the same region of the DNA (Figs. 21–18*d* and 21–18*e*; see also Fig. 22–8).

Five different **RNA polymerases** have been identified in eukaryotic cells. Of these, three are localized in the nucleus, one is present in mitochondria, and one is present in chloroplasts. Mitochondrial and chloroplast *RNA polymerases* are presumed to function during the expression of the limited amount of DNA present in these organelles. The nuclear RNA polymerases are called *RNA polymerase I*, *RNA polymerase II*, and *RNA polymerase III*. *RNA polymerase I* is concentrated in the nucleolus where it synthesizes the various ribosomal RNAs (except 5 S rRNA). *RNA polymerase II* is responsible for the synthesis of transcripts destined to become messenger RNA, and

RNA polymerase III transcribes DNA into a variety of low molecular weight RNAs, including transfer RNA and 5 S ribosomal RNA (see Table 22–5).

To initiate transcription, the *RNA polymerases* bind to specific regions of the DNA called **promoters**, each promoter consisting of about 40 nucleotide pairs. Synthesis of the primary transcript continues until a **terminating signal** is encountered in the sense strand. In a number of prokaryotes, the termination signal is a nucleotide sequence that gives rise to a *hairpin loop* in the elongating RNA; that is, the base sequence of the growing RNA molecule in this region is *self-complementary* and results in spontaneous folding into the double-helix conformation. Typically, the hairpin loop is followed by a poly U tail. Completion of the

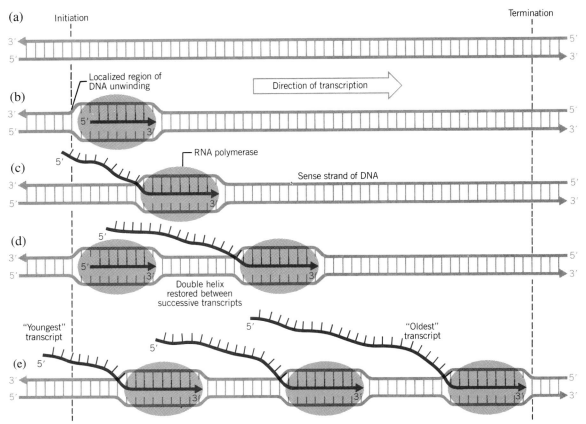

FIGURE 21–18 Transcription of DNA. In eukaryotic cells, DNA is transcribed into ribosumal RNA, messenger RNA, and transfer RNA by different RNA polymerases.

hairpin loop and poly U tail is followed by dissociation of the *RNA polymerase* from the template and release of the primary transcript.

BACTERIAL RECOMBINATION

Recombination is the exchange of genetic material between chromosomes that results in new combinations of genes in successive generations of cells. In eukaryotic cells, recombination occurs during **crossing-over** in meiosis (see Chapter 20). Although crossing-over cannot occur in haploid organisms, genetic recombinations in prokaryotic cells do result from transduction, conjugation, and cellular transformation.

In Chapter 20 we saw that bacterial plasmids can be transferred from one cell to another during conjugation and may then be integrated into the cell's main chromosome. The integration of plasmids is depen-

dent on *IS elements*—stretches of DNA (750–1400 base pairs long) found in both plasmids and the main chromosomes of bacteria. These IS elements are *transposable* in that they can move from one position to another in a chromosome.

GENETIC ENGINEERING

For many years, the gene has been thought of as a unit of *structure*, its definition based on the results of recombination experiments and the analysis of mutant cells. According to this older view, a gene was the smallest unit of inheritance that was not subdivisible by recombination and the smallest unit capable of independent mutation. Today, we recognize that a nucleotide pair is the smallest structural unit that fits the recombination and mutation definitions, although it certainly is not reasonable to call a nucleotide pair a gene. Therefore, a *functional* definition is more appropriate, and the gene is better defined as a unit of inheritance that encodes one polypeptide chain or one RNA molecule.

The complete analysis of the nucleotide sequence of a gene is a substantial undertaking. Nevertheless, the primary structures of a number of genes are now known, including most of the genes encoding the various globin chains of human hemoglobin (see Chapter 22). Even small chromosomes, such as the chromosome of the virus ϕX174, have been fully sequenced. This chromosome, which consists of 5,375 nucleotides, contains nine genes whose starting and ending locations are known. One of the more remarkable findings of recent years is that the nine genes do not occupy nine separate segments of the chromosome. Instead, two of the genes are located entirely within the coding sequences of two other genes. As will be seen in Chapter 22, the chromosomes of eukaryotic cells contain noncoding nucleotide sequences *between* and also *within* coding sequences.

Genetic engineering may be defined as the deliberate manipulation or alteration of the genetic information of an organism. Among the various forms that genetic engineering takes is the removal of genes from one organism followed by their insertion into the genome of another. Examples of the benefits to be derived through such undertakings include (1) the development of new strains of crop plants that are more resistant to adverse environmental conditions or to pests and (2) the modification of bacteria so that they produce certain proteins (e.g., human insulin, interferon, etc.) and other substances of value to humankind.

The major stages involved in genetically engineering an organism for purposes such as those described above are outlined in Figure 21–19, which depicts the alteration of the genome of a bacterium so that it can produce human proteins. In the first stage of the process, human cells are disrupted and the chromosomes containing the gene to be transferred are isolated and cut into small pieces using enzymes called **restriction endonucleases**. Bacterial cells are then disrupted to obtain "cloning vehicles"; the latter usually take the form of small, circular bacterial chromosomes called **plasmids**. Opened plasmids and pieces of the human chromosome are mixed in the presence of *ligases*, with the result that the human genes become spliced into the plasmid forming a **chimeric** or **recombinant** DNA molecule.

Once formed, the recombinant DNA can be transferred back into the same strain of bacteria from which the original plasmids were obtained. The bacterial *host* cells are then cultured, producing a **clone** consisting of millions of identical cells. During the growth of the clone, each bacterium produces a new copy of the recombinant DNA in addition to its normal chromosomal material. The result is a vast population of bacteria containing specific human genes. The expression of the human genes (as well as the host's native genes) results in the production of large quantities of human proteins.

We will now examine each of these stages in greater detail.

Preparing Donor DNA

Depending on the genes that are to be cloned, a tissue source is selected. Because nearly all cells contain a full complement of the organism's genes, literally any tissue could be selected as a source of any gene. However, the tissue that is usually chosen is the one that serves as the major source of the gene product in that organism. Accordingly, insulin genes are isolated from pancreas tissue and interferon genes are isolated from white blood cells. Although hemoglobin is produced by maturing red blood cells, the genetic engineering studies conducted with hemoglobin have employed liver tissue as a source of the globin genes. Either fresh or previously frozen tissue may be used as a source of donor DNA.

The tissues are vigorously homogenized to free the nuclear material and a detergent solution containing specific enzymes is added to free the DNA from attached proteins and inactivate nucleases that might otherwise begin to degrade the DNA. After the homogenate is deproteinated using phenol, the DNA is isolated by isopycnic density gradient centrifugation (see Chapter 12). The DNA, which forms a single band in the density gradient, is then withdrawn. The isolated DNA contains the genes that are to be cloned but it also contains many thousands of other genes, whose cloning may be of no interest.

Preparing the Cloning Vehicle

The Plasmid Method Cloning vehicles may take the form of small circular bacterial chromosomes called plasmids (as described in the example presented earlier) or phage DNA. Although plasmids account for only a small percentage of the total DNA of a bacterium, they contain a number of vital genes including those that confer antibiotic resistance to the cell (e.g.,

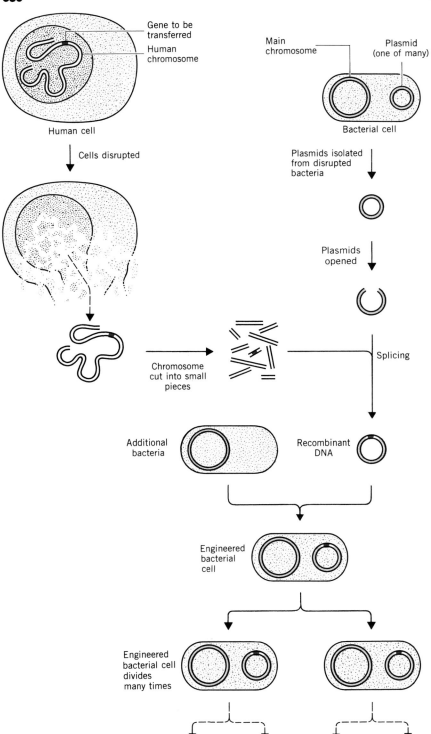

FIGURE 21–19 Major stages of gene cloning employing bacterial plasmids as the cloning vehicles.

the gene for *penicillinase*). The presence of these genes is essential in the subsequent search for cells containing the donor DNA. Because of their small size, plasmids are readily separated from the main chromosomes of disrupted bacteria by centrifugation.

Isolated donor DNA is treated with one or more of the **restriction endonucleases** (Table 21–3) that cleave the DNA into specific fragments. Restriction endonucleases possess site specificity, cleaving the DNA wherever certain nucleotide sequences, called **restriction sites**, occur. Some restriction endonucleases make "staggered cuts" in the DNA so that the exposed cut ends have a number of unpaired bases (Fig. 21–20). Because the nucleotide sequences that

are recognized by the restriction endonucleases are **palindromes** (i.e., base sequences that read the same in the forward direction along one chain and the backward direction along the other), the single-stranded ends are complementary.

When treated with the same restriction enzymes, plasmids also produce molecules having complementary, single-stranded ends. The donor DNA fragments and the plasmid DNA are then mixed in the presence of DNA ligase under conditions that permit the DNAs to be spliced together. The *recombinant DNA* that results has variable composition depending on which donor DNA fragments combined with the plasmid DNA. (Remember, the donor DNA consists of a mix-

TABLE 21–3   RECOGNITION SEQUENCES AND CLEAVAGE SITES OF REPRESENTATIVE RESTRICTION ENDONUCLEASES

Enzyme	Source	Recognition Sequence[a] and Cleavage Sites[b]	Number of Recognition Sequences per Chromosome of:	
			Phage λ	SV-40 Virus
EcoR1	*Escherichia coli*	↓ GAA TTC ● CTT AAG ↑	5	1
HindII	*Hemophilus influenzae*	↓ GTPy PuAC[c] ● CAPu PyTG ↑	34	5
HindIII	*Hemophilus influenzae*	↓ AAG CTT ● TTC GAA ↑	6	6
HpaI	*Hemophilus parainfluenzae*	↓ GTT AAC ● CAA TTG ↑	11	3
HpaII	*Hemophilus parainfluenzae*	↓ CC GG ● GG CC ↑	750	1

[a]The axis of symmetry in each palindromic recognition sequence is indicated by the colored dot.

[b]The position of each bond cleaved is indicated by a colored arrow. Note that the cuts are staggered (at different positions in the two complementary strands) with some restriction nucleases.

[c]Pu indicates that either purine (adenine or guanine) may be present at this position; Py indicates that either pyrimidine (thymine or cytosine) may be present.

Source: From E. J. Gardner and D. P. Snustad, *Principles of Genetics*, Wiley, New York, 1981.

FIGURE 21–20 Cleavage of DNA by a *restriction endonuclease* that makes "staggered cuts." The newly exposed ends of the DNA fragments have unpaired bases. (Modified from E. J. Gardner and D. P. Snustad, *Principles of Genetics* (7th ed.), John Wiley & Sons, 1984, p. 356.)

ture of DNA fragments only some of which contain the genes to be cloned!)

Recombinant DNA can be taken up by bacterial cells that are made transiently permeable to such macromolecules. Not all of the bacteria mixed with recombinant DNA will incorporate the modified plasmids but those that do can be identified if the bacteria used are antibiotic-sensitive. This is because incorporation of the recombinant DNA simultaneously confers antibiotic resistance to the bacteria. Therefore, only these bacteria will proliferate when cultured in the presence of antibiotic. (Those that do not take up the recombinant DNA, and therefore do not simultaneously acquire the antibiotic genes, die). Once incorporated by the bacteria, the recombinant DNA is replicated along with the remaining cellular DNA during growth.

Identification of the particular bacterial cells that contain the specific genes to be cloned demands specific assay procedures. The procedure (which is sometimes referred to as the "shotgun" approach) may be illustrated using as an example the gene that encodes an enzyme involved in the synthesis of the amino acid histidine (i.e., the *his* gene). The various forms of recombinant DNA produced by the procedure described above would be introduced into a *mutant* variety of *E. coli* cells that are unable to synthesize histidine. After transfer of the recombinant DNA to the cells, they would be cultured using a medium that lacked histidine. In this way, the cells that had not incorporated recombinant DNA containing the *his* gene would be unable to produce the necessary enzyme for histidine synthesis and would not survive in the histidine-free medium. Only the cells acquiring the *his* gene would survive; thus, the appropriate cells are selected and the gene is cloned as these cells grow and divide. A

different selection method is used for each gene to be cloned.

The Bacteriophage Method A bacteriophage (or phage) is a virus that infects bacteria. The initial stage of the bacteriophage method involves the isolation of a small number of phage from a bacterial culture. Bacteria can be cultured on the surface of a shallow dish containing nutrient medium (e.g., an agar plate), and on such a medium the bacteria grow until they cover the dish's surface. Because phage-infected bacteria eventually lyse, releasing new viruses that begin another cycle of infection, clear areas containing lysed cells and freed bacteriophage are soon seen on the surface of the nutrient medium; these areas are called *plaques*.

Phage particles removed from a plaque are mixed with a large culture of bacteria whose infection quickly yields much more phage. Phage and bacteria are separated by centrifugation and the phage is treated with enzymes that remove the protein coat; the phage DNA is then purified by density gradient centrifugation. Isolated phage DNA and donor DNA are then treated in a manner similar to that described earlier for plasmids with the result that the donor DNA is integrated into the phage DNA, thereby forming a cloning vehicle.

Gene Cloning Methods Using Reverse Transcription

In some instances it is possible to isolate from donor cells the mRNA for specific proteins. For example, globin chain mRNA can be isolated from reticulocytes, albumin mRNA can be isolated from chicken oviduct cells, proinsulin mRNA can be isolated from pancreas cells, and interferon mRNA can be isolated from white blood cells. The isolated mRNAs can then be used as template for the synthesis of complemen-tary DNA strands (called *cDNA*) using **reverse transcriptase**, an enzyme encoded in the genome of certain RNA viruses (Fig. 21–21; see also Chapter 7).

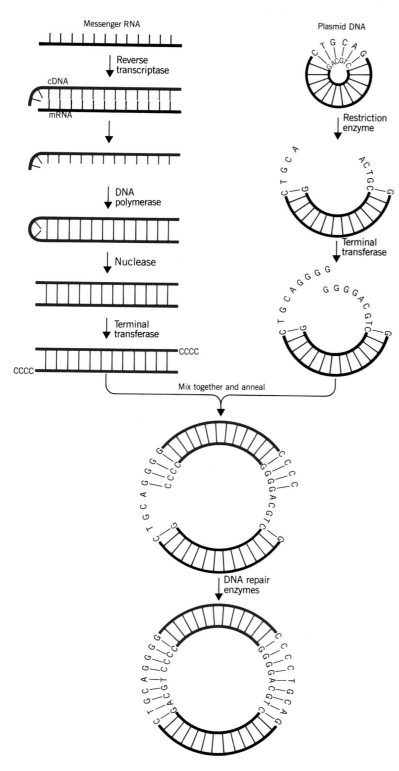

FIGURE 21–21   Gene cloning employing *reverse transcription*. Messenger RNA for a specific protein is isolated and *reverse transcriptase* is used to make a complementary DNA strand (called cDNA or copy DNA). *DNA polymerase* is then used to convert the cDNA to the double-stranded form. A *terminal transferase* then adds specific nucleotides, in this case CCCC, to each 3' end of the molecule. Plasmid DNA is opened using restriction enzymes and the sequence GGGG is added to the free ends by a terminal transferase. Finally, the plasmid and DNA are mixed together and allowed to anneal, following which gaps in the recombinant DNA are filled by repair enzymes. The recombinant DNA may now be incorporated into bacteria and cloned.

Single-stranded DNA produced by reverse transcription can be used as template for the synthesis of a complementary DNA strand. The resulting double-stranded DNA can then be inserted into an opened plasmid or phage DNA and cloned in bacteria as described above. Because the recombinant DNA used in this approach is specific (not random recombinant DNA molecules as are produced in the shotgun method), the necessity of using special *E. coli* mutants for cell selection is precluded. The incorporation of DNA molecules produced by reverse transcription into plasmids requires the use of enzymes called **terminal transferases** instead of restriction endonucleases, because reverse transcription does not produce DNA molecules that have offset complementary nucleotide chains at their ends (i.e., the enzymes do not produce staggered cuts). Terminal transferases add single-stranded tails to the 3' ends of the DNA strands.

Genetic Engineering of Mammalian Cells

Genes from a number of different eukaryotic cells have been excised and then cloned in bacterial cells using the restriction endonuclease and reverse transcription techniques. The bacteria containing the recombinant DNA are thus able to produce a number of eukaryotic proteins. The hormone *proinsulin* was the first such eukaryotic protein whose production was achieved in cultures of bacteria transformed using recombinant DNA. The gene encoding this protein was synthesized using reverse transcriptase and then transferred to the bacterial cells. Human *interferons* (proteins that act as antiviral agents) are also being made by applying the recombinant DNA technology to bacterial cells. In this instance, interferon mRNA is isolated from human white blood cells.

Interferon   The production of human interferon using recombinant DNA methods has attracted a great deal of public and scientific attention in recent years because interferon has also been found to attenuate the growth of transformed human cells. In a number of specific instances, cancerous tumors have been shown to go into remission after patients have been treated with interferon. Interferon was discovered in the 1950s by A. Isaacs and J. Lindenmann, who were studying a phenomenon called "virus interference" in which cells treated with one type of virus resisted infection by other types. Isaacs and Lindenmann found that virus interference was not due to one virus blocking the growth of another, rather, the infection by one virus was shown to induce a cell to produce and *secrete* a substance that conferred viral resistance to other cells. Because the substance interfered with the capacity for viruses to produce an infection, they called the substance "interferon."

Human interferons are now known to be a class of glycoproteins produced by a number of different cells including lymphocytes and fibroblasts. Synthesis of the interferons by these cells is triggered by the derepression of the interferon genes of the cell's genome (see Chapter 11). Once released from the cells, the interferons react with receptors in the plasma membranes of other cells and in some yet uncertain way block viral infection of these cells. Because interferon is produced and secreted in very small quantities, intensive study of the action of interferon has been hampered for decades as a result of the difficulty of isolating sufficient quantities of this material.

Because certain cancers are believed to be transmitted by viruses, and interferons have been shown to diminish the growth of tumors, it is tempting to suggest that interferon administration could prevent or cure virus-mediated cancers. However, it has *not* been shown that the effects of interferon on tumor growth result directly from antiviral activity. Rather, the interferon may act indirectly by stimulating other body defense mechanisms. Application of the recombinant DNA technology for large-scale interferon production should soon provide the scientific community with enough of this most interesting material to allow definitive tests of its clinical value. At the present we are seeing only the "tip of the iceberg" regarding the benefits that may be derived from the use of bacteria for cloning human genes and producing human proteins.

The genes of eukaryotic cells can also be cloned in other eukaryotic cells, but this is a much more difficult task. This is because the membranes of mammalian cells are not so readily made permeable to recombinant DNA. One method for transferring genes to mammalian cells that has had some success is the use of a whole virus as a vector for such transfer. The gene to be transferred is first linked to the viral DNA and the recombinant DNA-containing virus is used to infect the animal cells. The gene encoding the beta chain of rabbit hemoglobin has been successfully incorporated into the genome of the Simian virus (SV-40). When these viruses are then allowed to infect mammalian cells, the beta globin chain gene is transferred along with the native viral genes and rabbit beta globin chains are subsequently synthesized in these cells. Some success has also been achieved using mechanical microinjection procedures to transfer recombinant DNA-containing plasmids directly into the nuclei of mammalian cells.

SUMMARY

In all cells, DNA replication is **semiconservative**. In prokaryotic cells, replication may be **unidirectional** or **bidirectional**, but in eukaryotes it is always bidirectional. The overall growth of the two new complementary DNA strands during replication proceeds in the 5'→3' direction. The elongation of one strand is continuous and the other grows as a series of **Okazaki**

fragments that are eventually linked together. Many of the enzymes involved in the replication process have now been identified.

Viruses are incapable of directly replicating their nucleic acids. Instead, they employ the metabolic machinery of a host cell. Replication proceeds in any of a variety of ways depending on whether the viral nucleic acid is double-stranded DNA, single-stranded DNA, or RNA.

Although **genetic recombination** is a natural phenomenon in prokaryotes and eukaryotes, recombination can also be performed experimentally. Of special significance is the use of **recombinant DNA** methods for transferring human genes to and cloning them in bacteria.

REFERENCES AND SUGGESTED READING

Articles and Reviews

Alberts, B. M., Protein machines mediate the basic genetic processes. *Trends in Genetics* **1**, 26 (Jan. 1985).

Anderson, W. F., and Diacumkos, E. G., Genetic engineering in mammalian cells. *Sci. Am.* **245** (1), 106 (July 1981).

Brown, D. D., Gene expression in eukaryotes. *Science* **211**, 667 (1981).

Chambon, P., Split genes. *Sci. Am.* **244** (5), 60 (May 1981).

Chilton, M.–D., A vector for introducing new genes into plants. *Sci. Am.* **248** (6), 51 (June 1983).

Cohen, S. N., and Shapiro, J. A., Transposable genetic elements. *Sci. Am.* **242** (2), 40 (Feb. 1980).

DePamphilis, M. L., and Wasserman, P. M., Replication of eukaryotic chromosomes: A close-up of the replication fork. *Ann. Rev. Biochem.* **49**, 627 (1980).

Dickerson, R. E., The DNA helix and how it is read. *Sci. Am.* **249** (6), 94 (Dec. 1983).

Dickerson, R. E., Drew, H. R., Conner, B. N., Wing, R. M., Fratini, A. V., and Kopka, M. L., The anatomy of A–, B–, and Z–DNA. *Science* **216**, 475 (1982).

Felsenfeld, G., DNA. *Sci. Am.* **253** (4), 58 (Oct. 1985).

Gilbert, W., and Villa-Komoroff, L., Useful proteins from recombinant bacteria. *Sci. Am.* **242** (4), 74 (April 1980).

Grobstein, C., The recombinant-DNA debate. *Sci. Am.* **237** (1), 22 (July 1977).

Howard-Flanders, P., Inducible repair of DNA. *Sci. Am.* **245** (5), 62 (Nov. 1981).

Hubscher, U., DNA polymerase holoenzymes. *Trends in Biochem. Sci.* **9** (9), 390 (Sep. 1984).

Khorana, H. G., Total synthesis of a gene. *Science* **203**, 614 (1979).

Kolata, G., Z-DNA moves toward "real biology." *Science* **222**, 495 (1983).

Kornberg, A., DNA replication. *Trends in Biochem. Sci.* **9** (4), 122 April 1984).

Landy, A., and Ross, W., Viral integration and excision: Structure of the lambda *"att"* site. *Science* **197**, 1147 (1977).

Lehman, A. R., and Karran, P., DNA repair. *Int. Rev. Cytol.* **72**, 101 (1981).

Liu, L. F., DNA topoisomerases—Enzymes that catalyse the breaking and rejoining of DNA. *Crit. Revs. Biochem.* **15** (1), 1 (1983).

Moore, G. P., The C-value paradox. *Bioscience* **34**, 425 (1984).

Novick, R. P., Plasmids. *Sci. Am.* **243** (6), 102 (Dec. 1980).

Pestka, S., The purification and manufacture of human interferon. *Sci. Am.* **249** (2), 36 (Aug. 1983).

Schimke, R. T., Gene amplification and drug resistance. *Sci. Am.* **243** (5), 60 (Nov. 1980).

Sinsheimer, R. L., Recombinant DNA. *Ann. Rev. Biochem.* **46**, 415 (1977).

Takeda, Y., Ohlendorf, D. H., Anderson, W. F., and Mathews, B. W., DNA-binding proteins. *Science* **221**, 1020 (1983).

Walker, G. C., Inducible DNA repair systems. *Ann. Rev. Biochem.* **54**, 425 (1985).

Wang, J. C., DNA topoisomerases. *Ann. Rev. Biochem.* **54**, 665 (1985).

Wang, J. C., DNA topoisomerases. *Sci. Am.* **247** (1), 94 (July 1982).

Books, Monographs, and Symposia

Beljanski, M., *The Regulation of DNA Replication and Transcription*, Karger Press, Basel, 1983.

Bonner, J., and Varner, J. E., *Plant Biochemistry*, Academic, New York, 1976.

Drilica, K., *Understanding DNA and Gene Cloning*, Wiley, New York, 1984.

Freifelder, D., *Recombinant DNA*, Freeman, San Francisco, 1978.

Gardner, E. J., and Snustad, D. P., *Principles of Genetics* (7th ed.), Wiley, New York, 1984.

Hunt, T., Prentis, S., and Tooze, J., *DNA Makes RNA Makes Protein*, Elsevier, Amsterdam, 1983.

Kornberg, A., *DNA Replication*, Freeman, San Francisco, 1980.

Kornberg, A., *1982 Supplement to DNA Replication*, Freeman, San Francisco, 1982.

Lewin, K., *Genes II*, Wiley, New York, 1985.

Old, R. W., and Primrose, S. B., *Principles of Gene Manipulation*, University of California Press, Berkeley, 1980.

Swanson, C. P., Merz, T., and Young, W. J., *Cytogenetics*, Prentice–Hall, Englewood Cliffs, N.J., 1981.

CHAPTER 22

RIBOSOMES AND THE SYNTHESIS OF PROTEINS

Before the 1930s, it was the prevailing view that DNA was found only in animal cells and RNA only in plant cells. This notion was dispelled by a number of findings in the 1930s that definitively established that both DNA and RNA are present in animal and plant cells. Moreover, J. Brachet and T. Caspersson showed that the bulk of the RNA was present in the cytoplasm and that cells actively engaged in the synthesis of secretory protein (such as pancreas cells and the silk-gland cells of silkworms) contain greater amounts of RNA than cells that do not actively secrete protein. Albert Claude showed in the 1940s that the cytoplasmic RNA was included in tiny particles of ribonucleoprotein later to be called "ribosomes."

PROTEIN TURNOVER IN CELLS

The rate of breakdown and replacement of protein in cells was badly misunderstood prior to 1939. In growing animals generally and in secretory tissues in particular (liver, pancreas, endocrine glands, etc.), active synthesis of protein was known. However, the amount of protein synthesis taking place in other tissues of the adult was believed to be very low and confined to that necessary to replace protein lost in damaged or dying cells. These small protein losses, together with the catabolism of dietary amino acids, were believed to be responsible for the urea and ammonia measurable in the urine excreted from the body. Proteins were thus regarded as highly stable constituents lasting virtually the entire lifetime of the cell.

The first serious challenge to the "wear and tear" view of protein turnover came as a result of the work of R. Schoenheimer in 1938. Schoenheimer synthesized a number of amino acids in which the ^{15}N content of the alpha-amino nitrogen was considerably increased over the naturally occurring amount of this isotope. Schoenheimer then injected ^{15}N-containing glycine and leucine into rats and noted that these labeled amino acids were quickly incorporated into the proteins of many tissues. Although ^{15}N is not a radioactive isotope of nitrogen, it may nonetheless be distinguished chemically from the more common ^{14}N form and is called a "heavy" isotope of nitrogen. The results clearly indicated that protein synthesis in adult animals is not restricted to growing and secretory tissues but occurs in nearly all cells and that tissue proteins are in a continuous state of metabolic flux, being broken down and replaced by newly synthesized molecules.

Although the radioactive isotope of carbon, ^{14}C, was produced in the Berkeley cyclotron in 1940, it was not until 1947 that ^{14}C-labeled amino acids became available for use as biological tracers. The availability of radioactive amino acids was followed by a series of classic tracer experiments by H. Borsook, T. Hultin, P. Zamecnik, and P. Siekevitz, which verified the findings of Schoenheimer that most tissues readily incorporate amino acids into protein and also added crucial details to the newly emerging view of protein synthesis and metabolic turnover.

The first attempts to determine the subcellular site of amino acid incorporation into protein were carried out in 1950 by Borsook. Minutes after injecting ^{14}C-labeled amino acids into the bloodstreams of guinea pigs, Borsook removed the animals' livers and, using the technique of differential centrifugation (see Chapter 12), prepared subcellular fractions of the tissue. Borsook showed that it was the **microsomal fraction** that contained the greatest amount of radioactivity and suggested that the microsomes were the repository of the cell's protein-synthesizing apparatus. In the same year, Hultin demonstrated that it was the microsomal fraction of chick liver tissue homogenates that incorporate intravenously injected ^{15}N-glycine into protein.

By 1952, Siekevitz and Zamecnik had been able to demonstrate the *in vitro* incorporation of ^{14}C-labeled amino acids into liver cell proteins by both tissue slices and tissue homogenates. By measuring and comparing protein synthetic activity in cell-free whole homogenates, individual cell subfractions, and various combinations of subfractions, Siekevitz showed that the incorporation of amino acids into proteins by microsomes was dependent on an energy source provided by the mitochondrial subfraction and required enzymes and other factors present in the cytosol. The demonstration that amino acid incorporation into protein required metabolic energy laid to rest a view popular in the 1940s that polypeptide synthesis might be brought about by the reversal of the enzymatic reactions in which proteins are degraded. It is especially interesting to note that Siekevitz demonstrated the existence in the cytosol of a $MgCl_2$-precipitable factor required for protein synthesis. As $MgCl_2$ was known to precipitate RNA, Siekevitz suggested that RNA might somehow be involved in protein synthesis, a fact not fully recognized until many years later.

The studies described above established the general cytological and chemical basis of protein biosynthesis. Exhaustive research since the 1950s by dozens of groups of investigators has revealed the step-by-step, reaction-by-reaction details of the process and has given us an astounding insight into the molecular organization of the cell's protein-synthesizing apparatus. Although most of this chapter is devoted to the examination of ribosome assembly and structure and to the chemical events that accompany protein synthesis, much of what is to be presented is better understood if first placed in perspective with a brief, preliminary overview of the subject; this will set the foundation for the more comprehensive study that follows. For simplicity, this synopsis will be concerned only with cytoplasmic protein synthesis in eukaryotic cells.

A PRELIMINARY OVERVIEW OF PROTEIN BIOSYNTHESIS

The variety of proteins synthesized by a cell and the specific amino acid compositions of each protein are ultimately governed by the cellular DNA. This DNA is enzymatically transcribed in the cell nucleus to produce a host of RNAs (Chapter 21), including **ribosomal RNA** (*rRNA*), **messenger RNA** (*mRNA*), and **transfer RNA** (*tRNA*). Different RNA polymerases are responsible for the synthesis of each family of RNAs (see Chapter 21). The base sequences of these RNAs are *complementary* to the base sequences of the DNA molecules transcribed. rRNA is ultimately incorporated into the cytoplasmic ribosomes, which may be *free* in the cytosol or *attached* to the surface of the endoplasmic reticulum that faces the cytosol. Each ribosome consist of two parts or *subunits*—a *small* subunit and a *large* subunit. The small subunit binds mRNA entering the cytosol from the nucleus, and the functional complex is completed with the subsequent addition of the large subunit. Attached ribosomes are linked to the endoplasmic reticulum by their large subunits.

The nucleotides of mRNA are arranged as a linear sequence of **codons** (also known as triplets), each codon consisting of three successive nitrogenous bases. The codon sequence of each mRNA molecule contains all the information necessary to (1) properly *initiate* polypeptide synthesis on the ribosome, (2) designate

TABLE 22-1   THE GENETIC CODE

		SECOND BASE			
	U	**C**	**A**	**G**	
U	UUU ⎱ Phe UUC ⎰ UUA ⎱ Leu UUG ⎰	UCU ⎱ UCC ⎬ Ser UCA ⎰ UCG	UAU ⎱ Tyr UAC ⎰ UAA (terminator) UAG (terminator)	UGU ⎱ Cys UGC ⎰ UGA (terminator) UGG Try	U C A G
C	CUU ⎱ CUC ⎬ Leu CUA ⎰ CUG	CCU ⎱ CCC ⎬ Pro CCA ⎰ CCG	CAU ⎱ His CAC ⎰ CAA ⎱ Gln CAG ⎰	CGU ⎱ CGC ⎬ Arg CGA ⎰ CGG	U C A G
A	AUU ⎱ AUC ⎬ Ile AUA ⎰ AUG Met (initiator)	ACU ⎱ ACC ⎬ Thr ACA ⎰ ACG	AAU ⎱ Asn AAC ⎰ AAA ⎱ Lys AAG ⎰	AGU ⎱ Ser AGC ⎰ AGA ⎱ Arg AGG ⎰	U C A G
G	GUU ⎱ GUC ⎬ Val GUA ⎰ GUG (initiator)	GCU ⎱ GCC ⎬ Ala GCA ⎰ GCG	GAU ⎱ Asp GAC ⎰ GAA ⎱ Glu GAG ⎰	GGU ⎱ GGC ⎬ Gly GGA ⎰ GGG	U C A G

FIRST BASE · THIRD BASE

the specific *sequence* of amino acids to be incorporated (i.e., the primary structure of the polypeptide), and (3) *terminate* polypeptide synthesis and *release* the completed polypeptide. Table 22–1 shows the various codons of mRNA and their meanings in protein synthesis. This is called the "genetic code." The code is said to be *degenerate* because in certain instances a single amino acid may be coded for by more than one codon.

Molecules of tRNA entering the cytosol from the nucleus combine with amino acids; this is a molecule-specific association in that each amino acid species is enzymatically combined with a particular type (or species) of tRNA. The products, called **aminoacyl-tRNA**, represent the form in which amino acids are incorporated into newly synthesized protein. Each species of tRNA contains, among other functional groups, an **anticodon** (sequence of three bases) that is recognized by a corresponding (complementary) codon of mRNA and ensures that the correct amino acid will be incorporated into its proper position in the primary structure of the polypeptide being synthesized.

Once the mRNA–ribosome complex has been formed, amino acids bound to their specific tRNA molecules are sequentially brought to the ribosome and incorporated into the growing polypeptide chain. This process, called **translation**, is believed to take place by an orderly and linear movement of the mRNA along the ribosome (or vice versa) so that each codon is translated in sequence. The elongation of the polypeptide chain takes place by a series of enzyme-catalyzed reactions occurring at three adjacent sites of the ribosome; these are (1) the **A site** (i.e., *amino acid* or *acceptor site*), (2) the **P site** (i.e., *peptide site*), and (3) the **E site** (i.e., *exit site*). To understand the process of elongation, consider an intermediate stage in the synthesis of a polypeptide. At this time, the growing polypeptide chain is attached to the P site of the ribosome by a molecule of tRNA and is termed **peptidyl-tRNA**. The mRNA codon located at the adjacent amino acid site specifies the particular aminoacyl-tRNA that can be bound there. With a new aminoacyl-tRNA in position in the A site, the bond linking the growing polypeptide to its tRNA is broken and replaced by a peptide bond with the amino acid of aminoacyl-tRNA. This leaves the peptidyl-tRNA (which is now one amino acid longer) temporarily in the A site. The tRNA molecule freed in the process is moved to the E site, while at the same time the peptidyl-tRNA is moved from the A site to the P site; these movements constitute **translocation**. Binding of another aminoacyl-tRNA at the now vacant A site is accompanied by the release of the tRNA molecule from the E site. The released tRNA reenters the cytosol where it may combine with another amino acid to be used in protein synthesis. Translocation is accompanied by the movement of the ribosome and/or mRNA so that the next codon is in position at the amino acid site. Thus, tRNA molecules employed in bringing amino acids to the ribosome are transiently bound first to the A site, then to the P site, and finally to the E site before returning to the cytosol.

Amino acids are sequentially added to the growing polypeptide until its primary structure is complete. Once the end of the message encoded in the strand of mRNA is reached, the completed protein is released from the ribosome. The ribosome separates from the mRNA and dissociates into its two subunits; these may be used again in another round of protein synthesis.

Many mRNA molecules are large enough to be simultaneously translated by a number of ribosomes. These ribosomes move in series along the mRNA (or the mRNA is moved along the ribosomes), the net result being that its coded message is translated into a number of *identical* proteins. The release of one ribosome at the end of the message is accompanied by the attachment of a new ribosome at the beginning of the message. Such strings of ribosomes are called **polyribosomes** or **polysomes**, and most cellular protein synthesis takes place on these structures. Although each mRNA molecule may be attached to several ribosomes, each ribosome synthesizes but a single protein chain before dissociating into its subunits. The mechanics of protein synthesis described briefly here is treated in detail in later sections of this chapter.

STRUCTURE, COMPOSITION, AND ASSEMBLY OF RIBOSOMES

In this section, we will be concerned with the organization, composition, and assembly of the cytoplasmic ribosomes of prokaryotic and eukaryotic cells. The ribosomes present in cytoplasmic organelles (e.g., chloroplast and mitochondrial ribosomes) will be considered separately later in the chapter. Although functionally analogous, many differences exist between the ribosomes of prokaryotic and eukaryotic cells (Table 22–2). Considerably more is known about the structure and composition of bacterial ribosomes than ribosomes of eukaryotic cells, as will become evident during the discussion that follows. Most of the work on prokaryotic ribosomes has been carried out using *Escherichia coli*. Although some variations are observed among the prokaryotes, findings using *E. coli* are generally representative.

Ribosomes in the cytoplasm of eukaryotic cells have a sedimentation coefficient of about 80 S (MW about 4.5×10^6) and are composed of 40 S and 60 S subunits. In prokaryotic cells, ribosomes are typically about 70 S (MW about 2.7×10^6) and are formed from 30 S and 50 S subunits. The complete ribosome formed by combination of the subunits is also referred to as a **monomer**. Although ribosomes from both prokaryotic and eukaryotic sources are about 30 to 45% protein (by weight), with the remainder being ribonucleic acid, the specific protein and RNA components of these two major classes of ribosomes differ (Table 22–

TABLE 22–2 PROPERTIES AND COMPOSITION OF EUKARYOTIC AND PROKARYOTIC RIBOSOMES

	Eukaryotes	Prokaryotes
Monomers		
Sedimentation coefficient	80 S	70 S
Number RNAs	4	3
Number of proteins	70	55
Small subunit		
Sedimentation coefficient	40 S	30 S
RNAs present	18 S	16 S
Number of proteins	30	21
Large subunit		
Sedimentation coefficient	60 S	50 S
RNAs present	5 S	5 S
	5.8 S	23 S
	28 S	
Number of proteins	40	34

2 and Fig. 22–1); carbohydrate and lipid are virtually absent. Magnesium ions (and perhaps other cations) play an important role in maintaining the structure of the ribosome. Dissociation into subunits occurs when Mg^{2+} is removed. The precise role (or roles) of Mg^{2+} remains uncertain, although interaction with ionized phosphate of subunit RNA is presumed.

Prokaryotic Ribosomes

RNA Content The small subunit of prokaryote ribosomes contains one molecule of an RNA called 16 S RNA (MW 0.6×10^6), and the large subunit contains two RNA molecules, a 23 S RNA (MW 1.1×10^6) and a 5 S RNA (MW 3.2×10^4) (see Table 22–2). All three rRNAs are products of closely linked genes transcribed by *RNA polymerase* in the sequence 16 S → 23 S → 5 S. This assures an equal proportion of each RNA. The rRNA operon also contains genes for some tRNAs (Fig. 22–2). The transcription product of the rRNA operon consists of a 30 S RNA; this transcript is successively cleaved and trimmed to produce the final 16 S, 23 S, and 5 S RNAs that are incorporated into the small and large ribosomal subunits. Figure 22–2 presents the scheme of maturation of the prokaryotic rRNAs. For clarity, the incorporation of the ribosomal proteins is not shown. Ribosomal proteins combine with the rRNAs at various stages of subunit assembly: some are incorporated during transcription, others following release of the primary transcript and during processing, and still others once the mature rRNA products are formed. Certain proteins bind to the rRNAs only transiently and are not found in the fully assembled subunits.

Multiple copies of the rRNA genes occur in the genomes of prokaryotic (and eukaryotic) cells (Table 22–3); this is known as **reiteration**. In *E. coli* the number of rRNA genes is estimated to be between 5 and 10 and accounts for about 0.4% of the cell's total DNA. The primary structures of the three prokaryotic rRNAs have been extensively studied. 5 S RNA was identified in 1963 and, being the smallest of the three rRNAs (about 120 nucleotides), was sequenced first (in 1967).

Analyses of the nucleotide sequence of the *E. coli* 16 S RNA (1542 nucleotides) and 23 S RNA (2904 nucleotides) have recently been completed and the secondary structure of the 16 S RNA is shown in Figure 22–3. Methylation of certain bases in the sequence of 16 S RNA (and also in the sequence of 23 S RNA) occurs while transcription is taking place. No methylation of 5 S RNA nucleotides occurs. Unlike 5 S RNA in which duplication of certain sequences occurs, no repeated sequences are found in 16 S and 23 S RNA. The rRNAs contain a number of double-helical regions that are stabilized by conventional, complementary base pairing (Chapter 7). In *E. coli* 16 S rRNA, about half of all nucleotides present are involved in base pairing. Several *palindromes* (base sequences reading the same from either the 5' or 3' ends) exist in 16 S RNA, and these may play a role in restricting the formation of the double-helical regions. In 16 S RNA, a seven-nucleotide segment of the chain at the 3' end is

FIGURE 22–1 Scale representation of the relative sizes and RNA and protein complements of eukaryotic and prokaryotic ribosomes. Each area corresponds to the total molecular weight of that component. (Numbers in parentheses are the weight percentages of the components.)

FIGURE 22–2 Transcription and processing of prokaryotic rRNAs (see text for details). The ribosomal RNA operon contains genes for 16 S, 23 S, and 5 S RNAs, two tRNAs, and intervening spacer regions. During processing of the 30 S primary transcription product, the spacer is removed and the rRNAs and tRNAs trimmed to their final lengths.

FIGURE 22–3 Secondary structure of the 16 S RNA component of the small subunit of *E. coli* ribosomes. The double-helical regions (shaded) are stabilized by conventional base pairing.

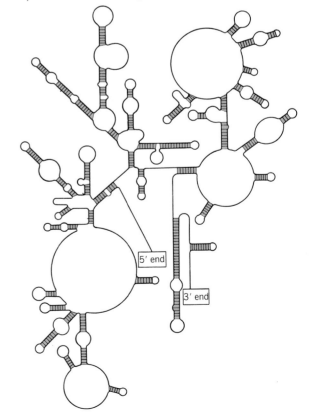

TABLE 22-3 REITERATION OF rRNA IN VARIOUS CELLS

Cell or Tissue	Number of Genes per Genome
Liver	750
HeLa cells	1100
Xenopus (toad)	900
Drosophila melanogaster	260
Tobacco leaves	1500
Saccharomyces cerevisiae	140
E. coli	5–10
Bacillus subtilis	9–10
B. megaterium	35–45

TABLE 22–4 AVERAGE MOLECULAR WEIGHTS OF
PROKARYOTIC AND EUKARYOTIC RIBOSOME PROTEINS

	Prokaryote	Eukaryote
Small subunit	18,900	25,300
Large subunit	16,400	28,100

FIGURE 22–4 Areas of *E. coli* 16 S RNA with which various small subunit proteins associate.

believed to interact with mRNA, leading to its binding during the initiation of translation. The 5 S and 23 S RNAs interact with one another in the large subunit, and both appear to be involved in aminoacyl-tRNA and peptidyl-tRNA binding during polypeptide chain elongation. Because the 16 S RNA as well as several proteins of the small subunit interact with 23 S RNA, the latter RNA may also have a role in subunit association.

Ribosomal RNA transcripts are not translated into proteins (i.e., rRNAs cannot serve as messengers), however, ribosomal proteins are the products of a typical transcription–translation process.

Protein Content Nomura, Kurland, and others have established that the small prokaryotic ribosomal subunit contains 21 different protein molecules; these are identified as S1, S2, S3 . . . S21. The large subunit contains 34 proteins (L1, L2, L3 . . . L34) but only 31 are different, that is, four copies of one of the proteins are present. The genes for the 52 different ribosomal proteins together with those for the three RNAs constitute about 5% of the genome of the bacterial cell. All the ribosomal proteins have been isolated and characterized, and nearly all have been fully sequenced. The small subunit proteins range in molecular weight from 10,900 to 65,000 and the large subunit proteins vary in molecular weight from 9600 to 31,500 (Table 22–4). Most of the ribosomal proteins are basic in nature, being rich in basic amino acids and having isoelectric points around pH 10 or higher. An exhaustive analysis of the primary structures of prokaryotic ribosomal proteins done in order to evaluate their degree of *homology* (see Chapter 4) indicates that these proteins did *not* have a common evolutionary ancestor. Homologies among them do not occur more often than would be expected on a random basis.

tRNA–Protein Interaction Lake, Nomura, Wittmann, Traut, Stoffler, Kurland, and others have studied the relationships between the three rRNAs and the ribosomal proteins and have shown that about 30 proteins bind specifically and *directly* to the rRNAs; these are the *primary* binding proteins. Those proteins that do

not bind directly to rRNA (i.e., the *secondary* binding proteins) interact with the primary binding proteins in the assembled ribosome. Figure 22–4 shows that approximate regions of 16 S RNA with which the small subunit proteins associate. It is believed that the backbone of the 16 S RNA polynucleotide winds its way among the proteins, with interactions occurring between hairpin turns of the RNA and surface residues of the protein molecules.

Assembly of Prokaryotic Ribosomes

Because all of the proteins and RNAs of the prokaryotic ribosome subunits may be isolated, it is possible through recombination studies to examine the assembly process. Nomura and others have shown that the assembly of individual subunits and their association to form functional ribosomes (i.e., ribosomes capable of translating mRNA into protein) occur *spontaneously in vitro* when all the individual rRNAs and protein components are available. Thus the ribosome is capable of *self-assembly*, and this is believed to be the mechanism *in situ*. The assembly is promoted by the unique and complementary structures of the ribosomal protein and RNA molecules and proceeds through the formation of hydrogen bonds and hydrophobic interactions. There is *order* to the assembly in that certain proteins combine with the rRNAs prior to the addition of others. Cooperativity also exists, because addition of certain proteins to the growing subunit facilitates addition and binding of others.

No self-assembly takes place when L proteins are added to 16 S RNA or when S proteins are added to 5 S and 23 S RNA. However, it is interesting to note that RNA from the 30 S subunit of one prokaryotic species will combine with the S proteins of another prokaryote to form functional subunits. The same is true for 50 S subunit proteins and RNAs from different prokaryotes. Assembly of *hybrid subunits* and formation of functional monomers from these occur in spite of the fact that ribosomal proteins and RNAs from different prokaryotes have different primary structures. It is clear that their secondary and tertiary structures, which are very similar, are more important in guiding rRNA–protein interactions. Although some proteins from yeast, reticulocyte, and rat liver cell ribosomes can be replaced by *E. coli* ribosomal proteins, *hybrid monomers* formed from these prokaryotic-eukaryotic subunits will not function in protein synthesis.

Model of the Prokaryotic Ribosome

Based on the available electron-microscopic data, results of small-angle X-ray analysis, and, of course, chemical studies, several proposals can be made about the structure of the ribosome monomer and its subunits. The 30 S subunit approximates a *prolate* ellipsoid of revolution (Fig. 22–5a). A transverse partition or groove encircles the long axis of the subunit, dividing it into segments of one-third (i.e., the **head**) and two-thirds (i.e., the **base**). A small protuberance called a **platform** extends from the base. The 50 S subunit is somewhat more spherical and possesses a flattened region on one surface (Fig. 22–5a). Extending from the main body of the large subunit are a **stalk** and **central protuberance**. Association of the subunits to form the 70 S monomer is depicted in Figure 22–5b. There is considerable morphological and biochemical evidence supporting the idea that the small tunnel formed between the two subunits upon their association (Fig. 22–5b) is the site of mRNA and aminoacyl-tRNA binding during protein synthesis (Fig. 22–5c).

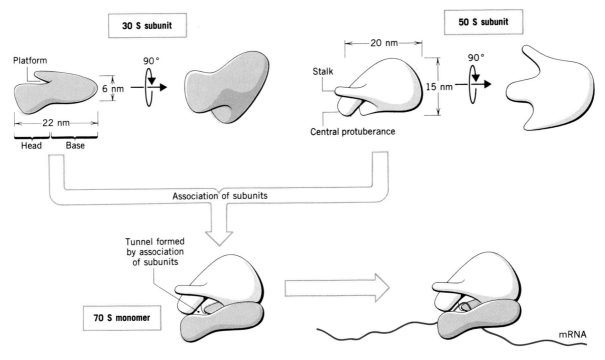

FIGURE 22–5 Model of the prokaryote ribosome.

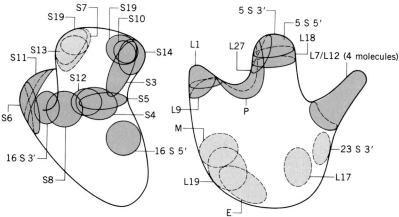

FIGURE 22–6 Distribution of proteins in the small (*left*) and large (right) subunits of the ribosome. Areas with heavy striping are near the surface facing the reader; areas with light striping are near the opposite surface. Also shown are the locations of the ends of the 5 S RNA molecule of the large subunit (labeled 5 S 3' and 5 S 5'), the end of the 16 S RNA molecule of the small subunit (labeled 16 S 3' and 16 S 5'), the site at which amino acids are linked to the growing polypetide (labeled P), the site at which the large subunit may be attached to a membrane (labeled M), and the site at which the growing polypeptide chain emerges from the ribosome (labeled E). (Courtesy of Dr. J. A. Lake; modified from J. A. Lake, The ribosome *Sci. Am.* **245** (2), 89 (1981).)

For example, (1) in many electron photomicrographs of polyribosomes, the thin mRNA strand seems to "disappear" into the ribosomes; (2) *in vitro* experiments have shown that when the synthetic messenger polyU is associated with the 70 S monomer, the polynucleotide is protected from ribonuclease attack over a length of about 70 to 120 nucleotides; and (3) transfer RNA is protected from cleavage by nuclease when associated with the ribosome. The observation that small nascent (i.e., growing) polypeptides are protected from proteolysis suggests that a stretch of the polypeptide chain may be enclosed within a second tunnel formed in the large subunit (Fig. 22–5c).

Genes for Ribosomal RNA and Protein

The genome of *E. coli* and other prokaryotes consists of a single, long, circular DNA molecule supercoiled and packed into the "nuclear" region of the cell. The *E. coli* chromosome is about 1100 nm long and appears to contain at least three separate regions coding for rRNA. Each region contains closely linked 5 S, 23 S, and 16 S rDNA genes. Because some 5 to 10 copies of each gene occur in the genome, more than one

copy of each gene is likely present in each rDNA region. Genes coding for ribosomal proteins are present in at least two separate regions of the *E. coli* chromosome. The same regions also appear to contain genes for *RNA polymerase*, some *transfer RNAs*, and the *elongation factors* required for protein biosynthesis (see later).

Lake, Nomura, and others have employed the techniques of immune electron microscopy and neutron diffraction to map the surface of the ribosomal subunits and identify the positions of the ribosomal proteins (Fig. 22–6).

Eukaryotic Ribosomes

The cytoplasmic ribosomes of eukaryotic cells differ from those of prokaryotes in both size and chemical composition (Table 22–2, Fig. 22–1). The monomer has a sedimentation coefficient of 80 S and is formed from 40 S and 60 S subunits. In addition, ribosomes occur in two states in the cytoplasm. They may be associated with cellular membranes such as those of the endoplasmic reticulum (i.e., "attached" ribosomes) and engaged in the synthesis of secretory, lysosomal, or membrane proteins or they may be freely distrib-

FIGURE 22–7   Processing of eukaryotic rRNAs (see text for explanation). During processing of the 45 S primary transcription product, the spacers are removed and the rRNAs and tRNAs trimmed to their final lengths. 5.8 S RNA becomes hydrogen bonded to 28 S RNA at the conclusion of processing.

uted in the cytosol. The functional differences between attached and free ribosomes will be pursued later, but let us turn first to a consideration of the chemical and morphological characteristics of eukaryotic ribosomes.

RNA Content   The small subunit of the eukaryotic ribosome contains one molecule of 18 S RNA (MW 0.7×10^6) and large subunit contains 28 S (MW 1.7×10^6), 5 S (MW 3.2×10^4), and 5.8 S (MW 5×10^4) RNAs. Hence, in addition to molecular weight or size differences, a major distinction between the RNA complements of prokaryotic and eukaryotic ribosomes

is the presence of an additional molecule of RNA (i.e., 5.8 S RNA) in the large subunit of eukaryotes.

18 S, 5.8 S, and 28 S rRNAs are the transcription products of closely linked genes in the chromosomes of the **nucleolar organizing region** (*NOR*) of the cell nucleus. Considerable redundancy exists as hundreds, perhaps even thousands, of copies of these rRNA genes are believed to be present (see Table 22–3). The genes for 5 S RNA are *not* present in the NOR but occur elsewhere in the nucleus. Consequently, unlike prokaryotes in which the 5 S RNA genes are linked to the genes for other rRNAs, the 5 S RNA genes of eukaryotes occur separately in the nucleus.

TABLE 22-5 RNA POLYMERASES

	Transcription Products
Prokaryotes RNA polymerase	tRNA, mRNA, and rRNA
Eukaryotes RNA polymerase I	5.8 S, 18 S, and 28 S rRNA
RNA polymerase II	mRNA
RNA polymerase III	5 S rRNA, tRNA

Figure 22-7 depicts the transcription and post-transcriptional processing of eukaryotic rRNAs. As in prokaryotes, transcription of DNA is mediated by the enzyme *RNA polymerase*; however, in eukaryotes there are three forms of this enzyme, each producing different transcription products. The *RNA polymerases* of prokaryotes and eukaryotes are compared in Table 22-5. The high-molecular-weight primary transcript, 45 S RNA, contains the precursors of 18 S, 5.8 S, and 28 S rRNAs. About half of the 45 S RNA molecule is represented by *spacer* sequences at the 5' end of the transcript and between the presumptive rRNAs. The spacers are removed during processing. As shown in Figure 22-7, the 5.8 S RNA produced during processing becomes hydrogen bonded to the 28 S RNA and the complex is incorporated into the 60 S subunit. Not shown in Figure 22-7 but discussed later is the incorporation of the ribosomal proteins. It should be noted that 5 S RNA is a *primary* transcription product and is not the product of posttranscriptional trimming.

Protein Content Various studies have established that the small subunits of eukaryotic ribosomes contain about 33 proteins (S1, S2, S3, etc.) and the large subunits 45 proteins (L1, L2, L3, etc.). The proteins of eukaryotic ribosomes are not only more numerous but also have greater average molecular weights than those of prokaryotic ribosomes (Table 22-4). From a chemical standpoint, eukaryotic ribosomal proteins have similar general properties as those in prokaryotes (e.g., rich in basic amino acids, high isoelectric point, etc.). Certain eukaryotic and prokaryotic ribosomal proteins reveal homologous regions, and these homologous proteins appear also to be functionally similar.

Nucleolar Organizing Region Eukaryotic cells contain several hundred copies of the genes encoding rRNA. These genes are arranged in a tandem fashion on one or more chromosomes of the nucleus. The DNA sequences between successive rDNA regions are not transcribed and represent **spacer DNA**. The rRNA genes and the spacer segments are usually looped off the main axis of the chromosome and are referred to as the *nucleolar organizing region*. It is here that most of the rRNA is synthesized. The NOR coalesces with nuclear proteins and forms visible bodies known as *nucleoli*. Most eukaryotic cells contain one or a few nucleoli, but certain egg cells are a striking exception. The oocytes of amphibians (e.g., the clawed toad, *Xenopus laevis*) are extremely large cells and are engaged in the synthesis of especially large quantities of cellular protein. These cells produce large numbers of ribosomes in order to provide the means to sustain such quantitative protein synthesis. Accordingly, it is not unusual to find hundreds or thousands of nucleoli (and NORs) in the nuclei of these cells. Such large numbers of nucleoli are the result of gene amplification—the differential replication of the rRNA genes of the genome. The ribosomes produced in the oocyte serve its needs for protein synthesis from the period prior to fertilization through the first few weeks of embryonic development.

By gently dispersing nuclear fractions isolated from oocytes of the amphibian *Triturus viridescens* and "spreading" the material on grids, O. L. Miller and B. R. Beatty in 1969 were able to obtain photomicrographs showing transcription *in progress*. Since then, a number of other investigators have extended the same approach to mammalian oocytes and to spermatocytes and embryo cells from various organisms. The visualization of transcriptional activity is achieved most easily with spread nucleoli because of the high degree of rDNA gene amplification (Fig. 22-8). The tandem rDNA genes are serially transcribed by RNA polymerases to produce 45 S rRNA. The rRNA (apparently complexed with protein) appears as a series of fibrils of varying length extending radially from an axial, linear DNA fiber (Fig. 22-9). These feather-shaped or "Christmas tree" regions are called **matrix units**. The spaces between successive matrix units are nontranscribed spacer DNA segments. The ribonucleoprotein (RNP) fibrils are seen to be in various stages of completion. The short fibrils near the tip of each feather are RNA molecules whose synthesis has only just begun and the longest fibrils represent RNA molecules whose synthesis is almost complete. Hence, *the direction of rDNA transcription is apparent in the* *photomicrograph.* In high magnification views (Fig. 22–9), even the *RNA polymerase* enzyme molecules carrying out the transcription of the DNA are visible along the axial DNA fiber.

Success in visualizing transcription has not been restricted to nucleolar genes. Almost identical results have been obtained with nonnucleolar chromatin. Here, however, the RNA transcripts represent messenger RNA.

Dispersed and spread nuclear fractions contain nontranscribing DNA as well as matrix units (Fig. 22–9).

The succession of *nucleosomes* (Chapter 20) reveals itself as a series of beadlike structures along the DNA fiber. Regions in which DNA is undergoing replication (called *replicons*) can also be seen (Fig. 22–10). S. L. McKnight and O. L. Miller have shown that DNA of homologous "daughter" fibers of the replicon also occurs as chains of nucleosomes, suggesting that replication may not require dissociation of nucleosomes or that nucleosomes are almost immediately reformed (see Chapter 21). Transcriptional activity can be identified within a replicon (Fig. 22–11), indicating that

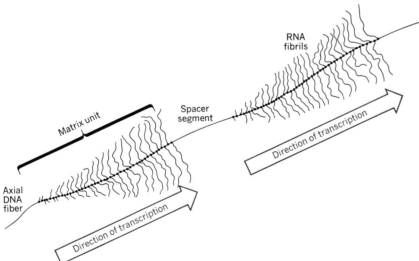

FIGURE 22–8 Visualization of transcription of rDNA genes. The thin axial DNA fiber is being transcribed simultaneously by a number of RNA polymerase enzymes (small black dots; see also Fig. 22–9). The transcripts (ribonucleoprotein complexes) appear as fine fibrils extending radially away from the DNA axis. (Electron photomicrograph courtesy of Dr. M. F. Trendelberg; from *Biol. Cellulaire* **42**, 1, 1981.)

the newly synthesized DNA is almost immediately available for transcription. The growing RNA fibrils are seen in homologous regions of both chromatid arms of the replicon.

Assembly of Eukaryotic Ribosomes The assembly of eukaryotic ribosomes is more complex than that of prokaryotic ribosomes. The principal stages of the process are outlined in Figure 22–12. Transcribed 45 S RNA combines with proteins in the nucleolus to form ribonucleoprotein complexes. However, not all the protein molecules of the complex become a part of the completed ribosomal subunit. Instead, certain proteins are released as RNA processing ensues; these "nucleolar proteins" return to a nucleolar pool and are reutilized. Those proteins that are retained during processing and become part of the completed subunits are, of course, legitimately called "ribosomal proteins." In the same sense, not all of the RNA of the complex becomes part of the ribsomal subunits, for the spacer is also processed out. (It should be noted that the spacer RNA is produced by transcription of

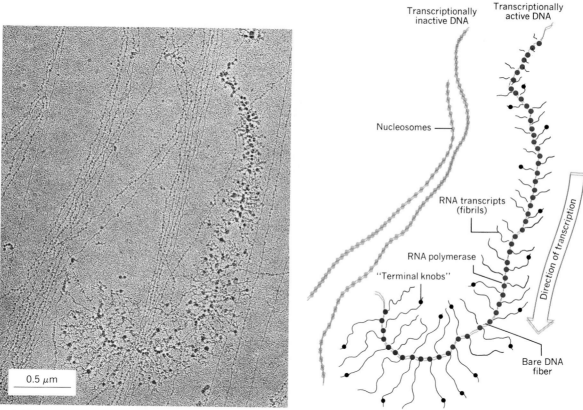

FIGURE 22–9 High-magnification electron photomicrograph of matrix unit and neighboring nontranscribed DNA fibers (see legend of Fig. 22–8 and text for explanation). (Photo courtesy of Dr. F. Puvion-Dutilleul.)

rDNA and *not* the spacer DNA between genes.) Processing produces three classes of fragments. One class contains spacer RNA and nucleolar proteins. The spacer RNA is hydrolyzed and the free nucleolar proteins return to the pool. A second class of RNP fragments contains a complex of 18 S RNA and certain ribosomal proteins that give rise to 40 S subunits. The third class of RNP fragments, which contains 38 S and 5.8 S RNA and ribosomal proteins, combines with 5 S RNA transcribed from extranucleolar rRNA genes and gives rise to 60 S subunits. Like the genes

for 45 S RNA, the extranucleolar 5 S RNA genes occur in multiple tandem copies. Among the various proteins synthesized in the cytoplasm using ribosome subunits derived from the nucleus are the ribosomal proteins themselves. These apparently reenter the nucleus for incorporation into new RNP complexes.

Model of the Eukaryotic Cytoplasmic Ribosome

In spite of the differences in overall sizes (as manifested in the greater molecular weights, sedimenta-

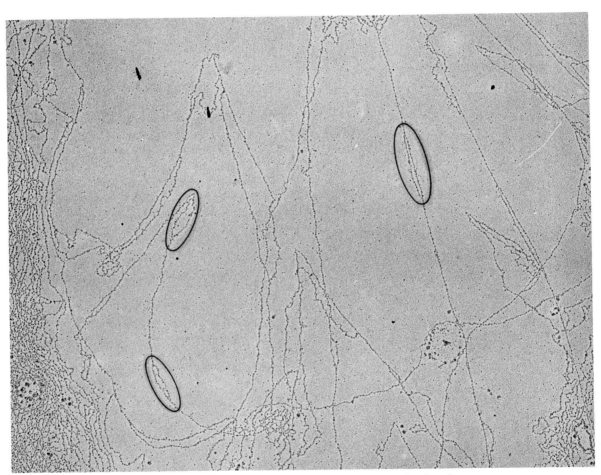

FIGURE 22–10   DNA fibers from dispersed nuclei exhibiting regions of replication (i.e., *replicons*). Three replicons are specifically identified in the figure. Magnification, 26,000X. (Electron photomicrograph courtesy of Dr. S. L. McKnight. Copyright © 1977 by MIT Press, *Cell* **12**, 795.)

tion constants, sizes, and numbers of RNAs and proteins), the cytoplasmic ribosomes of eukaryotes are remarkably similar in morphology to those of prokaryotes. As in 20 S subunits of prokaryote ribosomes, the 40 S eukaryote subunit is divided into head and base segments by a transverse groove (Fig. 22–13). The 60 S subunit is generally rounder in shape than the small subunit, although one side is flattened; this is the side that becomes confluent with the small subunit during the formation of the monomer (Fig. 22–13). The synthesis of proteins that are to be dis-

patched into the intracisternal space of the endoplasmic reticulum (ER) is carried out by ribosomes that attach to the membranes of the ER. As seen in Figure 22–13, attachment occurs via the large subunit.

Free and Attached Ribosomes

The cytoplasmic ribosomes of eukaryotic cells can be divided into two classes: (1) *attached* ribosomes and (2) *free* ribosomes (Fig. 22–14). Attached ribosomes are ribosomes associated with intracellular membranes,

FIGURE 22–11 Visualization of homologous transcriptional activity in the two daughter chromatids of a replicon. (Upper photomicrograph courtesy of Drs. A. L. Beyer, A. H. Bouton, and O. L. Miller; copyright © 1981 by MIT Press, *Cell* **26**, 155. Lower photomicrograph courtesy of Dr. S. L. McKnight; copyright © 1977 MIT Press, *Cell* **12**, 795.)

primarily the endoplasmic reticulum, whereas free ribosomes are distributed through the hyaloplasm or cytosol. Attached and free ribosomes are chemically the same. Although all animal and plant cells contain both attached and free ribosomes, the proportion of each varies from one tissue to another and can be caused to shift within a tissue in response to the administration of certain substances, notably hormones and growth factors.

Membranes of the endoplasmic reticulum that contain attached ribosomes constitute what is called "rough" ER (or RER) and membranes that are devoid of ribosomes are called "smooth" ER (SER) (see Chapters 1 and 15).

For many years, there has been considerable controversy about the functions of attached and free ribosomes. The currently accepted view suggests that proteins destined to be secreted from the cell or to be incorporated into such intracellular structures as lysosomes (which may or may not release their contents to the cell exterior) are synthesized on attached ribosomes. For example, many of the proteins circulating in

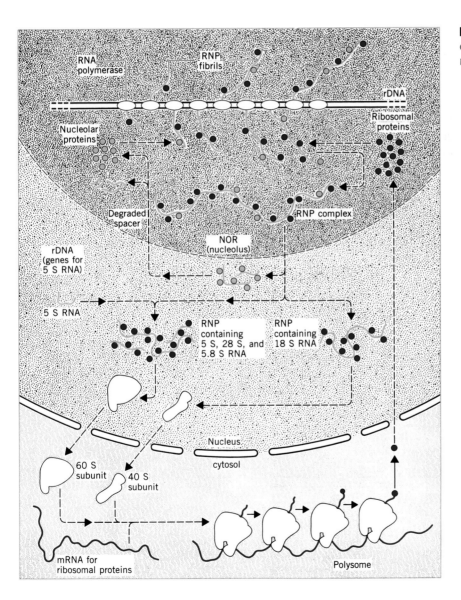

FIGURE 22–12 Synthesis and assembly of the components of eukaryotic ribosomes.

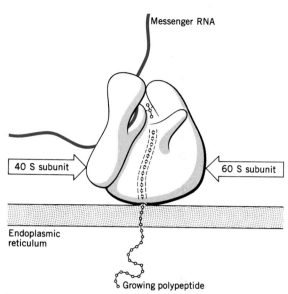

FIGURE 22–13 Model of the eukaryotic cytoplasmic ribosome (see text for details).

the blood plasma are derived via secretion by the liver, and these plasma proteins are known to be synthesized exclusively by the attached ribosomes of the liver cells. Most proteins destined to become constituents of the ER membranes or the plasma membrane are also synthesized by attached ribosomes (see also Chapter 15). Most, *but not all*, proteins destined for use in the cytosol are synthesized by free ribosomes. Exceptions include certain hormones like thyroglobulin, which is secreted by the thyroid gland and is synthesized by free ribosomes. Milk proteins produced by mammary gland cells are also synthesized by free ribosomes.

VECTORIAL SYNTHESIS

When cells are disrupted, the sheets of endoplasmic reticulum are broken into small vesicular fragments (Fig. 22–15a), which may be isolated with the *microsomal phase* by centrifugation. As noted in Chapter 12, this centrifugal fraction is quite heterogeneous and contains a variety of small particles in addition to fragmented endoplasmic reticulum. Fragmentation of the endoplasmic reticulum produces two kinds of vesicles; fragments of RER form vesicles whose outer surface is studded with ribosomes and SER vesicles are

free of ribosomes. The volume within the vesicle corresponds to the lumenal or cisternal phase of the cell.

In a series of elegant experiments, D. Sabatini and C. M. Redman examined protein synthesis *in vitro* by ribosome-laden vesicles. They were able to demonstrate that radioactive amino acids incorporated into protein by the ribosomes did not appear in the suspending medium but were recovered instead within the vesicles (Fig. 22–15b). These results argue strongly in favor of the proposal that attached ribosomes synthesize proteins for secretion, for the interior of the vesicles corresponds to the cisternal phase of the cell. Sabatini and Redman called this directional synthesis of protein **vectorial synthesis**. Once the protein is released into the cisternae, it is transported to the Golgi apparatus (see Chapter 19).

Although it is widely accepted that proteins destined to be secreted from the cell are synthesized on membrane-bound ribosomes, there are several pieces of evidence that indicate that membrane-bound ribosomes may have other functions as well. For example, J. R. Tata and others, working with muscle and nerve tissue, have observed the synthesis by RER ribosomes of small quantities of proteins for intracellular utilization. The rapid proliferation of RER during the period of active cell growth also suggests a nonsecretory function for attached ribosomes. The enzyme *serine dehydrogenase* has been shown to be specifically synthesized by attached ribosomes in liver cells, and yet this is an intracellular enzyme. Preliminary evidence suggests that some mitochondrial enzymes may be differentially synthesized by RER ribosomes.

Principally as a result of the studies of G. Blobel, D. D. Sabatini, C. M., Redman, D. I. Meyer, H. Lodish, and others, the mechanism that routes ribosomes to the endoplasmic reticulum has been elucidated. A major contribution to this scheme has been the confirmation of Blobel and Sabatini's **signal hypothesis** (Fig. 22–16; see also Chapter 15). Accordingly, proteins that are to be synthesized by attached ribosomes are encoded by mRNA molecules that contain a special nucleotide sequence near their 5' end called a "signal." The signal encodes a sequence of 16–26 amino acids. When a free ribosome translates such an mRNA molecule, the **signal peptide** emerging from the large subunit is recognized by a ribonucleoprotein complex in the cytosol called a **signal recognition particle** (SRP). SRP which consists of 6 polypeptides and 7 S RNA binds to the signal sequence, temporarily halting protein synthesis by that ribosome. Typi-

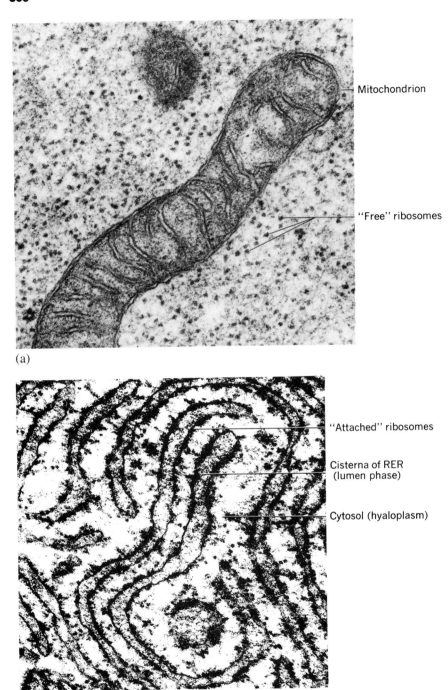

Mitochondrion

"Free" ribosomes

(a)

"Attached" ribosomes

Cisterna of RER
(lumen phase)

Cytosol (hyaloplasm)

(b)

(c)

(d)

FIGURE 22–14   Free (*a*) and attached (*b, c, d*) ribosomes. The SEM views seen in parts *c* and *d* are particularly unusual. In *c*, several neighboring membranous layers of the rough endoplasmic reticulum are seen. In *d*, the attached ribosomes of one layer are seen from the hyaloplasmic side; for some of these ribosomes, the subunits may be discerned. (*a* and *b* courtesy of R. Chao; *c* and *d*, courtesy of Dr. K. Tanaka. *Int. Rev. Cytol.* **68**, 102.)

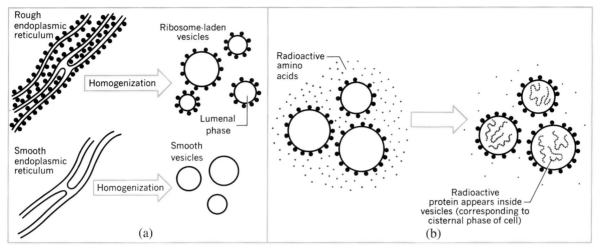

FIGURE 22–15   (a) Production of microsomal vesicles during homogenization of endoplasmic reticulum. (b) Vectorial synthesis of protein by ribosome-laden vesicles.

FIGURE 22–16   Synthesis of secretory proteins by attached ribosomes.

cally, chain length is about 80 residues at the time elongation is arrested. Protein synthesis by the ribosome is resumed only after attachment to the endoplasmic reticulum. Attachment occurs at specific membrane sites occupied by **docking proteins**. Once docking has occurred, SRP is released and protein synthesis by the newly attached ribosome is resumed, the elongating polypeptide progressively dispatched through the ER and into the intracisternal space. Prior to completion of synthesis, the signal peptide is removed by *signal peptidase*, an ER enzyme having its active site at the lumenal surface and the ER membranes.

RIBOSOMES OF ORGANELLES

The mitochondria and chloroplasts of eukaryotic cells contain their own DNA and protein-synthesizing apparatus. Although reports of the presence of DNA in mitochondria and chloroplasts appeared periodically in the scientific literature since the 1920s, such an astounding proposition was not generally accepted until

FIGURE 22–17 DNA fibrils in mitochondria. (Courtesy of Dr. E. G. Pollock.)

the 1960s, when H. Ris and M. Nass independently verified the existence of DNA fibrils in these organelles in animal and plant cells using electronmicroscopic methods originally developed to visualize DNA in prokaryotic cells (Fig. 22–17). Little attention was given to the possibility that mitochondria and chloroplasts might contain ribosomes and other elements involved in protein synthesis until the presence of DNA in these organelles was established. With such an impetus provided, it did not take long before ribosomes were indeed identified in and isolated from chloroplasts and mitochondria.

Chloroplast Ribosomes

Chloroplast ribosomes have a sedimentation coefficient of 70 S and consist of 50 S and 33 S subunits. In this respect, chloroplast ribosomes are similar to those of prokaryotic cells but distinct from eukaryotic cytosol ribosomes. The large subunit of the chloroplast ribosome contains 5 S and 23 S RNAs and the small subunit 16 S RNA (Table 22–6).

Mitochondrial Ribosomes

Unlike chloroplast ribosomes, which are similar in all groups of organisms studied, mitochondrial ribosomes are quite heterogeneous. With respect to their disposition within the mitochondrion, the ribosomes occur either free in the *matrix* of the organelle or are associated with the cristael membranes. (Ribosomes attached to the cytosol side of the outer mitochondrial membrane are cytoplasmic ribosomes possibly engaged in vectorial synthesis of certain mitochondrial proteins; see below.)

Although mitochondrial ribosomes from all sources studied consist of two subunits, there is considerable variation in the sizes of the subunits and the monomers formed from them (Table 22–6). Mitochondrial ribosomes of yeast, fungi, protists, and higher plants are characterized by a sedimentation coefficient of 70 to 80 S, whereas those of animal cells have a sedimentation coefficient of only 50 to 60 S. Unlike the ribosomes of chloroplasts, prokaryotic cells, and eukaryotic cytoplasm, mitochondrial ribosomes contain only *two* species of rRNA.

Protein Synthesis in Chloroplasts and Mitochondria

The amounts of DNA present in chloroplasts and mitochondria are only about 10 to 15% of that necessary

TABLE 22–6 PROPERTIES OF CHLOROPLAST AND
MITOCHONDRIAL RIBOSOMES

	Sedimentation Coefficient	
	Particle	RNAs
Chloroplast ribosomes		
Monomer	70 S	
Large subunit	50 S	5 S and 23 S
Small subunit	33 S	16 S
Mitochondrial ribosomes		
Animal cells		
Monomer	50–60 S	
Large subunit	40–45 S	16–18 S
Small subunit	30–35 S	12–13 S
Yeast, fungi, and protists		
Monomer	70–80 S	
Large subunit	50–55 S	21–24 S
Small subunit	32–38 S	14–16 S
Higher plant cells		
Monomer	70–80 S	
Large subunit	50–60 S	>23 S
Small subunit	40–44 S	>16 S

to encode the hundreds of different proteins present in these organelles. Therefore, most of the proteins of chloroplasts and mitochondria are the products of genetic information in the cell nucleus. Experimentally, ribosomal RNAs of chloroplasts and mitochondria can be shown to hybridize with their organellar DNA, and it is generally agreed that these RNAs are synthesized in the organelles. The origins of only a small number of chloroplast and mitochondrial proteins have been established to date, but the results are interesting and perhaps also surprising. Most, if not all, of the proteins that make up the organelles' ribosomes are synthesized in the cytoplasm of the cell and apparently make their way into the organelle for assembly (along with rRNA) into ribosomes. Entry into the organelle may be via vectorial discharge. This notion is supported by the observation that many mitochondrial proteins are synthesized with a transient *leader peptide*. The leader is removed after the protein is imported into the mitochondrion. The synthesis of several organelle enzymes, including *ribulosediphosphate carboxylase* (of chloroplasts), *cytochrome oxidase*, and an *ATPase* (of mitochondria), appears to be a "joint operation" of the organelle and the cytoplasm. For example, of the seven polypeptides that make up *cytochrome oxidase*, four are synthesized on cytosol ribosomes and three on mitochondrial ribosomes. There is

some evidence that a few of the chloroplast membrane proteins are synthesized on chloroplast ribosomes.

Much remains to be learned about the total function of organelle ribosomes; however, it is clear that though most chloroplast and mitochondrial proteins are the products of the nucleocytoplasmic protein-synthesizing machinery of the cell, a small number of proteins and portions of certain multisubunit enzymes are the products of genetic material intrinsic to the organelle and the result of translation on organelle ribosomes.

MECHANISM OF PROTEIN SYNTHESIS

Thousands of experiments carried out during the past 25 years involving hundreds of scientists have slowly and painstakingly revealed the intricate details of the mechanism by which proteins are synthesized in prokaryotic and eukaryotic cells. The overwhelming majority of these studies were conducted using two particular kinds of cells, namely, the bacterium *E. coli* and the mammalian *reticulocyte*. Bacterial cells represent highly desirable sources for studying protein synthesis because the cells themselves are readily obtained and conveniently cultured in the laboratory. Moreover, because the ribosomes of bacteria are not attached to intracellular membranes, they are readily isolated from disrupted cells.

The mammalian immature red blood cell or reticulocyte has been the overwhelming favorite among scientists studying protein synthesis in eukaryotic cells for a number of very important reasons; these are best appreciated by briefly considering the origin and features of this unusual cell.

In mammals, red blood cells are produced in the bone marrow and pass through a number of characteristic developmental stages before the mature red cell or *erythrocyte* enters the circulating blood (see Chapter 24). In its early stages of development, the red blood cell possesses most of the structural elements that characterize typical animal cells (nucleus, mitochondria, lysosomes, endoplasmic reticulum, etc.), but nearly all of these structures are lost by the time the reticulocyte stage is reached. The reticulocyte has only a small number of ribosomes, soluble enzymes, and other soluble constituents with which it completes the synthesis of hemoglobin begun at earlier stages. Indeed, hemoglobin accounts for nearly all the protein being synthesized by the cell, and in this respect the reticulocyte is a more desirable source for studying protein synthesis than bacteria, which synthesize many different proteins. Because the reticulocyte contains no organelles other than ribosomes, the latter are readily isolated following lysis of the cell. The cell is called a reticulocyte because its cytoplasm displays a fine reticulum when stained with certain basic dyes (such as methylene blue), the reticulum being formed in part by precipitation of the residual cytoplasmic RNA and ribosomes.

In their final stages of maturation, reticulocytes are released into the bloodstream from the bone marrow and complete their maturation during the ensuing few hours. In normal individuals, reticulocytes rarely account for more than about 2% of all circulating red blood cells.

The separation of reticulocytes from other bone marrow cells or from erythrocytes in order to follow protein (i.e., hemoglobin) synthesis is no simple matter, and for this reason marrow tissue and normal peripheral blood are rarely used as the source of reticulocytes. Instead, reticulocytes are obtained using the following procedure. The experimental animal (usually a rabbit or rat) is rendered severely anemic either by removing a large portion of its blood or by introducing a hemolytic agent (typically, phenylhydrazine) into its bloodstream. The hemolytic agent quickly produces an extensive intravascular hemolysis. In both instances, the resulting anemia is followed within several days by a marked increase in red blood cell production in the bone marrow and the premature release of reticulocytes into the circulating blood. The bloodstream becomes literally flooded with reticulocytes, which may account for more than 90% of all circulating red blood cells in severely anemic animals. Thus, large numbers of reticulocytes can easily be obtained by removing a blood sample from these anemic animals.

Protein synthesis can be studied using either intact cells or a "cell-free" system. That is, under appropriate experimental conditions, not only do *whole* cells incorporate amino acids into new protein but disrupted cells or simply isolated ribosomes supplemented with all the requisite soluble components (e.g., amino acids, tRNA, mRNA, enzymes, cofactors, etc.) also carry out protein synthesis. In recent years, other cells, including HeLa cells, liver cells, and yeast cells, have been employed to study protein synthesis, and these studies have confirmed most of the observations originally made using *E. coli* and reticulocytes. Some important differences do exist in the mechanisms of protein synthesis in prokaryotic and eukaryotic cells, but the overall process is fundamentally the same. Those differences that do exist will be noted below as we consider the details of this process.

Protein synthesis involves a number of distinct and sequential steps including (1) **activation** of amino acids, (2) formation of an **initiation complex** between messenger RNA and the ribosomal subunits, (3) polypeptide chain **initiation**, (4) chain **elongation**, (5) chain **termination** and **release** of the completed polypeptide, and (6) **dissociation** of the messenger RNA–ribosome complex. Before we consider each of these stages individually, it is worthwhile for perspective to discuss first the pioneering studies of H. M. Dintzis, who in the early 1960s established the *linearity* of chain elongation and determined the *direction* in which polypeptide chain assembly takes place. His work not only provided an insight into the complexity of the process yet to be revealed, but his brilliant selection of methodology served as a model and guide for many of the studies subsequently carried out by other scientists investigating the mechanics of protein synthesis.

Linearity and Direction of Polypeptide Chain Assembly—The Experiments of H. M. Dintzis

Dintzis' experiments were carried out to determine whether assembly of a polypeptide chain (1) began at one end and proceeded sequentially toward the other (and if so, which end was synthesized first), (2) began near the middle and then grew toward both ends simultaneously, or (3) occurred simultaneously at several (or many) points along the polypeptide chain with the eventual linking up of all segments. For his studies, Dintzis used reticulocytes isolated from the blood of rabbits that had been made anemic by phenylhydrazine injection.

To follow hemoglobin synthesis in these cells, Dintzis used radioisotopically labeled leucine. Leucine was selected because this particular amino acid is more or less uniformly distributed through the primary structure of human alpha and beta globin chains (see Chapter 4), and although the primary structure of rabbit hemoglobin had not been worked out at the time, there was good reason to suspect that it would be quite similar to that of human hemoglobin. This supposition turned out to be correct.

As a starting point, Dintzis proposed a model for protein synthesis according to which the polypeptide chains are assembled in sequence beginning at one end. Therefore, at any arbitrarily selected instant in time, say, t_0, we would expect to find polypeptide chains of various lengths (i.e., varying degrees of completion) attached to the mRNA–ribosome complexes in the cell. Such partially completed polypeptide chains are called *nascent* chains. If at t_0 cells are briefly incubated in a medium that permits continued nascent chain growth, and if radioactive amino acids are included in that medium, then we would expect that at time t_1 each nascent chain would have increased in length by adding a section containing the radioactive label. If the time period in which the cells are incubated in labeled medium is sufficiently short in comparison to the time required for synthesis of a whole polypeptide, then only a few chains will be completed in this interval and released from the ribosome (i.e., those nascent polypeptide chains that were already near completion at the time the incubation was begun). The remaining chains will still be attached to their respective mRNA–ribosome complexes. For short incubations, the radioactivity present in completed and released chains should therefore be confined to those regions of the chain synthesized *last*.

It should be clear from the preceding discussion that the longer the time interval of incubation in labeled medium (i.e., from t_0 to t_2 or t_3, etc.), the more chains will be completed and released and therefore the further "back" along the polypeptide's primary structure will radioactive label be found. That is, for any period of incubation equal to or shorter than the time required for complete assembly of whole polypeptide chains, one should find a "gradient of radioactivity" among those chains that were completed in that time interval, such that the highest radioactivity is toward the end of the chain synthesized last and the lowest radioactivity is toward the end synthesized first. Figure 22–18 depicts this concept in diagrammatic terms.

If incubation in the presence of radioactive amino acids is carried out for a period of time greater than that required for the assembly of a whole chain, then not only will all those chains already begun at t_0 be completed using the label but *new* chains containing the label throughout will also be synthesized and released. Therefore, after prolonged incubation in the presence of labeled amino acids, we would expect to find *no* "gradient of radioactivity" in completed chains. Instead, the radioactive amino acids would be more or less distributed uniformly through the entire length of most of the released polypeptides.

FIGURE 22–18 Model according to Dintzis for the linear growth of nascent polypeptide chains. Nascent (N), unlabeled chains at the time t_0 are shown in black. Radioactive segments added in separate experiments by times t_1, t_2, etc., are shown in color. Polypeptide chains completed and released from ribosomes are shown to the right and those remaining nascent are indicated by N. In this hypothetical polypeptide, trypsin digestion produces five polypeptide segments (see text for further explanation).

To test this model, it is necessary to isolate the polypeptide chains synthesized during the periods of incubation and to determine and compare the amounts of radioactive amino acids in various segments along their total lengths.

Dintzis incubated the rabbit reticulocytes at 15°C in a medium that included ^{3}H-labeled leucine, as well as all the other materials necessary for continued hemoglobin synthesis. He selected 15°C rather 37°C (the normal environmental temperature of these mammalian cells) because at this lower temperature protein synthesis is sufficiently slowed to permit these experiments to be carried out more easily. After incubation for varying periods of short duration, the reticulocytes were removed, washed, and lysed and the lysate separated into three fractions by centrifugation. A low-speed centrifugation removed plasma membranes and other large particulate debris from the lysate and a high-speed centrifugation of this supernatant provided a pellet containing ribosomes with nascent poly-peptides and a second supernatant containing hemoglobin—including those molecules nascent at t_0 but completed and released during the period of incubation.

The hemoglobin from these experiments was dissociated into its constituent heme and globin parts and the globin separated into alpha and beta chains by ion-exchange column chromatography. The alpha and beta chains were then treated with the proteolytic enzyme *trypsin*, which cleaves peptide bonds on the α-carboxyl side of lysine and arginine residues. Therefore, each chain is split into a specific number of peptide segments. For clarity, only five such segments are depicted in Figure 22–18, although rabbit alpha and beta globin chains regularly provided 35 segments. These peptides were then separated from one another by a technique called "fingerprinting," which combines paper electrophoresis with paper chromatography to produce a two-dimensional distribution of separate peptides across the sheet of filter paper. Having separated the peptide fragments from one another, the next task was to determine their specific ^{3}H-labeled leucine contents and to compare the results after varying periods of incubation in labeled medium. Not all peptide fragments produced by trypsin digestion of globin chains would be expected to contain leucine residues, and in fact only nine fragments in each chain did. Obtaining quantitative data on the specific radioactivity in each peptide fragment, comparing the results for several experiments, and determining whether these results support or contradict the proposed model for chain assembly required that two major problems be solved. One problem is that the total yield of peptide fragments unavoidably varied from one experiment to the other as a result of differential losses at each stage in the isolation procedures. The other problem is that if the model is correct, the radioactivity of an ^{3}H-leucine-containing peptide fragment would vary not only as a function of its position along the primary structure of the globin chain but would depend also on the number of amino acid positions in that fragment normally occupied by leucine. That is, it is necessary to *compensate* in some manner for the differential numbers of leucine residues in each peptide fragment.

Dintzis solved these problems by using what he called an "internal standard." At the end of each incubation interval, he added hemoglobin that was uniformly labeled with ^{14}C-leucine to the ^{3}H-leucine-containing samples, and the *mixture* was then carried

through the stages of digestion and fingerprinting. The uniformly labeled hemoglobin was prepared by long-term incubation of reticulocytes in medium containing ^{14}C-leucine; this yielded a preparation in which, for each hemoglobin molecule, either *all* the leucine positions were occupied by the radioactive form of the amino acid or *none* of them was labeled. (This would depend on whether the hemoglobin molecule was synthesized and released before or after the reticulocytes were placed in the labeled medium.) By expressing the radioactivity of each resulting peptide fragment as the ratio ^{3}H-leucine:^{14}C-leucine, Dintzis simultaneously circumvents both the problems of *differential losses* of material during the course of the experiment and the *differential numbers* of leucine residues in the peptide fragments.

The kinds of results obtained by Dintzis are shown in Figure 22–19. The numbers assigned to each peptide to identify and compare them in different "fingerprints" are more or less arbitrary. Dintzis arranged the data for each of the labeled peptides from very short (e.g., 4 minute) incubations to yield a curve showing increasing radioactivity as a function a selected peptide sequence. This in itself is not significant, as any collection of numerical data can be arranged in order of increasing value. What was important was that if the selected peptide sequence was then held constant when plotting the data from a series of incubations of longer duration, the resulting curves constituted a family having diminishing slopes (Figure 22–19). Examining the graph for alpha chains in Figure 22–19, it can be seen that after a 4-minute incubation, appreciable radioactivity was recovered in only four peptide fragments (i.e., numbers 14, 31, 22, and 16). As a result of longer incubations, proportionately more radioactivity was found in these peptides and in others that were not labeled at 4 minutes. Even after 60 minutes of incubation, a gradient of radioactivity still persisted for the peptide sequence, although by this time the slope was approaching zero. These data are consistent with the model of linear growth proposed by Dintzis and are not consistent with the other alternatives. Accordingly, peptide fragment 16 would be near the end of the polypeptide chain synthesized last and peptide fragment 21 would be near the end of the chain synthesized first. It can also be seen from the data in Figure 22–19 that after a 7-minute incubation in radioactive medium, some labeled leucine was found in all the peptides, indicating that only 7 minutes are required for the synthesis of a complete alpha or beta globin chain at 15°C. An ex-

FIGURE 22–19 Distributions of radioactivity among the peptides produced by tryptic digestion of alpha and beta chains of rabbit hemoglobin released from reticulocyte ribosomes after various periods of incubation at 15°C (see text for explanation).

trapolation of this to the normal environmental temperature for these cells would suggest that about 1.5 minutes is required for complete chain assembly at 37°C. Because the alpha and beta globin chains of rabbit hemoglobin each contain about 150 amino acids, this corresponds to the elongation of the chain at an average rate of about two amino acids per second.

It was subsequently found that peptide fragment 16 from the alpha globin chain digest contained the *C-terminal* amino acid. This indicated that it is the C-terminus that is synthesized last and that synthesis must therefore begin with the *N-terminal* amino acid of the polypeptide.

At about the same time that Dintzis was carrying out his experiments, J. Bishop, J. Leahy, and R. Schweet, also working with rabbit reticulocytes, reached similar conclusions about the direction of polypeptide chain assembly. The N-terminal amino acid of the globin chains is valine. Bishop, Leahy, and Schweet incubated reticulocytes in ^{14}C-labeled valine and then isolated the reticulocyte ribosomes together with their nascent globin chains. The ribosomes were then incubated in an *in vitro* system providing for continued growth of the nascent chains but containing ^{12}C-valine (i.e., ordinary valine). After incubation, the amount of N-terminal ^{14}C-valine was compared with that in other regions of the globin chains completed and released from the ribosomes and was found to be significantly higher. These results supported the concept that chain growth began at the N-terminus. A. Yoshida and T. Tobita, studying the synthesis of an amylase from bacteria, and R. E. Canfield and C. B. Anfinsen, studying egg white lysozyme synthesis,

also reached similar conclusions about the direction of protein synthesis.

In addition to showing the regular progression of polypeptide addition to growing globin chains, the data of Dintzis (and later Naughton and Dintzis, and others) also reveal differences in the *instantaneous* rates of chain elongation along the polypeptide. This was first suggested by S. W. Englander and L. A. Page, who proposed that curves such as those obtained in the pulse-label experiments of Dintzis were also *profiles of nascent chain lengths at t_0*. They noted that the increment of radioactivity between one leucine-containing peptide fragment and another reflected the number of nascent chains having their *growing ends between these two leucines at t_0* and that the slopes at various points along the curve are *inversely proportional* to the rates of chain growth through these points. Consider as an example the 7-minute curve for beta chains shown in Figure 22–19. According to Englander and Page, the rate of chain elongation through the region of the chain containing peptide 1 would be considerably faster than the rate of chain growth through the region containing peptide 12. (Compare the slopes of the curve in these two regions.) If a uniform (i.e., constant) rate of growth occurred over the entire length of the polypeptide, the data resulting from pulse-label experiments would yield a family of straight lines. Using bone marrow cells and carrying out pulse-label experiments similar to those of Dintzis, R. M. Winslow and V. M. Ingram obtained similar findings for the synthesis of alpha and beta chains of human hemoglobin A, namely, that the rate of chain growth is greater during the synthesis of the first half of the polypeptide than during the synthesis of the second half. Some hypothetical curves based on pulse-label experiments of the type conducted by Dintzis, Winslow and Ingram, and others are shown in Figure 22–20, along with an explanation of what these curves imply about the rates of chain growth and the instantaneous distribution of nascent

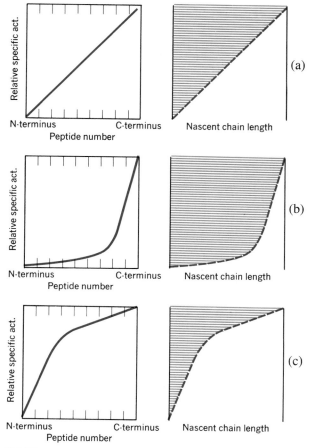

FIGURE 22–20 Hypothetical curves from pulse-label experiments similar to those of Dintzis showing implications of these curves regarding the rates of chain growth and nascent peptide chain lengths at t_0. Curve *a* would result from a situation in which there were equal numbers of nascent chains of all lengths at t_0, the growth rate being constant. Curve *b* (which is similar to Dintzis' results with hemoglobin) would result from a case in which there were more long nascent chains than short nascent chains, the growth rate being rapid at first but slowing down toward the end of the chain. Curve *c* would result from a case in which there were more short chains than long chains, the growth rate being slow at first but increasing toward the end of the chain.

chain lengths on the cell's ribosomes. A. J. Morris later showed that the decreases in the rate of assembly of globin chains occur specifically in the vicinity of amino acids 40, 57, 89, and 120–145 of the primary structure. Because it is clear from the hemoglobin data that polypeptide chain growth does not necessarily proceed at a constant rate, we would expect that in polysomes the distances between successive ribosomes along the strand of messenger RNA would not be equal but would reflect the differential rates of translation of the mRNA code; electron-microscopic studies indicated that this is, in fact, the case.

There are several possible explanations for these findings. For example, it could be that not all mRNA codons are translated at the same rate and that the rate of chain elongation over a given region of the polypeptide's primary structure depends on the types and amounts of amino acids and tRNAs present in the cell. It is also believed that for most proteins the assumption of tertiary structure begins during the course of chain growth, and this, too, might influence the rate of chain elongation.

Processing and Structure of the Transfer RNA

The first stage in the incorporation of an amino acid into a growing polypeptide chain involves the "activation" of the amino acid, that is, the enzymatic attachment of the amino acid to a specific transfer RNA molecule capable of inserting that amino acid into its appropriate position in the polypeptide chain being assembled on the mRNA–ribosome complex. Each tRNA molecule is specific for a particular amino acid. In a given tissue or cell, each amino acid-specific tRNA can exist in multiple forms called **isoaccepting species.** For example, *E. coli* contains five different (i.e., "isoaccepting") tRNAs capable of combining with leucine. Altogether there may be as many as 50 different tRNAs in a cell or tissue.

tRNA Processing   Transfer RNA is produced in precursor form by *RNA polymerase* transcription of DNA (see Table 22–5). Primary transcripts typically consist of about 100 nucleotides, 20 to 30 of which are removed from the 3' and 5' ends of the transcript during processing. In some cases, a single primary transcript contains the sequences of more than one tRNA molecule (as many as six) and these are separated during processing. All mature tRNAs contain the sequence C-C-A at the 3' end of the molecule. In some bacteria, processing at the 3' end of the primary tran-

TABLE 22–7   NUCLEOSIDES OF tRNA

Nucleoside	Symbol
Unmodified	
Adenosine	A
Uridine	U
Guanosine	G
Cytidine	C
Modified	
Inosine	I
Methylinosine	I^{Me}
Methylguanosine	G^{Me}
Dimethylguanosine	G^{DiMe}
2'-O-Methylguanosine	$G^{2'-OMe}$
Thymine riboside	T
Pseudouridine	Ψ
Dihydrouridine	D
Acetylcytidine	C^{Ac}
Methylcytidine	C^{Me}
Methyladenosine	A^{Me}

script involves the enzymatic removal of nucleotides one at a time until the C-C-A sequence is reached. However, the primary transcripts produced in eukaryotic cells lack the C-C-A sequence; in these cells, C-C-A is *added* to the 3' end of the processed molecule by the enzyme *tRNA nucleotidyl transferase*. Nucleosides at various positions in the primary structure may be modified enzymatically (see Table 22–7) to produce the final tRNA product capable of aminoacylation. In some species of tRNA, as many as 16% of the bases may be modified.

Structure of tRNA   The first successful purification of a tRNA species was achieved by R. W. Holley using countercurrent distribution; in 1965 Holley reported the primary structure of yeast alanine tRNA. The primary structure was determined using small polynucleotide fragments produced by enzymatic digestion of the isolated tRNA by pancreatic *ribonuclease* and *phosphodiesterase*. Since Holley's pioneering work, more than 75 tRNAs have been fully sequenced and all exhibit similar primary, secondary, and tertiary structures. Holley was awarded the Nobel Prize in 1968 for his pioneering studies.

The tRNAs contain a linear sequence of 70 to 80 nucleotides that can be arranged to form the classical "cloverleaf" pattern shown in Figure 22–21a and originally proposed by Holley. There are five folded regions: the **amino acid (or acceptor) arm**, the **dihydrouridine (DHU) arm**, the **anticodon arm**, the **TψC arm**, and the **extra (or variable) arm**. Each arm con-

sists of a double-helical *stem* stabilized by base pairing. All except the amino acid arm posses a *loop* region containing unpaired bases. With only a few exceptions, the base pairing that creates the secondary structure of the helical regions is of the conventional Watson–Crick type involving hydrogen bonds between A and U and between G and C.

As seen in Figure 22–21a, certain positions are invariant or semi-invariant among all tRNAs sequenced. (The term "semi-invariant" is used to denote a position invariably occupied by the same *type* of base, purine or pyrimidine.) For example, the four unpaired bases that terminate the sequence at the 3' end of the molecule are always purine-C-C-A; the last of these bases (i.e., adenine) forms the bond with the amino acid (see later). Most of the invariant and semi-invariant positions are found in the DHU loop and in the TψC loop. The invariant residues form hydrogen bonds with one another that are crucial to the maintenance of the characteristic tertiary structure of the tRNAs and also provide recognition sites for interactions with enzymes and with the ribosome. The loop of the anticodon arm contains seven bases, three of which form the anticodon.

One of the characteristic features of tRNA is that a large proportion of the nucleosides are modified. Table 22–7 lists some of the more than 40 modified nucleosides regularly occurring in the tRNA. Most modifications involve methylation of the regular base (i.e., A, U, G, and C) or methylation of the 2' hydroxyl oxygen of the riboses. The role (or roles) of the modified bases is not known with certainty, but suggested roles include the *prevention* of base pairing (1) with the tRNA molecule in order to provide a characteristic tertiary structure and (2) between tRNA and mRNA during translation. It is interesting that the base at

FIGURE 22–21   Structure of tRNA. (a) Cloverleaf pattern showing secondary structure resulting from hydrogen bonding (i.e., •—•) in the helical stems of each arm. Invariant and semi-invariant positions are indicated with the nucleoside symbol (see Table 22–7). Pu, purine, Py, pyrimidine; G*, guanosine or 2'-O-methylguanosine; A*, adenosine or 1-methyladenosine; α and β are variable regions containing up to four nucleosides. (b) Tertiary structure of tRNA proposed by Kim. (c) Rearrangement of the cloverleaf secondary structure to more clearly show the L-shaped tertiary structure; ribbonlike regions form helical segments through hydrogen bonding. (d) Stereo pair of RNA (by permission of Dr. S. H. Kim).

TABLE 22–8   DEGENERACY IN CODON–ANTICODON
RECOGNITION

I	U, C, or A
A	U only
G	U or C
C	G only
U	A or G

the 3' side of the anticodon is nearly always a modified purine when the first base of the corresponding codon is either A or U.

During translation, the three bases of the anticodon form hydrogen bonds with the corresponding codon bases of mRNA. An examination of Table 22–1 reveals that a single amino acid may be encoded by two or more codon sequences. For example, the codon for alanine may be GCU, GCC, GCA, or GCG—the *third* base being seemingly unimportant. Keeping in mind that during translation tRNA and mRNA molecules are antiparallel, it would appear that the base occupy-

ing the *first* position of the anticodon may recognize one or more different bases occupying the *third* position in the codon (Table 22–8). (Remember that the numbering of the polynucleotide chain begins at the 5' end and finishes at the 3' end.) Francis Crick refers to this as the "wobble" base, implying that it may orient in different ways in order to accommodate the appropriate base pairing. Recent studies suggest the possibility that the coding performance of the anticodon is enhanced by the specific sequences of nucleotides adjacent to the anticodon and in the nearby anticodon stem.

The most widely accepted model for the tertiary structure of tRNA is that proposed by S. H. Kim and is based principally on X-ray crystallographic studies of phenylalanine tRNA from yeast cells (Fig. 22–21*b*). The molecule has an L shape, with all double-helical regions being right-handed and antiparallel. The amino acid and TψC stems form one continuous double helix, and the DHU and anticodon stems form another. The two helices are perpendicular to each other, thereby forming the L, with the anticodon and C-C-A termini at opposite ends. The molecule is about 20 Å thick, which corresponds to the diameter of the RNA double helix. Figure 22–21*c* relates the tertiary struc-

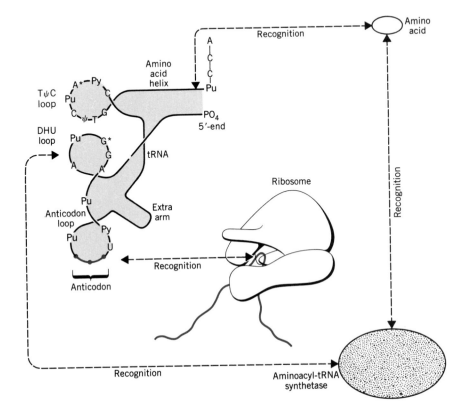

FIGURE 22–22   Relationship between various regions of tRNA and the enzymes, ribosomal proteins, and RNAs with which tRNA interacts during protein synthesis.

ture of tRNA to the cloverleaf secondary structure. The three-dimensional appearance of the molecules is presented in Figure 22–21d.

Different regions of the tRNA molecule appear to serve as recognition and binding sites for various enzymes, ribosomal proteins and RNA, and mRNA during protein synthesis (Fig. 22–22). Although the amino acid is bound to the adenosine nucleoside at the 3' end of the tRNA molecule, this region appears to have little if anything to do with codon–anticodon recognition or binding. This was elegantly demonstrated in a classic experiment by F. Lipmann and F. Chapeville, who prepared a tRNA specific for cysteine (abbreviated tRNA^Cys) that they then enzymatically combined with ^14C-cysteine to form cysteinyl-tRNA^Cys (or cys-tRNA^Cys). They treated the cys-tRNA^Cys with Raney nickel, which removes the sulfhydryl group from the cysteine residue, leaving ala-tRNA^Cys (i.e., a tRNA molecule specific for cysteine but containing bound ^14C-labeled alanine instead). When the radioactive ala-tRNA^Cys was employed in an *in vitro* protein-synthesizing system using a synthetic messenger RNA that coded for the production of a polypeptide rich in cysteine, the radioactive alanine residues were incorporated in place of cysteine. This showed that the specificity for codon recognition resided with the tRNA molecule and not with the amino acid.

Activation of Amino Acids

Amino acid activation involves two major steps (Fig. 22–23). In the first, the α-carboxyl group of the amino acid reacts with ATP to form an *aminoacyl-adenylate* and pyrophosphate. For each species of amino acid, there is at least one specific enzyme called an *aminoacyl-tRNA synthetase* that catalyzes the reaction. The aminoacyl-adenylate formed is not released from the enzyme but remains complexed to it, presumably by a linkage between the enzyme and the R-group of the amino acid. In the second step, the aminoacyl–AMP complex recognizes and reacts with a molecule of tRNA specific for that amino acid to form *aminoacyl-tRNA*, and the enzyme and AMP are released. The reaction between tRNA and the amino acid involves esterification to the 2' or 3' hydroxyl group of the ribose in the terminal adenosine unit of tRNA by the α-carboxyl carbon atom of the amino acid. These reactions occur in the cytosol. The aminoacyl-tRNA thus formed can now participate in protein synthesis on an mRNA–ribosome complex. A number of investigators contributed to the elucidation

of the above reaction sequences, but most notable among them are P. Zamecnik, M. B. Hoagland, P. Berg, and E. J. Ofengand.

Processing and Structure of Messenger RNA

Originally, it was believed that no special mechanism was required to initiate the synthesis of a polypeptide chain. It was supposed that a ribosome would simply attach to the 5' end of an mRNA and proceed to translate successive codons into the polypeptide's primary structure. It is clear now that this is not the case and that a specific mechanism exists for initiating translation and that also prevents *out-of-phase* translation of the mRNA.

The problem of out-of-phase translation warrants further consideration. Suppose that a section of mRNA contains the following codon sequence:

$$\ldots U\,G\,U_{\lfloor}A\,A\,G_{\lfloor}G\,C\,U_{\lfloor}A\,G\,A \ldots$$

This section of mRNA would therefore code for the addition of the amino acid sequence cys-lys-ala-arg to the growing polypeptide chain (see Table 22–1 for the genetic code). However, suppose that this section of mRNA was translated out-of-phase as follows:

$$\ldots U_{\lfloor}G\,U\,A_{\lfloor}A\,G\,G_{\lfloor}C\,U\,A_{\lfloor}G\,A \ldots$$

In this case, the sequence val—arg—leu would be incorrectly incorporated into the primary structure of the polypeptide. The need for a mechanism that ensures that translation is begun in-phase and carried out in-phase is apparent. Such a mechanism relies in part on the unique structure of mRNA.

Messenger RNA molecules isolated from the cytosol of eukaryotic cells are typically about 1500 nucleotides long and consist of both translated and untranslated regions. Because mRNA of this size could at most encode a polypeptide about 500 amino acids long, it is most likely that eukaryotic mRNA is **monocistronic** (or **monogenic**); that is, the mRNA contains the codon sequence for no more than one polypeptide chain. By way of contrast, the mRNAs of prokaryotic cells are quite variable in length, because most mRNA primary transcripts are products of two or more closely linked genes; these are termed **polycistronic** (or **polygenic**) mRNAs. For example, five of the enzymes involved in the metabolic pathway leading to the synthesis of tryptophan in *E. coli* cells are encoded in a single polycistronic mRNA produced by the transcription of 5 closely linked genes (the mRNA con-

FIGURE 22–23 The reaction sequence of amino acid activation. Although in the final product shown the amino acid is esterified to the 3′ hydroxyl position of the terminal ribosome of tRNA, an equilibrium mixture of 2′ and 3′ esters may be formed.

tains more than 7000 nucleotides). Interestingly, the various enzymes that are encoded in a polycistronic message are part of the *same* metabolic pathway and constitute an operon. Neighboring coding regions of a polycistronic message are separated by noncoding *intercistronic* regions containing up to about 40 nucleotides.

The 5' phosphate ends of all eukaryotic mRNAs contain the sequence m7Gpppn'$_m$N"$_m$p . . . , referred to as the "cap" (Fig. 22–24). The 3' ends of most eukaryotic mRNAs contain a "polyA" sequence that is 20 to 250 nucleotides long and is called the "tail." Not all of the remainder of the mRNA molecule may be translated. For example, in addition to the cap and polyA tail, the mRNAs for the globin chains of hemoglobin contain a 150-nucleotide *trailer* segment near the polyA tail that is not translated into globin. Hemoglobin alpha and beta chain mRNAs have an estimated molecular weight of 200,000 to 220,000 and contain 650 to 670 nucleotides. The alpha and beta globin chains are encoded by 423 and 444 nucleotides, respectively; the remainder of the mRNA contains a polyA tail 50 to 75 nucleotides long, a nontranslated sequence of 150 to 175 nucleotides, and, of course, a cap segment. The special chemical nature of the cap region of mRNA is apparently involved in ensuring the proper, in-phase initiation of translation. It is now fairly well established that a *leader* sequence also exists just following the cap region and just to the 5' side of the start codon.

Processing of mRNA Prokaryotic cells such as *E. coli* contain a single chromosome about 3,000,000 base pairs long. This is sufficiently long to encode the primary structures of about 3000 average-size proteins, and this number agrees fairly well with the number of different proteins believed to be present in this bacterium. Thus, most of the DNA of the bacterial cell constitutes structural genes. Eukaryotic cells contain far greater quantities of DNA. For example, the human genome is believed to contain as many as 4,000,000,000 base pairs apportioned among each cell's 46 chromosomes. This amount of DNA is enough to encode as many as ten million polypeptide chains—a number far greater than is believed to be present in human cells. The actual number of different polypeptides is more likely to be from about 30,000 to 150,000. The apparent discrepancy between the num-

7-Methylguanosine triphosphate
(M^{7}Gppp)

FIGURE 22–24 Cap region of mRNA. The 5' end of the molecule contains 7-methyl guanosine triphosphate followed by one or two nucleosides in which the sugar and/or base is methylated.

Chromosome 16　　　Chromosome 11

Exon-1 (93bp)
IVS-1 (95bp)
Exon-2 (204bp)　ζ2
IVS-2 (128bp)
Exon-3 (126bp)

ψβ2

Exon-1 (93bp)
IVS-1 (95bp)
Exon-2 (204bp)　ζ1
IVS-2 (128bp)
Exon-3 (126bp)

ε
Exon-1 (90bp)
IVS-1 (122bp)
Exon-2 (222bp)
IVS-2 (854bp)
Exon-3 (126bp)

ψα1

Gγ
Exon-1 (90bp)
IVS-1 (122bp)
Exon-2 (222bp)
IVS-2 (904bp)
Exon-3 (126bp)

Exon-1 (93bp)
IVS-1 (117bp)
Exon-2 (204bp)　α2
IVS-2 (149bp)
Exon-3 (126bp)

Aγ
Exon-1 (90bp)
IVS-1 (122bp)
Exon-2 (222bp)
IVS-2 (876bp)
Exon-3 (126bp)

Exon-1 (93bp)
IVS-1 (117bp)
Exon-2 (204bp)　α1
IVS-2 (149bp)
Exon-3 (126bp)

ψβ1

δ
Exon-1 (90bp)
IVS-1 (128bp)
Exon-2 (222bp)
IVS-2 (889bp)
Exon-3 (126bp)

β
Exon-1 (90bp)
IVS-1 (130bp)
Exon-2 (222bp)
IVS-2 (850bp)
Exon-3 (126bp)

FIGURE 22–25　Introns and exons of the globin chain genes. The α and ζ chain genes are located on chromosome 16 and are present as duplicates. A "psuedoalpha chain gene" (ψα1) is also present on this chromosome but no transcription or translation products of this gene have been found (i.e., the genes are "silent"). β,δ, ^{A}γ, ^{G}γ, and ε chain genes are on chromosome 11. Also present on this chromosome are silient beta genes ψβ1 and ψβ2. All of the coding sequences or exons of the globin chain genes are interrupted by noncoding intervening sequences or introns. The numbers of base pairs comprising each of the introns and exons are also shown.

ber of structural gene base pairs that "should" be present in a human cell to encode its proteins and the number that *are* in fact present can in part be accounted for by the fact that much of the RNA produced by the transcription of a structural gene does not end up in the mRNA to be translated. That is, much of the structural gene DNA is represented by base-pair sequences that encode RNA that fails to become part of mature mRNA. For example, the genes that encode the beta chains of human hemoglobin have more than four times the number of base pairs as are needed to specify the primary structure of this polypeptide. The extra DNA is represented by two **intervening sequences** or **introns** consisting of 130 and 850 base pairs. The coding sequences, called **exons**, are not only interrupted by introns, but they are also flanked by base sequences that do not encode amino acids for the globin chain. Thus the total coding sequence represents only a small portion of the total gene. Similar findings have been made for the alpha, gamma, delta, epsilon, eta, and zeta globin chains (Fig. 22–25) of human hemoglobin and the various globin chains of other mammalian hemoglobins. Intervening sequences are transcribed into RNA but intron transcripts do not end up in the message because they are removed during *processing* (see below).

Intervening sequences in the structural genes of eukaryotic cells are not at all uncommon. Introns have been identified in the genes for ovalbumin, conalbumin, ovomucoid protein, lysozyme, thyroglobulin, tubulin, albumin, and various immunoglobulins. The ovalbumin genes of chickens are formed from eight exons, each separated by an intron. Beginning at the position of the gene encoding the 5' end of the mRNA molecule and ending with the region encoding the 3' end, the gene contains 7500 base pairs. This length is four times greater than the final mRNA, which is only 1872 base pairs long. If we exclude the cap, the untranslated 3' segment, and the polyA tail, then the gene is seven times longer than the part of the message that is translated into polypeptide.

Whereas the genes of vertebrates have been found to contain as many as 50 introns, many genes lack introns altogether. For example, the genes for the histones have no introns and neither does the gene for interferon. Apparently, introns are not essential constituents of eukaryotic genes.

Introns often separate exons that encode specific functional segments of a protein. This is dramatically illustrated in the case of the globin chains of hemoglobin (Fig. 22–26). For the beta chains, exon-1 encodes

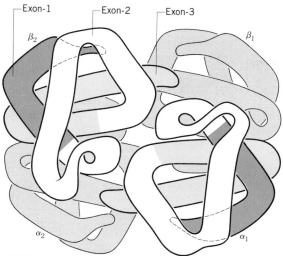

FIGURE 22-26 Model of the hemoglobin molecule showing the sections of each globin chain encoded by exons.

amino acids 1–31 and these form the A and B helices that support the polypeptide's heme pocket (see also Chapter 4). Exon-2 encodes amino acids 32–99; this segment forms the E and F helices that create the heme pocket. Finally, exon-3 encodes amino acids 100–141 and these form the G and H helices. It has been suggested that by moving the exons apart, introns increase recombination frequency and thereby hasten evolution. Indeed, as you move up the evolutionary ladder, structural genes have greater numbers of introns.

At the present time, the synthesis and processing of eukaryotic mRNA is believed to take the following form (Fig. 22–27). The enzyme *RNA polymerase* produces continuous *primary transcripts* of all structural genes to be expressed in the cell. After the primary transcript is about 20 nucleotides long, its free 5' end is modified by addition of the cap. Transcription proceeds through all exon and intron sequences and into the trailer region. For mRNAs that are to possess a polyA tail, transcription proceeds through the trailer to a point about 20 nucleotides past a signal that includes the sequence AAUAAA. The transcript is then released and the polyA tail is added to the free 3' end. Processing is completed as the introns are excised and the remaining exons are ligated to produce the mature messenger RNA. All of these mRNA processing events appear to take place within the nucleus of the cell, as neither the primary transcripts nor the processing intermediates are found in the cytosol. Each

intron begins with the sequence GT and ends with AG, which suggests that these base sequences have something to do with specifying the loci at which transcript cleavage is to occur. All intron base sequences appear to be unique, that is, they are not repeated elsewhere in the cell's genome, although the intron sequences of related genes (e.g., the globin chain genes) are quite similar.

The polyA tails of eukaryotic mRNAs appear to be involved in protecting the mRNAs from enzymatic degradation in the cytosol. mRNAs that lack a polyA tail (such as those for the histones) are stable for only a few minutes, whereas those with polyA tails are stable for many hours.

In Figure 22–27, the processed mRNA molecule contains an untranslated region between the stop codon and the polyA tail. In the case of the abnormal hemoglobin called "Hb constant spring," the alpha chains contain an extra amino acid segment at the C-terminal end of the chain. This is apparently due to a mutation that alters the stop codon with the result that during translation, the ribosome continues into the normally untranslated trailer region of the mRNA until it eventually encounters another stop codon.

Heterogeneous Nuclear RNA The nucleus of a cell contains a broad spectrum of RNAs including unprocessed primary transcripts, partially processed transcripts, and completely processed transcripts. The spectrum of mRNAs found in the nucleus is often referred to as **heterogeneous nuclear RNA** or **hnRNA** (Fig. 22–27), the size of the RNA being related in part to the size of the encoded polypeptide and the numbers and sizes of intron segments. During transcription, hnRNA associates with nuclear proteins to form ribonucleoprotein complexes called **heterogeneous nuclear ribonucleoprotein (hnRNP) particles** or *informofers*. The association of proteins at specific hnRNA sites may be related to the posttranscriptional processing that ensues. Associations between mRNA and proteins also occur in the cytosol; such complexes have been called **informosomes**.

Initiation of Polypeptide Chain Synthesis

Role of Formylmethionine In 1963, J. P. Waller reported the amazing finding that nearly one-half of all proteins in *E. coli* cells have the amino acid methionine in the N-terminal position. (Remember that protein synthesis begins at the N-terminus!) Then, in 1964, K. A. Marcker and F. Sanger discov-

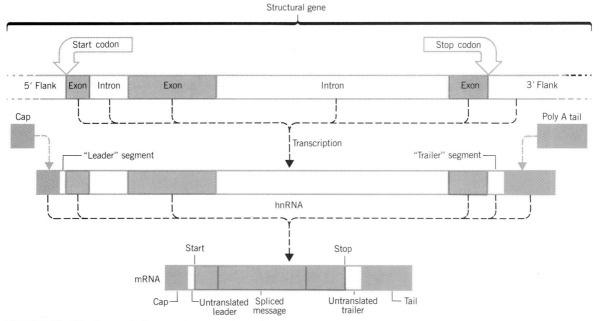

FIGURE 22-27 Transcription and processing of mRNA (see text for details).

ered an unusual species of aminoacyl-tRNA in *E. coli*—**N-formylmethionyl-tRNA**—and suggested that this molecule may play a role in the special mechanism of chain elongation because the presence of the *N*-formyl group in the amino acid (leaving only the α-carboxyl group available for peptide bond formation) would restrict this residue to the N-terminus. The structural formulae of *N*-formylmethionyl-tRNA and methionyl-tRNA are shown in Figure 22–28.

Two transfer RNA molecules specific for methionine are present in *E. coli*, but only one of these can participate in the subsequent enzymatic formylation of the methionine residue. These tRNAs may be denoted as tRNA^{Met} and tRNA$_f^{\text{Met}}$. The formation of *N*-formylmethionyl-tRNA occurs as follows:

$$(1) \text{ met} + \text{tRNA}_f^{\text{Met}} \xrightarrow[\substack{\text{specific aminoacyl-} \\ \text{tRNA synthetase}}]{} \text{met-tRNA}_f^{\text{Met}}$$

$$(2) \substack{\text{met-tRNA}_f^{\text{Met}} \\ +\text{ formate}} \xrightarrow[\substack{\text{formylating} \\ \text{enzyme}}]{} N\text{-formylmet-tRNA}_f^{\text{Met}}$$

The codon for methionine is AUG (Table 22–1), and when this codon occurs anywhere except at the beginning of mRNA, it codes for met-tRNA^{Met}. However, when AUG occurs at the beginning of the message, it

codes for *N*-formylmet-tRNA$_f^{\text{Met}}$ and chain initiation. For this reason, the AUG codon is also called the **initiator** or **start** codon. The picture is somewhat complicated by the fact that the codon GUG, which is a codon for valine, also codes for *N*-formylmet-tRNA$_f^{\text{Met}}$ when GUG occurs at the *beginning* of the message; anywhere else in the mRNA molecule, GUG codes for valine. The interaction that takes place between the aminoacyl-tRNA anticodon and the mRNA codon thus depends on both base pairing and the location of the codon in mRNA. It is now clear that *N*-formylmethionine's role as the initiating amino acid in protein synthesis is not restricted to *E. coli* but is a characteristic of prokaryotes in general.

The process of initiation in eukaryotes is fundamentally similar to that in prokaryotes. As in prokaryotes, there are at least two methionine tRNAs that recognize the AUG codon of mRNA, however, only one these tRNAs can participate in chain initiation. The initiator methionyl-tRNA (met-tRNA$_i^{\text{Met}}$) can be enzymatically formylated *in vitro*, although this does *not* appear to take place under native circumstances, and there are no formylating enzymes present in the eukaryotic cytosol. The other methionyl-tRNA (met-tRNA^{Met}) recognizes AUG codons located internally in mRNA. Of special interest is the observation that the initiation of protein synthesis in mitochondria and

FIGURE 22-28 Chemical structures of methionyl-tRNA and formylmethionyl-tRNA.

chloroplasts takes place in much the same manner as in prokaryotes. The initiating aminoacyl-tRNa is formylated using *formylase*, which is present in the organelles but absent from the cytosol.

Initiation Factors   A number of factors present in the *soluble* phase of the cell are required to *initiate* protein synthesis; others, to be discussed more fully later, are needed for polypeptide *elongation* and for *termination* (Table 22–9). Prokaryote initiation factors were discovered by the observation that washed *E. coli* ribosomes could not translate natural messengers unless supplemented with the wash. Eukaryote initiation factors were discovered by the similar observation that washed reticulocyte ribosomes would not initiate globin synthesis unless the wash was added back.

In eukaryotes, polypeptide synthesis is initiated by a series of steps beginning with the formation of a *ternary complex* between met-tRNA$_i^{Met}$, eIF-2, and GTP, which then attaches to the peptide site (P site) of the small ribosomal subunit. This is followed by attachment of mRNA to the small subunit, with the initiator AUG codon near the 5' end of the mRNA molecule aligned at the peptide site (Fig. 22–29). The next mRNA codon aligns at the amino acid site (A site). Association of mRNA with the small subunit requires eIF-1, eIF-3, eIF-4A, eIF-4B, and eIF-4C and is accompanied by the hydrolysis of one molecule of ATP. eIF-3 appears to remain transiently bound to the small subunit (evidence exists that eIF-4C may also remain bound). Addition of the large subunit requires

eIF-5 and is accompanied by the release of eIF-3 and eIF-2 and the hydrolysis of GTP. The final product, called the 80 S **initiation complex**, may now proceed to translate the remainder of the message. Figure 22–29 serves also to describe initiation in prokaryotic cells if (1) IF-1, IF-2, and IF-3 are substituted for eIF-1, eIF-2, and eIF-3 and (2) fmet-tRNA$_f^{Met}$ is substituted for met-tRNA$_i^{Met}$.

In certain cells, additional factors may be required for the formation of an initiation complex. For example, the initiation of globin synthesis in reticulocytes

TABLE 22-9   INITIATION, ELONGATION, AND TERMINATION FACTORS

	Prokaryotes	Eukaryotes
Initiation factors	IF-1 IF-2 IF-3	eIF-1 eIF-2 eIF-3 eIF-4A eIF-4B eIF-4C eIF-5 eIF-6
Elongation factors	EF-Tu EF-Ts EF-G	EF-1 EF-2
Termination (release) factors	RF-1 RF-2 RF-3	RF

FIGURE 22–29 Stages in the formation of the initiation complex in eukaryotic cells (see text for explanation). For diagrammatic simplicity, the small and large subunits of the ribosome are shown as disklike surfaces. The second codon, GCC (which codes for ala), is shown for illustrative purposes only.

depends on the availability of heme; consequently, the synthesis of globin chains and heme are tightly coordinated.

As noted earlier, nearly half of all proteins in *E. coli* have a methionine residue in the N-terminal position. Because it is not a formylated residue, the formyl group must be removed from the methionine either during or immediately following polypeptide synthesis. Indeed, enzymes that are able to remove formate from formylmethionine residues of polypeptides are present in *E. coli* and other prokaryotic cells. For those proteins that have amino acids other than methionine in the N-terminus (i.e., the majority of cell proteins), a mechanism must exist for removing methionine from the end of the polypeptide. Accordingly, aminopeptidases, which specifically cleave the peptide bond between methionine and the second

amino acid of the polypeptide chain, have been identified. Consequently, for most *E. coli* proteins, formylmethionine is removed from the end of each polypeptide chain by enzymatically cleaving first the formyl group and subsequently the methionine residue itself.

Only a small percentage of eukaryotic proteins have methionine as the N-terminal amino acid. Like prokaryotes, eukaryotes possess an aminopeptidase that removes methionine from the N-terminus of growing polypeptides. Studies with hemoglobin in which valine is the N-terminal residue of both the alpha and beta chains have revealed that the methionine that initially occupies the N-terminal position is removed only after the polypeptide is about 30 residues long. Until that point, the peptide bond linking methionine to valine is apparently protected in some fashion from cleavage by the enzyme. Indeed, other proteolytic enzymes, including papain, trypsin, chymotrypsin, and pronase, are unable to hydrolyze the peptide bonds of short nascent chains but can act on the outer segments of longer chains. This observation (and similar observations made for other proteins) is in accord with the current model of ribosomal structure in which the growing polypeptide chain is protected over that portion of its length residing in the interior of the ribosome.

Chain Elongation

Once the initiator aminoacyl-tRNA is located in the peptide site of the eukaryotic ribosome, chain *elongation* ensues. Addition of the second and subsequent aminoacyl-tRNAs follows a similar pattern. GTP reacts with a soluble-phase elongation factor, EF-1, to form a *binary complex*, which then combines with aminoacyl-tRNA to form a *ternary complex*. The ter-

nary complex interacts with the ribosome so that the aminoacyl-tRNA becomes bound to the vacant amino acid site. This step is accompanied by hydrolysis of GTP and the release of phosphate and an EF-1–GDP complex (the latter can be recycled to EF-1–GTP using additional GTP) (Fig. 22–30). Occupation of both the P and A sites of the ribosome is followed by the formation of a peptide bond between the amino acid bound to tRNA in the P site and the amino acid that just entered the A site. In forming this bond, the α-carboxyl group of the amino acid attached to the terminal adenosine unit of tRNA in the P site is transferred to the free α-amino group of the amino acid held by its tRNA in the A site (Fig. 22–31). The formation of the peptide bond is catalyzed by the enzyme *peptide synthetase*, one of the proteins of the large subunit, and temporarily leaves polypeptidyl-tRNA in the A site. A second elongation factor, EF-2 (also known as *translocase*), catalyzes a complex rearrange-

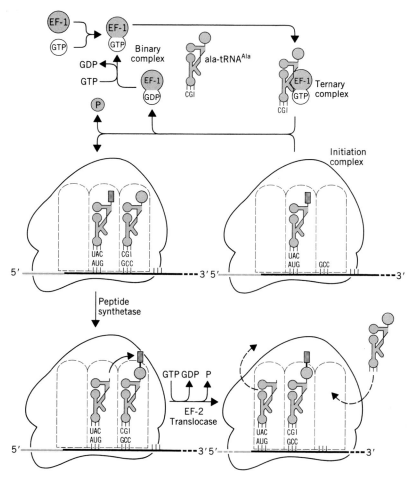

FIGURE 22–30 Chain elongation in eukaryotic cells. Elongation begins with addition of aminoacyl-tRNA to the initiation complex, but the reactions are the same whenever peptidyl-tRNA occupies the P site. For illustrative purposes only, GCC is shown as the second codon of the messenger RNA.

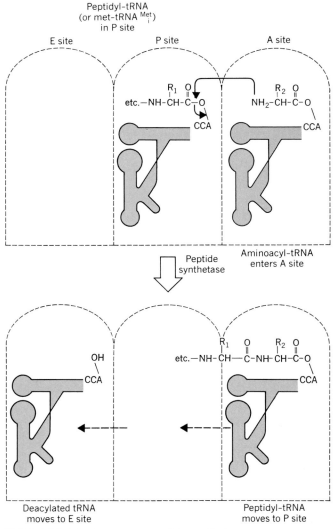

FIGURE 22–31 Action of *peptide synthetase.*

FIGURE 22–32 Utilization (and restoration) of EF-Tu·GTP complex during (and following) aminoacyl-tRNA binding to the A site in prokaryotic cells.

ment of the ribosome in which the deacylated tRNA at the P site is shifted to the exit site (i.e., the E site), the peptidyl-tRNA is shifted to the peptide site, and the messenger RNA molecule moves one codon further along the ribosome. As a result, a new codon appears at the vacant A site. Attachment of the next aminoacyl-tRNA molecule to the A site is accompanied by the release of deacylated tRNA from the E site (Fig. 22–30). The translocation step is accompanied by the hydrolysis of another molecule of GTP. These steps of the elongation reactions are repeated for each new codon of the message entering the A site.

Chain elongation in prokaryotic cells involves three soluble elongation factors: EF-Ts, EF-Tu, and EF-G. Binding of aminoacyl-tRNA to the A site requires an EF-Tu–GTP complex, and binding is followed by release of EF-Tu–GDP and inorganic phosphate. EF-Tu–GTP is replenished by EF-Ts-catalyzed transphosphorylation using GTP as substrate (Fig. 22–32). EF-G of prokaryotes functions in the same manner as EF-2 of eukaryotes.

The enzymatic cleavage of met and/or formate from the N-terminus of prokaryote proteins and met from the N-terminus of eukaryotic proteins takes place after a number of rounds of elongation have already been completed (Fig. 22–33). This leaves the amino acid coded for by the *second* mRNA codon in the "new" N-terminus.

Chain Termination

Chain *termination,* like initiation, involves a specific mechanism and does not occur automatically once the ribosome reaches the end of the message. An examination of Table 22–1 reveals that there are three triplets (sometimes referred to as the "nonsense" triplets) that do not code for any amino acid; these are UAG, UAA, and UGA. Studies with both prokaryotic and eukaryotic cells have implicated these codons in the process of chain termination.

The RNA from certain viruses that infect *E. coli* undergoes a mutation, with the result that virus coat protein is synthesized as incomplete polypeptide chains in an *E. coli in vitro* cell-free system. The incomplete polypeptides contain the N-terminus but not the C-terminus, suggesting that normal synthesis was interrupted and the partially completed chains released from the ribosomes. These mutations are suppressible, that is, *E. coli* mutants can be found that

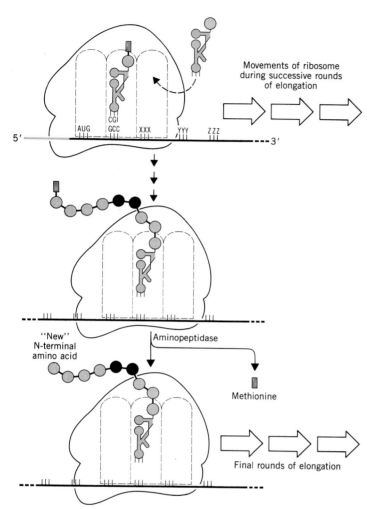

5'

3'

FIGURE 22-33 Removal of methionine from the N-terminal position of a growing polypeptide chain during elongation rounds in eukaryotic cells.

Movements of ribosome during successive rounds of elongation

"New" N-terminal amino acid

Aminopeptidase

Methionine

Final rounds of elongation

support the continued elongation of these polypeptides. However, in the resulting polypeptides, serine (code word UCG) or tryptophan (code word UGG) replaces glutamine (code word CAG). This has been interpreted to mean that the phage mutation involved the change of codon CAG to UAG (i.e., C was mutated to U), which in normal *E. coli* cells resulted in incomplete chains, whereas in the mutant *E. coli* strain the UAG was being read as the codon for serine or tryptophan. Observations of this sort indicated that the normal role for the UAG codon is chain termination. Other studies with *E. coli* suggest similar roles for the UAA and UGA codons.

In humans, a point mutation in which the first position of the UAA codon at the end of the translated region of alpha globin chain mRNA is altered results in

the translation of a major segment of the normally untranslated region (i.e., the ribosome continues past the terminator into the region near the 3' end of the message that normally remains untranslated). The result is the production of alpha chains containing 31 extra amino acids at the C-terminal end of the polypeptide. Hemoglobins formed using these mutant alpha chains (called Hb Constant Spring) function abnormally. Similar point mutations have recently been identified for beta globin chains.

Once the C-terminal amino acid has been added to the end of the polypeptide, the polypeptidyl-tRNA is translocated from the A site to the P site of the ribosome, as described earlier. This moves one of the nonsense or terminator codons into position in the A site (Fig. 22–34). The terminator is not recognized by a particular tRNA or other RNA species. Instead, release of the completed polypeptide from its tRNA requires participation of soluble proteins called **release factors** (RF). One release factor has been identified in eukaryotic cells and three in prokaryotes.

Binding of release factor and GTP to the free A site is followed by the activation of the peptidyl synthetase and translocase systems. The bond linking the completed polypeptide to tRNA is hydrolyzed, the polypeptide and tRNA released from the ribosome, and RF and GTP moved into the P site. Hydrolysis of GTP is followed by release of RF, GDP, and inorganic phosphate. At this time, the ribosome dissociates into its subunits, freeing mRNA. Factor eIF-6 has been implicated in chain termination and dissociation into subunits, however, the role of eIF-6 may be more closely related to *preventing* association of subunits (i.e., eIF-6 may be an "anti-association" factor).

For many proteins, the release of the completed polypeptide chain is followed by the spontaneous assumption of its functional secondary and tertiary structure. For others, all or part of the final secondary and tertiary structure is assumed *as the primary structure is being laid down*. It is conceivable that for some proteins the tertiary structure that is most favored thermodynamically is *not* the functional structure. Therefore, the *progressive assumption* of tertiary structure *during* elongation effectively reduces the number of possible alternative shapes that could be assumed by the polypeptide following release.

Polyribosomes (Polysomes)

Because the globin chains of hemoglobin each contain about 150 amino acids, their messenger RNAs must contain at least 450 nucleotides. Each nucleotide

FIGURE 22–34 Stages of chain termination in eukaryotic cells. Serine is shown as the C-terminal amino acid for illustrative purposes only.

yields a linear translation of 3.4 Å (Chapter 7), so that the mRNA would be at least 1500 Å long. In contrast, the diameter of a ribosome is only about 240–400 Å. These observations led Rich, Warner, Knopf, and Hall in 1962 to propose that the translation of a single mRNA might be carried out simultaneously by several ribosomes (i.e., *polyribosomes*) attached to and moving in succession along the message. For example, in the case of globin chain synthesis, four or more ribosomes could be attached to the mRNA.

Rich and his co-workers incubated rabbit reticulocytes for short periods in a medium containing ^{14}C-labeled amino acids. During this brief incubation, radioactive segments were added to each nascent chain. Following this, the cells were lysed and the lysate fractionated by centrifugation through a sucrose density gradient. Fractions collected from the gradient at the conclusion of centrifugation were examined in two ways: (1) the distribution of ribosomes through the gradient was determined by measuring the ultraviolet light absorption of the ribosomal RNA and (2) the distribution of nascent polypeptides was determined from the radioactivity of the collected fractions. Typical results are shown in Figure 22–35. Two UVL-absorbing regions were identified in the density gradient. The first (i.e., least rapidly sedimenting) peak (fractions 24 to 29), which corresponded to particles

of about 80 S and represented *single* ribosomes, had no radioactivity associated with it. Instead, the radioactivity was distributed over a region of the gradient containing more rapidly sedimenting (i.e., larger) particles (i.e., fractions 10 to 20). This indicated that protein synthesis in reticulocytes took place on structures that were larger than individual ribosomes, and Rich suggested that these were groups of ribosomes held together by mRNA.

When the enzyme *ribonuclease* was added to the lysate prior to centrifugation, the rapidly sedimenting peak disappeared and the first peak increased in size and was now associated with the radioactivity. This result, of course, supported Rich's proposal. Further confirmation came from electron-microscopic examination of the fractions, which revealed that the first peak contained single ribosomes and the rapidly sedimenting peak contained clusters of ribosomes—the further down the gradient the sample was withdrawn for microscopic examination, the larger was the observed cluster size. The predominant size cluster contained five ribosomes (called a *pentamer*), with smaller numbers of clusters containing six ribosomes (*hexamers*) and four ribosomes (*tetramers*).

Subsequently, electron-microscopic studies using negative staining techniques showed that the ribosomes were connected by a thin thread about 10–15 Å thick and about 1500 Å long and were separated by gaps varying from 50 to 150 Å. This corresponded to the diameter of an RNA molecule and the approximate length predicted for the globin messenger. Some polysomes are shown in Figure 22–36.

In the model for polysome function originally proposed by Rich, the several ribosomes move along the mRNA strand, each synthesizing a polypeptide chain. When a ribosome reaches the end of the message, it detaches, while at the other end another ribosome attaches to the mRNA.

Although it is now clear that polysome function in globin chain synthesis is not precisely as Rich predicted, the fundamentals of his model remain valid. The size of a polysome depends on both the length of the mRNA being translated and the amount of time required for initiation, elongation, and termination. For example, S. H. Boyer has shown that polysomes engaged in the synthesis of alpha globin contain an average of four ribosomes, whereas beta globin chain polysomes contain an average of six ribosomes. This occurs despite the fact that both globin chains are about the same size. The difference is due to the lower frequency with which initiation occurs on alpha globin

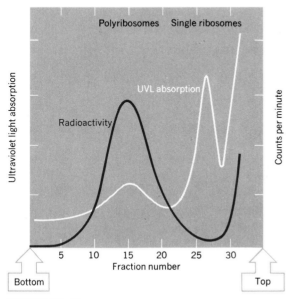

FIGURE 22–35 Distribution of ^{14}C-labeled amino acids and UVL-absorbing material in reticulocyte lysates subjected to density gradient centrifugation (see text for explanation).

FIGURE 22–36 Polysomes. The messenger RNA strands linking neighboring ribosomes of each polysome are indicated by arrows. Magnification, 200,000X. (Courtesy of Dr. O. L. Miller, from O. L. Miller, B. A. Hamkalo, and C. A. Thomas, *Science* **169**, 394 (1970). Copyright © by the American Association for the Advancement of Science.)

chain mRNA (αmRNA), which in the case of globin chain synthesis is the rate-limiting step. Elongation and termination occur at about the same rates for both globin chains. It was noted earlier that equal amounts of alpha and beta chains are produced in normal erythrocytes. If this is so, one might ask how such a balance is maintained in view of the polysomal differences just noted. The balance results in part from the presence in these cells of larger quantities of αmRNA than βmRNA, which thus compensates for differences in the frequency of initiation. The larger quantities of αmRNA result, in turn, from the presence of two pairs of alpha chain structural genes and only one pair of beta chain structural genes (see Fig. 22–25 and Chapter 4).

The current model for polysome function in eukaryotic cells is shown in Figure 22–37 and does not differ dramatically from that originally proposed by Rich. The detachment of the ribosome from mRNA is accompanied by the release of the completed (and probably folded) polypeptide chain. The ribosome immediately dissociates into its small and large subunits, which then enter a common cell subunit pool. During periods of reduced protein synthesis, subunits combine to form a pool of inactive but intact ribosomes (monosomes), but during periods of active cellular protein synthesis, these ribosomes again dissociate into subunits. As noted earlier, the small subunit attaches to the 5' end of the mRNA before the large subunit. There is some evidence that during periods of active protein synthesis, subunits recently released from

FIGURE 22–37 The polyribosome cycle.

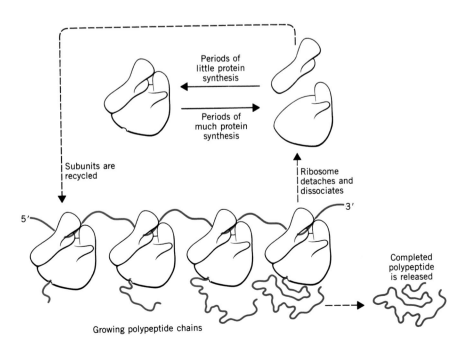

the messenger are *preferentially* reused for translation because of their closer proximity to the 5' end of mRNA than other subunits randomly spread through the cell.

Kinetics of Transcription and Translation in Maturing Erythrocytes.

Because so much is known about hemoglobin and the development of erythrocytes, it is possible to carry out a number of calculations that give us some insight into the frequency and rate of transcription and translation. Because there are about 5.4×10^9 erythrocytes and 0.16 g of hemoglobin in each 1 cm³ of human blood (these values are readily determined in the laboratory), this implies that a single red blood cell contains

$$0.16 \text{ g of hemoglobin per } 5.4 \times 10^9 \text{ cells} = 3 \times 10^{-11} \text{ g of hemoglobin} \quad (22\text{--}1)$$

The molecular weight of human hemoglobin A is about 64,500, and one gram-molecular weight (gMW) would contain 6.02×10^{23} molecules of hemoglobin (i.e., Avogadro's number); hence a single red blood cell contains

$$\frac{(3 \times 10^{-11} \text{ g Hb/cell}) (6.02 \times 10^{23} \text{ Hb molecules/gMW})}{(6.45 \times 10^4 \text{ g Hb/gMW})}$$
$$= 2.8 \times 10^8 \text{ molecules of hemoglobin} \quad (22\text{--}2)$$

It has already been noted that the synthesis of hemoglobin in the maturing red blood cell occurs over a period of 3 days. Knowing this, and using the result of equation 22–2 above, it is possible to calculate the *average* number of hemoglobin molecules whose synthesis is completed *each second* during this period of differentiation. It would be

$$\frac{2.8 \times 10^8 \text{ molecules of Hb/cell}}{(3 \text{ days}) (24 \text{ hr/day}) (60 \text{ min/hr}) (60 \text{ sec/min})}$$
$$= 1.1 \times 10^3 \text{ molecules of hemoglobin completed per second per cell} \quad (22\text{--}3)$$

Therefore, an average of 1100 molecules of hemoglobin are completed in each second of the 3-day maturation period. This, of course, assumes that synthesis is continuous and also uniform over the entire 3 days, which is not actually the case. Therefore, the real value would be greater than this during the period of peak synthesis and lower at other times. However, for purposes of this discussion we can assume continuous and uniform synthesis.

Because there are four globin chains in each hemoglobin molecule, there would be $4(1.1 \times 10^3)$ or 4400 globin chains completed per second in a single red blood cell. The experiments of Dintzis and others indicate that between 60 and 90 seconds are required for the synthesis of a single globin chain. If for convenience we use the larger value, then during any given second there would be

$$(90 \text{ sec}) (4400 \text{ chains/sec}) = 396,000 \text{ chains} \quad (22\text{--}4)$$

of globin in production in the cell. Because the most common form of alpha chain polysome in the cell is the tetramer, this means that there must be 198,000/4 or 49,500 molecules of αmRNA in the cell. The most common beta chain polysome is the hexamer, implying that there are 198,000/6 or 33,000 molecules of βmRNA. This assumes that each molecule of mRNA is stable for the whole 3-day maturation period and is available at the outset of hemoglobin synthesis. As neither assumption is entirely valid, the actual amounts of the globin mRNAs in the cell are most likely considerably higher during peak hemoglobin synthesis. Each erythroblast contains four structural genes for alpha chains and two structural genes for beta chains; therefore, each alpha chain structural gene would have to be transcribed into αmRNA 12,375 times (i.e., 49,500/4) and each beta chain structural gene would have to be transcribed into βmRNA 16,500 times (i.e., 33,000/2).

If the developing red blood cell contains an average of 49,500 molecules of αmRNA, and these are used to produce sufficient alpha globin chains for 2.8×10^8 molecules of hemoglobin, then the average αmRNA is translated 11,313 times, that is,

$$\frac{(2.8 \times 10^8 \text{ molecules Hb}) (2 \, \alpha \text{ globin chains/Hb})}{(49,500 \text{ molecules } \alpha\text{mRNA})} \quad (22\text{--}5)$$
$$= 11,313 \text{ alpha globin chains per } \alpha\text{mRNA}$$

The corresponding translation frequency for βmRNA would be

$$\frac{(2.8 \times 10^8 \text{ molecules Hb}) (2 \, \beta \text{ globin chains/Hb})}{(33,000 \text{ molecules } \beta\text{mRNA})} \quad (22\text{--}6)$$
$$= 16,969 \text{ times}$$

These figures indicate that the globin mRNAs are extremely stable (recent experimental evidence points to a half-life of at least several hours), especially in comparison with prokaryotic mRNAs, which may be translated only once.

In all of the above calculations, we have been considering *averages* only. The synthesis of hemoglobin in the maturing erythrocyte is *not* uniform or continuous throughout development. Instead, hemoglobin synthesis reaches a maximum in early development. Therefore, during the period of maximum synthesis more molecules of mRNA would be required, and the number of globin chains produced from a single mRNA molecule would be lower. Despite this, the high frequency of transcription and translation in this cell is clearly apparent.

Differences between Translation in Prokaryotic and Eukaryotic Cells

In eukaryotic cells, transcription occurs in the cell nucleus and translation occurs later in the extranuclear hyaloplasm. In prokaryotes, which have no nucleus, not only do transcription and translation occur in the same region of the cell but they also occur at the same time.

In bacteria such as *E. coli*, the transcription of DNA is accompanied by the translation of the *nascent messenger RNA*. The mRNA is synthesized beginning with its 5' end, and as soon as the mRNA strand is long enough, a ribosome attaches to the messenger and begins translation. As mRNA synthesis proceeds, more ribosomes attach to the elongating strand to form a polysome. Whereas in eukaryotes mRNA contains the codon sequence for a single polypeptide, prokaryotic mRNAs may be polycistronic (see earlier). Polycistronic mRNAs have multiple initiation codons and ribosomes may attach to any of these independently.

O. L. Miller and others have provided elegant proof of this in the form of electron micrographs such as the one shown in Figure 22–38. In this figure, a portion of the *E. coli* chromosomal DNA appears as a thin filament being actively transcribed by a number of RNA polymerase molecules into mRNA strands. To each of these mRNA strands, ribosomes have attached to form polysomes. In Figure 22–38, it appears that transcription proceeds from left to right along the DNA, as the mRNA length (and therefore polysome size) exhibits a general increase in that direction. Unfortunately, electron micrographs do not reveal the growing nascent polypeptide chains on the ribosomes, because the amino acids that make up the polypeptides are too small to be resolved.

Cotranslational and Posttranslational Protein Modification

Cotranslational Modifications In many cases, a number of changes are made in the structure and organization of polypeptide chains *during* their synthesis; these are called **cotranslational modifications** and include (1) **deformylation**, (2) **amino acid cleavage**, (3) **side chain alteration**, (4) **disulfide bridge formation**, (5) **sugar addition**, and (6) **tertiary folding**.

Deformylation. In prokaryotes and in eukaryotic mitochondria and chloroplasts, the formyl group of the N-terminal methionine of the growing polypeptide chains is enzymatically cleaved (see earlier).

Amino Acid Cleavage. In both prokaryotes and eukaryotes, N-terminal methionine and occasionally other amino acids as well are enzymatically cleaved from the free N-terminus by an aminopeptidase.

Side Chain Alteration. The R-groups of certain amino acids are often altered following inclusion of the amino acid into the growing polypeptide chain. For example, during the synthesis of collagen, certain proline and lysine residues are hydroxylated (to form hydroxyproline and hydroxylysine, respectively). Other amino acid side chains may be phosphorylated (e.g., serine).

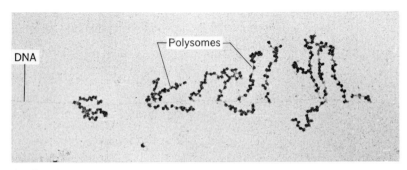

FIGURE 22–38 Visualization of the simultaneous transcription and translation of *E. coli* chromosomal DNA. Magnification, 43,000X. (Courtesy of Drs. O. L. Miller and B. A. Hamkalo, from O. L. Miller, B. A. Hamkalo, and C. A. Thomas, *Science* **169**, 394 (1970). Copyright © by the American Association for the Advancement of Science.)

Disulfide Bridge Formation. Juxtaposed sulfhydryl groups of cysteine residues may be oxidized to form disulfide bonds. This normally occurs after tertiary folding (see below) orients these R-groups into the necessary steric positions.

Sugar Addition. Sugars may be enzymatically attached to certain amino acids during the synthesis and completion of various glycoproteins.

Tertiary Folding. Although some proteins may spontaneously fold to form their biologically active tertiary structure following completion and release of the polypeptide from the ribosome (e.g., *ribonuclease*, see Chapter 4), others undergo tertiary folding during translation. In *E. coli*, tertiary folding of enzymatic polypeptides during their synthesis endows these nascent proteins with catalytic properties prior to termination and release.

Posttranslational Modifications *Posttranslational modifications* are changes that occur in protein structure *after* completion and release of the polypeptide have taken place. Some of the modifications already described as cotranslational may also occur following translation. For example, enzymatic hydroxylation and phosphorylation of amino acid side chains, the formation of disulfide bridges, and the addition of sugars to certain residues may occur following release of the completed polypeptide. Moreover, tertiary folding, although begun during translation, is completed following polypeptide release. Some modifications, however, are characteristically posttranslational; included in

this category are (1) **peptide cleavage**, (2) **quaternary association**, and (3) **addition of prosthetic groups**.

Peptide Cleavage. For some proteins, major changes in structure in the form of cleavage of specific bonds and removal of sections of the polypeptide occur following translation. For example, the A and B polypeptide chains that comprise the insulin molecule (Chapter 4) are produced by posttranslational cleavage of a single translation product (Fig. 22–39) called proinsulin, which has no hormonal activity.

The activation of the zymogen *chymotrypsinogen* to form the digestive enzyme *chymotrypsin* (see Chapter 8) serves to illustrate the level of complexity that posttranslational peptide cleavage can assume. *Chymotrypsinogen* is broken into two polypeptides by the enzyme *trypsin*, the product (which is still linked by disulfide bridges) being π *chymotrypsin* (Fig. 22–40). π *chymotrypsin* acts to catalyze its own conversion to the active digestive enzyme α *chymotrypsin*. This activation involves the removal of two dipeptides of π *chymotrypsin*, producing a product consisting of three interconnected polypeptide chains. Activation of *chymotrypsinogen* occurs in the small intestine, and this is where protein digestion takes place, whereas the pancreas, which is the site of production of the enzyme, secretes it in the zymogen form.

The posttranslational modifications that produce insulin and chymotrypsin also demonstrate that *a protein consisting of more than one polypeptide chain may not be encoded by a corresponding number of mRNAs (or genes!) but may be encoded by a single mRNA (or gene)*. Posttranslational peptide cleavage may produce a series of separate polypeptide chains that ultimately make up the final protein product. Indeed, evidence is at hand that indicates in some prokaryotes and in the case of certain viruses, a single polypeptide chain may be cleaved to produce several individual proteins.

Quaternary Association. As noted in Chapter 4 and in this chapter, some proteins that possess quaternary structure are assembled by the spontaneous interaction of individual polypeptide chains. In the case of hemoglobin, for example, separate alpha and beta chains spontaneously combine to form asymmetric dimers, and these combine to form the functional tetramer. Assumption of quaternary structure is accompanied by the formation of stabilizing bonds between neighboring protein subunits and modification of the individual tertiary structures they previously possessed.

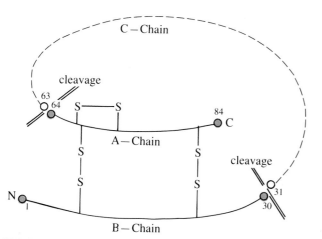

FIGURE 22–39 Posttranslational cleavage of *proinsulin* to form *insulin*.

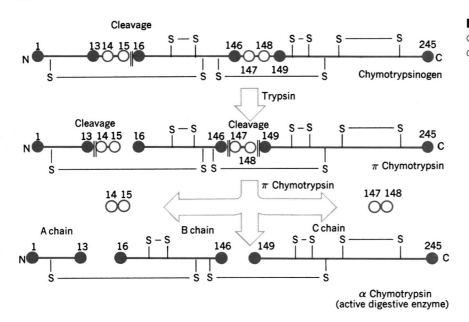

FIGURE 22–40 Posttranslational cleavage of *chymotrypsinogen* to form *chymotrypsin*.

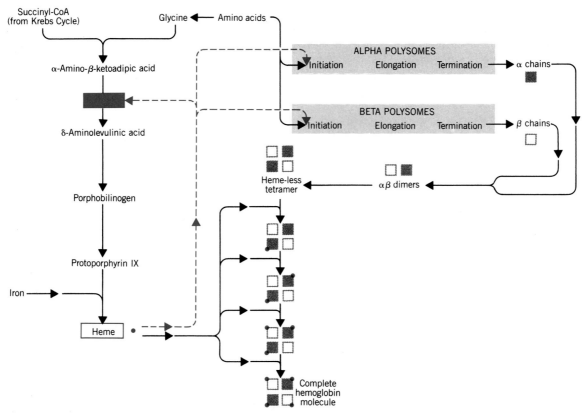

FIGURE 22–41 Synthesis of the hemoglobin molecule (solid arrows) and its regulation (dashed arrows).

Addition of Prosthetic Groups. The prosthetic groups of enzymes and other proteins are attached following release of the completed polypeptide chains, and attachment may be spontaneous or catalyzed enzymatically. In the case of hemoglobin, the insertion of the heme groups occurs *after quaternary association is complete* and begins with alpha globin chains.

The completion of the hemoglobin molecule by the successive attachment of its four heme groups brings to a conclusion a synthetic process about which more is known than for any other complex protein. In this chapter, we have seen that (1) heme regulates the initiation of globin chain synthesis; (2) globin chain elongation does not proceed at a uniform rate but that alpha and beta chains are produced in equal amounts; (3) asymmetric dimers spontaneously associate to form tetramers; and (4) heme insertion is sequential. To this must be added the long-known fact that heme acts as a negative effector of its own synthesis through feedback inhibition of an enzyme catalyzing an initial step in the heme biosynthetic pathway. In this fashion, the production of more heme than can be utilized for assembly of hemoglobin is avoided. Acting in concert, all of these mechanisms provide for the 1:1:1 ratio of alpha chain, beta chain, and heme group synthesis in the maturing red blood cell. Figure 22–41 summarizes our existing knowledge of the regulation of the synthesis of the hemoglobin protein.

Transfer RNA Specialization

When W. F. Anderson and J. M. Gilbert reported in 1969 that addition of isolated tRNA fractions to a reticulocyte cell-free system synthesizing globin chains altered the balance of alpha and beta globin chain production, the question was raised whether cells might have a rather specialized complement of tRNA. That this is indeed the case was borne out by the extensive studies of D. W. E. Smith on the amounts and types of tRNAs present in the reticulocyte. As Figure 22–42 shows, a direct relationship exists between the frequency with which each type of amino acid occurs in the hemoglobin molecule and the abundance in the cell of the tRNAs specific for that amino acid species. This implies that cellular mechanisms exist that coordinate the production of various tRNAs according to the types and amounts of different amino acids present in the proteins of that cell—a most striking implication! This notion is supported by additional evidence of tRNA specialization in other cells, including silk-gland cells, lymphocytes, and cells of the pancreas and liver.

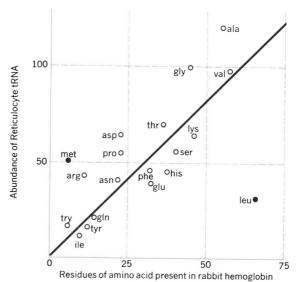

FIGURE 22–42 Transfer RNA specialization in the reticulocyte.

In Figure 22–42, the coordinates for met and leu appear to be exceptions to the linear distribution. For met, there appears to be an excess of tRNA^{Met} (and tRNA$_i^{Met}$) (the additional met residues involved in chain initiation are already taken into account in the data), and for leu, there is a shortage of tRNA^{Leu}. The shortage of leucine tRNA in reticulocytes is believed to be yet another rate-limiting control factor for the production of globin chains in the maturing cell.

INHIBITORS OF PROTEIN SYNTHESIS

Many substances are known to act as inhibitors of various stages of protein synthesis. Included among these are a number of *antibiotics* produced by one strain of microorganism and lethal to other strains of the same or a different species. Some of the best understood inhibitors of protein synthesis are listed in Table 22–10. Because the actions of many of these inhibitors are quite specific, they have proved to be extremely useful tools in the step-by-step elucidation of the mechanism of protein synthesis.

Inhibitors of Both Prokaryotic and Eukaryotic Protein Synthesis

Aurintricarboxylic acid inhibits formation of the initiation complex by preventing the association of mRNA with the small ribosomal subunit. Inhibitors of initiation are readily distinguished from inhibitors blocking other stages of protein synthesis because of the *delay effect* that follows their administration, that is, protein synthesis continues for a short time after administration of the inhibitor, because peptide chains whose growth had already begun are unaffected and grow to completion. *Edeine*, a polypeptide isolated from *Bacillus brevis*, inhibits the binding of aminoacyl-tRNA and N-formylmet-tRNA$_f^{Met}$ (in prokaryotes) to the small subunit. *Fusidic acid* is a steroidal antibiotic; in prokaryotes, it inhibits the binding of aminoacyl-tRNA to the ribosome, whereas in eukaryotes, it inhibits translocation by reacting with elongation factor.

Puromycin was one of the first inhibitors of protein synthesis to have its specific effect determined. This antibiotic mimics aminoacyl-tRNA and binds to the free A site of ribosomes engaged in protein synthesis. Catalytic formation of a bond between the nascent polypeptide and puromycin is followed by the release of the peptidyl-puromycin from the ribosome, as no further elongation is possible. The specific effects of puromycin have been used to advantage for studies of nascent chain length, the kinetics of chain elongation, and the identification of the effects of other antibiotics. *Tetracycline* inhibits protein synthesis by blocking aminoacyl-tRNA binding to the small subunit.

TABLE 22–10 INHIBITORS OF PROTEIN SYNTHESIS IN PROKARYOTIC AND EUKARYOTIC CELLS

	Effective in	
Inhibitor	Prokaryotes	Eukaryotes
Anisomycin	−	+
Aurintricarboxylic acid	+	+
Chloramphenicol	+	−
Colicin E3	+	−
Cycloheximide	−	+
Diphtheria toxin	−	+
Edeine	+	+
Erythromycin	+	−
Fusidic acid	+	+
Pactamycin	−	+
Puromycin	+	+
Ricin	−	+
Sodium fluoride	−	+
Sparsomycin	−	+
Streptomycin	+	−
Tetracycline	+	+
Trichodermin	−	+

Inhibitors Specific for Prokaryotes

Chloramphenicol (chloromycetin) binds to the large subunit of prokaryotic ribosomes and interferes with the functioning of *peptide synthetase*, thereby inhibiting chain elongation. *Colicin E3* inhibits protein synthesis in prokaryotes by interfering in some manner with the functioning of the small subunit. *Erythromycin* binds to ribosomes that are *not* engaged in protein synthesis, preventing their potential participation, but does not bind to ribosomes containing nascent chains (i.e., ribosomes that are part of a functioning polysome). *Streptomycin* was one of the earliest discovered antibiotics and was employed as an agent against bacterial infection for many years before its specific chemical actions were known. Streptomycin binds to protein S12 of the small ribosome subunit, causing release of N-formylmet-tRNA$_f^{Met}$ from initiation complexes (thereby preventing initiation of chain growth) and also causing misreading of the codons of mRNA by ribosomes already involved in chain elongation.

Inhibitors Specific for Eukaryotes

Anisomycin is an antibiotic produced by *Streptomyces* that inhibits peptide bond formation when bound to the small ribosomal subunit. *Cycloheximide* binds to the large subunit, preventing the translocation of tRNA in the A site to the P site. *Diphtheria toxin* (produced by a strain of *Corynebacterium diphtheriae*) inhibits protein synthesis through its action on EF-2 (*translocase*). EF-2 exists in cells in two forms—ribosome-bound and free. Diphtheria toxin acts enzymatically to alter free EF-2, rendering the factor inactive. Ribosome-bound EF-2 is not susceptible to inactivation by the toxin.

Pactamycin (produced by a strain of *Streptomyces*) binds to free small subunits (not to small subunits already part of polysomes), where it prevents initiation by inhibiting binding of met-tRNA$_i^{Met}$ and formation of the initiation complex. The toxic effects of *ricin* (a protein present in the castor bean) have been known for nearly a century. Ricin consists of two polypeptide chains (linked by disulfide bridges), one of which acts as the inhibitor once incorporated into the cell. Ricin acts on the large subunit, preventing formation of the 80 S initiation complex. Like ricin, *sodium fluoride* acts as an inhibitor or initiation; NaF blocks addition of the large subunit to mRNA.

Sparsomycin, another antibiotic produced by *Strep-*

tomyces, inhibits the association of the amino acid moiety of aminoacyl-tRNA from binding to the large subunit and, in so doing, blocks peptide synthetase. *Trichodermin* is the only chemical compound so far identified as a specific inhibitor of the termination stage of polypeptide synthesis.

Inhibitors of Organellar Protein Synthesis

Protein synthesis by mitochondrial and chloroplast ribosomes is also subject to inhibition by certain antibiotics and other chemicals. Shortly after the initial demonstration of organellar protein synthesis, it was found that chloramphenicol, a strong inhibitor of prokaryote protein synthesis, blocks synthesis in mitochondria and chloroplasts, whereas cycloheximide, which blocks eukaryote cytoplasmic ribosomal protein synthesis, is without effect on mitochondrial and chloroplast synthesis. These observations provided added credence for the notion that prokaryotic cells, mitochondria, and chloroplasts have a common evolutionary origin. It is now clear, however, that the picture is considerably more complex. For example, streptomycin, which inhibits prokaryotic protein synthesis, fails to inhibit mitochondrial protein synthesis in yeast cells. Other antibiotics inhibit mitochondrial protein synthesis but have no effect on prokaryotes. Erythromycin inhibits the synthesis of proteins in prokaryotes, yeast mitochondria, and chloroplasts but fails to block protein synthesis in mammaliam mitochondria. The latter observation suggests that the nature of mitochondrial protein synthesis varies among different groups of eukaryotes. In general, mitochondria from higher eukaryotes are more resistant to inhibitors of prokaryotic protein synthesis than are mitochondria from lower eukaryotes. In chloroplasts, protein synthesis is inhibited by the same agents that block this process in prokaryotic cells.

The differential sensitivity of eukaryotic cytoplasmic and mitochondrial ribosomes to specific inhibitors provides a means for examining the sources of certain mitochondrial proteins. The synthesis of a mitochondrial protein in the presence of cycloheximide (a cytoplasmic inhibitor) indicates that the mitochondria are the source of the protein, whereas synthesis of the protein in the presence of chloramphenicol indicates that the mitochondrial protein is produced in the cytoplasm and then moves to the mitochondria. Experimentally, determinations of this sort are carried out by incubating cells in media containing both radioactively labeled amino acids and inhibitor. The synthesis of the mitochondrial protein is manifested by the appearance of radioactivity in the protein later isolated from the mitochondria.

Using this approach, it has been possible to show that perhaps 85 to 90% of all mitochondrial ribosomal proteins are synthesized in the cytoplasm and then are imported into the mitochondria, where, together with mitochondrial rRNA, they are assembled into ribosomes. Of the seven polypeptides that make up the enzyme *cytochrome oxidase*, four are synthesized in the cytoplasm and three are synthesized in the mitochondria.

SUMMARY

Using isotopes of nitrogen and carbon as tracers of amino acid metabolism, a number of investigators working in the 1940s and early 1950s showed that protein turnover in tissues was not simply a function of "wear and tear." Instead, cell protein undergoes continuous breakdown and synthesis, the latter being a property of the **microsomal fraction** of tissue homogenates. It is the **ribosomes** in the microsomal fraction that carry out the assembly of the polypeptide chains that make up proteins, and these tiny organelles are currently the object of intensive biochemical and electron-microscopic analyses. The ribosomes of prokaryotic and eukaryotic cells are made up of two **subunits**, each subunit containing a specific combination of proteins and **ribosomal ribonucleic acids** (*rRNA*). Association of these subunits with **messenger** RNA (mRNA) is followed by **translation** of the mRNA base sequence (i.e., its **codons**) into the primary structure of polypeptides. **Transfer** RNAs (tRNA) combine with specific amino acids and enter the ribosome–mRNA complex. The succession of **aminoacyl-tRNA** species entering the complex and donating amino acid residues to the growing polypeptide is prescribed by the mRNA codon sequence. All the RNAs are **transcribed** from DNA and **processed** (cleaved, trimmed, or otherwise modified) prior to becoming functionally active.

Ribosomes may be **free** in the cytoplasm or **attached** to the endoplasmic reticulum membranes, each variety functioning in the assembly of different proteins. Individual polypeptides are assembled beginning at the N-terminus and proceeding toward the

C-terminus. The rate of chain elongation may vary in different regions of the polypeptide. In most cases, a single message is translated by several ribosomes traveling in close succession along the mRNA molecule; such complexes are called **polysomes**. Initiation and termination of chain growth, as well as the elongation cycle, require specific factors and are catalyzed by specific enzymes that are either constituents of the ribosomes or dissolved in the cytoplasm. Modifications of a protein may occur during or following translation. **Cotranslational** modifications include

deformylation at the N-terminus, amino acid **cleavage, side chain alteration**, formation of **disulfide bridges, addition of sugars**, and **tertiary folding**. Modifications that are characteristically **post translational** include **peptide cleavage, quaternary association**, and **addition of prosthetic groups**.

Many substances, including a number of **antibiotics**, act as inhibitors of protein synthesis. Some inhibitors are effective only in prokaryotes and some only in eukaryotes. Still others inhibit protein synthesis in both prokaryotic and eukaryotic cells.

REFERENCES AND SUGGESTED READING

Articles and Reviews

Baldwin, R. L., Seeding protein folding. *Trends in Biochem. Sci.* **11** (1), 6 (Jan. 1986).

Boyer, S. H., Smith, K. D., and Noyes, A. N., Immunological purification and characterization of hemoglobin chain-synthesizing polysomes. *Hemoglobin: Comparative Molecular Biology Models for the Study of Disease, Ann. N.Y. Acad. Sci.* **241**, 204 (1974).

Brawerman, G., Characteristics and significance of the polyadenylate sequence in mammalian messenger RNA. In *Progress in Nucleic Acid Research and Molecular Biology* (W. E. Cohn, Ed.), Vol. 17, Academic, New York, 1976, p. 118.

Brimacomb, R., Stoffler, G., and Wittmann, H. G., Ribosome structure. In *Annual Review of Biochemistry*, (E. E. Snell *et al.*, Eds.), Vol. 47, Annual Reviews, Inc., Palo Alto, Calif., 1978, p. 217.

Caskey, C. T., The universal RNA genetic code. *Q. Rev. Biophys.* **3**, 295 (1970).

Clare, B. F. C., and Marcker, K. A., How proteins start. *Sci. Am.* **218** (1), 36 (Jan. 1968).

Collins, F. S., and Weissman, S. M., The molecular genetics of human hemoglobin. In *Progress in Nucleic Acid Research and Molecular Biology* (W. E. Cohn and K. Moldave, Eds.), Vol. 25, Academic, Orlando, 1984, p. 317.

Crick, F., Split genes and RNA splicing. *Science* **204**, 264 (1979).

Darnell, J. E., mRNA structure and function. In *Progress in Nucleic Acid Research and Molecular Biology*, (W. E. Cohn and E. Volkin, Eds.), Vol. 19, Academic, New York, 1976, p. 493.

Darnell, J. E., Implications of RNA–RNA splicing in evolution of eukaryotic cells. *Science* **202**, 1257 (1978).

Darnell, J. E., The processing of RNA. *Sci. Am.* **249** (4), 90 (Oct. 1983).

Darnell, J. E., RNA. *Sci. Am.* **253** (4), 68 (Oct. 1985).

Davis, B. D., and Tai, P. C., The mechanism of protein secretion across membranes. *Nature* **283**, 433 (1980).

Delihas, N., Andersen, J., and Singhal, R. P., Structure, function, and evolution of 5-S ribosomal RNA. In *Progress in Nucleic Acid Research and Molecular Biology* (W. E. Cohn and K. Moldave, Eds.), Academic, Orlando, 1984, p. 161.

Dintzis, H. M., Assembly of the peptide chains of hemoglobin. *Proc. Nat. Acad. Sci.* **47**, 247 (1961).

Efstratiadis, A., Posakony, J. W., Maniatis, T., Lawn, R. M., O'Connell, C., Spritz, R. A., DeRiel, J. K., Forget, B. G., Weissman, S. M., Slighton, J. L., Blechl, A. E., Smithies, O., Baralle, F. E., Shoulders, C. C., and Proudfoot, J. J., The structure and evolution of the human beta-globin gene family. *Cell* **21**, 653 (1980).

Emr, S. D., Hall., M. N., and Silhavy, T. J., A mechanism of protein localization: The signal hypothesis in bacteria. *J. Cell Biol.* **86**, 701 (1980).

Englander, S. W., and Page, L. A., Interpretation of data on sequential labeling of growing polypeptides. *Biochem. Biophys. Res. Commun.* **19**, 565 (1965).

Erdmann, V. A., Structure and function of 5 S and 5.8 S RNA. In *Progress in Nucleic Acid Research and Molecular Biology* (W. E. Cohn, Ed.), Vol. 18, Academic, New York, 1976, p. 45.

Furuichi, Y., Muthukrishnan, J. T., and Shatkin, A. J., Caps in eukaryotic mRNAs. In *Progress in Nucleic Acid Research and Molecular Biology* (W. E. Cohn and E. Volkin, Eds.), Vol. 19, Academic, New York, 1976, p. 3.

Goldberg, M. E., The second translation of the genetic message: protein folding and assembly. *Trends in Biochem. Sci.* **10** (10), 388 (Oct. 1985).

Holley, R. W., The nucleotide sequence of a nucleic acid. *Sci. Am.* **214** (2), 30 (Feb. 1966).

Hunt, T., The control of globin synthesis in rabbit reticulocytes. *Hemoglobin: Comparative Molecular Biology Models for the study of Disease, Ann. N.Y. Acad. Sci.* **241**, 223 (1974).

Hunt, T., Control of globin synthesis. Haemoglobin: Structure, function and synthesis. *Br. Med. Bull.* **32**, 257 (1976).

Itano, H. A., Genetic regulation of peptide synthesis in hemoglobin, *J. Cell. Physiol.* **67** (Suppl. 1), 65 (1966).

Karlsson, S., and Nienhuis, A. W., Developmental regulation of human globin genes. *Ann. Rev. Biochem.* **54**, 1071 (1985).

Kearns, D. R., High-resolution nuclear magnetic resonance investigations of the structure of tRNA in solution. In *Progress in Nucleic Acid Research and Molecular Biology* (W. E. Cohn, Ed.), Vol. 18, Academic, New York, 1976, p. 91.

Kim, S. H., Three-dimensional structure of transfer RNA. In *Progress in Nucleic Acid Research and Molecular Biology* (W. E. Cohn, Ed.), Vol. 17, Academic, New York, 1976, p. 182.

Lake, J. A., The ribosome. *Sci. Am.* **245** (2), 84 (Aug. 1981).

Lake, J. A., Evolving ribosome structure: Domains in archaebacteria, eubacteria, eocytes and eukaryotes. *Ann. Rev. Biochem.* **54**, 507 (1985).

Lawn, R. M., Efstratiadis, A., O'Connell, C., and Maniatis, T., The nucleotide sequence of the human beta-globin gene. *Cell* **21**, 647 (1980).

Lodish, H. F., Translation control of protein synthesis. In *Annual Review of Biochemistry* (E. E. Snell *et al.*, Eds.), Vol. 45, Annual Reviews, Inc., Palo Alto, Calif., 1976, p. 39.

Mandal, K., The organization and transcription of eukaryotic ribosomal RNA genes. In *Progress in Nucleic Acid Research and Molecular Biology* (W. E. Cohn and K. Moldave, Eds.), Vol. 25, Academic, Orlando, 1984, p. 117.

Marcker, K., and Sanger, F., *N*-Formyl-methionyl-s-RNA. *J. Mol. Biol.* **8**, 835 (1964).

McKnight, S. L., and Miller, O. L., Electron miscroscopic analysis of chromatin regulation in the cellular blastoderm of *Drosophila melanogaster* embryo. *Cell* **12**, 795 (1977).

Miller, O. L., and Beatty, B. R., Visualization of nucleolar genes. *Science* **164**, 955 (1969).

Miller, O. L., Hamkalo, B. A., and Thomas, C. A., Visualization of bacterial genes in action. *Science* **169**, 392 (1970).

Moldave, K., Eukaryotic protein synthesis. *Ann. Rev. Biochem.* **54**, 1109 (1985).

Morris, A. J., Slabaugh, R. C., and Protzel, A., Size characteristics of nascent globin chains (peptidyl tRNA) in the reticulocyte. *Hemoglobin: Comparative Molecular Biology Models for the Study of Disease, Ann. N.Y. Acad. Sci.* **241**, 310 (1974).

Nierhaus, K. H., and Rheinberger, H. -J., An alternative model for the elongation cycle of protein synthesis. *Trends in Biochem. Sci.* **9** (10), 428 (Oct. 1984).

Noller, H. F., and Woese, C. R., Secondary structure of 16 S ribosomal RNA. *Science* **212**, 402 (1981).

Nomura, M., Ribosomes, *Sci. Am.* **221** (4), 28 (Oct. 1969).

Nomura, M., The control of ribosome synthesis. *Sci. Am.* **250** (1), 102 (Jan. 1984).

Ochoa, S., Regulation of protein synthesis. *Crit. Revs. Biochem.* **7**, 7 (1979).

Pestka, S., Insights into protein biosynthesis and ribosome function through inhibitors. In *Progress in Nucleic Acid Research and Molecular Biology* (W. E. Cohn, Ed.), Vol. 17, Academic, New York, 1976, p. 217.

Poyton, D. O., Cooperative interaction between mitochondrial and nuclear genomes: Cytochrome *c* oxidase assembly as a model. In *Current Topics in Cellular Regulation*, (B. L. Horecker and E. R. Stadtman, Eds.), Vol. 17, Academic, New York, 1980.

Proudfoot, N. J., and Brownlee, G. G., Nucleotide sequences of globin messenger RNA. Haemoglobin: Structure, functions and synthesis. *Br. Med. Bull.* **32**, 251 (1976).

Puvion-Dutilleul, F., Bachellerie, J-P., Zalta, J. P., and Bernhard, W., Morphology of ribosomal transcription units in isolated subnuclear fractions of mammalian cells. *Rev. Biol. Cell.* **30**, 183 (1977).

Rich, A., Polyribosomes. *Sci. Am.* **209** (6), 44 (Dec. 1963).

Rich, A., and Kim, S. H., The three dimensional structure of tRNA. *Sci. Am.* **238** (1), 52 (Jan. 1978).

Rich, A., and RajBhandary, U. L., Transfer RNA: Molecular structure, sequence and properties. In *Annual Review of Biochemistry* (E. E. Snell *et al.*, Eds.), Vol. 45, Annual Reviews, Inc., Palo Alto, Calif., 1976, p. 805.

Rich, A., Warner, J. R., and Goodman, H. M., The structure and function of polyribosomes. *Cold Spring Harbor Symp. Quant. Biol.* **28**, 269 (1963).

Safer, B., and Anderson, W. F., The molecular mechanism of hemoglobin synthesis and its regulation in the reticulocyte. *Crit. Revs. Biochem.* **5**, 261 (1978).

Sharp, P. A., On the origin of RNA splicing and introns. *Cell* **42**, 397 (1985).

Smith, D. W. E., Reticulocyte transfer RNA and hemoglobin synthesis. *Science* **190**, 529 (1975).

Weissbach, H., and Ochoa, S., Soluble factors required for eukaryotic protein synthesis. In *Annual Review of Biochemistry* (E. E. Snell *et al.*, Eds.), Vol. 45, Annual Reviews Inc., Palo Alto, Calif. 1976, p. 191.

Williamson, R., The processing of hnRNA and its relation to mRNA. In *Cell Biology: A Comprehensive Treatise*, (L. Goldstein and D. M. Prescott, Eds.), Vol. 3, Academic, New York, 1980.

Winslow, R. M., and Ingram, V. M., Peptide chain synthesis of human hemoglobins A and A$_2$. *J. Biol. Chem.* **241**, 1144 (1966).

Zamecnik, P., The machinery of protein synthesis. *Trends in Biochem. Sci.* **9** (11), 464 (Oct. 1984).

Books, Monographs, and Symposia

Dickerson, R. E., and Geis, I., *Hemoglobin: Structure, Function, Evolution, and Pathology*, Benjamin/Cummings, Menlo Park, Calif., 1983.

Hunt, T., Prentis, S., and Tooze, J. (Eds.), *DNA Makes RNA Makes Protein*, Elsevier, Amsterdam, 1983.

Lewin, B. M., *The Molecular Basis of Gene Expression*, Vol. 2, Wiley, New York, 1980.

Lewin, B., *Genes* (2nd Ed.), Wiley, New York, 1985.

Nomura, M., Tissieres, A., and Lengyel, P. (Eds.), *Ribosomes*, Cold Spring Harbor Monograph Series, Cold Spring Harbor Laboratory, N.Y., 1974.

McConkey, E. H. (Ed.), *Protein Synthesis*, Vol. 2, Dekker, New York, 1976.

Tedeschi, H., *Mitochondria: Structure Biogenesis and Transducing Function. Cell Biology Monographs*, Vol. 4, Springer-Verlag, Vienna, 1976.

CHAPTER 23

THE CYTOPLASMIC MATRIX

Although the cytoplasm of cells has been known since the 1950s to contain a variety of fiberlike structures called **cytoplasmic filaments** (**microfilaments** and **intermediate filaments**) and **microtubules**, the cytoplasm or **ground substance** itself had until recently been thought to be a structureless fluid. This view prevailed for many years despite the fact that the cytoplasm exhibits viscoelastic properties that are more like a gel than a liquid. However, the development of *high-voltage electron microscopy* in the 1970s provided the breakthrough that revealed that the ground substance is pervaded by a three-dimensional network of fine, gossamer threads, called the **microtrabecular lattice**. This lattice serves to anchor many of the cell's organelles in position and appears also to interconnect the microtubules and cytoplasmic filaments. Collectively, the microtrabecular lattice, cytoplasmic filaments, and microtubules form the **cytoplasmic (or cell) matrix**, which gives shape and support to the cell and appears also to be involved in such cellular activities as cell division, ciliary and flagellar movement, and cellular locomotion. In this chapter, we will examine the structure and functions of the microtrabecular lattice, the relationship of the lattice to the cytoplasmic filaments and microtubules, and the special roles played in cells by all of these structures.

THE GROUND SUBSTANCE AND THE MICROTRABECULAR LATTICE

Structure of the Lattice

Conventional transmission electron microscopes have accelerating potentials of several thousand volts and produce an electron beam that penetrates tissue and cell slices (i.e., sections) having thicknesses of about 0.2 μm. The development in the 1970s of *high-voltage electron microscopes* capable of accelerating electrons over a potential of *one million* volts has made it possible to study specimens having the thickness of whole cells. Such microscopes are quite spectacular, usually occupying two floors of a laboratory (they are about 30 feet tall) and weighing about 20 tons. Because the electron beam traverses the thickness of an entire cell, the interior of the cell is revealed *in depth*.

Principally as a result of the high-voltage electron-microscopic studies of K. R. Porter, M. Schliwa, J. J. Wolosewick, and J. B. Tucker, the cytoplasm or ground substance of eukaryotic cells has been shown to be divided into two major phases: the **microtrabecular lattice** and the **intertrabecular spaces**. A portion of the lattice is seen in the high-voltage photomicrograph of Figure 23–1 and is depicted diagrammatically in Figure 23–2. The lattice is not apparent

597

in electron photomicrographs of conventional thin sections because the sections lack the depth necessary to reveal the network. The microtrabeculae are rich in protein, whereas the intertrabecular space is aqueous and serves to dissolve or suspend the great variety of small molecules involved in cellular metabolism (e.g., glucose, amino acids, oxygen, and inorganic salts). At its margins, the lattice is attached to the plasma membrane of the cell, but it also interconnects many of the cytoplasmic organelles (e.g., mitochondria, endoplasmic reticulum, ribosomes, polysomes, etc.) as well as the ctyoplasmic filaments and microtubules that comprise the cytoskeleton (Fig. 23–2).

The microtrabecular lattice is not a rigid or static structure. Rather, it varies in response to changes in cell activity and shape. Its disposition in a cell also varies according to the cellular environment. For ex-

ample, when cells are cultured at a low temperature (e.g., 4°C), they round up, becoming spherical. The change in cell shape is due to a change in the cytoskeleton. First, the microtubules disassemble; then the cytoplasmic filaments disappear; and, finally, the microtrabeculae are modified although they do not decompose entirely. Most notable is the separation of some of the microtrabeculae from the lattice and their aggregation into **gobbets** (Fig. 23–2). The loss of parts of the lattice produces gaps that allow the cell organelles to move about more freely and also exhibit Brownian motion. If the cells are returned to their normal (higher) temperature, the microtrabecular lattice and the cytoskeleton reform, and the organelles are again confined in their movement.

Changes in the chemical composition of the environment of a cell also produce reversible changes in the

FIGURE 23–1   The *microtrabecular lattice* of a kidney cell. The three-dimensional network of filaments that forms the lattice is revealed by high-voltage transmission electron microscopy of whole cells. The round bodies seen at a few of the intersections of the lattice are ribosomes. (Courtesy of Dr. M. Schliwa.)

FIGURE 23–2 Model of the microtrabecular lattice.

Polysome · Ribosome · Plasma membrane · Endoplasmic reticulum

Gobbet

Microtubule · Mitochondrion · Microfibrils

Microfilaments

Microtrabecular strand

lattice. Cytochalasin B (Fig. 23–3), a drug obtained from the mold *Helminthosporium dematoiderum*, causes the microtrabeculae to thicken. High or low osmotic pressure, changes in the concentration of ions (e.g., Mg^{2+} and Ca^{2+}), and the presence of certain metabolic inhibitors also cause reversible changes in the lattice.

FIGURE 23–3 Structure of cytochalasin B.

Chemistry of the Lattice

Chemically, the most important constituents of the lattice are proteins. Two-dimensional electrophoresis of extracts of the lattice suggest that more than 100 different proteins comprise this polymeric network. Among the specific proteins that are thought to be present are actin, myosin, and tubulin. These proteins are also major constituents of cytoplasmic filaments and microtubules (see later). A basic difference in the composition of microtrabeculae and the cytoskeletal elements (i.e., microtubules and cytoplasmic filaments) is revealed when both are treated with organic detergents such as Triton X-100. Cells treated with Triton X-100 lose the microtrabecular lattice along with membranous structures such as mitochondria, endoplasmic reticulum, plasma membrane, and nuclear envelope, but they retain their cytoskeleton (Fig. 23–4). This behavior suggests that microtrabeculae have certain physical properties (and therefore a chemical composition) that are similar to those of membranes but unlike those of the cytoskeletal elements.

The observation that nonionic detergent extraction of cells leaves behind a residue consisting of microtubule and cytoplasmic filament proteins of the cytoskeleton has made it possible to study the cytoskeleton using conventional transmission electron microscopes. When the cytoskeletons of extracted cells are freeze-dried and replicas are produced using platinum, the cytoskeleton is revealed in considerable detail (Fig. 23–5).

Functions of the Microtrabecular Lattice

The microtrabecular lattice apparently serves as an intracellular scaffolding that helps suspend and organize the diverse structural components of the cytoplasm, including many of the cellular organelles. Acting in concert with the cytoplasmic filaments and microtubules, the lattice plays an important role in maintaining cell shape and in cellular movements. Evidence also exists that suggests that a number of the enzymes of intermediary metabolism including those of glycolysis are bound to the lattice. It is possible that these enzymes are associated with the lattice in an *or-*

dered fashion; for example, the enzymes of a particular metabolic pathway may be bound to the lattice in such a manner that successive reactions of the pathway are spatially coordinated.

CYTOPLASMIC FILAMENTS

Composition, Distribution, and Functions of Cytoplasmic Filaments

Cytoplasmic filaments are elongate, unbranched, proteinaceous strands that consist of bundles or groups of protein molecules sometimes wound into a helical shape. Although organized arrays of cytoplasmic filaments were first described for muscle cells and have been most extensively studied in this tissue, it is now apparent that essentially all eukaryotic cells contain

FIGURE 23–4 High-voltage transmission electron photomicrograph of a kidney cell *cytoskeleton* left behind as a residue after the microtrabecular lattice was extracted with the detergent Triton X-100. This detergent removes not only the lattice but also membranous organelles. (Courtesy of Dr. M. Schliwa.)

FIGURE 23–5 Platinum replica of a freeze-dried cytoskeleton of a fibroblast cell extracted with Triton X-100. Most of the filaments seen in this view are clustered together to form bundles running diagonally across the photo. Clusters of ribosomes are seen in areas where the filaments form a more diffuse network. (Courtesy of Drs. J. E. Heuser and M. W. Kirschner. Copyright © 1980 by the Rockefeller University Press; *J. Cell Biol.* **86**, 215.)

these filaments. Cytoplasmic filaments can be subdivided into three major classes based on filament size; these are (1) *microfilaments* (also called *thin filaments*), which have a diameter of about 6 nm and are composed primarily of the globular protein *actin*; (2) *intermediate filaments*, which have diameters between 7 and 11 nm and, depending on the tissue source, are formed from five different classes of proteins (see below); and (3) *myosin filaments* (also called *thick filaments*), which have diameters up to 22 nm and are rich in the fibrous protein *myosin*. Muscle cells, which are especially differentiated for contrac-

tion (discussed at length in Chapter 24), are rich in microfilaments and myosin filaments, but all cells contain cytoplasmic filaments and it has been shown that actin can account for as much as 20% of the total cell protein in some nonmuscle cells.

Microfilaments   Although actin molecules are globular proteins (ie., *G-actin*), in microfilaments the actin molecules are arranged to form two interwoven helical chains, each of which is called a **protofilament** (Fig. 23–6); in this polymeric form, the protein is called *F-actin*. Microfilaments are intimately associated with

FIGURE 23–6 Arrangement of polymerized G-actin protofilaments to form a microfilament. The two protofilaments form a right-handed double helix.

all cellular activities that involve movement. This is vividly demonstrated when cells are treated with cytochalasin B. In the presence of this substance, microfilaments dissociate and the loss of the microfilaments is accompanied by a loss of certain cellular functions. Some of the more common processes sensitive to cytochalasin B are *phagocytosis, pinocytosis*, and *exocytosis* (see Chapter 15), *cytokinesis* (discussed later in this chapter), *cytoplasmic streaming* (in plant cells), movements of *microvilli, cilia*, and *flagella*, movements of the *cytoskeleton*, and of course *muscle contraction* (Chapter 24).

Intermediate Filaments Five different classes of proteins comprise the intermediate filaments of cells and tissues; these are *desmin* in muscle cells (one type of protein of MW 52,000), *vimentin* in mesenchymal cells (one protein of MW 53,000), *cytokeratins* in epithelial cells (more than a dozen different proteins having MWs ranging from 40,000 to 70,000), *neurofilament proteins* in nerve cells (three proteins having MWs of 65,000, 105,000, and 135,000), and *glial fibrillary acidic protein* in astrocytes (one protein of MW 50,000). Although a dynamic role cannot be excluded for the intermediate filaments, in most cells they appear to play a structural role. For example, in epidermal and mesenchymal cells, bundles of intermediate filaments form a basketlike weave that encircles the nucleus and is intimately associated with the nuclear envelope at the nuclear pores. From there the filaments appear to radiate to the margins of the cells, where they terminate at desmosomes (junctions between neighboring cells in a tissue; see Chapter 15). Because of their intimate association with the nuclear pores, a role in mediating the transport of materials between the nucleus and the cyotplasm cannot be precluded.

In nerve cells, the long axes of the intermediate filaments are oriented parallel to the long axes of the cells' "long processes" (e.g., axons and dendrites; see Chapter 24) and are believed to provide these processes with a structural framework.

Myosin (Thick) Filaments *Striated, smooth*, and *cardiac* muscle cells contain vast numbers of cytoplasmic filaments that function during the contraction of these cells. Most of these filaments are microfilaments (i.e., thin filaments formed primarily from F-actin but also containing the proteins **tropomyosin** and **troponin**) and thick filaments (formed from myosin). The two types of cytoplasmic filaments are arranged in parallel rows and interact with each other through *cross-bridges* that enable the filaments to slide past one another and effect a shortening (i.e., contraction) of the cell. The contraction of striated muscle cells shortens the muscle, which then moves the skeleton. It is in striated muscle cells that the number of filaments is greatest and their geometric distribution most highly ordered. As seen in Figure 23–7, equally spaced about each thick filament are six thin filaments. Units of several hundred thick and thin filaments are bundled together to form **myofibrils**, and each cell contains many hundreds of myofibrils. Further discussion of the structure of muscle and the biochemistry and mechanics of contraction is deferred until Chapter 24.

Cytokinesis

Cell division in animal cells involves two separate mechanisms: **mitosis** (see Chapter 20) and **cytoplasmic cleavage**, or **cytokinesis**. In animal cells, cytokinesis begins toward the end of anaphase and is characterized by the appearance of a constriction about the midline of the spindle. The constriction deepens as the plasma membrane moves inward at the **cleavage furrow**. Material collecting at the midline of the spindle becomes quite dense, forming a structure known as the **midbody** (Fig. 23–8). Just before the infolding edges of the plasma membrane meet and fuse, the midbody disappears.

The furrowing or pinching-in of the plasma membrane at the cleavage furrow is reminiscent of the action of a purse string or of a rubber band tightening about a soft object. The process clearly involves the action of microfilaments and these are seen in abundance in the area of the cleavage furrow. Microfilament involvement is also supported by the observation that cytochalasin B inhibits the process. The presence of a band of microfilaments, the *contractile ring*, just underneath the plasma membrane in the area of constriction can be seen in the electron photomicrographs of Figure 23–9. Here *Arbacia* sea urchin eggs are observed at various stages of cleavage. Prior to the on-

FIGURE 23–7 Diagram (left) and photomicrograph (right) of a cross section of a striated muscle myofibril showing geometric arrangement of thick and thin filaments.

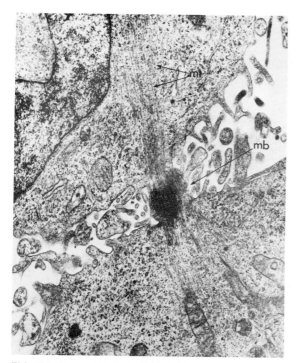

FIGURE 23–8 Final phase of cytokinesis in an animal cell. Only a few microtubules (m) of the spindle remain. Note the midbody (mb). (Courtesy of B. Byers. Copyright © 1968 Springer-Verlag, *Protoplasma* **66**, 423.)

set of cleavage, no microfilaments are seen in the area that is about to constrict (Figs. 23–9a and 23–9d). However, once cleavage begins, microfilaments quickly appear about the area of constriction, forming the contractile ring (compare Figs. 23–9d and 23–9e). Because the bundles of microfilaments are arranged in a circle just below the cleavage furrow, their long axes run perpendicular to the plane of the sections seen in Figures 23–9b and 23–9e. As cleavage is completed, the spindle fibers and midbody fade and the contractile ring and microfilaments disappear (Fig. 23–9c). The molecular basis of the constriction that characterizes cytokinesis in animal cells remains uncertain, but the clear presence of actin and the suspected presence of myosin has led to the speculation that the mechanism involves filament sliding similar to that which occurs during muscle contraction.

Plasma Membrane Movement

Intestinal epithelial cells have thousands of small fingerlike projections called *microvilli* (see Chapter 15, especially Figs. 15–28 and 15–29) that cyclically shorten and extend into the lumen of the intestine. This action facilitates the absorption of digested food by greatly increasing the surface area of the cells. Mi-

FIGURE 23–9 Cytokinesis in an *Arbacia* sea urchin egg (see text for details). (Photomicrographs courtesy of T. E. Schroeder. Copyright © 1976 by Cold Spring Harbor Laboratory, *Cell Motility*, Book A, p. 268.)

crofilaments in the intestinal microvilli are bundled together to form a core by the interaction of F-actin with the two proteins *villin* and *fimbrin*. The microfilament bundle runs parallel to the length of the microvillus and is anchored to the plasma membrane at the microvillus' apical tip. No myosin appears to be present in the microvilli. At the base of the microvilli is another microfilament network called the **terminal web**. Myosin has been localized in the terminal web and its interaction there with actin is believed to support the microvilli and perhaps also create a contractile mechanism that lowers the microvilli into the cell. A second possibility is that in the presence of high Ca^{2+}, villin can act to cleave F-actin into short fragments. This would cause the microvilli to collapse into the cell in a mechanism that is independent of myosin. Calmodulin has also been implicated in the regulation of movements of microvilli.

Amoeboid Movement

The locomotion of amoebae, slime molds, white blood cells, and a number of other cells involves the formation of **pseudopodia**—large cytoplasmic extensions from the main body of the cell and into which the remaining cytoplasm subsequently streams. Most amoeboid cells form more than one pseudopod at a time, but continued locomotion in one direction requires the reversal of the process in the nondominant pseudopodia. In pseudopodia-forming cells, the outer margin of the cytoplasm, called **ectoplasm**, is very viscous or *gel-like* and is generally free of granules and other cytoplasmic inclusions; the remaining cytoplasm, called **endoplasm**, is more fluid or *sol-like*. During locomotion, the fluid endoplasm flows forward into the advancing pseudopod, and as it reaches the anterior end of the pseudopod, it flows laterally and is converted into gel, thereby forming new ectoplasm. At the rear of the moving cell, in a region called the **uroid** (see Fig. 15–46), the ectoplasm *solates*, streams deeper into the cell, and becomes endoplasm (Fig. 23–10).

Two different views have dominated attempts to explain amoeboid movement. According to the "front-zone contraction" theory of R. D. Allen and others, the endoplasm in the anterior region of a pseudopod undergoes contraction as it is transformed into ectoplasm. In effect, this contraction *pulls* the endoplasm forward and into the pseudopod. An older view advanced in the 1920s by S. O. Mast and called the "ectoplasmic tube contraction" theory suggests that the ectoplasm toward the rear of the cell contracts, *pushing* the endoplasm forward.

Regardless of the site of origin of the force that causes the cytoplasm to stream forward, overwhelming evidence implicates actin microfilaments in the process. For example, the actin-binding drug cytochalasin B inhibits amoeboid movement. The presence of myosin filaments in amoeboid cells has also been reported as has an ATPase similar to one known to be important in the chemistry of muscle contraction. Contractions of the ectoplasm of amoeboid cells can be induced by adding ATP and Ca^{2+} to the cells. Although not fully understood, amoeboid movement appears to use an actin–myosin contractile system similar to that of muscle cells.

An amoeboidlike but much slower form of locomotion is also exhibited by mammalian connective tissue cells (i.e., *fibroblasts*) grown in culture. As a fibroblast moves across a surface (e.g., the culture dish), the cell continuously forms sheetlike extensions, called **lamellipodia**, and **microspikes** at its advancing edge (Figs. 23–11 and 23–12; see also Fig. 15–28). Some of the lamellipodia attach to the surface (at anchor points called **adhesion plaques**) and others sweep backward over the cell's upper surface in a wavelike motion called **surface ruffling**. Electron-microscopic studies

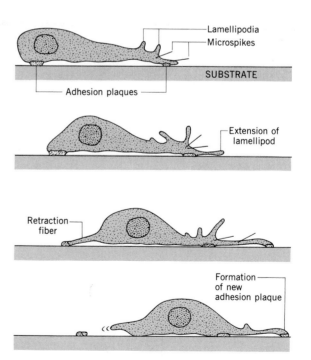

FIGURE 23–11 Formation of lamellipodia and microspikes during locomotion in cutltured fibroblasts (see text for details).

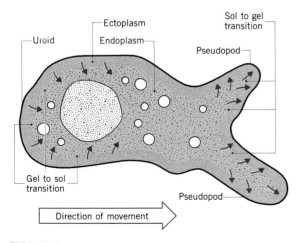

FIGURE 23–10 Sol-gel and gel-sol transitions inside the cell during amoeboid movement (see text for details).

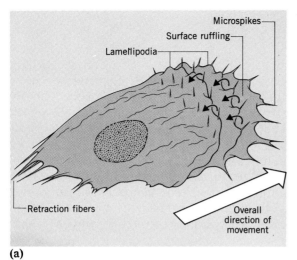

(a)

FIGURE 23–12 *Surface ruffling.* In the scanning electron photomicrograph (part *b*), the cell is moving from left to right. The surface ruffles along the margins of the cell's advancing edge are apparent. Also apparent are the microspikes and retraction fibers (pieces of which appear to have been left attached to the substratum as the cell moved forward). (Photomicrograph courtesy of Drs. D. Henderson and K. Weber.) From Electron Microscopy in Biology (J. D. Griffith, ed.). Copyright © 1982 John Wiley and Sons, Inc.)

(b)

reveal that the lamellipodia contain large numbers of actin microfilaments, and like true amoeboid motion, the movements of fibroblasts are inhibited by cytochalasin B. As the leading edge of the cell stretches forward, the trailing edge is drawn out into long *retraction fibers* that are eventually pulled free from their points of attachment to the underlying surface. The latter action may leave behind small pieces of plasma membrane and cytoplasm.

Capping, Membrane Flow, and Cell Locomotion
As discussed in Chapter 15, substances bound to mo-

bile plasma membrane receptors migrate laterally within the membrane and are concentrated in clathrin-coated pits prior to endocytosis. During endocytosis, these receptors and the ligands that they have bound are internalized along with small pieces of the plasma membrane. The receptors are eventually returned to the plasma membrane during exocytosis, which also restores membrane surface area lost during endocytosis. In nonmotile cells, endocytosis and exocytosis occur at positions scattered over much of the cell surface.

Many membrane proteins avoid internalization during endocytosis because their rates of diffusion within the membrane preclude their concentration within the coated pits. These membrane proteins are referred to as "noncycling proteins." When cells are treated with antibodies against noncycling membrane proteins, the proteins are cross-linked to form large aggregates. Because the diffusion rates of cross-linked proteins are so much lower than the diffusion rates of individual proteins, the aggregate can be clustered together. In motile cells, cross-linked proteins cluster at one point on the cell surface called a **cap**, and the process is called *capping*. Capping is a general phenomenon displayed by aggregates in the plasma membrane and is not peculiar to certain proteins; nonmotile cells, however, do not cap cross-linked surface proteins. Cytochalasin B and another drug called colchicine prevent capping, suggesting that membrane flow involves the actions of microfilaments (and perhaps also microtubules).

When fibroblasts are allowed to migrate over a substrate littered with tiny carbon particles, the particles become attached to the upper surface of extended lamellipodia and are slowly swept backward toward the trailing edge of the cell. This implies that the membrane as a whole is flowing backward. It has therefore been suggested that a moving fibroblast internalizes portions of the plasma membrane at the rear of the cell and reinserts this membrane at the leading edge of the cell (Fig. 23–13). This would provide a mechanism by which the cell could move itself forward along a substrate, for if the cell forms transient connections with the substrate at its anterior end, the forces that cause the membrane to flow backward would simultaneously propel the cell forward. The action would be similar to that of a tractor or tank track.

FIGURE 23–13 Locomotor activity in cultured cells is accompanied by membrane flow toward the rear of the cell. Membranous material enters the cell by endocytosis and is added to the advancing membrane by exocytosis at the cell's leading edge. Internally, membranous material and cytoplasm flow forward.

MICROTUBULES

Microtubules have many features that distinguish them from microfilaments and intermediate filaments. To begin with, the outside diameter of a microtubule (usually about 25 nm) is much greater than that of microfilaments. Furthermore, microtubules are *hollow*, containing a central lumen about 15 nm in diameter. Microtubule length is quite variable. Some microtubules are less than 200 nm long, but in the long processes of nerve cells their lengths may be as great as 25 μm (i.e., 25,000 nm). Microtubules can also be distinguished from microfilaments chemically. Microtubules contain two major proteins called α *tubulin* and β *tubulin*. Each protein consists of a single polypeptide chain about 500 amino acids long (MW 55,000) and both are similar in primary structure, indicating that they are probably derived from a common ances-

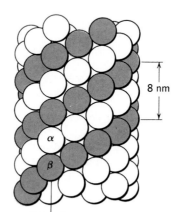

Heterodimer

FIGURE 23–14 Model of a microtubule as seen in cross section and laterally. The circumference of the microtubule has 13 subunits. Although subunits appear to be linear chains that are parallel to the axis of the microtubule, paired subunits (α and β heterodimers) form a helix about the central, hollow core. (Courtesy of J. Bryan. Copyright © 1974 by *Fed. Proc.* **33**, 156.)

FIGURE 23–15   Assembly of microtubules. In *in vitro* studies, α and β tubulin molecules (*a*) combine to form *heterodimers* (*b*). The heterodimers associate to form rings, spirals and other intermediate structures (*c, d, e*), which eventually open up to form strands or *protofilaments* (*f*). Side-by-side assembly of protofilaments creates sheetlike structures that curl to form a tube (*g*). Elongation of the tube occurs by direct addition of new heterodimers (*h*). (Modified from P. Dustin, Microtubules. *Sci. Am.* **243** (6), 66 (1980).)

tral protein. Not only are the α and β tubulins nearly identical but tubulins from diverse species of cells are very similar, suggesting either that they have hardly changed since they first appeared in eukaryotic organisms or that tubulin is a highly conserved protein.

α and β tubulin molecules combine to form **heterodimers** and these serve as the basic building blocks of microtubules. The model of heterodimer organization shown in Figure 23–14 is based on both chemical studies of microtubules and transmission electron microscopy. The microtubule is formed from a helical array of heterodimers with 13 subunits per turn of the helix. Neighboring heterodimers are linked to one another not only longitudinally but laterally as well.

Assembly and Functions of Microtubules

The current model for the manner in which tubulin subunits are assembled into a microtubule is based on *in vitro* studies. Under carefully controlled conditions (e.g., the appropriate concentration of tubulin and the absence of calcium), alpha and beta subunits spontaneously form dimers that when present in high concentrations assemble into chains (Fig. 23–15); the chains then form a variety of intermediate structures including single and double rings, spirals, and stacked rings. The rings eventually open up to form linear chains or **protofilaments** that associate side-by-side to form sheets. When a sheet is sufficiently wide, it is curled to form a tube. The end result is the formation of short cylinders of dimers (Fig. 23–14). After a short cylinder is formed, continued growth occurs by the direct addition of more dimers. Growth occurs primarily by addition of dimers at one end of the tubule. It is believed that during certain microtubular functions (such as the operation of the anaphase spindle) the addition of dimers to one end of a microtubule is accompanied by the loss of dimers from the other end.

Assembly of tubulin into dimers requires that these polypeptides bind GTP. GTP-activated dimers can then combine with other dimers or with the growing microtubule. Attachment of a dimer to the microtubule is accompanied by the hydrolysis of the GTP, but the resulting GDP and phosphate remain bound to the tubule. The tubulin dimer also has sites that can bind the drugs *colchicine, vincristine,* and *vinblastine* (Fig. 23–16) and these substances inhibit microtubule assembly. Tubules that are already present at the time of addition of these inhibitors disassemble. Calcium has long been recognized as an important ion in the

FIGURE 23–16 Microtubule assembly inhibitors colchicine (a) and vincristine (b). Vinblastine, another inhibitor, is similar to vincristine but has a methyl group in place of the formyl group.

microtubule assembly and disassembly process. Calcium may influence microtubules either directly or in association with the regulatory protein *calmodulin*.

Recently, a number of proteins have been identified that associate with the surface of microtubules (Fig. 23–17); these proteins are called **microtubule-associated proteins**, or *MAPs*. Two families of MAPs have been separated by polyacrylamide gel electrophoresis: MAP-1 and MAP-2. The MAPs facilitate microtubule assembly; that is, microtubules are formed considerably faster and at lower tubulin concentrations in the presence of MAPs. MAPs also protect microtubules from disassembly by colchicine and low temperatures. Interactions among microtubules and

FIGURE 23–17 Replica of fractured and etched microtubules. Portions of three of the microtubules seen in this view were fractured open, revealing their inner walls. The reticulum surrounding the microtubules is believed to consist of unpolymerized tubulin and *microtubule-associated proteins* (*MAPs*). (Courtesy of Drs. J. Heuser and M. W. Kirschner. Copyright © 1980 by the Rockefeller University Press, *J. Cell Biol.* **86**, 215.)

between microtubules and other cell components also involve MAPs. For example, microtubules may be interconnected via the microtrabecular lattice; in these regions the lattice is rich in both MAP-1 and MAP-2. MAP-2 appears to be involved in cross-linking microfilaments and intermediate filaments with microtubules.

Although the extensive interactions between microtubules, cytoplasmic filaments, and the microtrabecular lattice give support and shape to cells (see, for example, Fig. 1–3), microtubules play more than a supportive role for they are also intimately involved with cell motility, endocytosis and exocytosis, chromosome movements during mitosis, and the actions of cilia and flagella (see below).

Centrioles

Centrioles and **basal bodies** belong to a group of cell structures referred to as **microtubule-organizing centers**. These centers are involved in the elaboration of microtubules. Whereas the basal bodies are located at the bases of cilia and flagella, centrioles are usually found near the cell nucleus and occur in pairs; structurally, both organelles are identical. The typical centriole is composed of nine sets of *triplets*, each triplet consisting of one complete microtubule and two incomplete, *C-shaped* ones. The triplets are arranged parallel to one another and create a cylindrical body having a diameter of 150 to 250 nm. Although rigorous proof is still lacking, it is generally believed that centrioles are involved in the production of the microtubules that form the **spindle** of a dividing cell. However, not all cells that form a spindle during nuclear division have centrioles; for example, cone-bearing and flowering plants do not. In cells that do have paired

centrioles, the centrioles separate at the onset of nuclear division and move to diametrically opposite positions around the nucleus. Subsequently, as the chromatin condenses to form chromosomes and the nuclear envelope disappears (Chapter 20), fibers of the spindle make their appearance, extending from an area adjacent to one centriole through the cell to the other centriole (Fig. 23–18). As division proceeds, a new centriole appears near each original one; growth of the new centriole is always *perpendicular* to the long axis of the original centriole. By the time division is complete, each daughter cell has two complete centrioles.

Centrioles also play a role in the formation of the microtubules present in flagella and cilia. Here the centrioles are more often referred to as *basal bodies* or *kinetosomes* (also *blepharoplasts, basal granules,* or *basal corpuscles*).

Structure of the Centriole Though not all cells contain centrioles, in those that do the structure of the centriole is the same. Most algal cells (but not red algae), moss cells, some fern cells, and most animal cells have centrioles, but cone-bearing and flowering plants, red algae, and some nonflagellated or nonciliated protozoans (like amoebae) do not. Some species of amoebae have a flagellated stage as well as an amoeboid stage; a centriole develops during the flagellated stage but disappears during the amoeboid stage.

The most notable structural characteristic of a centriole is its nine sets of triplets. Each triplet contains three microtubules, which in cross section appear to be arranged like the vanes on a "pinwheel" (Fig. 23–19). Although there is no surrounding membrane, the nine triplets appear to be embedded in an electron-dense material.

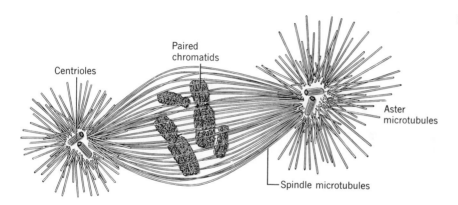

Centrioles

Paired chromatids

Aster microtubules

Spindle microtubules

FIGURE 23–18 Appearance of the centriole, aster, and spindle microtubules during nuclear division.

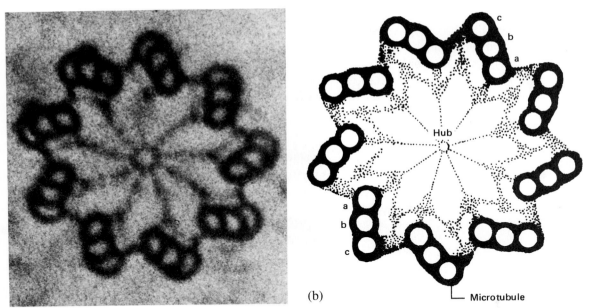

FIGURE 23–19 (a) Cross section of a centriole as revealed by electron microscopy. The spokes of the "pinwheel" turn inward in a clockwise manner, indicating that this view is from the proximal end of the centriole. (Courtesy of Dr. I. R. Gibbons.) (b) Diagram of a centriole showing the nine triplets that form the "pinwheel" and the electron-dense granular connectives.

The nine triplets are identical. The innermost (or a) microtubule of each triplet is a complete, round microtubule, but the middle (b) and outer (c) microtubules are incomplete, C-shaped, and share the wall of the neighboring microtubule. Also, the outermost (i.e., c) microtubules may not run the full length of the centriole. The triplets, although generally parallel to each other, may be closer together at the proximal end of the centriole (that end when observed "end-on" that has the triplets tilted inward in a clockwise direction, as shown in Fig. 23–19). The triplets may also spiral somewhat about the central axis of the centriole. Strands of material extend inward from each a tubule and join together at the central *hub*. These strands, when seen in cross section, give the centriole the appearance of a cartwheel (Fig. 23–19b).

The idea prevalent years ago that new centrioles arise by the division of existing centrioles is no longer accepted. Rather it appears that new centrioles are either produced *de novo* or are synthesized using an existing centriole as some form of template. In the latter case, growth of the new centriole occurs at right angles to the long axis of the existing centriole, the two organelles separated from each other by a distance of 50 to 100 nm.

Basal body (i.e., centriole) development has been studied in the ciliates *Paramecium* and *Tetrahymena* and in tracheal epithelium of *Xenopus* and chicks. The stages of development are virtually the same in all of these. Development of the basal body begins with the formation of a single microtubule in an amorphous mass. Microtubules are added one at time until there is an equally spaced ring of nine (Fig. 23–20). As the microtubules appear the amorphous mass is lost, as though it were being consumed in the production of the microtubules. There is some evidence that "connectives" exist between the microtubules, which could act to set the distance between them. Each of the nine microtubules in the ring are a microtubules. The b microtubules develop next and, finally, the c microtubules. Before the microtubules reach the doublet stage, the cylinder is rarely longer than 70 nm, but after this stage, the microtubules elongate. At the same time, the hub and "cartwheel" are added in the center (Fig. 23–20b). The a–c links are not formed until the end of development.

Basal bodies act as organizing centers for the development of the microtubules of cilia and flagella. New basal bodies form adjacent to centrioles. While still not associated with a flagellum or cilium, the basal

FIGURE 23-20 Development of *a basal body* in *Paramecium.* (*a*) The first microtubles form; (*b*) a developing ring of seven microtubules; (*c*) ring containing nine a microtubules, with some b microtubules forming; (*d*) complete set of nine triplets; (*e*) mature basal body with new basal body (nbb) seen in longitudinal section above. (Courtesy of R. Dippell. *Proc. Nat. Acad. Sci.* **61**, 461 (1968).)

body is more properly called a centriole, but after it migrates to a position just underneath the plasma membrane and acts as a center for flagellum or cilium development, it is called a basal body.

The synthetic functions of centrioles and basal bodies are not clear. It has been suggested that these bodies may contain DNA and carry out transcription, but so far only RNA has been reported to be present.

Cilia and Flagella

Cilia and **flagella** are organelles that project from the surface of certain cells and beat back and forth or create a corkscrew action (Fig. 23–21). In many instances, ciliary or flagellar movements propel cells through their environment. In other cases, the cell remains stationary and the surrounding medium is moved past the cell by the beating of its cilia (as in the layer of epithelial cells that lines the trachea or the collar cells lining the internal chambers of sponges).

Cilia are generally shorter than flagella (i.e., 5–10 μm versus 150 μm or longer) and are present in far larger numbers per cell. Flagella usually occur alone or in small groups; occasionally they are present in large numbers, as in a few protozoa and the sperm

cells of more advanced plants. The distinction between cilia and flagella is somewhat arbitrary, because other than differences in their lengths, the structure and action of cilia and flagella of eukaryotic cells are identical. (Bacterial flagella differ in structure and action; see below.)

A eukaryotic cilium or flagellum is composed of three major parts: a central **axoneme** or shaft, the surrounding plasma membrane, and the interposed cytoplasmic matrix (Fig. 23–22). The axonemal elements of nearly all cilia and flagella (as well as the tails of sperm cells) contain the same "9+2" arrangement of microtubules. In the center of the axoneme are two **singlet microtubules** that run the length of the cilium (Fig. 23–23). Projections from the central microtubules occurring periodically along their length form what appears to be an enclosing **sheath**. Each of the central microtubules is composed of 13 **protofilaments**.

Nine **doublet microtubules** surround the central sheath. One microtubule of each doublet (i.e., the A *subfiber*) is composed of 13 protofilaments. The adjoining B *subfiber* is "incomplete," consisting of 11 protofilaments (Fig. 23–23). **Radial spokes** having a *periodicity* of 24, 32, and 40 nm extend from each A

(a)

FIGURE 23-21 (a) Scanning electron photomicrograph of cilia on the dorsal surface of *Parmecium*. The cilia beat rhythmically creating a wavelike appearance (a *metachronal* wave). (Courtesy of E. Vivier. Copyright © 1974 by Elsevier Publishing Co., *Paramecium*, p. 14.) (b) Cilia at the lumenal surface of epithelial cells lining the oviduct. (Courtesy of J. Rhodin, *Histology—A Text and Atlas*, Oxford University Press. Copyright © 1974, New York.) In *Paramecium*, the beating cilia propel the cell, whereas the cilia of the oviduct epithelium serve to push the egg released from the ovary toward the uterus.

(b)

subfiber inward to the central sheath (i.e., they occur at repeating 24-, 32-, and 40-nm intervals along the axoneme's length). Adjacent doublets are joined by **nexin** or **interdoublet links**; the nexin links have a periodicity of 86 nm. Extending from each A subfiber are two **dynein arms**—an "outer" arm and an "inner" arm (see Fig. 23–23). Thin projections from the ends of the dynein arms touch the B subfibers of the neighboring doublets. The outer dynein arms occur with a periodicity of 24 nm, whereas the inner arms have a periodicity of 24, 32, and 40 nm (the same as the radial spokes). The cross section of the cilium depicted in Figure 23–23 is at a level that includes radial spokes, nexin links, and both dynein arms.

Each beat of a cilium or flagellum involves the same pattern of microtubule movement. The beat may be divided into two phases, the *power* or *effective* stroke and the *recovery* stroke (Fig. 23–24). The power stroke

(a)

FIGURE 23–22 *Cilia.* The cilia have been fractured (*a*) and sectioned (*b*) in a plane that runs parallel to the long axis of the organelles. C, cilium; mv, microvilli; L, lumen; k, kinetochore or basal body; cn, ciliary necklace. (Courtesy of Dr. E. Boisvieux-Ulrich. Copyright © 1977 by *Biol. Cellulaire* **30**, 245.)

(b)

occurs in a single plane, but recovery may not occur in the same plane as the power stroke.

The *sliding* microtubule model of ciliary movement is accepted by most investigators. In this model, the doublet microtubules retain a constant length and slide past one another in such a manner as to produce localized bending of the cilium. This activity is powered by ATP hydrolysis and the outer and inner dynein arms have been shown to contain most of the cilium's ATPase activity. The localized bending takes the

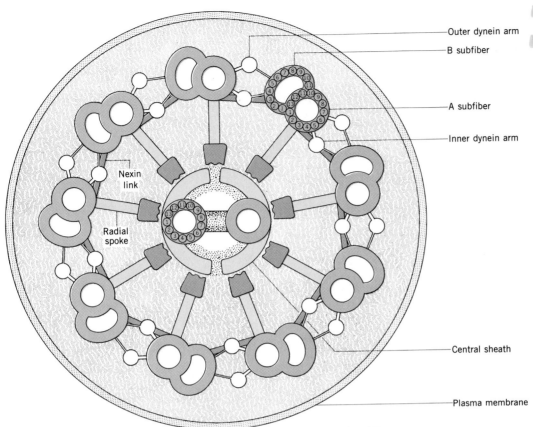

Outer dynein arm

B subfiber

A subfiber

Inner dynein arm

Nexin link

Radial spoke

Central sheath

Plasma membrane

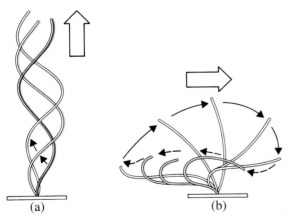

FIGURE 23–24   Two typical patterns of flagellar and ciliary motion. (*a*) Successive waves move toward the tip of the flagellum, propelling the cell (e.g., sperm) in the opposite direction. (*b*) The power stroke of a cilium is similar to the action of an oar in a rowboat; the recovery stroke (color) may not take place in the same plane as the power stroke. The large arrows show the direction of movement of the surrounding liquid.

form of a wave that begins at one end of the organelle and proceeds toward the other (usually, but not always, from base to tip). The localized bending is produced through the cyclic formation and breakage of links between the dynein arms of one doublet and the neighboring doublet. The protein filaments that make up each doublet are rows of tubulin molecules that apparently contain the sites to which the dynein binds. The fact that the sliding of microtubules past one another results in bending of the cilium may be explained by the behavior of the radial spokes that connect the outer nine doublets to the central sheath. In straight regions of the axoneme, the radial spokes are aligned perpendicular to the doublets from which they arise, whereas in the bent regions they are tilted and stretched (Fig. 23–25). Firm attachment of the radial spokes at both ends provides the resistance necessary to translate the sliding of the doublets into a bending action. Indeed, if the radial spokes and nexin links of sperm tails are destroyed by exposure to trypsin, addition of ATP results in the axonemes becoming *longer* and *thinner*, for microtubule sliding is no longer resisted. In effect, sliding is *uncoupled* from bending by elimination of the connections between the doublets and the central sheath.

Analogies are evident between the dynein–tubulin system of cilia and flagella and the actin–myosin system of muscle. However, whereas muscle fibers can only shorten and relax, cilia are capable of a much larger variety of movements. Ciliary and flagellar activity can be in a single plane, in three-dimensional or helical strokes, and can move the cell forward or backward using waves that are propagated from base to tip or from tip to base. At the chemical level, whereas Ca^{2+} appears to activate the actin–myosin system, these ions have the opposite effect on the dynein–tubulin system. The regulation of the Ca^{2+} level in a cilium or flagellum probably involves the plasma membrane surrounding the axoneme. Under normal circumstances (i.e., during periods of continuous beating), the internal Ca^{2+} level is low (about $0.1\mu m$)

FIGURE 23–25 Median longitudinal section through the basal end of beating cilia. Successive radial spokes occur in groups of three (numbered 1, 2, 3, 4, etc.). The spokes in each group are directly opposite one another in a straight cilium and are perpendicular to the doublet microtubules. In a bending cilium, the radial spokes are tilted and stretched. In the cilium to the right, displacement of the spokes results in group 4 being opposite group 5. Δl is a measure of the amount of displacement taking place as diametrically opposite doublets slide tipward on the inner concave side and toward the base on the outer convex side. (Courtesy of R. Warner and P. Satir, *J. Cell Biol.* **63**, 25. Copyright © 1974 by The Rockefeller Press.)

FIGURE 23-26 Diagram of a mature human sperm cell.

whereas Mg^{2+} (necessary to stimulate the ATPase of the membrane) remains in the millimole range. When the membrane is depolarized, the Ca^{2+} level inside the cilium increases and beating ceases. ATP is clearly the source of energy for movement and is produced by cellular respiration. In many cells, mitochondria are located adjacent to the basal body of the cilium or flagellum, and ATP diffuses toward the tip of the organelle. In sperm, a large mitochondrion is an integral part of the tail (Fig. 23-26) and is wrapped in a spiral about the middle piece of the axoneme.

Bacterial Flagella Flagella of bacteria are different from cilia and flagella of eukaryotic cells. The bacterial flagellum is not covered by the plasma membrane, rather it consists of a naked **spiral filament** about 13.5 nm in diameter and 10–15 μm long. The filament is composed of a chain of repeating protein subunits called *flagellin*. At its basal end, the spiral filament is attached to a "hook," which in turn is connected to a **rod** that penetrates the bacterial cell wall and plasma membrane (Fig. 23-27). A number of rings connect the rod with the membrane and wall layers.

Bacterial flagella work by rotation of the rod and

FIGURE 23-27 Diagram of the basal region of a bacterial flagellum. (Courtesy of J. Adler. Copyright © 1971 by American Society of Microbiology, *J. Bact.* **105**, 395.)

hook, which causes the filament to spin. When the filament spins in a counterclockwise direction, the cell is propelled smoothly and in a straight line, but when the spin is clockwise, a chaotic tumbling motion of the cell is observed. A period of counterclockwise rotation is followed by a burst of clockwise rotation, so that the bacterium is continuously set off along new linear paths. The source of energy for the rotation (believed to be generated at the cell membrane) is an electrochemical gradient established by an electron transport system that acts across the plasma membrane.

The Mitotic Spindle

Most studies of chromosome movement during mitosis have focused on the role played by the microtubules that make up the mitotic spindle fibers. The spindle fibers cause three distinct chromosome movements during mitosis (Chapter 20): (1) orientation of sister chromatids, (2) alignment of the centromeres on the metaphase plate, and (3) separation of centromeres and movement of sister chromatids (segregation) to opposite poles of the spindle. The microtubules that occur in the spindle include (1) the **centromere microtubules**, which terminate in a centromere; (2) the **polar microtubules**, which terminate at the poles; and (3) the **free microtubules**, which do not terminate in either a pole or a centromere. All three types can be dissociated into tubulin subunits by colchicine or cold temperatures.

Over the years, several models have been proposed to account for the movements of the chromosomes during anaphase. For example, it has been suggested that the chromosomes are *pushed* apart by spindle fibers developing between centromeres, that they are *pulled* apart by spindle fibers extending between the centromeres and the poles of the spindle, and that chromosomes *migrate* along spindle fibers. The various models are discussed in detail in Chapter 20. Although individual spindle fibers do not stretch or contract per se, they do change in length through either addition or removal of subunits. S. Inoue has shown that free microtubules alternately grow and decrease in length. His *in vitro* studies indicate that the microtubules assemble at one end and disassemble at the other.

Other Cell Movements

The references at the end of this chapter may be consulted for additional examples of the diverse forms of cell and cytoplasmic movement. Regardless of whether the movements are "internal" (such as cyclosis in plant cells) or result in a major change in shape or position of the cell, the present evidence indicates that microfilaments and/or microtubules are fundamental to these activities. In most cases, an interaction between proteins—such as in the actin–myosin or dynein–tubulin systems—with the simultaneous involvement of an ATPase is the underlying biochemical phenomenon.

SUMMARY

The **cytoplasmic matrix** is composed of **cytoplasmic filaments (microfilaments** and **intermediate filaments), microtubules, microtrabeculae,** and the **intertrabecular space**. These structures give shape to the cell, position and move the intracellular organelles, and move the cell as a whole. Cytoplasmic filaments are formed by a linear array of protein subunits (e.g., **actin**), whereas microtubules consist of an array of protein dimers (α and β **tubulin**) forming a hollow cylinder. Cytoplasmic filaments and microtubules interact with each other during a number of cellular activities including muscle contraction, amoeboid movement, and cytokinesis. Assemblies of microtubules form the structural basis of **centrioles** and **cilia** (and **flagella**) and as such are involved in chromosome movements and in ciliary (flagellar) locomotion of cells.

REFERENCES AND SUGGESTED READING

Articles and Reviews

Birchmeier, W., Cytoskeleton structure and function. *Trends in Biochem. Sci.* **9** (4), 192 (1984).

Bretscher, M. S., Endocytosis: Relation to capping and cell locomotion. *Science* **224**, 681 (1984).

Dentler, W. L., Microtubule–membrane interactions in cilia and flagella. *Int. Rev. Cytol.* **72**, 1 (1981).

Dustin, P., Microtubules. *Sci. Am.* **243** (6), 66 (Aug. 1980).

Goldman, R., Berg, G., Bushnell, A., Chang, C-M., Dickerman, L., Hopkins, N., Miller, M., Pollack, R., and Wang, E., Fibrillar systems in cell motility. In *Locomotion of Tissue Cells, Ciba Foundation Symposium*, Vol. 14, Associated Scientific Publishers, New York, 1973, p. 83.

Goodenough, U. W., and Heuser, J. E., Outer and inner dynein arms of cilia and flagella. *Cell* **41**, 341 (1985).

Inoue, S., Fussler, E., Salmon, E., and Ellis, G., Functional organization of mitotic microtubules. *Biophys. J.* **15**, 725 (1975).

Marx, J. L., Organizing the cytoplasm. *Science* **222**, 1109 (1983).

Pollard, T. F., Selden, S. C., and Maupin, P., Interaction of actin filaments with microtubules. *J. Cell Biol.* **99**, 33s (1984).

Porter, K. R., The cytomatrix: A short history of its study. *J. Cell Biol.* **99**, 3s (1984).

Porter, K. R., and Tucker, J. B., The ground substance of the living cell. *Sci. Am.* **244** (3), 56 (Mar. 1981).

Pratt, M. M., ATPases in mitotic spindles. *Int. Rev. Cytol.* **87**, 83 (1984).

Satir, P., How cilia move. *Sci. Am.* **231** (4), 44 (Apr. 1974).

Steinert, P. M., Jones, J. C. R., and Goldman, R. D., Intermediate filaments. *J. Cell Biol.* **99**, 22s (1984).

Steinert, P. M., Steven, A. C., and Roop, D. R., The molecular biology of intermediate filaments, *Cell* **41**, 411 (1985).

Weber, K., and Osborn, M., The molecules of the cell matrix. *Sci. Am.* **253** (4), 110 (Oct. 1985).

Books, Monographs, and Symposia

Borgers, M., and DeBrabander, M., *Microtubules and Microtubule Inhibitors*, North-Holland, New York, 1975.

Goldman, R., Pollard, T., and Rosenbaum, J., *Cell Motility.* Book A: *Motility, Muscle and Non-muscle Cells.* Book B: *Actin, Myosin and Associated Proteins.* Book C: *Microtubules and Related Proteins*, Cold Spring Harbor Laboratory, Cold Spring Harbor, N.Y., 1976.

Reinert, J., and Ursprung, H., *Origin and Continuity of Cell Organelles*. Springer-Verlag, Berlin, 1971.

Schweiger, H. G. (Ed.), *International Cell Biology 1980–81*, Springer-Verlag, Berlin, 1981.

CHAPTER 24

CELLULAR DIFFERENTIATION AND SPECIALIZATION

The development of a complex, multicellular organism such as a human being is accompanied by changes in which a single, unspecialized cell (the fertilized egg or **zygote**) ultimately gives rise to cells and tissues that are highly specialized in structure and function. This progressive transition is characterized by four component processes: **determination, differentiation, growth**, and **morphogenesis**.

New cells are usually not "committed" to a specific function. For example, following fertilization of an egg cell or **ovum** by a **sperm** cell, the resulting zygote divides mitotically many items to produce a ball of cells called a **blastula**. Each cell or group of cells in the blastula subsequently gives rise to specific tissues and organs of the developing embryo. Although a blastula gives rise to a single embryo, should one cell separate from the others at an early blastula stage, that cell may also develop into a complete embryo. In humans, it is this phenomenon that results in identical twins, identical triplets, and so on. At later stages of embryo development, cells that normally become epidermal tissue (i.e., **ectoderm**) can be surgically transplanted to another part of the embryo and there develop into **mesodermal** tissue (e.g., muscle) or

endodermal tissue (e.g., intestinal epithelium). At some early stages of an organism's development, all cells have the potential to develop into any of the variety of different tissue and cell types that characterize that organism. This potential is called **totipotency**. Generally, at some point in development, cells become committed to a specific course of **differentiation**, that is, they differentiate into a specific type of cell or tissue. The process that establishes the fate of a cell is called **determination**. During determination, certain genes become permanently "turned off" and others are sequentially expressed, further and further narrowing the course of differentiation of the cell.

During differentiation cells take on new and specific properties. These can be *structural* (such as the formation of organized arrays of actin and myosin filaments in muscle cells) or *biochemical* (as in the appearance of enzymes of a new metabolic pathway). Differentiation can also take the form of *loss* of existing structures or biochemical processes. For example, in the differentiation of mammalian red blood cells, the nucleus and other cellular organelles are lost, together with the biochemical processes that were provided by these structures.

The determination and differentiation of the cells of a developing organism are accompanied by *growth*, that is, an increase in the size and number of cells comprising the organism. In humans, for example, embryonic and fetal development proceeds from the hundreds of cells comprising the small blastula to a fetus weighing several kilograms and containing hundreds of millions of cells. Growth, in turn, is accompanied by **morphogenesis**—the generation of form and shape in the developing organism. In this process, differentiating and growing cells give rise to the characteristic organizational pattern of the organism. Small masses of cells take on the form and shape of specific and identifiable structures, such as bones, appendages, the brain, and other organs.

The pattern of differentiation of a cell is founded on the nature of the DNA in the cell nucleus. Before a cell can develop into a hair cell of a mammal, a feather cell of a bird, or a scale cell of a reptile, its nucleus must contain a genome whose transcription and translation into enzymes and other proteins allow the cell to differentiate in a specific direction. Moreover, given the appropriate genetic complement, conditions must allow these genes to be expressed. Gene expression is regulated at three levels. The first level involves mo-

lecular or metabolic interactions, such as mass action and allosteric enzyme function. These regulatory mechanisms were discussed in Chapter 11 and will not be pursued here. The second level of control is effected through the interaction between the cell nucleus, the cytoplasm, and the cytoplasmic organelles. The third level of control involves the interactions between the cell as a whole and its environment.

INTRINSIC (NUCLEOCYTOPLASMIC) INTERACTIONS

Transplantation of Cell Nuclei

In 1892, August Weissman proposed that cellular differentiation and specialization might be the consequence of the progressive *loss* of certain genes from a cell's genome during successive mitotic divisions. This hypothesis retained widespread acceptance for more than 50 years until a series of experiments carried out with amphibian egg cells indicated otherwise. The egg cells of amphibians (e.g., frogs and toads) are large enough to be manipulated using microsurgery. In the 1950s, R. W. Briggs and T. H. King enucleated unfertilized egg cells of *Rana pipiens* and replaced the hap-

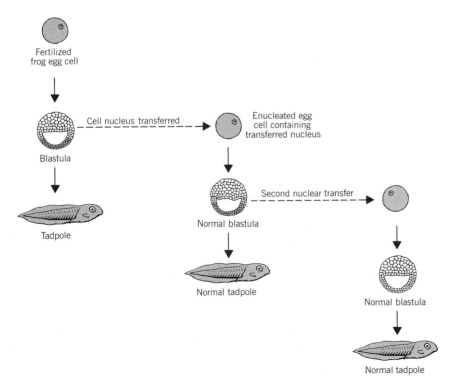

FIGURE 24–1 Effects of serial transplantation of nuclei from blastula stage cells to enucleated egg cells. The transplanted nuclei retain the capacity to express all genes and as a result, normal embryos and tadpoles develop from these eggs.

loid nucleus with a diploid nucleus taken from a cell at the blastula stage of development that would have developed into specific embryonic tissues. The cells produced by the nuclear transfer developed into normal frog embryos and normal tadpoles. This demonstrates that determination had not yet taken place in the nucleus of the blastula stage cell. Moreover, the cell still contains all the genes necessary to produce the different types of cells of the complete embryo. The totipotency of the blastula cell nucleus can be perpetuated even over a series of experimental transfers (Fig. 24–1), that is, if a nucleus is withdrawn from a blastula stage cell following the first transfer and transferred to a second anucleate egg, the embryonic development of that egg proceeds normally; if a nucleus is removed from a cell at this blastula stage and transferred to a third anucleate egg, normal embryonic development again occurs, and so on.

Whereas the observations of Briggs and King demonstrated that nuclear totipotency exists in cells at the blastula stage of embryonic growth, a series of experiments carried out a decade later by J. B. Gurdon

showed that totipotency is retained through much later stages of development. Gurdon removed nuclei from intestinal cells of *Xenopus laevis* tadpoles and implanted them in enucleated unfertilized egg cells. Some of the resulting cells developed into normal tadpoles and normal, fertile adults (Fig. 24–2). Thus, even the nuclei of differentiated intestinal cells contain the full complement of genes essential to the production of a complete organism. The experiments with amphibian cells indicate that there is neither a loss of genes during the many mitoses that lead from the undifferentiated fertilized egg cell to the highly differentiated cells of later developmental stages nor an irreversible loss of the capacity to express these genes.

Experiments with *Acetabularia*

Classic experiments with the unicellular alga *Acetabularia* demonstrated that the cytoplasm has a direct effect on gene expression. During the life cycle of *Acetabularia* (Fig. 24–3), newly formed zygotes attach to the sea bottom by developing rootlike structures called **rhizoids**. A slender **stalk** several centimeters long grows upward from the rhizoid and develops an umbrellalike **cap**. During growth of the cell, which usually takes several weeks, the nucleus remains in the rhizoid. When growth is complete, the nucleus divides mitotically many times and the resulting daughter nuclei migrate up the stalk into the cap; there meiosis produces haploid nuclei that become encysted. Eventually, the cysts are released, break open, and the gametes swim out. Fusion of two gametes to form a new zygote initiates a new cycle.

Acetabularia is capable of regenerating lost parts. For example, if the cap is cut off, the cell grows a new cap. If the new cap is removed, another cap is grown, and this cycle can be repeated many times with the same individual. If the cap is cut off and the cell nucleus then removed by microsurgery, a new cap is regenerated, but regeneration will not occur if the experiment is repeated with the same anucleate organism. Regeneration of the second cap in the absence of a nucleus indicates that nuclear information had previously been transferred to the cytoplasm and persisted there, at least through the time required to form the new cap. The information in the cytoplasm for regenerating the cap takes the form of messenger RNA.

Different-shaped caps are formed by different species of *Acetabularia*. *Acetabularia mediterranea* has a

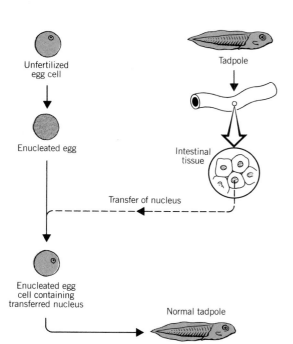

FIGURE 24–2 Effects of transplantation of nuclei from intestinal epithelial cells to enucleated egg cells. Normal embryos and tadpoles develop, indicating that even the differentiated intestinal cell's nucleus contains a full set of genes. Moreover, the capacity to express these genes was not permanently lost during differentiation into intestinal tissue.

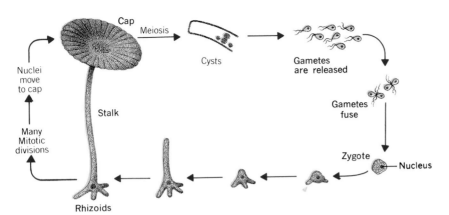

FIGURE 24–3 The life cycle of *Acetabularia* (see text for details).

complete cap and *A. crenulata* has a fingerlike cap. The interactions of the cytoplasm and nucleus are strikingly illustrated by grafting experiments using these two species (Fig. 24–4). If the foot of *A. crenulata* (which contains the nucleus) is joined to an anucleate piece of the stalk of *A. mediterranea*, the two fuse and a cap develops at the top of the stalk. The cap assumes characteristics of both species, presumably resulting from the combined influences of the cytoplasmic mRNA that was present in the *A. mediterranea* stalk and the nucleus and cytoplasmic mRNA that were in the *A. crenulata* foot. If the cap produced by the grafted cell parts is removed and a second cap allowed to form, the new cap has the *A. crenulata* shape. Thus, the cap information (mRNA) in the original cytoplasm of the stalk from *A. mediterranea* was

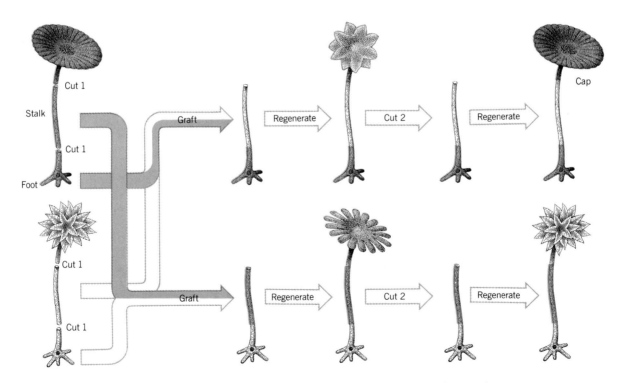

FIGURE 24–4 Results of grafting experiments using nucleated and anucleate segments of *Acetabularia mediterranea* (top) and *Acetabularia crenulata* (bottom). (See the text for description of the experiments.)

"lost" after one regeneration, whereas the cap information present in the *A. crenulata* nucleus is expressed.

EXTRINSIC (ENVIRONMENTAL) EFFECTS ON DIFFERENTIATION OF CELLS

A cell's differentiation is directly affected by neighboring cells and the surrounding environment. The fates of the late blastula–early gastrula stage cells of frog embryos (as well as embryos of a number of other species) have been determined and *fate maps* have been made of the types of tissues formed from each group of cells (Fig. 24–5). Even by the late blastula stage, cell determination has not yet taken place. For example, if the cells that normally give rise to eye lens tissue are surgically exchanged with cells that form gut tissue, embryonic development is still normal. The transplanted presumptive lens cells are influenced by their new position in the embryo and develop into gut tissue. Likewise, the transplanted presumptive gut cells develop into lens tissue. Apparently, influenced in some manner by their new surroundings, the transplanted cells follow an altered course of differentiation.

The direct effects of environmental factors on simple plants like *Fucus*, a brown marine alga, can be vividly demonstrated. The fertilized *Fucus* egg cell has no apparent topological polarity. Fifteen hours after fertilization a rhizoid develops, and this is followed by nuclear division and formation of a wall that separates

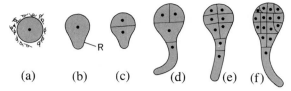

FIGURE 24–6 Early stages of development of the *Fucus* embryo.

the rhizoid from the *apical* cell. This first division establishes a polarity for the embryo (Fig. 24–6). Environmental factors can be shown to influence *where* on the fertilized egg the rhizoid will develop. In a cluster of fertilized *Fucus* eggs, the rhizoid develops on that side of each egg that is closest to the center of the group. When cells are placed in a temperature gradient, the rhizoids develop on the warm side of each egg. In a pH gradient, they develop on the more acid side, and in white light, they form on the shaded side.

As we have seen, the expression of a cell's genes is influenced by its cytoplasm, the adjacent cells, and the environment. The cellular differentiation that results from these influences produces a number of uniquely specialized cells. Specialized cell structure and organization is an overt reflection of specialized cell functions. Specialized cellular functions are in turn founded on the differentiation of a specialized structure. Among specialized cell types, muscle cells, nerve cells, and blood cells have been more extensively studied than any other form. Some of their specializations are considered in the following sections.

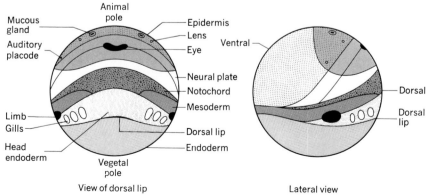

FIGURE 24–5 Fate map of the different cell areas of the frog blastula. The cells in each area of the blastula are associated with the development of specific types of tissues and organs.

RED BLOOD CELLS

The differentiation of mammalian red blood cells in bone marrow (a process called **erythropoiesis**) is one of the most striking examples of specialization occurring in nature. In adults, erythropoiesis begins with a *pluripotent* (i.e., multipotent) **stem cell** that gives rise in 7 to 10 days to mature, hemoglobin-filled erythrocytes. As noted earlier in the chapter, differentiation and specialization of cells may be accompanied not only by the acquisition and development of specialized structures but also by the *loss* of internal structures or physiological properties. The latter is the case during erythrocyte differentiation, for mature red blood cells lack nuclei, mitochondria, endoplasmic reticulum, Golgi bodies, ribosomes, and most other typical cell organelles.

The mature erythrocyte is a relatively simple cell, delimited at its periphery by a plasma membrane and containing internally a highly concentrated, paracrystalline array of hemoglobin molecules used for oxygen transport. The apparent simplicity of the maturing erythrocyte is the principal reason for its selection over most other kinds of cells as the preferred object for studying plasma membrane structure and protein (e.g., hemoglobin) structure and biosynthesis. These subjects, together with the contributions made by intensive study of the erythrocyte, are treated in some depth in Chapters 4, 15, and 22 and will not be dealt with here.

Erythropoiesis

The development and differentiation of the mammalian red blood cell is depicted in Figure 24–7. De-velopment takes place in the extrasinusoidal stroma of the bone marrow and begins with pluripotent stem cells capable of proliferating leukocytes (white blood cells) as well as erythrocytes. When primitive stem cells undergo division, one of the daughter cells remains undifferentiated and pluripotent, so that depletion of marrow stem cells does not normally take place. The erythropoietic activity of the bone marrow is under hormonal influence, increasing or decreasing according to the level of circulating *erythropoietin* (produced in the cortical region of the kidneys and secreted into the bloodstream). The erythrocyte progenitors show an increasing sensitivity to erythropoietin through the proerythroblast stage and a parallel, ever-decreasing proliferative potential (Fig. 24–8). As a result, by the proerythroblast stage, the cells are irreversibly committed to the maturation sequence leading to erythrocytes.

The mRNAs for the various globin chains of hemoglobin appear and increase in quantity in the proerythroblast and erythroblast stages, and this is followed by the synthesis and accumulation of hemoglobin. By the late erythroblast and normoblast stages, the synthesis of hemoglobin accounts for more than 95% of all protein synthesis occurring in the cells. Hemoglobin synthesis is concluded in the reticulocyte stage and is accompanied by the progressive dissolution of internal cell structure (including the loss of the nucleus) and residual nucleic acids. Late reticulocytes leave the bone marrow and enter the circulating blood, where in the ensuing hours they lose their granulation and are transformed into the biconcave disks typical of mature erythrocytes. Complete differentiation and maturation from stem cell to the erythrocyte takes from 7 to 10 days.

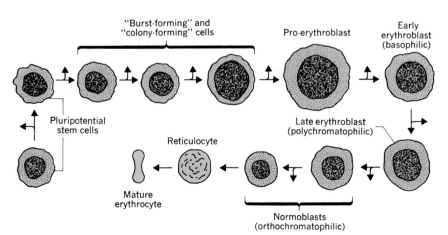

FIGURE 24–7 Differentiation and maturation scheme of erythrocytes of mammals. Differentiation begins with pluripotent stem cells in the bone marrow and ends 7 to 10 days later with the release of reticulocytes or mature erythrocytes into the bloodstream. Each double-headed arrow represents a mitotic division into two daughter cells only one of which is shown. Normoblasts become reticulocytes and then erythrocytes *without* division (the cell nucleus and other organelles are lost at the normoblast stage).

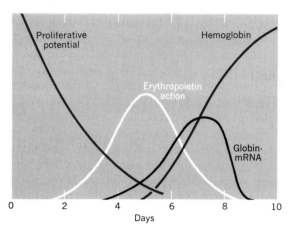

FIGURE 24–8 Relationship between the time course of differentiation and maturation of erythrocytes in the bone marrow and the proliferative potential of progenitor cells, sensitivity to erythropoietin influence, accumulation of mRNA for globin chains, and the buildup of cellular hemoglobin.

Because of its highly differentiated state, the mature erythrocyte is incapable of further proliferation. In humans, the average life span is 120 days. As there are 5 billion red blood cells in each cubic centimeter of blood, a few simple calculations* quickly reveal that in a typical adult the differentiation and maturation of 3 million erythrocytes is completed *each second*! Obviously, an appreciable proportion of the body's energy and resources is continuously consumed to support erythropoietic activity. This is in stark contrast with other highly differentiated cells such as those of muscle and nerve, whose proliferation ceases shortly after birth.

Genetic and Molecular Basis of Erythrocyte Differentiation

Because of the intensity with which the red blood cell has been studied by biochemists, cell biologists, molecular geneticists, physiologists, and others, the differentiation and specialization of this cell are better understood in molecular terms than any other; accordingly, erythrocyte differentiation may serve as a model for the genetic and molecular events that underlie cell differentiation generally.

As discussed in Chapter 4, development in humans is accompanied by the sequential appearance of six different hemoglobins composed of various combinations of seven different globin chains. The structural genes coding for these chains represent only a minuscule fraction of the total genetic complement of the pluripotent progenitor cells. In humans, there are two pairs of alleles (i.e., four genes) for alpha globin chains, one pair for beta chains, two pairs for gamma chains, one pair for delta chains, one pair for epsilon chains, and two pairs for zeta chains. These genes account for less than 0.0002% of the total DNA content of the nucleus, and yet globin chain synthesis accounts for more than 95% of all protein synthesis taking place in the red cell by the later stages of maturation. It is clear that during differentiation, the *progeny of the stem cells are committed to the highly selective expression of only a small number of genes.*

The globin chain genes are distributed between two pairs of homologous chromosomes—alpha and zeta chain genes on one chromosome pair and beta, gamma, delta, and epsilon chain genes on the other pair (see Fig. 22–25). The zeta, epsilon, and gamma chain genes are expressed only during embryonic and fetal development, whereas the beta and delta chain genes are expressed from midfetal stages through adulthood. Alpha chain genes are expressed throughout life beginning at an early embryonic stage. Therefore, superimposed on the highly selective expression of globin chain genes in maturing erythrocytes is the *temporal* expression of certain members of this family of genes during organismal development. Expression of these genes is also coordinated between the two different chromosomes bearing the members of this gene family. A most interesting finding of recent years is that the order in which the globin chain genes are arranged on their respective chromosomes parallels the order in which the various genes are expressed during embryonic and fetal development. Also present in the chromosome regions containing these genes are nucleotide sequences encoding polypeptide chains that are very similar in primary structure to the globin chains, but that are never expressed. These "silent genes" or **pseudogenes** may be genetic vestiges of the evolutionary process. Certainly, the inclination to draw a parallel between phylogeny and ontogeny at the organismal and cellular levels is unavoidable.

---

*Because about 8% of body weight is blood and blood contains 5 billion red cells per cubic centimeter, a person weighing 78 kg (172 lb) contains 6.24 kg or about 6.3 liters of blood. Therefore, altogether there would be 3.15×10^{13} circulating erythrocytes. With all of these turning over in an interval of 120 days or 1.04×10^7 seconds, this means that 3.03×10^6 red blood cells reach full maturity and are released into the bloodstream each second.

Cellular DNA includes not only structural genes but also sequences vital to DNA organization within the chromosomes and the coordination of gene expression during differentiation. About 60% of all the nucleotide sequences in the DNA of stem cells belong to the "unique sequence" class in which only a few copies of each sequence are present. The structural genes for the globin chains belong to this class. Highly repetitive DNA accounts for about 10% and consists of short sequences repeated in tandem many thousands of times. These regions are not transcribed and are believed to be located in condensed chromatin (see Chapter 20). The balance of the DNA—about 30%—is moderately repetitive (repeated up to several hundred times) and includes the genes for histones and the ribosomal RNAs (see Chapter 22).

Some moderately repetitive DNA sequences are located adjacent to the structural genes for globin and other proteins and may be involved in the coordination of gene activity. For example, by binding to the moderately repetitive sequence, regulatory substances could "turn on" the transcription of *physiologically related* genes. In the case of the red blood cell, such a gene set would be represented by (1) the globin chain genes, (2) the genes for the heme synthetic enzymes, and (3) the genes for the cell surface antigens (which serve as the basis for blood typing). Models accounting for gene regulation at this level were considered in Chapter 11.

As noted in Chapter 20, the genes of eukaryotic cells are associated with histones and other basic proteins to form *chromatin*, which is composed of repeating *nucleosome* units. Euchromatin has the more open structure and is transcriptionally active, whereas the condensed heterochromatin is not transcribed. Experiments with developing erythrocytes indicate that the globin chain genes are included in nucleosomes of condensed chromatin during the early stages of differentiation and that these are converted to the open form just before globin mRNA synthesis begins. The transcription of globin DNA sequences by *RNA polymerase II* that occurs when these genes are "turned on" can be traced to a modification of the associated histones and an interaction with nonhistone proteins. A selective expression of globin mRNA genes occurs in erythroblasts, with the result that globin mRNA is produced in amounts 100 times greater than expected on the basis of the proportion of total template DNA represented by the globin genes. Globin chain synthesis parallels the appearance of globin mRNAs in erythroblasts.

Although globin chain synthesis accounts for the overwhelming majority of all protein synthetic activity in the maturing erythrocyte, a number of other structural genes are expressed (e.g., those for the enzymes of glycolysis, enzymes for initiation, elongation, and termination of globin chain assembly, enzymes of the metabolic pathway for heme synthesis, enzymes and structural proteins of the plasma membrane, and so on). Although all protein synthesis comes to a halt during reticulocyte maturation, a red cell retains a limited metabolic capacity after release from the bone marrow. For example, glycolytic activity provides the ATP needed to maintain the sodium and potassium pumps of the erythrocyte membrane (see Chapter 15) and for other energy-requiring processes. What limited metabolism is retained by the mature cell serves to sustain it during its 120-day and 700-mile journey through the circulatory system.

The progressively selective expression of genes in the maturing red blood cell is accompanied by the loss of internal structure so that in the mature state the only remaining organelle is the plasma membrane itself. Internally, the cell consists primarily of a highly concentrated (30% by weight) crystallike arrangement of hemoglobin molecules suited to the cell's principal function—oxygen transport.

Morphological and Physiological Specialization of Red Blood Cells

Little cell growth occurs during the periods between the successive mitotic divisions of erythropoiesis. As a result, the mature erythrocyte is among the smallest cells of the body. In humans, the average red blood cell has a volume of only about 100 μm^3. The natural shape of the cell is that of a biconcave disk (Fig. 24–9), although the cell is flexible and its shape is easily distorted when the cell passes through narrow capillaries. The factors that maintain this rather unusual cell shape have fascinated scientists for more than two centuries. Prior to the discovery of the existence of cytoskeletons, the prevailing hypothesis suggested that the biconcave shape resulted from the chemical nature, arrangement, and interaction of the protein and lipid molecules that comprised the plasma membrane. However, it is now apparent that the biconcave shape is due to the modest cytoskeleton lying just beneath the plasma membrane (see Chapter 15). Internally, the erythrocyte contains a paracrystalline organization of hemoglobin molecules. Erythrocytes of individ-

FIGURE 24–9 Scanning electron photomicrograph of erythrocytes showing their biconcave disk shape. (Courtesy of Dr. M. A. Lichtman.)

DIFFERENTIATION OF B LYMPHOCYTES

A dramatic example of the degree of specialization that can accompany cellular differentiation is provided by a family of white blood cells called **lymphocytes**. There are two major types of lymphocytes: **T lymphocytes** and **B lymphocytes**. Both originate from pluripotent stem cells in the hemopoietic tissues of the embryo (i.e., the liver and bone marrow). From there the stem cells migrate to and colonize the *primary lymphoid tissues*; these tissues include the **thymus** (found in both mammals and birds) and the **bursa of Fabricius** (found only in birds, there is no mammalian equivalent). The stem cells that colonize the thymus gland (i.e., *prothymocytes*) give rise there to the T lymphocytes ("T" = "thymus-derived"). Many T lymphocytes leave the thymus and seed *secondary* lymphoid tissues such as lymph nodes, spleen, tonsils, and adenoids. Stem cells colonizing the bursa of Fabricius give rise to the B lymphocytes ("B" = "bursa-derived"). In mammals, which have no bursa, B lymphocytes are derived from colonized secondary lymphoid tissues and from the bone marrow itself.

Both kinds of lymphocytes protect the body against infection by microorganisms, parasites, and other agents and are responsible for the body's capacity to develop an **immunity** against certain diseases. In the case of the B lymphocytes, immunity stems from the production and secretion of **antibodies** or **immunoglobulins** into the bloodstream by these cells. The immunoglobulins secreted by the B lymphocytes combine with **antigens** in the surfaces of viruses, bacteria, and other agents of infection, ultimately leading to the destruction of the antigen-bearing material (see Chapter 4 for a discussion of immunoglobulin structure and interaction with antigens). The body's lymphoid tissues can give rise to more than a million different lines of B lymphocytes, each line developing an independent capacity for producing antibodies against a specific **antigenic determinant**.

The antibodies secreted by lymphocytes are the products of the transcription and translation of two types of genes called V and C genes. Despite the presence of many hundreds of different V and C genes in each lymphocyte, only one pair of V and C genes is expressed by each line of lymphocytes. The specific V and C genes expressed by an antibody-producing lymphocyte are not only expressed to the exclusion of other V and C genes but also to the exclusion of most other structural genes in the cell.

Like the B lymphocytes, T lymphocytes also protect

uals with *abnormal* hemoglobins (hemoglobins containing one or more amino acid substitutions) often lack the normal biconcave shape. The most notorious example is the "sickle" or crescent shape of erythrocytes in individuals with *sickle-cell anemia*. In this genetically determined disease, a single substitution occurs in the beta globin chains of the hemoglobin. Under conditions of oxygen deprivation or shortage, the hemoglobin molecules aggregate to form long fibers and these progressively group together into bundles. The bundles of hemoglobin molecules make the cells less flexible and eventually overcome the constraints of the cytoskeleton, deforming the cells and causing their sickling.

Oxygen enters and leaves the erythrocyte by diffusion and the biconcave shape facilitates oxygen flux by increasing the surface area-to-volume ratio of the cell. Oxygen inside the cell forms a reversible combination with hemoglobin (see Chapter 4), binding to hemoglobin as the blood circulates through the capillary networks of the lungs (where the net oxygen flux is directed into the erythrocyte) and being released as the blood circulates through oxygen-deficient tissues. The disk shape of the erythrocyte induces the formation of **rouleaux** or long *stacks* of cells, with the result that far greater numbers of cells can pass through the narrow capillaries when arranged in such a regimented manner than when the cells are freely and independently suspended. Experimental elimination of the biconcave shape using hypotonic solutions eliminates the capability of erythrocytes to form rouleaux.

the body against infection. Whereas the immunity conferred by the B lymphocytes is *humorally mediated* (i.e., it stems from antibodies secreted into the blood plasma), T lymphocytes provide *cell-mediated* immunity, that is, T lymphocytes interact directly with bacteria, viruses, or parasitic cells. An extended discussion of humoral- and cell-mediated immunity is included in Chapter 25.

Hybridomas and Monoclonal Antibodies Because of the highly specific nature of the antibodies produced by differentiated B lymphocytes, cultures of the many lines of these cells have long been contemplated as a source of antibodies for clinical application. Whereas the highly differentiated B lymphocytes cannot be cultured, hybrid cells formed by fusing malignant *myeloma* cells with B lymphocytes do proliferate in culture and synthesize their respective antibodies. These **hybridomas** can be separately cloned to yield large cultures of cells producing specific antibodies. Antibodies produced by this process are called **monoclonal antibodies** (see also Chapter 21).

MUSCLE CELLS

Vertebrates possess three types of muscle tissue: (1) **smooth** muscle, the contractile elements of most of the digestive system and most visceral organs; (2) **cardiac** muscle, found only in the heart; and (3) **striated** (or **skeletal**) muscle, responsible for most of the gross movements of the body, especially the skeleton. Each of these tissues is composed of cells called **muscle fibers**, which contain cytoplasmic filaments of *actin*, *myosin*, and other proteins responsible for the contractile nature of muscle. The filaments are arranged differently in each of the tissue types. The arrangement is most highly organized in striated muscle, and it is this type of muscle tissue that has been most extensively studied.

The organization of striated muscle is shown diagrammatically in Figure 24–10. During embryonic development, the muscle fiber is formed by the end-to-end fusion of many cells into a continuous tubelike structure. This explains why striated muscle fibers are such long cells and are *multinucleate*. The plasma membrane of the muscle fiber is called the **sarcolemma**; in addition to its exceptional length, the sarcolemma is characterized by numerous porelike invaginations that extend into the **sarcoplasm** (cytoplasm)

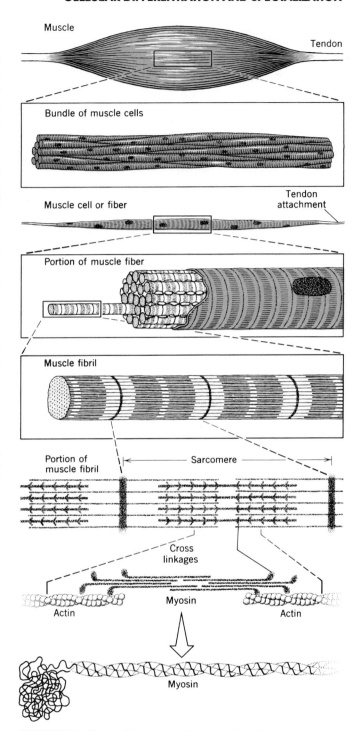

FIGURE 24–10 Cellular, subcellular, and molecular organization of striated muscle tissue.

FIGURE 24–11 The sarcoplasmic reticulum and T system of striated muscle cells.

at right angles to the long axis of the cell. These transverse or **T system** extensions of the sarcolemma (Fig. 24–11) make contact with most of the internal **myofibrils**—the contractile units of the cell. Each myofibril contains a large number of *thick* (15-nm diameter) and *thin* (7-nm diameter) filaments called

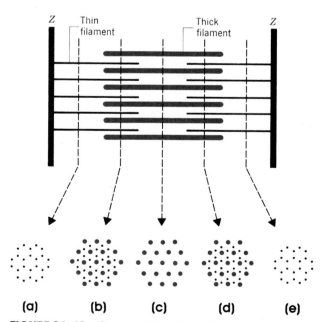

FIGURE 24–12 Cross sections through various positions of a sarcomere. The thin (actin) filaments are arranged parallel to the thick (myosin) filaments and form an hexagonal array in regions of overlap (i.e., regions *b* and *d*).

myofilaments that run parallel to each other and to the long axis of the cell. Seen in cross section (Figs. 23–7 and 24–12), the filaments are arranged in a repeating geometric pattern. The thick filaments are equidistant from each other with each surrounded by six thin filaments in a hexagonal array (Fig. 24–12*b*).

During the contraction of the cell, the thick and thin filaments slide past one another. When examined microscopically in longitudinal section, each myofibril reveals areas of different density (Fig. 24–13). The alternating light and dark areas are respectively called the **I** (i.e., **isotropic**) and **A** (i.e., **anisotropic**) bands. At the center of each I band there is a dark line called a **Z line**, and those portions of a myofibril that extend from one Z line to the next are termed **sarcomeres**. One end of each thin filament is anchored in the Z line and the other end projects toward the center of the sarcomere. The thick filaments are sandwiched between the thin filaments but are not attached to the Z lines. Within each sarcomere, the A bands extend from one end of a stack of thick filaments to the other. As seen in Figures 24–12 and 24–13, there is a region within each A band that is devoid of actin filaments; this region appears somewhat less dense than the remainder of the A band and is called an **H zone**. Finally, at the center of the H zone is the somewhat darker **M line**. As contraction occurs, the thick and thin filaments slide past each other, so that the H zone disappears and the I band becomes much narrower (Fig. 24–14). The Z lines are drawn closer together and the myofibril as a whole becomes thicker.

The myosin monomers that comprise the thick fila-

FIGURE 24–13 Relationship between the light and dark zones seen in electron photomicrographs of striated muscle tissue and the arrangements of thick and thin filaments. (Photomicrograph courtesy of R. Chao.)

FIGURE 24–14 Changes in the relative positions of the thick and thin filaments in neighboring sarcomeres during contraction and relaxation.

ments are fibrous proteins containing a "head" and "tail" portion (Fig. 24–15). Each myosin molecule is composed of two *heavy* polypeptide chains of MW 200,000 and four *light* chains of MW 20,000. Myosin can be cleaved into three pieces using proteolytic enzymes. Treatment with *trypsin* releases part of the tail (i.e., *light meromyosin*, **LMM**) from the remainder of the molecule. Light meromyosin has no ATPase ac-

tivity and cannot combine with actin. The remaining portion of the molecule, called *heavy meromyosin* (**HMM**), contains ATPase activity and binds actin. Treatment of HMM with the enzyme *papain* severs the head (the portion containing the ATPase activity) from the tail.

In thick filaments, the myosin molecules are arranged with their tails parallel to each other and their

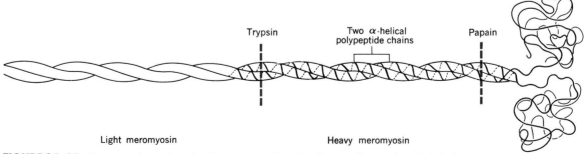

FIGURE 24–15 The myosin molecule. The enzyme trypsin cleaves the molecule into two pieces called light meromyosin and heavy meromyosin. The enzyme papain severs the head region (i.e., the region containing ATPase activity) from the tail. (Redrawn with permission. Copyright © 1979 by Worth Publishing, Inc.; A. L. Lehninger, *Biochemistry*, p. 512.)

heads projecting away from the long axis of the filament at intervals (Fig. 24–16). No heads are present in the center of the filament, this region coinciding with the H zone. The heads of the myosin molecules form cross-bridges with adjacent actin filaments. Two functional domains can be identified in each head. One domain contains the two cross-bridges that link the myosin to actin, and the other domain provides a flexible connection or hinge between the cross-bridges and the remainder of the myosin molecule.

Thin filaments are composed primarily of actin, but also present are small amounts of *tropomyosin, troponin, α actinin*, and *β actinin* (see Fig. 24–17). In the ionic environment of the cell, the actin exist in the fibrous (i.e., F-actin) form consisting of two chains of actin monomers coiled about each other. When extracted from muscle tissue and dialyzed, actin becomes globular (i.e., G-actin) as the monomers sepa-

rate. Each actin monomer is a single polypeptide of MW 43,000. The proteins tropomyosin and troponin are arranged at intervals along the actin filament and serve as regulatory agents in contraction. Actinin may function in polymerization of the actin monomers.

Muscle Contraction

When mixed together in solution purified myosin and actin molecules will bind to each other by forming cross-bridges. Experimentally, this is noted by an increase in the viscosity of the solution. Cross-bridges between actin and myosin filaments also exist in intact muscle fibers. It is the oar-like action of these cross-bridges accompanied by filament sliding that is the basis of contraction.

It has been known for some time that the cyclic con-

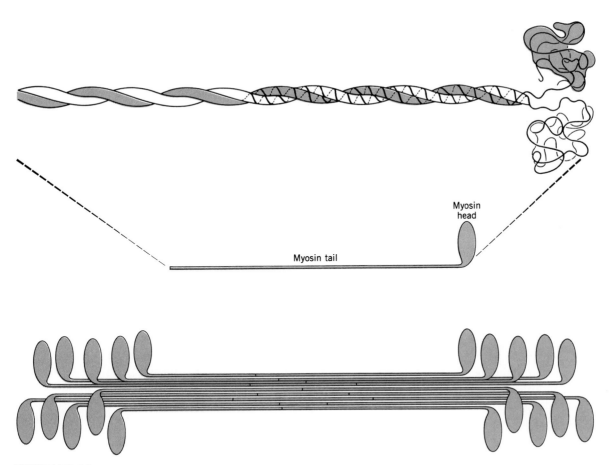

Myosin head

Myosin tail

FIGURE 24–16 The thick filament. In thick filaments, the myosin molecules are arranged with their tails parallel to each other and their heads projecting away from the long axis of the filament.

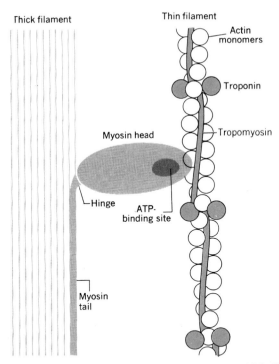

FIGURE 24–17 Association of the thick and thin filaments in striated muscle. The head of the myosin molecule is oriented at an angle to the long axis of the filament and forms a cross-bridge with the thin filament. The thin filaments are composed of two helical rows of actin monomers. The regulatory proteins tropomyosin and troponin occur in regular association with the actin filament. (Redrawn with permission. Copyright © 1979 by Worth Publishing, Inc.; A. L. Lehninger, *Biochemestry*, p. 397.)

the "power stroke" (i.e., contraction per se), (3) relaxation, and (4) the generation of energy for the process.

Triggering (Initiating) Contraction Prior to contraction the following conditions exist: (1) the cross-bridges between the myosin heads and the actin filaments are in a weakly interacting state, with each myosin head oriented at right angles to the filament's long axis; (2) the ATP and Mg^{2+} concentrations of the sarcoplasm are high, whereas the concentration of Ca^{2+} is low (Fig. 24–18, step 1); and (3) the myosin heads contain tightly bound ADP and phosphate molecules. Strong cross-bridge interaction between myosin and actin is prevented by the regulatory proteins troponin and tropomyosin.

Normally, contraction of a muscle cell is initiated when a nerve impulse causes depolarization of the sarcolemma at the *myoneural junction* (i.e., the junction between a muscle and nerve cell). The depolarization quickly spreads across the sarcolemma and down through the complex system of invaginations forming the sarcoplasmic reticulum or T system. Depolarization is accompanied by a change in the permeability of the sarcoplasmic reticulum membranes to ions. Of special importance are the resulting movements of calcium ions, which are rapidly released into the sarcoplasm from storage vesicles elaborated by the sarcoplasmic reticulum. Calcium ions bind to troponin causing a conformational change in the complex and a shift in the position of tropomyosin. As a result, the binding sites on actin become available for further and stronger interaction with the myosin.

traction and relaxation of muscle cells is intimately tied to the metabolism of ATP. For example, when ATP and Mg^{2+} are added to an actin/myosin mixture, the solution's viscosity rapidly diminishes and the ATP is hydrolyzed. This is interpreted as an ATP-elicited disengagement of the cross-bridges between actin and myosin. The cyclic breakage and reformation of these cross-bridges accompanied by their oar-like movement is believed to be the *in situ* source of filament sliding. The ATP hydrolysis that accompanies this action is catalyzed by an ATPase present in the head of the myosin molecule. The exact role played by ATP hydrolysis during filament sliding has been a subject of controversy for many years and still remains elusive. A full cycle from relaxation through contraction and back to relaxation involves four phases: (1) triggering (or initiating) contraction, (2)

The Power Stroke As just noted, calcium ions entering the sarcoplasm bind to troponin, causing the troponin and tropomyosin to undergo a conformational change. As a result, a binding site on a G-actin monomer is exposed and reacts with the actin binding site of the myosin head (Fig. 24–18, step 2). The myosin head immediately undergoes a conformational change, altering to 45° its angle to the tail of the myosin molecule. Consequently, the actin filament is caused to slide about 12 nm. This movement is the power stroke (Fig. 24–18, step 3) and occurs at almost the same moment for thousands of pairs of actin and myosin molecules, thereby positioning a new set of myosin heads and actin binding sites near one another. The process is repeated again and again, causing each actin filament to move a total distance of about 250 nm.

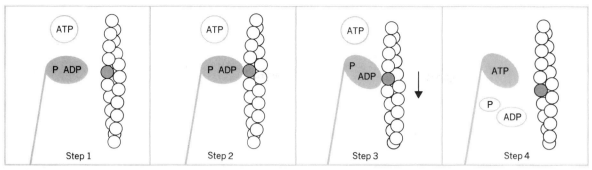

FIGURE 24–18 Steps in the contraction process. For easy reference, the specific actin monomer involved in the round of contraction is shown in color (see text for additional details). (Courtesy of K. C. Holmes. Copyright © 1973 by Nature; *Nature New Biol.* **241**, 226.)

The transition between the 90° and 45° conformation is accompanied first by the release of the bound phosphate from the myosin head and then by the release of ADP (Fig. 24–18, step 3). It is to be noted that no ATP is consumed during the power stroke. ATP binds to the myosin head, thereby occupying the site just vacated by ADP (Fig. 24–18, step 4). Binding of the ATP is accompanied by a transient detachment of the cross-bridges and transition back to the 90° conformation. With the myosin head once again oriented at right angles to the filament's long axis, the ATP is hydrolyzed (Fig. 24–18, step 1). Thus, the filament system is primed for another round of sliding activity as the cycle is ready to be repeated.

Relaxation The cessation of nervous stimulation allows the sarcolemma of the muscle cell to reestablish its polarity and regain its normal permeability. As the membrane repolarizes, calcium ions that entered the sarcoplasm are actively transported back into vesicles of the sarcoplasmic reticulum. The transport of Ca^{2+} is effected by an enzyme of the sarcoplasmic reticulum (i.e., a Ca^{2+}-dependent ATPase); this enzyme constitutes a large proportion of the sarcoplasmic reticulum protein.

With the removal of Ca^{2+} from the sarcoplasm, the conformation of troponin changes so that the binding site on the actin is again shielded from the myosin head. The thick and thin filaments slide back to their former positions, their movement being passive (i.e., resulting from the natural elasticity of the muscle tissue or the effects of gravity) or as a result of the contraction of *antagonistic* muscles (muscles pulling in the opposite direction).

ATP Consumption During Muscle Activity Large amounts of ATP are consumed during muscle activity. At least two ATP molecules are consumed for each interaction between a myosin molecule and actin (i.e., there are two ATP-binding sites per myosin molecule), and one ATP is required for each calcium ion returned to the sarcoplasmic reticulum (i.e., two-thirds of ATP consumption is attributed to contraction and one-third to relaxation).

The ultimate source of ATP for the muscle fiber is the phosphorylation of ADP during cellular glycolysis and respiration. However, the *immediate* source is the phosphorylation of ADP using *phosphocreatine* (Fig. 24–19), a substance uniquely present in high concentration in muscle cells. Phosphate is rapidly transferred to ADP through the action of *creatine kinase*, which maintains an almost constant level of ATP in the muscle fiber during normal levels of contractile activity. (Experimentally it is possible to deplete the phosphocreatine supply by causing repeated contractions while blocking cellular glycolysis and respiration.) The restoration of the depleted phosphocreatine is brought about first by the synthesis of ATP during glycolysis and/or oxidative respiration and then through the action of creatine kinase, which now catalyzes the transfer of phosphate in the opposite direction (Fig. 24–20).

FIGURE 24–19 Chemical structure of phosphocreatine.

FIGURE 24–20   Action of *phosphocreatine kinase.*

There are two types of striated muscle tissue—*red* and *white*—and the relative amounts of ATP contributed in each tissue by glycolysis and respiration vary with each type. The red (or slow) fibers are rich in mitochondria and cytochromes and employ oxidative phosphorylation for ATP production. Their demand for oxygen is satisfied in part by the presence in these cells of large quantities of the oxygen-storing protein *myoglobin* (Chapter 4). Fatty acids are a major substrate of these cells and are converted to acetyl-CoA. Acetyl-CoA enters the Krebs cycle, producing CO_2 and the reducing power for oxidative phosphorylation. White fibers contain few mitochondria and little myoglobin. In these fibers, glycolysis is the primary source of ATP.

Red fibers are found in greater number in muscles with slow, rhythmic contraction properties such as the flight muscles of birds. They are slow contracting because they have low myosin-ATPase activity. They resist fatigue because ATP production via oxidative phosphorylation is able to keep pace with ATP hydrolysis. White fibers predominate in quickly contracting muscles, such as the jumping muscles of frogs. These fibers are fast contracting because they have high myosin-ATPase activity. They fatigue quickly because ATP production via glycolysis does not keep pace with ATP breakdown. The proportion of red to white fibers in the leg muscles of humans appears to be determined genetically and is believed to account for individual differences between long distance and sprint runners.

During vigorous activity, blood supplying oxygen to and removing carbon dioxide from the muscles may be inadequate. Oxygen depletion of muscle promotes glycolysis and leads to the accumulation of lactic acid. On relaxation, oxygen reaches the muscle in adequate amounts and the lactic acid is removed by oxidation. Following a lengthy period of muscle exertion one's breathing rate may remain above normal for several minutes even though exercise is stopped. The extra oxygen consumed during this period is used to oxidize the accumulated lactic acid and is called the *oxygen debt.* Excess lactic acid is transported to the liver, where it is "recycled" first into glucose and then into glycogen.

NERVE CELLS (NEURONS)

The tissues of the nervous system contain a variety of cells, but one of the most highly differentiated and specialized is the neuron or nerve fiber itself. The primary and specialized functions of these cells are the **conduction** and **transmission** of impulses from one part of an organism to another. In some instances, the impulse may travel several feet, and a single neuron may bridge the entire distance.

In certain respects, neurons are similar to most other cells (Fig. 24–21) and the *cell body* contains the typical spectrum of organelles. It is the *processes* or extensions that make neurons easily distinguishable from other cells. Processes that receive transmitted impulses and conduct them toward the cell body are called **dendrites** and processes that conduct impulses away from the cell body are called **axons**. The axons terminate in a special structure called the **end-plate**, which is responsible for initiating the transmission of the impulse to the next cell. The junctions between successive nerve cells or between a nerve cell and an *effector* cell (such as a gland or muscle cell) are called **synapses**. At each synapse, the two cells are separated by a narrow gap called a **synaptic cleft**. A **nerve** is formed by a bundle of many neurons.

Neurons develop from the **neural tube** of the early embryo and in many animals continue development until some time after birth. The development of motor neurons that have elongated axons is best understood. These cells are derived from tissue on the ventral side of the neural tube. In an action reminiscent of amoeboid movement, the cytoplasm of these cells flows outward in a strandlike manner with progressive and continuous amoeboidlike activity at the end of the strand. The long, thin strand becomes the axon of the neuron and may give rise to several smaller branches. By the time growth is complete, the axon may contain thousands of times as much cytoplasmic material as the cell body and may extend for thousands of cell body diameters.

Growth of an axon is maintained by the continous flow of cytoplasm from the nerve cell body. Even after growth ceases, **axoplasmic flow** continues and serves to carry substances produced in the cell body toward the axonal endings. Mature nerve cells also exhibit a second form of intracellular transport called **axonal transport**. Axonal transport is more rapid than axo-

plasmic flow and conveys materials in *both* directions along the nerve cell process. Indeed, it is the *retrograde* movement by which herpes and rabies viruses make their way to the cell bodies of nerve cells. Axonal transport carries membranous elements and small vesicles filled with *neurotransmitters* or their precursors toward the nerve endings (see below). Much of the membranous material is incorporated into the plasma membrane or **axolemma** along the way, but the remainder reaches the nerve ending. The axoplasm of nerve cells contains large numbers of actin and myosin filaments as well as microtubules (Fig. 24–21). These are believed to play a role in both axoplasmic flow and axonal transport.

Conduction of Nerve Impulses

Neurons conduct signals or *impulses* from one part of the body to another. Under normal circumstances

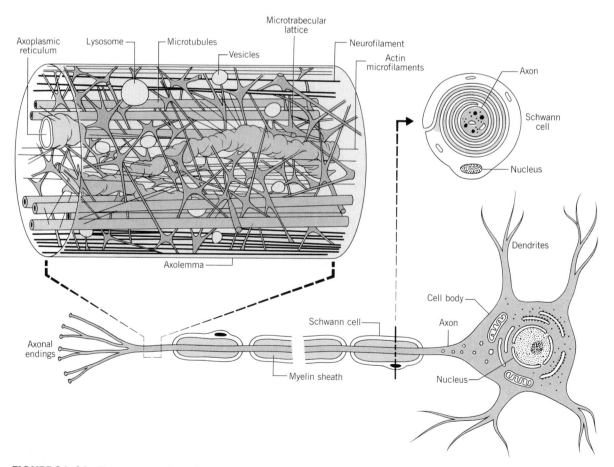

FIGURE 24–21   The nerve cell and associated Schwann cells.

each impulse begins at the dendrites (occasionally at the cell body) and spreads across the cell to the axonal endings. Experimentally, an impulse can be initiated anywhere on the cell surface and can be elicited by applying a variety of stimuli including electrical shock, pressure (pinching), heat, cold, and pH changes. The impulse results from transitory physicochemical changes occurring in the cell's plasma membrane, and once initiated it is propagated along the membrane without dependence on a continuing stimulus. The velocity with which the impulse travels along the fiber is not dependent on the strength of the stimulus, that is, it does not travel faster if initiated by a stronger stimulus. The rate of movement of an impulse varies according to neuron type but is in the range of 2–100 m/sec, a speed far too slow to compare the impulse to the movement of an electron through a wire during electrical flow. As the impulse passes along the plasma membrane, two major changes occur. One is a change in electrical potential across the membrane and the second is a change in membrane permeability.

Resting Potentials and Action Potentials The cytoplasm or **axoplasm** within the axon and the extracellular fluid contain a number of ions and are good electrical conductors. On the other hand, the axolemma acts as an insulator, albeit a weak one. When one member of a pair of microelectrodes connected to a

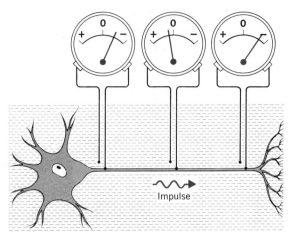

FIGURE 24–23 The transitory change in electrical potential that occurs across the axolemma as an impulse is conducted by the fiber can be measured using a series of recording electrodes. At several points along the fiber, one member of each electrode pair is inserted into the axon and the other is placed at the axon's surface. As the impulse passes each pair of electrodes, there is a transient reversal of the membrane's polarity.

voltmeter is inserted into the axoplasm and the other electrode is placed in the extracellular fluid, an electrical potential across the membrane is measured (Fig. 24–22). When the cell is not conducting an impulse, the potential is called the **resting potential** and maintains a constant value. For most nerve cells, the resting potential is 50–90 mV, with the inner (axoplasmic) surface of the membrane negative relative to the outside. When a neuron is stimulated and the impulse passes the region of the electrodes, a brief change in potential is recorded; this transitory change in potential is called an **action potential**. By placing electrodes at several points along the axon, it is possible to follow the action potential as it travels along the process. As the action potential passes each recording electrode (Fig. 24–23), the internal voltage rapidly changes from −90 mV through 0 mV to +20 or +30 mV. Within 0.5 to 1.0 msec, the membrane recovers, and the −90 mV resting potential is restored. The impulse sweeping along the axon represents a net transitory change in potential of 110–120 mV.

Ion Gradients across the Membrane

The resting potential of the nerve cell membrane stems from an unequal distribution of ions between the axoplasm and the extracellular fluid. The fluid

FIGURE 24–22 Measuring the resting potential of a nerve cell. One electrode is placed in the layer of liquid bathing the axon's surface and the other electrode is inserted through the axolemma into the axoplasm. The resting potential is about −90 mV.

bathing the membrane's surface contains Na^+, Cl^-, and HCO_3^- in higher concentrations than the axoplasm, whereas the axoplasm contains higher concentrations of K^+ and organic anions than the extracellular fluid. In addition to concentration differences across the membrane for individual ionic species, there is also a relatively higher concentration of negative ions inside the cell than outside (see the Gibbs–Donnan effect discussed in Chapter 15). As a result, the outside surface of the axolemma is positive relative to the inside surface.

The nerve cell membrane is permeable to sodium and potassium ions and these ions are continuously diffusing from the side of the membrane where they are at higher concentration to the side where they are at lower concentration. To maintain the concentration differences of these two cations, metabolic energy is used to pump inwardly diffusing Na^+ out of the cell and outwardly diffusing K^+ into the cell. The mechanism proposed to account for such a sodium/potassium pump is depicted in Figure 15–40 and involves cyclic changes in the tertiary structure of an enzyme carrier located in the axolemma. Sodium ions together with ATP are bound to an enzyme site exposed at the inside membrane surface and K^+ is bound to the outer surface. The binding causes a conformational change in the carrier that moves the ions through the membrane. The ATP is hydrolyzed and the ions are released.

Initiation of the Action Potential

Much of what we know today about the molecular basis of impulse conduction by nerve cells is based on the pioneering studies of A. Hodgkin, A. Huxley, and J. Eccles, who received the Nobel Prize in 1963 for their work. According to the presently accepted model, application of a stimulus to the neuron is followed by the rapid diffusion of Na^+ across the axolemma from the exterior into the axoplasm. The unusually rapid inward diffusion of Na^+ is apparently due to a transient increase in the size of the pores or "gates" in the plasma membrane permeable to Na^+. The inward rush of Na^+ reverses the electrical potential across the membrane. This is followed almost immediately by an opening of the membrane's potassium gates and a rapid efflux of K^+; the latter flow of ions restores the former potential. Only a small percentage of the Na^+ and K^+ intially present outside and inside the cell need traverse the membrane to cause the **reverse polarization** followed by the **repolarization**. Indeed, the

nerve fiber can conduct an impulse many times in succession without appreciably diminishing the concentration gradients across the membrane. Between successive conductions the Na^+/K^+ gates close and the Na^+/K^+ pump restores the original ion concentrations. It is estimated that the normal distribution of Na^+ and K^+ across the membrane can be restored in less than 5×10^{-3} sec.

If the stimulus that is applied reaches the neuron **threshold** level, an action potential is initiated. Any stimulus above the threshold level also initiates an action potential, but if the stimulus intensity is below threshold (i.e., **subthreshold**) the membrane recovers from its transitory depolarization without initiating the action potential.

Conduction of the Action Potential

Once the polarity at a given point on the nerve cell membrane has been reversed, current flows between this and the adjacent regions of the membrane. The flow of current to the neighboring region serves to open the Na^+ gates there, reversing the polarity in that region. The cycle is repeated as the action potential travels further and further along the nerve cell membrane.

The speed at which the impulse is propagated along the nerve fiber is directly dependent on the diameter of the axon (Fig. 24–24). An axon with a large diame-

FIGURE 24–24 The velocity of a nerve impulse is related to the diameter of the axon; however, large-diameter unmyelinated fibers (b) and narrow myelinated fibers (c) conduct equally rapidly because in myelinated fibers the impulse skips from one node of Ranvier to the next. This form of conduction is called saltatory conduction.

ter offers less *electrical resistance* than a smaller one and therefore depolarizes more readily and conducts faster. The rapid conduction speed observed in squid axons (25 m/sec) is due to the large axon diameter (1 mm or more!). The rate of impulse conduction in mammalian nerve fibers exceeds that in the squid, but the increased speed is gained by a change in *capacitance* and not by an increase in axon diameter. The capacitance is changed by the presence of myelin (a good insulator) in cells called **Schwann cells**, which are wrapped around the axon forming a sheath (Fig. 24–25). The outer surface of the axon membrane is exposed to the extracellular fluid only at the **nodes of Ranvier** (Figs. 24–21 and 24–24), which are about 2 mm apart. In effect, the impulse skips from node to node, a phenomenon called **saltatory conduction**, and travels considerably faster.

Synaptic Transmission

Transmission of an impulse from one neuron to another or to an effector cell may be mediated electrically or chemically. Electrical transmission occurs when the axonal endings of the neuron form *gap junctions* with the next cell (i.e., the plasma membranes of the two cells are bridged by connexons, see Chapter 15). In such an arrangement, the action potential passes directly and without delay from one cell to the next. More common is chemical transmission, which involves communication between cells that are separated by a narrow, fluid-filled space. When an impulse or action potential reaches the axonal endings of a neuron, neurotransmitters are released into this space (the synaptic cleft) and diffuse across the cleft to the next cell (e.g., the next neuron or an effector

FIGURE 24–25 Myelin-filled Schwann cells form a spiral envelope around the long processes of many nerve cells. As the axon develops (i.e., A_1, A_2, A_3, A_4), the Schwann cells form increasing numbers of turns around the axon as myelin synthesis proceeds. (Courtesy of Dr. H. Webster; copyright © 1971 by W. B. Saunders Co., *The Fine Structure of the Nervous System.*)

cell such as a muscle cell or gland cell). The axon terminals, the synaptic cleft, and the specialized region of the cell responding to the neurotransmitter comprise a **synapse**. The most common neurotransmitter is *acetylcholine*; others include *norepinephrine, dopamine*, and *seratonin*. Neurotransmitters are synapse-specific, that is, they are released at some synapses but not at others. Moreover, certain neurotransmitters act to stimulate the postsynaptic cell and others are inhibitory (some may seven be stimulatory when released at one synapse and inhibitory at another).

Best understood are synapses at which acetylcholine is the neurotransmitter. Until quite recently it was believed that acetylcholine is released into the synaptic cleft from acetylcholine-rich axoplasmic vesi-

cles (called **synaptic vesicles**) that populate the neuron's axonal endings. These vesicles were thought to fuse with the axolemma, thereby discharging their contents. However, this now appears *not* to be the case and the synaptic vesicles play a different role (see below). Acetylcholine is released instead from the **axoplasm** of the presynaptic cell through small channels formed in the membrane in response to the arrival of the action potential and the entry of calcium ions into the axoplasm from the synaptic cleft (Fig. 24–26). The released acetylcholine diffuses across the synaptic cleft and momentarily attaches to **acetylcholine receptors** in the membrane of the postsynaptic cell, eliciting a response by that cell (e.g., conduction of another action potential in nerve–nerve junctions).

All the acetylcholine released into the synaptic cleft

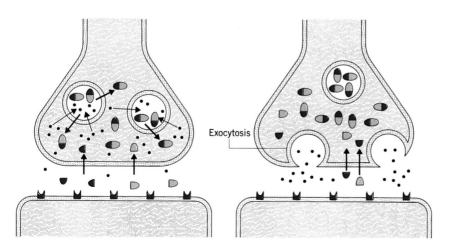

FIGURE 24–26 Chemical transmission of an impulse across the synaptic cleft from the presynaptic fiber to the postsynaptic cell. (*a,b*) When the action potential reaches the axonal ending, Ca^{2+} enters the axoplasm and is followed by the release of acetylcholine into the cleft (apparently through channels in the axolemma formed in response to the entry of Ca^{2+}). The acetylcholine binds to receptors in the membrane of the postsynaptic cell and is subsequently degraded to acetate and choline. Acetylcholine-rich synaptic vesicles do not release acetylcholine into the axoplasm until after prolonged conduction; their release of acetylcholine is accompanied by the sequestration of Ca^{2+} (*c*). Acetate and choline reenter the axoplasm of the presynaptic fiber and are recycled (*c*). Ca^{2+}-rich synaptic vesicles fuse with the axolemma and release their accumulated Ca^{2+} into the synaptic cleft (*d*).

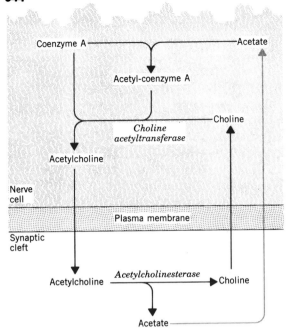

FIGURE 24–27 Synthesis and breakdown of acetylcholine.

by the presynaptic cell is broken down into choline
and acetate by the enzyme *acetylcholinesterase* found
on the surface of the postsynaptic cell and released
into the cleft. New acetylcholine is synthesized in the
presynaptic cell by the transfer of acetate from acetyl-
CoA to choline by the cytoplasmic enzyme *choline ace-
tyltransferase* (Fig. 24–27). Both the acetate and the
choline produced in the synaptic cleft by the degradation of acetylcholine may be transferred back into the
cytoplasm of the presynaptic cell and recycled (Fig.
24–27).

What then is the role of the acetylcholine-rich synaptic vesicles? During prolonged stimulation of a
nerve, rapidly diminishing axoplasmic acetylcholine is
replaced by the release of needed neurotransmitter
into the cytoplasm from the synaptic vesicles. As depicted in Figure 24–26, the release of acetylcholine is
accompanied by the sequestering of the calcium ions
that entered the cell from the synaptic cleft (i.e., actyl-
choline leaves the vesicles and Ca^{2+} enters them). The
calcium is then expelled into the synaptic cleft when
the synaptic vesicles fuse with the axolemma. The
synthesis of acetylcholine by the action of choline ace-
tyltransferase is accompanied by the formation of new
synaptic vesicles.

SUMMARY

Structural and biochemical changes in the cells of multicellular organisms leading to specialized cell roles
constitute **differentiation**. The potential for cellular
differentiation depends on the conditions for the expression (i.e., transcription and translation) of the
cell's genetic information. Interactions between the
cytoplasm and the nucleus (**intrinsic factors**) and environmental factors outside the cell (**extrinsic factors**) influence the course of differentiation.

The development of red blood cells illustrates the
sequential cause and effect mechanisms of differentiation. Stem cells in the bone marrow may give rise to
red or white blood cells. In the case of red blood cells,
increasing levels of the hormone **erythropoietin** traveling in the bloodstream to the marrow cause the stem
cells to form mRNAs for the globin chains of hemoglobin. The cells become committed to be red blood cells,
as 95% of all protein synthesis is directed to hemoglobin. The buildup of hemoglobin is then followed by the
dissolution or loss of cell structures such as the nucleus and residual nucleic acids. The cells are irreversibly differentiated into red blood cells.

Although lymphocytes are capable of producing and
secreting thousands of different **immunoglobulins**
(*antibodies*), each of which is specific to a single antigenic determinant, a single antibody is produced and
secreted by each fully differentiated cell.

Muscle cells develop a complex mechanism for contraction. Myofibrils containing **actin** and **myosin** filaments are geometrically differentiated parallel to the
long axis of the cell and are spatially arranged about
one another with a number of cross-bridges formed
between them. When stimulation is applied or ATP
and Mg^{2+} are added, the cross-bridges between the
filaments progressively break and reform as the filaments slide past one another and the myofibril (and
cell) shortens.

Nerve cell differentiation is also an excellent example of the degree of specialization that may accompany cell development. Neurons develop processes
many thousands of times the diameter of the cell body.
Impulses are **conducted** between remote parts of the
body along these processes and are **transmitted** from
one cell to the next as a result of specialized properties of the nerve cell membrane and the axoplasm.

REFERENCES AND SUGGESTED READING

Articles and Reviews

Buchthal, F., and Schmalbruch, H., Motor unit of mammalian muscle. *Physiol. Rev.* **60**, 90 (1980).

Carafoli, E., and Penniston, J. T., The Calcium Signal. *Sci. Amer.* **253** (5), 70 (Nov. 1985).

Dunant, Y., and Israel, M., The release of acetylcholine. *Sci. Am.* **252** (4), 58 (Apr. 1985).

Eisenberg, E., and Greene, L. E., The relation of muscle biochemistry to muscle physiology. *Ann. Rev. Physiol.* **42**, 293 (1980).

Eisenberg, E., and Hill, T. L., Muscle contraction and free energy transduction in biological systems. *Science* **227**, 999 (1985).

Huxley, H. E., Muscle contraction and cell motility. *Nature* **243**, 445 (1973).

King, T. J., and Briggs, R., Changes in the nuclei of differentiating gastrula cells as demonstrated by nuclear transplantation. *Proc. Nat. Acad. Sci.* **41**, 321 (1955).

Linas, R. R., Calcium in synaptic transmission. *Sci. Am.* **247** (4), 56 (Oct. 1982).

Nienhuis, A. W., and Benz, E. J., Regulation of hemoglobin synthesis during the development of the red cell. *N. Engl. J. Med.* **297**, 1318 (1977).

Patterson, P. H., Potter, D. D., and Furshpan, E. J., The chemical differentiation of nerve cells. *Sci. Am.* **239** (1), 50 (July 1978).

Rogart, P. H., Sodium channels in nerve and muscle membranes. *Ann. Rev. Physiol.* **43**, 711 (1981).

Schwartz, J. H., The transport of substances in nerve cells. *Sci. Am.* **242** (4), 152 (Apr. 1980).

Wallick, E. T., Lane, L. K., and Schwartz, A., Biochemical mechanism of the sodium pump. *Ann. Rev. Physiol.* **41**, 397 (1979).

Weber, A., and Murray, J. M., Molecular control mechanism in muscle contraction. *Physiol. Rev.* **53**, 612 (1973).

Werz, G., Determination and realization of morphogenesis in *Acetabularia*. *Brookhaven Symp. Biol.*, Upton, N.Y. (1965).

Books, Monographs, and Symposia

Aidley, D. J., *The Physiology of Excitable Cells*, Cambridge University Press, Cambridge, 1971.

Bessis, M., Weed, R. I., and Leblond, P. F. (Eds.), *Red Cell Shape*, Springer-Verlag, New York, 1973.

Browder, L. W., *Developmental Biology*, Saunders, Philadelphia, 1980.

Goldman, R., Pollard, T., and Rosenbaum, J., *Cell Motility* (Books A and B.), Cold Spring Harbor Conferences on Cell Proliferation, Vol. 3, Cold Spring Harbor Laboratory, Cold Spring Harbor, N.Y., 1976.

Hodgkin, A. L., *The Conduction of the Nervous Impulse*, Charles C. Thomas, Publishers, Springfield, Ill., 1964.

Snyder, S. H., The molecular basis of communication between cells. *Sci. Am.* **253** (4), 132 (Oct. 1985).

Stein, R. B., *Nerve and Muscle—Membranes, Cells and Tissues*, Plenum, New York, 1980.

CHAPTER 25

IMMUNITY AND CANCER

In this chapter, we deal with the subjects of **immunity** and **cancer**, and it might be well to ask at the outset what these subjects have in common. Immunity is a state attainable in higher vertebrates in which the body is protected against the harmful effects of certain disease-causing (i.e., *pathogenic*) agents such as parasites, fungi, bacteria, and viruses. Cancer, on the other hand, is a condition characterized by the abnormal and rapid growth of tissue cells to form **tumors** that in many instances interfere with normal tissue functions and often lead to death. In the past decade it has become clear that the body's system of defense against viruses, bacteria, and other infectious agents (i.e., the *immune system*) also serves as the principal line of defense against cancer. Moreover, it is now apparent that some cancers are intimately associated with viral infections, that is, certain viruses are *oncogenic* (cancer-creating). As we explore the subjects of immunity and cancer, a number of other interrelationships will also become apparent.

NONADAPTIVE AND ADAPTIVE SYSTEMS OF DEFENSE AGAINST INFECTION

A nonadaptive system of defense is one that protects the body against infection in a *fixed* way. Nonadaptive mechanisms are usually *nonspecific* in that the defense they provide is against many of a variety of different pathogens. For example, the *skin* serves as a nonspecific barrier against the penetration of a host of different toxic chemicals and most microorganisms. Internally, the *blood* and other body fluids contain several classes of leukocytes (e.g., **monocytes** and **neutrophils**) that serve to scavenge foreign particles and cancer cells by phagocytosis. Although the action of these scavengers may not be very specific (i.e., they may phagocytose any of a variety of foreign agents of disease), their interaction with products of the immune system (see below) increases their specificity and efficiency.

The **immune system** is an adaptive system of de-

647

fense and is *highly specific*. Through what is called an **immune response**, agents of infection are specifically recognized and selectively destroyed by the actions of white blood cells called **lymphocytes**. Moreover, the immune system exhibits long-term *memory* of its encounters with these agents. **Immunologic memory** permits more rapid response to a second, third, or subsequent attack by a given agent.

DUALITY OF THE IMMUNE SYSTEM

There are *two* ways in which the immune system responds to infection. In the **cellular immune response**, there is a direct interaction between lymphocytes and invading fungi, bacteria, or virus-infected cells. In contrast, the **humoral immune response** acts principally against *extracellular* phases of infection, that is, through the secretion of proteins called **antibodies** or **immunoglobulins** into the blood plasma, lymph, and tissue fluid. As we shall see, these antibodies combine with foreign substances called **antigens** that are carried in the surface of or are released by the pathogen. The combination of antibody with antigen then initiates a response leading to the elimination of the antigen and, more importantly, its source.

T Lymphocytes and B Lymphocytes

Cellular immunity is mediated by a class of lymphocytes called **T lymphocytes** or **T cells**. These cells are derived from pluripotent **stem cells** in the hemopoietic tissues of the embryo (i.e., the liver and bone marrow). From there they migrate to and colonize the *thymus gland* (hence, "T" implies "thymus-derived"). Many T lymphocytes migrate from the thymus to other lymphoid tissues such as the spleen and lymph nodes and colonize these tissues. Humoral immunity is mediated by **B lymphocytes** or **B cells**. In birds, B lymphocytes are derived from the *bursa of Fabricius*, an outpocketing of the hindgut that is colonized by hemopoietic stem cells early in embryonic development (thus, "B" implies "bursa-derived"). In mammals, which have no bursa, the origin of B lymphocytes remains uncertain, although many are derived from the bone marrow itself.

The activation of T and B lymphocytes by the presence of a foreign agent leads to the production of the *effector cells* that combat the infection. The T lymphocytes give rise to **cytotoxic, helper** and **suppressor cells**, whereas the B lymphocytes give rise to the

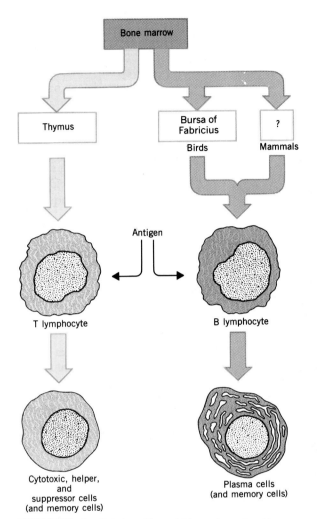

FIGURE 25–1 Origin of T and B lymphocytes. T lymphocytes are derived from the thymus and in response to the presence of antigen terminally differentiate into *cytotoxic, helper,* and *suppressor T cells.* In birds, B lymphocytes are derived from the bursa of Fabricius (for which there is no mammalian equivalent) and differentiate into *plasma cells.* Both T and B lymphocytes also produce *memory cells.*

plasma cells that secrete antibodies (Fig. 25–1). Both T and B lymphocytes also give rise to **memory cells**—cells responsible for immunologic memory.

Antigens of the foreign agent are recognized by two families of proteins peculiar to the immune system. The best understood of these proteins are the antibodies. Antibodies are found in the plasma membranes of B cells, where they act as antigen receptors, and are also secreted by the plasma cells produced by proliferation of B lymphocytes during an immune response. Because they are produced in large quantities during infection, antibodies are readily isolated for study and much is known about their structure and action. Less well understood are the T lymphocyte *receptor proteins*—proteins associated with the plasma membranes of these cells. These proteins are not secreted by T cells and are isolated with much greater difficulty. T-cell receptors provide T lymphocytes with the capability to attach directly to foreign cells and to host cells that have been infected by viruses. Despite the disproportion in our knowledge of these two classes of proteins, it is clear that antibodies and T-cell receptors are structurally similar and have a common evolutionary origin.

ANTIBODIES (IMMUNOGLOBULINS) AND THEIR DIVERSITY

Antibodies (or *immunoglobulins*; we will use the two terms interchangeably) are proteins secreted by B lymphocyte-derived plasma cells in response to the appearance of infectious agents in the body's tissues. Any substance capable of triggering the production of antibodies is called an *antigen*, and although antigens can take a variety of different chemical forms, they usually are proteins, polysaccharides, or nucleic acids. Antigenic proteins are frequently constituents of the surfaces of parasitic cells, fungi, bacteria, and viruses. Many different antigens may be associated with a single infectious agent. Even particles as small as viruses may carry several antigenic sources.

As noted earlier, it is the reaction that takes place between antibodies and the foreign antigens that leads to the elimination of the antigen and its source. This reaction is highly specific, that is, a particular antibody usually reacts with only one type of antigen. Despite its high degree of molecular specificity, the human immune system provides protection against a vast spectrum of antigen sources, for the body's lymphocytes are capable of making *many millions of different kinds of antibodies*. The antibody molecules do not usually destroy the infectious agent directly; instead, they "tag" the agent for destruction by other components of the immune system.

Antibodies account for about 20% of the protein present in blood plasma, the most common form being **immunoglobulin G** or **IgG** (see Table 25–1). IgG molecules are composed of four polypeptide chains of two different sizes; these are a pair of identical high-molecular-weight chains called **heavy chains** (or H chains) and a pair of identical low-molecular-weight chains called **light chains** (or L chains). The H chains have a molecular weight of 50,000 to 55,000 and contain about 450 amino acids, whereas the L chains have a molecular weight of 20,000 to 25,000 and consist of about 214 amino acids. Each L chain is covalently linked to an H chain by a disulfide bridge, and two light chain–heavy chain pairs are covalently linked by two disulfide bridges (Fig. 25–2). There are also 12 intrachain disulfide bridges, four in each H chain and two in each L chain. An asparagine residue in each H chain is bonded to carbohydrate, so immunoglobulins are also glycoproteins.

There are two different types of light chains found in immunoglobulins; they are designated *kappa* (κ) *chains* and *lambda* (λ) *chains*. The heavy chains of immunoglobulins belonging to the IgG class are of the *gamma* (γ) type, so that an IgG molecule may be represented as either $\kappa_2\gamma_2$ or $\lambda_2\gamma_2$, depending on the light chains that are present. Like IgG molecules, human immunoglobulins belonging to one of the other four classes also possess κ or λ light chains. However, the heavy chains of *immunoglobulin A* molecules (IgA) are *alpha* (α) *chains*; in *IgD*, they are *delta* (δ) *chains*; in *IgE*, they are *epsilon* (ϵ) *chains*; and in *IgM*, they are *mu* (μ) *chains* (Table 25–1). IgG, IgD, and IgE occur as monomers, but IgA molecules may occur as *monomers, dimers, or trimers*. IgA dimers are held to-

TABLE 25–1 THE MAJOR CLASSES OF HUMAN ANTIBODIES

	IgA	IgD	IgE	IgG	IgM
Light chain	κ or λ	κ or λ	κ or λ	κ or λ	κ or λ
Heavy chain	α	δ	ϵ	γ	μ
Percentage of total blood Ig	15	<1	<1	80	5
Number of H, L pairs	1, 2, or 3	1	1	1	5
Sedimentation coefficient	7, 10, 13 S	7 S	8 S	7 S	20 S

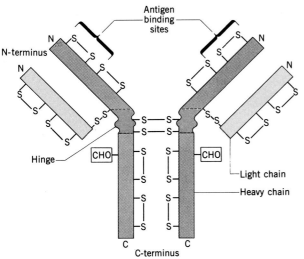

FIGURE 25–2 Generalized chemical structure of an antibody molecule. Each molecule contains a pair of identical heavy chains (i.e., H chains) and a pair of identical light chains (i.e., L chains). Each polypeptide is characterized by a number of intrachain disulfide bridges and polypeptides are interconnected by additional disulfide bridges. The H chains may contain some covalently bound carbohydrate (CHO). The antigens with which the antibodies react are bound near the N-terminals of each pair of L and H chains.

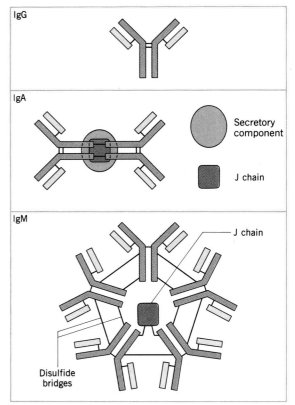

FIGURE 25–3 Schematic illustration of IgG, IgA, and IgM antibodies. IgG (and IgD and IgE) molecules occur as monomers only, but IgA may form dimers and trimers, and IgM may form pentamers. See text for additional details.

gether by another polypeptide called a *J chain* ("J" for "joining"). IgG molecules may also be linked by a second component called *secretory component* (Fig. 25–3). IgM molecules occur as pentamers in which the monomeric units are held together by disulfide bridges and by a J chain.

Constant and Variable Domains of an Immunoglobulin

An analysis of the primary structures of isolated immunoglobulins has revealed that they have certain amino acid sequences in common and certain sequences that differ. Amino acid sequences that are common are called **constant domains**, whereas sequences that vary are called **variable domains**. Each L chain has one constant and one variable domain, respectively designated C_L and V_L (Fig. 25–4). Because there are two forms of L chains, namely, κ and λ, there are also two different constant domains (i.e., constant regions common to κ chains and constant regions common to λ chains). Each H chain has one variable se-

quence (V_H) and three constant regions (C_H1, C_H2, and C_H3). It is now known that the three constant regions of the H chains and the hinge are encoded by separate exons (Fig. 25–4). By comparing Figures 25–4 and 25–2, it may be seen that each domain (constant and variable) contains one intrachain disulfide bridge.

At this point it might be best to take stock of what we have said concerning the diversity inherent in the immunoglobulins. First, there are five different classes of immunoglobulins (i.e., IgA, IgD, etc.) differing in the nature of their heavy chains (i.e., α, δ, ϵ, γ, and μ; second within each class, the light chains may be of two different types (i.e., κ and λ); and third, some of the immunoglobulins can occur as polymers. Despite the variation that all of this infers, it still falls far short of the variability that is actually known to exist. As you might suspect at this point, the added variability is due to the differential nature of the variable domains of each L and H chain.

FIGURE 25–4 The L and H chains of an immunoglobulin contain *variable* and *constant* domains. Each L chain has one variable (V_L) and one constant (C_L) domain. Each H chain has one variable domain (V_H) and three constant domains (C_H1, C_H2, and C_H3). The three constant domains of an H chain and its *hinge* region (Hg) are encoded by separate exons. Compare with Figure 25–2.

The variable domains of each L and H chain are found near their N-terminals, and this is where the antibody attaches to the antigen. Within each variable region there are three **hypervariable regions**, each containing 5 to 10 amino acids. The amino acid sequences in the hypervariable regions vary more than neighboring sequences in the variable domain (which are called **framework regions**) and form the antigen-binding sites (Fig. 25–5).

Antibody Diversity and the Genome

κ and λ light chains and α, δ, ϵ, γ and μ heavy chains are encoded in separate gene pools, each gene pool containing sets of different **C genes, V gene seg-ments**, and **J gene segments** ("J" = "joining"). The κ gene pool contains a C gene (C_κ) and a large number (perhaps several hundred) of V genes ($V_\kappa1$, $V_\kappa2$, $V_\kappa3$ . . . $V_\kappa n$). The λ gene pool also contains a C gene (C_λ) and several V genes ($V_\lambda1$, $V_\lambda2$, etc.). The heavy chain gene pool contains many V genes (V_H1, V_H2, V_H3 . . . $V_H n$) and a sequence of C genes (C_μ, C_δ, C_γ, C_ϵ, and C_α) (Fig. 25–6). Each V gene is more appropriately called a **V gene segment**, because it does not encode an entire variable domain of a chain; this is because there is a small "missing" stretch of DNA called a **J gene segment** that must be connected to a V gene segment to form a complete V gene. Each light and heavy chain gene pool contains four J gene segments. The V gene segments are located hundreds of thousands of nucle-

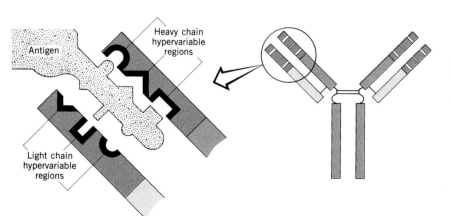

FIGURE 25–5 Each variable domain of an antibody contains *hypervariable regions* in which the amino acids vary more than in neighboring sequences. The hypervariable regions create the specificity of the antigen-binding site.

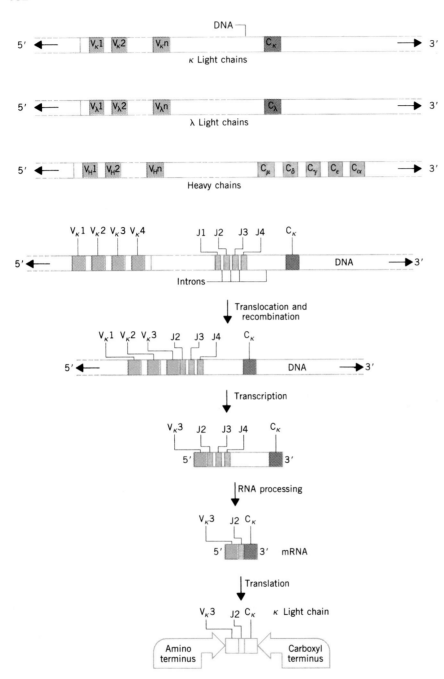

FIGURE 25–6 Antibody genes. The light chains of an antibody are encoded by a constant domain gene (e.g., C_κ and C_λ) and a number of variable domain genes (e.g., $V_\kappa 1$, $V_\kappa 2$, etc.). The constant domains of heavy chains are encoded by separate μ, δ, γ, ϵ, and α genes (i.e., C_μ, C_δ, etc.) and the variable domains encoded by separate variable domain genes (i.e., $V_H 1$, $V_H 2$, etc.).

FIGURE 25–7 Assembly of genes expressible as a complete kappa light chain. V gene segments are translocated so that they come to lie next to a J gene segment. Primary transcripts include a number of introns that are excised during RNA processing. See text for details.

otides upstream (i.e., on the 5′ side) of the C gene, whereas the J gene segments are located only a short distance upstream of the C gene and are separated from one another by introns (Fig. 25–7). The assembly of genes expressible as complete variable domains of L chains occurs during the differentiation of lympho-

cytes into antibody-producing cells and involves the translocation of a V gene segment so that it comes to lie next to a J gene segment. Any V gene segment can be connected to any J gene segment. Primary transcripts may include more than one J gene segment and intron. However, during RNA processing all introns and all J gene segments other than the one whose 5' side contains the translocated V gene segment are removed (Fig. 25–7).

The genetic basis for the diversity of heavy chain variable domains is somewhat more complex and involves yet another gene segment called a **D gene segment** ("D" = "diversity"). The formation of an expressible heavy chain variable domain gene requires translocational events that link any of the 20 D gene segments to both a V gene segment and a J gene segment. In light of the numbers of V, J, and D gene segments and the variety of combinations of these that are expressible as light and heavy chain variable domains, it is easier to comprehend the enormous potential for immunoglobulin diversity. This diversity is amplified by a lack of precision in the machinery for splicing the DNA and an unexpectedly high mutation rate in the variable domains.

T-CELL RECEPTORS

Like the plasma membranes of B cells, T-cell surfaces also contain antigen-binding receptors. Although **T-cell receptors** have many properties in common with antibodies, they are a distinct class of proteins (Fig. 25–8). The T-cell receptor is about two-thirds the size of an antibody and consists of two subunit polypeptides connected by a disulfide bridge. One polypeptide is designated an *alpha chain* and the other a *beta chain*. Each chain is composed of a constant domain and a variable domain. The disulfide bridge that connects alpha and beta chains occurs between constant domains. At the end of each constant domain, there is a region rich in hydrophobic amino acids; this region anchors the receptor in the plasma membrane. The gene pools for T-cell receptors are similar to those for antibodies. The variable regions of the alpha and beta chains are encoded by V, D, and J gene segments that are recombined in various ways to produce an expressible variable domain.

CLONAL SELECTION

The **clonal selection theory** explains the mechanism by which the body's immune system responds to the appearance of antigens present in or released from an agent of infection. The immune system includes many millions of different clones of B and T lymphocytes,

FIGURE 25–8 The T-cell receptor. Each receptor is about two-thirds the size of an antibody and consists of two polypeptides (alpha and beta) interconnected by disulfide bridges. Like the chains of the antibodies, the alpha and beta polypeptides contain variable and constant regions. Each chain also has a hydrophobic amino acid sequence by which it is anchored in the membrane.

each clone consisting of cells that are capable of responding to the same form of antigen. Thus, the body's immune system is continuously "on alert" in anticipation of the appearance of any of more than a million different antigens. The immune response is triggered when antigens of the infecting agent become bound to immunoglobulin molecules in the plasma membranes of B cells or to T-cell receptors. Usually, a given antigen can be bound by the surface receptors of a number of different clones, but this is a tiny percentage of the total number of different clones present. Thus, the clones that respond to the presence of the antigen are, in effect, *selected* by the antigen, and this is what is meant by "clonal selection."

Binding of antigen to the surface receptors of the cells of a particular B- or T-lymphocyte clone serves to activate these cells causing a succession of rounds of cell division that greatly increases the numbers of cells of that clone that are present in the body (Fig. 25–9). In the case of B cells, some of the progeny become **plasma cells** that synthesize and secrete quantities of antibody that will react with additional antigen; other progeny become **memory cells** that are reserved for a response to a subsequent exposure to the same antigen. Some of the T-cell progeny also become memory cells, but most develop into the **cytotoxic**, **helper**, and **suppressor cells** responsible for the *cell-mediated* immune response. In Figure 25–9, which illustrates the principles of clonal selection, the antigen is shown to activate only one of the millions of B and T-cell clones present in the body.

Activated B cells secrete antibodies, whereas activated T cells and unactivated B and T cells do not. It is therefore possible to distinguish activated B cells cytologically from activated T cells and virgin lympho-

cytes because activated B cells possess a well-developed rough endoplasmic reticulum (Fig. 25–10).

Antigenic Determinants and Haptens

As noted earlier, most antigens are either proteins, polysaccharides, or nucleic acids, although it is likely that many other naturally occurring substances (especially those of high molecular weight) and even synthetic (i.e., man-made) materials can also act as antigens. The specific region of the antigen that binds

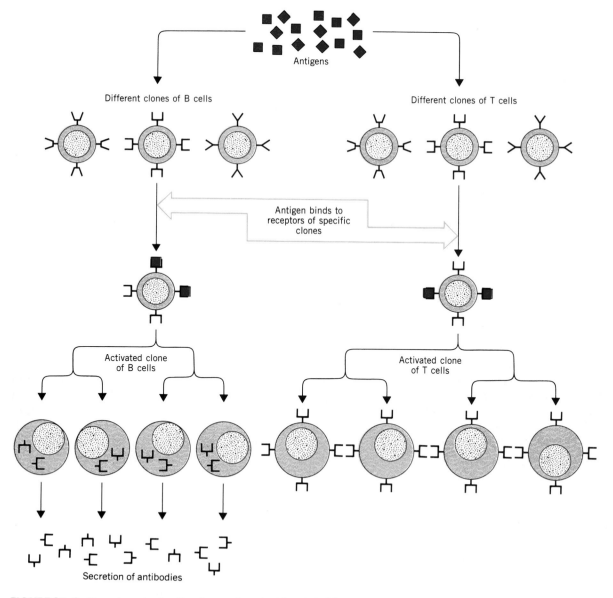

FIGURE 25–9 The clonal selection theory (see text for details).

FIGURE 25–10 Scanning (*a*) and transmission (*b*) electron photomicrographs of plasma cells. (*a* courtesy of John Wiley photo library; *b* courtesy of R. Chao.)

to a lymphocyte surface receptor or to an antibody is called the **antigenic determinant**. Large antigens may have several antigenic determinants. Some foreign substances can bind to an antibody or surface receptor and *not* elicit an immune response; such molecules are called **haptens**. Haptens can be made antigenic by chemically coupling them to a *carrier*, such as a protein or polysaccharide. When haptens combine with some of the body's own proteins, they may become antigenic and elicit an immune response that produces antibodies against the hapten.

Because a single antigenic determinant can react with more than one antibody, and because one antigen may possess more than one antigenic determinant, it follows that an antigen may activate several different lymphocyte clones. When a number of clones are activated, the response is said to be **polyclonal**. When only a few clones are activated, the response is said to be **oligoclonal**. When a single clone is activated, the response is **monoclonal**.

MONOCLONAL ANTIBODIES AND HYBRIDOMAS

If cells taken from various clones of normal antibody-producing and secreting lymphocytes could be grown in culture, it would be possible to have ready sources of various antibodies at one's disposal. Unfortunately,

laboratory cultures of normal lymphocytes grow very slowly or die out quickly. There is, however, an abnormal and cancerous form of lymphocyte called a *myeloma cell* that can be cultured in the laboratory. Myeloma tumors can be produced in experimental animals by intraperitoneal injection of mineral oil. In 1975, C. Milstein and G. Koehler were able to fuse myeloma cells with normal B lymphocytes, thereby producing a hybrid cell that could be grown in culture. Moreover, the hybrid cell, called a **hybridoma**, produces and secretes antibodies characteristic of the clone from which the normal lymphocyte is taken.

The procedure for producing a hybridoma is depicted in Figure 25–11. First, an immune response is induced in a normal animal by exposure to a specific antigen (either by injecting the purified antigen or by injecting bacteria or viruses). Lymphocytes are then obtained from the animal's spleen or other lymphoid tissue. Among the cells that are removed will be some from clones that were activated by exposure to the antigen (many, perhaps most, of the cells will be from clones not selected by the antigen). The normal lymphocytes are then mixed with myeloma cells in a solution of polyethylene glycol. The myeloma cells to be used lack the enzyme *hypoxanthine-quanine phosphoribosyl transferase (HGPRT)*, which catalyzes the synthesis of inosine monophosphate (IMP) and guanosine monophosphate (GMP). Polyethylene glycol induces fusion of the two families of cells, thereby form-

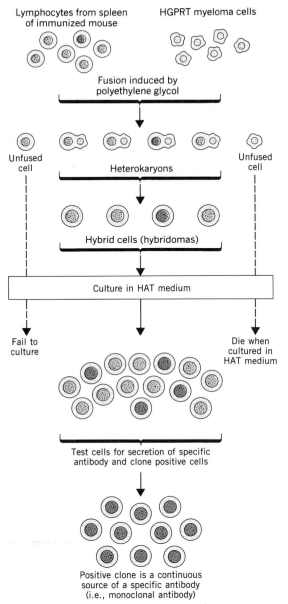

Lymphocytes from spleen of immunized mouse

HGPRT myeloma cells

Fusion induced by polyethylene glycol

Unfused cell   Heterokaryons   Unfused cell

Hybrid cells (hybridomas)

Culture in HAT medium

Fail to culture   Die when cultured in HAT medium

Test cells for secretion of specific antibody and clone positive cells

Positive clone is a continuous source of a specific antibody (i.e., monoclonal antibody)

FIGURE 25–11   Procedure for producing hybridoma cells (see text for details). The clone of hybridoma cells represents a continuous source of antibodies called monoclonal antibodies.

ing hybridomas. Some of the hybridomas that are formed will be hybrids of myeloma cells and the antigen-activated normal lymphocytes. Of course, the polyethylene glycol solution will also contain a number of unfused myeloma cells and normal lymphocytes. The unfused lymphocytes will fail to grow (or

grow so slowly that they produce insignificant numbers of progeny) when the cells are subcultured. If subculturing is carried out in a medium containing hypoxanthine, aminopterin, and thymidine (i.e., HAT medium), unfused myeloma cells will also die out because they cannot produce HGPRT. The hybridomas *will* proliferate because the lymphocyte nuclei contribute functional HGPRT genes to the heterokaryon. The particular hybridomas producing the desired antibody can be identified using a suitable assay procedure, and these can be selected and separately cloned. Thus, the end result is a culture of cells producing a single type of antibody—a **monoclonal antibody**. Such cultures can be maintained indefinitely and be used as a continuous source of antibody.

B LYMPHOCYTES AND THE IMMUNE RESPONSE

To understand how B lymphocytes are caused to secrete antibodies during an immune response, let's consider a case in which a person acquires either a bacterial or viral infection. Two events must generally occur if B lymphocytes are to be activated (Fig. 25–12). First, antigens present on the surface of (or released by) the pathogen become bound to antibodies in the plasma membranes of one or more of the millions of clones of B lymphocytes. Binding of the antigen to the surface of the B lymphocytes does not by itself cause activation of the clone. Instead, antigens must also be taken up during nonspecific phagocytosis of antigen-bearing particles by macrophages (i.e., phagocytic cells that act as scavengers in the body's tissues). The antigens taken up by the macrophages are degraded or "processed" and fragments containing antigenic determinants are then displayed at the cell surface. Macrophages that carry out this process are referred to as **antigen-presenting cells**. The antigenic determinant is then recognized by one or more clones of T cells possessing T-cell receptors for the antigen. T cells that recognize and are activated by antigen-presenting cells are called **T helper cells**. Activated T helper cells then interact with the B lymphocytes to which antigen had already been bound. The interaction between T helper cells and B lymphocytes serves to activate the B lymphocytes causing the rapid proliferation of the clone, thereby yielding plasma cells and memory cells (Fig. 25–12). Only the plasma cells produce and secrete antibodies. The memory cells are kept in reserve and will be called on to re-

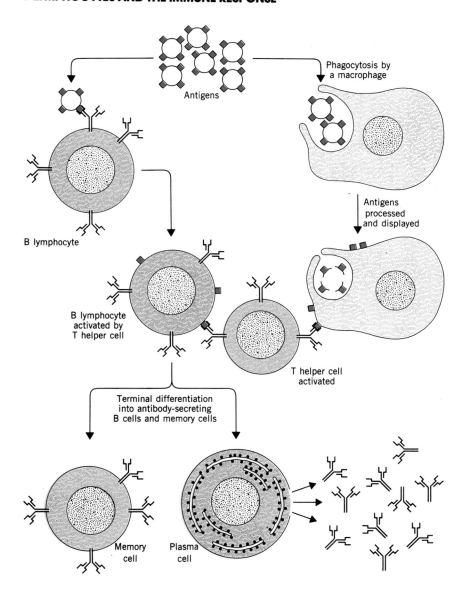

FIGURE 25–12 Role of the B lymphocytes during an immune response (see text for details).

spond during a second (or subsequent) infection by the same antigen-bearing pathogen.

Antibodies secreted by plasma cells may have several different effects: (1) they may interact with free (i.e., soluble) antigens causing **precipitation**; (2) they may interact with surface antigens of the pathogen (i.e., particulate antigens) causing **agglutination**; or (3) they may promote **complement fixation**.

Precipitation of Soluble Antigens

Antigens may have one or more antigenic determinants (Fig. 25–13). If one antigenic determinant is

present, the antigen is said to be *monodeterminant*; if two are present, the antigen is *bideterminant*, and so on. Most antibodies are *bivalent*, meaning that they can simultaneously combine with up to two antigenic determinants. As Figure 25–13 illustrates, the products formed by interaction of immunoglobulin and antigen depend on the number of antigenic determinants that are present. Two monodeterminant antigens can be cross-linked by a single antibody (Fig. 25–13a), but the product is not usually insoluble unless the antigen itself is very large. However, if two antigenic determinants are present, cross-linking by the antibody can produce chains of antigens that are in-

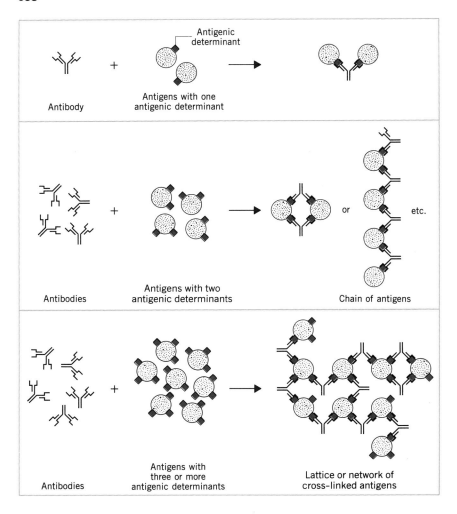

soluble and form precipitates (Fig. 25–13b). Multi-determinant antigens react with antibody to produce cross-linked networks or lattices that are insoluble (Fig. 25–13c).

Interactions between antibodies and free antigens can be considerably more complex than those illustrated in Figure 25–13. For example, some antibodies may exist as dimers (e.g., IgA) or pentamers (e.g., IgM) (see Fig. 25–3); these antibodies can simultaneously bind four or more antigenic determinants. Moreover, antigens may possess more than one kind of antigenic determinant, each determinant capable of reacting with a different antibody. Finally, the predominant form of interaction that takes place between antibodies and antigens is influenced by the respective concentrations of the interacting species. Small soluble complexes are favored when there is an excess of antibody; chains of cross-linked antigens are fa-

vored when there is an antigen excess; and cross-linked lattices are favored by nearly equal amounts of antibody and antigen. Regardless of the nature of the products formed, antigen–antibody complexes are eventually eliminated by the phagocytic action of macrophages (see below).

Agglutination

Antibodies that interact with antigens present in the surfaces of invading microorganisms or other foreign particles cause *agglutination* (Fig. 25–14). During agglutination the particles become cross-linked to form small masses, and the masses are eliminated by the phagocytic action of macrophages.

As illustrated in Figure 25–14, the plasma membranes of macrophages possess receptors that recognize and bind the C-terminal or F_c *regions* of

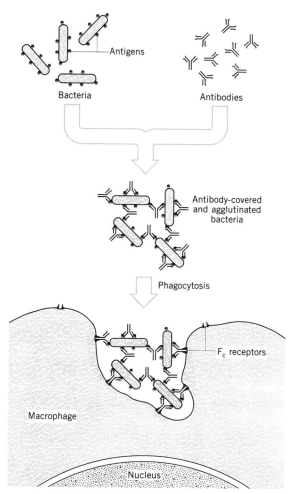

FIGURE 25–14 Agglutination by antibodies and phagocytosis by macrophages. Antibodies that interact with antigens in the surfaces of invading microorganisms can bring about cross-linking that agglutinates the invading cells. The agglutinated cells are then phagocytosed by macrophages whose F_c receptors form complexes with the antibodies.

immunoglobulin heavy chains (see Fig. 4–35). Consequently, the macrophage receptors are called F_c **receptors**. Because the F_c regions of the immunoglobulins include constant domains, macrophage F_c receptors can bind a variety of different antibodies. Interaction between a macrophage and a mass of agglutinated cells is followed by phagocytosis.

Although the mechanism is not fully understood, foreign cells that have attached antibodies can also be destroyed by **K** (or **killer) cells**. Killer cells bind the agglutinated mass by interacting with the F_c regions of antibodies but do not internalize it. Instead, it is thought that there is the transfer of toxic substances from the K cell to the pathogen.

Complement Fixation

The *complement system* is part of still another mechanism by which antibodies defend the body against invasion by pathogens. Complement consists of more than a dozen proteins that circulate in the blood. The binding of antibodies to a cluster of antigenic determinants in the surfaces of bacteria triggers a cascade of reactions in which the complement proteins (many of which are proenzymes) are sequentially activated. The cascade is initiated by the binding of a small complex of the complement proteins to the constant regions of antibodies that are bound to the bacterial antigens. In the ensuing reactions, additional complement proteins are bound and activated, eventually forming a *lytic complex* that creates an open channel through the bacterial surface. By disorganizing the bacterium's plasma membrane and allowing water to enter the cell by osmosis, the bacterium is killed. Complement fixation by antibody-coated bacteria and the lysis of the invading cells that follows is the most common defense mechanism attributable to B-cell-secreted antibodies.

Immunologic Memory

Figure 25–15 shows the relationship between time and the appearance of antibodies in response to a *first* exposure to a given antigen. Following a short lag period, antibodies begin to appear in the blood, rising to and maintaining a plateau level for some time before falling again. This characteristic response curve is called a **primary immune response**. As long as the antibody content of the blood remains at its plateau level, a condition of **active immunity** exists. The response to a second exposure to the same antigen—the **secondary immune response**—is much more dramatic. The lag period is shorter, the response is more intense (i.e., greater quantities of antibody are produced) and the elevated antibody level is maintained for a longer period of time. The difference between the two responses indicates that the body has "remembered" its earlier exposure to the antigen.

Immunologic memory may be explained in the following way. The initial exposure to antigen causes differentiation of B lymphocytes into *memory cells* as well as plasma cells. Whereas the plasma cells have a relatively short life span in which they are actively en-

FIGURE 25-15 Immunologic memory. During a primary immune response (i.e., a response to the first exposure to a particular antigen), the antibody level of the blood rises to a modest plateau and eventually declines. A second exposure to the same antigen (i.e., antigen A in the diagram) results in a secondary immune response that is quicker and in which greater quantities of antibody are made available. Simultaneous exposure to a different antigen (i.e., antigen B in the diagram) elicits a primary immune response.

gaged in antibody secretion, memory cells do not secrete antibody and continue to circulate in the blood and lymph for months or years. These memory cells are able to respond more quickly to the reappearance of the same antigen than undifferentiated B lymphocytes. Memory cells are also produced by the multiplication and differentiation of T lymphocytes (see below).

Autoimmune Diseases

The immune system normally produces antibodies against foreign proteins but not against the native proteins of the body, that is, the immune system can distinguish between "self" and "non-self." Yet one's own proteins will readily be regarded as antigens by the immune system of another organism. Thus, each individual's tissues possess a myriad of proteins (and other chemical substances) that are *potential* antigens.

The capability to distinguish self from non-self develops very early in life. In the 1950s, P. B. Medawar carried out a series of elegant experiments that bear on this concept. Adult mice from one strain reject skin grafts from another strain; that is, the recipient's immune system produces antibodies against antigens in the donor's tissue and this leads to the destruction of the donor's cells. However, when living spleen cells

(which carry the same antigens as skin cells) from one strain of mice were injected into *newborn* mice of a different strain and the skin graft experiments repeated when the newborn mice reached adulthood, the results were entirely different. Newborn mice that had been exposed to the spleen cells of another strain accepted skin grafts from that strain later in life. This is interpreted to mean that the spleen cells had been transferred to the new born mice while the mice were at an early enough stage of development to accept the spleen cells as "self" by the maturing mouse immune system.

In rare cases, individuals begin to produce antibodies against their own antigens. These antibodies are called **autoantibodies** and the diseases resulting from their presence are the **autoimmune diseases**. Among these diseases are *paroxysmal cold hemoglobinuria* (antibodies against one's own red blood cells), *myasthenia gravis* (antibodies against one's own muscle cell acetylcholine receptors), and *systemic lupus erythematosus* (antibodies against one's own nuclear DNA).

The causes of autoimmune diseases are not entirely clear and several different mechanisms seem to be involved. Clones of lymphocytes prepared to respond to a non-self (i.e., foreign) antigen that is structurally similar to self may undergo mutation during clonal expansion, thereby producing cells that now respond to self. It has recently become clear that T and B cells reactive to self antigens are present even in normal individuals. However, in normal individuals **T suppressor cells** serve to suppress the activity of these cells and thereby prevent autoimmune diseases.

MAJOR HISTOCOMPATIBILITY COMPLEXES

As we have already seen, B and T cells contain membrane receptors that react with specific antigenic determinants. B cells are activated by interaction with the antigen either in its dissolved form or while it is still a part of the surface of a pathogen. In contrast, the T cells of an animal are only activated when the antigen is displayed on the surface of a cell that *also* possesses markers of the animal's own identity. These markers are proteins encoded by a cluster of genes called the **major histocompatibility complex (MHC)** and are called the **MHC glycoproteins.**

The MHC glycoproteins were discovered as a result of tissue-transplantation or tissue-grafting experiments. Grafts involving donors and recipients from the *same strain* of experimental animals (i.e., the equivalent of identical twins) are usually accepted by the donor's body. However, when the donor and recipient are not genetically identical, the graft is rejected because the recipient mobilizes an immune response to the transplanted cells and destroys them.

It appears that different (i.e., unrelated) individuals express different sets of MHC genes and, like the antibodies and T-cell receptors, these proteins are incredibly diverse. However, unlike antibodies and T-cell receptors, which differ among the millions of different clones of cells of an individual, the MHC proteins differ only among individuals, that is, all cells of a single individual bear the same MHC proteins. Human MHC proteins are called **human leukocyte-associated antigens** or **HLA antigens** (they were first identified in leukocytes) and can be divided into two major classes: *class I* MHC antigens and *class II* MHC antigens (Fig. 25–16).

Class I antigens are found in the surfaces of nearly all cells. They consist of two polypeptide chains: a large *A chain* that is similar in size and organization to an immunoglobulin heavy chain and a smaller chain called a β_2-microglobulin (Fig. 25–16). Amino acid sequence homologies between these polypeptides and immunoglobulins suggest that an evolutionary link exists among them.

Class II antigens are not found in the surfaces of all types of cells; rather, they are limited to a few types of cells that play a role in the immune response. For example, they are found in most B cells, some T cells, and some antigen-presenting macrophages (see Fig. 15–12). Class II antigens are composed of two polypeptide chains: an *alpha chain* and a *beta chain*, each about the same size as an immunoglobulin light chain (Fig. 25–16). Like the class I antigens, the class II antigens exhibit some degree of homology with the immunoglobulins. The MHC antigens play a role in determining the actions of T cells.

T LYMPHOCYTES AND THE IMMUNE RESPONSE

T cells do not interact with free antigens or with antigenic sites in the surfaces of foreign microorganisms. Instead, T cells respond only to cells bearing *both* a self MHC antigen and an antigenic determinant from

FIGURE 25–16 Class I and class II major histocompatibility antigens. The A chains of class I antigens and the alpha and beta chains of class II antigens contain hydrophobic amino acid sequences by which they are anchored in the plasma membrane.

a foreign source (i.e., from bacteria, viruses, etc.). Thus, two stimuli are needed to trigger the proliferation and terminal differentiation of the required T-cell clones. **Cytotoxic T cells** respond to the combination of foreign antigen and class I MHC antigens, whereas **helper T cells** respond to foreign antigen and class II MHC antigens. Thus, the activities of these T cells are directed toward the body's own cells and not to free pathogens.

Cytotoxic T Cells

The actions of cytotoxic T cells may be illustrated by the body's response to viral infection (Fig. 25–17). When a virus attacks a cell, the viral nucleic acid enters the host cell and the proteins of the viral coat remain at the cell surface. Consequently, the infected cell has the proper combination of surface antigens to be recognized by cytotoxic T cells namely, class I MHC antigens and viral antigens.

Cytotoxic T cells attach to newly infected host cells, killing them (presumably through the transfer of cytotoxins) before virus replication occurs. A number of host cells are necessarily killed by this process before the virus infection is attenuated. A single cytotoxic T cell may kill several host cells. Both the viral antigen and the class I MHC antigen are involved in attachment of the T cell. Still unclear is whether one T-cell receptor combines with both antigens or whether two T-cell receptors mediate the process. Figure 25–17 depicts the attachment of the two cells via a single T-cell receptor.

Helper T Cells

As noted earlier in the chapter (see Fig. 25–12), macrophages and other scavengers ingest and degrade foreign antigens, ultimately displaying their antigenic determinants at the cell surface. The combination of antigenic determinant and class II MHC antigens at the surfaces of these cells leads to the attachment and activation of helper T cells (Fig. 25–18). Helper T cells can activate B lymphocytes and other T cells (e.g., cytotoxic and suppressor T cells). Certain helper T cells secrete **lymphokines** or **interleukins**, substances that activate macrophages and other lymphocytes. Some lymphokines attract macrophages to the site of infection. Other lymphokines prevent migration of macrophages away from the site of infection. Still other lymphokines stimulate T-cell proliferation. The net effect

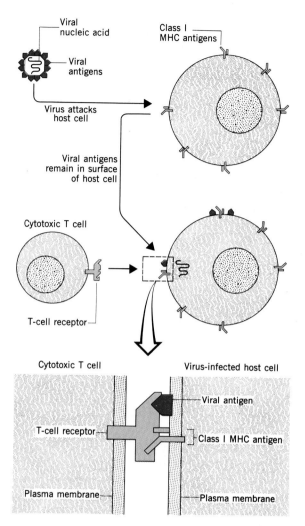

FIGURE 25–17 Action of cytotoxic T cells during a response to viral infection (see text for details).

FIGURE 25–18 Activation of helper T cells by antigen-presenting macrophages involves an interaction between the T-cell receptor and both foreign antigens and class II MHC antigens.

of lymphokine secretion is the accumulation of macrophages and lymphocytes in the region of an infection and is characterized by the inflammation that typically exists there.

Acquired Immune Deficiency Syndrome (AIDS)

No disease in recent memory has attracted so much public concern and fear as **acquired immune deficiency syndrome** or **AIDS**. The disease is caused by a virus called **human T-cell lymphotropic virus-III** or **HTLV-III** (HTLV-I and HTLV-II viruses are discussed later in the chapter in connection with cancers of the immune system). The HTLVs are **retroviruses**, that is, they are RNA viruses that induce their host cells to proliferate new viral genes by reverse transcription (see Chapter 21).

HTLV-III critically injures the immune system by infecting and eventually killing T cells. As a result of the progressive destruction of its T cells, the body is easily ravaged by a number of common infectious agents. In many instances, these infections would have caused little injury were there functional T-cell clones available. Unable to battle infections in the normal manner, victims that develop a "full-blown" case of AIDS eventually succumb. In AIDS patients the HTLV-III virus has been shown to be present in semen as well as in the blood. Though the virus is believed to be transmitted principally through sexual contact, a number of hemophiliacs have contracted the disease as a result of receiving transfusions of infected blood. A number of infants have developed AIDS by transplacental transmission from infected mothers.

CANCER

Cancer is a disease characterized by the excessive and abnormal growth of certain tissue cells. In the healthy tissues of an adult, growth is regulated such that cell proliferation exactly balances the rate of cell loss. As a result, once one reaches adult age, the sizes and cellular contents of the various body organs remain more or less constant. Various chemical, physical, and viral agents can cause the loss of normal control of cell growth; as a result, a normal cell may be **transformed** into a cancerous one. The unregulated cell growth that accompanies cellular transformation produces **tumors** or **neoplasms**, each tumor being the product of proliferation of a single aberrant cell. *Malignant* tumor cells are cancer cells that spread to and take up residence in neighboring tissues—a condition called **metastasis**. On the other hand, *benign* neoplasms consist of cancer cells that do not spread to distant sites and do not pose so great a threat to life. The body's major line of defense against cancer cells is the immune system, which recognizes the changes in cell surface antigens that frequently accompany cellular transformation. As a result, nearly all cancer cells are eliminated by the immune system before causing serious harm.

As discussed in Chapter 2, cellular transformation can occur *in vitro* (i.e., in cultures of cells). Cultures of normal mammalian cells such as fibroblasts have a finite life span. Usually such cultures can be maintained for no more than 40 or 50 generations, after which the cells become senescent and the culture dies out. *In vitro* transformation may convert a normal culture into an "immortal" one—one that can be indefinitely subcultured. The relative ease with which cancer cells can be grown and maintained in culture has been a boon to the study of cellular transformation.

Transformed fibroblasts, like cancer cells *in vivo*, exhibit abnormal (i.e., excessive) growth. Normal fibroblasts in culture are flattened cells that exhibit contact-inhibited cell growth once they form a single layer. Cancerous fibroblasts grow more rapidly, form clumps of cells, do not exhibit contact inhibition, and will grow on top of one another. In normal cultures, the cells are anchorage-dependent; cancerous cells are anchorage-independent.

When embryonic kidney or liver cells are grown in culture, they adhere to one another. It is as though they recognize their neighbors as cells of the same type and form the intimate associations characteristic of kidney and liver tissue. When kidney and liver cells are intermixed, they sort themselves out into regions that consist of one or the other cell type. Transformed cells exhibit diminished adhesion properties (apparently as the result of a change in the glycoprotein and glycolipid content of the plasma membrane). Such changes explain why cancer cells dissociate from the tissues from which they originate and are able to spread (i.e., metastasize) to other areas of the body and grow.

Mutagenesis, Proto-oncogenes and Oncogenes

The growth of a cancer stems from changes that occur in a single (and previously normal) cell. The transformed cell escapes immune detection and undergoes uncontrolled growth and division producing a large clone of transformed cells that is now identified as a tumor. Among the agents that have been shown to be *carcinogenic* (i.e., "cancer-causing") are various chemicals, radiation, and infection by certain viruses. Carcinogens transform normal cells by directly or indirectly causing changes in the cell's DNA. These changes are called **mutations**, and the carcinogen is thus said to be **mutagenic**.

It is now clear that most (perhaps all) cancers stem from the translocation, alteration, or activation of specific genes called **proto-oncogenes**. Proto-oncogenes are believed to be present in all normal cells, where they exist either in an inactive state or function in the production of normal quantities of transcription and translation products. Mutations change or activate proto-oncogenes and convert them to **oncogenes**, whose translation products (i.e., **onco-proteins**) are intimately associated with the cancerous state.

Cancer of the Immune System—Burkitt's Lymphoma

Some cancers are clearly associated with changes in the numbers of chromosomes in the nucleus. Normal human somatic cells contain 46 chromosomes (i.e., 23 homologous pairs), whereas cancer cells have irregular numbers—a condition termed *aneuploidy*. For example, HeLa cells (a cancer cell line maintained in subculture for many years and originally derived from a woman with cervical cancer) contain 70–80 chromosomes. Indeed, it is typical for there to be *more* than the normal chromosome number in cancer cells, but it is generally agreed that the aberrant number is the *result* of uncontrolled cell growth and *not* the cause of it. More subtle chromosome changes can be linked to cancer causation. One of the best illustrations of this is the human leukemia called *Burkitt's lymphoma*.

Earlier in the chapter, we saw that the remarkable diversity of antibodies produced by B cells results from the translocation of gene segments during cellular proliferation. In the most common forms of Burkitt's lymphoma, a *reciprocal translocation* occurs between chromosomes 8 and 14 (Fig. 25–19). A segment at the end of chromosome 8 is translocated to chromosome 14 and is replaced by a segment from the end of chromosome 14. As a result, a proto-oncogene from chromosome 8 ends up near genes of chromosome 14 that encode part of an antibody molecule. This translocation serves to convert the proto-

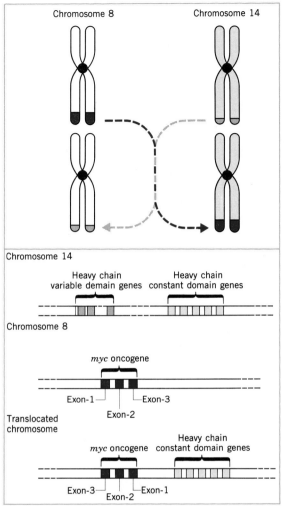

FIGURE 25–19 Burkitt's lymphoma results from a reciprocal translocation between chromosomes 8 and 14 (a). As a result the *myc* proto-oncogene of chromosome 8 ends up near heavy chain constant domain genes (b). This changes the *myc* gene's expression and results in cell transformation.

oncogene to an oncogene. The oncogene is called *myc* because the gene itself was first identified in a virus known as MC29 (myelocytomatosis virus).

The *myc* gene consists of three exons and two introns (Fig. 25–19). The introns consist of DNA that is not transcribed into mRNA and is not expressed as protein. However, the first exon (which is about 650 base pairs long) also does not code for protein; rather, protein synthesis begins with the second exon.

In the most common forms of Burkitt's lymphoma, either all three exons are translocated from chromosomes 8 to chromosome 14 or exons 2 and 3 are translocated (leaving exon-1 on chromosome 8). Translocation places the *myc* gene in a region of chromosome 14 that is adjacent to the genes encoding the constant regions of immunoglobulin heavy chains.

In rarer forms of the lymphoma, the *myc* gene remains on chromosome 8 but it is joined there by genes translocated from either chromosome 22 or chromosome 2. Translocation involving chromosome 2 brings κ light chain constant domain genes next to the *myc* gene; translocation involving chromosome 22 brings λ light chain constant domain genes next to the *myc* gene. Apparently, translocations that result in the *myc* gene being next to antibody encoding genes causes a change in the *myc* gene's expression. The nature of the change is not known with certainty, but it could be excessive production of the *myc* gene product or production at the wrong time in the cell cycle. The nature of the *myc* gene's protein product is still uncertain. However, the viral *myc* gene product is a protein that binds to DNA.

Naming oncogenes   Oncogenes are usually identified using a three-letter name that is related in some fashion either to the gene's source or to its cause. To distinguish between viral oncogenes and their native cell counterparts, the viral gene name is preceded by a "v" (for virus), whereas the cellular oncogene name is preceded by a "c" (for cell).

Cancer of the Immune System—T-Cell Leukemia

It is now clear that a number of cancers result from infection by *retroviruses*. Two general mechanisms have been proposed to account for the capability of retroviruses to transform cells: (1) the transfer of oncogenes from the retrovirus to the infected host cell and (2) the activation during infection of oncogenes already present in the host. Not all retroviruses contain oncogenes in their genome, but those that do are believed to have acquired them from the cellular DNA of a host cell during a previous infection. That is, the oncogene was a normal constituent of the host cell's genome but was picked up by the virus during its assembly within the host. Transfer of the oncogene from the virus to a new cell during a subsequent infection leads to the transformation of that cell. Retroviruses contain repeating nucleotide sequences at the ends of their genomes; these sequences are known as **long terminal repeats (LTRs)**. LTRs contain regulatory segments known as **promoters** and **enhancers**; these serve to activate transcription of viral genes into messenger RNA. Retroviral transformation of host cells by activation of oncogenes that are already present in the host cell may involve insertion of an LTR near one of the cellular oncogenes.

Two viruses, called **HTLV-I** and **HTLV-II**, are the cause of **T-cell leukemia** in humans. These leukemias are characterized by the excessive and abnormal production of T cells. The HTLV genome does not contain any oncogenes and appears *not* to transform cells by either of the mechanisms described above. One model proposed to explain the actions of the HTLVs postulates that the virus directs the infected host cell to produce a protein that stimulates not only production of new viruses but also production of new host cells. The viral promoter has been shown to be present in the LTR and when integrated into the host genome increases the transcription of genes that turn on cell division. This then leads to the abnormal and uncontrolled growth of the transformed cells.

A close relative of HTLV-I and HTLV-II, called HTLV-III, causes acquired immune deficiency syndrome (AIDS) (see above). In contrast to the effects of HTLV-I and HTLV-II, HTLV-III causes T-cell death rather than excessive proliferation.

Viruses and Cancer

That a link exists between tumors and viruses has been known since the beginning of the century. As early as 1910, P. Rous was able to show that an avian leukemia could be transmitted from one infected fowl to another using viral particle-containing extracts of the cancerous cells. The virus causing this cancer is known as the *Rous sarcoma virus* (RSV). In the 1960s, R. Dulbecco and M. Vogt were able to demonstrate virus induced cellular transformation *in vitro*. Dulbecco and Vogt worked with hamster embryo fibroblasts, which they cultured in Petri dishes containing agar growth medium. In normal cultures, cell proliferation produces a monolayer that eventually covers the surface of the dish. However, addition either *polyoma virus* or *Simian virus 40* (SV-40) to the culture transforms the cells. Growth becomes irregular and clumps of cells raised well above the surface of the agar quickly appear. When transferred to young animals, the transformed cells gave rise to tumors.

The polyoma and SV-40 viruses are *DNA tumor viruses*, that is, the viral genome consists of DNA. Included in the DNA is an oncogene that becomes integrated into the DNA of the host cell and that when expressed causes cellular transformation.

More extensively studied and better understood are the *RNA tumor viruses*. All of the RNA tumor viruses are *retroviruses*. When the RNA of a retrovirus is inserted into the host cell, it is transcribed into DNA (the process is usually referred to as *reverse transcription*). The DNA produced by reverse transcription of tumor virus RNA includes oncogenes. Integration of an oncogene into the host cell's genome and its eventual expression causes cellular transformation.

Oncogenes and Their Protein Products

The first viral oncogene was specifically identified in the 1970s in the Rous sarcoma virus and was named *v-src* (for sarcoma). As noted above, this virus causes a leukemia in chickens. The *v-src* oncogene encodes a protein called *p60src* having a molecular weight of 60,000. In 1975 it was found that the *src* gene is nearly an exact copy of a gene that is found in all chickens. Apparently, the chicken proto-oncogene was picked up by a retrovirus during infection, became a part of the viral genome, and changed into an oncogene that causes cancer when reintegrated into the genome of a newly infected host cell. About 20 oncogenes have now been identified in retroviruses. When transferred to host cells by infection, they cause a number of cancers (e.g., carcinomas, sarcomas, and lymphomas) in birds, mice, rats, cats, and primates. In each instance, the oncogene has proven to be closely related to a normal gene of the host and to encode a protein that is similar to a normal host cell protein.

Not all tumors result from viral infection, but like virus-induced cancers, the cells of tumors of nonviral origin also contain oncogenes. The oncogenes of such tumor cells are called **tumor oncogenes**. DNA that contains tumor oncogenes may be isolated and introduced into normal cells. When this is done, the normal cells are transformed into cancerous ones. The phenomenon is called **transfection** and is a valuable tool for studying cellular transformation.

As you may have inferred by now, tumor oncogenes and viral oncogenes appear to be derived from a common gene pool. For example, the retroviral oncogene *v-ras* carried by the murine sarcoma virus is homologous to a tumor oncogene. The *c-ras* tumor oncogene is derived from the proto-oncogene *proto-ras*, which is present in animal species ranging from yeasts to humans. This high degree of conservation suggests that in addition to being a source of oncogenes, the proto-

oncogenes play important roles in the untransformed cell.

Conversion of a proto-oncogene to an oncogene may occur in several ways. One mechanism believed to be important in the activation of oncogenes is *mutation* induced by a carcinogen. In 1982, R. A. Weinberg and M. Barbacid (working separately) showed that the nucleotide sequence in the T24 human bladder carcinoma oncogene *c-ras* differed at only *one* position from the nucleotide sequence of the normal, proto-oncogene. A single point mutation in which guanosine occurs in place of thymidine results in the incorporation of valine instead of glycine at the twelfth amino acid position of the oncogene-encoded protein. This solitary substitution appears to be all that is necessary to confer cell-transforming properties to the gene product. Another indication of the cancerous state of a cell is indicated by alteration of *protein kinases* in the membranes of certain cells.

Protein Kinases   Protein kinases are enzymes that phosphorylate proteins by adding phosphate to hydroxyl-containing R-groups of certain amino acids (Fig. 25–20). In normal cells, about 90% of the phosphate coupled to protein is coupled to serine and 10% is coupled to threonine. Phosphorylation of the hydroxyl group of tyrosine is very rare (about 0.02%). In 1978, it was found that the protein encoded by the *v-src* oncogene (i.e., *p60src*) is a plasma membrane-bound *protein kinase* that phosphorylates tyrosine, and shortly thereafter the *src* proto-oncogene was also shown to encode a tyrosine-phosphorylating protein kinase. These findings indicated that the products of viral oncogenes mimic the functions of their normal cellular counterparts. However, in transformed cells there is a 10-fold increase in the phosphotyrosine level. It appears, therefore, that an abnormally high phosphorylation of tyrosine may be related to cell transformation.

Among the cellular proteins that are acted upon by protein kinases is *vinculin*, a cytoskeletal protein. In cells transformed by the Rous sarcoma virus, there is a 20-fold increase in the amount of phosphorylated tyrosine present in vinculin. Because vinculin is believed to play a role in determining cell shape, there may be a connection between the unusual shape of transformed cells and the excessive phosphorylation of cellular vinculin.

Certain metobolic pathways can also be altered in cells by changes in protein kinase levels. For instance,

FIGURE 25–20 Among the products of oncogenes are enzymes called protein kinases. These enzymes phosphorylate hydroxyl-containing R-groups of serine, threonine, and tyrosine residues in specific intracellular proteins.

it has been known for many years that tumor cells produce large amounts of lactic acid as a result of increased glucose catabolism by glycolysis (see Chapter 10). Though a small fraction of three of the enzymes of glycolysis contain phosphotyrosine (e.g., *enolase*, *phosphoglycerate mutase*, and *lactate dehydrogenase*), the percentage is so small as to raise considerable doubt regarding a link to transformation. Moreover, the rate-limiting enzyme for glycolysis (*phosphofructokinase*) does not appear to contain phosphorylated tyrosine.

Growth Factors and Growth Factor Receptors
Among other things, the growth of cells in culture can be regulated by addition of a number of secretory polypeptides called **growth factors**. It is believed that *in vivo* these factors are secreted into the blood by the cells of one tissue and interact with specific plasma membrane receptors, called **growth factor receptors**, in other tissues. Binding of growth factors to growth factor receptors triggers the production of intracellular signals that ultimately cause DNA synthesis and mitosis. Several oncogenes have been identified whose protein products are similar to growth factors and to growth factor receptors.

Among the known growth factors is **platelet-derived growth factor (PDGF)**, a protein having a molecular weight of about 30,000. It has recently been found that the PDGF is nearly identical to a portion of

the protein product of a viral oncogene—the *v-sis* oncogene of Simian sarcoma virus (SSV). It is believed that in SSV-transformed fibroblast cells, the *v-sis* oncogene is expressed as a secretory oncoprotein that interacts with PDGF receptors in the fibroblast plasma membrane. Binding of the oncogene product to the PDGF receptors acts to stimulate tyrosine phosphorylation of intracellular proteins and causes unrestricted growth of the cells. *In situ*, normal cells bearing the PDGF receptor respond to PDGF from *other* cell sources. Transformed cells produce and release their own PDGF, which then acts to promote cell proliferation by binding to the cells' PDGF receptors.

Another growth factor called **epidermal growth factor (EGF)** stimulates cell proliferation following binding to cellular EGF receptors. EGF receptors are large glycoproteins (MW 175,000) having a tyrosine kinase activity that is stimulated severalfold on binding EGF. EGF receptors have amino acid sequences that are closely related to the product of the viral oncogene *v-erb-B*. This oncogene is present in the genome of *avian erythroblastosis virus (AEV)*, a virus that attacks chickens and other birds. Apparently, the viral gene was derived from the normal EGF receptor gene that is present in avian cells.

Results of studies on PDGF and EGF, their receptors, and the corresponding oncogene products again implicate protein kinases with cellular transformation. One of the models proposed to account for the ef-

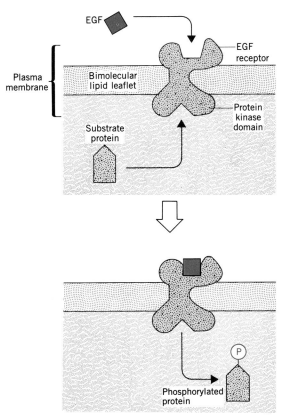

FIGURE 25–21 Growth factor receptors (e.g., epidermal growth factor receptor or EGF receptor) bind growth factors (e.g., EGF). Binding of the growth factor activates the receptor's protein kinase domain and results in the phosphorylation of tyrosine residues of specific target proteins. Protein phosphorylation somehow acts as the trigger for DNA synthesis and mitosis. The products of certain oncogenes are believed to mimic growth factors and growth factor receptors resulting in untimely and excessive cell proliferation.

fects of the growth factor–receptor system is shown in Figure 25–21. The growth factor receptors are intrinsic membrane proteins that span the lipid bilayer. The growth factor receptor site is located in that portion of the protein that faces the outside of the cell, whereas the site for tyrosine phosphorylation faces the interior. Binding of the growth factor to the receptor site serves to activate the tyrosine-specific protein kinase site, and as a result intracellular target proteins are phosphorylated. It is believed that the tyrosine-phosphorylated proteins then act in some way to promote cell proliferation.

Nuclear Oncoproteins

Nuclear Oncoproteins The products of some oncogenes are proteins that have an effect directly within the cell nucleus. Included in this group is the *myc* gene discussed earlier in connection with Burkitt's lymphoma. The *c-myc* oncoprotein binds to nuclear DNA and in some manner makes replication of the DNA a continuous process. Other oncogenes whose products act within the nucleus are the *c-myb* and *c-fos* oncogenes. The *c-myb* gene is associated with a leukemia in birds and *c-fos* produces a sarcoma in rats. At this time, little is known about their oncogenic protein products or their modes of action within the nucleus.

Figure 25–22 summarizes the sites of action of the oncogenes discussed in the chapter and Table 25–2 lists the genes, their retroviral sources, and their supposed action. Because more than 20 oncogenes have been discovered so far, Figure 25–22 and Table 25–2 are by no means complete.

Multiple Oncogene Activation and Cancer

The link between cancer and either the activation of cellular proto-oncogenes or the acquisition of viral oncogenes is irrefutable. Activation of cellular oncogenes may stem from exposure either to chemical mutagens or to physical mutagens (e.g., radiation). Uncertain at the present time is whether activation or acquisition of a single oncogene is sufficient to fully transform normal cells into cancerous ones. R. A. Weinberg has shown that cells cultured immediately

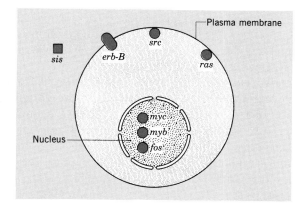

FIGURE 25–22 Some of the known oncogenes. Expression of the *sis* oncogene results in production of an extracellular growth factor. The *erb-B*, *src*, and *ras* oncogene products are plasma membrane proteins. Products of the *myc*, *myb*, and *fos* oncogenes act directly on the nuclear DNA.

TABLE 25-2 ONCOGENES AND THEIR EFFECTS

Oncogene	Retroviral Source	Cell Locus of Oncoprotein	Function of Oncoprotein
src	Rous sarcoma virus	Plasma membrane	Protein kinase
erb-B	Avian erythroblastosis virus	Plasma membrane	Protein kinase
sis	Simian sarcoma virus	Extracellular (i.e., secreted)	PDGF-like growth factor
ras	Rat sarcoma virus	Plasma membrane	Protein kinase (?)
myc	Myelocytomatosis virus	Nucleus	DNA binding
fos	Mouse sarcoma virus	Nucleus	?
myb	Avian leukemia virus	Nucleus	?

after removal from an animal are not easily transformed by a *single* oncogene, whereas culture cells from established cell lines are. Pairs of oncogenes, such as *myc* and *ras*, appear to cooperate in some manner such that they readily transform cells cultured immediately after removal from an animal. It has been believed for some time that the development of a cancer is a multistep process. What we have learned in recent years about oncogenes and oncoproteins places us at the threshold of understanding these steps at the molecular level.

THE IMMUNE SYSTEM AND CANCER

As noted at the beginning of the chapter, in addition to protecting the body against infection, the immune system serves as a principal line of defense against cancer. Therefore, it is not surprising that suppression of the immune system often results in cellular transformation and the appearance of tumors. For example, it is not uncommon for individuals with acquired immune deficiency syndrome (AIDS) to develop a cancer called *Kaposi's sarcoma.*

Tumor cells are not only functionally different from their normal predecessors but they may also be antigenically different. Tumor cells may lose the major histocompatibility antigens (see above), may produce new antigens, or may produce antigens normally found only at much earlier (e.g., embryonic) stages of development. Tumors that stem from infection by oncogenic viruses may display antigens that are characteristic of the viral genome. If the antigens present in the surfaces of a tumor cell are sufficiently different from those of a normal cell, the cell will be regarded as "non-self" and will be attacked by the immune system.

Natural Killer Cells

Foreign cells that have attached antibody and virus-infected cells can be destroyed by a type of lymphocyte called a *natural killer* cell or *NK* cell (see above). The NK cell is believed to function by transferring cytotoxic substances to the target cell. The interaction between NK cell and target cell is accompanied by the production of **interferon**, which serves to promote rapid proliferation of additional NK cells. *In vitro* studies using cultures of transformed cells indicate that NK cells can also destroy many kinds of tumor cells (e.g., leukemia, myeloma, sarcoma, and carcinoma cells).

Tumor Detection and Tumor Immunotherapy

The growth of tumors *in vivo* may be modified by **immunotherapy**. In its simplest form, immunotherapy involves stimulating the immune system so that it may respond more vigorously to the presence of transformed cells. Immunostimulants that have met with some success are thymic hormones and doses of attenuated bacteria. One of the most promising areas of tumor diagnosis and tumor therapy involves the use of **monoclonal antibodies** that have been produced against specific tumor antigens. Monoclonal antibodies that have been complexed with radioactive ligands may be injected intravenously and used to detect and localize cancerous tissue by "radioimaging."

"Magic Bullets" A most spectacular application of monoclonal antibodies for cancer therapy involves complexing tumor-specific antibodies with cytotoxic agents. These complexes would then specifically seek out and associate with tumor cells, leading to the tumor's death. Cytotoxic agents that may be complexed with monoclonal antibodies to form such "magic bullets" include radioisotopes (whose radiation destroys the tumor cell), bacterial toxins, and plant lectins. This form of therapy is presently in a very early stage of development but holds great promise for the future. The significant advantage of this technique over less specific chemotherapeutic measures and radiation treatment are clear.

SUMMARY

The **immune system** is the body's system of defense against viruses, bacteria, and other pathogenic agents and also serves as the principal line of defense against cancer. The immune system responds to infection in two ways: (1) the **cellular immune response**, in which there is direct interaction between **lymphocytes** and invading pathogens; and (2) the **humoral immune response**, which acts primarily against extracellular phases of infection and in which proteins called **antibodies** or **immunoglobulins** are secreted into the body fluids.

Two types of lymphocytes mediate the immune response; these are the **T lymphocytes** or **T cells** (which mediate cellular immunity) and the **B lymphocytes** or **B cells** (which mediate humoral immunity). In response to the presence of foreign substances called **antigens**, B lymphocytes give rise to antibody-secreting **plasma cells** and T lymphocytes give rise to **cytotoxic**, **helper**, and **suppressor** T cells. Both B and T lymphocytes also produce **memory cells**, which are reserved by the body and respond during later exposures to the same source of antigen.

Several different classes of antibodies are secreted by plasma cells, each antibody molecule containing a pair of identical **heavy chains** and a pair of identical **light chains**. The heavy and light chains are encoded by hundreds of genes and **gene segments**. The recombination of genes and gene segments that takes place during **terminal differentiation** of lymphocytes results in millions of cell clones, each capable of producing a different antibody. Hybrid cells called **hybridomas** can be produced *in vitro* by fusing B cells with tumor cells. These cells can be grown in culture, where they provide a continuous source of **monoclonal antibodies**. Antibodies secreted by plasma cells protect the body by **precipitating** free antigens, **agglutinating** foreign cells, and triggering **complement fixation**.

Antibodylike proteins present in the plasma membranes of differentiated T cells endow T cells with their capacity to interact directly with the body's own virus-infected cells and with tissue macrophages. This interaction is founded on the presence in the membranes of target cells of foreign antigens and proteins of the **major histocompatibility complex**.

Cancer is a disease characterized by excessive and abnormal tissue growth. **Transformation** of a normal cell into a cancerous one may result from exposure to carcinogenic chemicals or radiation; these cause translocation, alteration, or activation of specific genes, called **proto-oncogenes**, that are normal constituents of the genome. Proto-oncogenes are converted to **oncogenes** and these encode **oncoproteins**. Cell transformation may result from (1) excessive expression of an oncogene, resulting in too much product, (2) production of an altered product, or (3) production at the wrong time of the cell cycle. Transformation also results from viral infection, during which an oncogene carried by the virus is transferred to the host cell's genome.

DNA tumor viruses and **RNA tumor viruses** (all of which are **retroviruses**) have been identified. Viral oncogenes are similar to cellular oncogenes and are believed to have been derived from a common gene pool, the viral oncogenes having been picked up from a host cell's genome during a previous cycle of infection. Among the oncoprotein products of the oncogenes are a number of **protein kinases**, **growth factors**, and **growth factor receptors**.

REFERENCES AND SUGGESTED READING

Articles and Reviews

Abelson, P. H., Biotechnology: An overview. *Science* **219**, 611 (1983).

Bishop, J. M., Oncogenes. *Sci. Am.* **246** (3), 80 (Mar. 1982).

Capra, J. D., and Edmundson, A. B., The antibody combining site. *Sci. Am.* **236** (1), 50 (Jan. 1977).

Croce, C. M., and Klein, G., Chromosome translocations and human cancer. *Sci. Am.* **252** (3), 54 (Mar. 1985).

Croce, C. M., and Koprowski, H., The genetics of human cancer. *Sci. Am.* **238** (2), 117 (Feb. 1978).

Dulbecco, R., The induction of cancer by viruses. *Sci. Am.* **216** (4), 28 (Apr. 1967).

Fauci, A. S., Masur, H., Gelmann, E. P., Markham, P. D., Hahn, B. H., and Lane, H. C., The acquired immune deficiency syndrome: An update. *Annals Int. Med.* **102**, 800 (1985).

Hunter, T., The proteins of oncogenes. *Sci. Am.* **251** (2), 70 (Aug. 1984).

Hunter, T., Oncogenes and growth control. *Trends in Biochem. Sci.* **10** (7), 275 (July 1985).

Hunter T., and Cooper, J. A., Protein–tyrosine kinases. *Ann. Rev. Biochem.* **54**, 897 (1985).

Kennel, S. J., Flynn, K., Foote, L., and Lankford, T., Monoclonal antibodies in cancer detection and therapy. *Bioscience* **34** (3), 150 (Mar. 1984).

Land, H., Parada, L. F., and Weinberg, R. A., Cellular oncogenes and multistep carcinogenesis. *Science* **222**, 771 (1983).

Laurence, J., The immune system in AIDS. *Sci. Am.* **253** (6), 84 (Dec. 1985).

Leal, F., Williams, L. T., and Robbins, K. C., Evidence that the *v-sis* gene product transforms by interaction with the receptor for platelet-derived growth factor. *Science* **230**, 327 (1985).

Leder, P., The genetics of antibody diversity. *Sci. Am.* **246** (5), 102 (May 1982).

Lerner, R. A., and Dixon, F. J., The human lymphocyte as an experimental animal. *Sci. Am.* **228** (6), 82 (June 1973).

Marx, J. L., Chemical signals in the immune system. *Science* **221**, 1362 (1983).

Marx, J. L., What do oncogenes do? *Sci. Am.* **223**, 673 (1984).

Marx, J. L., The immune system "belongs in the body." *Science* **227**, 1190 (1985).

Nicolson, G. L., Cancer metastasis. *Sci. Am.* **240** (3), 66 (Mar. 1979).

Oppenheimer, S. B., Causes of cancer: Gene alteration vs. gene activation. *Am. Lab.* 40 (Nov. 1982).

Oppenheimer, S. B., Control of the spread of cancer. *Am. Lab.* 12 (Feb. 1983).

Oppenheimer, S. B., Prevention of cancer. *Am. Lab.* 66 (Feb. 1983).

Rose, N. R., Autoimmune diseases. *Sci. Am.* **244** (2), 80 (Feb. 1981).

Silverstein, A. M., Development of the concept of immunologic specificity. *Cell Immunol.* **67**, 396 (1982).

Tabin, C. J., Bradley, S. M., Bargmann, C. I., Weinberg, R. A., Papageorge, A. G., Scolnick, E.M., Dhar, R., Lowy, D. R., and Chang, E. H., Mechanism of activation of a human oncogene. *Nature* **300**, 143 (1982).

Teillaud, J. L., Desaymard, C., Giusti, A. M., Haseltine, B., Pollock, R. R., Yelton, D. E., Zack, D. J., and Scharff, M. D., Monoclonal antibodies reveal the structural basis of antibody diversity. *Science* **222**, 721 (1983).

Tonegawa, S., The molecules of the immune system. *Sci. Am.* **253** (4), 122 (Oct. 1985)

Weinberg, R. A., Oncogenes of spontaneous and chemically induced tumors. *Adv. Cancer Res.* **36**, 149 (1982).

Weinberg, R. A., Oncogenes and the molecular biology of cancer. *J. Cell Biol.* **97**, 1661 (1983).

Weinberg, R. A., A molecular basis of cancer. *Sci. Am.* **249** (5), 126 (Nov. 1983).

Weinberg, R. A., Cellular oncogenes. *Trends in Biochem. Sci.* **9** (4), 131 (Apr. 1984).

Witkowski, J., The myth of cell immortality. *Trends in Biochem. Sci.* **10** (7), 258 (July 1985).

Books, Monographs, and Symposia

Bier, O. G., da Silva, W. D., Gotze, D., and Mota, I., *Fundamentals of Immunology*, Springer-Verlag, New York, 1981.

Cooper, E. L., *General Immunology*, Pergamon, Oxford, 1982.

Hood, L. E., Weissman, I. L., Wood, W. B., and Wilson, J. H., *Immunology* (2nd ed.), Benjamin/Cummings, Menlo Park, Calif., 1984.

Oppenheimer, S. B., *Cancer*, Allyn & Bacon, Boston, 1982.

Paul, W. E., *Fundamentals of Immunology*, Raven, New York, 1984.

GLOSSARY

Acrosome An organelle of animal sperm having lysosomelike activity.

Actin Protein present in thin filaments of cells.

Actinomycin D An antibiotic that inhibits the elongation of RNA molecules.

Activation energy Energy required to initiate a chemical reaction.

Active site Region of the enzyme's surface that binds and alters the substrate molecule(s).

Active transport The movement of molecules through a membrane against a concentration gradient and at a cost of cellular energy.

Adenosine triphosphatase ATPase; an enzyme that hydrolyzes ATP to form ADP and inorganic phosphate.

Adenosine triphosphate ATP; one of the nucleoside triphosphates; an intermediate in many energy transfer reactions.

Adenylcyclase (adenylate cyclase) An enzyme that catalyzes production of cyclic AMP from ATP.

Aerobes Cells that live in and utilize oxygen.

Affinity chromatography A technique for separating macromolecules based on the biological affinity of the molecule for a matrix-bound ligand.

AIA (Anti-Immunoglobulin Antibodies) Antibodies produced in response to the presence of foreign antigens in an organism

AIDS (Acquired Immune Deficiency Syndrome) A viral disease in which the immune system's T cells are progressively destroyed.

Allele One of the alternative forms of a gene.

Allosteric effectors Small molecules, usually metabolites, that bind to allosteric proteins at a site other than the active site and cause a change in protein shape and activity.

Allosteric enzymes (allosteric proteins) Enzymes whose activity is modulated by the binding of allosteric effectors at sites other than the active site.

Alpha actinin A protein found in the Z line of muscle fibers.

Aminoacyl adenylate In protein synthesis, an intermediate in the formation of a covalent bond between an amino acid and its tRNA.

Amphipathic Molecules that have both hydrophilic and hydrophobic regions.

Amplification The production of additional copies of a gene.

Anabolism Energy-requiring synthesis of more complex molecules from simpler ones.

Anaerobes Cells that can live in the absence of oxygen.

Anaphase Stage of mitosis or meiosis in which the chromosomes move toward the poles of the spindle.

Anaplerotic reactions Reactions that replenish intermediates depleted by other metabolic pathways.

Aneuploidy A chromosome number that is not an exact multiple of the haploid number (a condition common in transformed cells).

Angstrom (Å) A unit of length usually used to describe molecular dimensions; equal to 10^{-8} cm.

Antibiotic A drug that kills bacteria.

Antibody Protein formed and secreted by B lymphocyte-derived plasma cells in response to the presence of a foreign substance (antigen).

Anticodon The three-base group of a tRNA molecule that recognizes and pairs with the three-base codon of mRNA.

Antigen A substance, often a protein, that upon injection into a vertebrate stimulates the production of antibodies.

Antigenic determinant The specific region of an antigen that binds to an antibody or to a T-cell receptor.

Apoenzyme The protein component of an enzyme; apoenzyme + coenzyme = holoenzyme.

Aromatic amino acids Amino acids whose side chains include a derivative of a phenyl group. In proteins, the aromatic amino acids are phenylalanine, tyrosine, and tryptophan.

Aster The region at the poles of a dividing cell, composed of microtubules and a pair of centrioles.

ATP See *adenosine triphosphate.*

ATPase See *adenosine triphosphatase.*

Attenuator region A region of DNA within an operon at which most *RNA polymerase* molecules stop transcription. Receipt of a specific antitermination factor will cause transcription to proceed.

Autoantibodies Antibodies produced against one's own antigens.

Autoimmune disease An abnormality in which the immune system produces antibodies against the body's own, native tissue cells.

Autophagic vacuoles Membrane-lined vacuoles containing morphologically recognizable cytoplasmic components. They include autolysosomes (which are secondary lysosomes) and autophagosomes (which are vesicles sequestering cytoplasmic organelles).

Autophagy The process of sequestration of intracellular components in vacuoles.

Autoradiography Determination of the location of radioactive tracers introduced into cells by exposure of photographic emulsions placed in contact with the cells.

Autosomes Chromosomes other than the sex chromosomes.

Autotrophic cells Cells that can synthesize macromolecules from simple nutrient molecules, such as carbon dioxide, ammonia, and water.

Axoneme The microtubular organization of a cilium or flagellum.

B-DNA Double-helical DNA in which the helices are right-handed (see also *Z-DNA*).

B lymphocyte A cell that gives rise through division to antibody-secreting plasma cells.

Bacteriophage A virus for which the host cell is a bacterium.

Basal body An organelle located at the base of cilia and believed to be involved in the organization of ciliary microtubules.

Base pair Two nitrogenous bases, one in each strand of double-helical DNA, that interact through hydrogen bonds. The base pairs are adenine and thymine (i.e., A and T) and guanine and cytosine (i.e., G and C).

Benign tumor A tumor whose cells do not spread to other tissues.

Beta-galactosidase An enzyme that catalyzes the hydrolysis of lactose, producing glucose and galactose.

Bimolecular leaflet The two monomolecular layers of lipid that form the basis of membrane structure.

Bioenergetics Thermodynamics as applied to living cells or whole organisms.

Bivalent A synapsed pair of homologous chromosomes.

Bursa of Fabricius An outpocketing of the hindgut of birds that is the source of B lymphocytes.

C value The total amount of DNA in a haploid genome.

C_4 cycle (C_4 photosynthesis) A photosynthetic CO_2-reducing pathway in some green plants; also known as the Hatch–Slack pathway.

Calmodulin A protein that binds four calcium ions and regulates various metabolic processes in eukaryotic cells.

Calorie A unit of energy; the amount of heat required to raise the temperature of 1.0 g of water from 14.5° to 15.5°C.

Cancer   A disease characterized by the excessive and abnormal growth of certain tissue cells.

Capsid   A shell-like structure, composed of aggregated protein subunits, arranged in a regular pattern about the nucleic acid component of viruses.

Capsomeres   An individual protein subunit of a capsid.

Carcinogen   An agent that causes cancer.

Carcinoma   A tumor that arises from the abnormal growth of epithelial cells.

Carotenoids   Accessory photosynthetic pigments.

Catabolite Activator Protein (CAP)   A protein that regulates glucose-sensitive operons.

Catabolite repression   In bacteria, the decreased synthesis of specific enzymes concerned with degradation of glucose or similar catabolites. Caused by low levels of cyclic AMP in such cells.

Catalyst   An agent that increases the rate of a chemical reaction without altering the reaction's equilibrium.

cDNA   Single-stranded DNA synthesized from RNA by reverse transcription.

Cell culture   A population of cells grown *in vitro*.

Cell cycle   The cyclic sequence of events in dividing cells including the G_1, S, G_2, and M periods.

Cell division   Formation of two daughter cells from a parent cell by enclosure of the two new nuclei in separate cell compartments.

Cell fusion   Formation of a hybrid cell (or *heterokaryon*) from the nuclei and cytoplasm of different cells. Induced by treatment of a mixed cell culture with Sendai virus.

Cell plate   New cell wall material laid down in the center of a dividing plant cell.

Cellular immunity   Immunity provided by the actions of T lymphocytes.

Cell wall   A rigid or semirigid structure enclosing the protoplast of most plant and prokaryotic cells.

Central dogma   The relationship between DNA, RNA, and protein: DNA serves as a template for both its own replication and the synthesis of RNA; RNA, in turn, serves in protein synthesis.

Centriole   Microtubule-containing structures formed near the nucleus of many cells, at the spindle poles in most dividing cells, or forming the basal portion of a cilium or flagellum.

Centromere   The primary constriction of a chromosome to which the spindle fibers attach.

Chemiosmotic theory   A hypothesis that accounts for the coupling of electron transfer across a membrane and ATP formation by postulating the formation of H^+ gradients across the membrane.

Chiasma   Site of DNA exchange between two chromatids of a bivalent.

Chimeric DNA   A DNA molecule containing two or more regions of different origin.

Chlorophyll   Light-capturing pigment located in chloroplast thylakoids of plant cells or in prokaryotic cells.

Chloroplasts   Membranous organelles of plant cells containing chlorophyll and the sites of photosynthesis.

Chromatids   Identical members of a replicated chromosome, attached to one another at the centromere(s).

Chromatin   The nuclear material consisting of DNA, RNA, and proteins.

Chromosome   The gene-containing structure(s) in the nucleus or nucleoid of a cell.

Cilia (singular **cilium**)   Locomotor organelles located at the cell surface.

Cisterna (pl. **cisternae**)   A flattened, membrane-bordered channel.

Cistron   A unit of DNA representing a polypeptide chain or protein (equivalent to a structural gene).

Clone   A group of identical cells descended from one cell by mitosis and cytokinesis.

Codon   A sequence of three nucleotides of messenger RNA that codes for an amino acid or for chain termination.

Coenzyme   A small organic molecule weakly bound at or near the active site of the enzyme and required for enzyme activity.

Colchicine   A drug that binds to tubulin and causes the breakdown of microtubules; colchicine also prevents formation of spindle fibers and is used to prevent anaphase separation of chromosomes during mitosis.

Complement   A class of blood serum proteins that can lyse cells to which antibodies have attached.

Complementary base pairing   Hydrogen bond formation between a particular purine and a particular pyrimidine in nucleic acids; for example, guanine and cytosine and adenine and thymine.

Concanavalin A   A lectin used to study the surface properties of cells.

Conjugate base   A base derived from an acid through the removal of one or more hydrogen ions.

Connexon   See *gap junction*.

Constant domains   Constant amino acid sequences in the light and heavy chains of antibodies.

Constitutive enzyme An enzyme maintained at a constant level because the structural gene(s) for this enzyme are continuously expressed.

Contact inhibition A halt in cell division that occurs when freely growing cells from a multicellular organism physically contact one another.

Copolymer A polymeric molecule containing more than one kind of monomer unit.

Core particle Result of partial digestion of a nucleosome leaving the histone octamer and 146 base pairs of DNA.

Corepressor A metabolite that combines with repressor protein and blocks transcription of messenger RNA.

Coupled reactions Two chemical reactions that have a common intermediate through which energy is transferred from one reaction to the other.

Coupling factor F_1 factor; the headpiece of the mitochondrial inner membrane sphere that has ATPase activity.

Covalent bond Bond between atoms formed by sharing electrons.

Cristae Infoldings of the mitochondrial inner membrane and the site of enzymes of oxidative phosphorylation and electron transport.

Crossing-over Exchange of homologous chromosome segments leading to genetic recombination.

CURL Compartment for uncoupling ligand and receptor formed during concentrative receptor-mediated endocytosis.

Cyclic adenosine monophosphate Cyclic AMP; adenosine monophosphate in which the phosphate group is bonded between the 3′ and 5′ carbon atoms to form a cyclic molecule; this nucleotide is active in regulating certain reactions in cells.

Cycloheximide An inhibitor of protein biosynthesis.

Cytochromes Electron transport intermediates containing heme (or related prosthetic groups) in which the iron undergoes valency changes during electron transfer.

Cytoplasm The protoplasmic contents of the cell, exclusive of the nucleus.

Cytoskeleton An intracellular framework composed of filaments and microtubules.

Cytosol The fluid portion of the cytoplasm in which the organelles are suspended.

Cytosome A vacuole or vesicle of unknown function in the cytosol of eukaryotic cells.

Dark reactions Photosynthetic reactions that can proceed in the absence of light. Carbon dioxide is reduced ("fixed") to form sugars and other carbohydrates, using ATP and NADPH produced in the light reactions.

Deletion A loss of part of a chromosome or DNA molecule from the genome.

Denaturation Change in the native configuration of a macromolecule resulting, for example, from heat treatment, extreme pH changes, or chemical treatment. Denaturation is usually accompanied by the loss of the molecule's biological activity.

Density-dependent inhibition A halt to cell division in a layer of cells when their edges touch one another.

Density gradient A solution of changing density usually used in centrifugation to separate different kinds of cells, particles, or macromolecules.

Deoxyribonucleic acid (DNA) The genetic material of all cells and many viruses.

Desmosome One of several different kinds of plasma membrane specializations that occurs where adjacent cells in a tissue contact one another.

Dictyosome A stack of cisternae that forms part of the Golgi apparatus. In plant cells, the term is often used for the entire Golgi apparatus.

Diffusion The net overall movement of molecules in the direction of a lesser concentration.

Dimer Structure resulting from association of two identical subunits.

Diploid A cell or individual or species having two sets of homologous chromosomes in the nucleus of somatic cells.

Disulfide bond (disulfide bridge) A covalent bond between two sulfur atoms in separate amino acids of a protein.

DNA ligase An enzyme that links (ligates) the ends of two DNA molecules.

DNA polymerase An enzyme that synthesizes new DNA using the information in an existing DNA molecule.

DNA–RNA hybrid A double helix in which one polynucleotide is DNA and the other is RNA.

Doubling time The amount of time required for the number of cells in a culture to double during the exponential phase of the culture's growth curve.

Dynein Protein component of microtubules.

Effector A regulatory metabolite that activates or inhibits an enzyme by binding to its allosteric site.

Electron carriers Intermediates such as flavoproteins and cytochromes that reversibly gain or lose electrons.

Electron transport The movement of electrons from substrates to oxygen via the intermediates of the oxidative respiratory chain.

Electrophoresis A method of separating macromolecules or particles according to their charge, size, and shape as they migrate through a gel or other medium in an electrical field.

Elongation factor Proteins that cyclically associate with the ribosome during the elongation of the nascent polypeptide chain.

Endergonic reaction A chemical reaction with a positive standard free energy change; an energy-consuming reaction.

Endocytosis Intake by a cell of solutes or particles by enclosing them in an infolding of the plasma membrane.

Endonuclease An enzyme that makes internal cuts in a polynucleotide.

Endoplasmic reticulum ER; membrane-lined tubes and cisternae distributed within the cytosol of eukaryotic cells; the ER frequently has attached ribosomes (rough ER).

Endosome An endocytic vesicle derived from the plasma membrane; probably the same as a *receptosome*.

Endothermic process A process that absorbs heat.

End product repression (or **inhibition**) A control mechanism in which the production of an enzyme required in a metabolic pathway is inhibited by the final product of that metabolic pathway.

Entropy The randomness or disorder of a system.

Episome A genetic element that can exist either free or as part of the normal cellular chromosome. Examples of episomes are the F factors (plasmids) of *Escherichia coli*.

Euchromatin Noncondensed, active regions of chromosomes in the interphase nucleus of a cell.

Eukaryotic cells Cells in which the nucleus is separated from the cytoplasm by a pair of membranes (i.e., the nuclear envelope).

Exergonic reaction A reaction with a negative standard free energy change; an energy-releasing reaction.

Exocytosis A mode of transport of substances out of the cell by enclosure in a vesicle, fusion of the vesicle with the plasma membrane, and subsequent expulsion of the vesicle's contents.

Exon Polypeptide encoding portion(s) of structural genes that are separated from each other by intervening sequences. See also *intron*.

Exonuclease An enzyme that cleaves polynucleotide chains at their ends.

Exothermic process A process in which heat is evolved.

Facilitated diffusion Assisted transport of molecules through the plasma membrane along a concentration gradient.

Feedback (end product) inhibition Inhibition of the activity of the first enzyme in a (branch of a) metabolic pathway by the end product of that pathway.

Fermentation Oxidation of carbohydrate in non-oxygen-requiring pathways such as glycolysis.

Ferritin An iron-rich protein found in the liver, spleen, and bone marrow.

Fertility factor A form of plasmid present in bacteria.

Feulgen reaction Staining reaction for DNA used in microscopy.

Fingerprinting A technique used to separate peptides that combines paper chromatography with electrophoresis.

First law of thermodynamics Energy cannot be created or destroyed; statement of the principle of the conservation of energy.

Flagellum (pl. **flagella**) A cell organelle ultrastructurally similar to a cilium but usually longer and present in smaller numbers per cell and used for locomotion.

Flavin adenine dinucleotide (FAD) An electron carrier molecule that acts as a carrier in energy transfer reactions.

Fluid-mosaic membrane Model of the plasma membrane in which proteins are distributed in a phospholipid bilayer.

Fluorescent antibody technique Detection of selected antigens in cells by staining with a specific antibody that has been conjugated with a fluorescent dye.

Formylmethionyl-tRNA fmet-tRNA; the initial aminoacyl–transfer RNA complex that reacts with the small ribosomal subunit at the beginning of polypeptide chain synthesis (in prokaryotes).

Free energy A component of the total energy of a system that can do work under conditions of constant temperature and pressure.

Freeze-fracture Procedure for preparing materials for electron microscopy by rapid freezing and fracturing of the tissues; the exposed fracture faces are used to create a replica that is observed and photographed in the electron microscope; the fracture faces may or may not be further sublimed before the replica is made.

Furrowing A cell division mechanism that involves a pinching-in, or cleavage, to form two daughter cells from the parent cell.

Galactosidase β (beta) See *beta galactosidase*.

Gametes Haploid cells (e.g., ova and sperm) that unite at fertilization to produce the diploid zygote.

Gap junction Portions of the plasma membranes of adjacent cells that are bridged by narrow, tubuler channels that permit cell-to-cell communication. Also called a *nexus*. Each tubular channel is called a *connexon*.

Gene A section of the DNA that encodes a single polypeptide chain or one or more molecules of tRNA, mRNA, or rRNA.

Gene amplification Differential replication of specific genes.

Gene cloning A technique used to produce millions of copies of a particular gene.

Generation time The interval of time in which growing cells double their numbers or mass (also called the *doubling time*).

Gene segment Light (and heavy) polypeptide chains of immunoglobulins are encoded by genes whose DNA occurs in widely separated segments; each segment encoding a portion of the chain is a gene segment.

Genome The genes of a haploid set of chromosomes.

GERL Golgi-associated endoplasmic reticulum involved in production of lysosomes.

Gluconeogenesis Synthesis of carbohydrates from noncarbohydrate precursors such as fats or proteins.

Glycogen A polysaccharide formed by polymerization of glucose units through 1→4 (and occasional 1→6) glycosidic linkages between adjacent sugar residues. The most common form in which carbohydrate is stored in animal cells.

Glycolipids Lipids that contain polar, hydrophilic carbohydrate groups.

Glycolysis A pathway for glucose catabolism that does not require oxygen.

Glycoprotein A protein containing one or more chains of sugar residues.

Glyoxylate cycle An anaplerotic metabolic pathway replenishing two-carbon metabolites; the pathway is associated with glyoxysomes.

Group-transfer reactions Reactions (excluding oxidations or reductions) in which molecules exchange functional groups.

Growth curve The change in the number or mass of cells in a growing culture as a function of time.

Growth factor A specific substance that must be present in the growth medium to permit cells to grow and divide.

Hairpin loops A folded region of single-stranded DNA or RNA formed by the pairing of two contiguous complementary stretches of bases.

Haploid Cell or individual having but one copy of each homologous pair of chromosomes.

Haptens Small nonantigenic molecules that are capable of stimulating specific antibody synthesis when chemically coupled to a larger molecule.

Heavy chain The heavier (larger) of the polypeptides comprising an antibody.

Heavy isotope Form of atoms containing greater than the common number of neutrons and thus heavier and more dense than the commonly observed isotope (e.g., 15N, 14C).

HeLa cells An established line of human cervical carcinoma (cancer) cells derived initially from the patient Henrietta Lacks.

Helix A spiral structure having a repeating pattern described by two simultaneous operations—rotation and translation. It is the natural conformation of many regular biological polymers.

Helix-destabilizing proteins Proteins that bind to and stabilize the unwound DNA strands during replication.

Heme An iron-containing porphyrin that serves as the prosthetic group of the hemoglobins and enzymes such as *catalase* and cytochromes.

Hemoglobin Protein carrier of oxygen found in red blood cells; composed of two pairs of polypeptide chains and iron-containing heme groups.

Hemolysis Rupture of a red blood cell and release of its hemoglobin.

Heterochromatin Highly condensed chromatin regions of the interphase cell nucleus.

Heteroduplex Double-stranded DNA molecule in which the two strands are not entirely complementary; produced by mutation, recombination, or by annealing DNA single strands *in vitro*.

Heterogeneous nuclear RNA (hnRNA) Gene transcripts having a very wide size distribution and often precursors of messenger RNA.

Heterokaryon See *cell fusion*.

Heterophagic vacuoles Membrane-bordered vacuoles containing material that has been endocytosed. A form of lysosome.

Heterophagy The process of forming a heterophagic vacuole.

Heterotrophic cells Cells requiring complex nutrient molecules (e.g., glucose, amino acids) for energy and for the synthesis of their own macromolecules.

High-energy bond A bond that yields a large amount of free energy when the bond is hydrolyzed.

High-energy phosphate compound A phosphorylated compound having a highly negative standard free energy of hydrolysis.

Histone A protein component of the chromosome having a high content of basic amino acids. Eight histones comprise the core of a *nucleosome*.

Holoenzyme The complete form of an enzyme (coenzyme + apoenzyme = holoenzyme).

Homologous Having the same or similar gene content.

Homologous chromosomes Chromosomes that pair during meiosis, have the same morphology, and contain genes governing the same characteristics.

Hormone A chemical substance released into the bloodstream by one tissue or organ that modulates biochemical functions in the cells of another tissue or organ.

Humoral immunity Immunity provided by the secretion of antibodies by plasma cells.

Hybridoma A hybrid cell formed by fusion of a normal B lymphocyte with a cancerous myeloma cell; hybridomas are the sources of *monoclonal antibodies*.

Hydogen bond A weak electrostatic force between one electronegative atom and a hydrogen atom covalently linked to a second electronegative atom.

Hydrolysis The cleavage of a molecule into two or more molecules by the addition of a water molecule.

Hydrophilic Molecules or parts of molecules that readily associate with water; usually containing polar groups that form hydrogen bonds in water.

Hydrophobic Molecules or groups in molecules that are poorly soluble in water because of their nonpolar nature.

Hydrophobic bond The associations formed by hydrophobic groups when surrounded by water.

Hyperchromic shift An increase in UVL absorption by a solution of DNA caused by disruption of hydrogen bonds of a DNA duplex to yield the single-stranded structure.

Hypervariable region A short stretch of amino acids in a variable domain of an antibody molecule that exhibits even greater variation in amino acid sequences.

Immunity Built-in resistance to an infection that results from an earlier exposure to (and bout with) the infectious agent.

Immunoglobulin See *antibody*.

Immunologic memory Long-term memory exhibited by the immune system enabling rapid response to a second exposure to a given antigen.

Inducer A molecule that by binding to a regulator protein induces gene transcription.

Inducible enzymes Enzymes whose structural genes can be activated by the presence of specific inducer molecules in the cell.

Initiation factors Proteins required for the initiation of protein synthesis in prokaryotic and eukaryotic cells.

Inner membrane sphere Spherical particles attached to the matrical surface of the mitochondrial inner membrane.

Integral protein Protein molecule embedded in the lipid bilayer of the plasma membrane (or other cellular membranes).

Interleukin See *lymphokine*.

Intermediate filaments Cytoplasmic filaments having a diameter (about 10 nm) between that of thin and thick filaments; components of the cytoskeleton.

Interphase The state of the eukaryotic nucleus when it is not engaged in mitosis or meiosis; consists of G_1, A, and G_2 periods of the cell cycle.

Intervening sequences See *introns*.

Intron DNA sequences within a structural gene that are transcribed into RNA but do not give rise to amino acid sequences in the protein. Intron transcripts are excised during RNA processing.

In vitro (Latin: "in glass") Experiments carried out with isolated cells, tissues, or cell-free extracts rather than *in situ*, in place within the organism.

In vivo (Latin: "in life") Experiments carried out using the intact organism.

IS element (insertion sequence) A short sequence of nucleotide pairs in DNA that facilitates transposition of itself and/or other genetic elements in the genome.

Isoaccepting tRNAs Different tRNAs that activate the same amino acid during protein synthesis.

Isoenzymes See *isozymes*.

Isomers Alternative molecular forms of a chemical compound.

Isoosmotic Having the same salt concentration or exerting the same osmotic pressure.

Isopycnic density gradient centrifugation A centrifugal method used to separate particles in density gradients on the basis of differences in particle density.

Isotonic Producing equivalent osmotic pressures, usually by containing equivalent concentrations of salts.

Isotopes Alternative nuclear forms of an atom, all having the same atomic number (proton number) but different atomic weights (i.e., the number of neutrons varies).

Isozymes Alternative molecular forms of the same enzyme within a single cell or organism.

Joule (J) A unit of work or energy; 1 joule = 1 newton-meter = 0.239 calories = 10^7 ergs.

Karyotype A photograph or diagram of a complete complement of chromosomes from a cell or individual.

Kbp Abbreviation for kilobase pairs (i.e., 1000 base pairs of DNA).

Kinetochore Site of attachment of a spindle fiber to the centromere of a chromosome.

Krebs cycle Most common pathway for oxidative metabolism of pyruvic acid; as the citric acid cycle or the tricarboxylic acid cycle.

Label (radioactive) A radioactive atom introduced into a molecule to facilitate observation of the molecule's metabolic transformation.

Lambda phage Bacteriophage that is used extensively in molecular genetics research and in genetic engineering.

Lamellipodia Sheetlike extensions of the plasma membranes characteristic of motile cells (especially in culture).

Lagging strand DNA strand formed by ligating the smaller fragments produced by discontinuous synthesis.

Leader The nontranslated sequence of nucleotides at the 5′ end of mRNA (upstream of the initiation codon).

Leading strand DNA strand formed by continuous synthesis.

Lectins Cell-agglutinating proteins. Most lectins are isolated from plant seeds.

Ligase Enzyme that joins together the parts of single strands of DNA between the 5′ end of one strand and the 3′ end of another.

Light chain The lighter (smaller) of the polypeptide chains of an antibody.

Light reactions Photosynthetic reactions in which light energy is converted to chemical energy in the form of ATP and NADPH.

Lipid Class of organic compounds that are poorly soluble or insoluble in water but soluble in nonaqueous (organic) solvents such as ether and acetone.

Lipid bilayer Long-standing model for the structure of cell membranes based on the hydrophobic interactions between phospholipids. The polar heads face outward and the hydrophobic tails are clustered in the interior.

Liposomes Single or multilamellar membrane-bordered vesicles formed by vigorously dispersing lipids in water.

Loculus The interior of a thylakoid.

Long terminal repeats (LTRs) Nucleotide sequences at the ends of RNA molecules in which the same base appears repeatedly (as in the polyA tail of eukaryotic messenger RNA).

Lymphocyte A class of leukocytes that protects the body against infection.

Lymphokine A substance secreted by a lymphocyte that activates macrophages and other lymphocytes.

Lysis The bursting of a cell by the rupture of its plasma membrane.

Lysogenic bacterium A bacterium that contains a prophage.

Lysogenic viruses Viruses that can become prophages.

Lysosomes Intracellular organelles that contain a large variety of hydrolytic enzymes; lysosomes fuse with vacuoles formed during endocytosis and break down their contents.

Lysozyme Enzyme that degrades the polysaccharides found in the cell walls of certain bacteria.

Lytic infection Viral infection leading to the lysis of the host cell.

Lytic viruses Viruses whose proliferation within the host cell leads to the cell's lysis.

Macromolecules Molecules having molecular weights in the range of a few thousand to hundreds of millions of molecular weight units.

Macrophage   A large, phagocytic white blood cell.

Major histocompatibility antigens   Glycoproteins in the plasma membranes of cells that act as markers of "self" (i.e., serve to distinguish one's own cells from foreign cells).

Malignant tumor   A tumor whose cells spread (metastasize) to and proliferate in other tissues of the body.

MAPs   Proteins associated with the surfaces of microtubules (i.e., microtubule-associated proteins).

Meiosis   The reduction division of the cell nucleus in sexually-reproducing organisms; produces daughter nuclei having half the number of chromosomes as the original nucleus.

Melting   The separation of the two strands of duplex DNA to form single strands by disruption of the hydrogen bonds that hold the two strands together.

Memory cells   Cells produced by the immune system in response to exposure to an antigen (i.e., a foreign agent) but that are stored in the body for use during a subsequent exposure.

Mesosome   An extensively infolded portion of the prokaryotic cell's plasma membrane that is believed to function in cell division.

Messenger RNA (mRNA)   The complementary copy of the sense strand of DNA that is made during transcription and that codes for protein during translation.

Metabolic pathway   A set of consecutive enzymatic reactions that converts one molecule to another.

Metabolism   The type and diversity of various chemical reactions occurring in a living cell; the sum total of these reactions.

Metaphase   The stage of mitosis or meiosis when chromosomes are aligned along the equatorial plane of the spindle.

Metastasis   The separation of malignant tumor cells from a tumor followed by their spread to other tissues.

Microbody   A membrane-bounded cytoplasmic organelle with varied enzyme content and functions; e.g., peroxisomes and glyoxysomes.

Microfilaments   Narrow intracellular fibers that contain polymerized actin and that are thought to function in the maintenance of cell structure and movement.

Micron (micrometer, μ)   A unit of length used for describing cellular dimensions; it is equal to 10^{-4} cm or 10^4 Å.

Microsomes   A membrane-rich fraction of a tissue homogenate produced during high-speed centrifugation of the mitochondrial supernatant.

Microtrabecular lattice   A gossamer network of filaments radiating through the cytosol and anchored to the plasma membrane and organelles.

Microtubule   An unbranched cylindrical assembly of protofilaments involved in cell movement; microtubules are present in spindle fibers, cilia, and centrioles.

Microvilli   Fingerlike projections of the plasma membranes of certain cells.

Mitosis   The division of the nucleus that produces two daughter nuclei exactly like the original parental nucleus; somatic nuclear division.

Mole   A gram-molecule weight of a chemical substance.

Molecular weight   The sum of the atomic weights of all the atoms that comprise a molecule.

Monoclonal antibody   Antibodies secreted by hybridoma cells.

Monolayer   A single layer of cells, molecules, or other particles.

Monomer   The basic subunit from which, by repetition of a single reaction, polymers are made. For example, amino acids (monomers) yield polypeptides (polymers).

Monosome   The monomeric form of a ribosome that may contain no mRNA, tRNA, or nascent polypeptide chain.

Multienzyme system   A group of enzymes that catalyze sequential steps of a metabolic pathway and are in physical proximity to one another.

Mutagens   Agents, such as radiation and certain chemicals, that raise the natural frequency of mutation.

Mutation   A change in gene structure.

Myofibril   Parallel units of a muscle cell composed of bundles of myofilaments.

Myofilament   Individual thick (myosin) and thin (actin) filaments of the myofibril.

Myosin   Protein molecules that form a thick filament.

NAD, NADP   Nicotinamide adenine dinucleotide and nicotinamide adenine dinucleotide phosphate; acceptors of hydrogen in many enzyme-catalyzed oxidation–reduction reactions.

Natural killer cells   Lymphocytes that can destroy foreign or transformed cells by secretion of cytotoxic substances.

Neoplasm A tumor formed by abnormal cell proliferation.

Neutral fats Glycerides; fatty acid esters of glycerol; a major storage form of fats.

Nexus See gap junction.

Nuclear envelope The two membranes enclosing the eukaryotic cell nucleus.

Nucleases Enzymes that cleave the phosphodiester bonds linking neighboring nucleotides in a nucleic acid molecule.

Nucleic acid Polymer of nucleotides in an unbranched chain; DNA and RNA.

Nucleoid A region in the cytosol of prokaryotic cells that contains nuclear material but is not segregated by membranes.

Nucleolar organizing region (NOR) The specific part(s) of the genome containing genes for ribosomal RNA, usually clustered together on one (or more) chromosome.

Nucleolus Structure found in nucleus of eukaryotic cells. Involved in rRNA synthesis and ribosome precursor formation.

Nucleoplasm The unstructured portion of the nucleus in which the chromosomes and nucleoli are suspended.

Nucleoside Molecule containing a purine or pyrimidine linked to a pentose sugar.

Nucleosomes Spherical particles seen along decondensed chromatin; each nucleosome is composed of eight histones around which there are nearly two turns of DNA.

Nucleotide A phosphorylated nucleoside.

Okazaki fragment Short stretches of nucleotides produced during discontinuous synthesis of DNA and eventually linked together to become part of the lagging strand.

Oncogenes Genes whose activation can cause cellular transformation and lead to a cancerous state.

Oncoprotein The protein product of an oncogene.

Operator A specific nucleotide sequence in the operon that binds the repressor and exerts control over transcription of adjacent structural gene(s).

Operon A cluster of associated genes and recognition sites on the chromosome that participates in regulating transcription; including regulatory gene, promotor site, operator site, and structural gene(s).

Organelle A subcellular component; a discrete structural differentiation of the cell containing particular enzymes and performing particular functions for the whole cell; e.g., mitochondria, ribosomes, etc.

Oxidant An oxidizing agent; a substance that accepts electrons (or hydrogen) from a reducing agent or reductant.

Oxidation The loss of electrons or hydrogen from an atom, ion, or a compound.

Oxidation–reduction reactions Reactions that involve electron transfer from a reductant to an oxidant. The reductant is said to be oxidized and the oxidant reduced as a result of the transfer.

Oxidative phosphorylation In mitochondria, the enzymatic phosphorylation of ADP to form ATP that is coupled to electron transport along the respiratory chain to oxygen.

Palindrome A stretch of DNA in which the base sequences read the same from the 3′ or 5′ end.

Peptide bond A covalent bond between two amino acids in which the α-amino group of one amino acid is bonded to the α-carboxyl group of the other.

Peptidoglycan Macromolecule formed from both protein and carbohydrate parts; abundant in the cell walls of bacteria.

Peripheral protein Protein molecule loosely attached to the lipid bilayer of the plasma membrane.

Permease A carrier protein in the plasma membrane that is involved in the transport of specific substrate molecules across that membrane.

Peroxisomes Intracellular organelles that contain a fine granular matrix and often crystallike cores. They contain enzymes involved in hydrogen peroxide metabolism, including catalase.

pH A measure of the hydrogen ion concentration of an aqueous solution.

Phage See *bacteriophage*.

Phagocytosis A form of endocytosis in which large amounts of particulate material, even whole cells, are enclosed in endocytic vesicles.

Phase-contrast microscope An instrument that translates differences in the phases of transmitted or reflected light into gradations of contrast.

Phosphodiester linkage A covalent linkage involving esterification to phosphoric acid.

Phospholipids Lipids that contain charged, hydrophilic phosphate groups; primary components of cell membranes.

Photophosphorylation Process of formation of ATP from ADP and inorganic phosphate in the light reactions of photosynthesis; occurs by a cyclic or noncyclic pathway involving photosystems I and II.

Photorespiration Uptake of oxygen and release of carbon dioxide by photosynthetic cells or whole plants in the light.

Photosynthesis The enzymatic conversion of light energy into chemical energy by forming carbohydrates and oxygen from CO_2 and H_2O in green plant cells.

Photosynthetic phosphorylation The enzymatic formation of ATP from ADP in green plants coupled to light-dependent transport of electrons from excited chlorophyll molecules.

Photosystem I (PS-I) A photochemical reaction system in photosynthesis that is not associated with the evolution of molecular oxygen; is coupled to photosystem II.

Photosystem II (PS-II) A photochemical reaction system in photosynthesis that transfers electrons from water to photosystem I in the course of which molecular oxygen is evolved.

Phycobilin An accessory photosynthetic pigment present in red and blue-green algae (cyanobacteria).

Pinocytosis Endocytosis of small molecules together with aqueous medium.

Plaque Round, clear areas in a confluent sheet of cells; results from the killing or lysis of clusters of cells by several cycles of virus growth.

Plasmalemma See *plasma membrane*.

Plasma membrane The membrane that surrounds the cell and encloses the cytoplasm. This membrane is semipermeable and largely composed of lipid and protein.

Plasmids Small, circular chromosomes in bacteria.

Plasmodesmata Cytoplasmic channels through plant cell walls that connect the protoplasts of the adjacent cells.

Plastid Eukaryotic cell organelle that stores pigments or carbohydrates.

Polyacrylamide gel electrophoresis (PAGE) A method of molecular separation that relies on the differential migration of molecules, usually proteins or polynucleotides, through a polyacrylamide matrix on application of an electrical potential.

Polyadenylation The addition of a sequence of polyA to the 3' end of eukaryotic RNAs *after* transcription.

Polymer An association of monomer units to form a large molecule.

Polymerase An enzyme catalyzing the synthesis of DNA or RNA from nucleoside triphosphate precursors.

Polynucleotide A linear sequence of nucleotides in which the sugar of one nucleotide is linked through a phosphate group to the sugar on the adjacent nucleotide.

Polynucleotide ligase Enzyme that covalently links free 3' and 5' ends of polynucleotide chains.

Polypeptide A long, unbranched polymer (chain) of amino acids.

Polyploid Cell or individual having one or more extra sets of chromosomes.

Polyribosome Complex of ribosomes joined together by a messenger RNA molecule (number of ribosomes depending on size of mRNA); polyribosomes are engaged in polypeptide synthesis.

Polysomes See *polyribosome*.

Polytene chromosome Giant chromosome arising from successive rounds of chromatid duplication.

Pore An opening in a membrane or other structure; often referring to the nuclear pore complex of the nuclear envelope.

Primary constriction Location of a centromere on a chromosome.

Primary protein structure The number of polypeptide chains in a protein, their sequences of amino acids, and the location of interchain and intrachain disulfide bridges.

Primer A structure that serves as a growing point for polymerization.

Prions Infectious agents that appear to contain little or no nucleic acid.

Procentriole An immature centriole.

Proenzyme See *zymogen*.

Prokaryotes Organisms such as bacteria, cyanobacteria (blue-green algae), and mycoplasmas in which the nucleus is not separated from the cytoplasm by membranes.

Promoter A specific nucleotide sequence in the operon in which *RNA polymerase* binds.

Prophage A phage genome that has become integrated into a bacterial chromosome.

Prophase The first stage of mitosis or meiosis, after DNA replication and before chromosomes align on the equatorial plane of the spindle.

Proplastid An immature plastid.

Prosthetic groups Conenzymes that are more or less permanently bound to their enzymes.

Protamines A class of proteins rich in the basic amino acid arginine. They are found complexed to the DNA of sperm in many invertebrates and fish.

Protease An enzymatic protein that is capable of degrading other proteins.

Protein kinase An enzyme that phosphorylates the

hydroxyl-containing R-groups of specific amino acids (e.g., serine, threonine, and tyrosine); a number of oncoproteins appear to be protein–tyrosine kinases.

Protist   Unicellular eukaryotic organisms such as protozoa and algae.

Proto-oncogene   A gene whose product is a normal cell constituent but that can, if expressed at the wrong time of the cell cycle or if altered in some way, lead to abnormal cell growth and proliferation; the inactive (or precursor) form of an oncogene.

Protoplasm   The substance of a cell.

Protoplast   A plant cell (or bacterium), but *not* including the cell wall.

Provirus   The state of a virus in which it is integrated into the genome of a host cell and is transmitted from one cell generation to another.

Pseudogene   Inactive gene that may be a remnant of an ancestral gene.

Puff   A region of a chromosome that is undergoing active transcription, usually observed in giant polytene chromosomes.

Pulse–chase experiment   A radioactively labeled compound is added to living cells and a short time later an excess of unlabeled compound is added. Samples are then taken at periods after the pulse to follow the course of the label as the compound is metabolized (chase).

Purine   Parent compound of the nitrogen-containing bases adenine and quanine.

Puromycin   Antibiotic that inhibits polypeptide synthesis by competing with aminoacyl-tRNAs for the A site of the ribosome.

Pyrimidine   Parent compound of the nitrogen-containing bases cytosine, thymine, and uracil.

Quantum   The energy of a photon.

Quaternary structure   The manner in which the separate polypeptide chains of a protein are held together and oriented in space with respect to one another.

Radioactive isotope   Isotope with an unstable nucleus that emits ionizing radiation; important as labels or tracers in biology.

Reannealing   Renaturation; specifically, the restoration of duplex DNA regions through complementary base pairing of single-stranded DNA molecules.

Receptosome   A cytoplasmic vesicle that delivers the products of receptor-mediated endocytosis to specific intracellular sites; may be the same as *endosome.*

Recombinant DNA technology   Methods used to transfer genes from one organism to another usually employing a vector such as a bacterial plasmid or viral nucleic acid.

Redox couple   Compounds that occur in both the oxidized and reduced forms and that are participants in oxidation–reduction reactions (e.g., NAD^+/NADH).

Reductant   A reducing agent that donates electrons or hydrogens in oxidation–reduction reactions.

Reduction   The gain of electrons or hydrogens by an atom or molecule.

Regulatory genes   Genes whose primary function is to control the rate of synthesis of other genes; control is effected through either an RNA product or protein product of the regulatory gene.

Release (termination) factor   Specific macromolecule involved in the reading of the "stop" signal during protein synthesis.

Renaturation   The return of a protein or nucleic acid from a denatured and nonfunctioning state to its "native," functioning configuration.

Repetitive DNA   Repeated sequences of DNA nucleotides in the chromosomes of eukaryotic cells.

Replication forms (RF)   The structure of a nucleic acid during its replication; most frequently used to refer to double-helical intermediates in the replication of single-stranded DNA and RNA viruses.

Replication fork   The point at which the individual strands of the DNA duplex are separating during DNA replication.

Replicon   A unit of DNA replication; contains one origin and two replication forks or termination sites.

Repressor   A protein product of the regulator gene of the operon that binds to the operator site and prevents transcription of structural genes.

Residual bodies   Secondary lysosomes containing undigested residues, membrane fragments, and whorls.

Respiration   The oxidative breakdown and release of energy from molecules by reaction with oxygen in aerobic cells.

Restriction enzyme (restriction endonuclease)   Enzyme that cleaves double-stranded DNA at specific sequences.

Reticulocyte Immature red blood cell still capable of limited hemoglobin synthesis.

Retrovirus An RNA virus that on infecting a host cell induces *reverse transcription* of the viral RNA into DNA; many retroviruses cause transformation of normal cells.

Reverse transcriptase An enzyme encoded by the RNA of certain viruses that is able to make complementary single-stranded DNA chains from RNA templates.

Reverse transcription Synthesis of DNA using RNA as template.

RIA (Radioimmunoassay) A highly sensitive method employing radioactive isotopes to assay antibodies.

Ribonuclease An enzyme that cleaves the phosphodiester bonds of RNA.

Ribonucleic acid (RNA) Nucleic acids that function in transcription and translation. The genetic material of certain viruses.

Ribosomal DNA (rDNA) The genes of the nucleolar organizing region that code for ribosomal RNA.

Ribosomal proteins A group of proteins that combine with rRNA and give the ribosome its three-dimensional structure.

Ribosomal RNA (rRNA) Ribonucleic acids that are part of the ribosome structure and that function in protein synthesis.

Ribosomes Small cellular particles made up of rRNA and protein. Ribosomes are the site of protein synthesis; in eukaryotic cells, they are often attached to the endoplasmic reticulum.

RNA See *ribonucleic acid.*

RNA polymerase Enzyme that catalyzes the formation of RNA from ribonucleoside triphosphates, using DNA as a template.

Rough ER (RER) Portion of the endoplasmic reticulum bearing ribosomes.

S period The period of the cell cycle (eukaryotes) in which the nuclear DNA is replicated.

Saltatory conduction Rapid conduction of nerve impulses characteristic of myelinated nerve fibers; the impulse skips from one node of Ranvier to the next.

Sarcolemma The plasma membrane of a muscle cell or fiber.

Sarcoma A tumor that arises from the abnormal proliferation of mesenchymal cells.

Sarcomere The contractile unit of a muscle fiber, extending from one Z line to an adjacent Z line.

Sarcoplasm The cytoplasm of a muscle cell or fiber.

Sarcoplasmic reticulum Endoplasmic reticulum of a muscle cell or fiber.

Scanning electron microscope (SEM) Electron microscope that permits observation of a specimen's surface structure; the beam is not transmitted through the sample but causes the release of secondary electrons from the metal-coated surface of the specimen.

Secondary constriction Any constriction site along a chromosome other than the primary constriction at the centromere.

Secondary structure Structure of a polypeptide chain describing the location, extent, and types of helices (as well as nonhelical regions).

Second law of thermodynamics The principle that all physical and chemical changes proceed in a direction such that the entropy of the universe increases.

Sedimentation coefficient A quantitative measure of the rate of sedimentation of a given substance through water at 20°C in a unit centrifugal field; expressed in *Svedberg units* (S).

Semiconservative replication The usual mode of duplex DNA synthesis resulting in daughter duplex molecules that contain one parental DNA strand and one newly formed strand.

Sex chromosome Any chromosome involved in sex determination, such as the X and Y chromosomes.

Smooth ER (SER) Portion of the endoplasmic reticulum devoid of ribosomes.

Spindle Aggregation of microtubules during nuclear division that functions in the alignment and movement of chromosomes at anaphase.

Spindle fiber A microtubule in mitotically or meiotically dividing cells that extends from one pole to the centromere of a chromosome or that extends from pole to pole.

Spontaneous process A process accompanied by a decrease in free energy.

Standard electrode potential E'; the oxidation–reduction potential of a substance relative to a hydrogen electrode; expressed in volts.

Standard free energy change $\Delta G°$; a thermodynamic constant representing the difference between the free energy of the reactants and the free energy of the products of a reaction.

Standard state Most stable form of a pure substance at 1.0 atmosphere pressure and 25°C (298° K). For reactions occurring in solution, the standard state of a solute is a 1.0 M solution.

Stem cell A cell from which other cells arise by differentiation.

Stereoisomers Molecules that have the same structural formula but different spatial arrangement of dissimilar groups bounded to a common atom. Stereoisomers have differences in their crystal structures and differ in the direction in which they rotate polarized light; they also differ in their capacity to be used as a substrate in an enzyme-catalyzed reaction.

Steric (stereochemical) Relating to the arrangement in space of the atoms in molecules.

Steroids Compounds that are derivatives of a tetracyclic structure composed of a cyclopentane ring fused to a substituted phenanthrene nucleus.

Streptomycin An antibiotic isolated from *Streptomyces griseus* (a soil bacterium) that binds specifically to bacterial 30 S ribosomal subunits, thereby blocking protein biosynthesis.

Stroma Unstructured matrix of the chloroplast in which the grana and stoma thylakoids are suspended.

Structural gene A gene that encodes (or is expressed as) a polypeptide.

Substrate Specific compound acted upon at the active site of an enzyme.

Supercoiled DNA Twisted forms taken by double-stranded DNA molecules.

Suppressor gene A gene that can reverse the phenotypic effect of a variety of other genes.

Svedberg unit The unit of sedimentation equal to 10^{-13} seconds. The number of S units of a molecule or particle in a given centrifugal field is related to the weight, shape, and density of the molecule or particle.

Synapsis Specific pairing of homologous chromosomes, typically during zygotene of prophase I in meiosis.

Synaptonemal complex A complex structural component linking a pair of synapsed homologous chromosomes.

T lymphocyte A cell derived from the thymus gland that gives rise to cytotoxic, helper, and suppressor cells during an immune response.

T_m Abbreviation for the midpoint melting temperature of DNA; temperature at the midpoint of the

transition of duplex DNA molecules to single strands during melting.

Telophase Last stage of mitosis in which each daughter nucleus reestablishes its interphase structure.

Termination codon One of three triplet sequences that causes termination of protein synthesis.

Tertiary structure The three-dimensional folding of a macromolecule into a compact structural form; in proteins this is brought about by interactions among side chains of amino acids.

Tetracycline An antibiotic that acts by inhibiting protein synthesis.

Thermodynamics The science that deals with exchanges of energy.

Thylakoid A membranous sac present in chloroplasts that may be disk-shaped (in the grana) or elongated; it is the site of the light-requiring reactions of photosynthesis.

Tight junction A type of junction between neighboring cells in a tissue; in the tight junctions, neighboring plasma membranes fuse.

Topoisomerases Enzymes that relax DNA supercoils by nicking and then resealing one of the DNA strands.

Totipotent Capable of giving rise to all of the various cell types that comprise the tissues of an organism.

Tracer A radioactive form of an atom or compound whose metabolism can be followed by virtue of the detectable radiation that it emits.

Trailer The nontranslated sequence of nucleotides at the 3' end of a mRNA molecule, immediately following the termination codon but not including the polyA tail.

Transcription Process by which the base sequence of DNA is used to synthesize a complementary RNA molecule.

Transduction The transfer of bacterial genes from one bacterium to another.

Transfection The transfer of genetic material from transformed cells to normal cells, resulting in the transformation of the normal cells.

Transferases Enzymes that catalyze the exchange of functional groups between molecules.

Transfer RNA (tRNA) The RNA molecule that carries an amino acid to a specific codon on the messenger RNA–ribosome complex during translation.

Transformation The genetic modification induced by the incorporation into one cell of DNA from another cell (or a virus).

Translation Process by which amino acids are assembled into a polypeptide on the ribosome under the direction of the base sequence transcribed from DNA into messenger RNA.

Translocation A structural rearrangement involving parts of entire nonhomologous chromosomes.

Tritium ^{3}H; a radioactive isotope of hydrogen; extremely useful in tracer studies.

Tropomyosin A muscle protein that associates with actin and troponin and is involved in regulating contraction.

Troponin Protein associated with actin filaments in muscle cells that binds calcium ions, thereby effecting contraction.

T system Invaginations of the sarcolemma in muscle fibers, producing a system of transverse tubular infoldings.

Tubulin Globular protein subunits whose regular helical packing forms microtubules.

Ultracentrifuge Centrifuge capable of producing rotor speeds up to 100,000 rpm and able to rapidly sediment tiny particles and macromolecules.

Ultraviolet light Electromagnetic radiation having a wavelength shorter than that of visible light (390–200 nm). Causes DNA base pair mutations and chromosome breaks.

Uncoupling agent A substance (e.g., 2, 4-dinitrophenol) that can uncouple phosphorylation of ADP from electron transport.

Unit membrane Membrane showing a three-layer or dark-light-dark pattern of electron density in the electron microscope; a former incorrect model of membrane structure proposing that phospholipid bilayer is coated on its outer and inner surfaces by proteins.

Vacuole A membrane-enclosed sac in the cell cytoplasm filled with molecules and particles in a watery medium; common in plant cells.

van der Waals force A weak, attractive force between atoms; particularly important in hydrophobic bonding of amino acids in proteins.

Variable domain Amino acid sequences in the light (and heavy) chains of antibodies in which the sequence varies from one antibody to another.

Vesicle A small, spherical, membrane-bordered intracellular chamber.

Viroids Pathogenic agents that consist only of short RNA molecules.

Viruses Infectious, disease-causing particles that require a host cell for replication and that contain either DNA or RNA as their genetic material.

Wobble Capability of the third base in the tRNA anticodon (5′ end) to form a hydrogen bond with any two or three bases at 3′ end of the messenger RNA's codon. Thus, a single tRNA species can recognize several different codons.

X-ray crystallography The use of X-ray scattering by crystals to determine the three-dimensional structure of molecules, especially proteins and nucleic acids.

Z-DNA Double-helical DNA in which the helices are left-handed (see also *B-DNA*).

Zygote The product of fusion of two gametes during fertilization; the cell from which a new diploid individual develops.

Zymogen A digestive enzyme precursor lacking catalytic activity but that can be converted to the active form.

Zymogen granule Membrane-enclosed packets of enzymes produced and released by secretory cells such as pancreas acinar cells.

INDEX

The book's Glossary (page 673) contains the definitions of a number of terms that are not listed in the Index. Students should consult the Glossary for additional terms and their definitions.

Rothman, J.,   455
Rouleaux,   350
Rous, P.,   665
Rous sarcoma virus,   665
Ruben, S.,   433
Ruffling (surface),   605

S

Sabatini, D.,   344
Sachs, J.,   430
Salt bridges,   101
Salting in,   284
Salting out,   284
Sanger, F.,   99
Saponification (of fats),   140
Scanning electron microscope,   20
Schachter, H.,   455
Schleiden, M.,   4
Schliwa, M.,   597
Schneider, W.,   465
Schoenheimer, R.,   543
Schwann, T.,   4
Schwann cells,   642
Schweet, R.,   568
Scrapie,   51
SDS-PAGE,   292
Sealing strands,   351
Second messengers,   253
Secondary lysosomes,   468
Sedimentation coefficients,   253–265
Selective proteolysis,   445
Semiconservative replication (of DNA),   515, 524
Senescence,   59
Shadowing-angle,   17, 19
Sickle-cell anemia,   631
Siekevitz, P.,   544
Signal hypothesis,   344, 559
Signal peptide,   345, 469, 559
Signal recognition particle,   345
Signal sequence,   469, 559
Simonsiella,   21, 42, 44
Singer, S.,   333
Sis oncogene,   669
Sinha, N.,   523
Sjöstrand, F.,   387
Soaps,   75
Spacer DNA,   553
Sparsomycin,   592
Specific activity,   310

Spectrin,   349
Spheres of hydration,   77, 363
Sphingolipids,   144
Sphingomyelin,   144, 146
Sphingosine,   144, 146
Spindle,   484, 611
Spiral filaments,   618
Spongy mesophyll,   419
Spot desmosomes,   353
Src oncogene,   666, 669
SRP,   345, 346
SRP receptors,   345
Sta-Put chamber,   279
Standard free energy,   202
Standard units of measure,   74
Steck, T.,   337
Stem cells,   628, 647
Stereo microscopy,   21
Stereo viewers,   24
Steroids,   139, 147, 254
Stokes' law,   264
Streptomyces,   405, 592
Streptomycin,   592
Stringent control,   249
Stroma,   39, 423, 425
Substrate cycles,   193
Sucrose,   129
Supercoils (DNA),   491
Suppressor cells,   648, 654
Surface ruffles,   26, 354, 606
Svedberg unit,   265
Svensson, H.,   294
Synapses,   638, 642, 643
Synaptic body,   209
Synchronous cultures,   67
Szent-Gyorgyi, A.,   402

T

Tay–Sachs disease,   144
Taylor, J.,   524
TCA cycle,   222, 257, 393, 395, 411
T cells:
    leukemia,   665
    lymphotropic viruses,   663
Telolysosomes,   468
Temperature cycles,   68
Temperature shocks,   68
Terminal transferases,   540
Terminal web,   354, 604

COMMON ABBREVIATIONS

A	Adenine or adenosine
Å	Angstrom
ACTH	Adrenocorticotrophic hormone
AMP, ADP, ATP	Adenosine 5'-mono-, di-, and triphosphate
cAMP	Cyclic AMP
ala	Alanine
arg	Arginine
asn	Asparagine
asp	Aspartic acid
ATPase	Adenosine triphosphatase
C	Cytosine
c	Centi, 10^{-2}
c	Velocity of light (vacuum), 2.997×10^{10} cm sec^{-1}
^{14}C	Carbon 14
cal	Calorie
cm	Centimeter
CMP, CDP, CTP	Cytidine mono-, di-, and triphosphate
CoA, CoA-SH, acyl-CoA, acyl-S-CoA	Coenzyme A and its acyl derivatives
CoQ	Coenzyme Q; ubiquinone
CPM	Counts per minute
CsCl	Cesium Chloride
cys	Cysteine
cyt.	Cytochrome
d	deci, 10^{-1}
DNA	Deoxyribonucleic acid
DNAase	Deoxyribonuclease
DPN+, DPNH	Same as NAD+, NADH
E_o	Standard electrode potential
e$^-$	Electron
EDTA	Ethylenediaminetetraacetic acid
ER	Endoplasmic reticulum
ESR	Electron spin resonance
ETP	Electron transfer particle (from mitochondrial membrane)
$\mathscr{F}$	Faraday
f	femto, 10^{-15}
FAD, FADH$_2$	Flavin adenine dinucleotide, oxidized and reduced forms
FDP	Fructose-1, 6-diphosphate
fmet	N-formylmethionine
FMN, FMNH$_2$	Flavin mononucleotide and its reduced form
FP	Flavoprotein
g	Gram, gravity
G	Gauss, guanine, or guanosine
G	Giga, 10^9
$\Delta G^{o'}$	Standard free-energy change at pH 7
Gal	D-Galactose
GDH	Glutamate dehydrogenase
GLC	Gas-liquid chromatography
glc	Glucosamine
GlcNAc	N-Acetyl-D-glucosamine
gln	Glutamine
glu	Glutamic acid
gly	Glycine
GMP, GDP, GTP	Guanosine mono-, di-, and triphosphate
G3P	Glyceraldehyde-3-phosphate (3-phosphoglyceraldehyde)
G6P	Glucose-6-phosphate
hr	Hour
h	Planck's constant, 6.624×10^{-34} joule-sec or 1.584×10^{-34} cal-sec
h	Hecto, 10^2
H+	Hydrogen ion
^3H	Tritium
Hb	Hemoglobin
his	Histidine
ile	Isoleucine
IMP, IDP, ITP	Inosine mono-, di-, and triphosphate
J	Joule (1.0 J $= 0.239$ cal $= 10^7$ erg)
K	Kelvin (degrees)
k	Kilo, 10^3
kcal	Kilocalorie
l	Liter
λ	Lambda phage or wavelength of radiation
LDH	Lactate dehydrogenase
leu	Leucine
LH	Luteinizing hormone
ln	Logarithm to the base e